Food, Fiber, and the Arid Lands

FOOD, FIBER
and the
ARID LANDS

Editors:

WILLIAM G. McGINNIES
BRAM J. GOLDMAN
PATRICIA PAYLORE

THE UNIVERSITY OF ARIZONA PRESS
Tucson, Arizona

THE UNIVERSITY OF ARIZONA PRESS

Copyright © 1971
The Arizona Board of Regents
All Rights Reserved
Manufactured in the U.S.A.

I.S.B.N.-0-8165-0299-4
L.C. No. 75-152038

Foreword

Food, Fiber, and the Arid Lands is an especially appropriate volume at a time when considerable apprehension prevails about sufficient foods for hungry nations. With a few notable exceptions, the arid lands of the world still are sparsely populated and generally low in agricultural productivity. For centuries the inhabitants of the desert and semidesert regions have eked out their subsistence from the parched land through limited irrigation of the soil or by nomadic herding of cattle, sheep, and goats. Today the situation in most of the arid parts of the world is little different from what it has been for hundreds of years; for most of the vast regions of the earth characterized by high temperatures and low rainfall, there has been little reason to believe that the future holds much in the way of increased productivity of food and fiber.

Under the pressures of increasing population and urbanization, however, this view is beginning to be challenged. The world's dramatic awakening to the impact of ecology on the quality of human life, and a new awareness of the potential for exploitation of the desert environment — in which our particular interest lies — give rise to more thoughtful, even sober, attention to the problems faced by those aspiring only to bigger yields from crops and more protein from animals. To these very basic needs of food and fiber, we are, therefore, addressing ourselves further to a consideration of housing, educational programs, and nonagrarian resource evolution, to the end that the planners and the decision makers may have the benefit of the most intelligent and contemporary understanding possible of the cultural and social aspects of arid lands as well as the latest technological advances on a wide-ranging spectrum of scientific developments.

JAMES H. ZUMBERGE, *Dean*
College of Earth Sciences
University of Arizona

A Note From the Editors

In building this book and its predecessor (*Arid Lands in Perspective,* 1969), we have sought to serve those with broad interests in the arid lands. Scientific specialists have found that they cannot achieve mastery in a narrow field without understanding better the neighboring disciplines, the whole system that constitutes a desert environment; authorities such as government officials who have broad responsibilities for desertic regions have found that they need more specific information than the generalizations so often served up about the nature and the future of the arid lands. We have tried to provide for the enthusiasms, professional concerns, and planning requirements of several classes of potential readers — the highly technical as well as the nonspecialized readers.

An acknowledgment of extraordinary service, above even the contributions of many others, must be made to three individuals. Even so, no matter how laudatory, it cannot do justice to these people for their essential contributions: Mary Lu Moore (bibliographical and translation), Anne Stamper Haney (editorial), and Rosemary Quiroz Santa Cruz (executive and translation).

WILLIAM G. MCGINNIES
BRAM J. GOLDMAN
PATRICIA PAYLORE

Office of Arid Lands Studies
The University of Arizona

Contents

PART ONE
CULTURAL AND SOCIAL PROBLEMS

Social and Institutional Obstacles to Agricultural Development
    of Arid Lands in the Middle East   3
    Marion Clawson

Resource-Allocation Problems for the Development of Arid Zones   17
    Manfred A. Max-Neef

Agrarian Capitalists and Urban Proletariat: The Policy of Alienation
    in American Agriculture   39
    Harland I. Padfield

Arid-Lands Agriculture and the Indians of the American Southwest   47
    James E. Officer

The Papago Cattle Economy: Implications for Economic and Community
    Development in Arid Lands   79
    Rolf W. Bauer

Building Design and Planning for Self-sustaining Communities in Remote
    Localities of Australia's Arid Zone   103
    Balwant Singh Saini

International Training: The Interamerican Center for Land and
    Water Resource Development: A Case Study   123
    Bruce H. Anderson

Organization of Educational Programs in Sparsely Settled Areas of the World   137
    Everett D. Eddington

Early Human Contacts from the Persian Gulf through Baluchistan and
    Southern Afghanistan   145
    George F. Dales

PART TWO
LAND AND RESOURCE USES

Applications of Remote Sensing to Arid-Lands Problems   173
    Robert N. Colwell and David M. Carneggie

The Gobi Desert of Mongolia: Geographic Description and Prospects for
    Land Use on the Basis of Soil, Vegetation, Hydrology, and Climate   187
    Jerzy Lefeld

Pastures and Sheep Production in the Arid Zone of the Soviet Union   211
    Nina T. Nechayeva

Range-Cattle Production Under Dry, Warm Conditions   217
    Carl B. Roubicek and Donald E. Ray

Shrub Productivity: A Reappraisal of Arid Lands   235
    Joe R. Goodin and Cyrus M. McKell

Economic Botany of Arid Regions   247
    Peter C. Duisberg and John L. Hay

Soil Management: Humid Versus Arid Areas   271
    Thomas C. Tucker and Wallace H. Fuller

PART THREE
WATER AND AGRICULTURE

Exploitation of Groundwater for Agricultural Production in Arid Zones   289
  David J. Burdon

Water Conservation For Food and Fiber Production in Arid Zones   301
  Lloyd E. Myers

Runoff Agriculture in the Negev Desert of Israel   311
  Michael Evenari, Leslie Shanan, and Naphtali H. Tadmor

Artificial Inducement of Runoff as a Potential Source of Water in Arid Lands   323
  Daniel Hillel

Physiological Basis for Plant Growth Inhibition Due to Salinity   331
  James W. O'Leary

The Protoplasmic Basis for Drought Resistance: A Quantitative   337
  Approach for Measuring Protoplasmic Properties
  Eduard J. Stadelmann

PART FOUR
ECOLOGY OF ARID REGIONS

Sand-Stabilization Methods in Arid Lands: Protection of Agricultural and   355
  Settlement Areas
  Mikhail P. Petrov

Stabilization of Sand Dunes in the Semiarid Argentine Pampas   369
  *Antonio J. Prego, Roberto A. Ruggiero, Florentino Rial Alberti,
  and Federico J. Prohaska*

Effect of Insecticides on an Ecosystem in the Northern Chihuahuan Desert   393
  Howard G. Applegate, Wayne Hanselka, and Dudley D. Culley, Jr.

PART FIVE
INFORMATION SOURCES

Arid-Lands Information in United States Government-Sponsored Indexing   407
  Tools: What? Where? When? How?
  Patricia Paylore

A Regional Bibliography of Calcrete   421
  Andrew S. Goudie

Part One
CULTURAL AND SOCIAL PROBLEMS

Social and Institutional Obstacles to Agricultural Development of Arid Lands in the Middle East

MARION CLAWSON

Land Use and Management Program, Resources for the Future, Washington, D.C.

ABSTRACT

THE LAND AND WATER RESOURCES of the Middle East have the potential for a greatly increased agricultural production; the chief obstacles to achievement of the potential are social and institutional. Irrigation has been practiced for thousands of years, but can be improved and extended; drainage is necessary for most of the irrigated area. Different land-use systems for rainfed croplands and for native grazing lands can increase output greatly in a relatively short time. If increased output is to be achieved, additional inputs of various kinds will be required; fertilizer application will have to be increased greatly, weeds will have to be controlled by use of chemicals or in other ways, and other materials combined in a proportioned package of productive inputs. By such means, crop yield could be doubled or more, and crop rotations intensified. Markets within the region can absorb large increases in output over the next two or three decades, including more cereals; the outlook for such export crops as cotton and citrus is less encouraging and less clear.

The major obstacles to achievement of these development potentials in agriculture are unstable and inefficient government, lack of transportation and marketing facilities, absence of technical knowledge about agriculture, entrenched customs (especially in livestock grazing), obsolete land-ownership patterns, and lack of investment capital. A strategy of providing a reasonably complete package of all needed inputs in a few selected localities, rather than scattering efforts widely and thus largely wasting such resources as are available, is suggested as the best means of overcoming the social and institutional obstacles to development.

SOCIAL AND INSTITUTIONAL OBSTACLES TO AGRICULTURAL DEVELOPMENT OF ARID LANDS IN THE MIDDLE EAST

Marion Clawson

THE ARABLE SOILS, the natural grazing lands, and the water resources of the Middle East (*1*) — ancient home of Man, where irrigation has been practiced for thousands of years — are physically and economically capable of greatly increased agricultural output. The major obstacles to early — or, indeed, to eventual — realization of the potentials are social and institutional situations and arrangements of various kinds. These can be stubborn, but progress is possible; and a frank analysis of the problems may be the first step in their solution (*2*).

We estimate that there are about 8 million hectares (20 million acres) of reasonably good land which are, or might be with tolerable costs, irrigated (*3–18*). This is only 5 percent of the gross area of these countries, but irrigable lands are typically only a minor fraction of total area in any country, state or province, or district where irrigation is necessary. The difficulty is not merely lack of water; extreme desert conditions often do not produce accumulations of suitable soil materials and rarely produce significant acreages of true soils. The Nile Valley includes about 3 million hectares (7½ million acres) of irrigable land; but, outside of the valley, there is only an extremely small area of reasonably good soils or soil materials. The deserts are simply too sandy and/or potential soil materials too thin over rock for successful agriculture. In the Mesopotamian Plain, perhaps 3 million hectares (7 million acres) could be irrigated successfully, if adequate dependable water supply is provided and if the land is fully drained. Elsewhere there are irrigable lands, locally of great importance, as in Syria, Israel, Jordan, Lebanon, and Saudi Arabia, but geographically sharply limited in extent.

We estimate that there are about 10 million hectares (23 or 24 million acres) of rainfed arable lands capable of economic crop production. Some of these, perhaps 2 million hectares (4 to 5 million acres), were included in the irrigable acreage; hence there is duplication to this extent. But the Nile Valley and much of the Mesopotamian Plain, and lesser areas elsewhere, are irrigable but not capable of successful farming without irrigation because precipitation is too limited. The rainfed arable lands differ considerably in soil type, soil depth, and water-holding capacity; and the amount of rainfall also varies greatly. We think crop production has been pushed too far, onto soils too poorly adapted to successful continuous cropping in view of their low and erratic rainfall, and that such marginal lands would be more productive if permanently seeded to grass for livestock grazing. Modern moisture-conservation techniques can greatly increase the output of this relatively large area of rainfed croplands.

Between the arable lands and the deserts lacking any significant vegetation, as well as on the higher hills and mountains, are extensive areas suitable only for grazing. Although considerable is known about such lands, no comprehensive survey of their extent, seasonal usefulness, carrying capacity, and other characteristics has been made (*19*). All reports indicate that they have been overgrazed severely and for long periods of time, far past the point of maximum output per acre (*20–24*). Information on the utilization of such lands is lacking or is unreliable; but something like 20 million sheep, 6 million goats, fewer than 1 million camels, and 3 million cattle, as well as smaller numbers of other livestock, may have received a substantial part, if not all, of their food from natural grazing areas of the seven countries listed (*25*). Historically for thousands of years, livestock have moved back and forth across this large area and surrounding territory, thus adding to the problems of trying to estimate the number of livestock and the capacity and usefulness of the grazing land.

These seven countries of the Middle East are, taken as a region, very dry; aridity is a general shortage of water, and this is surely an arid region (*26-29*). We estimate that the average precipitation for the region as a whole is in the general magnitude of 100 millimeters (4 inches); this average is strongly influenced by very extensive desert areas in Egypt, Jordan, Syria, Iraq, and Saudi Arabia where precipitation is extremely low. Annual precipitation at Cairo is about 40 millimeters (1½ inches); at Aswan, in Egypt, it is less than 25 millimeters (1 inch), for example. But the rainfed croplands have annual precipitation ranging from 250 to 450 millimeters (10 to 18 inches), and even higher; water is still a limiting factor in crop production, but successful cropping is possible. The higher mountains in the Judean Hills, Lebanon mountains, Anti-Lebanon mountains, and elsewhere receive much more; and the high mountains along Iraq's northern border get still more — up to a meter (40 inches) and more.

The Middle East, as we define it, has two great river systems, the Nile and the Tigris-Euphrates (*30-40*). These rivers bring great quantities of water into the region, from watersheds well outside of it. The Nile at Aswan has an average annual flow of 92 billion (milliard) cubic meters (nearly 75 million acre feet), all of which originates south of Egypt; the combined annual flow of the Tigris and Euphrates and their tributaries averages

4

nearly 80 billion (milliard) cubic meters (about 65 million acre feet), two-thirds of which originates in Turkey. The Nile has been a magnificent natural resource for millenia; its natural floods in late summer and autumn thoroughly saturated the valley soils with water, depositing thin layers of fertile sediments, and flushed out salt accumulations. The latter role has been less generally recognized but may have been decisive. Winter crops, especially wheat and other cereals, could then be grown. The river and the valley had a pronounced seasonal cycle, and Man adjusted to it.

For about a hundred years, summer or perennial irrigation has been practiced in the Nile Valley, on a growing scale. Low dams diverted low river flows of late spring and early summer to provide irrigation for summer crops, especially cotton. Completion of the High Aswan Dam has totally altered the natural regime of the Nile; floods can now be completely controlled (*41–42*). Storage capacity equal to 2 years' average flow can iron out year-to-year fluctuations in runoff, and irrigation water can be available in any reasonable quantity at any season. These gains come at a cost, in addition to the $700 million invested: evaporation from the reservoir will probably take 12 billion (milliard) cubic meters (10 million acre feet) of water annually, and continued year-round irrigation on an increased scale will exacerbate an already serious drainage problem. The ancient crops were mostly adequately drained in the unregulated streamflow era, partly because men stayed out of marshy and swampy lands, especially to the north, in the delta. But summer irrigation has raised the water table, to less than 2 meters (6 feet) in much of the valley, to less than 1 meter (3 feet) in areas of considerable size. Yields of summer crops, especially cotton, have been seriously and adversely affected. Salinity has not risen seriously *as yet*, but salt accumulations are beginning to show and will be increasingly serious unless much more and better drainage is provided.

The Mesopotamian Plain is another area of long continued irrigation — 7,000 years or more. While the Tigris and Euphrates rivers have brought irrigation water to this area, they have never been as neatly self-regulating as was the Nile. Floods were less predictable, more variable, possibly more damaging as a result. Siltation, not only of canals and drains, but deposits on the land, impaired the natural drainage; in some areas, well-developed soil horizons can be found, buried under several feet of sediments washing in from surrounding hills or deposited by the river waters. Although the quality of the water in these rivers is fairly good (mostly under 400 ppm), evaporation of so much water over many centuries has led to great accumulations of soluble salts in the topsoil — up to 600 tons and more per hectare (250 tons and more per acre). The surface of the land is white with salts, much of it sodium chloride. These lands can be drained today without unusual difficulty; later, we discuss their drainage more fully.

Space here does not permit full discussion of other water supplies in the region: the Jordan which, with all its tributaries, has only about 2 billion (milliard) cubic meters (1½ million acre feet) annually — a river the focus of much controversy, over little water (*43–44*); the Orontes, the Litani, and other still smaller rivers; numerous springs, as near Damascus and Jericho, oases in the desert where man has lived for many thousands of years. These water sources have been and are extremely important in the local economy; while they can be used much more effectively, they are still small (Fig. 1).

It is not necessary, nor possible, to consider here the climate of the Middle East in detail. Precipitation has been mentioned; it nearly always comes in the winter, when temperatures are moderate, and this greatly facilitates crop production. The region is characterized by lots of sunshine — well over 90 percent of potential in the Nile Valley. As would be expected, evaporation from water surfaces is very high, 2½ meters or more per year.

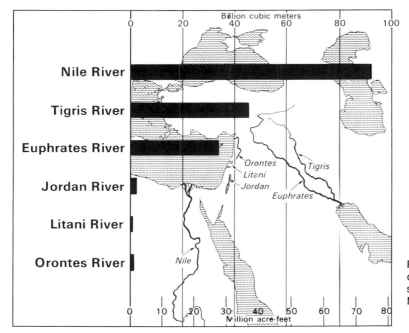

Fig. 1. A comparison of average annual streamflow from major Middle Eastern sources.

Much of the region, especially Iraq, has relatively high, hot, dry winds in the summer. Within this general pattern, considerable variation arises within short distances, primarily as a result of elevation. In the coastal plain of Israel, frost-free winters permit an extensive citrus industry, while in the hills only a few airline miles to the east winter temperatures drop enough to provide the necessary dormant stimulus to apples, pears, and stone fruits, for instance. Likewise, Lebanon produces citrus and bananas on the lower lands, and deciduous fruits at higher elevation.

DEVELOPMENT POTENTIAL OF MIDDLE EASTERN AGRICULTURE

To facilitate exposition, we describe a few major ways to improve resource use in this region, then consider how modern farm techniques developed elsewhere could be applied effectively in the region, and finally present some summary data to quantify the expected increased output. We necessarily must describe measures one by one; but it cannot be too strongly emphasized that it is a *package* of new productive inputs and practices, each properly proportioned and timed to complement the others, which is really productive (45, 46). It does little good to irrigate when drainage is defective; improved crop varieties are only modestly productive unless additional fertilizer is provided; there is no use producing crops if marketing and transportation facilities are lacking; and so on, and so on. Moreover, it is hard to say which measures are most important; all are essential and each adds to the productivity of all the others.

Nearly all the presently irrigated and potentially irrigable lands require drainage. Drainage is especially needed in much of the Mesopotamian Plain, where saline conditions are now so severe; see Buringh (14) and Poyck (46). On some lands, the farmers now apply a preplanting irrigation to flush soluble salts down into the soil; then a crop of winter barley is grown and sparingly irrigated. Yields often average no more than 450 to 550 kilograms per hectare (8 to 10 bushels per acre) under these conditions, for salt content is still high. After harvest, the lands are allowed to dry out and grow up to weeds for 1½ years, after which the cycle is repeated. A harvest of such a small yield of barley on an alternated crop-fallow basis, under irrigation, is about as marginal as crop production can get, and still continue. Fortunately, sufficient research and mapping has been done to demonstrate clearly that the technical problems of draining these soils can be met at reasonable costs. Drainage of the Mesopotamian Plain would cost approximately ¾ billion dollars, but could be carried out by stages over a period of years. It would take considerable water and a year or two to flush salts out to the point where reasonable crop yields would be possible, once the drains had been installed, and a longer period of years before salt content would be reduced to an economically and technically optimum level.

While present drainage in the Nile Valley is not as obviously and impressively deficient as in the Mesopotamian Plain, yet it is now a major barrier to higher crop yields and will become more serious in the future unless substantial efforts are made to improve it (47, 48). Internal soil drainage in the Nile Valley, as in the Mesopotamian Plain, ranges from moderate to good. A substantial mileage of drainage canals now exists in Egypt, but often these drains are not properly maintained. Buried drains, tile or plastic, would do a far better job, with less maintenance; moreover, they would reclaim a substantial acreage of good cropland (up to 13 percent of total land area) now made unusable for crops by the open drains. Research, field tests, and mapping have demonstrated that such buried drains are wholly practical, at a cost of about $220 per hectare ($85 per acre), or about ½ to ¾ billion dollars for total area needing drainage.

The other, smaller, irrigated and irrigable areas of the Middle East often have a proportionally small demand for drainage, but we shall not try to cover them in detail.

For the rainfed croplands, a different system of land use than now prevails would be more productive. Most rainfed lands are farmed on a crop-fallow rotation, but weeds are allowed to grow almost unchecked in the fallow years and provide a significant amount of low-grade forage for nomadic flocks and herds. The fallow thus accomplishes little in saving moisture for the crop year; the weeds take what otherwise would be saved for the next crop. The fallow may add a little nitrogen, in the form of ammonia washed out of the air, and may provide a little control over crop diseases and pests; but there are far better ways of meeting these objectives today. We propose the following changes in use of rainfed croplands.

a) Where rainfall is reasonably good, abandon fallow; grow a grain or legume or fodder crop every year; practice moisture conservation; add fertilizer; control weeds by cultivation and chemicals.

b) Where rainfall is less but still passable, and soils are too shallow to store a season's rainfall for the next crop, follow same practices as above, but expect lower crop yields.

c) Only under exceptional conditions, where rainfall is seriously deficient *and* soils are deep, practice crop-fallow rotation; use fertilizer as necessary, and control weeds.

d) Where rainfall is too low and/or too variable, or where soils are too thin, abandon annual cultivated crops, seed to perennial grasses, and manage effectively for grazing. The livestock feed thus produced would fully offset, or more, the feed lost by eliminating the weeds in the fallow year on the lands remaining in cereals.

The natural ranges of the region could be made much more productive than they now are (20–24). Reseeding of marginal croplands, noted above, and possibly some reseeding of depleted unplowed grazing lands where soils and moisture conditions are favorable, would add greatly to forage production. Some range improvements, such as water-spreading and livestock water development would also be desirable. By far the most important measure needed, however, is restriction of numbers of livestock and of season grazed. Forage is literally eaten into the ground, in a manner which far exceeds overgrazing on western United States ranges. Vastly more forage would be produced, over a period of years, if less of that which grows were harvested each year; the institutional barriers to improvement will be considered later. Far more livestock output would be produced if the available forage were put through fewer, healthier, better-bred animals.

The forage output of grazing lands varies sharply with the seasons, reflecting the seasonal character of the precipitation. These grazing lands have amazing capacity to withstand severe overuse and to respond when given the opportunity. When Arab-owned herds and flocks were withdrawn in 1948 from what is now northern Israel, in 3 to 5 years without grazing the lands gained two whole range condition classes, and grasses previously believed extinct were flourishing. If modern range management could be applied to extensive range areas, substantial increases in output could be achieved in a few years without reseeding.

Increased usable irrigation water supply would also be helpful, especially from the Tigris and Euphrates *(30–40)*. Storage dams would reduce flood hazard, produce electric power, and provide water for irrigation. One major problem is the lack of treaties specifying the water rights of Turkey, Syria, and Iraq. Total present claims exceed total river flow, but each country may be exaggerating for bargaining purposes. Numerous possible storage sites exist; it is beyond the scope of this chapter to list and analyze them. Development by stages is possible, especially since there are several smaller tributaries to the Tigris river in Iraq. If Turkey develops primary power dams on the upper Euphrates, this may provide streamflow regulation to benefit Syria and Iraq. The usual complexities of multiple purpose development on the two rivers (which join only near their mouths) are exacerbated by the international character of the streams.

With the High Aswan Dam completed, both opportunity and need for development of additional irrigation water from the Nile is very limited. The Jordan and its tributaries, such as the Yarmuk, are also about fully developed or planned for development — perhaps not efficiently so, since sharing by hostile countries has made coordinated river system development impossible. Elsewhere, on the Orontes, Litani, and other smaller streams and springs, some additional water, but not large quantities, might be developed.

POTENTIALS OF IMPROVED FARM PRACTICES

Improved natural resource use gradually merges into utilization of modern farm science and practices. Farm production practices might be improved in numerous

ways; we shall discuss only three major ones: increased use of fertilizer; introduction, development, and use of improved crop varieties; and better control of weeds *(49-60)*.

The use of chemical fertilizers has been growing rapidly in the Middle East since World War II; even if present use is optimum for present farm practices and crop yields, improved farm practices in the future will require much more fertilizer for higher crop yields. Little or no fertilizer is now used on rainfed croplands; if our other proposals for management of these lands are followed, then substantial applications of fertilizer will be required for optimum results. Present crop varieties must be replaced by improved varieties which can efficiently utilize more fertilizer, especially more nitrogen. Irrigated lands will require more nitrogen, phosphate, and potash fertilizer for continued high yields. Our estimates of fertilizer use for the region are shown in Table 1.

TABLE 1

Estimates of Fertilizer Use for the Middle East

| Fertilizer | Estimate (in million short tons) | |
	For Present	For Future (if yield potential is to be realized)
N	0.3	2.1
P_2O_5	0.1	1.4
K_2O	<0.05	1.4

Present crop varieties, especially cereals, do not respond well to increased fertilizer application. Newly developed short-straw varieties (Mexi-pak, for example) of wheat and of other cereals have shown remarkable yields in the Middle East. The same is true of imported rice varieties. Maize varieties can be imported or developed within the region. New and improved varieties of various other crops can be imported or developed. Given favorable soil, water, and fertilizer conditions, new varieties can often produce several times as much per acre as do present varieties under existing conditions.

Many croplands of the Middle East are severely weed-infested. Weeds utilize a large part of the available water and fertilizer, and otherwise interfere with high crop yields. Reduction or eradication of weed infestations may be achieved by cultivation, chemical sprays, or combinations of these practices. The technical problems, while not simple, are capable of solution. Costs which may be excessive with present crop yields are reasonable when crop yields are increased as much as we think possible. Weed control is in part a capital outlay; once weed populations have been reduced to relatively much lower levels than at present, the cost of keeping them in control is less.

The foregoing has sketched the more important measures of improved resource use and of modern farm practice; other complementary measures would be involved, as well as numerous important details. If all of the measures were adopted *as a package,* what would be the effect on crop yields and on total crop and livestock output?

POTENTIAL CROP OUTPUT (61)

Present and potential yields per acre and present and potential total production of a few of the main crops for a few countries are shown in Table 2. Wheat yield in Egypt on irrigated land, about 2.7 metric tons per hectare (40 bushels per acre), could be increased to 6.7 metric tons (100 bushels) on the average. If this seems high, it should be pointed out that wheat yields on irrigated land on the Columbia Basin irrigation project in the United States already average nearly this high; and further increases in the next few years are probable. A similar yield of irrigated wheat in Iraq seems possible; for unirrigated wheat in Syria, on the other hand, the potential yield is less than 3 metric tons per hectare (40 bushels per acre), on the average. Other substantial increases in grain yield are possible; likewise, some additional acreages of grain could be grown. Compared with present total production of grain of about 11 million metric tons, total potential grain production is nearly 50 million metric tons; this is three times present grain consumption in the region, and substantially more than grain consumption will likely be in 25 years or so when the potential output might be achieved, as we discuss later. Thus, it seems unlikely that both grain yields and grain acreage will actually be pushed to this level; some land will likely be diverted to other uses, or a somewhat less intensive production of grain carried out. However, this comparison does show what is possible.

Cotton yields could also be increased materially. Although cotton yield in Egypt now somewhat exceeds a half ton per hectare (one bale per acre), it is double that on comparable irrigated lands in Israel and in California. Cotton yields could also be increased greatly in Syria and Iraq.

For the crops using large areas of land, such as the grains and cotton, potential increases in total output are largely limited by potential increases in yield per acre; some additional land can be cropped as irrigation is extended, as presently fallow lands are cropped, and as other shifts are made, but the extent of such increases in crop area are limited. When vegetables and fruits are considered, large increases in crop yields are also possible, and also large increases in crop acreage are possible by shifting land from the staple crops. For instance, onions are an important crop in Egypt, including for export; crop acreage could be expanded to a point where Egypt would be producing a significant portion of total world production of onions, if market demand permitted. Hence, a consideration of potential production for some crops must consider market possibilities as well as physical potentials. Increases in output of livestock products are also possible.

We estimate that the use of improved resource utilization and of improved agricultural practices could result in an increase in total agricultural output of about 2½ times in perhaps 25 years. This includes the combined effect of higher crop yields, some increased crop acreage, and a considerable increase in livestock production. It is physically possible and, if markets permit, could be economically sound. The extent to which this potential will actually be realized depends primarily upon overcoming institutional and cultural barriers. At best, some years would be required to achieve it; this is why we consider it as a 25-year outlook, even at the best. While

TABLE 2

Present and Potential Yield per Unit of Cropped Area and Total Production, Some Major Middle Eastern Crops

Country and Crop	Yield per Hectare (metric tons)		Yield per Acre			Total Production (1,000 metric tons)	
	Present	Potential	Units	Present	Potential	Present	Potential
Egypt:							
wheat	2.7	6.7	bushels	40	100	1,620	5,400
maize	3.6	6.3	bushels	57	100	2,360	4,400
rice	4.1	8.0	pounds	3,650	7,100	2,000	5,600
Syria:							
wheat	0.9	2.8	bushels	13	42	1,049	5,600
barley	0.9	2.6	bushels	17	48	589	1,300
Iraq:							
wheat							
rainfed	——	2.7	bushels	——	40	——	2,700
irrigated	——	6.7	bushels	——	100	——	13,400
Six countries, all grains	——	——		——	——	10,948	49,592
Egypt, cotton	0.59	1.2	pounds	525	1,070	462	1,200
Syria, cotton	0.5	1.0	pounds	445	890	123	300
Iraq, cotton	0.25	1.0	pounds	222	890	10	100

Source: Calculated from data in (2)

an estimate of this kind may be unreal as a forecast of what is most likely to happen, it does have great utility in suggesting what is wholly possible.

MARKET POSSIBILITIES

The possibilities of marketing additional agricultural output, assuming that it can be achieved, divide into domestic and export considerations *(62-65)*.

On the domestic side, the Middle East as a whole is a region of rather rapid population increase. Food and Agricultural Organization, in its Indicative World Plan, estimates that total population of the region we consider will about double between 1962, the latest year for which data were available when that plan was made up, and 1985; while it does not explicitly consider population change after 1985, there is no reason to expect that population growth will cease then, even if population control methods should be widespread and effective *(66)*. The same plan envisages an increase in per capita income of substantial proportions, although somewhat less than doubling, from 1962 to 1985. Food consumption per capita is now relatively low, more so for the foods of higher cost than for the staple grains. A combination of rising population and rising consumption per capita will greatly increase the total effective demand for food, more so for some foods than for others. Total demand for all cereals will slightly more than double, for vegetables it will go up by 2½ times, for fruits it will increase even more than for vegetables, and still larger increases will occur in total consumption of meat and eggs *(67–74)*.

Moreover, this region as a whole was a net importer of several important foods in 1962, and its import deficit had been growing. This is true for cereals, especially for wheat in Egypt, but for all grains into Israel, and for most grains in nearly all the countries to a lesser degree. The region also imports about 6 percent of its meat supply at present. All of these imports could be replaced by domestic production, if the latter should rise sufficiently; the question as to the extent, if any, that domestic prices should be held above world prices to stimulate such domestic production is another matter.

Thus, on the whole, the domestic market for agricultural production in these countries can absorb a substantial increased agricultural output over the next 25 years. The outlook of agricultural exports is less clear and, on the whole, less bright. Two of the major agricultural exports from the region now are cotton from Egypt and citrus from Israel; there are, of course, other exports from these and other countries, but these two dominate the trade-income statistics. Cotton exports generally, not only from Egypt but also from Syria and Iraq if the productive potential is realized — and, for that matter, from other countries in the world — may encounter increasingly rough weather in the years ahead. For one thing, the production of man-made fibers of various kinds is increasing so fast, and they are so steadily encroaching on cotton, that the outlook for cotton is necessarily pessimistic. In 1968, for instance, cotton lost five percentage points in the total fiber picture — its largest single loss in one year, to date. The picture is further complicated by the fact that Egypt produces a long-staple cotton, whose price margin over ordinary cotton has also shrunk. Increased exports of cotton may be possible, but the trade obstacles are considerable and the prices probably cannot be highly remunerative.

Citrus is exported from Israel in relatively large volume, and from other countries in smaller volume. Its postwar market has been primarily western Europe; its rivals in that market are other Mediterranean countries, as well as others. In recent years, Israel has exported some oranges to the United States, where they compete with oranges and orange juice from California, Texas, and Florida; it seems most unlikely that this is a highly profitable export outlet for Israeli citrus, although it may be true that a given volume of oranges brings more money in the United States than would be realized by adding that supply to present shipments to Europe. If Russia and other eastern European countries would open up to Middle East citrus, output could expand greatly without any adverse effect on prices.

There is much interest in the Middle East and elsewhere in the possibilities of exports from the region of vegetables and fruits out of the main season, and of specialty products such as flowers. Some such exports now take place. But there are serious difficulties in expanding their volume much or rapidly. Egypt has attempted export of winter tomatoes, for instance, to take advantage of its winter sunshine and its adequate water supply and cropland acreage, but the volume of such exports has not risen in recent years — indeed, it has fallen in the past half-dozen years — and is still miniscule in relation to productive capacity. The markets for such products are mostly severely limited, and small additional supplies often result in substantial decreases in price. Successful export of such products requires a closely articulated and highly efficient production, shipment, and marketing organization, geared to meet the demands of the export market. The problems of producing for export are, by and large, more serious than those of producing for the domestic market. One should not dismiss the possibilities of exports too easily, but the best present outlook is that they will never utilize a significant fraction of either the available cropland or of the available water supply, although they might make a useful contribution to the foreign exchange earnings of some countries.

OBSTACLES TO AGRICULTURAL DEVELOPMENT

We come now, finally, to the focal point: What are the social and institutional obstacles to achievement of the agricultural productive potential of the Middle East? Here, as on the technical side, a *package* of improved measures is necessary; it does relatively little good to carry out agricultural research if there is no agricultural extension or other mechanism to carry the results to the farmer, but extension without something to extend is unproductive too, and so on, for many other institutional inputs. In a situation of this kind, it is difficult indeed to say which is most important or which should come first — all are important, and all are needed if each is to be reasonably productive. As a simple expository matter, one must deal with problems and opportunities one at a time. But, with the possible exception of the first topic we discuss, each is equally important. In each case, there is an obstacle or a problem, and its elimination or conquering provides an opportunity.

A basic difficulty is a general lack of stable and efficient government. The countries of the Middle East differ considerably in this respect. Egypt has had a degree of stability in government since its 1952 revolution, in the sense that Nasser was the effective power throughout the period, though there were some changes among his chief lieutenants. But, by most accounts, Egyptian bureaucracy is very inefficient, both in the sense of costing too much in relation to output and in the sense of impeding economic progress. Moreover, Egypt has yet to demonstrate that political power can be transferred peacefully and efficiently. Lebanon has had a precarious balance of power, neatly balancing competing religions and other groups in a reasonably stable and reasonably efficient government. It may not be pure chance that Lebanon has the highest per capita income of all Arab countries, excluding those depending upon oil.

In Jordan, the king has provided a continuing and stabilizing influence. From 1948 to 1967, Jordan made extraordinary economic progress, in spite of several severe handicaps. Although it enjoyed some financial help from other countries, there was nothing like the inflow of funds that Israel and the oil-rich Arab countries had. In 1967, Jordan suffered most severely from the war — loss of territory, and another flood of displaced refugees, crowding into a severely disrupted economy, lacking in easily exploitable natural resources. Saudi Arabia has had its own particular brand of political stability, built in large part upon a powerful monarchy and lots of oil money.

Each of the foregoing countries has had its problems and has been far from perfect, in spite of which each has achieved substantial economic progress. But the situation has been far different in Iraq and Syria. In each, there have been many changes in government, some peaceful, some not, in the 1950s and 1960s. No government seems to have had a firm base of political power. Turnovers have been rapid, and no government could afford to undertake programs with long payoff periods — nor would it have done much good if they had begun something which the next government would feel compelled to change, if only to demonstrate its uniqueness.

Some of the necessary improvements in agriculture require efforts continued over many years, to be really successful. Unless, or until, more stable and more efficient government comes to the Middle East, specifically agricultural programs will be severely hampered.

In looking at the present generally unsatisfactory situation as to government, some historical perspective is necessary (75). Present countries of the Middle East emerged after World War II; in the interwar years, they had been colonies of Britain and France; before World War I they were part of the Ottoman Empire. Egypt and Saudi Arabia have a somewhat different history, but indigenous and reasonably independent government is recent for them too. The Arab culture, with its strong family loyalties and feuds and its lack of nationalism, except in recent years, has constituted an obstacle to stable and efficient national government. Some progress has been made in the 1950s and 1960s; more can be hoped for in the next generation. Will it come fast enough?

The Middle Eastern countries have generally lacked infrastructure. Transportation facilities have been deficient, for example. Many agricultural villages or local areas lacked roads by which inputs could be brought in and surplus agricultural commodities moved out. Even where major arterial highways existed, local roads were often lacking. Transport has been relatively costly; trucks, tractors, and other equipment were imported and hence costly; they were often poorly maintained, owing in large part to a lack of trained mechanics, and this added to their depreciation; spare parts have often been lacking or available only after long delay, resulting in much downtime; and so on. This is a familiar list to anyone with experience in developing countries, but it is nonetheless serious. Some progress has been made in 20 years and more will be made in the next 20 years.

Marketing facilities and organized markets have often been seriously deficient (76–81). Produce often moved slowly from producing area to consumer; storage facilities for nonperishables such as grain were often inadequate or poorly located, with consequent crop loss. Efforts to develop export outlets or to produce new commodities have frequently been unsuccessful for lack of adequate marketing mechanisms. This, too, is a familiar story in developing countries; while some progress has been made in the Middle East in the 1950s and 1960s, almost every agricultural report brings up marketing as one serious problem yet remaining.

Although electricity has been brought to rural areas in the Middle East at a great rate since World War II, many rural areas still lack it. Those who have it often use it primarily for lighting, but wider usage will surely come, if agriculture develops. Battery-operated transistor radios are now found in nearly all agricultural villages, though their ownership is by no means yet universal. Television has begun to make its appearance. In all of these ways, rural areas in the Middle East, as rural areas throughout most of the world, are in rapid change. But the gap between the isolated, custom-bound, poverty-stricken Middle Eastern villager of 20 years ago and the highly modern, urbanized farmer in the more developed countries today was immense. In 20 years it could not be, and has not been, fully bridged. Change in thought-patterns, in attitudes towards specialized knowledge, and in customary action, can take place only with some years of adjustment — sometimes very painful adjustments, resisted by many participants. These general social and intellectual changes will exert a marked influence on agriculture in the Middle East, even when their connection to agriculture is indirect rather than direct.

There has been, and is, a dearth of agricultural knowledge in the Middle East, and of ways to get to the farmer such information as did exist (*82–85*). Some excellent agricultural research has been carried out, but some research has been technically far less competent. In any event, the total research effort has been inadequate to meet the needs of an expanding agriculture. Similar things may be said about the agricultural extension and specialized agricultural college training — some good, some less so, but too little in any case. Many farmers have lacked really good technical advice that they could and would trust.

A lack of adequately trained agricultural specialists is in large part a result of the inadequate agricultural educational system. Some countries have sent young men (and, less frequently, young women) abroad for agricultural training. Even when these young people received suitable graduate training in the foreign country, some did not return and were lost to the region. The shortage of trained people has been most severe at the middle level — tractor management on rangelands would shortly pay off handneers, for instance.

Agricultural change in the Middle East has been hampered by deeply entrenched customs. Nomadic herdsmen have grazed livestock on the natural rangelands of the Middle East for many thousands of years. With limited exceptions in some areas, no individual or no small group owned these ranges, or could fully prevent their use by others; the first men or the first group there with its livestock took all the available feed (*86*). The result has been, naturally, severe overgrazing. The range forage is seasonal. At other seasons many livestock graze on the weeds that are allowed to grow on the fallow lands used for grain in alternate years. These weeds are poor livestock feed, and they rob the grain land of moisture which otherwise might be stored in the soil. Improved livestock management on rangelands would shortly pay off handsomely, but it is very difficult to institute without fundamental changes in land control and use.

Irrigators in the Middle East, like irrigators everywhere in the world, tend to use too much water (*87*). Water is wasted that might be used productively elsewhere, or salts accumulate, or the soil becomes waterlogged, or some of each of these. The situation is exacerbated by the common lack of water charges based on the amount of water the farmer uses. Research, teaching, and demonstration may well improve water use efficiency, but more stringent methods are likely to be necessary.

Individual farmers and farmers generally in the Middle East have typically lacked such production inputs as fertilizer, other chemicals, seeds of improved crop varieties, and the like, or have had to pay high prices for them. Where these inputs have been provided by government agencies, they have frequently been made available too late for really effective use. The most effective and versatile input of all — production credit — has been especially deficient or very costly or both. Annual interest rates of 25 percent and more are not uncommon. The money lender may have had his problems, in terms of costly supervision and high losses, but the overall result has been a severely restricted use of production credit and of the inputs that it could buy.

The Middle East had a highly concentrated land ownership 20 years ago — about 1950. Egypt has completed a truly major and generally satisfactory land reform (*88–90*). The very large estates have been broken up, and the stranglehold their former owners exerted over general government has been broken. Iraq initiated a land reform which has lagged badly (*91, 92*). Considerable areas of land have been taken from large landowners, but distribution of land to small peasants has lagged seriously: large acreages in government hands, unfarmed or farmed under short-term arrangements. Breaking up large irrigated areas in the highly saline irrigated areas has worsened the land-use situation because the soils in small fields dry out less thoroughly when the neighbor's small field is irrigated, and yields have declined in many areas.

The Middle East has generally lacked organizations of the kind which the irrigation and drainage districts have been in the United States. Decisions about when to make water available, on the conditions under which the farmer could get water, on how much water would be provided, on where drains were to be located and how they were to be maintained, and other similar matters have been made by government engineers and administrators, with farmers having virtually no voice in such decisions.

One method of trying to control water use in the Nile Valley has been to construct the irrigation canals so that the water level is frequently lower than the land surface. In some areas water is provided by a full canal for a five-day period, during which time surface-gravity diversions are possible; this is followed by five days of low water, when the farmer must raise the water two feet or more to get it onto his land; and then five days of no water. Although small diesel or electric pumps are now coming into use, in the past the raising of water has been primarily by human and animal power — extraordinarily expensive power, even when the source receives only the barest subsistence food.

The farmers of the Middle East have not been encouraged to participate in the planning and management of either irrigation or drainage systems; doubtless their capacity to do so is limited, but the result has been that the farmer has sought to find ways to counter or offset the administration of the government engineer and administrator, rather than to cooperate with him. A fully modern and progressive agriculture would seek ways to enlist the cooperation of its farmers in irrigation and drainage, and ultimately to turn over to them much of the administration.

The achievement of the agricultural potential we have

outlined will require investment of additional capital, and capital has been scarce in most of the Middle East in the past. Nevertheless, the shortage and cost of capital is only a lesser factor or problem, not a truly major one, in my judgment. First of all, some countries have had ample capital for agriculture; this is surely true of Israel. But Iraq has had a great deal of oil royalties to invest; it has allocated a substantial part of its annual development budget to agriculture, including irrigation and drainage, and for a considerable number of years the expenditures were less than 30 percent of the allocations. The shortage of trained people, lack of managerial experience, and the instability of government have been major causes for this situation. Saudi Arabia has also had ample capital for investment in agriculture; the major difficulty there has been a lack of natural resources readily developable for agriculture and a lack of knowledge about such resources as did exist.

While Egypt, Lebanon, Jordan, and Syria have had neither large oil royalties nor a large inflow of funds from other sources, yet loan funds have been available, at least in some volume. Egypt found the capital to build the High Aswan Dam, for instance. More serious than any lack of capital for investment in agriculture has been the lack of sound plans for its economic use, and, still more seriously, the lack of a comprehensive program for agricultural development in which large capital investment was only one part. There is grave danger today that, if some of these countries somehow obtain large amounts of capital for investment in agriculture, they would use them for spectacular but relatively unproductive projects — and they could well be aided and seduced by foreign consulting and engineering firms. It is so much easier to get a foreign firm to design and build a big dam than it is to provide the package of needed measures for a particular area and, by slow and continued painstaking work over a period of years, achieve the agricultural potential of the area.

The question of priorities and timing in agricultural development naturally arises. With so much needing to be done, more or less all at once, and with a shortage of

trained manpower, limited capital, and a limited overall capacity to mount an agricultural revolution everywhere at once, and yet with the great importance of employing all elements of the package of improved measures more or less at the same time, where does a country turn and where does it start? Faced with this situation, our conclusion is that a country would be well advised to apply a reasonably complete package of improved measures all at once for the development of a limited number of areas, rather than to apply a little of each measure everywhere — or, worst of all, to apply one measure in one locality, and another elsewhere, and so on.

If irrigation, drainage, fertilizer, improved crop varieties, weed control, better roads, better marketing, adequate production credit, and all other parts of the whole package of improved measures would be applied in reasonable proportions in one area, the results should be vastly greater, and more impressive, than if the same efforts were scattered. Likewise in the rainfed areas, if some lands could be cropped annually, with enough fertilizer, best crop varieties, weed control, and so on, while other intervening or adjacent areas were managed for best livestock production, the results should be much greater than from spreading a little of these measures over a much larger area. The total output would be greater from a whole package than from bits and pieces of the package widely scattered, and the psychological or demonstration effect would also be greater. Admittedly, there might be political or other problems arising out of choosing some localities for a full package while denying similar measures to other localities. But stimulation of unbalance, within politically and socially tolerable limits, may be an effective development strategy.

We close, as we began: The physical and economic potentials of agriculture in the Middle East are very great; the obstacles to their achievement are primarily social and institutional. While the obstacles should not be minimized, yet surely their removal lies within the power of the people and the governments of the Middle East.

REFERENCES AND NOTES

1. For present purposes, Middle East includes Egypt, Israel, Lebanon, Jordan, Syria, Iraq, and Saudi Arabia.

2. This chapter draws extensively on *Agricultural Development Potential of the Middle East,* by Marion Clawson, Hans H. Landsberg, and Lyle T. Alexander (N.Y., American Elsevier Co., 1971). The interpretations made herein are the responsibility of the author only. Many persons contributed to the book, and their assistance is gratefully acknowledged. In addition to the sources cited, the authors of the book had access to unpublished reports and other materials which cannot be cited; hence some of the statements herein are not based on any published sources.

3. The map delineations on which this discussion of soils is based were taken from *4.* In addition, maps *5-10* and reports *11-17* were made available.

4. U.S. DEPARTMENT OF AGRICULTURE, SOIL CONSERVATION SERVICE, WORLD SOIL GEOGRAPHY UNIT
    n.d.   1:1,000,000 maps. Unpublished.

5. YAHIA, H. M.
    n.d.   Map showing location of soil surveys in parts of Iraq. Iraq Ministry of Agriculture, Division of Soils and Agricultural Chemistry, Baghdad. Unpublished.

6. RAVIKOVITCH, S.
    n.d.   A soil map of Israel, scale 1:250,000, in two parts. Hebrew University, Rehovoth, Agricultural Experiment Station.

7. RAVIKOVITCH, S.
    n.d.   A salinity map of Israel, scale 1:750,000. Hebrew University. Rehovoth, Agricultural Experiment Station.

8. Working manual for a 1:20,000 scale detailed soil map of Lebanon. The manual and maps are probably inaccessible at this time.

9. Various maps of soils and soil capability for different parts of Lebanon.

10. Six sheets of the 1:20,000 scale detailed soil map for Lebanon *(8)*.

11. An unpublished report, Soils of Lebanon (in French) by Bernard Gèze.

12. BALBA, A. M., AND M. M. EL-GABALY
    1965 Soil and groundwater survey for agricultural purposes in the northwest coast of Egypt. Alexandria University Press.

13. BALL, J.
    1939 Contributions to the geography of Egypt. Government Press, Cairo, Egypt. 308 p.

14. BURINGH, P.
    1960 Soils and soil conditions in Iraq. Ministry of Agriculture, Republic of Iraq, Directorate General of Agricultural Research and Projects, Baghdad. 322 p.

15. FOOD AND AGRICULTURE ORGANIZATION
    1966? High Dam Soil Survey, United Arab Republic, General Report. United Nations Development Program and Food and Agriculture Organization FAO/SF:16:UAR.

16. MOORMANN, F.
    1959 Report to the Government of Jordan on the soils of East Jordan. FAO, Rome, EPTA Report 1132. 73 p., 6 maps.

17. KADDAH, M. T.
    1956 Soil Survey of the Northwest Sinai Project. Institut du Désert, Publication 9.

18. FOOD AND AGRICULTURE ORGANIZATION OF THE UNITED NATIONS
    1966 Indicative world plan for agricultural development, 1965–1985. Near East, Subregional Study 1: Vol 2: explanatory notes and statistical tables, 243 p., 200 tab., Job 50506. IWP-NE/66/Provisional.
    Data on area of arable land, irrigable and rainfed.

19. FAO (Food and Agriculture Organization) is now engaged in a comprehensive analysis of the natural grazing areas of the world; when completed, this study should provide useful data for this as well as other regions.

20. BAUMER, M., AND O. M. HACKETT
    1965 The development of natural resources in Jordan. Nature and Resources 1(3):16-29.

21. HEADY, H. F.
    1963 Report to the Government of Saudi Arabia on grazing resources and problems. FAO, Rome, EPTA Report 1614. 32 p.

22. IBRAHIM, K. M.
    1967 Summary report and bibliography, the pasture, range and fodder crop situation in the Near East. FAO, Rome, PL:PFC/1. 160 p., 1016 refs.

23. KLEMME, M.
    1965 Pasture development and range management, with respect to increasing livestock production: Report to the Government of Saudi Arabia. FAO, Rome, EPTA Report 1993. 31 p.

24. VAN DER VEEN, J. P. H.
    1967 Range management and fodder development. Report to the Government of Syria. FAO, Rome, UNDP Report TA2351. 78 p.

25. These statistics are taken from FAO statistical yearbooks.

26. This discussion is drawn from *27-29.*

27. BACON, L. B., AND OTHERS
    1948 Agricultural geography of Europe and the Near East. U.S. Department of Agriculture, Office of Foreign Agricultural Relations. 67 p. (USDA Miscellaneous Publication 665)

28. FOOD AND AGRICULTURE ORGANIZATION OF THE UNITED NATIONS
    1959 FAO Mediterranean Development Project; the integrated development of Mediterranean agriculture and forestry in relation to economic growth, a study and proposals for action. Rome. 213 p.

29. BRICHAMBAUT, G. P. DE, AND C. C. WALLÉN
    1962 A study of agroclimatology in semi-arid and arid zones of the Near East. Technical Report of FAO/UNESCO/WMO Interagency Project on Agroclimatology. FAO, Rome.

30. There is considerable literature that deals, in one way or another, with these rivers. The various studies of H. S. Hurst on the Nile are particularly useful; see *31–33.* There have been a number of engineering studies, in relatively recent times, which are not published in the ordinary sense of the word, yet copies have circulated to a considerable number of professional persons; see *34–40.*

31. HURST, H. E.
    1952 The Nile. Constable, London. 826 p.

32. HURST, H. E., R. P. BLACK, AND Y. M. SIMAIKA
    1946 The Nile basin. VII: The future conservation of the Nile. Egypt Ministry of Public Works, Physical Department Paper 51. 178 p.

33. HURST, H. E., R. P. BLACK, AND Y. M. SIMAIKA
    1959 The Nile basin. IX: The hydrology of the Blue Nile and Atbara and of the main Nile to Aswan, with some reference to projects. Egypt, Ministry of Public Works, Nile Control Department, Paper 12.

34. HARZA ENGINEERING COMPANY, CHICAGO, AND BINNIE AND PARTNERS, LONDON
    1963 Hydrological survey of Iraq. Iraq Ministry of Agriculture.

35. HATHAWAY, G. A., H. W. ADAMS, AND G. D. CLYDE
    1965 Report on international water problems: Keban Dam — Euphrates River, a report prepared for the International Bank for Reconstruction and Development.

36. INTERNATIONAL BANK FOR RECONSTRUCTION AND DEVELOPMENT
    1968 Irrigation development potentialities dependent upon Litani River flows.

37. KNAPPEN-TIPPETTS-ABBETT-MCCARTHY ENGINEERS
    1952 Report on the development of the Tigris and Euphrates River systems. Iraq Development Board.

38. NETHERLANDS ENGINEERING CONSULTANTS
    1963 Report on the investigation in the Euphrates Project area. Euphrates Project Authority. Syria.
    Also catalogued under Nederlands Advies Bureau voor Ingenieurswerken in net Buitenland.

39. NETHERLANDS ENGINEERING CONSULTANTS, THE HAGUE, AND DAR AL-HANDASAH, BEIRUT
    1967 Jordan Valley Project, agro and socio-economic study. I: Main report, the Hashemite Kingdom of Jordan. Jordan River and Tributaries Regional Corporation.
    See annotation to *38.*

40. MAIN, (CHARLES T.) INC.
    1953 The unified development of the water resources of the Jordan Valley region. Prepared at request of the United Nations, under the direction of TVA. Boston. 78 p.

41. LITTLE, T.
    1965 High Dam at Aswan, the subjugation of the Nile. Methuen, London. 242 p.

42. OWEN, W. F.
    1964 Land and water use in the Egyptian High Dam era. Land Economics 40(3):277–293.

43. DOHERTY, K. B.
    1965 Jordan waters conflict. Carnegie Endowment for International Peace, New York, International Conciliation 553. 66 p.

44. STEVENS, G. G.
    1965 Jordan River partition. Hoover Institution on War, Revolution, and Peace, Stanford University, Hoover Institution Studies 6. 90 p.

45. This point has been stated with unusual clarity and wisdom in:
    HOPPER, W. D.
    1968 Investment in agriculture: the essentials for payoff. *In* Strategy for the conquest of hunger, Proceedings of a Symposium convened by the Rockefeller Foundation, p. 102–113. Rockefeller Foundation, New York.

46. POYCK, A. P. G.
    1962 Farm studies in Iraq, an agro-economic study of the agriculture in the Hilla-Diwaniya area in Iraq. Landbouwhogeschool te Wageningen, Mededelingen 62(1). 99 p.

47. BALLS, W. L.
    1953 The yields of a crop, based on an analysis of cotton-growing by irrigation in Egypt. Spon, London. 144 p.

48. WEBSTER, I. M.
    1967 Pilot project for drainage of irrigated land, United Arab Republic, Final Report. FAO, Rome, UNDP/SF Report 30/UAR. 72 p.
    Fieldwork, October 1961 — November 1964. Apparently not released.

49. There is a very extensive literature which bears, in one way or another, on the agriculture of the Middle East; references *50* through *60* are some that I have found most useful.

50. ABDALLAH, H.
    1965 UAR agriculture. UAR Ministry of Agriculture, Foreign Relations Department, Cairo. 115 p.

51. AGRAWAL, B. L.
    1968 Prospects of foodgrain production in Iraq. UNSF/FAO Institute of Cooperation and Agricultural Extension, Baghdad.

52. ANONYMOUS
    1966 Iraq, country report. FAO, Rome. Animal husbandry, production, and health.

53. FOOD AND AGRICULTURE ORGANIZATION, MEDITERRANEAN DEVELOPMENT PROJECT
    1959 Lebanon, country report. FAO, Rome.

54. FOOD AND AGRICULTURE ORGANIZATION OF THE UNITED NATIONS
    1966 Indicative world plan for agricultural development, 1965–1985. Near East, Subregional Study 1: Vol. 1, text, 184 p., 93 tab., Job 51878. Vol. 2: explanatory notes and statistical tables, 243 p., 200 tab., Job 50506. IWP-NE/66/Provisional.

55. FALK PROJECT FOR ECONOMIC RESEARCH IN ISRAEL
    1964 Long-term projections of supply and demand for agricultural products in Israel, vol 1, by Yair Mundlak. Hebrew University, Jerusalem, Faculty of Agriculture.

56. MOE, L. E.
    1965 Israel: Projections of supply and demand for agricultural products to 1975. U.S. Department of Agriculture, Economic Research Service, ERS-Foreign 137. 41 p.

57. TREAKLE, H. C.
    1965 The agricultural economy of Iraq. U.S. Department of Agriculture, Economic Research Service, ERS-Foreign 125. 74 p.

58. WARREN, C. J.
    1964 The agricultural economy of the United Arab Republic (Egypt). U.S. Department of Agriculture, Economic Research Service, Foreign Agricultural Economic Report 21. 57 p.

59. WEITZ, R., AND A. ROKACH
    1968 Agricultural development: Planning and implementation, an Israeli case study. D. Reidel Publishing Company, Dordrecht, Holland, F. A. Praeger, New York. 405 p.

60. WORZELLA, W. W.
    1968 Cultural studies on yield and quality of field crops in Lebanon. American University of Beirut, Faculty of Agricultural Sciences, Publication 30.

61. This section is based on Clawson, Landsberg, and Alexander (*2*), an appendix of which includes statistical material on crop acreages, crop yields, crop production, livestock numbers and production, total agricultural input and output, and the like, drawn from all available sources, for as many years as data were available.

62. In addition to references previously cited, in particular FAO Indicative World Plan (*15*), *63–65* are directly relevant.

63. PIQUER, G.
    1968 Advisory service on production, marketing and export trade of fruits and vegetables in the Near East and North Africa. FAO Regional Seminar on Marketing Programmes, Procedures and Organizations, Near East Region, Beirut, 4–12 November 1968. 4 p.

64. FOOD AND AGRICULTURE ORGANIZATION OF THE UNITED NATIONS
    1967 Outlook for production and trade of selected horticultural products in Mediterranean countries. Summary of findings and conclusions, by J. Wolf and G. Coda-Nunziante. Conferenza Nazionale per l'Ortoflorofrutticolotura i Commissione di Studio, Sessione di lavoro di Bari, 29–30 settembre 1967. WS/65322.

65. ORGANIZATION FOR ECONOMIC COOPERATION AND DEVELOPMENT
    1968 Tomatoes: Production, consumption, and foreign trade of fruits and vegetables in OECD member countries. Present situation and 1970 projects. Paris.

66. See various sections of *54*.

67. Several recent publications (*68–74*) deal with the economy of Middle Eastern countries, especially Egypt.

68. ALNASRAWI, A.
    1967 Financing economic development in Iraq, the role of oil in Middle Eastern Economy. Frederick A. Praeger, New York. 188 p.

69. CAMPBELL, D. R.
   1967– Jordan: The economics of survival. Interna-
   1968 tional Journal (Canadian Institute of International Affairs) 23(1):109–123.

70. AMERICAN UNIVERSITY OF BEIRUT, ECONOMIC
   RESEARCH INSTITUTE
   1967 A selected and annotated bibliography of economic literature on the Arab countries of the
   Middle East, 1953–1965. Beirut. 458 p.

71. FOOD AND AGRICULTURE ORGANIZATION OF THE
   UNITED NATIONS, NEAR EAST COMMISSION ON
   AGRICULTURAL PLANNING
   1968 Papers presented at the Fourth Session, 23
   March–2 April. Baghdad, Iraq.

72. HANSEN, B., AND G. A. MARZOUK
   1965 Development and economic policy in the UAR
   (Egypt). North-Holland Publishing Company,
   Amsterdam. 333 p.

73. ISSAWI, C. P.
   1963 Egypt in revolution, an economic analysis.
   Oxford University Press, London, New York.
   343 p.

74. MEAD, D. C.
   1967 Growth and structural change in the Egyptian
   economy. Richard D. Irwin, Homewood, Illinois. 414 p.

75. WICKWAR, W. H.
   1962 Modernization of administration in the Near
   East. Khayats, Beirut. 201 p.
   A publication I found particularly informative.

76. Reports *77–81* catch some of the flavor of the agricultural marketing problem in the Middle East.

77. CLARKE, J. G., assisted by C. O. EMMRICH and I. LAHAM
   1967 A report on trial shipments to Aleppo in March
   1967. FAO, Amman, Jordan. UNDP/SF Project JOR/7, Research Report 3. 28 p.

78. EL SHERBINI, A. A.
   1968 Organization and establishment of wholesale
   markets in the Republic of Iraq, Interim Report
   to the Government of Iraq. FAO, Rome. 18 p.

79. FOOD AND AGRICULTURE ORGANIZATION OF THE
   UNITED NATIONS
   1965 FAO Near East Regional Conference on the
   Marketing and Refrigeration of Perishable Pro-
   duce, Beirut, Report. (mimeo).
   Probably circulated little beyond conference itself.

80. RANDHAWA, N. S.
   1967 Production and marketing of vegetables in Syrian Arab Republic. UNSF, Ghab Development
   Project, Damascus, Syrian Arab Republic,
   Land Use and Production Economics Series 8.

81. WOLF, J.
   1965 The citrus economy and the feasibility of international market arrangements. FAO Monthly
   Bulletin of Agricultural Economics and Statistics 14(9):1–15.

82. Reports *83–85* are on this point.

83. AL-DAHIRY, A. W. M.
   1967 The role of Baghdad University in agricultural
   policy. National Seminar on Agrarian Reform
   in Iraq, Baghdad.

84. EL-KOUSSY, A. A. H.
   1966 A survey of educational progress in the Arab
   States 1960–1965. UNESCO Conference of
   Arab Ministers of Education. Published by
   Regional Centre for the Advanced Training of
   Educational Personnel in the Arab States,
   Beirut.
   Probably circulated little beyond conference itself.

85. TAYLOR, D. C.
   1968 Research on agricultural development in
   selected Middle Eastern countries. Agricultural Development Council, New York. 166 p.

86. Where there was control over amount and season of
   range use in one area, this has been abandoned in recent
   years, for ideological reasons, with predictable adverse
   consequences to the range vegetation. See Klemme (*23*).

87. LEUENBERGER, R.
   1963 Report to the Government of Iraq on improvement of irrigation practice. FAO, Rome, EPTA
   Report 1709. 16 p.

88. BAER, G.
   1962 A history of landownership in modern Egypt,
   1800–1950. Oxford University Press, London,
   New York. 252 p.

89. OWEN, W. F.
   1964 Agrarian reform and economic development,
   with special reference to Egypt. Rocky Mountain Social Science Journal 1(2):62–76.

90. SAAB, G. S.
   1967 The Egyptian agrarian reform, 1952–1962.
   Issued under the auspices of The Royal Institute of International Affairs, by Oxford University Press, London, New York. (Middle
   Eastern Monographs, 8) 236 p.

91. LIBYA, NATIONAL AGRICULTURAL SETTLEMENT
   AUTHORITY
   1967 Land policy in the Near East, Proceedings of
   the Development Center on Land Policy and
   Settlement for the Near East, Tripoli, Libya,
   16–28 October 1965. FAO, Rome, 412 p.

92.   1967 National seminar on agrarian reforms in Iraq.
   Baghdad (in Arabic).

Resource-Allocation Problems for the Development of Arid Zones

MANFRED A. MAX-NEEF

Project Manager, Programmes of Modernization of Rural Life in the Andes
United Nations Development Programme

ABSTRACT

THE GREATEST FOOD POTENTIAL for the future will be made available to humanity as a consequence of the development and rational exploitation of the seas and the arid and semiarid lands. If we are now preoccupied with the development of dry lands we are undertaking nothing new. We are in fact only making an effort to recover a consciousness that existed long ago and which our forefathers conspicuously destroyed during colonial times.

The development of arid regions is a huge collective venture, and one of its fundamental features — as opposed to other types of regions — is that no isolated investment project can be contemplated without having to consider immediately all necessary interrelated service projects that are required for the basic "strategic" projects to be able to survive and prosper. Highest priority should be given investments promoting exports or import-substitutes to obtain the most telling impact on the balance of payments. A goal of investment projects should be the maximization of rate of growth over time. Continuity of this rate can be guaranteed only through increasing participation of the people. Investments should, therefore, be selected according to their capacity for generating a symmetric income distribution.

An aspect of great concern in arid-lands development is the requirement of new human settlements which is, in fact, a matter that must be handled with the utmost care. For successful modernization of a region, it is important that already-established groups of people be moved in, to preserve their sense of security. Technology should be adapted to the potential skills of these people, thus reducing training costs and minimizing tensions involved in adjusting to a new environment.

Three possible national consequences might be expected from the development of a desert or semiarid region: the arid zone will tend to attract select human resources with a pioneering spirit; the arid region may develop without the handicap inherent to areas that, though less hostile from the natural point of view, are affected by tradition and inertia; as a consequence of the previous factors, the arid region once in its full development process will most likely generate a high standard of living and per capita income.

RESOURCE-ALLOCATION PROBLEMS
FOR THE DEVELOPMENT OF ARID ZONES

Manfred A. Max-Neef

THIS CHAPTER is divided into three parts and one annex. Those readers interested exclusively in the main subject, that is, dry lands, can do without reading the second part and the annex and yet lose nothing of the central argumentations, since both are of specialized interest to economists, due to their technical implications and vocabulary. The second part is, in fact, a review of criteria proposed by different authors as to resource allocation for development. The annex presents the analytical details, concepts, and premises for the development formula presented in the third part. Whoever is not interested in such details may do well just to accept the formula's validity in good faith. The first part — of general interest — deals with the resource-allocation problem in general and puts the rest of the material into proper perspective. The third part — also of general interest — contains the criterion proposed by the author for the development of arid zones and is, thus, the nucleus of the whole work. As for the Introduction, it is — in the author's opinion — recommended reading, since it lines out for the reader the spirit and motivations of the writer vis-à-vis his product.

At the end a bibliography has been included. An explanation about it is required. The author was unable to find publications that dealt in a specific way with his subject matter. Either the publications on arid lands have no mention of the resource-allocation problems for arid-lands development, or the literature on resource allocation has no concrete references to the very special conditions of desertic regions. Therefore the bibliography is composed mainly of works that deal with resource allocation in general. So errors of this chapter, and any misconceptions, are the author's exclusive responsibility (1). The writings cited in the bibliography are all basic ones, regardless of publication date. It will be seen that the oldest is dated 1928 and the most recent one 1970.

Most of the proposals that appear in this paper are based on the field experience of the author as expert in regional development, first with the Pan American Union and then with the Food and Agriculture Organization of the United Nations.

INTRODUCTION

The greatest food potential for the future will be made available to humanity as a consequence of the development and rational exploitation of the seas and the arid and semiarid lands. Many of today's deserts were flourishing meadows during the life spans of ancient civilizations. A good deal of the Sahara desert fringe was once fertile ground, and so was the Tamarugal desert in northern Chile, where a great number of canchones (2) are still to be seen, only that today they are covered with a dramatic crust of hard and broken earth that gives us, with eloquence, a testimony about centuries of modern indifference. The irrigation systems of the old Incas and Mochicas in Peru and Bolivia are still examples of the solution of highly complicated hydraulic problems. In fact, many of the irrigation schemes now being developed follow the paths and patterns of the old designs which apparently, after many studies, still turn out to be the most appropriate.

It is quite notable that the great ancient civilizations of America and northern Africa — Egyptians, Incas, Mochicas, and Mesoamericans — had two things in common: water consciousness and a capacity to develop the potentials that were hidden in the moistureless grounds of the deserts. Those were civilizations of tremendous cultural achievements, and we are led to believe that a correlation must have existed between the capacity of those groups to provoke ecological changes of transcendence — like irrigating a desert — and their capacity to generate and irradiate a vast cultural influence upon less energetic human aggregates. Whether it was the degree of civilization that made such endeavors possible, or whether it was those undertakings that accounted for the formation of a great civilization, is something we shall probably never know. But there is a probability that it was a case of interacting forces, perhaps an example of Toynbee's concept of environmental challenge. In any case, we know at least one thing: those human groups that were unable to improve their environmental conditions remained culturally behind. The creation of technology is a consequence of a given cultural level, but is also the cause for higher cultural achievements. So, we might affirm that due to the technology created by these civilizations, and due to the successful application of it, they were able to reach their well-known degree of cultural influence and historical immortality.

While ancients were faced with tremendous environmental challenges, which, in the case of the mentioned groups, they were able to solve with amazing ability, later and more modern groups — like immigrants during and after the colonization period of America — settled in the midst of more pleasant and less challenging natural environments. Agriculture became mainly a rainfed activity, to the degree that still today Latin American countries in general are affected by an "irrigation fear." Thus agriculture developed in those places where the supply of water represented no basic problem. After all, immigrants were coming from a Mediterranean Europe where deserts were the image of something to run away from. So it happened that after four centuries of European colonization and following the forced integration of Indians as labor in mining and Mediterranean-type agriculture, all deserts that were once irrigated ended up drying out completely again. It is as the jurist and internationalist from Bolivia, Carlos Salamanca, has said in his Mission Reports to the United Nations about the possibilities and potentials of the Cuenca del Plata: "Only those who lack water are able to develop a water consciousness; those who have and had it in abundance have never been capable of grasping its true importance."

The first thing we must, thus, recognize is that if we are preoccupied with the development of dry lands, we are undertaking nothing new. We are, in fact, only making an effort to recover a consciousness that existed long ago and which our forefathers conspicuously destroyed during colonial times.

Transformation of desertic areas into agriculturally productive fields is always a huge collective venture. It is an endeavor of tremendous proportions, the success of which is strictly dependent on the degree of human solidarity that those who engage in such an enterprise are able to develop. Group conflicts and wars between nations, when it comes to the physical conquest of a desert territory for the satisfaction of power games, disappear inevitably after wisdom is recovered and the felt need arises to develop such a territory simply for the benefit of human beings. People working in a desert, for its development, have no choice other than to accentuate their own most positive human qualities. No other natural environment — not even the jungle — provokes in human beings the same necessity to the same degree. There is something unique about the way a desert affects man's soul and sentiments. Therefore, the qualitative conquest of a desert is a feat of solidarity. Efforts geared toward such a conquest not only help solve hunger problems but contribute, at least in part, to the slow formation of a more humane humanity.

The reader may, by now, be under the impression that my motivation to justify and persuade about the need for the development of deserts is somewhat idealistic and emotional. This I cannot deny, because early in my young adulthood, while living in northern Chile, I started a lifelong love affair with such dramatic and overwhelming lands. It was there that through watching and learning from the desert people, I began building up all that exists within myself in terms of human solidarity. There is no doubt that only those who have lived long enough in a desert; having seen the red moon; having suffered the frightful impact of the howl of absolute silence; having seen life emerging from death; having felt the unending and eternal vastness of solitude; having experienced the drama of human, animal, and plant survival; having witnessed the foolish yet glorious and disturbing heroism of the lonely miner wandering all of his life in pursuit of the impossible, with just a dream and his shadow — the

blackest of all shadows — as only companions; having learned, in such a manner, the true meaning of manhood; and finally having understood that it is that desert where man can best discover and find his inner self; such persons will be able to understand my thoughts.

But let me insist on historical examples, beyond my own emotional outbursts. Western historians — save some distinguished exceptions — are often tempted to present us with supposedly successful hegemonies such as the Pax Romana, the Pax Britannica and — if they are capable of overcoming the traditional intolerance implicit in the "missionary complex" of Christian culture — they may with some reluctance add the Pax Sinica to their list. Yet the most perfect of them all, the hegemony of the Inca Empire, has only very discreetly been advertised. After all, its destruction and annihilation represent one of the deep and inner shames of our "glorious" western "civilization." The Pax Incaica — I would like to call it thus — was consolidated more as a consequence of wide-ranging education and planning policies than through the institutionalization of power based on force.

It is well known that the leaders of newly integrated cultural groups, during the process of the Empire's expansion, were often sent for training and education to the School of the Curacas (Philosophers) mainly in Cuzco, before being reinstalled as political rulers of their own people. Such a policy, together with the notorious sense for planning that was intrinsic to the Inca nature as shown by the invention of dehydrated foods that could be stored without risk of deterioration for long periods, the establishment of extensive storage networks, the use of a highly efficient system of accounting and control, the well-decentralized political-administrative organization which left practically no citizen without a clear notion as to his rights and duties and the three commandments (not ten) that were the foundation of the juridical system: "Thou shalt not steal; Thou shalt not lie; Thou shalt not be lazy"; allowed the Incas to be able to exhibit for posterity one of the greatest human achievements the establishment of a society from which hunger was banished.

However, due to the ecological characteristics of the environment in the midst of which the Empire took its seat, none of the listed features would have accounted for a successful social organization had it not been for the additional capability of such a people to transform, through irrigation, the hostility of the deserts into a sort of bridge between separated cultural aggregates destined to become one great civilization. That, I insist, could only be achieved by people who had the group cohesion that solidarity alone can generate. Such facts confirm the opinion of the great German geographer Ratzel, who wrote: "Everything was adverse to the Inca Empire, except man." The Incas were well aware that social cohesion, cultural identity, and political hegemony are possible only if no empty and undeveloped spaces are left to separate one human community from another. Until we Latin-Americans recover that part of Inca wisdom —

the harm that empty spaces do to human feelings and progress — the much-talked-about regional integration shall remain as a vain dream or as a favorite subject for after-banquet discourses.

In present times we can mention the example of Egypt. A nation full of internal conflicts has solved a great number of them and has achieved a high level of internal peace after undertaking constructive enterprises of the magnitude of the Aswan Dam. The war of the Chaco between Bolivia and Paraguay was a territorial conflict about a good deal of desert land. The war was fought in 1932. Today and in the near future, that same land will become part of the greatest regional development program of modern history — the Cuenca del Plata Programme — but only due to the solidarity that is emerging now between those who killed each other four decades ago. Deserts can guarantee a horrible death to those who enter with hatreds, but they reserve the most generous prizes to those who enter with a spirit of sacrifice and human solidarity. Therefore, when I am asked how to educate people for desert development, I can only answer: "Let them become good technicians; as to the rest, it is the desert that will teach them how to become good men, capable of applying technical knowledge in a humane manner."

The rest of this chapter will deal with the more technical thoughts and principles that are necessary for the solution of problems of arid-zone development. I shall refer basically to investment criteria for regional development in general, a revision of different criteria according to several authors, and a proposed set of criteria for arid lands. In addition I will discuss types of training and education suitable for the purpose.

I. RESOURCE ALLOCATION — THE GENERAL PROBLEM

Resource allocation criteria are indispensable as tools for the planning of development, whether national, sectoral, or regional. Everyone would most certainly agree with such a statement, but unfortunately that is as far as any general agreement on the matter will go. Resource allocation, or, in a more limited sense, investment, "is a special case of the general problem of choice" (*3*, p. 366). As such, its solution requires the existence of selective norms, and it is precisely in the formulation of them that a wide-ranging literature of disagreements has already been generated (see *4, 5*).

The common aim of all those participating in the discussion, whether at the academic level or at the level of the planner, is to be able — through the application of certain selective norms — to choose only those investments that will produce an optimal impact on the development process. If such is the aim, the grounds for disagreement should be obvious to anyone. In the first place too many opinions still exist as to what a development process is really. Furthermore the whole concept of "optimum" still remains surrounded by an almost impenetrable cloud composed of generally misconceived

pieces of ethics, dogmas, and even superstitions. So, if no accord exists as to what development is, no unanimity could ever be expected as to the validity of any chosen selective norms. If the recognition of the validity of the norms is absent, there will then, finally, be no ground for agreement as to what an "optimal" impact would be.

We must realize our confrontation with a problem the solution of which will never be based upon general acceptance. We must recognize also that reasoning, in this case, must follow the lines of inference. This means both the unpleasant task of giving definitions, in order to overcome a semantic misunderstanding, and the acceptance of the fact that some value judgments are not only inevitable but must be accepted as valid premises without which no reasoning by inference could take place.

For definitions I shall constrain myself to only three. In the first place, I would like *development* to be understood here as "a process composed of economic growth and structural changes, meaning by the latter all and any changes that will facilitate the process to become continuous and self-sustaining, but in such a way that there be constantly increasing participation of both existing and potential resources in it, as a consequence of a widening spectrum of alternatives — cultural, social and economic — from which human beings can draw for the satisfaction of their needs." For *investment* should be understood — in the more general sense — any allocation of resources destined to produce an impact — positive or negative — on the growth of the product. We shall, however, herein confine the use of the term investment — unless otherwise specified — to capital investment. Finally, a set of *optimal investment criteria* will be one that permits allocation of resources to promote effective and positive results both for the growth of the product — either national or regional — and for the increase in the participation of existing and of new sources of economic and social vitality, as implicit in the definition given for development.

Value judgments are clearly present in the definitions, but they are inevitable. To consider structural changes as part of development is a value judgment, despite all historical and sociological evidence that may be shown to support it; so is the indication that an always increasing proportion of existing and potential resources — especially human — should be put into action if the process is to become self-sustaining. I, thus, propose that although my premises are deprived of the privilege of irrefutability, they be accepted in their role of starting propositions for further inference. I shall, therefore, not discuss, nor shall I advise discussion, as to the merit of the value judgments implicit in my propositions. I propose, in turn, that the consistency of the criteria I intend to present to the reader be evaluated — as is often necessary in scientific inquiries — within the framework of limitations established by its own premises. I must make it clear, though, that I am not offering a generalization that would give way to an unjustifiable immunity for all and any premises. That would, in fact, be an alibi for an eventual glorification of nonsense. I am bestowing protection — with all due intellectual humility — only on premises which, according to my experience and common sense, represent necessary value judgments.

If, for example, for the establishment of an investment criterion in a desert zone, one premise is the existence of perfect competition, a situation in which factor prices will equal marginal product, that premise can legitimately be attacked. If the premise is that investments be rated as to their ability to solve human problems, or to promote a more symmetric income distribution, for it is considered desirable that more people participate in the development process and effort, then this premise cannot be attacked on scientific grounds for it does not obey a scientific choice alone but an ethical one as well. Discussion as to the validity of the proposition will be of doubtful fruitfulness. One side might argue that under the latter conditions the impact of investments on the growth of the product will not be maximal and that, since maximization should be granted first priority, preference ought to be given to those investments that maximize individual profit. The other side might argue that to maximize the immediate impact on growth of the product is not only senseless but harmful, if the price is an increasing tendency for the income distribution to polarize and to leave relatively poorer areas to remain undeveloped, and furthermore that what is desirable is not to maximize the immediate impact on the growth of the product but to assure its maximum *rate of growth* within a relative equilibrium between all regions. The discussion could go on and on. However, the value of a final decision will be a matter of technical considerations and common sense. One thing is true, though. Were we to follow orthodox investment and economic-growth rules, no justification could ever be found for the *development* of a desert, except in a country that is all desert, thus not offering any other choice.

The degree of sophistication that has been achieved so far in the advancement of investment criteria for the development of scarcely developed areas is such that a great deal of prior development seems to be required in order to put the criteria into practice. There are some honorable exceptions; this I cannot deny. But at any rate, I believe that just as water is to be added to a moistureless ground, or spice is to be added to a flat taste, common sense should be added to an excess of professional sophistication, because "it would be a mistake to identify complexity with completeness and sophistication with wisdom" (*6*, p. 3).

It is often believed that once an investment criterion has been established, it should be applied for all investment decisions in the country, regardless of the type of regions — arid or not — where the decisions are to be carried out. In fact, the criterion itself — the argument will continue — is not only a way for determining in *what* to invest, but also *where* to invest. This is, or at least should be, true as long as no factors intervene in the decision-making process other than the principles outlined by the general planning policy. This is, however, not the case. The planner has no easy job, for he must

fight not only to demonstrate the importance of his work, but must also defend the country's future from the nonsenses of the politician, who is a man, in turn, without an easy job, because he not only has to fight to demonstrate that the votes he got were well deserved, but must also defend the country's future from the nonsenses of the planner. It is under these prevailing emotional conditions that development must be promoted. It is also under these conditions that, after all the necessary dueling, regional investment priorities are normally established.

What region is to be developed in a country will normally not be a decision resulting from an exclusively rational choice between technically tabulated alternatives (fortunately). It will rather be the result of intuition and political considerations with a degree of technical backing. This not because of simple capriciousness, but due to the understandable and legitimate impulse governments of developing nations have to engage in experimentation. The expert must recognize this, respect it, and live with it. Technical considerations will, no doubt, be given their true importance for the development of an already chosen region. It is, thus, the intention of my essay simply to point out the basic deliberations one has to go through in order to select a set of adequate resource allocation criteria for arid zones, once the development of such a zone has been decided upon by decision-makers, whoever they may be.

In order to put all that has been said so far into one all-encompassing conceptual framework, what is really intended in this written enterprise is to offer — partially and with obvious limitations — a *planning methodology* for arid lands. In fact a set of resource allocation criteria is paramount within a general or integral planning effort. No planning can fructify in the absence of such basic criteria, and no such criteria, in turn, make any sense when isolated from the corresponding planning policy.

Natural and spontaneous human behavior follows an inductive tendency. Experiences in and about the particular lead to generalizations. Planning, on the contrary, is characterized by a deductive process. Generalizations (general goals) are intended, and particular forms of behavior are then induced (or imposed) to achieve the aims desired. The sociological premise on which the justification of planning is based is that unrestricted spontaneous human economic behavior — in my opinion only under the presently main dominating sociopolitical systems — tends to generate social and economic distortions detrimental to the maintenance of a social system that intends to advertise equal opportunities for all, or at least for the majority.

This is, no doubt, true as far as the main reigning systems of our world today are concerned, which are nothing but "variations" on the themes of centralized power such as capitalism and socialism or communism, all of which are, after all, close kin in their essence, though official enemies in their form. In other words, given the dominant socioeconomic systems of our present world — with all their interpretational differences originated as a consequence of diverse cultural and historical backgrounds of the interpreters — human economic behavior in organized civilization has an egotistical tendency that is motivated more by felt needs of self-protection than by the unperceived convenience of a more generalized human cooperation and welfare. So, as a sensible palliative, people recur to planning. I am, thus, advancing the notion that planning — *as applied with all its formal variations, in our present world* — represents simply a rational effort to avoid increasing economic imbalances and social injustices, through the promotion of individual and/or group incentives and restrictions, even granting and allowing for some so-called structural changes, while maintaining by all other possible means the foundations of the reigning sociopolitical and economic system.

Development can be given a push through the use of certain techniques implicit in *planning* and *programming.* Both refer, in a general theoretical sense, to "deliberate and rational efforts that tend to accelerate and orient the development process through a complete and detailed selection of objectives, as well as through the determination and allocation of those resources and means necessary for the achievement of the objectives." All objectives — I strongly believe — must, in the end, be geared toward the satisfaction of human needs. Man is, and must continue being, the beginning and the end of all and every developmental effort of a rational type. But let me make it very clear that planning in itself, and as such, does not promote development automatically. Planning and programming are *tools,* and tools can be used for many purposes. They can with equal efficiency be used for the advancement of human aggregates or for their enslavement. Development must, therefore, be a high-level decision that must *precede* the use of planning and programming techniques. Planning that does not follow a developmental decision will be nothing but a tool working in a vacuum, thus lacking wide-ranging effects. Although this may sound as a subtlety, it is of paramount importance, since it implies that planners require first of all a clear knowledge as to the true intentions of the decision-makers in the field of development and social change.

Planning, as practiced today, is not intended as a method for changing the existing fundamental values of any system within which it has been allowed as a working technique. It has, on the contrary, been applied — by the champions of any existing order — only as a palliative against the "accepted" imperfections of the particular system, and in proper congruence with its "officially" perceived environmental defects. Planning, as already stated, is nothing but a tool that follows the general principles and rules of a deductive process. It can, thus, be adapted to *any system,* since all systems are composed of a set of fundamental premises (values); and this must be understood as a realistic statement of the utmost importance. Planning is deprived of any preconceived sign of its own since its sign will always result from the way in which it is used and applied. Planning will normally result in improvements *within* a system (any system), but it can hardly be expected to lead passage from one social system to another. The rational elimination of a system (revolution) or its natural decay (evolution), correspond to the sphere — and are as well the consequence — of political action engraved into historical tendencies and perspectives. Planning comes afterward, as a means for the establishment of more efficient ways of socioeconomic interaction and group behavior.

In this chapter I am dealing only with one aspect intrinsic to planning: resource allocation. A presentation about integral planning techniques and methods for arid zone development would merit at least a book. So, despite my natural temptation as a planner, I shall from now on constrain my comments to only those chosen aspects of the general problem. This is why I have considered it necessary to make, at least, some gross general remarks about a framework that will not be commented on in detail, but into which the rest of my proposals are to be fitted. I shall neither describe nor defend my preferences as to what system should command the social organization within which the development of deserts or other arid or semiarid regions will be granted due priority. Some of my proposals will appear quite unorthodox, so I leave it best to the reader's imagination to detect my position if he finds it at all of any interest. Yet I should make it very clear that the system that might be perceived as implicit in my suggestions does not completely satisfy my true ideals and expectations either. I have, in fact, made an effort to adjust my proposals — though allowing myself a minimal personal degree of ideological and ontological hope, ambition, and satisfaction — to the factual possibilities dominating the parts of the present world having dry lands.

Desert development is an undertaking of such a unique nature that it can give origin to the formation of socially innovating personalities, politically well qualified to promote deep ulterior changes — even of system — of vast societal effects. Although the sway desert development exerts on the people engaged in it was only superficially treated in the Introduction, when referring to some ancient civilizations, these psychological impacts and influences, as well as characterological mutations, would well merit analysis by a specialist.

II. A REVIEW OF DIFFERENT CRITERIA

A century and a half ago, Jean-Baptiste Say formulated the law that made him dominate a great deal of the unfolding classical economic thought, almost until the great depression of the early 1930s. He stated: "Supply (production) creates its own Demand." His conception, as to the effects production has on the size of the market, remained unchallenged and "was not given serious theoretical consideration until Young revived it" (7). Ever since Young's paper was published, an increasing number of scholars have realized the importance of the subject, especially inasmuch as its understanding was perceived as fundamental to the comprehension of the mechanisms of economic change and, eventually, development. "The emphasis on the interdependence between size of the total production and the size of the market (in an inclusive sense), and on the idea of balance, has been developed more fully in recent years under the name of balanced growth" (5).

Young implied in his writing that ". . . an increase in the supply of one commodity is an increase in the demand for other commodities and . . . [that] every increase in demand will evoke an increase in supply." But even he realized that the imperfections in the market system make the working of his proposed automatism difficult. The many market defects are not taken into account by the orthodox theory of resource allocation.

The orthodox theory, which assumes perfect competition and perfect mobility of capital and labor, thus making optimal factor combinations possible, prescribes the selection of investments according to their ability to maximize individual profit. The underlying assumption is that these types of investment will maximize both the welfare of the individual and that of the society as a whole. This may well be so *if* perfect competition is the case, since the perfect factor mobility that follows makes the social-opportunity cost of a good or service equal to its market price. Otto Eckstein, using some additional aspects, makes a similar statement. He is of the opinion that "under conditions of perfect competition and in the absence of external economies, each enterprise will undertake all those investments for which the internal rate of return exceeds the interest rate. Every factor will be hired up to the point where the value of its marginal product is equal to its market price" (8). The final conclusion within this framework of classical thought is that without external economies or diseconomies — they could not exist under the described conditions by definition (9, see also 10) — the market price of factors, as stated above, will equal its social opportunity cost, and this being the case, will make private profit to be equal, in turn, to social profit. The whole classical approach is characterized by its great ingenuity, and it allows for extremely elegant demonstrations. This is probably its greatest intrinsic danger, since elegance makes for the delight of students and professionals without field experience, with the result that a particular model turns into a sort of article of faith.

The conditions under which the orthodox theory is supposed to perform certainly do not exist in reality. The market imperfections are just too many to grant the principle of *ceteris paribus* — in this case — any real validity. To begin with, the mobility of factors, especially under conditions of development, seems to work in a proportion inverse to its availability. We thus find that the mobility of capital (scarce resource in a developing country) is by far greater than that of labor (normally abundant resource in those countries). From this fact we can infer that the difference between social opportunity cost and market price will tend to be greater in the case of labor than of capital allocation. These differences tend to generate external economies, thus creating an additional problem for the planners, who will necessarily be confronted with the need not only of estimating their relative importance, but also of orienting investments into interrelated projects, whenever these can adequately profit from the external economies generated by the original project or set of projects. The orthodox approach is an impractical one for planning purposes.

Hollis Chenery (11) is one of the few — as is correctly pointed out by Michael Belshaw (12) — "to take marginal theory seriously as a useful device for dynamic capital allocation," when he writes that "economic theory tells us that an efficient allocation of investment resources is achieved by equating the social marginal productivity of capital in its various uses." For Chenery any project will affect the national income, the balance of payments, and the distribution of income. These effects, he states, "must be reduced to a common measure . . . through a welfare function containing an indefinite number of variables characteristic of a particular investment" (11).

Chenery establishes the concept of Social Marginal Productivity and proposes that "the formulation of an optimum investment program does not require an accurate measurement of the marginal productivity of each investment. It is sufficient to rank the projects in order of their social value, determine the marginal project from the total funds available, and exclude all lower ranking projects" (11). But here, by ranking projects in order of their social value, one would conclude that he is really not making use of marginal analysis. As Belshaw says: 'it would seem instead that Chenery is presenting us with a rather sophisticated form of capital/output ratio' (12). The social value of the project will be higher, the higher the proportion of the output to the initial investment. This in itself should, obviously, not be considered as sufficient evidence for determining the social value of a project. This is an important criticism to the approach offered by Chenery, although some of his statements — like the impact of investments — make a lot of sense and are truly practical. Another big problem for the application of the marginal theory of resource allocation is the extremely unrealistic assumption that projects are independent. Project interrelationship happens to be precisely one of the most crucial aspects in the whole decision-making process implicit in any investment policy.

Leibenstein and Galenson suggest several corrections of Chenery's formula, by considering the effects of external economies and the inclusion of "extra social overhead" required to maintain the labor force, such as schools, hospitals, and housing. They propose also to include all the extra costs of raising nutritional standards to the level required by the aims of the project. Other corrections proposed by the authors relate to the interdependence of projects (13).

The model implicit in Galenson and Leibenstein's paper is well summarized by Eckstein (8). It goes basically along the following line: (a) There are two types of income in the economy: wages and profits. (b) Wage earners save nothing, while profit recipients make their entire income available for investment. (*c*) One production function, which makes output per worker purely a function of the amount of capital per worker, applies throughout the economy. In Eckstein's own terms "it can then be concluded that maximization of per capita output at a specific future date requires that the amount of capital then available be maximized; this requires that the amount of investment in each preceding period be maximized, which, in turn, requires that profits be maximized and wages minimized." This set of criteria favors those projects having the greatest capital intensity, that is, projects where the capital/labor ratio is the highest. In Galenson and Leibenstein's opinion, this should be the policy even in countries where there is an excess in the labor supply. The underlying assumption is that if efforts are concentrated in capital-intensive investments, the rate of savings will increase more than with labor-intensive investments, thus creating a multiplier effect that will help to generate new employment.

One criticism of Galenson and Leibenstein's approach is that, considering that any project has significant effects on the distribution of income, it may become necessary to create special fiscal measures in order to maintain a satisfactory distribution. It can easily occur that fiscal means become ineffective to cope with the income-polarization problem, and, since the "reinvestment coefficient" proposed by the authors favors high profits and low wages, the final result will be — in Eckstein's terms — "at odds with most standards of equity." The greatest risk involved, therefore, in Galenson and Leibenstein's set of criteria is the natural tendency the method shows to polarize income distribution. Furthermore, even if through fiscal means income distribution could be maintained in a satisfactory manner, it is still doubtful whether a government should be given a contradictory role, in which on the one hand it stimulates investments to take place under conditions that maximize profits and minimize wages and on the other adopts fiscal measures that tend to correct the imbalances that come about as a consequence of the investment policy it has favored. I do not believe that governments could nor should be sold ideas that appear to be evidently inconsistent.

So far I have mentioned, apart from the classical approach, two sets of criteria, both of which try to define a measure of "social productivity." There are others along this line, such as those proposed by Kahn (*14*), Tinbergen (*15*), and the Sub-Committee on Benefit Cost Analysis (*16*). The objective of all these is to maximize social welfare, but there is, however, no unanimity as to what the concept really means.

There are other sets of criteria, some less sophisticated than those already analyzed. Such would be the case of the position adopted by Polak and Buchanan (*17*, see also *18*), for whom the investment criterion is to select projects the type and size of which are such that no major imbalance will occur in the balance of payments.

Another group of opinions is the one that includes the United Nations Economic Commission for Latin America (ECLA) and the Russian planning system as described by Grossman. Both take a "general equilibrium" view of the problem. Grossman points out that "under socialism the law of planning replaces the anarchic law of market value and private profitability" (*19*). He quotes a statement of the socialist criterion that runs as follows: "The rate of development of one or another branch of the national economy is determined in the USSR not by the profitability which it yields, but by the national economic need it satisfies and by its role in the resolution of the political and economic tasks of socialist reconstruction."

Finally there is the group represented by Hagen and Hirschman, who take an interdisciplinary approach. Hagen (*20*) considers, apart from the strictly economic aspects, the importance of factors such as social acceptability of the program, development of resistances, and the effect of alternative programs on the socioemotional stability of the people. He considers the fact that human adjustment to new values and a new environment generates considerable psychological pressures, thus giving origin to insecurity, restlessness, and dissatisfaction.

Hirschman demonstrates himself basically skeptical as to the contributions the economist can make for the solution of the problem. He stresses that "the economist does not *invent* an investment pattern [inasmuch as the] principal areas of needed development are usually fairly obvious." He also states that there is no way of asserting that "the last million pesos of scheduled expenditure on education will have approximately the same impact on the growth of the national income over the years to come as the last million pesos to be spent on transportation." He concludes that "if there is at all a reasoned decision . . . it will have to be based on a fundamental conviction as to whether the change of environment intended to change the people should take precedence over attempts to change the people directly — a question to which economists can supply no answer" (*21*). Hirschman also criticizes strongly the practice of attacking a particular use of resources in terms of some unidentified superior alternative uses. Even if "the economist is convinced that [a] particular project . . . does not have the highest priority in terms of the economic development of the country, unless he can say in detail what the alternatives are, unless he can produce projects in a state of readiness to be undertaken similar to the one he attacks, he will not and should not be listened to" (quoted from *5*).

Hirschman is absolutely right in his arguments, especially in the last one. Frequently in developing countries, projects in readiness are indefinitely postponed on the advice of consultants of "great reputation" to the effect that "more important projects" must be taken care of. Often there is not even consensus among the experts involved as to what those other "important projects" are. Here the advice of Galbraith should be remembered: "A bad decision made on time will not usually be as costly as a good decision made too late" (*6*, p. 64).

III. A PROPOSED SET OF CRITERIA
FOR ARID ZONES

Let us first of all assume that we are dealing with an arid or semiarid region with a low population density — in other words, a zone for the development of which the allocation of human resources is *conditio sine qua non*. We shall, furthermore, assume that, according to the region's potentials, a number of activities could be developed, thus confronting us with a problem of choice between alternatives. Finally, we will assume that the resources necessary for carrying out an investment program are limited. Some of them are limited because of natural reasons and others because of the limitations of interrelated resources that would otherwise help to increase their individual availability. The latter would be the case of water in a desert. Perhaps not more than a limited quantity of water can — through a plan in its initial steps — be made available, because increase of its supply beyond a certain limit would have to be accomplished through further capital allocation for that specific purpose; capital being a scarce resource, its hyperproportional use in water development alone might leave an insufficient amount to be allocated to the basic projects for which that water is ultimately going to be used.

Interrelation of Investments: First Step

A fundamental characteristic of an arid zone is that *no isolated investment project can be contemplated without having to consider immediately all the necessary interrelated service projects that are required for the basic or "strategic" project to be able to survive and prosper.* This fact makes the arid zone one that can only be treated within the framework of an *integral program.* A consequence is that very often the service projects that are an indispensable complement to the "strategic projects" are considerably larger in size and cost than the latter to be served by them (*22*). Moreover, service projects have an additional peculiarity in that they are limited by problems of scale, that is, limited minimal sizes. A power plant to serve a specific regional industry must have a minimal size, regardless of whether the whole power capacity it can generate is used or a percentage of it is wasted. The same occurs with dams, the sizes of which depend partly on topography, and with infrastructure investments. A given economic activity requires a road of communication in order to survive. That road is limited by a minimal size no matter how intensively it is initially used. The analysis of the interrelationships between strategic projects and their service projects, plus the limitations of minimal size of the latter, belongs then to the very first fundamental aspect to be taken into consideration for the development program of an arid zone.

If the excess capacity generated by the service projects makes them operate under unprofitable conditions, additional investments that would use part of that surplus should be considered, as long as there exist also other *raisons d'etre* for them to justify the additional capital allocation. If increasing capital allocations are made only

in order to transform a service project that yields no direct profit into one that does, a big mistake will be made, since the cost of that capital, in terms of an alternative investment the society will now have to do without, may be very high. It should be pointed out that the construction of a service project — let us say a dam — during a development process, and especially in the case of a region like a desert, should be justified not in terms of its ability to generate direct profit but in terms of its capacity to satisfy the economic and social needs of the region as a whole and of the specific strategic projects it serves, say, new agricultural activities. That is the reason such projects are established — as *services*. This is of fundamental importance, since the time span between the creation of a service project and the full use of its output capacity in terms of inputs of strategic projects will probably be much longer in an arid zone than in metropolitan centers or other relatively more advanced and densely populated regions. It is, therefore, important not to base investments in service projects (in an arid zone of the type described) according to a strict cost/benefit analysis; in so doing, the investment program as a whole — including strategic projects which do yield direct profit after maturity — may be discouraged. Undeniably it would be agreeable if service projects were profitable in themselves from the beginning, but if such is not the case, as it seldom is, this should not worry planners and policy-makers to the extent that, in order to overcome a situation misperceived as undesirable, more important overall tasks of an integral nature are postponed or forgotten. After all, what should be understood by profit? Must it be only economic or financial profit? Can there be no economic loss in one particular service unit that, nevertheless, yields — as part of a whole — social profit?

The problem of interrelationships not only affects strategic projects and their service projects but also appears in a more subtle way when so-called external economies are generated as a consequence of an investment program. The concept of external economies implies that the output of an economic unit is not only a function of its own inputs but also of the factor input that takes place in other economic units as well. It would be a function of the following type:

$$O_1 = F(T_1, L_1, K_1, C_1; T_2, L_2, K_2, C_2, \ldots T_n, L_n, K_n, C_n),$$

in which T represents technological inputs; L, labor inputs; K, land and natural resources input; and C, capital inputs. The output of the economic unit O_1 is determined not only by its own inputs, but by the inputs of the units $O_2, O_3, \ldots O_n$. The external economies are represented after the semicolon in the function.

In the terms of Kumagai (*9*), the concept implies: (*a*) the existence of some economic effects exerted by activities of any economic unit upon those of another, and (*b*) the alleged existence of institutional or technical defects inherent to the pricing mechanism which impede payments to be made for those economic effects, positive or negative. These external economies, which express nothing else but the degree of interrelationship and interdependence that exists between different investments, are to be found in any economy, although their relative importance will be considerably higher the less mature and diversified the regional economy is. In a highly industrialized environment, although the interdependence still persists, its effects tend to be minimal. The same is the case, the closer the system is to one of perfect competition. Under such situations — as Kumagai points out — "the position of a single firm must be like that of a drop in the ocean" (9).

The external economies as described by the above function should be identified as "technological external economies," for they affect the production function of a given economic unit. It has been pointed out in the literature that, although important, these external economies are difficult to detect and measure. It is even difficult to give adequate examples of them. Two good examples are, however, the case in which a firm "benefits from the labour market created by the establishment of other firms, and that in which several firms use a resource which is free but limited in supply" (23). An example of the latter case is given by Scitovsky (23), that of an oil well the output of which depends on the number of other oil wells being in operation in the same field. In our case it would be the use of underground water for several agricultural units in a desertic area.

Another type of external economy is that in which the profits of one economic unit are affected by the actions of other units. This could be represented by a function of the following type:

$$P_1 = G(O_1, L_1, K_1, T_1, C_1; O_2, L_2, K_2, T_2,$$
$$C_2, \ldots O_n, L_n, K_n, T_n, C_n),$$

which shows that the profit of the unit 1 depends not only on its own output and factor input but on that of units 2, 3, ... n as well. These external economies, which should also be estimated in any program, are generally referred to as "pecuniary external economies."

The effects of external economies in the development of an arid zone will be considerable, since the lower the economic diversification, the greater will be their impact. But, on the other hand, the arid region may prove to be ideal for accumulating valuable empirical information about the effects of external economies at the regional level, mainly because it will be a region relatively "undisturbed" from the multiple and simultaneous economic processes that characterize other areas. Measurement will be easy, since mainly the simultaneousness and multiplicity of investments — for instance, in metropolitan areas — have made it difficult to estimate the real impact each new activity has on other economic units.

Investment Effects: Second Step

Every investment produces, apart from others, an impact on the national or regional income, on the balance of payments, and on the distribution of income. Since the development of an arid zone will normally carry a considerable weight for the economy as a whole, for it will be a large undertaking, it should be stressed that each particular investment project should be chosen according to the desired impact on the three above-mentioned fields.

As far as national or regional income is concerned, the tendency to select projects that would maximize output at a point of time should be avoided by any possible means, since the goal should be the maximization of the *rate of growth over time*. This means that coefficients such as the capital/output ratio, or embellishments of it, should be avoided for the determination of investment *priorities* (24). Capital/output ratios are only helpful in the determination of the amount of capital (or investment, to use a simpler term) to be allocated once projects have already been selected according to their possible contribution to the rate of growth over a crucial period, and to the solution of fundamental regional problems. This is particularly important for an arid zone that is supposed, through an initial investment program of high cost, to generate attraction for further investment initiatives, both private (or cooperative) and public. These latter investments will be of a derivative type. When I referred to the question of interrelationships, strategic and service projects were mentioned. A strategic project must have at least two characteristics: (a) it must be able to yield, after maturity, direct profits, and (b) it must be able to generate a multiplier effect in the sense of attracting investments in derivative projects, that is, projects that will use as inputs part of the production of the strategic unit. A case in point might be a cement factory (strategic) that stimulates the appearance of construction and similar industries.

As far as the balance of payments is concerned, the arid zone can be of crucial importance. From the general point of view, we should agree with Polak who, when considering the balance of payments problem, classifies investments into three types: (a) export-promoting or import-substituting investments; (b) investments that produce goods sold in the home market, replacing similar goods sold in the home market (this mainly due to technological improvements); (c) investments that produce goods sold in the home market in addition to those already sold (17).

The first type of investment has a positive impact on the balance of payments, the second is neutral, and the third one can either be neutral or have a negative effect. The first type should be given, whenever possible, the highest priority, since an increase in the reserves of foreign exchange is desirable from every point of view, especially as a means to avoid unnecessary devaluations. Arid zones in particular tend to have a potential for the exploitation of export-promoting investments — mainly mining and special agricultural products like fibers, for instance. Therefore, even if other alternatives of investment exist, preference should be given to those that promote exports, unless the balance-of-payments situation is already such that other investments can be afforded, or unless it is a heavily populated arid zone (which is quite improbable to begin with).

With regard to income distribution, investments should be selected according to their capacity for generating a more-or-less symmetric type of distribution (25). I do not believe, nor do I favor, investments that have a tendency to polarize the distribution of the newly generated income, even when monetary and fiscal devices can be applied in order to cope with the problem later. It is not wise to create consciously an anomalous condition and then generate a special means to correct the situation.

If the immediate maximum growth of the product were the important aim, the distribution of the income would not be as crucial a matter as if the greatest importance were given to both a maximization of the *rate of growth* over time and an increasing participation of human resources in the development process. I am convinced that the continuity of a process of growth can be guaranteed only through that increasing participation of the people. The fewer the people sustaining the process of growth, the greater the risk of collapse.

I have defined development as being composed of economic growth (growth of the national income) and of structural changes. These structural changes are difficult to define, although they imply reforms that allow growing participation of human resources in the internal market. I would, therefore, like to propose that every structural change will be reflected in and by the income distribution, and that no fundamental redistribution of income can take place without implying some degree of structural change. If this proposition is acceptable, then a measurement for development can be proposed that includes both economic growth and structural changes — the latter measured indirectly — through a simple equation. The proposed equation would basically represent the *quality* of development being obtained under a certain program, thus being a test for the investment policy's results.

Economic growth per capita can be measured as:

$$EG = \frac{\Delta Y_1}{Y_o} - r, \tag{1}$$

where ΔY is the increase in the total income, Y is the total income of the previous year, their quotient is equal to the product of net savings multiplied by the capital/output ratio, and r is the rate of population growth. This formula shows the rate of growth of income per capita, but since (according to my definition) growth alone does not represent development, a corrective factor (coefficient) should be considered. The proposed coefficient will represent the income distribution at a given period and will be identified as the "distribution coefficient." It is obtained by dividing the modal income of the region by the per-capita income of the region. The coefficient would take the form:

$$\frac{Y_m}{Y_p} = S, \tag{2}$$

where Y_m is the modal income and Y_p the per-capita income. It must be remembered that the modal income (mode, in statistics) is the value that appears more often

in a frequency distribution; per-capita income (arithmetic mean, in statistics) is simply the total income divided by the total population. So, then, the more symmetric the income distribution, the closer will be the value of the coefficient to that of unity. The more polarized the income distribution, the closer will be its value to that of zero. The formula for the measurement of development would then be as follows (26):

$$D = \left(\frac{\Delta Y_1}{Y_o} - r\right)\frac{Y_m}{Y_p}, \tag{3}$$

or in short:

$$D = EG \cdot S. \tag{4}$$

This formula represents to some degree a value judgment as referred to in the introductory remarks, for it considers development not to be optimal until a maximum number of people are participating in it, both in terms of benefits and efforts. In the proposed formula, the distribution coefficient will show the proportion in which the increase of the income is being wasted because of maldistribution. What is meant by waste, in this case, is the fact that maldistribution generates "conspicuous consumption" on one extreme and, on the other, minimizes purchasing power for goods that could most probably be produced internally. Conspicuous consumption practically never implies production of import-substituting goods; quite to the contrary. A more symmetric distribution, on the other hand, can be much more stimulating for import substitution or for export increase. Finally, a relatively symmetric distribution of the product will most probably act as an incentive to save. Polarization inevitably represents waste.

Under a more-or-less symmetric type of distribution (when S tends toward 1), the income generated will be utilized in a way closer to the optimum, and development will from then on depend mainly on the rate of economic growth. In other words, if distribution of income tends toward relative symmetry, most efforts can be concentrated on techniques to bring about further product growth allowing maximum participation of the people in regional and national economic processes.

In the case of an arid region with low population density, the control of the distribution coefficient will be relatively easy, thus making the arid zone of the described characteristics an ideal test case.

Factor Intensity: Third Step

The third step in the process of decision-making will include all questions related to the factor intensity of the proposed investments. This means passing judgment on the volume and type of labor required for the projects and their execution. Furthermore, it will mean a decision as to the labor/capital ratios to be established in order to optimize the conditions set forward in the program, considering, of course, all limitations. This is one of the most difficult and delicate decisions in the entire process, since it involves the already-old discussion as to whether labor-intensive or capital-intensive techniques should be favored. This discussion adopts a very special proportion, as shall be seen, when referred to a desertic region.

A region like the one with which we are concerned can be approached in two ways. One alternative is to exploit some abundant natural resources — say, petroleum or other minerals — for the benefit of other regions. This would be a case in which no particular interest exists as to the development of the region in an integral sense. The region is to remain simply as an exploitable "hinterland" instead of an inclusive land to be developed. The other alternative is the opposite; promote the region's integral development. In the first case, human settlement must be minimal and the use of capital must be maximal. In the second case, human settlement and even resettlement is an essential feature of the development program. In this chapter, attention should center on the second case. If so, the decisions adopted as to human settlement are crucial, mainly because whenever such an activity gets under way, there is little margin left for late readjustments or amendments. Moving human beings is not easy, and having to move them back again because of oversights in the programming can be absolutely disastrous both socially and economically.

Many economists tend to defend labor-intensive projects in underdeveloped regions on the grounds that the national rate of unemployment is normally high, and that even among the employed human resources a quite considerable margin is underemployed, thus having a marginal productivity equal to zero. Another defense for labor-intensive projects is based upon the belief that the utilization of unemployed labor in one investment program does not reduce output in other areas and that, therefore, it represents no social opportunity cost (27). The same would hold true for underemployed labor, since the transfer of it from a place in which its marginal productivity is zero to a place in which it will be greater than zero would imply no social opportunity cost either.

Although the above arguments may seem logical, they can be refuted with some solid arguments. In the first place — as already stated — *the mobility of the factors tends to be in inverse proportion to their availability.* Labor has a very low mobility, and, therefore, the price elasticity of its supply will most likely be very low, or, to put it in other words, as mentioned by Raj, "its supply price is likely to be quite high" (28). This holds true, in practice, even for unemployed or underemployed labor. Any regional planner with field experience could testify to this end. In the case of an arid zone, the supply price of labor is, no doubt, going to be very high, thus representing a social cost to the community. The social cost in this case will be at least a part of the supply price of that labor. On the other hand, the increase in consumption that has to be expected as a consequence of the newly generated employment will represent, for some time, a social cost, especially to the degree that it affects consumption somewhere else. The increase in consumption will represent a social cost as long as it takes for the expected increase in output generated by the new program to take care of it. It could then be concluded that

the lower the wage elasticity of supply of labor, and the higher the propensity to consume out of additional income, the larger the social cost of employing additional labor in underdeveloped economies (regions, in our case) (28). It should be added that the supply price of labor is determined not only by its low mobility but also by the characteristics of the new environment into which it is supposed to move and settle. Climate is probably one of the most important environmental conditions that will have a direct effect on the supply price of labor. It is thus important to keep in mind that the already high price of the supply of labor will be increased still further because of the environmental conditions of an arid-zone program.

The increase that will be generated in consumption, in an arid region, by the new labor force employed will most probably not only affect in a negative way the possibilities of consumer expansion in other nearby regions but will also in all likelihood generate the need to draw upon the foreign-exchange reserves in order to supplement partly with imports the initial increase in regional consumption. The increase in consumption will eventually be met by the increase in income generated by the concrete projects carried out within the regional program. This, however, may take quite some time. The time lag between the initial investment and the maturity of the project is, therefore, a determining factor for the right decisions to be taken. If the time lag is long, one would be tempted to advise a reduced amount of labor employed, especially in an arid region where consumer goods have to be imported anyway, either from another region or from abroad. On the other hand, if the time lag is short, conditions might favor a higher labor-intensive program.

When considering as a social cost the increase in consumption generated by the arid region, it is convenient to point out that the part of that new consumption that is covered by resources with no alternative uses anywhere in the economy at that time should be excluded as part of the social cost. In other words, what represents social cost, without any doubt, is the increase in consumption met by resources with alternative uses and by resources obtained through the use of foreign exchange for import purposes or through a reduction of exports.

So far for technical argumentations. Now I should like to look at the problem using a more realistic outlook and a necessary degree of common sense. No region can be developed without sufficient people. People are not only necessary for the construction of projects, but for the induced human settlement that must follow. People are needed above all — and let us keep this clear in our minds — to give life to a region. Villages, towns — and later even cities — must appear, and with them all types of human activity. The more life a region is capable of generating, the more it will be able to develop. So *people have a value in themselves and cannot be reduced to economic cost-benefit analysis.*

The main problem is, however, how to make people migrate into a region like a desert, still undeveloped or just in the initial stages of development. It can and must be done in an indirect way. There are two possibilities. First, thousands of laborers are required for the construction of dams and irrigation systems. Since these works take a long time, the laborers will most probably have to be transferred with their families. The tradition in such cases has been to build near the construction a campsite to be taken away after the work is finished. What should be done instead, in the case of a desert, is to initiate a properly located housing project together with the construction of the dam, allowing the families of the workers to become owners of houses once the project is terminated and as long as they decide to settle in the area and integrate into newly generated economic activities. The cost of such a housing project might partially be computed as a cost of the dam and partially as a cost of the National Housing Program, which, in one way or another, exists in practically every country. This strategy seems appropriate, since ownership of a house is a strong reason to settle down in an area.

Second, in view of the natural difficulty that exists in bringing people into a new region, even if they are to be drawn let us say from very poor urban dwellings, another aspect must be taken into consideration. People who live in very poor villages or suburban slums have an organized community that, despite their poverty, gives them a sense of security. Mutual protection, status, and leadership within such communities are of the utmost importance, and it is difficult to expect a family to abandon the communal environment just for the sake of a better income and a better house. The ties of communal security are just too strong. Therefore, the most appropriate way to solve the problem is to move, if not an entire community, at least a consistent part of it, so that its members can settle in a new place yet maintain their fundamental social ties. This has been done with amazing success in Egypt, where new towns have been organized as a consequence of the Aswan Project. Another important feature of this settlement strategy is that when a communal group changes its environment yet maintains its social ties, it adapts easily to new activities and becomes much more receptive to modernization and innovation than a group that remains in its old environment. The author was able to establish through research carried out in the fishing industry of Peru — a good deal of it in the desertic coastal area — that Indians coming from the Sierra in groups were much more capable of adapting to a modern industry than were coastal people engaged in traditional forms and ways of fishing. The change of environment is stimulating for people, as long as it does not mean the destruction or definite loss of basic communal social modes of interaction. This is seldom taken into account in development programs, yet it may signify success or failure from the human point of view.

Another aspect worth mentioning for the justification of the outlined settlement strategy is the fact that people living in urban marginal dwellings are normally considered to belong to the society's lower strata or class. Research the author was able to carry out in such places

— especially in Lima — led to the conclusion that such an assumption is quite misleading. As a matter of fact *marginal urban dwellings are not representative of a social class but of a satellite society with a complete social stratification system of its own.* It was even possible to detect conspicuous cases of what I have called in a yet unpublished essay "the oligarchies of misery." All ties and modes of social interaction that are found in the dominant society — even oligarchic groups — are *mutatis mutandi* also found in the satellite society. Therefore great care must be taken when turning to such human aggregates as a source of labor supply and settlement in developing arid regions.

The suggestions offered above must, of course, be accompanied with the creation of a labor market. This means that parallel to construction of irrigation systems, for example, future employment facilities must be developed. This is one of the most serious problems in an arid zone. Things cannot be done, as in other cases or under different circumstances, one after the other. Many activities must unavoidably be developed simultaneously. This is again an argument in favor of the *integral* approach that must be adopted for the development of an arid region.

It should by now seem quite clear to the reader that I favor, for the development of arid regions, the largest possible amount of labor, since it is this labor force, together with their families, that should turn into the initial immigrating force. All other argumentations of a cost-benefit nature are vain and byzantine. No development is costless or painless. However, after the region becomes relatively colonized and projects get under way and reach maturity, results will more than repay the original costs.

Regardless of the amount of labor employed in the construction of the projects, one aspect that, for technological reasons, must be taken into consideration is the intensity of labor, under given circumstances, vis-à-vis substituting machinery. Machinery, especially of the heavy type, is generally imported in developing countries, and the pressure these imports exercise on the foreign exchange is such that "the social cost of using machinery is higher than the private accounting cost, [while] the social cost of using labor is, on the other hand, lower than the private accounting cost" *(28).* The preceding does not mean, however, that the final social cost of labor will be lower than that of labor-substituting machinery, nor does it prove the opposite either. The relation may vary in each particular case, and that is why a comparative analysis is advisable in most instances.

After the considerations on the factor-intensity matter, outlined above, it may be wise to remember that although substitution between labor and machinery is a key issue, the rate of substitutability between the two is not unlimited. In fact, substitution of one for the other is many times limited by technological reasons. Furthermore, as pointed out by Kahn, the substitution of labor for capital is seldom costless and may necessitate a substantial complementary investment *(14).* This complementary investment may be especially high in the case of arid-zone development.

It might well be concluded that a development program for an arid region with a population density so low as to make human settlement indispensable should, whenever possible, favor labor-intensive investments in order to apply the settlement strategy outlined a few paragraphs earlier. Were we to deal with an arid region with an adequate population density, the reasoning might be different, although not necessarily so. If the human resources are on hand, the problem of factor mobility, the difficulties inherent to the low wage elasticity of the supply of labor, and the high cost of supply, will have little weight in the decision-making. In other words, if the population of the region allows the development program to draw, for its needs, from the existing human resources, one should be inclined — save in exceptional cases — to favor labor-intensive projects as well. So in one case labor-intensity is justified because of settlement needs, in the other because of the possibility of drawing from an already existing labor supply. The final conclusion should be that *the arid region has the characteristic — different from other regions — of always requiring labor-intensive projects if it is to develop in a balanced and appropriate way.*

The problem of labor is not to be reduced only to its scarcity or abundance, or to its proportion with respect to capital. An additional aspect leads us to the fourth step in the investment planning for arid zones; this is the problem of education and of skills of the human resources. The supply of labor may be high, and it can still be a scarce factor if the surplus of its supply is below the standards of needed education or skills.

Labor Skills: Fourth Step

It is probable that the wage elasticity of the supply of skilled labor in developing countries is higher than that of unskilled labor. This would make it less difficult for the settlement of this type of labor. But the problem will still be its overall scarcity, thus making each transfer of skilled labor represent quite a high social-opportunity cost which, in the long run, may adversely affect the economy as a whole.

In the case of an arid zone with low population density, it is highly improbable that the proportion of skilled labor required by the program will be available in the region. There is no choice other than to draw from a scarce supply somewhere else, thus deeply affecting other branches or regions of the economic activity, or to engage in investments that will require low skills, provided that the latter have other conditions to make them attractive.

A. J. Brown gives a good example of a division of investments according to their capital/skilled labor requirements (29). He classifies investments into four types: (*a*) low skill and capital light: textiles and footwear (we may add some forms of agriculture); (*b*) high skill and capital light: engineering and metal works; (*c*) high skill and capital heavy: petroleum refining; and (*d*) low skill and capital heavy: cement (we might add mining). It is

between these general types of investments that choices will have to be made. In the case of our arid zone, however, the choices will have to be confined to those investments that require low skill and light capital and those that require low skill and heavy capital. This holds valid, of course, only as long as the development program of the arid region has not been decided upon the grounds that the natural resources potentially available — such as petroleum — make all efforts in another direction worthwhile. But if the decision to develop the region stems from reasons other than the good luck of an oil discovery, choices should, at least at the beginning, be confined to low-skill alternatives.

Last, but not least, another human factor has to be taken into account. As Hagen has stressed, among the important factors to be considered for any investment criteria are the "social acceptability of the program, [the] development of resistances, and the effect of alternative programs" (*20,* see also *30*). The importance of the three factors enumerated by Hagen, plus others such as the frustration effect, the loss of the sense of security, and the increase in anxiety, will depend to a great extent on the strategy adopted for the introduction of modernization into a region not accustomed to it.

Education and Training: Fifth Step

As a general criterion it would be wise to accept that whenever a new region is to be affected by modernization, and human resources are drawn from the same region or from others, *technology should be adapted,* as possible, *to the potential skills of the people and not the other way around.* This criterion will be helpful not only as a means for reducing the cost of training, but also for minimizing the normal tensions involved in any forced or induced transition. This principle seems especially relevant in the case of arid zones, since it seems that the man who lives under adverse natural conditions (in this case aridity) is able to develop, when properly stimulated, amazing skills that are seldom properly exploited when modernization is introduced.

It may be proper to illustrate what I mean, with a theoretical example (*31*). Let us assume that some kind of inventory of natural resources of the country, divided by regions, exists. (If it does not exist, it will have to be made sooner or later.) According to the inventory, it is concluded that the country has potential capacity for producing products 1, 2, 3, 4, . . . 7. Some of them can potentially be produced in more than one region, others only in one. Let us further assume that the necessary studies have been made and it has been concluded that our arid zone should concentrate its productive efforts on the elaboration of 1, 4, and 7. So far it has been decided *what* to produce. Now it is of paramount importance to decide *who* should be engaged as labor in the production of the three types of goods, given the type of technology that will be used and which we accept to be the most modern available in the international market.

Such an inventory was made for natural resources; one should be made for human resources. This study should,

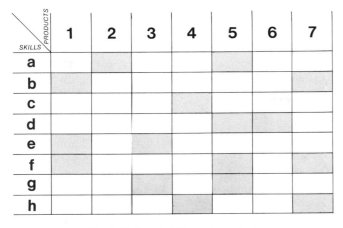

Fig. 1. National skill-product chart.

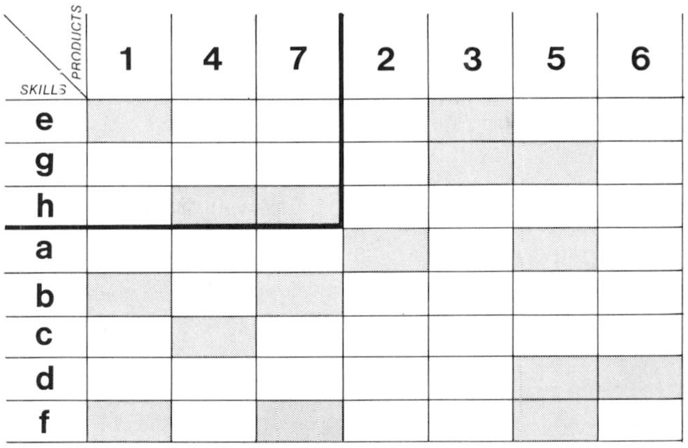

Fig. 2. Regional skill-product chart.

however, not be one of simple availability, but one of cultural-anthropological nature, so as to classify human resources by region according to their most outstanding communal skills whether originated in formal education or not. Once the census of skills is completed, an adequate interdisciplinary team should establish for what alternative applications each one of the skills is potentially useful (*32*). After this has been solved for each of the main existing skills, a skill-product chart should be prepared in order to have a visual image of the possible combinations between the two variables.

In Figure 1 we show a skill-product chart for the whole country. The columns represent products and the (horizontal) lines skills. It can then be seen that product *1* can be produced — with minimum training and minimum tensions of readaptation — by groups with skills *b, e* or *f*. Product 2 can be produced only by group *a*. On the other hand, the chart also shows that groups with skill *a* can be efficiently used for the manufacturing of products *2* and *5*. By the same token, skill *f* can be used in the production of *1, 5,* and *7*, thus representing the human group with the most versatile skill bank. The chart should be referred to, for short, as the national SP chart, and each possible combination of skill and product as an SP complex.

Returning now to the initial example, let us make the assumption that the cultural-anthropological study of the area has proven the existence of skills *e, g,* and *h*. A new SP chart can now be prepared for the region, in which the desired products are represented by the *n* first columns and the existing regional skills by the *m* first lines. The following columns and lines will be in their corresponding alphabetical and numerical order. A chart like the one in Figure 2 is obtained, in which the first three columns are *1, 4,* and *7* of the original national SP chart. The first three lines are *e, g,* and *h* of the original national SP chart. The heavy line shown in Figure 2 represents what should be called the "regional SP efficiency frontier." The chart shown in Figure 2 is the regional SP chart. What is represented inside the regional SP efficiency frontier are the only possible combinations of SP complexes inside the region. Thus we see that the possible complexes are *1e*, *4h*, and *7h*. There are no possible combinations for the use of skill *g*. If *4h* and *7h* are both possible at the same time, all products will efficiently be covered. Yet it may occur that *4h* and *7h* are alternative choices due to insufficient supply of *h* for both productive undertakings.

Should the latter be the case, we would have an unused skill and at the same time a product to which no skill can be applied. In this case *g* and *7* represent what can be identified as *X* complexes. One can be represented as *Xg* and the other as *7X*. We have then for our region one of two cases, depending on the availability of *h*. In the first case the final regional SP complex will be:

$$\text{RSP complex} = (1e) + (4h) + (7h) + Xg.$$

In the second case we may have:

$$\text{RSP complex} = (1e) + (4h) + Xg + 7X.$$

It should be pointed out that the above do not represent algebraic sums, but only combinations.

The *X* complexes can become SP complexes only if the corresponding reallocation of the affected human resources takes place. As seen in the regional SP chart (Fig. 2), the possible solutions for *Xg* are (*3g*) and (*5g*). Both imply moving group *g* to the region where *3* and/or *5* can be produced. On the other hand, the possible solutions for *7X* are (*7b*) and (*7f*). In this case it would imply moving groups *b* and/or *f* into our region where *7* is to be produced. Whether the solution for *X* complexes is one or the other will depend first of all on the other alternative SP complexes that can be formed with the group that would have to be moved; in other words, it will depend on the social-opportunity cost implicit in each *X* complex becoming a final SP complex. One additional alternative for the solution of *7X* may be (*7u*) in which *u* represents an unskilled group to be found in the region. An unsatisfactory solution (*7g*) represents the definite loss of a skill and it should only be adopted in the event that the cost involved in forming SP complex (*3g*) proves to be higher than (*3u*), and SP complex (*5g*) proves, in turn, to have a higher cost than SP complexes (*5a*), (*5d*), and (*5f*).

The advantages in using the proposed method of combining existing skills with production possibilities, given the technological levels to be applied, are several. In the first place the author is convinced that the psychological problems that arise when traditional labor is transferred into the industrial activity (frustration, anxiety, insecurity, and others) are not due simply to the change of activity, but mainly to the fact that the individual has to give up skills in order to learn new ones. The essence of the suggested method is based upon the evident fact that any skill a person has can be used for more than one purpose. There are, in fact, fewer basic human skills than alternative possibilities for the use of those skills. A person — say a peasant artisan — talented for making painted ceramics has sensibility in his fingers, good eyes, good pulse, and a sense of equilibrium. Can we possibly assume that those skills can only be applied to painted ceramics? Certainly not. They could be applied in laboratories, electronics industries, and many other undertakings.

Skilled labor scarcity is, in my opinion, to a great extent an overrated problem. Any person is potentially skilled or unskilled depending upon what activity is contemplated. If an individual has strong and calloused hands, he can be forced to become a pianist, but he will be a disaster as a pianist. A person with delicate hands can be trained to work with heavy tools in an iron foundry, but he will most probably be a frustrated worker.

Another advantage of the proposed system is the low cost of retraining. No skills have to be taught, only the readaptation of those skills to a new activity. The cost will be lowered even further because a person who need not be taught new skills can be retrained in a very short time. This is a concrete advantage for an arid zone. People working in modern activities, but according to their skills, will allow for a much faster and more effective development of the region.

Time Lag: Sixth Step

When discussing factor intensity, some considerations about the "time lag" of alternative investments were made. A more detailed analysis of time should be a basic step in the selective process.

When reference is made to the time lag of alternative investments, what is actually meant is the time of gestation of the project, or, in other words, the time that elapses between the initial investment and the moment at which the products become available. It is not only important to take the time lag into consideration for the reasons already outlined in the comments on factor intensity, but also for the social cost that should be imputed to it. In this sense, the social cost of time — as suggested by Raj — would be "the social opportunity cost to the society arising from the use of scarce resources . . . [in other words] the output that could alternatively be [generated] in the meanwhile with the same amount of investment" (28).

When considering time, one must not only estimate the implications of the gestation period, but also take into account the duration of the project once it reaches maturity. The larger the life expectancy, the smaller should be the cost imputed to time of gestation.

Social Investments: Seventh Step

The next step should be a detailed analysis of the social investments involved in the program, meaning education, housing, community welfare, health, and others. The problem here is that since these investments do not pro-duce marketable outputs, it seems difficult to evaluate their direct impact. Although some authors propose that price equivalents be imputed to their intangible outputs, one would probably be better off just using common sense. Nobody would ever deny that better social investments will have a positive impact on production and productivity. Social investments should, therefore (as in the case of service projects) not be evaluated according to the direct profits they generate, which cannot be calculated, but according to the economic and social needs they satisfy. In fact, in the case of arid-zone development, social investments should be considered in a light similar to that of the service projects that must accompany all strategic investments.

At the beginning of a development program the volume of social investments must be calculated with great care in order to avoid immoderate expenditures. Investments of the social type, beyond the minimum required for starting, should be financed with national capital resources and, if possible, through the new capital formation generated by the program. Their financing through foreign loans should be avoided, since the construction of social projects is composed almost exclusively of inputs of national origin. Loans in foreign currency should be used for investments in projects that will produce marketable outputs for final consumption or for intermediate demand.

Financial Policy: Eighth Step

Without going into technical details — because this chapter is not on strict economics nor is it for economists alone — a general financial rule is suggested, especially in the case of developing nations. Every project, let us say an irrigation scheme in the arid region, requires factor inputs of national origin (labor, food, shelter, local engineers, cement) and of imported origin (consultants, heavy machinery). Since most development projects in poorer countries must count on foreign loans, it should be stressed that these be used only for the payment of imported inputs. National inputs should by all means be financed with internal capital resources. Unfortunately this is not the case in most instances; many of the national inputs are paid for through foreign loans. The fact is, and the author has been able to demonstrate it in a concrete regional program under the auspices of the United Nations, that with proper internal reorganization of the monetary and banking policy, more financial resources can be made available than people are generally willing to believe (33).

Given a limited capacity for foreign indebtedness in a developing country, the suggested policy should be followed whenever possible. It will allow a maximum of projects to be financed.

Ponderations: Final Step

The final step would be to give weight to each one of the preceding elements. Through an adequate tabulation of the results, the projects should then be selected and their priorities and timings of construction established.

No prescription of a fixed type can or should be given as to the weight to be granted to each element. The particular circumstances will determine that.

It might as well be recognized and accepted, at this point, that there is no formula in existence that can efficiently replace common sense. Any formula or any criterion should be used only to refine, organize, and evaluate what must beforehand be consonant with common sense. There are many instances in developing countries in which resources have to be allocated for purposes that are contrary to what any formula or rigid set of criteria might advise. If people are starving or dying of infectious diseases, or are living under extremely adverse social conditions, all necessary capital allocations have to be made so as to remedy the situation, regardless of the ability of those investments to yield direct economic profit. If these obvious needs are not met in time, the social unrest that may result can easily destroy all efforts in other fields. In other words, it would be well to keep in mind that a developing country or region must use elaborate investment criteria only when it can afford to do so, and it can afford to do so only after having met the most obvious urgencies. It must be pointed out, moreover, that a set of investment criteria must not represent a complete hindrance to any legitimate and important need for experimentation, typical and necessary in every nation that is engaged in a process of development and modernization.

We economists too often behave in an almost obsessive manner when it comes to exact and precise proofs and demonstrations. We should try not to exaggerate, because our behavior can provoke frustration among people who really are interested in doing something.

Concluding Remarks

It might be proper to consider briefly some of the possible national consequences that might be expected from the development of a desert or semiarid region, apart from those that were already mentioned in the Introduction.

In the first place, because of the limitations imposed by adverse natural conditions of environment, the arid zone will tend to attract select human resources with a pioneering spirit that, in the long run, may gravitate there with great positive effects on the society as a whole. Secondly, the arid region may develop without the handicap inherent to areas that, though less hostile from the natural point of view, are affected by tradition and inertia. This means that the region's progress will be based, most probably, upon a natural acceptance of innovation on the part of the settlers to the new environment. Thirdly, and as a consequence of the previous factors, the arid region once in its full development process will most likely generate a high standard of living and per-capita income, as is the case with the northwest of Mexico. The area, due to the psychological effects it will have on its new settlers, may produce in the long run an interesting "demonstration effect" for the rest of the society.

There are, of course, many cases of arid regions that are already handicapped by tradition and rigidity. Such is probably the case with the northeast of Barazil, although I do not doubt for a minute that the potentials of the impoverished *nordestino* are among the best of the Brazilian population. The fact is that hardship bears the seed of progress. To make that seed flourish it is sufficient to provide the missing "strategical" requirements, which often are fewer and less costly than one would tend to believe at first. It is these additional requirements that must be allocated intelligently, putting into full use, for that purpose, the six fundamental qualities that will distinguish a good planner: academic or formal knowledge of the discipline of planning, environmental experience, common sense, political sense, a feeling for human solidarity, and as has been pointed out by the economist Hermann Max, the imagination of the poet.

ANNEX: METHOD FOR ESTIMATING INCOME DISTRIBUTION

Premises

1. For the same level of income — national or regional — the structure of consumption and demand changes according to the form in which that income is distributed.

2. Deriving a conclusion from Engels's Law, it can be said that the more polarized the income distribution, the less diversified will be the structures of consumption and demand.

3. If we define, in a limited sense, economic waste as that part of income that goes into forms of conspicuous consumption (in the sense of Veblen), it can be said that the waste potential grows in a direct proportion to the degree of polarization in income distribution.

4. The greater the probability of waste, the lower the probability for an increase in the national rate of investment.

5. Through several empirical studies it has been demonstrated — at least for some developing economies — that the propensity to save, while nil at very low income levels, increases at medium and higher income levels and decreases again at the top income levels. It can then be said that the propensity to save has the form of a logistic curve (or Pearl-Read curve), the equation of which is:

$$y = \frac{k}{1 + e^{a+bx}}, \qquad (1)$$

where $b < 0$.

The value of y at $x = 0$ is:

$$\frac{k}{1 + e^{a}}, \qquad (2)$$

and as $x \to \infty$, $y \to k$.

The increments in y, as x increases, are such that the difference of increments of $1/y$ is proportional to the corresponding difference in $1/y$. The described curve belongs to the type known also as Growth Curves and is often used to describe population trends.

6. From the above it follows that the more polarized the income distribution, the larger the proportion of that income in the hands of the top income groups. The consequence will be a lower propensity to save and, in turn, a lower propensity to invest, than would be the case with a less-acute degree of polarization.

7. Probably the highest propensity to save and to invest will be reached with a relatively symmetric income distribution. With such a type of distribution the structures of consumption and demand will most probably be more diversified, and finally the degree of waste will diminish considerably.

8. With a relatively symmetric income distribution, most probably the level of internal demand will also increase, resulting in stimuli for further internal investments. The reason is that conspicuous consumption is mainly satisfied through the import market, so, the lower the level of conspicuous imports, the higher the internal demand for nationally produced goods.

9. Although a process of income redistribution is not a structural change in itself, we can affirm that it is the consequence of economic and structural changes. Moreover we can say that no structural transformation can take place, in an economy or a society, without having an impact on the income distribution, and no basic change in the income distribution can occur marginal to a process of structural alteration. This is an important premise on which the author has based the validity of the formula that will be described. In fact, structural changes cannot be quantified, but income distribution can be quantified. So, in order to evaluate those structural modifications, we shall use a quantifiable value which we know is always and directly affected by changes in the internal economic and social structure. We thus establish an indirect way for estimating those changes.

Definition

10. In part one of this chapter we defined the concepts of economic growth and development. Although a quite large definition of development was given, I want to redefine it here simply as the combination of economic growth and structural changes of a positive nature.

Measurements

11. Per-capita economic growth can be defined simply by the following equation:

$$EG = \frac{\Delta Y}{Y} - r, \qquad (3)$$

where Y is total national income of a given year, ΔY is the income increase in the next year, and r is the rate of growth of the population. The increase in income is determined by the rate of investment and the capital/output ratio.

12. What we have above is simply the measurement of economic growth, and we know that such a quantification alone does not give us an idea about the rate of development. The formula needs, thus, to be corrected by a given ratio capable of indicating the way in which income is distributed, so that the rates of growth and of development can be compared.

13. The corrective factor I propose is a coefficient that will indicate the degree of asymmetry in the distribution of income. It should be identified as "distribution coefficient" and given the symbol S. The proposed coefficient is obtained by dividing the modal income by the per-capita income:

$$S = \frac{Y_m}{Y_p}, \qquad (4)$$

where Y_m is the modal income and Y_p the per-capita income.

14. We know from statistics that the mode is the most common value of a frequency distribution. Therefore if we refer to modal income, we mean the most frequent or typical income we find in the economy. On the other hand, the per-capita income is just a simple arithmetic mean obtained from dividing the total national income by the total population. We also know that in a normal distribution curve, that is, a symmetric curve, the mode and the arithmetic mean coincide. We can thus say that the value of the coefficient S will have limits ranging between 0 and 1.

15. The limits mentioned are based on the following reasoning:

a) Although theoretically there can exist a statistical distribution in which the mode may be greater than the arithmetic mean, it will occur only in very exceptional cases.

b) On the other hand, we know that as far as income is concerned we can never expect to find a case in reality in which the modal income will be greater than the per-capita income.

c) We can thus say that the upper limit of the coefficient will be 1 only in the case of a totally symmetric income distribution, a case we also cannot expect to find in any real situation.

16. We can now establish that the closer the value of S is to 1, the better the income distribution. The closer it gets to the value of 0, the greater the polarization.

17. On the other hand we have already established in the premises what can be added now to the characteristics of the coefficient S. The closer its value to 0, the higher the possibilities of economic waste due to conspicuous consumption at the top-level income groups. The inverse happens, the closer the value of S gets to 1.

The Development Formula

18. The formula that will allow us to evaluate the degree of development that is taking place at any given time can now be written in the following way:

$$D = \left(\frac{\Delta Y}{Y} - r \right) \frac{Y_m}{Y_p}, \qquad (5)$$

or, in a more simplified way:

$$D = EG \cdot S. \qquad (6)$$

Characteristics of the Formula

19. This formula, although using only quantitative components, is nevertheless of a qualitative nature. It will always give us a numerical result that will be less than that given by the formula for economic growth. We shall have to work then with three different values that seem contradictory or at least paradoxical: first the value obtained from the economic growth formula, second that obtained from the development formula, and third the value obtained by subtracting that of development from that of economic growth.

20. The truly important values are the first and the third. The value obtained from the economic growth formula shows us the *real* increase in income. It is based on statistics and it represents a true value which we cannot deny. The development formula gives us, instead, a value which is statistically not true. However, the difference between both values will give us a third figure which will indicate the part of the total income from which conspicuous consumption might be generated due to income maldistribution. It also indicates the part of total income that will have the lowest contribution to savings and investment.

21. The third value described is the one which is to be identified as waste potential. Its formula will be:

$$W_p = \left(\frac{\Delta Y}{Y} - r\right) - \left[\left(\frac{\Delta Y}{Y} - r\right)\frac{Y_m}{Y_p}\right] \quad (7)$$

or, in a more simplified manner:

$$W_p = EG - \left(EG \cdot S\right), \quad (8)$$

22. By so dividing total income into two parts and knowing, for instance, the investment coefficient that exists for a given period in the economy, we can assume that if the coefficient is, say 12 percent of total income, there is one segment of that total income that contributes more than the other to the investment rate. We can thus eventually establish something of great importance for planning. That is the investment coefficient not as a percentage of total income but as a percentage of the segment of income that contributes most to it.

Argumentations

23. It might be argued that the proposed formula for development is based on a correction coefficient S that does not denote in an absolutely precise way the form in which income is distributed. This is true. However, the formula has been based on operational possibilities in developing countries or developing regions. It is well known that there are few or no developing countries in which precise data on income distribution can be found. This is mainly due to the political implications such an information has, when made public. Therefore the proposed formula has practical application, not only in an economy with complete and reliable information but in a developing economy as well.

Methodological Comment

24. The two basic informations needed for the application of the proposed formula are the per-capita income that is known for any country or region, and the modal income that, although not obtainable from official statistics, can nevertheless be established through a sampling method.

25. The distribution coefficient and the waste potential are extremely valuable information for the evaluation of a regional development plan. In the case of an arid region, the estimate of the modal income is easy, and thus the application of the formula will be of great help for permanent evaluation.

Final Comments on Comparative Analysis

26. It might be well to call the attention of the reader to the fact that if comparative analyses are made between different countries using the suggested formula, it should not be surprising if the distribution coefficient for the United States is the same as, say, for Peru. This will in fact be the case when comparing several highly industrialized nations with developing countries. From such a fact one must be very careful in order not to conclude that the distribution coefficient is meaningless for development purposes. The conclusion to be drawn is quite different.

27. There is a dynamic relation between the coefficient and the absolute level of per-capita income. In other words, a nation with a high income level turns into a consumer society in which even conspicuous consumption is necessary to stimulate further growth. However, an economy where Engels's Law is fully at work (in short, an economy with very low income levels for the great majority of the population) cannot afford such a heavy increase in consumption. A good deal of consumption must, in fact, be postponed for purposes of increasing investment.

28. We can divide the process into three stages, each one with different characteristics for the distribution coefficient. First, in an economy with very low income levels, the coefficient will most probably be small: say less than ½. Second, in an economy of medium income levels, the coefficient must have previously increased, and may even require a further increase in order to stimulate investments and the internal market. The coefficient will have to rise to a level between ⅔ and ¾. Third, in an economy with high income levels, the coefficient not only can diminish again but most probably will. It may well go down to ½. But the important point is that a coefficient of ½ in a developing economy has a very different meaning from a coefficient of ½ in a highly industrialized society. In the first case it means that due to the low coefficient, very few people are really participating in the internal market. In the latter case it means that the majority of the people have already reached a level of income that allows them a good living standard and a full participation in the internal market. A polarized income distribution is in such a case considerably less harmful than in a poor country.

REFERENCES AND NOTES

1. The opinions expressed in this chapter are the author's own and do not necessarily represent those of the United Nations Food and Agriculture Organization, where he was formerly engaged as a general economist, or of the UN Development Programme, where he is now engaged as a Project Manager.

2. Canchones are irrigation systems composed of a series of parallel canals, about a meter deep, and separated by about 10 meters. Cultivation takes place between the canals.

3. AHUMADA, J.
   1961   Investment Priorities. *In* H. Ellis and H. Wallich, eds, Economic development for Latin America. Macmillan & Co, New York.

4. LEIBENSTEIN, H.
   1958   Why do we disagree on investment policies for development? Indian Economic Journal 4:369–386.

5. VAIDYANATHAN, A.
   1956   Survey of the literature on "Investment Criteria" for the development of underdeveloped countries. Indian Economic Journal 2:122–144.

6. GALBRAITH, J. K.
   1962   Economic development in perspective. Harvard University Press, Cambridge, Massachusetts.

7. A. Vaidyanathan (*5*) refers to:
   YOUNG, A.
   1928   Increasing returns and economic progress. Economic Journal 38(152):527–542.

8. ECKSTEIN, O.
   1957   Investment criteria for economic development and the theory of intertemporal welfare economics. Quarterly Journal of Economics 71(1):56–85.

9. KUMAGAI, H.
   1959   External economies and the problem of investment criteria. Osaka Economic Papers 7(2):13–20.

10. AGARWALA, A. N., AND S. P. SINGH, eds.
    1958   The economics of underdevelopment. Oxford University Press, Bombay. Part 4 is especially dedicated to external economies, including papers by Rosenstein-Rodan, Nurkse, Fleming and others.

11. CHENERY, H.
    1953   The application of investment criteria. Quarterly Journal of Economics 67(1):76–96.

12. BELSHAW, M.
    1958   Operational capital allocation for development planning. Economic Development and Cultural Change 6(3):191–203.

13. GALENSON, W., AND H. LEIBENSTEIN
    1955   Investment criteria, productivity and economic development. Quarterly Journal of Economics 69(3):343–370.

14. KAHN, A. E.
    1951   Investment criteria in development programmes. Quarterly Journal of Economics 65(1):38–61.

15. TINBERGEN, J.
    1961   The relevance of theoretical criteria in the selection of investment plans. *In* Massachusetts Institute of Technology, Center for International Studies. Investment criteria and economic growth, papers presented at a conference sponsored jointly by the Center for International Studies and the Social Science Research Council, October 15–17, 1954. Asia Publishing House, New York.

16. FEDERAL INTER-AGENCY RIVER BASIN COMMITTEE
    1950   Proposed practices for economic analysis of river basin projects, report by the Sub-Committee on Benefits and Costs.

17. POLAK, J. J.
    1943   Balance of payments of countries reconstructing with the aid of foreign loans. Quarterly Journal of Economics 57(2):208–240.

18. BUCHANAN, N. S.
    1945   International investment and domestic welfare. Holt, New York, 240 p.

19. GROSSMAN, G.
    1961   A suggested model of the distribution of Soviet investment. *In* Massachusetts Institute of Technology, Center for International Studies. Investment criteria and economic growth, papers presented at a conference sponsored jointly by the Center for International Studies and the Social Science Research Council, October 15–17, 1954. Asia Publishing House, New York.

20. HAGEN, E.
    1961   The allocation of investment in underdeveloped countries: the case of Burma. *In* Massachusetts Institute of Technology, Center for International Studies. Investment criteria and economic growth, papers presented at a conference sponsored jointly by the Center for International Studies and the Social Science Research Council, October 15–17, 1954. Asia Publishing House, New York.

21. HIRSCHMAN, A. O.
    1961   Economics and investment planning: some reflections based on the experience in Colombia. *In* Massachusetts Institute of Technology, Center for International Studies. Investment criteria and economic growth, papers presented at a conference sponsored jointly by the Center for International Studies and the Social Science Research Council, October 15–17, 1954. Asia Publishing House, New York.

22. A strategic project in a desert area may be of an agricultural type, for the existence and survival of which service projects such as dams and irrigation schemes must be constructed. Another strategic project might be a large factory — cement for instance — in which case water development is again a service project.

23. SCITOVSKY, T.
    1954   Two concepts of external economics. Journal of Political Economy 62(2):143–151.
           Also reprinted in Agarwala and Singh (*10*).

24. The capital/output ratio measures the productivity of capital. It indicates the amount of investment required in order to increase output in one unit. It is a type of incremental ratio.

25. By symmetric I mean a frequency distribution for which the variate-values equidistant from a central value are equally frequent. Although total symmetry cannot be expected in the case of income distribution, it seems important to me to get as close to it as possible.

26. For readers who might be especially interested, the Annex at the end of this chapter presents the detailed argumentations on which I have based my proposed formula.

27. The social opportunity cost is represented by what the society will have to do without in order to do something else.
28. RAJ, K. N.
    1956 Application of investment criteria in the choice between projects. Indian Economic Review 3(2):22–39.
29. BROWN, A. J.
    1943 Industrialization and trade: the changing world pattern and the position of Britain. Oxford University Press, New York. 71 p.
    Mentioned by Vaidyanathan, (5).
30. MAX-NEEF, M.A.
    1965 En torno a una sociología del desarrollo. Lima, Universidad de San Marcos, Departamento de Sociología.

31. The method that follows was devised by the author of this paper and published for the first time in 1967 in the Bulletin of the International Labour Organization in Montevideo, Uruguay (mimeo).
32. I remember having read — I do not remember where — that the people in a region in Japan had great skills with their fingers, which they had been using for centuries in calligraphic art. Because of their skill they were transferred to the electronics industry, in which the installation of transistors required the skill those people already had. The operation proved very successful.
33. Reference is made to the Brazilian-Uruguayan Project for the Development of the Merim Lagoon Basin, an area of 64,000 square kilometers.

BIBLIOGRAPHY

This bibliography is composed mainly of basic works that deal with resource-allocation problems in general. Items from the references and notes section that fit this criterion are repeated in the bibliography for the sake of completeness.

Books

AGARWALA, A. N., AND S. P. SINGH, eds.
1958 Economics of underdevelopment. Oxford University Press, Bombay. 510 p.

BETTELHEIM, C.
1962 Problemas teóricos y prácticos de la planificación. Editorial Tecnos S. A., Madrid. 426 p.

ELLIS, H. S., AND H. C. WALLICH
1961 Economic development for Latin America. Proceedings of a conference held by the International Economic Association. Macmillan, London. 478 p.

ENKE, S.
1963 Economics for development. Prentice-Hall, Englewood Cliffs, New Jersey. 616 p.

FURTADO, C.
1965 Dialéctica del desarrollo. Diagnóstico de la crisis del Brasil. Fondo de Cultura Económica, México. 160 p.

GALBRAITH, J. K.
1962 Economic development in perspective. Harvard University Press, Cambridge, Massachusetts. 76 p.

HACKETT, J., AND A. M. HACKETT
1963 Economic planning in France. Harvard University Press, Cambridge, Massachusetts. 418 p.

HAGEN, E .E., ed
1965 Planeación del desarrollo económico. Fondo de Cultura Económica. México. 462 p.

HIRSCHMAN, A. O.
1961 Strategy of economic development. Yale University Press, New Haven. Yale Studies in Economics 10. 217 p.

HOSELITZ, B., ed
1952 Progress of underdeveloped areas. University of Chicago Press. 296 p.

HOSELITZ, BERT F.
1960 Sociological aspects of economic growth. Free Press, Glencoe, Illinois. 250 p.

MAX, H.
1970 El valor de la moneda. Editorial Joaquín Almendros, Buenos Aires. 350 p. (See especially chapters on credit for the more dynamic role banks could play in the financing of development projects.)

MAX-NEEF, M. A.
1965 En torno a una sociología del desarrollo. Lima, Universidad de San Marcos, Departamento de Sociología.

SCIENTIFIC AMERICAN
1963 Technology and economic development. Alfred A. Knopf, New York. 205 p.

TINBERGEN, J.
1962 La planeación del desarrollo. Fondo de Cultura Económica, México, 2d Spanish edition. 107 p.

UNITED NATIONS, ECONOMIC COMMISSION FOR LATIN AMERICA
1958 Manual on Economic Development Projects. United Nations, New York. 242 p.

WEITZ, R.
1963 Agriculture and rural development in Israel: projection and planning. Israel Division of Publications in cooperation with the Settlement Department of the Jewish Agency, Rehovot, Israel. National and University Institute of Agriculture, Bulletin 68. 148 p.

Articles

BELSHAW, M.
1958 Operational capital allocation criteria for development planning. Economic Development and Cultural Change 6(3): 191–203.

CHENERY, H. B.
1953 Application of investment criteria. Quarterly Journal of Economics 67(1):76–96.

ECKSTEIN, O.
1957 Investment criteria for economic development and the theory of intertemporal welfare economics. Quarterly Journal of Economics 71(1):56–85.

GALENSON, W., AND H. LEIBENSTEIN
1955 Investment criteria, productivity and economic development. Quarterly Journal of Economics 69(3):343–370.

KAHN, A. E.
1951 Investment criteria in development programmes. Quarterly Journal of Economics 65(1):38–61.

KUMAGAI, H.
 1959   External economies and the problem of invest-
        ment criteria. Osaka Economic Papers 7(2):13–
        20.
LEIBENSTEIN, H.
 1958   Why do we disagree on investment policies for
        development? Indian Economic Journal 4:369–
        386.
MAX-NEEF, M. A.
 1964   El desarrollo de la comunidad y la programación
        nacional del desarrollo. Guatemala, C. A., Secre-
        taría de Bienestar Social de la República.
MAX-NEEF, M. A.
 1967   Planificación integral ante el reto del mundo del
        mañana. Comunidad (México) 2(10):601–611.
POLAK, J. J.
 1945   Balance of payments of countries reconstructing
        with the aid of foreign loans. Quarterly Journal of
        Economics 57(2):208–240.
RAJ, K. N.
 1956   Application of investment criteria in the choice
        between projects. Indian Economic Review 3(2):
        22–39.
SCITOVSKY, T.
 1954   Two concepts of external economies. Journal of
        Political Economy 62(2):143–151.
VAIDYANATHAN, A.
 1956   Survey of the literature on "investment criteria"
        for the development of underdeveloped countries.
        Indian Economic Journal 2:122–144.
YOUNG, A.
 1928   Increasing Returns and Economic Progress. Eco-
        nomic Journal 38(152):522–542.

Proceedings

MASSACHUSETTS INSTITUTE OF TECHNOLOGY,
CENTER FOR INTERNATIONAL STUDIES
 1961   Investment criteria and economic growth, papers
        presented at a conference sponsored jointly by the
        Center for International Studies and the Social
        Science Research Council, October 15–17, 1954.
        Asia Publishing House, New York. See especially:
        GROSSMAN, GREGORY — Suggestions for a theory
        of Soviet investment planning, p. 95–120.
        HAGEN, E. E. — Allocation of investment in under-
        developed countries: Observations based on the
        experience of Burma, p. 55–94.
        HIRSCHMAN, A. O. — Economics of investment
        planning: Some reflections based on experience
        in Colombia, p. 33–53.
        TINBERGEN, J. — Relevance of theoretical criteria
        in the selection of investment plans, p. 1–15.
MAX-NEEF, M. A.
 1963   Investment criteria for the development of arid
        zones in underdeveloped countries. American
        Association for the Advancement of Science,
        Committee on Desert and Arid Zones Research.
        Symposium on Arid Lands of Latin America:
        Their Problems and Approaches to Solution,
        Cleveland, Ohio, 1963. Paper presented.
MAX-NEEF, M. A.
 1967   Integral planning for the challenge of tomorrow's
        world. "Man and the City of Tomorrow." Con-
        ference sponsored by the Town Planning Institute
        of Canada and the University of Montreal, Mont-
        real, Canada, June 21–24, 1967. Paper presented.

Agrarian Capitalists and Urban Proletariat

The Policy of Alienation in American Agriculture

HARLAND I. PADFIELD

Department of Anthropology and Bureau of Ethnic Research
University of Arizona, Tucson, Arizona, U.S.A.

ABSTRACT

MODERN AGRICULTURE in the United States is a major contributor to the urban crisis. Its development should be regarded as a cause rather than a result of the massive population migrations of the twentieth century. Moreover, both the nature and the magnitude of the urban slum problem have been determined largely by the cooperation of agricultural policy and technology, especially arid-land technology. American agriculture is a unique product of (*a*) eighteenth century European capitalism, (*b*) an agrarian cultural system conditioned by two hundred years of frontier experience and a century more of frontier thinking, (*c*) an agricultural policy justified by an image of beneficiaries who work and live on the land, but designed in fact to assist owners of land and capital investors in the production process, and (*d*) the fact that American agriculture's technology developed under more or less permanent frontier conditions — abundant virgin land, low population densities, and an east-to-west encounter with North American ecology.

United States agriculture is presently and historically the source of two demographic streams flowing into the cities. One emanates from the population of marginal or surplus farmers and the other from the population of immigrant agricultural laborers. Each demographic phenomenon can be attributed in large measure to domestic policies vis-à-vis agriculture.

Commodity price supports, land diversion, and other such programs favor the large producer and accelerate the marginal farmer's transition to the urban sector. In addition, policies designed to protect agriculture against labor assist in the recruitment of unsophisticated or culturally alienated peoples and reinforce their economic and social alienation. When migratory workers move to the cities, they are singularly ill-equipped for employment in industry, and they settle into an adaptation to alienation in the black and brown ghettos. Government policies assure agriculture a steady supply of occupationally immobile workers alienated economically and culturally from the American mainstream. Government policies, influenced by agriculture, retard technological innovation and keep agriculture dependent upon that labor supply. Human obsolescence is probably the most costly consequence of this agricultural system.

Because of the decisive advantage that this or that program can afford a given industry or firm, public policy must be regarded as a prime area of competition for differential advantage. The concept of technology should be broadened to include the innovative response of industry to policy, including taking advantage of and influencing the formation of policy. Public scientists and scholars must increasingly turn their attention to this area, the arena of political power, because it is here that the rules are made determining what "externalities" the economy as a whole must absorb. Examining this arena requires a willingness to accept sacrifices, however, because the public scientist and his institution usually depend for appropriations upon the same legislative committees which private interests dominate.

AGRARIAN CAPITALISTS AND URBAN PROLETARIAT
The Policy of Alienation in American Agriculture

Harland I. Padfield

SOME HIGHLY REGARDED agricultural scientists in the United States have begun raising the issue of the connection between an efficient, highly productive agricultural system and the social ills besetting the nation (1–6). While the negative legacy of the United States plantation system is almost a cliché among social scientists, the assertion or implication that *modern* American agricultural policy and technology are contributing to the urban crisis is somewhat more controversial. I am pursuing this controversy with the following two theses:

a) Modern agriculture in the United States is a major contributor to the urban crisis. It is more correct to regard modern agriculture's development as a cause rather than a result of the massive population migrations of the twentieth century. Moreover, both the nature and magnitude of the urban slum problem have been determined largely by the cooperation of United States agricultural policy and technology vis-à-vis ecology — especially in arid lands.

b) United States agricultural policy as distinguished from its health and welfare policy or its urban policy has one implied goal, the same goal it has had for decades — increasing economic efficiency. This policy is a logical outgrowth of the definitional boundaries of what constitutes the agricultural sector. "Agriculture" means the producer — the private owners of land and capital devoted to natural food and fiber production. Given this definition, there has been no dilemma nor is there a dilemma for agriculture. There may be a dilemma for the President of the United States or for Congress. There may be a dilemma for scientists identified with agriculture whether to oppose or assist this policy. There may be confusion, intended or otherwise, as to the goals of that policy, but for agriculture there is no dilemma or equivocation. Its effective policy is deliberate, predictable, and logical.

Although the trend of United States agriculture in terms of end results is clear, how it got that way is not. The American scientific, industrialized food-and-fiber system is a unique product of four components:

a) The original institutional input of the eighteenth century European capitalistic model of private ownership of land and capital as the basis for the allocation of agricultural surplus. As agriculture became commercialized in the early nineteenth century and its production took the form of commodities and ultimately money, other capitalist concepts such as the self-adjusting market also constituted institutional foundations.

b) An agrarian cultural system conditioned by 200 years of frontier experience lingering almost a century beyond the closing of the frontier — consisting on the one hand of an outmoded rural political system and on the other hand of a rural value system and a nineteenth century image of agriculture that can be conveniently manipulated by politicians and private interests.

c) An agricultural policy justified by an image of beneficiaries who work and live on the land, but designed in fact to assist "producers" or owners of land and capital. This assistance includes direct subsidies to the producers as well as special protection from the external effects of their land-use and labor practices.

d) The development of agricultural technology under more or less permanent frontier conditions — the constant factors of abundant virgin land, low population densities, and a dynamic encounter with ecology created by the east-to-west movement of the frontier from temperate rain forests to humid prairies to semiarid plains to the arid lands of the West.

It is as if the course of United States agrarian development had been dictated by an agricultural czar, because it has moved unequivocally in one direction — to maximize production per unit of land and per unit of labor and concentrate production in fewer and fewer hands (7, p. 150–160). The longterm effects were inevitable — the wholesale transformation of peasant and rural populations into an urban proletariat, while commercial agriculture has become dominated by urban industrialists. The irony is that almost without exception, government intervention in agriculture is undertaken to promote the welfare of rural people and to alleviate rural poverty and the plight of the small farmer. In contrasting political economies, such as Mexico, where agriculture is subject to monolithic government control, the dual goals of developing the capacities of rural peoples and developing an efficient industrial machine for the exploitation of land are seen as a dilemma rather than as identical strategies. As America's so-called urban crisis becomes more visible each passing year, it becomes more plausible to argue that Mexican strategists are correct, and the urban slums of the United States are a case of her agrarian chickens come home to roost.

United States agriculture is presently and historically the source of two demographic streams flowing into the cities. One emanates from the population of marginal or surplus farmers and the other from the immigrant agricultural labor population. Each demographic phenomenon can be attributed in large measure to domestic policies vis-à-vis agriculture. Underlying both of these phenomena is the basic economic institution of private ownership. This sets United States agriculture apart in a fundamental way from agriculture in Mexico and other countries that have deep-rooted socialistic agrarian traditions to deal with. The American agrarian system was never peasant or feudal. These institutions were planted but never materialized, primarily because of the abundance of unsubjugated virgin land (*8*, p. 175–176). Although desire for religious freedom and other highly moralistic and altruistic motives are important, it would be more realistic to regard the European colonists to America as pioneer shareholders in huge land corporations (*9*).

Commercial agriculture was established early in the Southern colonies. Even in the New England colonies, where land allocation and town settlement were governed by communal principles, land was privately owned. Because they were self-sufficient and noncommercial, the early New England farm and the hill farm of the South were distinguishable from the commercial plantation. Nevertheless, private allocation of the public domain transformed colonists into private owners, and although these primitive rural capitalists were heavily constrained by communal institutions, egalitarian values, the surrounding wilderness, and hostile Indians, their microcosmic businesses contained all the potential dynamics for the changes that befell agriculture in the 1800s. Thus the colonial agricultural system, North and South, was inherently antagonistic to the polar type of economic institution it is supposed to epitomize — a "community" with maximum social integration and cooperation. It was only the temporary constraints imposed by the environmental and technological situation which made the early American farm seem to conform to the community image. Moreover, if the American Revolution is to be regarded as an agrarian movement, it should be regarded as an agrarian capitalist revolt against the political and economic domination of Eastern colonial absentee landlords and not as a movement aimed at transforming the prevailing economic institutions. The history of the American frontier is filled with the violence of the struggles between contending capitalists over control and ownership of resources. The degree of violence and bloodshed, or the fact that one set of contestants was composed of farmers, in no way sets apart private agriculture from private ownership of minerals, timber, or factories. Failure to recognize the basic capitalistic nature of farming early in the history of United States political formation is responsible for much of the self-delusion inherent in United States agricultural policy.

MARGINAL-FARMER STREAMS

The marginal-farmer population is a demographic and economic residue of the frontier phase of United States development. The free-land policy that prevailed throughout the settlement of the trans-Mississippi frontier is prerequisite to the marginal farmer. This land system led to the placement of more than 32,000,000 people, or one-third of the entire population of the United States, on farms from coast to coast (*8*, p. 196; *10*, p. 1046). Basic to the free-land policy as a cause is the limitation of size. Basic to size is technology, and basic to technology is the mobility of labor. Therefore, I consider the slavery issue to be primary to the land policies and the single most decisive agricultural policy in United States history.

After the War of Independence and during the years of government under the Articles of Confederation, slavery was systematically abolished in the states of the North. Its prohibition was extended into the Northwest Territory by the Northwest Ordinance of 1787. On the other hand, slavery in the states of the South was implicitly reinforced by the United States Constitution written that same year (*11*, p. 46). The right to chattel labor in the South and the prohibition of it in the Northwest led to enormous differences between agricultural practices of those two sections. The North's pattern of the family-sized farm was a function of this policy in combination with cheap land. With free land available to every man and no segment of the population legally immobilized for agricultural labor, farms tended to be the size a man and his family could farm, given the available technology.

American society in 1790 could truthfully be called agrarian. The first national census, in 1790, showed over 90 percent of the population engaged in agriculture (*12*, p. 1184). Opportunity, nineteenth century style, presented itself to the mass of people, not as jobs in industry, in government contracts, or advertising, but as land. So the majority of Americans became farmers, and American society entered the age of popular agriculture. In this way part of the stage was set for the elimination of the marginal farmer that began shortly after the free-land supply was cut off — referred to as "the closing of the frontier."

Commercialization set another part of the stage. This began soon after the opening of the Northwest Territory, where wheat, favored by new soils, became a principal cash crop. The transfer of manufacturing to industry (hence loss of "self-sufficiency") was part of this process. Changing ecology, free land, mobile labor, and commercial competition stimulated mechanization. The prairies and the semiarid plains became the cradle of farm technology. The mechanical systems developed here extended the workpower of the farm family and broke the 40-acre barrier. Farms of 80 and 160 acres became common (*13*, p. 385–452). Thus the setting for the exodus of marginal farmers was complete. It simply remained for the dole of free public land to be cut off for rural depopulation to reach the proportions it did in the 1920s and 1930s.

In terms of percentage of the population employed in agriculture, there was a steady decline from 90 percent in 1790 to 31 percent in 1910 to 6.6 percent in 1960. However, in terms of absolute numbers there was a constant increase until 1910, 20 years after the close of the frontier (*12*, p. 1184–1193; *14*, p. 108). The arid-land frontier, which did not really begin to absorb population until after 1900, undoubtedly had some holding effect on rural-to-urban migration.

Although popular agriculture is viewed traditionally as a persisting omnipresent institution, it was but an economic episode that waxed during the cheap- and free-land era of the 1800s. Its passing should have been expected from the start, but just the opposite prevailed. Naturally, agricultural policy's encounter with this dynamic process would be at the threshold of the transition from farmer to exfarmer, which is the marginal farmer. Early land policies set him up, and his appearance was speeded up by such other policy innovations as land-grant colleges and experiment stations (7, p.147–160). Despite this contradiction, the marginal farmer was a *cause célèbre* for policy makers in the 1930s (15–17). Policies like commodity price supports, liberal loans, land diversion, and payments in lieu of production were adopted. But these like other previous policies favored the large producer over the small farmer, widened the gap and produced more *marginal* farmers.

Like all human strategies, the era of popular agriculture reaped consequences never foreseen by those who sought it. One of these was its passing. However, during its reign it developed the arable lands of the continent while providing jobs for most of the population for more than a century. One unforeseen consequence came from the pervasiveness of the agricultural experience. Virtually a whole national population moved into and out of agriculture, working out their occupational destinies. Like veterans of wars, the veterans of agriculture remember it like it was. Moreover, since the agrarian frontier is where serious community settlement began, the country's political and legal institutions intially formed in this economic environment. This kind of powerful institutional underpinning and cultural reinforcement insures American society an enduring legacy of rural arrangements and agrarian values — a legacy that seems reinforced by its incongruity with existing realities. This consequence of the agricultural policies of an earlier era could well rank as one of the most serious problems modern agricultural policy has to deal with.

IMMIGRANT-LABOR STREAMS

Both southeastern and southwestern technological traditions in United States agriculture are predicated on immobilized labor. Both systems were bolstered and are bolstered by a network of policies. The United States Constitution itself ranks as one of the earliest and certainly the most auspicious documents affecting agricultural policy (see Article I, Section 2; Article IV, Section 2 regarding slaves).

Although the involuntary immigration of agricultural labor was made illegal in 1808, informal policy entered into the picture as it has in the Southwest with regard to illegal "wetback" Mexican immigration. The ban on African immigration simply was not enforced. This policy continued until the Civil War was imminent (11, p.36). The Civil War, revolving as it did around agricultural labor policy, naturally altered the status of slaves. However, in the long haul, the war did not fundamentally change either the technology of Southern agriculture or the institutions of its immigrant workers. Legal freedom from bondage meant little without the capability to function in some economic capacity other than what they were enslaved in. No economic resources were made available for this change. Emancipation of the South's chattel labor simply conferred horizontal mobility, the freedom to wander about destitute — the freedom to choose where to be exploited.

Refugees from southern agriculture poured into the cities of the North during the Civil War, after Reconstruction, during World War I, and again into North and West Coast cities during and after World War II. These facts are well known (18, p.75). The social and economic consequences of the great Negro migrations are also well known. These migrations represent the massive transfer of costs from the rural to the urban sector: the astronomical costs of the institution, operation, and modernization of southern agriculture (19;20, p.25–85; 21).

Two points should be emphasized about this all-too-familiar phenomenon. One, the negative social and economic capacities of the Negroes of the South who moved to cities of the North emanated directly from the principle of immobile labor which was — and to a certain extent still is — basic to agricultural technology in the South. Therefore *maintenance* of these negative traits is related to only recently passing technologies. After the Civil War, the maintenance of immobility in Southern agriculture's labor supply simply took new forms. In this regard, the tenant and sharecropping arrangements in combination with segregationist practices affecting the legal standing, health, and education of Negroes must be regarded simply as an extension of plantation agricultural policy, aided in the main by United States agricultural policy (22). Second, the emigrant from southern agriculture swelled the migrant stream pouring into the new cotton systems of the Southwest and joined the ranks of semichattel labor in the process. There the foundations were laid for the impending coalition of southern agrarian interests with the arid-land farmers in that both systems were dependent upon immobilized labor.

Arid-land agriculture in the Southwest developed partly by accident. One can only speculate what kind of system would have developed had America been colonized from West to East, from arid to humid lands, instead of the other way around. When it arrived in time and space, arid-land agriculture was confronted with an awkward technological gap. Mechanization had developed under a peculiar set of ecological conditions on the prairies (Central Plains). The critical attributes were limited water in the form of natural rainfall, level land, and northern latitude. The modern grain complex arose — the iron plow, reaper, thresher, steel plow, and drill prior to the Civil War; and the cultivator, harrow and combine after the Civil War. Farm size was linked to this complex because it was an important variable by which the individual owner could increase his profits. The greater an area to which horse, steam, and gasoline power could be applied, the more benefit from this power the individual owner could capture for himself, thus giving him a competitive advantage in the market.

Farmers, further west in the Great Plains, encountering more difficult growing conditions, tended to increase the size of farms even more. This trend was stimulated further by high demand for commodities and labor caused by the Civil War. Finally the South's absence from Congress during the war paved the way for the Homestead Act. This gave legal sanction to a trend that had been going on in fact for decades. The point is simply this: the types of mechanical systems developed in American agriculture were *linked* to large-scale farming. Again this was a deliberate trend — a function of the rules of agrarian capitalism and public policy. It was inevitable only in that sense.

Once a technological complex gets established, however, its dictates are compounded. When irrigation enabled farmers to open the arid West, they brought the grain complex of the Midwest to the new region completely intact. By this time, scale economies dictating large units were already built into this technological complex. But the farmers encountered two different ecological conditions. Water, being subject to artificial control and subsidized by the sale of power generated by the dams, was practically unlimited and could be applied in quantities equivalent to or exceeding rainfall in humid lands. Also much of the reclamation land was in the southern latitudes, where longer growing seasons made possible a large array of crops besides forage and grains.

With these two ecological differences combined with the growth of nearby cities and the cotton boom of World War I, it did not take arid-land farmers long to begin growing fruits, vegetables, and cotton. But these are crops with technological requirements fundamentally different from grain crops. Farmers could plow and plant with grain mechanical systems but they could not harvest with the reaper and thresher or the combine. For instance, a farm family aided by a handful of permanent hired hands would do well to harvest 40 acres of cotton. Yet the size of his farm (160-acre homestead), was predicated on the technological dry-farming complex. In this Procrustean bed, the corrupting and chronic labor troubles of arid-land agriculture arose.

The heart of the labor problem for the Southwest lay in the fact that its agricultural systems were not consistently inefficient or labor-intensive as were the cotton systems in the South. Large-scale irrigation agriculture was labor-intensive only during harvest — perhaps two, three, at the most four months of the year. Modified chattel systems of the South were not the answer. This would obviate the use of gasoline tractors, four-bottom plows and four-row cultivators. Clearly, what Southwest agricultural systems needed was a moving supply of cheap labor. The overriding social attribute of this type of labor was identical to that of the cotton belt Negro — occupational immobility.

As in the South, government at the local and national level cooperates to assure agriculture a steady stream of migratory workers. Policy operates on two levels to achieve this end. First it assists farmers in recruiting occupationally immobile workers. When a "special type" of worker is said to be needed in agriculture, what is meant is a worker who is alienated economically and culturally from the mainstream of American society. Sometimes domestic economic conditions produce a sufficient supply of this kind of worker. Happily, this is not often the case. One notable exception is the dust-bowl and depression migration of the 1930s. Also through the years, the rural South has supplied harvest labor to the cotton fields of the Southwest. During World War II, prisoners of war were used.

By any criterion, however, Mexico ranks as the most reliable source of labor for Southwestern agriculture. Notable examples of public policy in this regard are the Agricultural Act of 1949 and Public Law 78 of 1951, amending this act, admitting Mexican aliens to the United States for labor in agriculture. The workers under the provisions of these laws did not remain in the United States, however, but were transported back and forth under the supervision of offices of the United States Department of Labor. This program reached a peak in the late 1950s, with nearly a half a million Mexican nationals used in 1959.

While the bracero program (P.L. 78) did not directly increase the population, other forms of Mexican migration are a different story. Public Law 414 has been used through the years to recruit agricultural workers by the expedient process of making them eligible for citizenship in the United States. It is impossible to indicate how many immigrants have entered the country under this agriculturally oriented program, but as of 1968 a total of 684,533 legal immigrants from Mexico eligible for United States citizenship reported under the annual alien registration program (*23*). This figure has climbed steadily through the years and reflected sharp increases during the waning years of the bracero program (P.L. 78) during 1959–1963. This figure does not include the unknown numbers of Mexicans working illegally in agriculture throughout the Southwest, nor does this figure include the unknown number of legally immigrated Mexicans ("green carders") who reside in Mexico but commute daily to work in the United States. An investigation I made in 1963 indicated 8,000 to 10,000 border crossings daily from Mexicali, just one of ten or more Mexican border cities where this sleeper population of semi-United States citizens resides. All these phenomena are related directly to agricultural demand for cheap labor and the responsiveness of public policy to supply it (*24*).

As immigrant labor streams into agriculture, public policy operates at a second level to afford agriculture protection against its workers by exempting migratory and other forms of farm labor from the benefits enjoyed by nonfarm labor. Agricultural laborers have been systematically excluded from the rights to collective bargaining under the National Labor Relations Act, Workmen's Compensation, Social Security, and minimum wage laws. In this manner public policy compounds the external effects of importing vast numbers of economically depressed and culturally separated people into the United States by reinforcing their isolation from the economy and society at large.

One of the effects of these policies has been to retard technological innovation and keep agriculture dependent upon massive seasonal labor inputs. Agriculture has traditionally tended to focus on public policy as opposed to technical innovation as the key element in control of labor and thence of the production process. While technology remained relatively static from 1915 until the mid-1950s, the demography of agriculture has been dynamic. Historically, migratory labor on industrial southwestern farms is a demographic episode in the rural-to-urban transition of peasants and obsolete farmers. Mexicans, Filipinos, southern Negroes, Indians, and alienated Anglos, all have functioned as pools for harvest labor and have undergone transformation in the process. Some have moved up in the occupational hierarchy of the farm industry itself. Most move out of agriculture. Some are pushed out by technological displacement, but given the relative stagnation of technological innovation in harvesting through the years it would be safe to say most workers moved out voluntarily seeking greater opportunity in the cities. If this were not so there would not be the perennial "labor shortage" in agriculture (24, p.282–292).

Thus the destinies of southeastern and southwestern arid lands merge in two ways. The agricultural interests of these two regions unite to promote protectionist policies vis-à-vis labor, thus collectively reinforcing the immobility of industrially incapacitated people. In addition, the destinies of their genuinely rural people merge demographically in the harvests and more recently in the urban ghettos of Dallas, El Paso, Phoenix, Los Angeles, and other cities, where the rural emigrant's adaptation to alienation becomes a way of life.

EXPLOITATION OF POLICY

Among the growing number of external effects of the American agricultural system, human obsolescence is probably adding up to more long-range costs to the American economy as a whole than all other effects combined. I have discussed briefly and inadequately two populations of obsolete humans — marginal farmers and immigrant workers. All of the institutional components of the American system of agriculture — capitalism, public policy, and technology — contribute to human obsolescence. Public policy is the most disturbing factor in the disturbing problem because, at this point, events in the system are administered by public office holders, public appointees, and public scientists. The absurdity, of course, is that the *public* takes the beating. It pays out billions in support payments annually while serving as the human scrap pile for the humanity no longer beneficial to the agrarian capitalist.

The basic absurdity of the price-support system, which is aimed at inequality of farm income and thus implicitly at helping the marginal farmer, is letting the same rules which established the inequality to begin with serve as the allocation model for the administration of benefits — namely size and efficiency. The larger the producer the more public assistance he receives, and the more efficient he is the greater his net benefit. In 1963, for example, approximately 11 percent of all farms had annual sales of $20,000 and over. These received payments of public tax monies averaging $2,391 per farm and accounted for 54.5 percent of the total payments. On the other hand, 42.5 percent of all farms averaged less than $2,500 in sales, and these averaged only $51 in public assistance (3, p.57). By what logic do public administrators seek to reduce inequality by doling out to the more prosperous farmer 46 times the amount of public assistance than is given to the poor farmer? The answer to this question, of course, is that they don't. The question as framed implies a false proposition. The true proposition, which makes the program explicitly logical, is that this stratum of large producers is 46 times more "valuable" than the marginal producer.

Another way to resolve the original absurdity would be to say that — like the income tax — farm policy, in this case the price-support program, is set up with the best of intentions, but that the normal operation of the capitalist system underlying all of our institutions means that the arena of public policy becomes a prime area of competition for differential advantage both in determination of policy and in the organizational and technological response to its rules. The arena of policy is a prime resource to be exploited just as natural resources and labor are exploited. The competitive advantage of one industry over another or one agricultural business over another in this arena is decisive. The underlying propositions of the capitalistic system enable economists to project behavior of firms in choosing the least-cost combinations of inputs from a finite array of production possibilities. It should come as no surprise to these supreme skeptics then, what the response of this same "animal" would be toward the profit implications of policy. The rich get richer and the poor farmer is boosted into the urban orbit even faster than if these policies had never existed. Why agricultural economists were so long getting around to looking at this would be an interesting, however digressive, inquiry. For instance, compare the *1940 Yearbook of Agriculture,* especially articles by Mario (17), Taylor (10), Alexander (5) and Tolley (25) with *Food and Fiber for the Future* (7).

"Mining" public policy appears nowhere more blatant nor costly to the society at large than in the area of immigrant agricultural labor. This problem has been set by some economists in the context of displacement costs of mechanization in agriculture or cushioning the external effects of mechanization (5, 6, 26). However, the underlying problem is the fundamental powerlessness of these populations. To agriculture this has been the criterial attribute upon which their roles in agriculture depend. Moreover, agriculture has reinforced the social, psychological, and economic alienation of its immigrant laborers through time — in the South by the tenant and sharecrop system and in both the South and Southwest by the seasonal migratory system. In both the recruitment of the powerless and the reinforcement of alienation, agriculture has been historically assisted by policy. The cotton-picking machine and the tomato harvester are simply mechanical illustrations of a social and economic reality implicit from the very beginning of immigrant labor's role in southern and southwestern agriculture. I stress redundantly the direct and continuing relationship between immigrant labor's basic immobility in agriculture and the culture of alienation in the urban ghettos. The National Advisory Commission on Food and Fiber has made recommendations bearing on this. One of these is to separate policies for poverty from policies for agriculture (7, p. 5). This distinction is a step in the right direction, since it recognizes the central hypocrisy in agricultural policy. However, more than this recognition must take place if we are to correct basic causes, some of which originate in the protectionist policies created specifically *for* agricultural business.

Because there is a direct relationship between agriculture and poverty, any policy bearing directly or indirectly upon the poor, especially the rural poor, will be naturally regarded as a domain of agricultural interest. The exploitation of the domain of public policy has two dimensions. One dimension of action is simply shrewd economic decision-making — capturing the maximum benefit and externalizing every conceivable cost implicit in policy. I think that the concept of technology should be broadened to incorporate this area of action. However, in the final analysis, the ultimate exploitation of policy lies in the political power wielded by the agricultural-industrial system. It is more direct and simple than scholars and scientists like to realize. Basically it boils down to the association and fusion of the bureaucracy in question with the groups it regulates, promotes, and serves. The seniority system in Congress and the long tenure of southern Congressmen are part of this network of power relationships (1, p. 183–236; 3, p. 9–10; 27; 28). Too much has been said on this subject in the scholarly and popular press for the scientist to be any longer naive about this.

The specter of direct policy determination and administration by private interests is a reality at the state as well as the national level. A recent study by E. B. Eiselein of the politics of vested interest in the regulation of water resources at state level revealed overlapping memberships of key legislative committees and private associations organized for the sole purpose of promoting the interest of a given industry. Further, his study revealed that these lobbying associations were invariably dominated by the richest and most powerful firms in that industry (29).

Universities, colleges of agriculture, and agricultural experiment stations inevitably find themselves locked up in this network of power relationships at the bottom of the structure, for the obvious reason that they depend for appropriations upon these very same legislative committees dominated by the private interests these colleges are supposed to be "serving." The results of this bread-and-butter system of power are coercive, pervasive, and corrupting. I contend this system of the bold, direct intervention of private interests in the political process is one of the chief contributors to every national crisis the United States faces today.

What is occasionally called a dilemma for agriculture (6) thus falls into proper perspective. The dilemma does not exist for the private sector. Its course is predictable and rational. Given the short-term goals it lives by, agrarian capitalism is successful. In fact the success of the agrarian capitalist system is simply the obverse side of the crisis. The dilemma does not exist for the alienated in the ghettos. Their choices are inevitable and linear. This fact is implicit in the theory of the culture of poverty or the culture of alienation (30, 31). Relegated to the human scrap pile by the American industrial system, the alienated of the ghettos no longer keep up the pretense of participation in that system. They profoundly reject the system and in so doing make theirs any way they can. At best they are a chronic drain on the economy and at the worst a direct threat to the stability of the society.

The dilemma exists for the public person — the scientist and practitioner suspended in the bread-and-butter power structure of the bureaucratic industrial complex. The dilemma is simply this — should he serve the private interests in proportion to their control on his purse strings, or should he serve the whole society and thereby threaten implicitly and explicitly the purse strings of his institution and hence his own status and capacity to do research?

Because the public people occupy the space between the powerful and the powerless, the dilemma is really theirs. Their willingness or unwillingness to accept the sacrifice of serving the public interest may well determine the final outcome of America's agriculturally compounded urban crisis.

REFERENCES AND NOTES

1. HATHAWAY, D. E.
   1963   Government and agriculture: public policy in a democratic society. The Macmillan Company, New York. 412 p.

2. BONNEN, J. T.
   1965   Present and prospective policy problems of U.S. agriculture: as viewed by an economist. Journal of Farm Economics 47(5):1116–1129.

3. HARDIN, C. M.
   1967   Food and fiber in the nation's politics. U.S. Government Printing Office, Washington, D.C. 236 p.

4. NATIONAL ADVISORY COMMISSION ON RURAL POVERTY
   1966   The people left behind. U.S. Government Printing Office, Washington, D.C. 160 p.

5. SHAFFER, J. D.
   1969   On institutional obsolescence and innovation — background for professional dialogue on public policy. American Journal of Agricultural Economics 51(2):245–267.

6. KELSO, M. M., AND J. S. HILLMAN
    1969   Social and political dilemma of the agricultural industry and agricultural institutions. University of Arizona, Arizona Agricultural Experiment Station, Tucson. 15 p. (mimeo).

7. NATIONAL ADVISORY COMMISSION ON FOOD AND FIBER
    1967   Food and fiber for the future. Report of the National Advisory Commission on food and fiber. U.S. Government Printing Office, Washington, D.C. 361 p.

8. EDWARDS, E. E.
    1940   American agriculture: the first 300 years. *In* G. Hambidge (ed), Farmers in a changing world, p. 171–276. U.S. Department of Agriculture, Washington, D.C.

9. WEBB, W.P.
    1951   Ended: 400 year boom: reflections on the age of the frontier. Harper's Magazine 203 (1217): 25–33.

10. TAYLOR, C. C.
    1940   The contribution of sociology to agriculture. *In* G. Hambidge (ed), Farmers in a changing world, p. 1042–1055. U.S. Department of Agriculture, Washington, D.C.

11. MEIER, A., AND E. M. RUDWICK
    1966   From plantation to ghetto: an interpretive history of American Negroes. Hill and Wang, New York. 280 p.

12. GOODWIN, D. C., AND P. H. JOHNSTONE
    1940   A brief chronology of American agricultural history. *In* G. Hambidge (ed), Farmers in a changing world, p. 1184–1196. U.S. Department of Agriculture, Washington, D.C.

13. WEBB, W. P.
    1931   The Great Plains. Ginn and Co., Boston. 345 p.

14. ROHRER, W. C., AND L. H. DOUGLAS
    1969   The agrarian transition in America; dualism and change. The Bobbs-Merrill Company, Inc, New York. 197 p.

15. ALEXANDER, W. W.
    1940   Overcrowded farmers. *In* G. Hambidge (ed), Farmers in a changing world, p. 870–886. U.S. Department of Agriculture, Washington, D.C.

16. DAVIS, C. C.
    1940   The development of agricultural policy since the end of the World War. *In* G. Hambidge (ed), Farmers in a changing world, p. 297–326. U.S. Department of Agriculture, Washington, D.C.

17. MARIS, P. V.
    1940   Farm tenancy. *In* G. Hambidge (ed), Farmers in a changing world, p. 887–906. U.S. Department of Agriculture, Washington, D.C.

18. HAUSER, P. M.
    1965   Demographic factors in the integration of the Negro. *In* T. Parsons and K. B. Clark (eds), The Negro American, p. 71–101. The Beacon Press, Boston.

19. FRAZIER, E. F.
    1939   The Negro family in the United States. The University of Chicago Press, Chicago. 372 p.

20. GLAZER, N., AND D. P. MOYNIHAN
    1963   Beyond the melting pot: the Negroes, Puerto Ricans, Jews, Italians, and Irish of New York City. The M.I.T. Press, Cambridge. 360 p.

21. MYRDAL, G.
    1962   American dilemma. Harper & Row, New York. 1483 p.

22. PERLO, V.
    1953   The Negro in southern agriculture. International Publishers Co, Inc, New York. 128 p.

23. U.S. DEPARTMENT OF JUSTICE
    1968   Annual report of the Immigration and Naturalization Service. Washington, D.C. 108 p.

24. PADFIELD, H., AND W. E. MARTIN
    1965   Farmers, workers and machines: technological and social change in farm industries of Arizona. The University of Arizona Press, Tucson. 325 p.
    A detailed study of Southwest agricultural systems vis-à-vis labor that bears directly on portions of this paper.

25. TOLLEY, H. R.
    1940   Some essentials of a good agricultural policy. *In* G. Hambidge (ed), Farmers in a changing world, p. 1159–1183. U.S. Department of Agriculture, Washington, D.C.

26. SECKLER, D. W., AND A. SCHMITZ
    1969   Mechanized agriculture and social welfare: the case of the tomato harvester. University of California, Berkeley. Unpublished paper. 18 p.
    Although the authors address themselves to the problem of the displaced migrant worker, their frame of reference is broadened in the analysis to recognize the *basic* immobility of the migrant worker. Moreover the authors recommend general solutions which would reduce the *need* for compensation.

27. BONNEN, J. T.
    1969   The crises in the traditional roles of agricultural institutions. *In* V. W. Ruttan, A. D. Waldo, and J. P. Houck (eds), Agricultural policy in an affluent society, p. 48–62. Norton & Co, New York.

28. HATHAWAY, D. E.
    1969   The implications of changing political power on agriculture. *In* V. W. Ruttan, A. D. Waldo, and J. P. Houck (eds), Agricultural policy in an affluent society, p. 63–68. Norton & Co, New York.

29. EISELEIN, E. B.
    1969   Weststate U.S.A.; the association in the politics of water resource development. University of Arizona, Department of Anthropology. 271 p. Unpublished PhD dissertation.

30. LEWIS, O.
    1966   The culture of poverty. Scientific American 214(4):19–25.

31. PADFIELD, H.
    1970   New industrial systems and cultural concepts of poverty. *In* Poverty and social disorder: a symposium. Human Organization 29 (1): 29–36.

Arid-Lands Agriculture
and the Indians of the American Southwest

JAMES E. OFFICER

Office of International Programs and Department of Anthropology
University of Arizona, Tucson, Arizona, U.S.A.

ABSTRACT

MORE THAN ONE-SIXTH of the land in the arid states of Arizona and New Mexico lies within the boundaries of Indian reservations. The Indians occupying these reservations were among the first north of Mexico to develop sedentary communities with an agricultural base. Certain of these reservations today include some of the best farming and grazing land in the United States, and, in a few cases, the Indian tribes have legal rights assuring them an adequate supply of water to irrigate large acreages.

In spite of the presence of an agriculture tradition among the Indian tribes and the availability of fertile land and water, Indian reservations have over the past half century contributed far below their potential to the food and fiber production of the southwestern United States.

Among the historical factors which account for the relatively low level of agricultural development on Indian reservations are the following: the long struggle of some tribes to obtain title to their lands, the battle over water rights, the lack of adequate credit and technology, the individualization of tribal holdings which has resulted in the creation of uneconomic farming units, the failure of the federal government to make good on its promises to construct irrigation projects and related works, and the failure of tribal governing bodies to resolve critical internal problems. Some of these factors are comparable to those which have inhibited agricultural development in other parts of the world, including the rural areas of Mexico and the Maori Reserves in New Zealand.

As other lands formerly devoted to agriculture are converted to urban and industrial uses in the American Southwest, Indian reservations may be called upon to play a more vital role in food and fiber production; and the elimination of the social, political, and economic barriers inhibiting full development of these lands may become a critical aspect of regional economic planning in the area.

ARID-LANDS AGRICULTURE
AND THE INDIANS OF THE AMERICAN SOUTHWEST

James E. Officer

MORE THAN ONE-SIXTH of the total land acreage of the arid states of Arizona and New Mexico lies within the boundaries of Indian reservations (Fig. 1). Since the ancestors of these southwestern tribes were among the first Indians north of Mexico to develop sedentary communities with an agricultural base, we might reasonably expect their lands to be major contributors to food and fiber production today. That they often are not is partially attributable to the fact that some are unsuited to the cultivation of crops and to animal husbandry. However, this is only part of the explanation, for several of the reservations are located in highly fertile areas where irrigation farming is not only feasible but has been shown to be highly profitable.

Anthropologists Courtland Smith and Harland Padfield (1, p. 327) have noted that "in part the problems of irrigated agriculture and the subsequent economic development are not of land and water alone, but of the relations of land, water and social institutions." This thesis is well supported by the story of the development of agriculture and livestock raising on the Indian reservations of Arizona and New Mexico, and we will have an opportunity to consider its relevance in the present chapter.

THE SETTING

Indian lands, taken as a whole, represent a good cross section of the geography of Arizona and New Mexico. They include areas of low elevation, as well as some of the highest mountains in the region. While the hot and barren lands of the Papagos and the Navajos in Arizona may seem disproportionately to weight the Indian areas on the side of nonarable desert, consideration of the Pueblo villages on the Rio Grande in New Mexico and the several reservations in Arizona located along such streams as the Gila, the Salt, and the Colorado indicates a better balance. Furthermore, if we take into account the mineral wealth of the Hopi, Papago, and Navajo reservations in Arizona, and the large supplies of harvestable timber on the Navajo, Hualapai, San Carlos, and Fort Apache reservations in Arizona and the Mescalero Reservation in New Mexico, it is clear that, in the aggregate, Indian lands compare favorably with those not owned by Indians.

Few of the southwestern reservations are located so as to have flowing streams traversing or bordering them. This is a fact of nature in the entire area, and, in this respect, Indian landowners are no worse off than others.

A number of the Arizona reservations not close to flowing streams have access to irrigation water from river storage and underground sources. (The latter, in recent years, have become much less reliable than formerly because of the drain on aquifers as a result of agricultural development and population increases in the areas surrounding the reservations.)

Although some of the southwestern tribes made use of floodwaters for dry-farming, and a few still do, no reservation is situated in an area where dry-farming on a commercial scale is feasible today. The Hopis of Arizona plant and harvest crops without irrigation in an area so arid that only the most persistent and industrious of farmers would even consider making the attempt. However, for the most part these plots are limited to subsistence uses.

THE BEGINNINGS OF INDIAN AGRICULTURE

The sedentary tribes of the arid Southwest have been cultivating crops for both food and fiber since at least the two centuries immediately before the birth of Christ. The common aboriginal crop complex consisted of maize, beans, pumpkins, and cotton. A unique feature of the agriculture of the tribes on the lower Gila River was the "semicultivation" of certain seed plants, often referred to as native millets, which constituted an important source of food (2, p. 97 and chapter 8). This practice was observed in 1540 by Alarcón, the first Spaniard to reach the area, and was still in existence at least until the middle of the past century (3, p. 34–58). Cotton was, for certain groups, both a food and a fiber crop. Castetter and Bell (4, p. 198) note that the Pimas ate cotton seeds in great quantity.

In general, maize was the principal food staple because of its storage qualities and the fact that it could be produced in fairly large amounts. By modern standards, the yields of 10 to 12 bushels per acre appear low, but they seem to have been adequate for the subsistence-oriented Indian farmers. When greater yields were obtained, the surpluses could easily be stored for later use or for trading purposes.

In the Gila Valley of central Arizona, the prehistoric Hohokam, considered by many authorities to have been ancestral to the modern Pima Indians, constructed extensive irrigation works as early as 300 B.C. (5). There is little evidence, however, to suggest that the Hohokam or the later Pimas constructed elaborate diversion or water-storage structures of permanent character. Dam building was principally necessary in poor water years when streamflow was not sufficiently high to reach the openings of the irrigation canals.

While the most spectacular development of irrigated agriculture took place in the Gila and Salt River valleys near present-day Phoenix, Arizona, examples of irrigation ditches are available from other areas of the American Southwest. Woodbury *(6,* p. 38–39) in summarizing the archaeological evidence, notes the presence of ditches in the Verde Valley of Arizona, extending as far north as Flagstaff, and in the Four Corners region where the state boundaries of Arizona, New Mexico, Colorado, and Utah intersect. At Mesa Verde, in southwestern Colorado, archaeologists uncovered a large ditch nearly 4 miles in length. They also found ditches of somewhat lesser length in the Chaco Canyon area of northwestern New Mexico. The builders of these ditches are assumed to have been ancestors of the modern Pueblo Indians of Arizona and New Mexico.

The preponderance of the evidence for ditch irrigation outside central Arizona dates from the tenth to the thir-

teenth centuries *(6,* p. 39), considerably later than the earliest — and more elaborate — canal building efforts of the Hohokam.

Apart from the irrigation methods of the ancestral Pimas and Pueblo Indians, the early agriculturalists of the Southwest depended primarily upon one or another variation of flood-farming to assure a proper supply of moisture for their crops. In the arid desert country of southern Arizona, the Papagos farmed at the mouths of dry washes, counting upon the flow from seasonal rains to carry them through the crop cycle. This method is often referred to as "Akchin farming," the name deriving from words in the Pima language meaning "at the mouth of the wash." The Papagos also constructed small diversion ditches to carry runoff water from the nearby mountains to their delta fields.

The Yuman-speaking peoples of the lower Colorado and Gila rivers planted their crops along the fertile floodplains, putting their faith in annual overflow to provide needed moisture. Early reports *(2,* p. 67), indicate "cultivated abundance" in these areas.

Hack *(7,* p. 26) reports a variety of farming practices

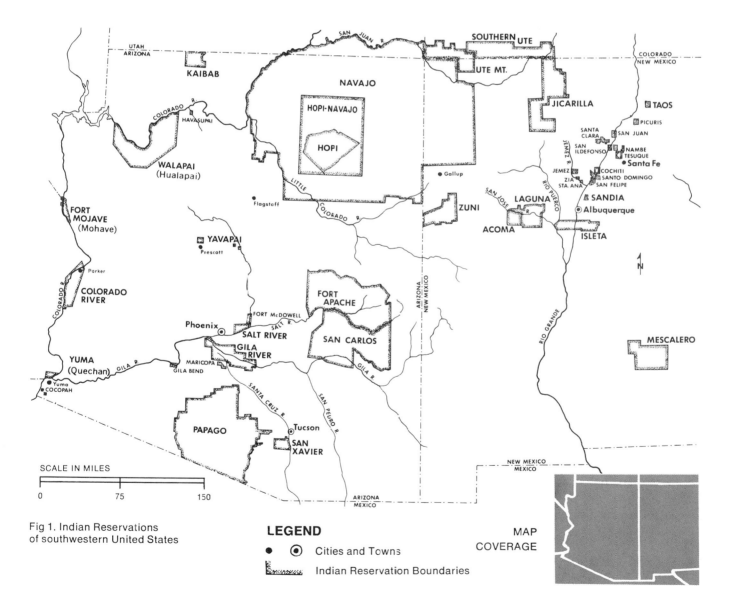

Fig 1. Indian Reservations of southwestern United States

LEGEND

● ◉ Cities and Towns

Indian Reservation Boundaries

MAP COVERAGE

among the Hopis of the high plateau of northern Arizona. Included were variations of flood-farming, planting in sand dunes, planting in areas with underground seepage, and limited irrigation, primarily using springflow. The Hopis, and to a lesser degree the Papagos, are among the few southwestern groups whose present-day farming techniques differ little from the ancestral ones.

The Navajos, neighbors of the Hopis, borrowed agricultural practices from them and from other Puebloan groups. The antiquity of farming among the Navajos, according to Young (8, p. 120), is indicated by the fact that the name applied to them by the Puebloans and later adapted into Spanish orthography means "cultivated fields." Spanish documents dating from the seventeenth and eighteenth centuries clearly distinguish the Navajos from other Apachean groups on the basis of their use of agriculture.

Young, in describing the rudimentary agricultural techniques of the Navajos, quotes from a manuscript by W. W. Hill of the University of New Mexico. According to Hill, the Navajos traditionally practiced three kinds of simple irrigation farming, all of which are presumed to have been borrowed from their Pueblo neighbors. These included (a) the interception of floodwaters on lands advantageously situated, (b) the diking of flatlands to catch and hold water from spring floods, and (c) limited ditch irrigation (in the Canyon de Chelly and Canyon del Muerto areas). The Navajo crop complex was the familiar one previously described, with melons being an important addition after the coming of the Spanish (8, p. 120–121).

In connection with aboriginal agriculture throughout the Southwest, it is of passing interest to note that each sedentary group devised its own schemes for harvesting the available water supply and putting it to use. From the variety of water-conservation methods described in this section, the reader may well infer that the early southwesterners displayed great ingenuity in getting the most from their arid environments.

SPANISH INFLUENCES

The first Indians of the Southwest to add significantly to their crop complex through diffusion from the Spanish were the Pueblo tribes of the Rio Grande valley of New Mexico, among whom Spanish colonization began in 1598. These Indians had for hundreds of years grown maize, pumpkins, beans, and cotton, primarily through flood-farming, but with some use of simple irrigation. The Pueblos north of the village of Cochiti apparently were not able to cultivate cotton because of climatic conditions (see Lange 9, p. 95). It may be presumed that irrigated agriculture became more prominent after the arrival of the Spanish, both because of the exposure of the Indians to new techniques and because the presence of Spanish settlers in the area provided new markets for disposing of surplus harvests. The Dominguez report of 1776 and the report of Friar Antonio Cavallero in 1779 (9, p. 80) mentioned "fields irrigated by deep and wide

ditches west of the [Rio Grande] river, above and below the pueblo [of Cochiti]."

The most important of the crops introduced by the Spanish in New Mexico was wheat. We may infer that the introduction of this grain provided an additional stimulus to expansion of the aboriginal irrigation systems, since nowhere in the Southwest except in areas of extensive irrigation did wheat apparently become a major crop (2, p. 123–124).

Among the other European introductions which attained importance within the early period of Spanish colonization were chilis, watermelons, muskmelons, onions, cabbages, beets, string beans, peas, *habas* and *garbanzos*. Fruit trees were also popular, peaches and apples being the most common, with plums, apricots, and cherries somewhat less numerous (9, p. 99).

Outside the present area of New Mexico the diffusion of the European crops mentioned above tended to precede the arrival of the first Spanish colonists, knowledge about new crops being passed along from tribe to tribe.

Watermelons were among the earliest European importations in the Gila and Colorado river areas of Arizona. Kino (10: v. 1, p. 249) found melons growing near the junction of the Gila and Colorado in 1700. Sedelmayr (11, p. 108, 110) also observed melons in cultivation along the lower Colorado River in 1744, and members of the Anza California Expedition (12: v. 3, p. 42, 51, 227, 228, 323) recorded that the Yuma Indians offered them an estimated 3,000 watermelons in 1775.

Castetter and Bell (4, p. 114) speculate that wheat had been received from the south by the Pimas of the Gila basin before the arrival of Father Kino in 1687. Kino himself distributed wheat seed in considerable quantity among these Indians, who quickly began to cultivate it to such an extent that it replaced maize as their principal crop.

Anza (12: v. 1, p. 184; v. 2, p. 240, 304), writing of a Pima village in 1774, said their fields of wheat were so large that, "Standing in the middle of them, one cannot see the ends, because they are so long. Their width is also great, embracing the whole width of the valley on either side. . . ."

Wheat had not become established in the lower Gila and Colorado river areas when Kino visited the region. This seems clear from his statement (10: v. 1, p. 373) that "I sent wheat to sow at the Colorado River and in the Yuma and Quiquima nations, grain and seed which had never been seen or known there, to see if it would yield there as well as in those other fertile new lands; and it did yield and does yield very well indeed."

In spite of Kino's enthusiasm, wheat did not catch hold among the Yuma-speaking people as it had among the Pimas of the Gila. Castetter and Bell (2, p. 123–124) relate this to the fact the Yumans were primarily flood-farmers, and on the Colorado there was sufficient autumn overflow only about every other year to make possible the germination of wheat. Also, the spring floods, coming usually in May and June, were too late to do the wheat crop much good. In contrast, the peak flow on the Gila occurred in January, February, and March, the perfect time for irrigating wheat.

The area in which the Hopis were living at the time of the coming of the Spanish was not suited to the growing of wheat. Large-scale irrigation was impossible in this arid plateau region. In spite of this fact, some wheat *was* grown by the Hopis at an early date. Archaeologist J. O. Brew found wheat straw embedded in the adobe walls of the old mission at Awatobi *(13)*. Far more successful as a Spanish import were peaches, which have constituted an important item of the Hopi diet since the seventeenth century *(14,* p. 9). Onions, chilis, and melons were other crops introduced by the Spanish which were added to the Hopi complex. Curiously, the modern Hopi word for watermelon is a combination of the corrupted Spanish word for horse and the Hopi word for squash *(14)*.

The Navajos, too, borrowed from the Spanish many of the vegetables which they had introduced elsewhere and in certain spots planted fruit trees. Wheat, however, failed to become a significant crop among these Indians because of the absence of large irrigable plots.

In summarizing the Spanish influence on southwestern Indian agriculture, it may be said that the principal contribution of these early European explorers and settlers was in adding to the existing crop complex and in diffusing certain new agricultural implements. While in some areas, principally the Pueblo region of New Mexico, the Spanish may have also introduced more sophisticated irrigation techniques, the real changes in aboriginal methods did not come about until after the Anglo-Americans entered the area in significant numbers and the region came under the control of the United States.

THE COMING OF THE ANGLO-AMERICANS

The Puebloans

In the period between the Pueblo Revolt of 1680 and the assumption of control over New Mexico by the United States government in 1846, we may presume that the Pueblo Indians of New Mexico were hard pressed even to retain their previous level of agricultural productivity and sophistication. Spicer *(15,* p. 162–169) has detailed the turbulence of these years, beginning with the breakdown of unity among the Puebloans, continuing with their constant strife with the Spaniards (who had regained control of the area by the late 1690s) and then worsening as a result of incessant raiding of their villages by nomadic, neighboring tribes. By the middle of the eighteenth century *(15,* p. 166) the Pueblo population had been reduced by at least half, a majority of the communities had been depopulated, and economic prosperity had severely declined. A heavy smallpox epidemic hit the Rio Grande valley in 1780–81, causing a further reduction in both population and prosperity.

Peace and prosperity did not come quickly to the Pueblo area with the assumption of Anglo-American control. Spicer *(15,* p. 169) notes, "When General Kearny marched into Santa Fe in 1846, he found that he had taken possession of a region which had become inured to unpredictable, periodic raiding by the Navajos and Apaches, and that . . . the Pueblo villages regarded themselves as in a perpetual state of siege. . . ." It was not until 1864 that the Navajos were thoroughly defeated and the Bureau of Indian Affairs of the federal government could carry out a program with any consistency among the Pueblos.

By 1870 *(15,* p. 171), the Pueblos were beginning to expand their farming operations with some assistance from the resident agents of the Bureau of Indian Affairs. However, a new crisis was brewing which shortly thereafter dealt them a severe setback. With the reduction of the danger from marauding tribes, the Mexican-American farmers along the Rio Grande also increased their farming effort, often encroaching on Indian lands in the process. As Anglo-American settlers poured into the region, additional pressure was exerted on both the Indians and their Spanish-speaking neighbors. As a protective measure, some Indians acquired land by purchase. However, the majority simply stood by and watched as others moved into their territory. The incoming Anglo-Americans not only appropriated lands which the Indians had long used — in some cases without benefit of title — but they moved in sheep and cattle in such quantities that thousands of acres were soon overgrazed and denuded of vegetation. One result of this overgrazing and the extensive timber cutting which accompanied it, was a great increase in water runoff. The floods which followed washed away Pueblo ditches and fields.

The combination of floods and occasional drought produced such an impact on Pueblo agriculture that in 1895 agent Dolores Romero *(16,* p. 158) called upon the federal Indian Commissioner's office for assistance. By 1901 *(17,* v. 1, p. 64–65), a limited amount of help was forthcoming, notably at San Ildefonso and Zuni, where ditches were widened and deepened at an expenditure of slightly more than $14,000. Subsequent reports of the Indian Commissioner reveal additional efforts of the Indian Bureau, mostly on a small scale. The largest project constructed during this period in the Pueblo area was a dam and accompanying ditches at Zuni, for which the Bureau spent $300,000 over a period of about 5 years *(18,* v. 2, p. 57).

In spite of Indian Bureau efforts to improve farming opportunities, the Pueblo situation continued to worsen. By 1913, 3,000 non-Indian families had moved onto Indian lands. Only two villages — Acoma and Zia — still held intact the lands originally granted to them *(15,* p. 172). Adding to the woes of the Pueblos was the fact that the Indian Bureau was not assisting them in protecting their property against outsiders. A decision of the United States Supreme Court in 1876 had held that, by virtue of having received their lands originally through Spanish grants, the Pueblo Indians could not be considered in the same class as others whose reservations had been created through action by the United States. Uncertain as to whether the Pueblos were legally Indians under their jurisdiction, Indian Bureau officials stood by while non-Indian settlers appropriated Pueblo lands.

In 1913 the Supreme Court reversed the 1876 decision in deciding the so-called Sandoval Case (231 U.S. 28). The issue in this lawsuit was whether Pueblo villages were subject to federal liquor laws applicable to "Indian country." The Court held that lands occupied by "distinctly Indian communities" dependent upon the federal government were a part of Indian country, and entitled to federal protection (19, p. 16).

The result of the Sandoval decision was a chaotic period during which the courts attempted to adjudicate village boundaries and determine the rights of squatters. Finding judicial settlement a slow and uncertain process, Congress in 1924 passed the Pueblo Lands Act, providing that all Spanish grant lands lost by the Indians would be restored to them. In the event this could not be done, provisions were made for payment of reparations. Even with this act, however, another decade and a half passed before all the disputed titles were cleared.

With the land question settled, the Pueblos were able to move forward with agricultural development. Help from the Indian Bureau between 1935 and 1945 made possible the cultivation of an additional 1,000 acres. The Indians began sowing new crops and adopting cooperative practices of buying and marketing (20, p. 21).

When the Middle Rio Grande Conservancy District was organized during the 1930s, lands of six of the Pueblo villages were included. This development has proved beneficial to the communities of Cochiti, Santo Domingo, San Felipe, Sandia, Santa Ana and Isleta. The District provides drainage and flood control and maintains a main canal system for all irrigable lands, Indian and non-Indian. The Bureau of Indian Affairs during the past 10 years has assisted the middle Rio Grande Pueblos to subjugate an additional 5,000 acres of land, and plans have been developed for turning the remaining 16,000 acres of irrigable land into more economic farm units.

Plans to increase the extent and effectiveness of Pueblo Indian agriculture may not yield significant results in the near future, however, because of uncertainties surrounding the water rights to which these Indians are entitled. The issue of Pueblo water rights has been pending for more than 100 years, but a new urgency has now developed to get the matter settled. This urgency stems from the initiation of the so-called San Juan-Chama Reclamation Project. Included as part of the same legislative package which produced the Navajo Irrigation Project (see below), the San Juan-Chama Project will, when completed, provide for the diversion of approximately 235,000 acre-feet of water from the headwaters of the San Juan River in sparsely settled northwestern New Mexico across the Continental Divide into the headwaters of the Chama River for use in the Rio Grande Valley.

The New Mexico State Engineer, who will administer the Project, feels that accounting for all the imported water will require knowledge on his part of the present water rights of all users in the watersheds to be benefitted by the importation. Since the Pueblo Indians own large portions of land within these watersheds, New Mexico has brought suit against them in several separate but related actions to adjudicate the waters of the Rio Grande tributaries, including the Chama River, which will be used to transport the imported water to the Rio Grande. The United States government has entered these suits in behalf of the Indians. The legal issues involved are extremely complex, and arguments advanced by the federal government to support the claims of certain Pueblos may, should they prevail, threaten the interests of others. The Indians, therefore, have not been able to speak with a unified voice in the matter. Settlement of the lawsuits may take many years.

While the Pueblos have suffered more than other southwestern tribes as a result of the encroachment of non-Indians onto lands formerly occupied by them, incidences of trespassing occurred elsewhere, notably among the Papagos of southern Arizona, for whom a large reservation was not established until 1917. Furthermore, such tribes as the Pimas were adversely affected in other ways by the movement of Anglo-Americans into the Southwest. Foremost among the problems of some of these other tribes has been that of water rights, with which the Pueblos also are now wrestling.

The Pimas

At the time the U.S. government assumed control over the area in which the Pimas lived (1854), these Indians were unquestionably the leading agriculturalists of the Southwest. Reference has already been made to their extensive wheat fields during the eighteenth century. In 1858, the first year of the Overland Mail Route, the Overland Company purchased a 100,000-pound surplus crop of Pima wheat. In succeeding years, the Pimas sold wheat in increasing amounts to the federal government for the use of troops operating in the territory. By 1862, the government was purchasing wheat at the rate of about 1,000,000 pounds per year from the Pimas (21, p. 110).

A flood in 1868 inaugurated a dry cycle, which lasted for 5 years on the Gila River. The absence of rainfall during this period, combined with increased diversions of river water by non-Indian farmers upstream, dealt a severe blow to Pima agriculture. (It is ironic that non-Indian settlement was made possible through subjugation of the hostile Apaches by military forces eating bread made from Pima wheat and who, often, were guided by Indian scouts from the Pima villages.) Because of the absence of river water for irrigation, an estimated 1,200 Pimas left the Gila River Reservation in 1872–73 and took up residence on the nearby Salt River (22). At this same period other Indians moved from farming locations near the Gila River to spots on the reservation where seepage water was available most of the year (23, p. 42).

The Bureau of Indian Affairs did make some token efforts to help the Pimas during the early 1870s. However, these efforts were directed more toward enabling the Indians to make full use of the limited water available to them than toward battling the neighboring non-Indian farmers over water rights in the river. The first Bureau allocation for irrigation on the Gila River Reservation came in 1873, when a small amount was used to repair ditches. Such expenditures continued until about 1903 *(24,* p. 3).

As an alternative to remaining on their water-scarce reservation, the Pimas, in 1874, were invited to consider relocation to Oklahoma territory *(23,* p. 45). When the Indians rejected this alternative, the federal government responded by enlarging the boundaries of the reservation. However, the new lands were no more suitable for irrigation agriculture than those which the Indians had previously farmed.

After 1880, the Pima situation became steadily worse. Many Mormons from Utah began about this time to found settlements on the upper Gila River near the modern community of Safford, Arizona. Some 3,000 Mormons entered the Safford area between 1880 and 1900 *(23,* p. 46). Through drilling wells and constructing irrigation ditches, they further deprived the Indians of water from the river. In 1887, the Florence Canal Company appropriated the entire flow of the Gila for the benefit of non-Indian farmers in the Florence area.

By 1895, the situation among the Pimas had deteriorated so badly that the government was obliged to purchase and distribute among them some 225,000 pounds of wheat as rations. Since the cost of these rations amounted to about $30,000 per year, federal officials began to talk of doing something to restore the reservation residents to their previous state of self-sufficiency. In 1896, the Indian Bureau conducted a study of the feasibility of reservoir storage, and in 1901 and 1902, legislation was introduced to construct a large dam *(23,* p. 54). Congress did not act on this legislation. The response of the Indian Bureau was to abandon the idea of a dam and recommend, instead, an irrigation project dependent upon electric pumps and a new canal system within the reservation. Beginning with the drilling of five experimental wells in 1905, the Bureau of Indian Affairs, with help from the Reclamation Service, constructed a series of wells, canals, and diversion dams over a period of nearly 20 years. However, without a major storage reservoir on the river, much of this effort was wasted and the Indians had little incentive to return to farming on the previous scale.

While this activity was taking place, the subject of a reservoir was still being discussed in various quarters. In 1905, the Reclamation Service reported that reservoirs sufficient to irrigate 40,000 acres could be constructed on the Upper Gila and San Francisco rivers. The Southern Pacific Railroad, whose tracks traversed a part of the area which would be covered by such a reservoir, objected, and the plan for a dam at San Carlos was laid aside.

By 1912, the Board of Army Engineers began a study of the feasibility of a San Carlos reservoir and continued to investigate the matter until 1915, when it recom-

mended that this project be deferred until diversion dams downstream could be completed. By 1924 these were nearly complete, and in that year Congress finally authorized the construction of a major storage reservoir near San Carlos *(24,* p. 6–8). The terms of the legislation provided for the establishment of an irrigation district of 100,000 acres, one-half of which was to fall inside the reservation for the benefit of the Indians; the other half would be located in the Florence / Coolidge / Casa Grande area for the benefit of the non-Indian farmers who had settled in the region. The act did not designate the Indian land to be included, nor did it apportion water among the Indians and the other users.

The decision to build the storage reservoir did not solve the agricultural problems of the Pima Indians. A decade before (during the period 1914–1921), the Indian Bureau had, in carrying out the terms of the Dawes Act of 1887, allotted parts of the reservation among Indian family heads. The standard allotment was 10 acres of irrigable land and 10 acres of desert. These small acreages were not patented to individual Indians with any consolidation for future water-resource development in mind *(23,* p. 74). The attempts to work out an allotment exchange program in order to achieve this consolidation were, to a large degree, unsuccessful, and until 1969 — when the tribes and the Indian Bureau finally reached agreement — no one could say precisely which reservation lands lay within the Project.

By the time the attempt at consolidation was made, two other events of significance had taken place which further affected the Pimas in reestablishing an agricultural economy. To begin with, some of the original allottees had died, and their small holdings had been inherited in common by large numbers of heirs, greatly complicating the management problem. This problem will be dealt with at greater length in a subsequent passage.

The second development that affected Pima agriculture was the shift from subsistence farming to cotton farming and mechanization of the areas surrounding the reservation. World War I was the principal stimulus for this development, the Goodyear Tire and Rubber Company initiating it on properties controlled by them near Phoenix. Prior to this time, the non-Indian farmers, like the Indians, had grown crops primarily for subsistence.

The struggles of the Pimas to obtain water and of the Pueblos (and to a lesser degree, the Papagos) to hold on to their land represent the fiercest battles between Indians and non-Indians following the subjugation of the hostile tribes in the 1880s. However, other tribes had problems of a different sort.

The Colorado River Tribes

The Colorado River Reservation, inhabited principally by Mohave and Chemehuevi Indians, was established in 1865. In comparison with all the other tribes of the Southwest, these Indians were very favorably situated. The Colorado River, a flowing stream of great size, lay alongside their lands, and, although the Mohaves and other Yumans had previously relied heavily on flood farming, the opportunities for irrigated agriculture were abundant. Taking note of this fact, Congress in 1867 appropriated $50,000 for the construction of a large canal. This was the first irrigation project on an Indian reservation, and employees of the Indian Bureau today like to point out that neither it, nor any other Indian irrigation project, has ever been completed (25).

The first canal and headgate were completed in 1869 and water was turned into the canal on July 4, 1870. To the great consternation of everyone, the water rushed forth in much greater quantities than had been expected. As a result, the banks washed away and floodwaters spread over the land. When the level of the river subsided, the headgate was repaired and the canal was used successfully for a brief period in the summer of 1871 (26, p. 4). The Bureau immediately began surveys to determine how the difficulties could be overcome and finally approved a recommendation that the canal be extended up the river (to gain elevation) and that heading be provided by tunneling through Headgate Rock. Tunnel construction began in 1872. The Indian laborers found the work difficult and demanded pay instead of the food and clothing rations they had formerly received. To quiet the ensuing disturbance and force the Indians to return to work, the Indian Agent, in August 1873, brought in a troop of soldiers. With their assistance, the work proceeded and the tunnels were dedicated in a memorable celebration on June 23, 1874.

From the beginning, the tunnels also caused trouble. Within a few days, one caved in so badly that it prevented any further use of the canal at that time.

Despite the excellent irrigation potential of the reservation, the Indian Bureau was not able for many years after 1874 to secure the necessary appropriations to make large-scale agriculture possible. Small amounts were allotted to repairing canals and to constructing irrigation facilities making use of steam pumps, water wheels, and windmills. When the river was high, the canals carried some water; but in other years they were dry. The reports during this period (26, p. 5) mention only one significant crop year (1884) when there was sufficient water available to produce a bountiful harvest.

More efficient pumps were placed into operation with the aid of larger allocations of funds in 1912 and 1918. In the latter year, Congress also appropriated $50,000 for a study which might lead to the subjugation and irrigation of up to 150,000 acres of reservation land. Using this appropriation and another made in 1919, the Indian Bureau undertook an extensive survey which resulted in a report (issued in 1920) setting forth a plan for the irrigation of 104,000 acres. The report called for a diversion dam at Headgate Rock, a main canal with distribution system, a protective levee, and drainage works. Fifteen years went by before Congress, by the Act of August 30, 1935 (49 Stat 1039), authorized the construction of Headgate Rock Dam across the Colorado River. Two more years passed before an appropriation was made for this purpose, and the dam itself was not completed until 1941, at a total cost of $4,632,775.

At no time prior to the beginning of work on Headgate Dam had the number of irrigated acres on the reservation ever exceeded 10,000, in spite of the fact that the Bureau had devoted more than $1,000,000 to this effort over the preceding three-quarters of a century (27, p. 29, 31).

Except for the Quechans (usually known as the Yuma Indians) at the junction of the Gila and Colorado rivers, and that segment of the Mohave and Chemehuevi tribes on the Fort Mohave Reservation near Needles, California, the other tribes of the Colorado River have always been less favorably situated for the development of agriculture than those we have been describing. The Havasupais engaged in subsistence farming in their small homeland at the bottom of the Grand Canyon, and their near relatives, the Hualapais, did some small-scale farming on the high plateau above the river. Neither group has any appreciable potential for agriculture today. The Hualapais do raise cattle, but their herds are small.

The Fort Mohave Reservation has a substantial water right, sufficient for the irrigation of nearly 19,000 acres, but no farming is being done there today. The richest of the bottomlands belonging to this reservation were lost in 1940 as a result of the construction of Parker Dam and the reservoir known as Havasu Lake.

Members of the Quechan tribe, who practiced flood-farming extensively prior to the coming of non-Indians, own slightly more than 7,000 acres of irrigable land, much of which is farmed under lease to non-Indians. A reservation for these Indians was established on both sides of the Colorado River near present-day Yuma in 1884. Following the passage of the Dawes Act (also known as the Allotment Act) in 1887, the federal government negotiated an agreement (on the order of the treaties of an earlier period) with the Quechans by the terms of which the Indians were to give up their reservation in exchange for individual allotments. The irrigable lands of the reservation remaining after all eligible Indians had received allotments were to be sold to non-Indians under the general land laws of the United States and the proceeds from such sales deposited in the United States Treasury to the credit of the Indians.

Both the Indians and local representatives of the Indian Bureau felt confident in 1894 that irrigated agriculture would soon come to the reservation and the surrounding area by virtue of the efforts of a private company which was to construct the necessary facilities. Unfortunately for all, the venture failed. Thereupon, Congress, in 1904, made provision for including the irrigable lands of the old reservation within the Yuma Federal Reclamation Project being undertaken pursuant to the Reclamation Act of 1902 (32 Stat 388). In taking this action and a subsequent one in 1911, Congress legislated that Indian allotments would be 10 acres in extent and that the cost of irrigating these allotments would be repaid to the United States government out of any funds which might be realized from the sale of excess lands. Furthermore, the *unpaid* costs of the Indian portion of the project were to constitute a lien against the allotments. (The 1893 agreement had provided that *non*irrigable lands within the reservation would revert to the public domain with no payment to the tribe, and Congress did not change this portion of the agreement.)

The allotment of Yuma lands took place in the early 1900s, following the act of 1904, but some of the irrigable lands of the old reservation were never sold to non-Indians as the act had provided. Instead, they were withdrawn for use by the Reclamation Service. The Indians were not compensated for these lands, nor have the lands been restored to them, although legislation to accomplish the latter has been proposed. The Indians have also recently sued the federal government, asking to be paid for all the irrigable lands not disposed of under the provisions of the 1904 act. The lawsuit over this issue is still pending before the Indian Claims Commission.

The Navajos

On the vast Navajo Reservation, which includes parts of the states of Arizona, New Mexico, and Utah, the federal government began to provide its first assistance to Indian farmers and stock raisers about 1880, with a small appropriation of $3,500, which was used to supply windmills and stock pumps for a limited irrigation system. An effort was made about this same time to build a dam at Blue Canyon, but floods washed it away. Over the ensuing half century, the Indians and the federal government individually or collectively completed about 67 small irrigation projects across the reservation, ranging in size from a few acres to several thousand. Within these projects, land was farmed by family groups. All the early efforts at agricultural improvement were related to subsistence farming (*8*, p.121–123).

In 1948 the Bureau of Indian Affairs reported that there were 78 small irrigated tracts on the Navajo Reservation, aggregating 23,000 acres and capable of supporting around 400 families on a subsistence basis. Furthermore, it was estimated that the reservation contained nearly 59,000 acres capable of agricultural development (exclusive of the Navajo/San Juan Project, which will be discussed later).

Congress, in 1950, passed the Navajo-Hopi Long Range Rehabilitation Act (Public Law 474 — 81st Congress) to make a concentrated attack on the social and economic problems of these two Indian tribes. By the terms of the act, the sum of $9,000,000 was authorized to provide maximum development of the irrigation farming potential of the reservation. During the decade between 1950 and 1960, 5,134 acres of new farmland were placed under irrigation. Furthermore, irrigation canals and other structures sufficient to serve a much larger acreage were completed (*8*, p. 123–125).

Of all the irrigated farmland developed on the Navajo Reservation, more than 50 percent lies in the Shiprock-Fruitland area, along the San Juan River in northwestern New Mexico. Even more intensive agricultural development of this comparatively well-watered river valley will take place when the 110,000-acre Navajo Project (see below) is built.

The Salt-River Pima-Maricopa Community

In an earlier passage, we mentioned that some of the Pima and Maricopa Indians from the Gila River Reservation had migrated from their traditional villages in the early 1870s and taken up residence along the Salt River near its junction with the Verde. Although they had done so at the invitation of non-Indian farmers in the area, within a short time the increasing numbers of non-Indian settlers entering the Salt River Valley came to constitute a threat to their remaining in the new location. Taking note of this fact the Bureau of Indian Affairs recommended the establishment of a reservation, and one was created in 1879.

The new lands that these Indians occupied were well suited for irrigated agriculture, and the federal government aided them in constructing ditches and diversion dams. Under the Kent Decree of 1910, the Indians received an adjudicated right to nearly 19,000 acre-feet of water, an amount sufficient for 3,434 acres. In the contract for the construction of Bartlett Dam on the Verde River — completed in 1939 — the federal government reserved 20 percent of the stored waters for reservation use, enough to irrigate an additional 6,310 acres. To supplement the water supply for agricultural use, the Indian Bureau drilled three deep wells, and non-Indian lessees have drilled ten more. Approximately 4,000 acres are served by these wells, making a total of nearly 14,000 irrigated acres within the reservation boundaries (*28*, p.68).

The Yavapais

Adjacent to the Salt River Pima-Maricopa Community is the Fort McDowell Reservation, occupied by members of the Yavapai Tribe (also known as the Mohave-Apache). Except for those lands lying along the banks of the Verde River, little of the acreage on this reservation is susceptible to irrigation agriculture, and farming has never been important either for subsistence or cash income. A similar situation prevails at the tiny Camp Verde Reservation, upstream from Fort McDowell, which is the home of another band of Yavapais. This reservation, which is actually divided into two parts, was purchased for Indian use by the federal government in 1916. Included in the purchase price were water rights and an interest in the existing system of irrigation canals, which have since served both Indian and non-Indian lands. Fewer than 100 acres of the reservation have been consistently cultivated (*28*, p. 58).

The Papagos

In an earlier passage, we indicated that the Papago Indians of southern Arizona had engaged in farming, primarily at the mouths of arroyos, prior to the coming of the Spanish. One band of Papagos residing at the village of San Xavier del Bac, south of Tucson, had also made considerable use of irrigation (*29*). By diverting water from the intermittently running Santa Cruz River, they were able to farm an area estimated at 1,500 acres

(*30;* chapter 3, p. 6). Although a reservation was established for these Indians in 1874, the Indian Bureau did little prior to 1914 to assist them with expanding or improving their farming operations. By the latter year, the river had cut such a deep channel that surface irrigation by diversion was no longer possible. The federal government then drilled a well, and in subsequent years added six more (*30;* chapter 3, p. 11). The reservation supply of groundwater has been consistently good in this area, making it possible to keep pumping costs low.

At another location on the Santa Cruz, downstream from San Xavier, the Papagos of the community of Chui Chuischu were diverting a small amount of water for their fields when the Indian Bureau first began providing services to them in 1912. Fewer than 200 acres were cultivated in this fashion. Beginning in 1914, the Bureau provided funds for drilling wells and lining the main canals. By 1937, the irrigable land had been expanded to 615 acres (*31,* p. 5–6). An estimated 1,028 acres were being farmed by irrigation in the Chui Chuischu area in 1959 (*30;* chapter 3, p. 11).

The Bureau of Indian Affairs, especially during the 1930s, also directed funds to the construction of limited irrigation facilities at other locations on the main Papago Reservation, as well as on the reservations at Gila Bend and Ak Chin (Maricopa), where small groups of Papagos reside. However, these projects were extremely small in scope.

The Papago Development Program, proposed in 1949, urged the expenditure of $4,500,000 for the irrigation of an additional 15,000 acres. Funds for this purpose were never appropriated, however. In 1957, the tribe leased 7,500 acres in the southern part of the reservation to a private corporation, which agreed to subjugate 5,600 acres of this land and drill a total of 18 wells over a period of 10 years. By 1960, they had drilled six wells and brought 1,920 acres under cultivation (*32,* p. 90). Unfortunately, this project failed and the lease was cancelled in 1961. Since then, neither the tribe nor the Indian Bureau has been able to develop a mutually satisfactory alternative proposal for the use of the land.

The Papagos (and some Pimas) who live on the small Ak Chin (Maricopa) Reservation south of Phoenix have had considerably more success with development leases like the one described above. By 1953 (*32,* p. 72), this group had leased out 6,400 acres. Under the terms of these contracts, the lessees were obliged to drill wells and subjugate land. Upon expiration of the leases, the land reverted to the Indians for management by a tribal farming corporation. By 1968, the tribe had taken over the management of an additional 3,500 acres of irrigated lands with a gross annual return in excess of $1,000,000.

The Apachean Tribes

The Apachean tribes of the Southwest, with the exception of the Navajos, were not agriculturalists by tradition. Nor have they taken up farming to any extent on the four reservations that they occupy in Arizona and New Mexico. The Indian Bureau has provided some irrigation facilities to serve these reservations, and there is still a slight potential to be exploited. However, the limitations on farm development among the Apacheans are manifest in estimates of the Bureau of Indian Affairs that the two large reservations in Arizona (San Carlos and Fort Apache) have a combined total of no more than 10,000 acres which could be converted to agricultural use through irrigation (*33*). On the other hand, all four reservations are well suited for stock raising, and the Apaches have gained considerable fame for the quality of their herds of cattle.

INDIAN LIVESTOCK DEVELOPMENT

The Spanish colonists introduced sheep and cattle in small numbers among the Pueblo villages at the end of the sixteenth century. Very gradually, these animals assumed a solid place in the Pueblo economy. Spicer (*15,* p. 546) notes that meat and wool were staple products in the Rio Grande area by the early 1700s, but in no Pueblo village had livestock assumed the place of more than an adjunct to the still-dominant small-scale farming.

The most far-reaching effects of the introduction of livestock were upon the less sedentary tribes, for within a century and a half this innovation changed their whole way of life. The Apaches and Navajos, who at best had depended only slightly on farming, began to turn to raiding as an economic pursuit, preying upon the herds and flocks of the Spanish colonies and sedentary Indian communities. Spicer (*15,* p. 547), observes:

[I]t is certain that they undertook to live to a large degree by raiding in the period marked by the beginning of the 1700's. They knew of and now wanted horses, and proceeded to acquire them in small numbers by raiding the Spanish settlements, ranches, and haciendas that had begun to appear in New Mexico, Sonora and Chihuahua. Those Athapaskans who became differentiated from the others as Navajos in the north also became interested in sheep and goats, as well as horses, sheep having thrived in the New Mexico environment since their introduction by the Spanish. On the other hand, the Athapaskans of the south, later called the Apaches, were raiding settlements which had begun some cattle, rather than sheep, raising.

As the Navajos became more accustomed to the use of horses as riding animals, they developed into highly skilled raiders. By about 1750, they were carrying off large numbers of sheep, which they used both as an immediate source of meat and as domestic animals. Steadily the size and numbers of their flocks increased, so that by the early nineteenth century, sheep herding had assumed a place in their lives of at least equal importance with farming (*15,* p. 547). The Apaches, on the other hand, did not develop an interest in herding. They kept strings of horses for raiding purposes, but did not themselves attempt raising either sheep or cattle. To the Indians of both groups, the horse became an important symbol of wealth. The Navajos used horses for both riding and subsistence.

The Jesuit priest Francisco Eusebio Kino introduced cattle raising among the Pimas and Papagos in the late eighteenth century. Prior to his coming, these animals were doubtless known to the Indians of southern Arizona, since they had been previously introduced into the north Mexican state of Sonora (*29*, p. 69). Most cattle in the eighteenth and early nineteenth centuries were probably semiwild and formed only a minor part of the subsistence base of the Pimas and Papagos.

Soon after the Gadsden Purchase of 1853, which brought the traditional homeland of the Pimas and Papagos into the United States, Anglo-American ranchers began moving into southern Arizona, invading the territory long occupied by the Papagos and appropriating the watering holes upon which Papago cattle depended. This undoubtedly resulted in diminished Papago herds. Finally, shortly after 1900, the Bureau of Indian Affairs began drilling wells for stock water in the Papago area; the modern period of cattle raising dates from this era.

The Papagos were not the only Indians who received help and encouragement from the Bureau of Indian Affairs to engage in stock raising. In the late 1860s, following their conquest and subsequent imprisonment at the hands of the Anglo-Americans, the Navajos were placed on a reservation and issued sheep with which to begin the establishment of new herds.

Sheep first became commercially important to the Navajos in the late nineteenth century with the development of a market for Navajo blankets, woven from wool. As their flocks increased, the Navajos also began to find markets for the sale of lambs, wool, and hides.

By the 1920s, Indian Bureau agents and others began to perceive that the Navajo flocks and herds of horses were increasing at a rate far faster than the available range could accommodate them. Within a decade thereafter, the Bureau and the Soil Conservation Service initiated a reservation-wide program of stock reduction. Using coercive methods which have left much bitterness among the Navajos, they succeeded by 1948 in reducing stock numbers to what the range technicians considered a proper carrying load. Navajo sheep raising was reportedly on a sound basis for continuance, although it was here, as elsewhere in the vicinity, a marginal enterprise (*15*, p. 548–549).

A similar, though somewhat less drastic, program for curtailment of excess stock numbers was undertaken on the Papago Reservation during the 1930s. As a result of the well-drilling efforts of the Indian Bureau, Papago herds had increased to the point where approximately 30,000 head of cattle and an equal number of horses were grazing on the large reservation. These numbers were far in excess of the carrying capacity of the range, much of which was undergoing serious erosion.

The Papagos did not willingly accept livestock reduction proposals. Many concealed the number of horses and cattle which they owned in order to avoid having to reduce their holdings. Most of the Papago families kept herds ranging in size from one to ten cows, and, although these numbers appear extremely small, they provided the subsistence-oriented Papagos with their principal source of cash income (*30;* chapter 4, p. 2).

During the 1930s, in spite of vigorous actions by the Indian Bureau, the Papagos were still grazing 27,000 head of cattle and 18,000 head of horses. Continued action, coupled with periods of drought during the 1940s, brought these numbers down significantly, so that by 1950 they were approaching the carrying capacity of the range. However, in the period since then, the cattle herds have been increasing at a faster rate than range conditions warrant.

On both the Papago and Navajo reservations, the programs of stock reduction were accompanied by efforts to improve the water supply and to rehabilitate the range. The federal government gave particular attention to these matters during the 1930s.

Stock raising among the Pueblo Indians has in only a few cases ever been as important as farming. As late as 1934, most of the Puebloans raised cattle and sheep for subsistence, with the excess being driven to the local traders in exchange for credit. Livestock were not weighed, but sold by the head (*20*, p. 17). Beginning about 1935, the Indian Bureau offered help to the Pueblos in improving their range and fencing it, constructing stock trails, corrals, and water tanks, and improving the quality of their animals. Furthermore, they assisted the Indians in developing new markets.

Aberle (*20*, p. 20–21) points out that as a result of this attention to livestock raising, the Pueblos were able in the four-year period between 1938 and 1942 to increase significantly the size of their calf and lamb crops, the weight of yearling lambs, and the production of wool. Only two of the Pueblos — Acoma and Laguna — were obliged to undertake stock reduction programs of any magnitude.

The raiding economy of the Apaches was cut off sharply with their confinement to reservations in the 1870s and 1880s. Although the federal government apparently made some attempt to persuade these Indians to take up cattle raising (*34*, p. 27) and farming (*15*, p. 549) shortly after 1880, these efforts met with little success. Water shortages on the Gila River contributed to the failure of the farm program among the Apaches of the San Carlos Reservation.

By the late 1920s, some of the Indians of the latter reservation were tending small herds of cattle, but most of the available range was being leased to outsiders. Nevertheless, the Indian Bureau superintendent who came to San Carlos at this time set to work to convert the Apaches into cattlemen. By not renewing leases, he slowly eliminated the non-Apache cattle owners and secured cattle which he urged the Apaches to manage. By the end of the 1930s, the San Carlos Apaches were well on their way to the development of a sound cattle business. These Indians have progressively improved their herds until today they bring some of the best prices of any cattle in the Southwest (*15*, p. 549–550).

Developments similar to those described above can be reported for the Apaches of the Mescalero Reservation in southern New Mexico and the Fort Apache Reservation in Arizona. The Jicarilla Apaches of northern New

Mexico have engaged much more extensively in sheep raising than the other Apacheans, and less extensively in herding cattle. The sheep industry was started among the Jicarillas in 1920 when the government issued 9,000 animals to Apache families. Sheep numbers reached their maximum in 1943, when 38,654 head were grazed. By 1959, they had declined to slightly under 16,000 (*35*, p. 189).

The Hualapais and Hopis of Arizona are also stock raisers, the latter having had sheep herds since Spanish times. More recently they have begun to raise cattle, although not in large numbers. Unlike the situation on most of the other reservations we have been discussing, the full range resources of the Hualapai have never been put to use. Indian Bureau estimates place the carrying capacity of the range in the neighborhood of 9,000 to 12,000 head (*28*, p. 52); the Hualapai herd has never exceeded 4,000 animals.

This lengthy recitation of the history of food and fiber production among the southwestern tribes sets the stage for a discussion of the modern situation, its potentials and problems. However, before proceeding thereto, we need to provide at least a brief statement on such matters as Indian concepts of land ownership and the relationship between agriculture and social organization.

INDIAN AGRICULTURE, LAND OWNERSHIP, AND SOCIAL ORGANIZATION

Among the Pima and Pueblo groups, both of whom made more extensive use of irrigation agriculture than other southwestern tribes, the principal social unit for agricultural purposes was the village. Bringing a new tract of Pima land under cultivation and irrigation required the cooperation of the men in the villages to be affected. All met together under the leadership of their headmen to plan the work to be done. Persons considered to have special knowledge about such matters selected the land to be served by the canals and determined the courses these canals would take. The actual work of building canals was a cooperative endeavor directed by the village headman. The latter also made work assignments with respect to canal maintenance.

Following the construction of a canal, the village headman, with the aid of an advisory council of community leaders, assigned farm plots to those who had participated in the work. Land thus assigned became the inalienable property of the assignee and his heirs. The patriarchal head of each extended family determined how land assignments would be used (*4*, p. 126–127).

The patriarchal family was also the center of the Papago agricultural system (*4*, p. 128). However, except in those few areas where the Papagos practiced irrigation, there was little need for the more sophisticated cooperative work units found among the Gila River Pimas.

As the Pima water supply began to fail in the 1870s, canals fell into disuse. We have already noted that during this time many Pima families moved to the area of the reservation where seepage water was available. Exploiting these water resources did not require any such elaborate social structure as was necessary for canal building and maintenance. Furthermore, when the Indian Bureau came to the rescue by drilling wells shortly after 1900, the construction and maintenance of these wells and related irrigation works were performed by wage laborers, rather than by cooperative effort on the part of the men in the villages.

As the Indian Bureau began to allot land in 1914, another significant change in Pima social organization occurred. As previously mentioned, the size of the Pima allotments was 10 acres of irrigable land and 10 acres of desert land. Prior to 1914, few Pima families had cultivated as many as 10 acres. Furthermore, families had lived together in close proximity within the villages, cultivating fields lying outside the village boundaries. After 1914, families came increasingly to build their homes on their allotments, thus separating themselves spatially from their neighbors. Hackenberg (*23*, p. 62) has observed that the allotment program ended the old Pima village.

In the era before allotment, Pima lands were considered as belonging to the tribe, although held individually (*4*, p. 127). A person had the right to loan land in case he was not using it himself, but sale and trade of land were generally unknown. The allotment program introduced the new concept of individual, as opposed to tribal, ownership, and also that of the right of the individual to dispose of his land through sale (few of the Indians on the reservation have chosen to exercise this latter right).

Among the Yuman flood-farmers of the lower Gila and Colorado rivers, clearing land for cultivation was an individual, rather than a community, effort (*2*, p. 140). Small family groups resided together near the more or less shifting agricultural sites scattered at random through the cultivable portions of tribal territory. The boundaries of such groups of relatives were determined by lines running to peaks in the mountains and other high landmarks.

Occasionally disputes arose concerning the boundaries of farmlands. Usually, settlement of such disputes was by conciliation between the interested parties, a third person sometimes being called to serve as referee. However, when the disputants failed to settle their differences through these means, they resorted to regulated combat. According to Castetter and Bell (*2*, p. 142):

[A] sort of pushing contest was resorted to in which one of the disputants was surrounded by his friends who sought to push, drag, or carry him across the disputed territory in face of the opposition of the other claimant. . . . The winner of the contest, of course, gained his point in retaining or recovering the land in question. If the losers were dissatisfied, a second battle was staged in which poles and sticks of limited size and weight were employed by both sides to determine supremacy. . . . The winning side obviously always fixed the boundary. One reliable Mohave informant said the last battle of this kind that he saw was at a time which, from related information, we were able to date as about 1885. The last Mohave dispute of this kind took place about 1887–88. . . .

There was no distribution of land by allotment among the Yumans as reported for the Pimas. The land an individual held was of his own choosing or was inherited from his father. If he wished to expand this plot, he obtained permission of his neighbors, not of the chief. If a son married, he either cultivated jointly the land held by his father, meanwhile enlarging the holding, or he cleared and brought an entirely new area under cultivation. That land was held individually is indicated by the fact that it could be traded (*2*, p. 142–143).

The allotment program carried out by the Indian Bureau among the Mohave, Chemehuevi, and Quechan tribes had far less impact on the social organization of these groups than was the case with the Gila River Pimans. To begin with, the concept of individual ownership of land was apparently well established prior to the allotment period. Furthermore, the three tribes did not dwell in compact villages. However, the idea of being able to sell land was new, as was the procedure of distributing ownership in common among the surviving relatives of an allottee. Traditionally, Yumans inherited land only through the male line (*2*, p. 143).

Probably the most experimental and industrious agriculturalists of all southwestern Indians are the Hopis, although they live in a land where irrigation agriculture is not generally practical. In view of the scarcity of water and arable land in the region, the Hopis have succeeded remarkably well as farmers. Hodge (*36;* vol. 1, p. 565) notes that very early in the twentieth century they were planting an average of 2,500 acres in corn. In addition, they had 1,000 acres of peach orchards and 1,500 acres devoted to such vegetables as melons, beans, pumpkins, and chilis. Except in years of extreme drought, Hopis always had surplus crops to trade with neighboring tribes or to sell.

Although the absence of extensive irrigation facilities reduced the need for communal labor on the order of that practiced among the Rio Grande Puebloans and the Pimas, the Hopis within each village did get together for such work as cleaning the springs upon which the community depended for its water supply and limited irrigation of nearby fields. Traditionally, the hereditary chief of each village directed this work (*36;* vol. 1, p. 565). Colton (*37*, p. 24) notes that in modern times communal tasks such as those described above have more often than not been undertaken by particular individuals with initiative, leadership ability, and public spirit, rather than by teams of persons working under the direction of the chief. However, some persistence of older patterns was observed at the village of Bakabi, where the leader of communal tasks during the 1930s was the village chief. When a man ordered by him to join a working party failed to do so, a masked figure (kachina) appeared at his door with a whip and demanded that he fulfill his community obligation. This method of stimulating the reluctant apparently was very effective (*37*, p. 24).

Farming was generally carried on by individual or extended families. A man lived in the home of his wife and confined his farming largely to fields which belonged to her lineage. In ancient times (*37*, p. 23), agricultural land was assigned to individual clans by the chief who founded the pueblo, and as new clans joined the village, his hereditary successors made assignments from the common land. A member of a clan had the right to cultivate any suitable unused land of his own clan or that of his wife. With respect to decisions about the use of these lands, the chief and the village council were without authority. Clan chiefs had the right to make such decisions.

Lands not assigned to a particular clan were held in common by all the villagers. Any person choosing to do so could farm those not already in use. However, such use did not confer any right to will or sell them. The introduction of peaches by the Spanish brought slight changes in this system; peach trees could be owned by the person who planted and cared for them. Orchards, being personal property, could be sold or willed to one's heirs.

The advent of the reservation system brought other changes. Today, a man can select a piece of common land and fence it for his own use. At his death, it reverts to his children. So long as the fence remains, the land is considered to "belong" to the individual who fenced it, whether or not he cultivates it regularly (*37*, p. 23).

Within clans, disputes sometimes arose over the question of which individual family would farm which plot. The clan chiefs, in such an instance, were expected to step in and settle the matter without calling upon outsiders. Colton (*37*, p. 23) notes that Hopi tradition records cases of disputes between villages over particular tracts of land. (Some have occurred in recent years over grazing lands. Prior to the formation of an overall Hopi Tribal Council in the late 1930s, such disputes were usually settled by the Indian Agent.)

The religion of the Hopis — as is true in the case of all the Puebloan tribes — is interwoven with the practice of agriculture, and the ceremonial cycle is closely tied with the crop year. However, among the Hopis, there are relatively few specific planting and harvesting rituals (*38*, p. 361).

Traditionally, among most of the other Pueblo groups, the war chief rather than the village chief was the person responsible for enforcing participation in communal labor. Even before the coming of the Spanish, the Rio Grande Pueblos had developed a high degree of organization with respect to the construction and maintenance of irrigation facilities. Speaking of Santo Domingo, White (*39*, p. 141–142) notes that the head war chief (tsiakiya) makes the decision as to when the time has come to open an irrigation ditch. He then summons all the village officers and informs them of his decision. The village governor (an official not present in the Pueblo villages prior to Spanish times) assumes the responsibility of notifying the workers when they are to report and also supervises the actual construction.

Throughout the life of the ditch, maintenance supervision is provided by an irrigation ditch chief called kokatc in Keresan, but more popularly known by the Spanish name of mayordomo.

Lange, who studied the community of Cochiti during the late 1940s, found no regular office of mayordomo there, although each year the village governor selected someone to serve in that capacity on a temporary basis. Under his direction, with authority delegated by the governor, the men of the village cleaned and repaired the ditches (9, p. 81). At Cochiti, as is commonly the case throughout the Pueblo area, community work also includes caring for the farmlands of the principal religious leader of the village who is known as the cacique (usually kasik in Keresan). The latter is not expected to engage in any secular activity. Lange (9, p. 83) notes that all adult males are expected to contribute to the community labor pool, but that "there is an increasing tendency for many to be 'busy' with other tasks." This situation is similar to that described by Colton (see above) for the Hopis.

Lange (9, p. 86–87) also reports the presence of a system of community farm labor performed cooperatively by family groups. The only compensation for this kind of work is a share of the harvest and reassurance that the relative who has been helped will reciprocate should there be need for him to do so. Formerly, this type of activity was more extensive, including members of the same clan.

Except as indicated above, farmlands among most of the Puebloans are used individually, although they are said to belong to the village. Aberle (20, p. 21) points out that families have no deeds nor written records that the land is theirs, but everyone in the community knows who is entitled to use each plot. When a man dies, the right to use his land is passed along to his wife. In the event she is no longer alive, the land is divided equally among the children, a method of inheritance similar to that employed by the Bureau of Indian Affairs with respect to titled allotments. The results are the same: fractionation of plots into uneconomic units. This, plus decreasing interest in farming on the part of the younger Indians, is contributing to a decline in agriculture among the Pueblos.

It should be noted in passing that in spite of the fact that Pueblo lands were not individually allotted according to the White Man's law — as was the case on the reservations of some other tribes — there seems always to have been a strong concept of private ownership, especially among the Rio Grande people. This is illustrated by an incident at Cochiti in the late 1940s. In the spring of 1948, all of the Cochiti men living in Santa Fe were called back to the village for a meeting. Since these individuals had not been participating in the community labor of ditch cleaning and repair, they were advised that they should either pay the village for having this work done or agree to come and do it themselves. Some agreed to one or the other alternative; others did not. Those who refused maintained that they should not have to contribute so long as their lands were lying idle. They also stated that they would be willing to permit others to use their lands in their absence, provided the tenants would perform the ditch work or pay for it. The governor then threatened to confiscate these tracts for the village. The Santa Fe men insisted this would be illegal because the land was "privately owned." Their opinion prevailed and the community council dropped the matter (9, p. 95).

Although, in terms of the actual acreage under cultivation, the Navajo reservation ranks high among those of the Southwest, the Navajos have always farmed principally for subsistence. Even the government help which they have received in bringing irrigation water to certain areas has been aimed at developing tracts for small family farming.

Like the neighboring Hopis, but to a lesser degree, the Navajos practiced matrilocal residence; that is, a man moved in with his wife's family. However, the Navajos differ notably from the Hopis in that farmlands are not considered to be owned by either clan or matrilineal extended family. Rather, they are regarded as individual property gained originally by the clearing of land and the planting of crops. Sometimes many fields are found inside one fence, each individual's field being marked by small posts. Should an individual lapse in his farming for a period of time, others have the right to move in, reclear and plant the field, and claim possession. Navajo farmlands may be held by either men or women (40, p. 22).

In localities where peach trees have been cultivated for many years, there is a general right by clan to the yield of each tree. Van Valkenburgh (40, p. 21) reports that many people from great distances come into the Canyon de Chelly during the peach harvest to claim their share by remote clan affiliation.

A spring, waterhole, or tank, when surrounded by a fence or after having been used over a long period of time, is claimed and respected as private property. However, when water is developed by a group of people with or without government assistance, it is regarded as communal.

With respect to matters of inheritance, farmlands create the greatest source of dissension among the Navajos (40, p. 23). Disagreements sometimes arise among kinsmen (but not on the basis of clan affiliation) and the immediate family of a deceased person. The remote kinsmen may claim that the land was cleared and improved by them or their immediate relatives. The family of the deceased will counter with the claim that *they* helped clear and operate the land. In most cases, the immediate heirs retain the land, and it is usually the son who took the most active part in the farm operation who assumes control.

Small individual or family herds are common among those tribes for whom stock raising is important. Although many Indian families have some sheep or cattle, few depend primarily upon them for either subsistence or a major part of their cash income. Among the Papagos, the overwhelming majority of cattle owners have fewer than a dozen cows (30; chapter 4, p. 6).

These animals are turned loose in the desert to search for food and are only casually tended by their owners. Formerly, each village chose a local stockman to direct village roundups and sales. His work, though voluntary and unpaid, conveyed considerable prestige (*32*, p. 71). Community participation took place under the stockman's supervision at roundup and sale time.

During the 1930s — when stock reduction and range conservation were burning issues on such reservations as the Navajo and Papago — the federal government encouraged tribal councils to divide the range into grazing districts, issue permits to stock owners limiting the size of their herds and flocks, and enact grazing regulations to control the use of the tribal rangelands. This was a new role for the tribal governments, which they assumed most reluctantly. On the Navajo Reservation, the council promulgated its first set of grazing regulations in 1937 and issued the first grazing permits three years later in 1940. The following year, however, the council petitioned the Bureau of Indian Affairs for the issuance of special permits which would allow stockowners to retain larger numbers of animals. Based in part on the fact that federal appropriations for range improvement had been cut back, thus preventing the Indian Bureau from fulfilling its commitment to increase range carrying capacity, and in part on the fact that rainfall that year was heavy, the bureau agreed to the special permits. These were extended the following year and remain in effect today, having never been rescinded by the Commissioner of Indian Affairs (*8*, p. 155–156).

Livestock raising among the Navajos is often a family or communal enterprise. However, actual title to the animals rests with individuals. The concept of individual ownership of particular animals has been obscured somewhat by the grazing permit system since all of the members of a particular extended family may graze their animals under a permit issued to only one of the family members. This individual may refer to the herd as his, but when pressed for details, will reveal the names of the owners of all the animals involved.

The Navajo sheep and goat herds usually consist of animals owned by the individuals of the matrilineal group; that is, a woman, her daughters and sons, and perhaps even her brothers who live elsewhere with their wives. A husband may add sheep to such a herd, in which instance they remain his property. During lambing season, children receive animals as gifts from their parents and other relatives. Thereafter, they have the right to say what is done with these animals. However, when older children go to off-reservation schools for long periods and cannot contribute to the care of the herd, such animals may be slaughtered for food.

Responsibility for tending sheep and goats is clearly that of the entire homestead group. Men do as much of this work as time and energy will permit, but women and children help, and even may assume the major responsibility. In spite of the total family effort involved in caring for the animals, decisions regarding slaughter and sale are reserved for the individual owner, who also controls the disposition of the carcass following slaughter, or the use of the money after sale (*41*, p. 63–64).

Among the Navajos, cattle are much less a group enterprise than are sheep and goats. Cattle tending is considered to be strictly a man's job. More often than not, the cattle belonging to a particular homestead are the property and concern of only one or two of the men in the group. Other members may assume general responsibility for the animals, but it remains up to the owner to find them if they are lost, to round them up and herd them to branding, and to assist in the branding. When others help with these tasks, they expect to be repaid in some way (*41*, p. 64).

The difference in concepts of ownership and responsibility with respect to sheep and cattle may produce conflicts within a Navajo family. Younger sons, aware that cattle raising is more profitable than sheepherding, may seek relief from responsibilities in connection with the latter, so as to permit them to build up cattle herds. Other family members are likely to resist because of the additional burden of work which would thus be shifted to them. Downs (*41*, p. 65) reports that women, especially, are antagonistic to cattle raising, since a shift to a cattle economy would weaken their economic role in the household.

With the limited grazing lands available to them in modern times, the Pueblo Indians have not engaged extensively in livestock operations. Only at four of the reservations are sheep herded in significant numbers, and only at four are more than 1,000 cattle grazed.

During the Spanish period, the Pueblos doubtless relied more heavily upon livestock than today, although they had no need for an elaborate social structure to manage their animals. Herds and flocks have always been individually owned (usually by family heads) and have grazed upon lands belonging to the villages (*20*, p. 19). At the present time, a new man wishing to enter the livestock business must secure permission to do so from the governing authorities of the pueblo. Officials of the Indian Bureau work closely with the village governors to keep them advised of the carrying capacities of the ranges, and permits are issued to family heads with due regard for these capacities. Families usually run their livestock in one group. When a woman is the head of a household, she may own sheep, cattle, or horses and graze them on the pueblo lands under the same kind of permit issued to a male family head.

In performing such chores as fence mending and the maintenance of stock water supplies, the Pueblos resort to cooperative labor. The participants are usually only those persons who own (or whose families own) livestock. Lange (*9*, p. 219) reports that at Cochiti, the council formerly chose a fence rider who kept the village leaders advised as to what work needed to be done in repairing the fence surrounding the pueblo. However, after 1952, as an economy measure, this position was eliminated and the responsibility shifted to the persons who kept stock on the range.

The Apaches of the San Carlos Reservation in Arizona

are today considered the most successful Indian stockmen in the Southwest, but they have achieved this status in very recent times. Prior to the establishment of their reservation in 1872, these Indians lived by raiding the cattle herds of the surrounding region. With some encouragement from the Bureau of Indian Affairs, a few Apaches had begun to build up cattle herds by the early 1890s. However, most of the Indians evinced no interest in cattle raising, and quickly slaughtered the animals issued them by the Bureau (*42,* p. 181).

With the confinement of the marauding Apaches, white ranchers were able to increase their cattle herds in areas adjacent to the Indian reservations. Since the lands of the Indians were unfenced, cattle in large numbers were soon grazing thereon. Rather than attempt the monumental job of fencing hundreds of acres to prevent cattle trespass, the Indian Bureau set up a system of grazing permits, charging the non-Indians for the use of the Apache rangelands. This system continued in effect until 1932, by which time most of the better grazing lands of the reservation were providing forage for cattle owned by outsiders.

As early as 1923, Superintendent James B. Kitch of the Indian Bureau formulated plans for expansion of the Indian cattle industry. He was opposed by the white ranchers and by many of the Indians, who felt that removal of the outsiders would decrease the income of the tribe. However, he persisted in his effort, refusing to renew grazing permits when they expired, and, simultaneously, prevailing upon small family groups to take up cattle raising. By 1938, a sufficient number of Indians were running cattle on the reservation to prompt the tribal council to approve the establishment of ten cattle associations. Four years later, the number was increased to eleven. Each association had its own board of directors, and there was an overall tribal board consisting of representatives from each separate association.

Once established, the associations quickly came to control the cattle industry on the reservation. A young man wishing to enter the livestock business was required to file an application with the association of his choice. If his application was approved, he was placed on a waiting list. When his turn came, he could select 20 head of breeding stock from either of the two tribal herds, with the understanding that he would pay back an equal number of animals, plus two more for interest, within eight years (*42,* p. 182).

The San Carlos Apaches reorganized their livestock industry in 1956 with a reduction in the number of associations from eleven to five. In spite of heavy opposition from many quarters, the council in that year established a grazing fee for each individually owned cow, in order to compensate the tribe for the use of the range. It also eliminated the previous practice of requiring the owners to participate in roundups. Since 1956, the work of rounding up and caring for the cattle has been performed by paid laborers. The Apaches have also hired an overall manager for their cattle operations.

In spite of the apparent success of the Apaches in raising cattle, only a small percentage of the Indians can be said to be truly ranchers in the usual sense of that term. As is the case on the Papago reservation, many Apache families have small herds of fewer than a dozen cows. In 1954, Getty (*42,* p. 185) found that only 15 Indians, out of a total of more than 700 cattleowners, were running over 100 animals. While the proportion of larger operators has increased since that time, there are still many small, inefficient cattle operations on the reservation today.

The cattle industry among the White Mountain Apaches of the Fort Apache Reservation in Arizona and among the Mescalero Apaches in New Mexico has developed in a fashion similar to that just described. Cattle associations are found at both locations. Ranching is today big business for a number of the Indians on the Fort Apache Reservation, and the number of cattle grazed there falls only slightly below that at San Carlos.

In recent years, the range management specialists of the Bureau of Indian Affairs and the Hualapai Tribal Council have devoted considerable attention to building up the livestock industry on the large but sparsely populated Hualapai Reservation in northwestern Arizona. The Hualapais, too, have a cattle association and a tribal herd. The Bureau has been attempting to create new range for the Hualapai cattle by clearing brush and removing juniper trees. At the present time, the Hualapai range is understocked.

THE MODERN PERIOD: PROBLEMS AND PERSPECTIVES

A glance at Tables 1 through 7 will provide the reader with some idea of the magnitude of farming and livestock operations on Indian reservations today in the arid states of Arizona and New Mexico. The total value of Indian livestock in 1968 was placed by the Indian Bureau at slightly more than $36,000,000. The total value of crops produced in the same year exceeded $30,000,000, approximately half of this total representing the worth of farm production on the Colorado River Reservation in Arizona. The latter (see Table 7) contributed substantially to the agricultural output of Yuma County, one of Arizona's principal farm areas.

As many of the prime agricultural lands of the arid Southwest pass from farm and range to urban uses, which has been the trend in recent years in Maricopa and Pinal counties in Arizona, and Bernalillo County in New Mexico, the agricultural and range lands of Indian reservations become more attractive as centers of food and fiber production. In 1968 Arizona Indian reservations contributed more than one-tenth of the total cultivated crop acreage in the state, yet only a little over one-third of the reservation potential was being utilized. In Yuma County alone full development of the Colorado River Reservation could add about 60,000 acres to the county's present agricultural production. Another 19,000 could be added to the farm production of Mohave County through development of the irrigable lands of the Fort

Mohave Reservation. Northwestern New Mexico could undergo a major agricultural boom through completion of the Navajo Irrigation Project, which would place over 100,000 acres in production.

On the other hand, as we indicated at the beginning of this chapter, the extent to which the potential of the Indian lands will be realized depends upon many factors other than the availability of land and water. In an attempt to put this situation into perspective, we have chosen in this final section to discuss the situation at four reservations: the Gila and Colorado River reservations in Arizona, where the available farmland is greatest; the Navajo Reservation in Arizona and New Mexico, which has for several years been awaiting the completion of a major irrigation project; and the Papago Reservation in Arizona, where livestock production has been limited because of culture conflicts and political considerations. The case studies of these four reservations cover the range of problems which inhibit the full development of Indian lands for the production of food and fiber.

The Gila River Reservation (Arizona)

In previous passages we have outlined the long struggle of the Pima and Maricopa Indians of the Gila River Reservation to reconstruct an agricultural economy in the period between 1870 and 1924. Many of the Indians (and others) felt this struggle had been concluded in the latter year with the authorization of the San Carlos Irrigation Project providing for the construction of Coolidge Dam and related irrigation works. However, the struggle did not end at that point and, in fact, continues to the present day.

The congressional act providing for the dam authorized the use of water by both Indians and non-Indians but did not establish priorities between the two groups of users. A previous act passed in 1917 had provided that water diverted from the Gila River should be distributed among Indians and others in accordance with their respective rights and priorities. It had provided further that such rights and priorities would be determined by the Secretary of the Interior or by a court of competent jurisdiction.

It was nearly 10 years after the passage of the act authorizing Coolidge Dam (1935) before the so-called Gila River Decree was issued in an attempt to define water rights. The Indian interest was represented before the court by lawyers from the Department of Interior, and much bitterness resulted when the tribe was denied the right to have its own legal counsel. The decree itself held that all Indian and white users were "entitled to share equally in all of the stored and pumped water" of the project. Thus, the Indians were not credited with a right predating that of the white settlers, as they had expected would be the case. A later tribal attorney has referred to the decree as "one of the greatest crimes of water law" (*43*, p. 41).

In connection with the construction of the dam and related works, the old irrigation canals of the Indians were renovated and new canals constructed. The Indian water users were called upon just as were the non-Indian

TABLE 1

Indian Reservation Farmlands — 1968
(Pueblo Indian Tribes) *

Reservation	Land Acreage	Irrigable Acreage	Irrigable Acreage Subjugated †	Irrigable Acreage Planted — 1968	Dry-Farm Acreage	Dry-Farm Acreage Planted — 1968
Acoma	243,752	2,500	1,800	847	None	None
Cochiti	28,139	1,864	880	502	None	None
Hopi	650,013	86	86	26	9,100	9,100
Isleta	210,948	6,195	4,570	3,971	None	None
Jemez	88,387	2,500	1,828	848	None	None
Laguna	417,453	5,000	1,690	403	None	None
Nambe	19,075	700	265	152	None	None
Picuris	14,947	350	215	188	None	None
Pojoaque	11,599	180	36	30	None	None
Sandia	22,885	3,418	1,760	1,542	None	None
San Felipe	48,930	3,808	1,670	540	None	None
San Ildefonso	26,192	1,200	500	414	None	None
San Juan	12,234	2,000	1,200	902	None	None
Santa Ana	42,085	1,150	1,150	392	None	None
Santa Clara	45,748	1,700	950	608	None	None
Santo Domingo	69,260	4,300	2,384	1,027	None	None
Taos	47,341	6,000	3,272	1,934	None	None
Tesuque	16,813	800	600	173	None	None
Zia	111,789	1,000	516	185	None	None
Zuni	407,247	8,750	2,600	1,458	500	156
Totals	2,534,837	53,501	27,972	16,142	9,600	9,256

* Information provided by the Bureau of Indian Affairs area offices in Phoenix and Albuquerque.
† Cleared lands under irrigation systems.

JAMES E. OFFICER

users, to pay operation and maintenance charges in return for receiving water. However, limited as they were to small allotments of 10 acres, the Indian users felt they could not afford these charges and resisted paying them. In an attempt to find a solution for the problem, the Bureau of Indian Affairs hit upon the idea of subjugating and developing a large tract within the project which could be farmed by the Indian Bureau, with the income used to pay operation and maintenance charges for the individual Indian farmers. Devoting this much land to such a purpose did, of course, deprive unallotted Pimas of any opportunity to gain ownership of land within the project, and the tribe agreed reluctantly to this proposal. In 1952 the so-called tribal farm was turned over to an Indian manager, and the tribe has continued to operate it to the present day.

The Coolidge Dam Act provided for a landowners' agreement between the Secretary of the Interior and the water users, both Indian and non-Indian. By the terms of this agreement, all wells drilled on project lands were to be brought under the supervision of the project. The purpose of this provision was to keep farmers within the district from depleting groundwater supplies at the expense of the other water users. Since Indian lands were served first by water from the storage reservoir behind the dam, and the reservoir did not fill as rapidly as had been expected, the project manager authorized the drilling of wells on non-Indian lands so that the non-Indian users would receive an amount of water equal to that going to the Indians. Many of the non-Indians also

acquired, through lease or purchase, lands lying just outside the project and drilled their own wells there, thus giving them extra water without violating the terms of the agreement. When the Indians tried to do the same thing, they were challenged by leaders of communities surrounding the reservation who maintained that the Indians were draining the underground aquifers upon which the surrounding cities and towns depended for their domestic and industrial water supplies. Nevertheless, in the mid-1950s, the Indians did drill a few wells of their own, which are still in use.

The construction of the dam was followed by a number of dry years, and the reservoir, upon which the Indian lands depended for irrigation, was never filled. In fact, it has to the present day never reached the level predicted for it. Wholly inadequate supplies of water for the reservation lands discouraged many Indians from farming altogether, whereas the neighboring non-Indian lands continued to receive water from wells.

Two other factors seriously inhibited the Pimas and Maricopas from making maximum use of their agricultural lands. The first of these is the heirship problem; the second is the difficulty of obtaining credit.

Long before the San Carlos Project was completed, many of the original Pima allottees had died. Ownership of their small plots was passed on in common to their heirs. Since these allotments were too small to be farmed successfully by many persons, two trends developed. In the first instance, one of the Indian owners might take it upon himself to farm the land, often without the consent

TABLE 2

Indian Reservation Farmlands — 1968
(Southwestern Tribes, Except Pueblos) *

Reservation	Land Acreage	Irrigable Acreage	Irrigable Acreage Subjugated†	Irrigable Acreage Planted — 1968	Dry-Farm Acreage	Dry-Farm Acreage Planted — 1968
Cocopah	527	431	236	191	None	None
Colorado River	264,091	122,259	55,500	48,019	None	None
Fort Apache	1,664,872	7,885	2,868	1,056	‡	None
Fort McDowell	24,680	1,300	600	323	None	None
Fort Mohave	38,384	18,974	None	None	None	None
Gila Bend	10,337	710	710	279	None	None
Gila River	371,933	132,146	67,464	36,909	None	None
Havasupai	3,058	204	204	147	None	None
Hualapai	993,173	625	432	432	‡	None
Jicarilla	742,303	None	None	None	1,090	None
Kaibab	120,413	68	30	30	65	None
Maricopa Ak Chin	21,840	10,852	10,852	3,479	None	None
Mescalero	460,255	600	506	102	610	None
Navajo	13,981,530	32,680	32,680	15,175	19,532	10,885
Papago	2,773,358	13,790	5,154	513	4,000	600
Quechan	8,801	7,645	7,645	3,921	None	None
Ramah (Navajo)	146,996	None	None	None	800	276
Salt River	46,619	25,849	12,389	10,175	None	None
San Carlos	1,877,216	2,326	724	475	‡	None
San Xavier	71,095	1,398	1,398	1,024	None	None
Totals	22,621,481	379,742	199,292	122,250	26,097	11,761

* Information provided by the Bureau of Indian Affairs area offices in Phoenix, Albuquerque, and Window Rock.
† Cleared lands under irrigation systems.
‡ Estimate not available.

of the other heirs. This led to family disputes. In the second instance, family members might disagree both on the question of who would farm the land and on the question of whether or to whom it would be leased. In such an instance, the land simply lay idle. Even where agreement could be reached regarding the leasing of allotments, the management problem proved to be enormously complicated. At present, the Bureau of Indian Affairs assumes the responsibility for contacting all the owners, approving such leases, and distributing the income thusly derived. The Indians pay nothing for this service, regarding it as their right since it was the federal government which forced the allotment policy on them in the first place.

During the past two decades, Congress has considered many proposals for solving the heirship problem, which is a common one on Indian reservations in most parts of the country, but no general legislation has been enacted on the subject. The Indians themselves have generally resisted such proposals, since they regard them as threats to their land ownership. They fear that if the common court procedures of partition and sale, which are available to non-Indians in similar situations, were to be applied to Indian lands (an element in all such proposals), they would quickly pass out of Indian ownership.

A second major inhibitor to the development of lands on the Pima Reservation has been the difficulty of obtaining credit. At present, the law does not permit the mortgaging of tribal land, and individual allotments can be mortgaged only with the approval of the Indian Bureau. The Pimas and Maricopas, like most other Indians in this country, fear mortgages since in the event of payment default the bank or other lending institution would be able to take over the mortgaged property. Thus, they have excluded themselves (or in the case of tribal lands, have *been* excluded) from resort to the credit source generally relied upon by non-Indian farmers.

The Bureau of Indian Affairs maintains a revolving loan fund from which Indians may borrow, but it has been inadequately funded by Congress, and little help has, therefore, been available from this quarter.

In 1936, when the land subjugation program related to the San Carlos Project was completed on the Pima Reservation, most of the Indians had their allotments planted in pasturage. The Indian Bureau employees working with the Pimas encouraged them to convert to other crops, especially cotton, which was by then a major crop in the area. However, credit money was not available with which to obtain the necessary equipment to convert their acreage to more profitable crops, and many lacked the necessary technical skills to grow a cash crop like cotton. As a result, the Indians followed the line of least resistance, leaving their land in pasturage which required little cultivation and "hiring out" wherever possible as agricultural laborers in the fields of non-Indian farmers in the surrounding area (*23*, p. 115). Many resorted to leasing.

In the period between 1952 and 1968, the Pima Tribal Farm was hard pressed to meet its obligations to pay operation and maintenance charges for the Indian allottees within the San Carlos Project. Farm income was barely adequate to make the necessary improvements to the land and maintain the irrigation system serving it. The principal benefit to the tribe from the farm during this period was in the employment of around 50 tribal members.

The tribal council in 1968 decided to change management and revise the farm operation. The Indian employee who had been in charge of the farm previously was dismissed and a new manager hired. In the first year of operations under a revised farm plan based on that employed by the Papagos of the neighboring Maricopa Ak Chin Reservation, the tribe was able to make a profit from the farm and to meet all its obligations. The long-range plan now in operation calls for the tribal farm to acquire, through lease, certain acreages outside and within the San Carlos Project as present development leases to non-Indians expire. Some of these acreages may be unitized under separate managers, but with help from the tribal farm.

While there is hope for the Gila River Reservation in the approach of tribal corporate farming recently adopted, many barriers remain before the full agricultural potential of the area can be realized. Of singular importance is the need for a solution to the heirship and credit problems. The impact of these inhibiting factors on reservation agriculture can be seen in the fact that during 1968 only about 37,000 acres, out of a total of more than 67,000 which have water delivery, were actually being cultivated.

It is unlikely that many individual Pima Indians will continue to farm. In fact, of the 28,909 allotted acres on the reservation which were cultivated in 1968, only about 6,000 were farmed by individual Indians.

The Colorado River Reservation

Favorably blessed with fertile river bottomlands and an adjudicated water right sufficient to irrigate more than 100,000 acres, the Colorado River Reservation has the greatest agricultural potential of any Indian reservation in the southwestern United States and probably in the whole country. Yet, through the years, the Mohave and Chemehuevi Indians, the principal tribes inhabiting this reservation, have been plagued by a series of problems, which, in spite of more than a century of federal assistance, have prevented them from developing more than half of the reservation's farm resources.

We have already outlined the history of agriculture on the Colorado River Reservation prior to the construction of Headgate Rock Dam in 1941. On the basis of plans conceived for this project, one might have assumed that within a few years the reservation would be in full production. However, at about the same time the construction got underway, the Bureau of Indian Affairs began to look to this reservation as a means of easing the economic situation on several other reservations in the area, and this produced a new set of complications. Indian Commissioner John Collier met with the Colorado River tribes in 1939 and discussed with them the status of the reservation lands. He pointed out, among other things, that the reservation had been established initially in 1865 for the use of "Mohaves and such other Indians of the Colorado basin as might be settled" there by the Secretary of the Interior. He told the tribes that the expenditure of public funds for subjugating 100,000 acres could be justified only by getting Indians from other tribes onto the project.

After much pressure from the Bureau of Indian Affairs, the tribes on March 25, 1944, passed a resolution opening up the southern half of the reservation to returning Indian military personnel regardless of tribal affiliation. Following the passage of this resolution, further meetings were held culminating in the enactment by the tribes in 1945 of Ordinance No. 5, setting forth the recommendations for colonization. Over the next three years, 32 families from other reservations — primarily Navajos and Hopis — were resettled at Colorado River.

By 1949, the Mohaves and Chemehuevis had changed

their position with respect to colonization and were prepared to repudiate Ordinance No. 5. Two more years of discussion and deliberation followed, however, before the tribes actually rescinded the ordinance late in 1951. This action required the approval of the Secretary of the Interior, who declined to grant it.

Failing in this action, the tribes in 1952 held a referendum of the existing members in which they proposed changes in the ordinance. This referendum passed but was also rejected by the Interior Department, which asserted its intentions to continue with the colonization program. The tribal claims attorney then filed suit against the United States before the Indian Claims Commission, asking for relief from the offending ordinance.

In 1954, in response to a request for clarification of the ownership status of the reservation, the Solicitor of the Department of the Interior issued an opinion in which he stated that, subject to judicial or legislative determination to the contrary, the lands of the Colorado River Reservation were held in trust for the benefit of the

TABLE 3

Crops Grown and Dollar Values — 1968
(Pueblo Indian Tribes) *

Reservation	Acres Farmed by Indians	Acres Farmed Under Leases	Crops Grown	Acreage	Gross Value	Return to Indians
Acoma	847	None	Row crops	276	$ —	$ —
			Small grains	6	—	—
			Hay, pasture, forage	471	—	—
			Produce, garden	94	104,602	104,602
Cochiti	383	109	Row crops	31	—	—
			Small grains	10	—	—
			Hay, pasture, forage	429	—	—
			Produce, garden	32	74,368	64,918
Hopi	9,126	None	Row crops	5,505	—	—
			Produce, garden	3,621	94,050	94,050
Isleta	3,971	None	Row crops	769	—	—
			Small grains	82	—	—
			Hay, pasture, forage	2,933	—	—
			Produce, garden	187	594,832	594,832
Jemez	848	None	Row crops	292	—	—
			Small grains	26	—	—
			Hay, pasture, forage	376	—	—
			Produce, garden	154	211,376	211,376
Laguna	403	None	Row crops	83	—	—
			Small grains	1	—	—
			Hay, pasture, forage	234	—	—
			Produce, garden	85	86,372	86,372
Nambe	152	None	Row crops	32	—	—
			Hay, pasture, forage	97	—	—
			Produce, garden	23	26,565	26,565
Picuris	47	141	Row crops	43	—	—
			Small grains	3	—	—
			Hay, pasture, forage	79	—	—
			Produce, garden	63	65,084	13,720
Pojoaque	30	None	Row crops	3	—	—
			Small grains	1	—	—
			Hay, pasture, forage	20	—	—
			Produce, garden	6	4,478	4,478
Sandia	1,072	470	Row crops	48	—	—
			Small grains	84	—	—
			Hay, pasture, forage	1,400	—	—
			Produce, garden	10	211,049	138,399
San Felipe	540	None	Row crops	234	—	—
			Small grains	8	—	—
			Hay, pasture, forage	210	—	—
			Produce, garden	88	110,872	110,872

members of all tribes of the Colorado River and its tributaries who might be placed there by federal authority. He further stated that Ordinance No. 5 constituted an enforceable agreement.

In spite of the controversy over colonization and ownership of the reservation, the subjugation of lands on the Colorado River Reservation proceeded at a rapid pace after 1941. In the following decade, approximately 20,000 acres were brought under irrigation works, and by 1953 the total for the whole reservation had reached nearly 34,000 (*27*, p. 31).

Because of their concern about the establishment of water rights on the reservation, which were the subject of a massive lawsuit involving the states of Arizona and California, the tribes of the Colorado River Reservation in 1953 requested the federal government to undertake no additional land subjugation. Only about 4,000 acres were brought under irrigation in the succeeding five-year period. Some of these lands were subjugated under development leases, the first to be undertaken on this reserva-

tion. Such leases were made possible by the passage of special legislation in 1955 (69 Stat 539).

The colonization program did not have the success hoped for by the Indian Bureau. Relatively few families from other reservations chose to change their surroundings. There was a large spurt in the number of colonists in 1950 as the result of a congressional act which appropriated $5,750,000 for relocation and resettlement of Navajos and Hopis. However, this spurt occasioned the tribal action to rescind Ordinance No. 5, and, in spite of the refusal of the Secretary of the Interior to approve this action, it effectively ended the relocation program. In recent years, the tribes have made provision for incorporating the remaining colonists into the organization as full members.

The question of reservation ownership was finally settled by congressional action in 1964. At that time, the beneficial owners of the reservation were determined to be those Indians eligible for membership in the original tribal organization as it was constituted in 1937 under the

TABLE 3 (Continued)

Crops Grown and Dollar Values — 1968
(Pueblo Indian Tribes)*

Reservation	Acres Farmed by Indians	Acres Farmed Under Leases	Crops Grown	Acreage	Gross Value	Return to Indians
San Ildefonso	226	188	Row crops	29	—	—
			Small grains	23	—	—
			Hay, pasture, forage	340	—	—
			Produce, garden	22	56,676	39,704
San Juan	902	None	Row crops	92	—	—
			Small grains	22	—	—
			Hay, pasture, forage	739	—	—
			Produce, garden	49	82,190	82,190
Santa Ana	367	25	Row crops	68	—	—
			Small grains	25	—	—
			Hay, pasture, forage	262	—	—
			Produce, garden	37	68,460	66,210
Santa Clara	568	40	Row crops	39	—	—
			Small grains	3	—	—
			Hay, pasture, forage	532	—	—
			Produce, garden	34	57,669	55,074
Santo Domingo	1,027	None	Row crops	398	—	—
			Small grains	40	—	—
			Hay, pasture, forage	441	—	—
			Produce, garden	148	245,932	245,932
Taos	1,934	None	Row crops	99	—	—
			Small grains	149	—	—
			Hay, pasture, forage	1,629	—	—
			Produce, garden	57	184,716	184,716
Tesuque	173	None	Row crops	45	—	—
			Small grains	4	—	—
			Hay, produce, forage	117	—	—
			Produce, garden	7	22,766	22,766
Zia	185	None	Row crops	38	—	—
			Small grains	7	—	—
			Hay, pasture, forage	109	—	—
			Produce, garden	31	35,620	35,620
Zuni	1,614	None	Row crops	157	—	—
			Small grains	156	—	—
			Hay, pasture, forage	670	—	—
			Produce, garden, orchard	157	109,185	109,185
Totals	24,415	973		24,924	$2,446,862	$2,291,581

* Information provided by the Bureau of Indian Affairs area offices in Albuquerque and Phoenix. Difference in total crop acreage and the combined total acres farmed by Indians and under leases is accounted for by the fact that in a given year several crops may be grown on the same land.

Indian Reorganization Act, together with those persons born subsequent to that time who could qualify for membership. The act of 1964 also made provisions for including the colonists from other reservations. The Indians, in accepting the 1964 act, were required to withdraw their claim against the federal government relating to the colonization of the southern reserve.

The question of ownership was not the only one settled in 1964. That same year the Solicitor of the Department of the Interior issued an opinion holding that the terms of the 1955 leasing act were broad enough to permit 25-year development leases for agricultural purposes. This opinion, coupled with the decision in the Arizona-California lawsuit vesting the Indians with an adjudicated right to Colorado River water sufficient to irrigate more than 100,000 acres, provided the impetus for a new period of land subjugation and development. The latter had proceeded slowly between 1953 and 1963, not only because of problems related to land ownership and colonization, but also because the Bureau of Indian Affairs during those years had lost ground in convincing Congress of the wisdom of appropriating federal funds for Indian irrigation projects. Money for this purpose was exceedingly scarce by 1964, and has become even more so in the interval since. The possibility of long-term agricultural leases was regarded by tribal leaders as their only opportunity to get more land into production, and they wasted no time in prodding the Indian Bureau into action.

The biggest of the development leases on the Colorado River Reservation was let in 1964 to Bruce Church, Inc., one of the largest firms in the West devoted to growing, packing, and shipping vegetables. The Church lease is 6,400 acres in extent, and the company was called upon to have the land completely developed within 5 years. Actual subjugation was finished a year ahead of schedule. Officials of the Church company regard the Colorado River development as their "special order" ranch and have invested heavily in preparing and cropping it. The crops grown there thus far include lettuce, cantaloupes, cotton, alfalfa, barley, wheat, and sorghum.

Since 1964, the Colorado River Tribes have leased more than 40,000 acres of land for development under the long-term agricultural leases. Some of the lands previously subjugated have fallen into disuse during the period, and others previously reported by the Indian Bureau as subjugated have since been determined to lack proper irrigation facilities. As a result, in spite of the recent efforts, the total amount of land under irrigation works at the end of 1968 was only slightly more than 55,000 acres, out of the estimated total of 120,000 on the reservation which might be irrigated.

The Bureau of Indian Affairs and the Department of the Interior have not been wholly enthusiastic about going the route of the development lease on the Colorado River Reservation. Recognizing that such a procedure ties up the land for long periods during which the Indians receive relatively little return, departmental officials, including a former Secretary of the Interior, have encour-

aged the tribes to try alternative approaches. One of these would be for the Indians to establish a farm corporation and lease land from the tribes. The corporation could then borrow money, pledging the leasehold interest as security. If a competent manager could be found for the corporation, and in view of the agricultural potential which the reservation offers this seems highly likely, the Indians could begin receiving much greater returns from their land than at present. For example, during 1968, the gross value of farm produce on the reservation was about $16,000,000, of which the Indians received only about one-eighth. The percentage of the return going to the Indians will increase as the leases mature, but the tribes would be better off financially to develop the land themselves. On the other hand, looking at the picture only from the point of agricultural production, it is probably true that in its early years a tribal farm corporation would get less yield from the land than a highly experienced, well-financed, well-managed private corporation such as Bruce Church, Inc., can obtain.

Complicating the development picture at the Colorado River Reservation is the matter of allotments and tribal assignments. A total of 841 allotments of 10 acres each were made to Indians on this reservation in the period after 1911. In 1940, the tribal council adopted a land code making it possible for Indians with allotments to exchange them for 40-acre assignments of tribal land. In 1954, the assignment program was changed to increase the size of the units from 40 to 80 acres. Two years later, the land code itself was amended to make the 80-acre assignment explicit (27, p. 26).

Of the original allottees or their descendants, 352 took advantage of the assignment program to deed to the tribe all or part of their allotments in exchange for tribal assignments. By 1953, slightly over 6,000 acres of the reservation were privately owned. More than twice this acreage (14,280) was tied up in 197 assignments to individuals.

While at least half of the remaining allotments are in heirship status, the actual amount of land involved is slight, as compared with the Gila River Reservation, and the heirship problem is much less of an inhibitor of agricultural development than in the case of the latter. However, the assignment problem is beginning to introduce some complications which may threaten the future subjugation and development of farmlands.

Many of the younger people at the Colorado River Reservation do not have land assignments at the present time. The assignment program was held up for a long period while other matters involving such critical items as water rights and the ownership of the reservation were being decided. Now, a great clamor has arisen to begin making assignments again. The council has responded to this appeal by calling upon the tribal attorney to draft a new land code.

The Bureau of Indian Affairs has attempted to get the council to consider alternatives to further assignments, but thus far the council has stuck to its guns. Bureau officials point out that the minimum of 200 acres is now

TABLE 4

Crops Grown and Dollar Values — 1968
(Southwestern Tribes, Except Pueblos) *

Reservation	Acres Farmed by Indians	Acres Farmed Under Leases	Crops Grown	Acreage †	Gross Value	Return to Indians
Cocopah	None	191	Row crops	45	$ —	$ —
			Small grains	131	—	—
			Produce, garden	15	27,822	2,465
Colorado River	5,650	42,369	Row crops	18,010	—	—
			Small grains	9,750	—	—
			Hay, pasture, forage	24,734	—	—
			Produce, garden	4,735	15,914,578	2,000,000
Fort Apache	1,056	None	Row crops	450	—	—
			Hay, pasture, forage	370	—	—
			Produce, garden, orchard	236	114,950	114,950
Fort McDowell	323	None	Small grains	25	—	—
			Hay, pasture, forage	285	—	—
			Produce, garden	13	1,300	1,300
Gila Bend	None	279	Row crops	199	—	—
			Hay, pasture, forage	80	53,960	4,000
Gila River	17,000	19,909	Row crops	11,796	—	—
			Small grains	9,887	—	—
			Hay, pasture, forage	14,332	—	—
			Produce, garden	894	4,290,187	1,729,430
Havasupai	147	None	Row crops	25	—	—
			Hay, pasture, forage	110	—	—
			Produce, garden	12	3,809	3,809
Hualapai	432	None	Row crops	85	—	—
			Hay, pasture, forage	110	—	—
			Produce, garden	12	29,309	29,309
Jicarilla	1,090	None	Hay, pasture, forage	1,090	$ 2,000	2,000
Kaibab	60	None	Row crops	8	—	—
			Small grains	30	—	—
			Hay, pasture, forage	12	—	—
			Produce, garden	10	2,744	2,744
Maricopa Ak Chin	3,479	None	Row crops	2,872	—	—
			Small grains	407	—	—
			Produce, garden	200	1,026,276	1,026,276
Mescalero	102	610	Row crops	10	—	—
			Small grains	5	—	—
			Hay, pasture, forage	660	—	—
			Produce, garden	37	59,870	29,400
Navajo	26,060	None	Row crops	10,848	—	—
			Small grains	1,189	—	—
			Hay, pasture, forage	8,224	—	—
			Produce, garden, orchard	4,631	2,077,453	2,077,453
Papago	1,113	None	Row crops	73	—	—
			Small grains	184	—	—
			Hay, pasture, forage	256	—	—
			Produce, garden	600	88,124	71,344
Quechan	None	3,921	Row crops	1,739	—	—
			Small grains	1,190	—	—
			Hay, pasture, forage	982	1,687,862	245,673
Ramah (Navajo)	276	None	Row crops	143	—	—
			Small grains	8	—	—
			Hay, pasture, forage	125	3,979	3,979
Salt River	114	10,061	Row crops	5,899	—	—
			Small grains	1,899	—	—
			Hay, pasture, forage	2,516	—	—
			Produce, garden	1,042	3,515,451	284,300
San Carlos	475	None	Row crops	151	—	—
			Hay, pasture, forage	292	—	—
			Produce, garden	32	55,136	55,136
San Xavier	76	948	Row crops	831	—	—
			Small grains	117	—	—
			No report	76	118,596	9,649
Totals	57,453	78,288		144,450	$29,073,406	$7,693,217

* Information provided by Bureau of Indian Affairs area offices in Phoenix, Albuquerque, and Window Rock.

† Where acreage devoted to crops exceeds the total of acres shown in columns 2 and 3 above, it reflects the fact that in a single calendar year more than one kind of crop is grown on the same land.

necessary for an economic unit. Furthermore, unless they are consolidated into larger tracts, assignments of 80 acres will be difficult to lease even if the lands are served by irrigation facilities, and impossible to lease if the lessee must undertake the subjugation and development of the land himself. Tribal officials have responded to these arguments by pointing out that some of the existing assignments *are* leased and produce more income per acre than any other lands on the reservation. (This response does not address itself to the case of *undeveloped* assignments. It is conceivable that those Indians who may receive such assignments will find it impossible to subjugate the land themselves and will be unable to lease it out for development. The result will be continued idleness of the land.)

For the time being, the tribes appear to be content to proceed under long-term development leases in order to get their land into production. They have not completely abandoned the idea of a tribal farm corporation and, as some of the existing development leases expire, may be motivated to move more positively in that direction. This would set the stage for the eventual assumption by the tribes of the responsibility for managing the largest consolidated farm acreage in single ownership in the entire Southwest. The possibilities are interesting to contemplate, although any development of this kind (because of the tenure of development leases) lies 25 to 50 years in the future.

The Navajo Irrigation Project

In 1962, after many years of discussion, Congress authorized the Navajo Irrigation Project, located in the extreme northwest corner of the state of New Mexico. As contemplated, the project would provide water for irrigating more than 110,000 acres of desert. Those who conceived the project estimated that, once completed, it would provide farms for 880 families and jobs for up to 2,500 laborers. However, in the interval since the early planning stages, the tribe has decided to use a corporate farming approach rather than to resettle families on individual farms.

The Navajo Project was part of a larger legislative package which included a proposal for diverting water from the San Juan River through the mountains to the growing city of Albuquerque. Since the irrigation project itself had a relatively low cost-benefit ratio (according to the formula employed by the Bureau of Reclamation to determine the feasibility of such projects), it is doubtful whether it could have gained the necessary support for passage had the Albuquerque proposal not been included in the same legislation.

Although the Albuquerque portion of the package has received consistent funding in general accordance with the original proposal, the Navajo Irrigation Project has fallen far behind schedule. It was originally expected to cost $135 million; the estimate was revised in 1966 to $175 million and again in 1969 to $195 million. The original date for the first water delivery has been moved back from 1970 to 1977. A large part of the total cost is involved in land acquisition. At present only about 13,000 acres of the required total are in Indian ownership (*44*).

The Office of Management and Budget, which must pass upon and approve the budget requests of all federal departments, has not been favorably disposed toward the Navajo Irrigation Project, both because of its cost and its low cost-benefit ratio, and has consistently refused to permit the Indian Bureau to seek from Congress the funds necessary to move the project along. For example,

TABLE 5

Pueblo Indians, 1968
(Livestock Numbers and Value) *

Reservation	Cattle	Estimated Value	Sheep	Estimated Value	Horses	Estimated Value
Acoma	768	$ 136,704	7,260	$101,640	240	$ 16,800
Cochiti	194	34,532	None	None	28	1,960
Hopi	3,638	727,600	5,940	83,160	505	35,350
Isleta	1,747	310,966	None	None	34	2,380
Jemez	453	80,634	None	None	20	1,400
Laguna	1,150	204,700	12,110	169,540	340	23,800
Nambe	94	16,732	None	None	8	560
Picuris	40	7,120	None	None	11	770
Pojoaque	63	11,214	None	None	7	490
Sandia	356	63,368	None	None	12	840
San Felipe	178	31,684	680	9,520	28	1,960
San Ildefonso	150	26,700	None	None	5	350
San Juan	67	11,926	None	None	34	2,380
Santa Ana	482	85,796	None	None	10	700
Santa Clara	275	48,950	None	None	6	420
Santo Domingo	1,234	219,652	None	None	36	2,520
Taos	710	126,380	None	None	300	21,000
Tesuque	55	9,790	None	None	50	3,500
Zia	700	124,600	None	None	66	4,620
Zuni	808	143,824	17,431	244,034	500	35,000
Totals	13,162	$2,422,872	43,421	$607,894	2,240	$156,800

* Information provided by Bureau of Indian Affairs area offices in Phoenix and Albuquerque.

in its request for funds to be spent during fiscal year 1970, the Bureau of Indian Affairs requested $15,000,000 for the Navajo Irrigation Project. The budget examiners recommended a reduction in this amount to $3,500,000. Congress added $2,000,000 for a total of $5,500,000, still only a little over a third of the funds originally sought (*45*).

The Navajo Tribe has recently been discussing the formation of an agricultural products industry which would assume management of the two existing major irrigation projects on the reservation, both located in the Shiprock, New Mexico, area. If this tribal corporation is formed, it will provide the Indians some of the management experience they will need when, and *if,* the larger project is completed.

Caught up as it is in regional and national politics, the Navajo Irrigation Project presents another example of the kinds of factors, other than land and water, which have influenced the course of agricultural development on Indian reservations in the arid lands of the American Southwest.

The Papago Indians and Their Livestock

Early in June of 1969, the newspapers throughout Arizona carried stories relating to a heavy loss of cattle on the Papago Reservation because of drought conditions resulting in insufficient water and forage. These stories, which continued through most of the summer, stimulated the people of Tucson to contribute time and money to helping the tribe get supplemental feed and water to the stricken animals. Few of those who contributed to this effort were aware of the fact that the Papago cattle owners suffer heavy losses every year during the same season, or that the problem itself is an exceedingly complicated one

for which no solution is likely to be found in the foreseeable future.

In an earlier passage, we indicated that cattle were introduced into southern Arizona by the Jesuit priest Francisco Eusebio Kino in the closing years of the eighteenth century. While the Papagos were pleased to acquire cattle, they did not take to herding with particular enthusiasm, and prior to the beginning of the present century were generally content to let the animals run wild on the desert. When Anglo-American cattlemen came to Arizona after the close of the American Civil War, they appropriated the waterholes upon which the Papago herds depended, and the Indians suffered accordingly. To overcome this problem, the Bureau of Indian Affairs began drilling wells and constructing stock tanks (charcos) on the lands which were later incorporated into the present reservation. With an assured water supply, the Indian cattle rapidly increased in numbers and the range was soon overstocked. During the 1930s, the Indian Bureau made a concerted, only partially successful, effort to persuade the Indians to reduce their herds.

Since the days of Father Kino, most of the Papagos have maintained small herds ranging in number from 2 to 12 cows. In general, the Indians have been quite casual about caring for these animals. Many Papagos have come to expect an annual loss of animals during the drought which occurs late in the spring, and to accept such losses as "natural" events in their lives.

A few of the Papagos are large ranchers with herds including many hundreds of animals. These cattle graze on the public domain of the tribe; but, unlike the situation outside the reservation, the owners do not pay grazing fees for this privilege. Through the years, these large cattle operators have developed a vested interest in main-

TABLE 6

Indian Livestock Numbers and Dollar Values, 1968
(Southwestern Tribes, Except Pueblos) *

Reservation	Cattle	Estimated Value	Sheep	Estimated Value	Horses	Estimated Value	Goats	Estimated Value
Colorado River	75	$ 13,875	—	$ —	—	$ —	—	$ —
Fort Apache	18,000	3,600,000	—	—	558	55,800	—	—
Fort McDowell	600	111,000	—	—	—	—	—	—
Fort Mohave	—	—	—	—	—	—	—	—
Gila Bend	30	4,500	—	—	5	350	—	—
Gila River	—	—	—	—	—	—	—	—
Havasupai	47	8,695	—	—	179	12,530	—	—
Hualapai	3,936	728,160	—	—	168	11,760	—	—
Jicarilla	2,245	399,610	13,242	185,388	329	23,030	—	—
Kaibab	657	131,400	—	—	18	1,260	—	—
Mescalero	6,555	1,166,790	—	—	592	41,400	—	—
Navajo	30,246	3,307,320	367,328	9,370,740	22,201	2,204,880	136,245	3,491,060
Papago	15,713	2,356,950	—	—	4,229	296,030	—	—
Ramah	510	90,780	5,140	71,960	422	29,540	—	—
Salt River	230	42,550	—	—	—	—	—	—
San Carlos	26,919	4,980,015	—	—	1,164	116,400	—	—
San Xavier	250	37,500	—	—	130	9,100	—	—
Totals	106,013	$16,979,145	385,710	$9,628,088	29,995	$2,802,080	136,245	$3,491,060

* Information provided by Bureau of Indian Affairs area offices in Phoenix and Albuquerque.

TABLE 7

A Comparison of Agricultural Production, 1968
Yuma County and the Colorado River Indian Reservation *

Crop	Total Acreage Yuma County	Total Acreage Colorado River Reservation	Percentage Colorado River Reservation
Alfalfa	51,000	12,361	24.2
Barley	25,000	8,468	33.8
Cotton	32,800	7,693	23.4
Peanuts	450	15	3.3
Produce	45,629	4,735	10.3
Sorghum (grain)	22,000	10,992	49.9
Wheat	13,800	1,282	9.2

* Figures provided by Colorado River Indian Agency and Yuma County Extension Service. In terms of the *value* of agricultural products, excluding citrus, the Colorado River Reservation is now contributing more than 25 percent of the Yuma County total.

taining the status quo with respect to grazing fees and in holding on to the large acreages of the reservation upon which they graze their animals. Some of these ranchers are said to deliberately overstock the range to keep other cattle owners out. They are willing to suffer heavy losses during the late spring drought, knowing from experience that a sufficient number of animals will survive to assure them an adequate income.

In several of the "districts" on the Papago Reservation, the large cattle owners are the wealthiest and most powerful persons. The district representatives to the tribal council are often individuals selected and supported by them. These representatives are not likely to stand idly by and permit the other council members to take actions to the detriment of their sponsors.

Tribal government among the Papagos is highly decentralized. The individual districts share almost equally with the overall tribal organization in terms of authority. The opposition of a single district is usually sufficient to doom a legislative proposal. In this context, it has been impossible for the tribal council to come to grips with the problem of overstocking. The formation of strong cattle associations within the districts and the imposition of grazing fees — both of which have been a part of the successful cattle program of the San Carlos Apache Reservation — would go a long way toward improving the quality of animals on the Papago Reservation, restoring the range to its full potential and preventing the annual loss of animals during the drought season. However, such action is not likely to be taken by the Papagos so long as the tribal council remains weak and the large cattle operators continue to dominate tribal politics.

Within this frame of reference, the Bureau of Indian Affairs has been attempting for more than 30 years to administer a program aimed at improving herd quality, rehabilitating the range, and assuring an adequate supply of stock water. The funds available for this work have usually been inadequate, which has been another factor inhibiting the success of the effort. As at San Carlos, Bureau officials have encouraged the formation of livestock associations and several have been organized. How-

ever, only one of these has functioned with any significant degree of effectiveness.

In the past, the Bureau has resisted the establishment of a supplemental feed program using federal funds, although officials of that agency have communicated their concern to individual cattle owners that if provisions for feeding are not made, the annual cattle die-off will continue. The range-management specialists on the Bureau staff, while disturbed by the losses of cattle during the spring drought, are fearful that if supplemental feeding *is* made available on a regular basis, the cattle owners will allow their herds to increase even more rapidly than at present, thus occasioning further deterioration of the range.

Because of pressures from the Indians and the general public in Tucson the Indian Bureau during the summer of 1969 did make supplemental feed available to cattle owners suffering heavy losses. As part of the bargain, the tribal council agreed to an independent study of the problems of livestock raising on the reservation.

Stated most extremely, the position of the Indian Bureau has been that if the Indians are unwilling to take action on their own to conserve the resources of the range, the federal government should not spend money that will only lead to further destruction of the reservation grasslands. The Indian position, also stated most extremely, has been that the Bureau, through improving water supplies at an earlier period, created the present problem and should, therefore, be willing to provide supplemental feed during the drought season so that Papago cattle will not die.

Range conservationists for the Bureau of Indian Affairs report that the resources of the reservation, if properly utilized, can support at least the number of cattle now grazing there and perhaps more. The reservation superintendent stated in 1969 that by constructing more stock water tanks in areas with good forage and by reseeding certain areas near existing water supplies, the present situation can be alleviated somewhat. However, officials of the Indian Bureau seemed pessimistic over the prospect of permanent solutions until tribal leaders acquire both the motivation and the authority to establish and enforce grazing regulations at the district level.

Late in the summer of 1969, it did not appear that, in spite of publicity given the Papago cattle issue in the Arizona press, the problem was any nearer a solution than it has been for the past several decades. Thus it seems that into the foreseeable future the Papagos will continue to produce a grade of cattle inferior to that found on other reservations in the surrounding area, parts of the range will continue to deteriorate, and relations between members of the Indian community and the Indian Bureau will continue to be strained.

INTERNATIONAL PERSPECTIVES

Certain of the problems that have inhibited more extensive agricultural development on Indian reservations in the arid Southwest are peculiar to the sociopolitical context of American Indian communities generally, and it is difficult to find their precise counterparts elsewhere. However, such problems as a shortage of capital, converting from subsistence to large-scale mechanized agriculture, consolidating uneconomic holdings, overcoming ownership fractionation, leasing, achieving equitable land distribution while maintaining or increasing productivity, and funding irrigation projects from the public treasury are to be encountered throughout the world, especially in the so-called emerging nations of Africa, Asia, the Middle East, and Latin America.

Mexico is the classic example of a nation that has been wrestling with the dilemma of land reform and increased farm production for nearly a century. Benito Juárez, President of Mexico and himself an Indian, believed that the traditional communal landholding system of the Mexican Indians was a serious barrier to agricultural development. He visualized a Mexico consisting of "a community of landed yeomen, of free, private landholders selling their produce and buying their wants in an open market" (*46,* p. 36, 49). Juárez himself did not accomplish the destruction of the communal system, but his successor Porfirio Díaz did. The resulting concentration of the most productive farmland in the hands of a few wealthy latifundistas was one of the factors leading to the Mexican Revolution of 1910 (*47,* p. 27–47).

Following the revolution, the Mexican government attempted to return to communal landholding systems (ejidos) in the rural areas, but the results have not been wholly successful from the point of view of the previously landless peasants, nor have they taken Mexico far down the road in converting to a more efficient and productive agricultural effort.

Federal Indian administrators in the United States, beginning in the 1850s, looked to the breakup of communal landholdings on Indian reservations as the means both to make Indians more self-sustaining and to integrate them into the fabric of the larger society. Such thinking led eventually to the passage of the Dawes Act of 1887, which provided for the allotment of reservations, with land titles passing from the tribes to individual Indians. The result in many parts of the country (*not* in the Southwest) was the transfer of a large portion of the reservation estate from Indian to non-Indian ownership. Where allotments remained in Indian hands, the owners were, and often still are, faced with the problems of obtaining the capital necessary for development and of increasing fractionation of their holdings through inheritance. Leasing their lands to others has often proved the most satisfactory way of obtaining income from them.

Mexico's experience under the ejido system has been similar. Ejidos generally are of two types: those in which the communal lands are assigned to individuals without an actual transfer of title, and those in which the lands are farmed collectively by the ejiditarios. The former is the more common of the two and approximates the situation on Indian reservations of the United States where lands have been individually allotted or assigned. Because of the geography of many areas of Mexico, assignments of individual ejiditarios have been quite small, often uneconomic. Unable to receive income from such lands commensurate with the labor expended in developing them, or unable to obtain the credit necessary to farm them efficiently, assignees have rented out their holdings, a procedure much practiced although forbidden by law. Erasmus has pointed out, "By and large, the practice of renting shifts land management from Indians to non-Indians while the work force remains the same" (*48,* p. 221). This situation is identical with that on many United States Indian reservations, where allottees or assignees may perform the manual labor on their own lands, while a majority of the farm profits go to those who lease the property from them!

By resorting to leasing or sharecropping arrangements, large corporate farmers in Mexico are able to put together substantial tracts of land that they can farm by modern mechanized means. Such individuals often wield great political power and are able to attract substantial public monies for irrigation development and land subjugation. A somewhat similar situation prevails in the United States. The author, while serving as Associate Commissioner of Indian Affairs during a 6-year period, observed that on Indian reservations many of the irrigation projects most successful in receiving continuing funds from the federal treasury are those whose principal beneficiaries are the large corporate farmers who lease the Indian lands.

The above is especially true in the instance of reservations that have been allotted and where the problem of offering large tracts for development leases is more complicated. At a reservation like the Colorado River in Arizona, where many of the best agricultural lands have never been allotted, the large farm corporations prefer the profitable development leases, which bring extensive acreages under their management for long periods at relatively low rental costs. The corporations, therefore, have been less motivated to lobby for extensive investment of public funds in land development.

Some of the Mexican ejidos that have used a collective-farm approach have been more successful than those dividing land among assignees, although this is not always the case. One ejido (actually a consolidation of three) frequently cited as among the most successful in Mexico is the Quechehueca collective in southern Sonora state, located within the Yaqui Indian irrigation zone. Erasmus (*48,* p. 223–224) attributes the success of the Quechehueca collective to the honesty and ability of its manager, but it is evident that other factors are also important, notably the fact that this enterprise is managed according to the corporate farm model. This has occasioned some dissension in the community, in that the managers, who are employed the year round, receive more income than do the ejidatarios, who work only seasonally. Similar complaints sometimes arise in the case of Indian tribal farm operations in the United States.

In many of the emerging nations, the desire of rural dwellers for lands of their own stems both from a thirst for ownership and a desire to have holdings upon which they can farm for subsistence. In the case of the Indians of the United States — as in the instance of those at the Colorado River Reservation — the hunger for land is primarily derived from a desire to have property that will produce income *through leasing* (49). Even those Indians whose ancestors were farmers have little inclination to farm today, and the number of Indian agriculturalists is rapidly declining on all reservations, just as it is generally in rural areas of the United States.

The ejido system of Mexico has also produced its own variety of the "heirship" problem. Descendants of deceased ejiditarios may wish to continue farming the lands of an original assignee or receiving income these lands produce through renting or sharecropping. However, as the number of interests increases, the income for each shareholder declines. In some ejido situations, lands lie idle today because of the inability of the numerous shareholders to agree as to land use, or because, through partition, individual holdings have been reduced in size to uneconomic levels (50).

The heirship problem is also found in some other parts of the world where lands are held in trust or are communally owned. A notable example is the case of the Maori of New Zealand. In a 1960 report prepared by J. K. Hunn, then the Acting Secretary of Maori Affairs, the magnitude of the problem was fully laid out (51, p. 52–59). The situations described by Hunn are almost identical with those found on United States Indian reservations. The government of New Zealand has attempted to solve the problem through purchase of fractionated interests and through partition. Both are approaches which have been advocated in the United States. In fact, during the entire period of the 1960s, the United States Congress debated various proposals for solving the problem in this country through partition and sale of fractionated holdings. The Indians themselves worked to defeat the legislation because of their fear that the suggested solutions would result in further alienation of Indian land.

The obsession of American Indians to hold on to all the land they presently have and, if possible, acquire more, affects also the matter of obtaining credit for agricultural and other forms of economic development. The

standard procedure for obtaining loans in the United States is by mortgaging landholdings. Indians view this practice through fearful eyes. Any lien against their land is a potential threat to continuing ownership. At present, the title to both tribal and individual lands on Indian reservations is held by the United States government *in trust* for the Indians. Individually held land (allotments) may not be mortgaged except with the consent of the Secretary of the Interior, and tribally held lands are not mortgageable at all. The original intent of these restrictions was to prevent land alienation, and Indians have opposed general extension of mortgaging authority to them because of their fear of foreclosure. Many would apparently rather see land idle than to risk it in the mortgage market.

In the above sense, the credit situation on Indian reservations differs from that in many emerging countries, where a shortage of development capital within society generally is the critical issue.

The lack of Indian farming skills has been another factor that has held back full agricultural development on United States Indian reservations. This, too, is a problem shared with other regions of the world. Relatively few Indians today have been trained as farm managers, and, in spite of the presence of 4H and other programs of similar nature on Indian reservations, the Bureau of Indian Affairs has rejected the idea of attempting to make small family farmers out of the Indians. In a country where the transition from family farm to corporate farm is far advanced, such a step would, they believe, be retrogressive and unfair to the Indians. Furthermore, the attempt to make small farmers of the Indians during the period between 1887 and 1934 largely failed.

In Mexico, the federal government has been more motivated to assist the small farmer. Mexican Indians have received help in this regard from the National Indian Institute (INI) and its centers throughout the country (51).

One problem for the Bureau of Indian Affairs in the United States is that of identifying and encouraging those tribes willing to try corporate (tribal) farming to do so and see that tribal members are provided opportunities to undertake management training. At the present time,

TABLE 8

Ownership Status of Farmlands on Allotted Reservations, 1968 *

Reservation	Tribal Irrigable Acreage	Allotted Irrigable Acreage	Tribal Sub-jugated Acreage	Allotted Sub-jugated Acreage	Tribal Acreage Planted	Allotted Acreage Planted	Tribal Dry-Farm Land	Allotted Dry-Farm Land	Tribal Dry-Farm Planted	Allotted Dry-Farm Planted
Colorado River	116,214	6,045	48,455	6,045	42,369	5,650	None	None	None	None
Gila River	82,060	50,083	17,378	50,083	8,000	28,909	None	None	None	None
Hualapai	318	307	125	307	125	307	None	None	None	None
Navajo	32,480	200	30,260	200	14,854	321	17,923	1,600	9,385	1,500
Quechan	150	7,495	150	7,495	145	3,776	None	None	None	None
Salt River	5,849	20,000	1,743	10,676	1,425	8,750	None	None	None	None
San Xavier	225	1,173	225	1,173	76	948	None	None	None	None
Totals	237,296	85,303	98,336	75,950	66,985	48,661	17,923	1,600	9,385	1,500

* Information provided by Bureau of Indian Affairs area offices in Phoenix and Albuquerque.

the two tribal farms on southwestern Indian reservations are both managed by non-Indians.

While we have alluded to parallels between the situation in the United States and elsewhere only through examples drawn from Mexico and New Zealand, we might well have selected examples from other countries. Latin America, especially, provides many comparable situations, although the sociopolitical complexes in which they are found may be quite different (53).

SUMMARY AND CONCLUSIONS

In the foregoing pages we have presented something of the history and current status of food and fiber production on the Indian reservations of the arid American Southwest, primarily the states of Arizona and New Mexico. It should be clear to the reader that these reservations, considered in the aggregate, not only have a high potential for agriculture and stock raising, but at the present time are contributing substantially to United States food and fiber production. It should be equally clear, however, that because of a combination of factors, some derived from Indian culture but most relating to the unique reservation system, the full potential of the Indian lands is a long way from being realized.

Among the factors that have inhibited the development of agriculture and stock raising on the reservations, we can cite the following: the struggle of the Indians to establish and protect their rights to land and water, the imposition on certain reservations of the allotment system and its attendant problem of fractionated ownership, the failure of the federal government to make good on its promises for the construction of irrigation projects and related facilities to serve Indian reservations, the slowness of the federal government to assist the Indians in resolving such basic questions as reservation ownership, the inability of the Indians to obtain credit, and the failure of tribal governing bodies to resolve critical internal problems. There are unquestionably other factors, but those immediately above are of obvious and major significance.

Although in the preceding part of the present chapter, we presented case studies of four particular situations, the reader should not infer that the problems outlined in these studies are confined to the tribes mentioned in connection therewith. The heirship problem, for example, is one of some degree of importance on seven reservations (see Table 8), and is critical in the case of the Gila River, Salt River, San Xavier (Papago), and Quechan (Yuma) reservations. The credit problem is universal. The failure of the federal government to complete irrigation projects has been serious for the Puebloans of New Mexico and for the Colorado River tribes. It is becoming serious for the Navajos. Nearly everywhere Indian resource development has been inhibited by weak or unimaginative tribal leadership.

Some of the tribes of the Southwest still have problems with respect to establishing their water rights, but for most, this is not a factor in determining the pace at which the agricultural resources of the reservations are being developed. A few tribes, notably those on the Colorado River and Fort Mohave reservations, are eager to make full use of the water rights previously granted them, fearful that otherwise these rights may be taken away. This preoccupation has been one of the basic motivating factors in persuading the Colorado River tribes to permit long-term development leases of large acreages on their reservation.

With the settlement of the question of land ownership at the Colorado River Reservation, there are no remaining major problems in this area for the southwestern tribes. However, there are a number of Indian groups—notably those living along the Colorado River in Arizona and California—who have disputes with the federal government or with private parties over the precise location of reservation boundaries. In the past five years, the Department of the Interior has taken a more aggressive stand than formerly in assisting the Indians in solving such problems, and in no instance today are these boundary disputes major factors in impeding the development of agriculture on Indian reservations.

REFERENCES AND NOTES

1. SMITH, C. L., AND H. I. PADFIELD
   1969 Land, water and social institutions. *In* W. G. McGinnies and B. J. Goldman (eds), Arid lands in perspective, p. 325–336. University of Arizona Press, Tucson.
2. CASTETTER, E. F., AND W. H. BELL
   1951 Yuman Indian agriculture. University of New Mexico Press, Albuquerque.
3. HEINTZELMAN, S. P.
   1857 Indian affairs on the Pacific. House Executive Document No. 76, 34th Congress, 3rd Session, Washington, D.C.
4. CASTETTER, E. F., AND W. H. BELL
   1942 Pima and Papago agriculture. University of New Mexico, School of Inter-American Affairs, Inter-Americana Studies 1. 245 p.
5. Personal communication from Emil W. Haury, August, 1969. Information is based on Haury's excavations at Snaketown, the best known Hohokam site.
6. WOODBURY, R. B.
   1961 Prehistoric agriculture at Point of Pines, Arizona. American Antiquity 26(3:2). Also issued as Society for American Archaeology Memoir 17. 48 p.
7. HACK, J. T.
   1942 The changing physical environment of the Hopi Indians of Arizona. Harvard University, Peabody Museum, Papers 35(1). 85 p.
8. YOUNG, R. W.
   1961 The Navajo yearbook, report VIII: 1951–1961, a decade of progress. Navajo Agency, Window Rock, Arizona.

9. LANGE, C. H.
   1959   Cochiti, a New Mexico pueblo, past and pres-
          ent. University of Texas Press, Austin.
10. KINO, E. F.
   1919   Kino's historical memoir of Pimería Alta.
          Translated and annotated by H. E. Bolton.
          Arthur H. Clark Co, Cleveland. 2 vols. Reis-
          sued, 1948, by University of California Press,
          Berkeley, in 1 vol.
11. SEDELMAYR, J.
   1939   Sedelmayr's relación of 1746. Translated and
          edited by R. L. Ives. U.S. Bureau of American
          Ethnology, Bulletin 123:97–117. (*Its* Anthro-
          pological Papers 9)
12. BOLTON, H. E.
   1930   Anza's California expeditions. University of
          California Press, Berkeley. 5 vols.
13. BREW, J. O.
   1937   The first two seasons at Awatovi. American
          Antiquity 3(2):122-137.

In modern times, wheat has been reintroduced among
the Hopis and is grown today in small quantities at the
village of Moencopi located about 40 miles west of the
other Hopi villages. Small-scale irrigation works supply
the needed moisture.

14. WHITING, A. F.
   1939   Ethnobotany of the Hopi. Museum of North-
          ern Arizona, Flagstaff, Bulletin 15. 120 p.
15. SPICER, E.
   1962   Cycles of conquest. University of Arizona
          Press, Tucson.
16. U.S. COMMISSIONER OF INDIAN AFFAIRS
   1895   Annual report, 64th. *In* U.S. Department of
          Interior, annual report of the secretary. Wash-
          ington, D.C.
17. U.S. DEPARTMENT OF THE INTERIOR
   1901   Annual report of the secretary. Vol. I, Wash-
          ington, D.C.
18. U.S. DEPARTMENT OF THE INTERIOR
   1908   Annual report of the secretary. Vol. II, Wash-
          ington, D.C.
19. U.S. DEPARTMENT OF THE INTERIOR
   1958   Federal Indian law. U.S. Government Print-
          ing Office, Washington, D.C.
20. ABERLE, S. D.
   1948   The Pueblo Indians of New Mexico, their land,
          economy and civil organization. American
          Anthropologist 50(4:2). Also issued as Ameri-
          can Anthropological Association Memoir 70.
          93 p.
21. BROWNE, J. R.
   1869   Adventures in the Apache country. Harper,
          New York. 535 p. Reissued, 1950, under title
          "A tour through Arizona, 1864, or Adventures
          in the Apache Country," by Arizona Silhou-
          ettes, Tucson. 292 p.

22. A reservation was ultimately established for these and
other Pimas and Maricopas who settled north of the Salt
River in the vicinity of present-day Scottsdale, Arizona.

23. HACKENBERG, R.
   1955   Economic and political change among the Gila
          River Pima Indians, a report to the John Hay
          Whitney Foundation. 177 p.
24. U.S. BUREAU OF INDIAN AFFAIRS
   n.d.   History of the Gila River Reservation San Car-
          los Irrigation Project. Unpublished; issued in
          mimeo form by Bureau of Indian Affairs Area
          Office, Phoenix, Arizona.

25. While this may be an exaggeration of the fact, it under-
scores the struggle that the Indian Bureau has always
undergone to obtain the support of the Bureau of the
Budget and Congress for irrigation projects whose beneficiaries have been primarily Indian. The Bureau has
had considerably more success in obtaining appropriations for projects that have benefited large numbers of
non-Indians as well as Indians.

26. U.S. BUREAU OF INDIAN AFFAIRS
   n.d.   History of the Colorado River Reservation,
          Colorado River Irrigation Project. Unpub-
          lished; issued in mimeo form by Bureau of
          Indian Affairs Area Office, Phoenix, Arizona.
27. UNIVERSITY OF ARIZONA, BUREAU OF ETHNIC RESEARCH
   1958   Social and economic studies: Colorado River
          Reservation. Series of three reports prepared
          for the Bureau of Indian Affairs. II: History
          of the Colorado River Reservation. 66 p.
28. KELLY, W. H.
   1953   Indians of the Southwest. University of Ari-
          zona Bureau of Ethnic Research, Tucson, An-
          nual Report, 1st. 129 p.

29. Other Pima-speaking Indians, known as the Sobaipuris,
who were closely related to the Papagos and Pimas, were
irrigating fields along the San Pedro River in Arizona
when the first Spanish soldiers and missionaries penetrated their area. During the eighteenth century, however, they were forced to abandon their villages because
of continuing Apache raids. Some joined their cousins
in the Pima communities along the Gila River; others
fled to join the Papagos at San Xavier near Tucson.

30. METZLER, W. H.
   n.d.   Papago economic studies. Draft copy of manu-
          script loaned to me by Dr. Bernard L. Fontana,
          Arizona State Museum, Tucson.
31. U.S. BUREAU OF INDIAN AFFAIRS
   n.d.   History of the Papago Reservation Chuichu
          Irrigation Project and development of water
          supply. Unpublished; issued in mimeo form
          by Bureau of Indian Affairs Area Office, Phoe-
          nix, Arizona.
32. KELLY, W. H.
   1963   Papago Indians of Arizona, a population and
          economic study. A report prepared for the
          Bureau of Indian Affairs, Sells Agency. Uni-
          versity of Arizona Bureau of Ethnic Research,
          Tucson. 129 p.

33. Personal communication from the Bureau of Indian
Affairs, Phoenix Area Office, June 1969.

34. GETTY, H. T.
   1963   San Carlos Indian cattle industry. University of
          Arizona Anthropological Papers 7. 87 p.
35. SASAKI, T. T., AND H. W. BASEHART
   1961– Sources of income among Many Farms /
   1962   Rough Rock Navajo and Jicarilla Apache:
          Some comparisons and comments. Human Or-
          ganization 20(4):187-190.
36. HODGE, F. W. (ed)
   1960   Handbook of American Indians north of Mex-
          ico. Pageant Books, New York.
37. COLTON, H. S.
   1934   A brief survey of Hopi common law. Museum
          of Northern Arizona, Flagstaff, Museum Notes
          7(6):22-24.
38. OSWALT, W. H.
   1966   This land was theirs, a study of the North
          American Indian. John Wiley and Sons, New
          York. 560 p.
39. WHITE, L. A.
   1935   Pueblo of Santo Domingo, New Mexico.
          American Anthropological Association, Mem-
          oir 43. 210 p.

40. VAN VALKENBURGH, R.
   1936 Navajo common law I. Museum of Northern Arizona, Flagstaff, Museum Notes 9(4):17-22.

41. DOWNS, J. F.
   1964 Animal husbandry in Navajo society and culture. University of California Press, Berkeley and Los Angeles.

42. GETTY, H. T.
   1961– San Carlos Apache cattle industry. Human
   1962 Organization 20(4):181-186.

43. FEY, H. E., AND D'A. MCNICKLE
   1959 Indians and other Americans. Harper and Brothers, New York.

44. Personal communication, Bureau of Indian Affairs Area Office for Navajos, Window Rock, Arizona, August 1969.

45. Information from files of the author and personal communications from the Bureau of Indian Affairs Area Office for Navajos, Window Rock, Arizona, August 1969.

46. VERNON, R.
   1963 The dilemma of Mexico's development. Harvard University Press, Cambridge, Massachusetts.

47. EDER, G. J.
   1965 Urban concentration, agriculture, and agrarian reform. American Academy of Political and Social Science, Annals 360:27-47.

48. ERASMUS, C.
   1961 Man takes control. University of Minnesota Press, Minneapolis.

49. Personal communication to the author from the Land Operations Officer, Bureau of Indian Affairs, Colorado River Agency, June 1969.

50. Personal communication to the author from Harland Padfield, University of Arizona anthropologist, who has studied the ejidos of northern Mexico in connection with his other research interests.

51. HUNN, J. K.
   1961 Report on Department of Maori Affairs, with statistical supplement. R. E. Owen, Government Printer, Wellington, New Zealand.

52. As the United States Representative to the Interamerican Indian Institute, which has its headquarters in Mexico City, the author has been privileged to visit several of the Centros Coordinadores for Indians which are operated by the Instituto Nacional Indigenista of Mexico.

53. For an excellent bibliography of published materials on land reform in Latin America and its relationship to food and fiber production, the reader is referred to:
   SCHAEDEL, R. P.
   1965 Land reform studies. University of Texas, Institute of Latin American Studies, Offprint Series 19. (Reprinted from Latin American Research Series 1(1):75-122.)
   Information current through 1965.

The Papago Cattle Economy

Implications for Economic and Community Development in Arid Lands

ROLF W. BAUER

Director, the Papago Office of Economic Opportunity
for The Papago Tribe of Arizona
Sells, Arizona, U.S.A.

ABSTRACT

THE UNITED STATES CONGRESS has established special federal agencies to carry out its treaty obligations to North America's Indian tribes. In contrast to non-Indian communities, where the emphasis has been on local self-government, Indian reservations became federally administered: schools, lands, finances, and economic development programs. National Indian policy has oscillated during the last five decades between federal administration and tribal self-government, between federally subsidized and managed economic development and tribal self-sufficiency. The pendulum is presently swinging in the tribal direction and is momentarily passing through a phase of "Indian involvement" enunciated by both the Bureau of Indian Affairs and the Division of Indian Health. The long-range goal of the federal government remains, however, Indian self-government and economic self-sufficiency.

A history and comparison of federal projects to develop a modern cattle industry for the Papago Tribe of Arizona points out important social factors affecting the transfer of responsibility for program planning and implementation from professional administrators and technicians to Indian communities. Conclusions are drawn concerning economic development in arid-land cattle economies, administrative relations in community development, and the relations between economic and community development. It is seen that development of the Papago cattle industry succeeded only when recognized techniques for developing community organization were employed. Papagos did not take responsibility until they could call a community organization their own. The author recommends, in similar situations, recognition of an existing or specially formed group widely viewed as belonging to and representative of the enclave group and interested in improving the cattle economy; investments into the enclave-controlled organization; maintenance (by the outside agency) of a geographical separation from the enclave community or policy-making body; and negotiation of economic-development plans with the enclave community organization under its control, at its request, and with accountability to that body.

A hypothesis incorporating these factors is presented in order to explain why some federally initiated livestock programs were successfully transferred to the Papagos while others were resisted; why some programs were maintained after the administrators left while others were abandoned. The hypothesis is that when similarly situated people describe a program as their own, they will continue it after the change agent leaves. To fulfill the conditions of the hypothesis, they must accept the initiator, help identify the problem, participate in planning the solution, believe that they will benefit, adapt the program to their sociocultural life, recognize and participate in the organization as their own, accept the major responsibilities, and/or recognize the consequences as their own, in some combination of the above-listed factors.

THE PAPAGO CATTLE ECONOMY
Implications for Economic and Community Development in Arid Lands

Rolf W. Bauer

INTRODUCTION AND PERSPECTIVE

AMERICAN INDIANS and other aboriginal peoples have usually been pushed by conquering Western peoples onto poor lands; frequently those lands are arid. Modern nation-states have recently attempted to integrate and develop these economically depressed groups, with the United States, Peru, Chile, Ecuador, Australia, the Soviet Union, the Arab states, and others employing a variety of approaches.

Strategists who seek to develop native economies should distinguish among the types of ecosystems already developed by these communities. "Ecosystem" refers to the orderly way in which living organisms in communities interact with their nonliving environment and with each other to maintain themselves, and specifically the systemic interaction of man's cultural and natural environment.

Recently immigrated aboriginal enclave peoples should be distinguished by their ecosystems from those living in that harsher climate for generations. Enclave groups that have controlled their adaptation to arid-land conditions over centuries should be distinguished from those groups whose adaptation was imposed by the intervention of a dominant cultural group. Development strategists should also distinguish those groups that view economic development and/or cultural integration as a threat to their ecology and identity rather than as a contribution. The purposes of economic development should be distinguished as well. Whose interests will be primarily served? The nation's? Private groups'? The enclave community's? Once initiated, is development responsibility to be carried on indefinitely by a governmental or private agency or is it to be transferred from the initiators to the community?

For the latter goal, a strategy is needed to develop the community's responsibility for its own improvement. It is now widely accepted that such responsibility does not evolve on its own. Special techniques must therefore be employed to engender some kind of group organization along with the techniques that are used for economic development.

If these distinctions among arid-land aboriginal enclave groups and development goals are not made, and if strategies are not tailored to these kinds of differences, progress can be consciously thwarted as has happened in the case of the Papago Indian Tribe.

The Papagos have lived in and adapted to the Sonoran Desert of North America during the past 10,000 to 25,000 years as hunters and gatherers, farmers, and cattle grazers (Fig. 1). Over the last few centuries Piman-speaking Indian groups, under pressure from the growth of the Mexican and American nations and Apache Indian raids, have been able to hold less and less land. In the late nineteenth and early twentieth centuries, the Papagos, along with other Southwest Indian groups, were established on reserves much reduced from the land they had previously occupied.

The Sonoran Desert occupied by the three Papago reserves is part of a series of deserts in the southwestern United States, northwest from Mexico. Average rainfall varies from less than 7 inches at the extreme north and increases from less than 10 inches as you proceed from the western edge to 12 inches at the lower elevations on the eastern edge. It increases to 20 inches in the highest eastern mountains. Average annual rainfall, however, is not an accurate indicator of aridity there, for precipitation is divided between a summer season of scattered, intense storms from July to mid-September, and a winter season of gentler, more widespread rains from early December to early March. The annual total varies tremendously for any given locality.

The reserves are drained by numerous intermittent streams or "washes," none of which flows the year round. Scattered mesquite trees occur throughout the reserves, apparently replacing formerly lush stands of grass. Most of the reservation is within the area of two major vegetation patterns; at higher elevations and on rougher terrain the dominant plants are paloverde, bur sage, and cacti; at lower elevations, creosotebush and desert saltbush. Many parts of the reservation typify what has been referred to as the "giant cactus forest," the saguaro towering above dense stands of paloverde, shrubs, and other cacti (1).

A large number of the common desert plants were traditionally used by the Papago, particularly as food, and also as raw material for houses, baskets, and numerous tools and utensils. Particularly important were the fruits of the saguaro and mesquite, seeds of sacaton grass, and amaranth. From spring through late summer, many plants were valued as greens, contributing significantly to the diet: lambsquarter, canaigre, the buds of the cholla, the mescal or century plant, the sotol, and others (1).

INDIO

PHOENIX

TUCSON

HERMOSILLO

GUAYMAS

LOS MOCHIS

N

LEGEND

● TOWNS

---------- UPPER PIMERIA

THE PAPAGO RESERVATION TODAY

Fig. 1. The Sonoran Desert and the Papago Indian Reservation.

In contrast to the variety and relative importance of wild-plant sources of food, the desert offered only a sparse animal life for economic exploitation. Hunting was always a relatively minor activity for the Papagos. Before the introduction of horses and cattle, the mule deer was much more abundant and widespread and was the chief game animal of the Papagos. Of less importance were the white-tailed deer, antelope, mountain sheep, and peccary or javelina. Small animals, as well as game birds, were hunted more systematically (1).

Unlike other aboriginal peoples whose ecosystems were rapidly destroyed or readapted through Western conquest, the Papago Indians continued to make modifications of their desert in the presence of Western society. The ecology of the Indian peoples of the American Plains was shattered as much by the destruction of their buffalo as by their military defeat. Confined on reservations, the buffalo and other game destroyed by white hunters, the Indians were provided with rations by a government which thereafter tried to remake them into farmers and cattlemen (2). In contrast to the Plains Indians, the Papago were not only undefeated but their subsistence hunting, gathering, and flood-farming economy was replenished and expanded by Western contact.

Piman agriculture at the close of the seventeenth century ... centered around the maize-bean-pumpkin complex.... Although maize was the most important Piman crop... among the Papago tepary bean approached it in value because of its tolerance to drought.... With the coming of the Spaniards, various Old World crops were introduced into Pimería Alta.... In order to furnish a more stable food supply for the Indians at the missions Kino distributed seed of wheat, chick peas, bastard chick peas, lentils, cowpeas, cabbages, lettuce, onions, leeks, garlic, anise, pepper mustard, mint, melons and cane, as well as grapevines, roses, lilies and trees of pear, apple, quince, mulberry, pecan, peach, apricot, plum, pomegranate and fig.... As the penetration of foreign animals into Pimería became very rapid soon after Kino's arrival, he quickly established a number of stock ranches with cattle, oxen, horses, mules, burros, goats, sheep and chickens.... The introduced crops which proved to be of greatest significance in early Piman economy were wheat, barley, watermelons and cowpeas, and in that order; the most important domesticated animals were horses and cattle. (3)

For nearly two centuries the Papagos lived more on the periphery of than enclosed by the northern reach of New Spain, and later the newly independent and expanding nations of Mexico and the United States. During this period, Papago communities could select those Western innovations that they wanted and fit them into

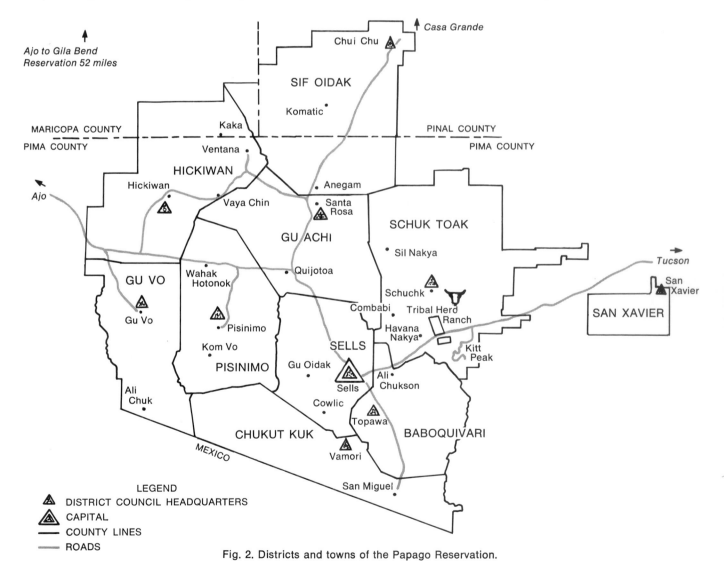

Fig. 2. Districts and towns of the Papago Reservation.

their ecosystem. Spanish mission communities were the scene for peaceful, nondisorganizing culture change, for they emphasized maintenance of the modified Indian community as a social, economic, and political unit with Indian leadership under one or a few Jesuits. To supplement hunting and gathering, the Papagos selected and adapted new Spanish crops along with cattle and horses into their economy. The Papagos developed both hostile and dependent relations with the Mexican settlers and ranchers. They were nowhere united in their resistance to the progressive land encroachment by Mexicans. Gradually crowded out of the river valleys of the Sonoran Desert, many emigrated northward into desert villages in the United States. But they also moved south to work on Mexican farms and cattle ranches seasonally for goods and money, learning cattle-ranching while maintaining their homes far from their employers. Apache raiding pressure, while it intensified from 1810 to the 1870s and pushed Papagos away from the river valleys west to desert defense villages, also brought Papagos and Anglo ranchers and military together in a defensive alliance to protect their crops, cattle, and horses. This affiliation with the American troops was strengthened when an Indian agent was appointed for

the Papagos and began his administration by giving out hoes, shovels, and picks a decade before the Apaches were defeated, and a decade after the United States purchased from Mexico the northern portion of Pimería Alta. But the alliance did not prevent American cattle ranchers and miners from encroaching into the heart of Papago country. An American mining town grew to 10,000 at Quijotoa by 1884 and a few years later disappeared as the ore ran out. Cattlemen with herds from Texas expanded from the Santa Cruz valley onto the eastern and central portions of Papago country, where there was good grass cover for their herds.

Ranchers moved steadily into these parts. In some instances they found water holes which they appropriated, driving off Papagos who customarily used the water. In other cases the new ranchers developed their own wells and water supply, excluding Papagos from what they developed. Small battles were fought. *(4)*

But hostilities lessened as the cattle market dropped and American ranches in Papago country receded, though not completely. Some Anglo-American ranchers remained in Papago territory until ejected by the Anglo courts in 1929 *(4)*.

Not having been defeated, the Papagos were only

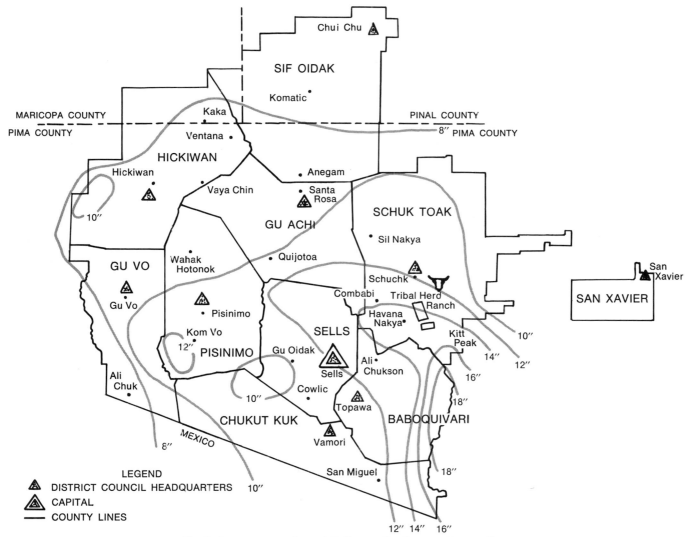

Fig. 3. Average annual precipitation on the Papago Reservation.

slowly established within the American reservation system. In 1874 and 1884 two small reserves were created adjacent to American communities at San Xavier and Gila Bend. The main Sells reservation was not established until 1917.

The Papagos thus became an enclave group surrounded by a rapidly developing nation, and within this context they encountered over the next half century a set of conditions that differed significantly from their previous contact experience with the Spanish, Mexicans, Apaches, or American allies, and which threatened their ecosystem and cultural identity as never before. Three generalizations can be drawn from the rough sketch of this enclave group's historical experience. *First,* their land base shrank under pressure from all four contact groups occupying much of Pimería Alta in 1700. Most Papagos now make a living on the less than 3,000,000 acres enclosed by the three reservations. *Second,* Papago communities of Upper Pimería were on the periphery of the three Western societies, not enclosed by them. When Western ranchers, settlers, or miners moved onto their customary land or Apache raids became too intense, Papago communities could move further away. Not until the late nineteenth century did large numbers of non-Papagos establish themselves among the villages, and then for less than ten years. These conditions allowed the Papagos maximum opportunity to select the Western innovations they felt would contribute to their ecology without being overwhelmed by or coerced into accepting them. And *third,* because the Papagos were never surrounded by a dominant society, they had two centuries before the reservations were established to select what they believed would contribute to their cultural life in the desert: to adapt, modify, and integrate new crops, a new religion, cattle, horses, wagons, firearms. Thus by alliances, migrations under pressure, and warfare when necessary, the Papagos controlled, defended and expanded their ecology. But in the twentieth century the federal government imposed Anglo ways and thus threatened Papago control of their ecology.

The American reservation system, that is, national government administration of Indian lands and Indian communities, had been developing for nearly a century before it was slowly applied to the largest portion of Papagos on the Sells reservation by 1917. With it and the nation's trust responsibilities over Indian lands and people, came a set of conditions that, taken together, constituted this threat.

a) The Papago homeland was restricted to its most reduced size, and title to the land was vested ultimately in the American national government.

a) Papago land was surrounded by expanding American communities and ranches and penetrated by American government, mining, ranching, and clerical interest groups.

c) The Papago communities were politically disenfranchised within the reservation system and within state and national politics.

The Papago land base was delimited by American law and surrounded by Anglo communities. Papago land was less their own than it had been. It was held in trust by the national government, whose title had been purchased from the Mexican government in 1853, without Indian approval. As trust land its use must ultimately be approved by the national government or its appointed trust agents. In principle and, at times, in practice, traditional Papago land use has not been approved by the trustees, thus allowing the government the authority to intervene unilaterally in the ecological adaptations made by Papagos to "their" land. Even mineral rights were denied the Papagos when the reservation was approved. The reservation remained open to "entry and location" for Anglo miners for nearly four decades before Congress established the tribe's rights (*5*).

No longer were Papagos at the periphery of the developing nations, but legally enclosed and physically surrounded by the Anglos. Their land and communities were penetrated by mining interests, a variety of churches, and government agents, all "selling" and imposing their interests, programs, and innovation. No longer could Papago communities, when threatened by these incursions, move elsewhere. And when they told the government agents, with their guns lowered, that their children would not go to the government school, armed troops soon appeared to escort them there, and to arrest the resistant parents. The Papago communities and later the tribal government were politically disenfranchised. The American reservation system did not recognize the existence of Indian political organizations until a century after the first American reservation was established. A pervasive authority to intervene in and administer Indian community affairs was vested in the Indian agent or superintendent, appointed by the national government and accountable only to it. The powers of the superintendent, without accountability to the Indian community, inevitably eroded whatever political authority these communities had traditionally held. Indian people became politically defenseless.

The role of the superintendent and the debilitating effects of authoritarian rule on Indian communities in the reservation system are discussed by E. H. Spicer (1962) and J. Van Willigen (1969):

The authority of the superintendents was based on the assumption that the Indians had no governing bodies which should exercise control of their actions. Compulsory education, for example, was a measure conceived as for the good of the Indians regardless of what they might think of it. Similarly, the steps to increase irrigable land ... were measures to be executed by the superintendents on orders from their superiors. It was apparently not believed that the Indian communities as they existed could or should participate in the formulation of such programs or organize their execution once they were formulated. Both the objectives and their execution were matters which concerned only the authority of the employees of the Indian Bureau.... The inevitable result was the breakdown of many aspects of Indian community organization, and a strong sense on the part of the Indians that what was being done was being imposed on them.

Since the superintendent's power was thus imposed, and there was no possibility of modifying his actions through any means except direct request to him, a strong sense of dependency on the superintendent also developed. The various aids which were available through him, food in hard times, medicine, tools, protection from encroaching Whites, etc., could be obtained not through the operation of any community organization, but only by going to the superintendent's office and making a personal appeal to him. He controlled the distribution of such benefits. He also acted as judge in matters of offense and punishment, so that his power extended to the social and political, as well as the economic, affairs of the Indian communities. . . . *(4)*

The effects of this relationship are insidious and devastating. With displacement of control to (an administrative) unit outside the community, traditional political processes and structures become debilitated and ineffectual through disuse. These institutions become less and less effective in managing their own affairs. Dislocation and dependency generates political ineffectiveness which encourages dislocation and dependency. This goes on until certain communities have totally lost their capacity for executing plans in concert.

One might argue that there are sectors of Indian community life which were not affected, through ignorance or indifference on the part of the Federal representative. This suggests, perhaps, that certain aspects of Indian community life were permitted to flourish, taking their own direction. Unfortunately the dislocation of political power had multiple effects. With deterioration of leadership roles you also find deterioration in other areas. In terms of social control, the weakened community organization could not muster enough legitimate power to maintain their standard of order. The law and order function was met more and more by extra-community roles which were activated more and more in terms of behavior standards, and punishments which were based on a White concept of an orderly universe. What used to mean a community council meeting and public chastisement now means a call to the Bureau of Indian Affairs Law and Order branch with their ubiquitous grey pickup trucks.

The pervasive and multiple effects of the process are seen in the high rate of economic dependency, suicide, violent crime, incarceration, and alcoholism. . . . Loss of control of basic decision-making processes is intimately related to the tragic inventory of what are carelessly called social problems on American Indian reservations. *(6)*

There you have a general picture of the system and the new conditions with which the Papago communities had to contend: the dislocation of community decision-making to the superintendent, the pervasive authoritarian powers of his office, and the resulting loss of Papago political identity. The reservation system was prepared to impose or introduce America's standards of (middle-class) community life: education, law and order, religion, land use, housing, business, economic development, health, sexual morality, marriage, political organization, boards, committees, standards of dress, restaurants, airports, eight-hour working days, national holidays, cocktails, the national sports, the Boy Scouts, bands and cheerleaders, motels, souvenirs, movies, late-model cars, public-nurse home visits, surveys, toothbrushes, and on and on the list goes. Without some means of keeping these alterations at a distance so that they might be considered, tested, selected, rejected, Papago community life could be overwhelmed, especially if the superintendent's authority pushed any of the

changes. Even the time needed to adapt and integrate selected changes into their ecosystem is not available in the reservation system, for it operates by a clock and calendar set in Washington, D.C. Promises are made and broken, programs initiated and then halted when funds run out, experienced employees are transferred, policies change with new administrations in Washington or new superintendents on the reservation, and all these changes occur as in some Kafkaesque novel, beyond the influence of the Papagos.

Two options have been open: either the Papagos could learn to defend themselves against such apparent irrationality, or they could become frustrated, apathetic, blindly bitter, and wholly dependent upon the personal opportunities available from time to time. The latter response is found among too many Plains Indian peoples who succumbed to the reservation system following their military defeats. Papago communities, however, learned to defend themselves, even at the cost of stalemating, slowing down, or rejecting economic development programs proposed by the superintendent. To protect themselves within the reservation system Papago communities and individuals have developed various techniques of passive resistance: (*a*) noncooperation, (*b*) cognitive boundaries between the dominant society and the threatened subsociety, (*c*) physical and psychological withdrawal, and (*d*) organized political opposition.

The authority and influence of the Papago Agency grew slowly after the main reservation was established. It has always been understaffed relative to the number of villages (40 to 65) scattered over nearly 1,200 square miles and connected primarily by rough dirt roads and some wagon trails. Improvement projects in range management, water sources, or agriculture for every one of the villages, or even a large portion of them, could not possibly be developed and maintained solely by the agency staff under these physical conditions. Some responsibilities, therefore, had to be delegated to Papagos. Those projects that threatened an aspect of Papago culture or ecology would simply not be maintained when the government agents were not around. Government engines broke down. Government irrigation ditches were not cleaned out. Accurate figures on the existing number of cattle were exceedingly difficult to obtain from Papago cattlemen when the agency staff attempted to enforce stock reduction.

Noncooperation with the government and those other interest groups of the dominant society that threatened Papagos was strengthened when Papagos began to distinguish elements in the reservation system which belonged to "them," government organizations and employees, as opposed to "us," the Papagos. Outsiders were tested, by close observation and questioning, to determine whether they could be trusted, whether they were part of "them" or "us." Range projects and agricultural and educational programs that belonged to the government or the white men were avoided, resisted, or not maintained by Papagos, while those belonging to the people had a much better chance for success. Under conditions of threat, the boundaries between Papago and Anglo society became clearer.

Passive resistance took other forms as well. Parents protected their children from being sent hundreds of miles away to military-like boarding schools. Absenteeism in the government Indian schools was high, and so was the dropout rate.

Even today, when the Papago expect a visitor from the Agency whom they fear or dislike, it not rarely happens that they leave their houses and hide in the mountains until they feel safe again. . . . For many years the government tried to stop the summer rain ceremonies because they involved wine and intoxication. With characteristic elasticity, the Papago merely went to the ceremony in another village when pressure on their own community became too severe. Such efforts on the part of the government were abandoned in 1933, when it was conceded that Indians like other citizens have a right to religious freedom. (7)

The instances cited above represent passive resistance through withdrawal in contrast to organized resistance through political confrontation.

Political opposition is the fourth form of defense. At least three different superintendents and a larger number of staff members have been removed from office as a result of Papago efforts. The Good Government League of Papago leaders sought government services for its people before the reservation was formed. About 1925, eight years later, the League split over allegations by some members that ". . . a small group of the League members 'ran the Agency' and 'the Superintendent always called them in and did what they said'." The opposition members organized the League of Papago Chiefs and gathered sufficient support to have the superintendent removed in 1929 (4). Following investigations by the Tribal Council and Chairman, two more superintendents were removed in the 1960s for having bypassed the authority of the Tribal Council in the planning and carrying out of economic development projects.

The most prominent Papago political organization has been the Papago (Tribal) Council with its elected officers and staff. Adopted by a two-to-one majority of only one-third of the eligible voters in 1937, the Papago Council has slowly gained the support of the people during periods of able leadership. Even the council has been, and for some villagers still is being, tested to decide whether it can be trusted as belonging to the people. During its early years, when it lacked sufficient revenue to support a full-time staff, the Chairman of the Papago Council was also employed by the government agency. Within a decade the Council elected a young man who quit this employment because he felt it compromised the independence of the Council. He became the first full-time and salaried tribal official. With his able leadership for five years, the Council became considerably more independent of the government's direction. Papago tribal politics, however, is still dominated by the issues dividing the Good Government League and the League of Papago Chiefs: close cooperation with the government agency or separation

and independence of it. The oscillation of tribal politics between these two polar positions can be related to the acceptance or resistance of government-initiated development programs. Generally speaking, development programs have been more successful when they encouraged tribal separation from and independence of government authority and have been resisted when too tightly controlled by the government agency. And since the Papago Council and eleven district councils have had the authority as well as the ability to reject or stalemate those programs that threatened their security, Papago cooperation is now an absolute requirement.

Thus the Papago Indians in the Sonoran Desert of southwestern United States are a member of the general class of economically underdeveloped arid-land enclave groups around the world. They may be distinguished for development purposes by the following characteristics:

a) They have lived in an arid-land ecosystem for over 10,000 years.

b) They had controlled the selection and adaptation of Western innovations while living in the presence of three versions of that society. However, in the twentieth century their control was threatened by new conditions brought on by the American reservation system.

c) They have viewed as threats to their security those economic development programs that they could not call their own.

d) The ultimate stated goals of the United States government's development efforts are Papago economic self-sufficiency and self-government.

We will find further ways to distinguish the Papago case as we take a close look at what has historically been the primary economic development venture of the federal government among these people: the improvement of their cattle industry. Two different approaches to this project and the simultaneous attempt to encourage community responsibility for its management will be compared, and their contrasting results evaluated. Finally, conclusions will be drawn from the material presented concerning economic development in arid-land cattle economies, administrative relations with the community, and the relations between economic and community development.

PRELUDE TO FEDERALLY ADMINISTERED DEVELOPMENT PROJECTS

Father Kino introduced the Papago Indians to cattle-raising practices between 1692 and 1710 with the creation of the mission chain north from New Spain (8). He established herds of cattle, horses, sheep, and goats at each of at least nine missions, and distributed as gifts about 2,000 cattle to the Indians along the Santa Cruz and San Pedro valleys of present-day Arizona.

He expected that the ranches he established would be kept up, and that others would finally be formed throughout the country. But the whole plan seems to have lapsed after his death, and though there were Jesuit missionaries until 1764, there were no further gift herds. All mission property was confiscated when the Jesuits left, and the Franciscans who followed had nothing to do with ranching. (9)

The introduction of cattle as gifts (furnishing some instruction in their care as part of mission life) was important to the Spanish Catholic effort to convert the Papagos to Christianity.

In listing the means by which the natives could be made to submit to conversion and the whole continent totally reduced, he [Father Kino] names cattle as the most important item: "for, although in the past year I have given more than 700 cattle to the four fathers who entered this Pimeria, I have for the new conversions and missions, which by the favor of heaven it may be desired to establish, more than 3,500 more cattle; and some of them are already far inland. . . ." (8)

But with the dwindling of the missions in the eighteenth and nineteenth centuries, the newly introduced practice of cattle-ranching may have lapsed as well.

Woodbury and Woodbury (10) believe that cattle formed only a minor part of the subsistence base of the Papago in these centuries, and that the animals were probably semiwild. James Simpson (11) suggests that the post-Civil War cattle boom in the Southwest and the vast herds brought to southern Arizona were probably more important to the development of the Papago cattle industry than were the stock introduced by Father Kino. Instead of complementing hunting as an important aspect of their economy, cattle began to replace it, he believes. Southern Papagos, close to the Anglo and the Mexican ranchers, were acquiring animals through the abandonment of ranches due to Apache raids and drought, and as pay for their services with the soldiers (12). There are some strong indications that herds had become in one century more than a minor portion of Papago subsistence before the Indian Service began introducing changes to develop cattle-ranching as a permanent Papago industry in the early twentieth century. Between 1711 and 1912, that is, between the time Father Kino's missions lapsed and the time the Indian Service made their first development effort, *there were two centuries during which time the Papago had the opportunity to adapt ranching practices to their subsistence economy.*

There is some indication that during the early portion of those 200 years, those stock that bred outside the influence of the missions were hunted like deer, or (9) caught as "Mustang" and "tamed and used . . . in the manner taught at the mission." During those two centuries there seems to have been a merging of hunting and cattle practices, the latter adapted from or fitted to the former, ending with the shift from emphasis on hunting to emphasis on grazing, especially in the southern district of the present-day reservation.

Most of the cattle were allowed to run wild at first, considered much as deer or other game, to be hunted and killed as the families needed food. According to Dr. Underhill the 1 or 2 appointed hunters of each family group ordinarily brought in 10 or 15 deer a year to provide for the family. As the settlers drifted into the country, taking up mines and ranchlands, and colonizing the valleys, the wild deer and antelope became scarcer, and the family hunter killed about the same number of half-wild cattle, to be cut up and distributed for food, gifts and barter, as they had distributed venison. (9)

Ruth Underhill cites how, as cattle-hunting became cattle-grazing, the roles changed, but the social structure remained.

Since Kino's time, the Papago have begun to own cattle, at first in the lands near Mexico, later in the North. Cattle are in the charge of the men, like house building and agriculture, and their care is directed by the family head. He usually delegates one of his sons as cowboy, just as in former times, one was the hunter. (12)

And. . .

Now that cattle raising has taken the place of hunting, its arrangements are very similar. A village unit used the neighboring range as it had formerly used the hunting, but that range has no definite limits and no objection is raised if the cattle of other units are found there. A few families have accumulated enough cattle so that their ownership ranks as a business, managed on modern lines. Many, however, sell only two or three a year to local buyers and butcher about as many for the family as they formerly butchered deer. They distribute gifts of beef as they once distributed venison. (12)

Thus, without outside direction for two centuries, Papagos were making social and cultural adjustments in their arid-land ecosystem to the gradually emerging emphasis on cattle: for hunting first, and grazing later. The increasing emphasis on grazing may be seen in Table 1 showing the number of cattle ranches from pre-1850 to 1960.

THE INDIAN SERVICE PROVIDES WATER

Referring to Table 1, we see in the period 1900-1925 an abrupt increase in the number of newly founded cattle ranches. This is the same period in which white ranchers and the Indian Service drilled wells for stock water on what is now the Sells Reservation. According to Metzler (14), by 1919 there were around 30,000 head of cattle and 30,000 horses.

TABLE 1

The Number of Papago Ranch and Grazing Locations Since Prior to 1850 *

	Total	Newly Founded
Before 1850	0	0
1850–1875	1	1
1876–1900	6	5
1901–1925	19	14
1926–1960	23	12

* From Mark (13)

That was about five to six times the estimated carrying capacity of the Papago rangeland for 1944. Such intensive overgrazing resulted in gullying, washing, and eroding of the pasture land. Three to nine inches of topsoil were destroyed on over 1,250,000 acres of land (*14*).

Attempts to persuade the Papagos to reduce the number of their cattle and horses were bitterly resisted. Reduction in the number of horses was especially important, since they did more to damage the range than did cattle. There was less economic need for horses than for cattle, the latter functioning as a checking account where one could obtain cash when they were sold to cattle buyers. About 1930, more drastic methods of stock reduction were employed by the Indian Service in order to save the range. The number of cattle a family could have was limited to 100, the number of horses to 50.

Stock-reduction efforts were assisted by periods of drought and disease. Between drought and a dourine infection among the horses, the 27,000 cattle and 18,000 horses estimated for 1939 were reduced substantially to 13,000 and 7,000 respectively by 1950 (*14*). By 1960, according to District records, the number of cattle had increased, and overgrazing was half again the 1944 estimated carrying capacity (*10*). The livestock figures are summarized in Table 2.

The rise and fall and rise again of livestock numbers on the three reservations should be kept in mind while stock water improvements are being discussed, in order to evaluate whether these improvements stimulated or tried to keep up with overgrazing of Papago lands.

There have been three periods of intensive stock water development since the Indian Service began directing the development of the Papago cattle economy: the early 1900s, when the first wells were drilled for human and stock consumption; the middle 1930s, when the Emergency Conservation Work Program was accepted by the people; 1953 to 1958, when $1,000,000 was spent to develop and distribute wells, more spreader dikes, and deep water charcos.

Woodbury and Woodbury (*10*) believe there was a tremendous increase in the number of cattle the Papagos ran after the government well-drilling program of the early 1900s, because lack of water as a barrier to herd size was broken. Metzler (*14*) agrees that after white ranchers and the Indian Service began drilling wells, cattle production increased to 30,000 by 1919 and so intensively overgrazed the limited range surrounding the wells that erosion set in. James Simpson, and Charles Whitfield, Land Operations Officer, believe that the size of Papago herds was already too large to keep from overgrazing the available range by 1913 when the first government wells were drilled. The Indian Service erred grievously, Simpson (*11*) believes, by "attempting to make range improvements to catch up with the numbers of cattle there, instead of going on the opposite approach and bringing the numbers down to fit the range needs." He suggests that the government, before drilling any wells, should have worked out a range-improvement and cattle-reduction plan with the Papagos in each affected area. Without Papago agreement to the plan, no well would have been provided in the area. Such a plan would

have required an extensive range and cattle educational effort with the Papago ranchers, but might have begun to accomplish earlier in the century what the government has still not accomplished: Papago understanding of and participation in a stock-reduction and range-management program that they recognize as necessary for a successful livestock industry.

Charles Whitfield, on the other hand, does not believe that the government's early stock water program was a major contributing factor to the livestock numbers of the year 1919. Furthermore, he does not think that the Indian Service tried to provide more water to keep up with herd size. It is his opinion that by 1920 the Indian Service had tried unsuccessfully to persuade Papagos not to increase their herds. Appeals to the Indians, however, should have been replaced with restrictions, he believes (*15*).

TABLE 2

Estimated Numbers of Livestock, 1914 to 1967

Year	Cattle	Horses	Estimator
1914	30[a]-40-50,000[b]	8-10,000	Castetter & Bell; Clotts
1919	30,000	30,000	Metzler
1939	27,000	18,000	Metzler
1950	13,000	7,000	Metzler
1960	15,000		Woodbury
1967 (High Estimate)	18,000	3,000	Simpson
1944	11,000	1,000	B.I.A. (Carrying Capacity)

a Castetter and Bell.
b Clotts.

The consequences of not having solved this overgrazing problem early in the century have haunted and discouraged the successive government administrations since. Millions of dollars have been spent trying to revive range forage. The Indian Service had to threaten to carry out a stock-reduction program against the bitter resistance of the Papagos, who viewed it as a direct threat to their way of life, rather than a boon as the government saw it. The bitterness remaining from this issue spilled over to many other government programs. As Metzler (*14*) says, "The Papagos relate all Bureau activities to livestock reduction, and have tended to oppose all measures in regard to livestock production."

THREE-PRONGED GOVERNMENT APPROACH

To make up for the consequences of overgrazing, there has been a three-pronged effort by the government from the 1930s to the present not only to persuade Papagos to reduce their cattle holdings, but also to rebuild the eroded range and to teach a quality concept of cattle-raising at the same time. By the 1940s the government had lost much of the leverage it once had, for New Deal funds had been promised the tribal council in return for adopting grazing districts. And these funds had gone to increase the water sources and spreaders, thus insuring greater forage, encouraging more cattle, and without restrictions or stock reduction, guaranteeing continued overgrazing. This vicious circle had to be broken, for each of the three factors was dependent upon the success of the other two. For example: if the quality concept of cattle was not learned by the Papago stockmen, the stock-reduction program would continue to be resisted, and thus range-improvement projects would make little headway against continued overgrazing. The circle was broken by the droughts and the dourine disease during the 1940s. However, such a break cannot be expected to continue for very long, unless the government is counting on continued natural disasters.

The crucial factor, then, is that of Papago attitudes toward large herds rather than smaller herds of quality cattle that bring an equal economic return. Since Papago attitudes did not change, except for those of a dozen or more families today raising cattle as an industry, we should try to understand why. We can gain some insight by looking at the way of life, developed over two centuries in the presence of cattle, which so strongly re-enforced those attitudes. Then we will consider the government's three approaches and evaluate their success or failure.

Papago Subsistence Attitudes Toward Cattle

To appreciate the strength with which Papago attitudes toward cattle could be held, we have to remember the two or more centuries they had to fit subsistence cattle-hunting and cattle-grazing into their social, cere-monial, and economic life. As late as 1937 there had still been little government influence upon their cattle activities.

The stock industry is as yet practically unorganized. It has just grown up of itself, handled by the Indians as best they could. The reservation is very large, and the Agency admin-istration has had neither money nor personnel to give much government aid or supervision. *(9)*

A gift-giving and barter economy was beginning to mix with an intervening cash economy. The Emer-gency Conservation Work Program (ECWP) of the late 1930s, with its work-for-wages programs, was a major thrust in this direction. A description of the Papago economy and its complementary social relations for the nineteenth century could still have remained largely intact in the 1930s. Ruth Underhill *(12)* described the importance of gift-giving in the Papago economy, and the importance of food as gifts. Gift-giving, barter, and gambling were the primary means to distribute economic wealth. Wealth was distributed throughout the year in response to social and religious obligations, not accord-ing to a fixed schedule. The need for food, crops or cat-tle, was continuous. Not to respond to these obligations was a disgrace. It was therefore necessary to be able to butcher a cow or sell it for cash at any time. Thus, any programs that might curtail this freedom to choose when to sell or butcher one's cattle would be resisted. Such a

program would not only threaten this liberty, but would appear as well to threaten the whole Papago way of life. The freedom to dispose of one's cattle whenever the need demands it is still defended today. According to mem-bers of the Gu Achi Stockmen's Association, the people need their stock on hand during the entire year. They listed the following pressures on small cattle owners that keep them from selling their head at a few large sales, or reducing the number of cattle they own, as those advo-cating a commercial cattle industry recommend. Ranked in order of most general occurrence, the uses (para-phrased from *16*) are:

a) Selling cattle for money, as when they return from the cotton camps, or just before they go to Magdalena, or in hard times.

b) Exchange of cattle with someone for beans, etc. That's been going on a long time.

c) For church feasts, saints' days, weddings, other fiestas. Catholic families have pictures of their saints. On the day of their saint, these people organize the fiesta and butcher cattle for the meals. But it's also a commu-nity fiesta, so some people will make donations of meat to help out.

The need for ready cash has been added to the need to barter for other foods and to the need to fulfill family and community religious obligations. Now cattle are in demand more often for cash than for the traditional requirements, which shows the increased dependence of the Papagos upon the wider cash economy.

Even in the late 1930s the Papagos had used their cattle as a bank reserve:

Cattle are the reserve fund for the people, the "Papago Bank." When there are no jobs and no crops many villages would sell cattle.
Therefore the advantages of the plan urged by the govern-ment, regular big cattle sales spaced periodically through the year in such a manner as to bring higher prices, do not appeal to the average Papago cattleman. Nor do cattle asso-ciations, for the same reason. There is a great advantage, to his mind, in being able to lay his hands on ready cash when he wants it; and he is ready to let the extra profit go for the sake of convenience or to supply immediate need. That is the way he has always been accustomed to handle things. . . . This deeply ingrained attitude presents a problem in Indian resistance to the government's attempts to reorganize the Papago stock industry even though the attempted reor-ganization be for better economic results through the devel-opment of cattle, and such things as supervised and organized meetings in large periodic cattle sales. The Superintendent and stockman are well aware of the attitude as a barrier to their efforts to bring the industry to a better organized and more profitable state, and are doing their best to over-come it by explanation and informal education. *(9)*

Not only Papago social organization and economy, but Papago values have been in contradiction to Western-type economic development. According to Metzler:

Anglo values emphasize individual initiative, respon-sibility, aggressiveness, and getting ahead. Papago ideals are cooperation, harmony, and doing your part. . . . This type of moral code created an individual who was highly subordinate and submissive.... Any desire to get ahead was outside his realm of thinking. This type of moral code created a society in which cooperation and sharing were the key values. Individual or family advancement was sinful, as it tended toward inequality and lack of unity.... This culture produced friendly, cooperative human beings, but had no goals in the way of material progress.... The whiteman, who simply assumes that the Papago can adopt his way of life, is ignoring the depth of the Papago system of life. *(14)*

(The superintendent *(17)* summed up these contradictions of Papago and Western values by commenting, "Their priorities are backwards.")

Some government programs to develop a cattle industry out of a subsistence cattle economy encountered resistance and an unwillingness to learn the Anglo ways because the whole society's culture was being questioned, not just its ranching techniques.

Stock Reduction

Now let us consider the government's efforts to develop the requirements for a Papago cattle industry from the 1920s to the present, with the previous discussion in mind. The three aspects of their general plan, stock reduction, quality cattle education, and range improvement, will be studied in that order.

Drought and disease have been the prime factors in the agency's stock reduction success. (See Table 2, years 1939 to 1950.) Attempts to persuade the Papagos to reduce their holdings have been unsuccessful and unenforceable. In the early 1940s a dourine infection spread among Papago horses. The state of Arizona required the disposal of thousands of the diseased animals, a ruling which the Indian Service used to carry on its horse-eradication program begun in 1934. Horses eat forage closer to the ground than cattle and as a result cause greater overgrazing. According to the Agency, there were far more horses on the rangelands than were economically required for roundups and transportation. But the Papagos held horses in higher esteem than their cattle and might have resisted the eradication program even more had not the dourine infection spread. Dobyns estimates that the extent of horse reduction was even greater than Table 2 indicates.

In 1941 the Extension Division estimated that 20,000 head of horses were on the reservation. After roundups had been carried out, and after horse losses from other causes... (by 1949) there are about 7,500 horses on the reservation.... *(18)*

During those nine years, thousands of horses were sold and disposed of in connection with state disease requirements, and hundreds more were rounded up and sold as part of the Indian Service's eradication program.

During the late 1930s and late 1940s, the Papago Reservation suffered severe droughts. Numbers of cattle-owners as well as numbers of cattle decreased sharply by 1950. In 1949, Thomas A. Segundo, Tribal Chairman, told Dobyns:

As water developments were made, cattle ownership grew. Then came drought.... It hit the small owner the hardest,

and forced him out. The larger ones survived, though they lost many cattle, they had some left. This last year many have gone out of the cattle business. *(19)*

Where the tribal council and the grazing districts were unwilling to limit the number of head per family, or even to pass and enforce grazing regulations before 1950, successive droughts had largely reduced cattle numbers from an estimated 27,000 in 1939 to 13,000 by 1950 (see Table 2).

Not only did Papagos resist the government's stock-reduction programs, but they resisted as well government programs they associated with stock reduction. Cattle associations to foster cooperative selling and to apply commercial range-management principles were suggested by the Indian Service in the 1930s. Papagos were skeptical of any government efforts to restrict or regulate their livelihood. District associations might be another way to effect stock reduction. It was not until the droughts of the next decade and a half had nearly fulfilled the government's stock reduction goals that Papagos could be persuaded of the benefits (especially range work and local control) of cooperative cattle associations.

Even though the Agency has been able to encourage cattlemen to associate, it has been less successful in convincing them that they should set limits to the number of cattle that can be grazed in the district range, or limit by permit the number of head a family can own. Some of the district associations have placed the recommended limit of 250 head per family, but most of them do not have any limits adopted. For instance, the most active and recently formed association, the Gu Achi Stockmen's Association, has still not adopted a permit system or attempted to limit the size of its members' herds, even though its officers recognize (verbally) that they are overgrazing their district's ranges *(16)*.

The Agency has found that it cannot force Papagos to limit the size of their cattle herds directly, because of the passive resistance and antagonism such an effort engendered. And Papagos have not been convinced "that fewer cattle and horses, maintained in much better condition, would be advantageous" *(10)*.

A near stalemate has resulted. The Indian Service and the agency have worked hard to make use of what drought and disease have accomplished for them. They have tried to demonstrate to Papago stockmen the economic advantages of quality cattle on properly grazed pastures, now that stock reduction has made that situation more likely of attainment than ever before. But, because of a variety of factors, some of which have already been discussed, Papagos (excluding the largest cattle-owning families) have not been convinced that the economic gains are yet worth the social losses they believe will result. The Agency has taken Papago stockmen on tours of modern Anglo cattle ranches and has shown them the best reservation pasture projects. The agency has bypassed a resistant tribal council and gone to the district and village councils to instill the need for cooperative livestock associations.

To what extent have these government educational and organizational efforts loosened this stalemate?

Quantity Versus Quality Concepts

The Papago Agency developed a modern, profit-making cattle ranch to demonstrate to Papago stockmen the benefits of a scientifically managed cattle industry.

The following history of the Agency's effort to establish the Tribal Herd Ranch, was compiled in 1964 by Charles Whitfield in the Bureau of Land Operations, Papago Agency:

1. The Tribal Herd has been developed from the offspring of New Mexico Rehabilitation's cattle shipped to the jurisdiction (of the Agency) in the Fall of 1935. In the shipment from New Mexico were a number of heifers to be issued for slaughter. On arrival many of these animals were so thin they were considered unfit for food and were turned out on the reservation. Later roundups showed that 12 head had survived. (*20*)

2. In addition to these 12 heifers, the herd was increased to over 1,000 head in 1941. In March of 1942, $10,000 was made available through a federal loan, and in June of 1942, $6,000 additional funds were provided for the purchase and establishment of a purebred herd. From this beginning the herd has grown and at one time amounted to 2,000 breeding cows. (*20*)

3. The Papago Agency maintained close supervision of the Tribal Herd activities until 1956 and assisted greatly in the location of pasture lines, fencing of these pastures with Tribal labor, and generally established the Tribal Herd operation as a first rate ranch enterprise. (*20*)

4. In 1956 the Agency decided that the Tribal Herd enterprise was ready to be turned over literally to the Papago Tribe, who the Agency felt, was ready to assume complete management of the operation with the Papago Agency playing a minor technical support role. (*20*)

5. In 1962 a letter from the Papago Tribal Chairman requested that the Agency provide additional technical support and supervision to the Tribal Herd operation. This support greatly increased and has been accelerated to the present date (1964). (*20*)

Subsequently:

6. In 1966 the Tribal Chairman and his supporters felt the Agency's increasing assistance given to the Tribal Herd Ranch had gone too far, and they accused the Agency of trying to take over the Tribal Herd again. Since then the Land Operations staff has just been "on call." (*21*)

7. From 1966 to the present, the Papago Tribal Herd Manager has guided the operations and management of the Ranch.

Why couldn't the Papago Tribe maintain the Tribal Herd Ranch when it was turned over to them? Why did they in 1966 reject any further assistance from the Land Operations staff? Were the Papago cattlemen learning and adopting the modern methods of cattle-ranching being demonstrated?

Papago cattlemen have traditionally owned and grazed as many head of cattle as the availability of water and customary land use would allow. Before the arrival of Anglo ranchers, the Indian Service and the water and forage developments of the Papago Agency, herd size was limited by the natural ecology of desert rangeland and by the limited number of cattle that Papagos could acquire or that were bred on the open range. Traditionally Papagos have held "monopolistic" attitudes toward their subsistence cattle economy. They held that an individual had a right to land only while he was using it. Each cattleman has as many cows as he can get, and grazes them just as hard as he can, because if he doesn't someone else is going to beat him out. This part of the Papago communal land use system encourages each cattle owner to view his operation as a monopoly rather than to view it as limited by every other cattle owner's sharing of an exhaustible natural resource. As in ocean fisheries, if everyone has a monopoly view of the resource and catches as many fish as possible, eventually the ocean will be depleted of that resource. This is what has happened on the limited Papago range. Combined with the monopolist's view, most of the Papago cattlemen believe that the more cattle you have, the more money you can make at sales, or the more reserve assets you have to draw on when you're in need. They don't realize that they can sell a heavier calf and still get the same return as from a number of poorer quality and lighter calves. And, still further, combined with these views are their residual attitudes from the stock-reduction program. Proposals to change from a quantity to a quality view of cattle ranching means stock reduction to the small owners who have continually resisted attempts at stock reduction (*11*).

The Tribal Herd Ranch, as developed by the Papago Agency, was managed by modern principles of animal husbandry and range conservation. The Ranch is composed of three distinct pasture areas with select perennial grasses. The stock are systematically rotated among the three pastures. Rainfall averages about 15 inches per year. Purebred bulls are the basis for a scientific breeding program. An inexpensive bull-rental program has meant that Papago cattlemen throughout the reservations could secure a purebred bull for four years at a minimal cost of $70. A carefully selective culling program has eliminated dwarfs from the Herd. A Livestock Board was established by the Agency to oversee the operation of the Ranch. It was composed predominantly of Papago cattlemen with large herds and supervised by Land Operations staff of the agency.

Range conditions and Papago cattle management throughout most of the reservations, however, contrast sharply with the Tribal Herd Ranch. Within each of the eleven fenced Districts, cattle range freely. Cattle breeding is not selective on the open range. Pastures were not developed outside the Tribal Herd Ranch until the 1950s. Rainfall diminishes from 15 to 18 inches around the

Baboquivari Mountains, where the Tribal Herd Ranch is located, down to 5 inches in the westernmost parts of the reservations. In contrast to the carefully grazed Tribal Herd pastures, all but two or three of the eleven districts have been overgrazed. Herds have not been carefully culled, so that dwarfism is a recurrent problem to Papago cattlemen.

The Livestock Board and those Papagos who worked at the Tribal Herd Ranch were taught the basic principles of animal husbandry and range management in preparation for the time when the Ranch would be transferred to the Papago Tribe. Apparently, the Livestock Board was not very active while the agency developed the ranch into a profit-making business. In 1949, Burton Ladd, the agency superintendent, remarked in a recorded interview that the Livestock Board had showed little interest in the Ranch's operation. He also commented upon the opposition toward the Ranch shown by the Board's large cattle owners (19). There is doubt whether the Livestock Board was actually making policy decisions while the Tribal Herd was being developed and before it was transferred to the tribe. Eugene Walker of the agency's extension division suggested this in an interview recorded in 1949.

The Tribal Herd is owned by the Tribe, and everything concerned with it is cleared through the Livestock Board of the Tribal Council. *In theory* it is under the Board. (19)

The composition of the Livestock Board, predominantly owners of large cattle herds, probably encouraged the Papago Agency to bypass the board in certain policy decisions, that is, those that might open the door for government or tribal regulation of their cattle interests or encourage government-subsidized competition, for these men had developed large and marketable herds and

preferred to manage their own range and herds without help or interference from outsiders. Through the years opposition to regulation of the cattle industry with reference to total tribal interests was maintained by these wealthiest large cattle owners among the Papagos. (4)

Under the impetus of the Indian Reorganization Act, 1934, "the reservation superintendent was instructed to persuade Papagos to organize a tribal council based on a constitution in the formulations of which Papagos were themselves to participate." The leadership in the tribal council was immediately assumed by persons from the southeastern area of the reservations, where most of the wealthy cattle owners lived (4).

Tribal and district government was entirely new to the Papago. There had only been prolonged intervillage organization, at best, for defense against Apache raiding parties, 1740 to 1875. Political organization was based on the autonomy of each village.

There was no immediate acceptance of tribal organization in the terms in which the government men saw it. There had never been a subordination of one village group's interests to another, nor was there now. Representatives were not regarded by their districts as empowered to enact legislation, but rather were thought of as "legs," to use the old Papago term; that is messengers and communicators of news to their district councils. (4)

But a tribal organization allowed the government agency the possibility of dealing with one group or one man instead of having to negotiate development programs individually with each of over 50 villages. From 1937 to 1948 the tribal chairmen were not salaried by the Tribe, but were either employed by the agency or on their own resources. In 1948 new precedents were established when a young tribal chairman resigned from his position in the agency and became salaried as a full-time chairman. He believed the tribal chairman and government could then be less dependent upon the agency's view of development. Politically, tribal chairmen have developed three different roles in their relations with the Papago Agency. Some have been allies sufficiently to be the agency liaison to the people. Some have been cooperative, but more independent. Others have been bitterly critical of the agency and skilled enough to have one or two agency superintendents removed. In 1966, when the agency was told to stop assisting the Tribal Herd Ranch, the tribal chairman's relations with the agency were quite strained.

The primary purpose of the Tribal Herd Ranch, as developed by the Papago Agency, was to "provide a demonstration area for the education of Papago stockmen in matters concerning range and livestock management." A secondary and complementary purpose was to "provide a service by producing registered bulls for use by individuals on the reservation" (20).

In order to accomplish both objectives, "the Tribal Herd enterprise must make a continuous profit. The profit must be held in reserve and then used for maintenance . . . and development of new facilities . . . necessary to improve the overall operation of the ranch" (20).

The demonstration value of the ranch was complicated in operation by these factors. The climatic and technical conditions of the Tribal Herd Ranch were in sharp contrast with the conditions existing throughout most of the reservation where the lessons of the ranch were to be applied. Though the Tribal Herd Ranch is an enclosed range allowing selective breeding, Papago cattle run in the open range within each district even though they are individually owned. About 90 percent of Papago cattle owners, all but the wealthiest cattlemen, lack the understanding or training necessary to apply the modern principles of animal husbandry and range management practiced at the Tribal Herd Ranch by the Papago Agency, even if conditions between the open range and the Ranch were comparable.

While developing the Tribal Herd Ranch, the agency took Papago ranchers off the reservation to tour successful Anglo ranches. They also visited pasture projects on the reservation. By these tours they hoped to show the Papago cattlemen the benefits of modern cattle ranching and encourage them to learn to do likewise. Trips were made to the Rocky Mountain Experimental Station and to the Vamori pasture.

The tours have worked out really well except for one problem. They're taking the wrong people. They're taking the people who volunteer to go on this kind of trip; people already fairly sophisticated. They're not the ones who need to volunteer. You can see this also in their annual on-reservation range field trip. The same people go on it every year. They're not getting any of the new people. *(11)*

The registered bull-rental program developed at the ranch was intended as a service and implicitly as an incentive to bring Papago cattlemen to the ranch for educational purposes. In actual operation, the incentive for Papagos to rent purebred bulls at less cost than they could find off the reservation has worked, but there has apparently been limited educational value to the program. The Tribal Herd Manager discussed the program in an interview. He was asked if the cattlemen who come to the ranch to rent bulls also learn some of the principles and practices of modern ranching and breeding that he had received from the Land Operations staff of the Agency. He replied,

Stockmen from the associations come out to look at the bulls to rent. They don't ask me many questions. They're most interested in selecting bulls to rent. *(21)*

He said they sometimes do ask questions, such as, how does he take care of dwarf problems, or treat "pink eye," or prevent "black leg." But he felt there was little educational value to the bull rental service, and that it was a real expense to treat and fatten up the bulls, which return in poor condition after four years on an overgrazed range.

After developing the Tribal Herd Ranch as a profit-making enterprise, the Papago Agency tried to transfer it to the Papago Tribal Government in 1956. Why couldn't the tribe maintain the successfully operating Ranch during the next six years? And after the tribe requested and received agency technical assistance for the ranch, why did the tribal chairman and council accuse the agency of trying to take over the Ranch again?

In order to prepare Papagos and the tribal government to manage the ranch, participants were taught "generalized concepts of animal husbandry." Selected Papagos were not trained for the specific positions previously filled by agency employees necessary to operate the Ranch. The Livestock Board, which was supposed to supervise the herd manager, "fell apart." Without its supervision and experience, the manager could not adequately carry out his responsibilities. The tribal council lacked sufficient capital of its own to keep the Ranch at the level of operation it had attained when it was developed and subsidized by the federal agency *(15)*.

Apparently, in the eyes of the Papago tribal chairman and council, the increasing technical assistance provided by the agency after the request of 1962 began to look by 1966 like planning, supervision, and repossession. According to the Tribal Herd Manager and the agency range management specialist, Papagos were accusing the agency of taking over the ranch again.

Range Improvements

"The foundation for a range management program was established in 1934 when the Papagos accepted the Emergency Conservation Work Program (ECWP). As the first step toward the systematic use of range resources, nine large range units or districts were to be established and separated by division fences." The Bureau of Indian Affairs, in consultation with the newly formed Tribal Council, made these district divisions. Brand owners were required to keep their stock within their own districts. In many cases, they felt that this requirement was a hardship *(14)*.

Stock water and range development projects have been carried out from the early 1900s (discussed earlier) through the 1960s. Carefully distributed stock water supplies are necessary to commercial cattle industry, but if not coordinated well enough with range conservation, they may also lead to the overgrazing that can destroy the basis for that same industry.

With the New Deal wage work programs, the Emergency Conservation Work Program, the Civilian Conservation Corps (Indian Division), and the Agricultural Adjustment Program, many water-supply and erosion-control projects were initiated in the 1930s and 1940s. In 1935, 44 earthen tanks, 16 dams, 11 storage tanks, and 22 wells were constructed; 9 springs were developed; and 1,465 pounds of seed were planted on the erosion areas. In 1936, 32 more charcos (earthen tanks) and 2 masonry dams were built in order to secure the correct spacing of watering places. Thirty large masonry storage tanks were erected; 31 wells were developed and 3 drilled. In addition, 9,000 pounds of Johnson grass and chamiza seed were gathered, and 5,000 pounds planted on the nearby San Xavier Reservation. Even with federal help, the Papagos have been unable to cope with overstocking and the deterioration of their ranges *(18)*.

Funds from the Agricultural Adjustment Administration, 1946 to 1949, offered the Papago Agency administrators an opportunity to plan and carry out stock-water and range-improvement projects with the people on district-wide and village bases. Three districts qualified for AAA funds, and the development approach in two of these varied enough to point out important lessons in development administration on the Papago reservations. They differed by the degree of Papago participation in the planning and execution of the projects, and by how strongly the Papagos felt the program was needed. The successes varied accordingly.

Schuk Toak and Gu Achi

In the Schuk Toak case the administrators went to the village councils and reconstituted the moribund district council with promises that Papago men would be paid for work done. They sold the program more on this point than on the benefits that would be derived from the work. Though intending to allow Papagos sufficient time for them to decide and take over the administration of AAA projects, the Agency men were soon pressed by Washington, D.C., time limits to get the funding agreement signed. As a result the projects carried out by Schuk Toak men were viewed by these men as an essentially government program.

During 1946 work was done on 527,842 acres of grazing lands which was approved for $7,918 in AAA benefit payments. A community pasture was built at Queen's Well, a truck trail constructed, a charco at Haivana Nakya was deepened, 15 miles of fence was repaired or built, and five corrals repaired, a new charco was built at Santa Rosa Ranch, brush was cleared from a Johnson grass meadow, and brush dams were put in gullies on the Tribal Herd Ranch. (22)

The people were dissatisified with receiving their benefit payments only at the end of the first year's projects, for they had equated the AAA work with CCC work for which they were paid regular wages. (This confusion was largely due to the rush to get the original funding agreement signed by Washington deadlines.) As a result, the work undertaken in the district under the second year's funding agreement was significantly less than the agency conservationist officer had hoped for, despite his having worked through the district political organization, and despite the fact that the projects had been selected by the villages (22).

There are two major factors here that may explain why the project work fell off the second year. First, "We told them that if they signed these papers the government would give them some money. The explanation was on that scale. They were only confused otherwise. We didn't want them to reject it; we had to sell them." So when the people learned that they would only receive payment after the year's work was complete instead of as weekly wages, they were less willing to undertake the same effort the second year. Implicit here is that weekly wages had priority over completed conservation projects. Second, because of the Washington, D.C., deadline for completing the contract for the year's work (set by the county Association),

The Indians felt pressed; they did not have time to discuss, understand, and adopt the plans as theirs. The program remained the property of the Agency, accepted for the money it would bring.... The amount [of money] was insufficient to produce continuation of the work once pressure relaxed. (21)

In the Gu Achi District case, the people went to the agency conservationist to get advice and help to halt the flooding that was threatening one village and eroding farm and grazing lands. Two successive drought years had severely depleted the number of cattle available, and thus the subsistence cattle owner's "bank reserve." The conservationist went out to Gu Achi and met with the village headman, who called a meeting. The local bus driver translated. With an extension employee whose home was in Gu Achi, the agency conservationist surveyed the extent of erosion. At another meeting he explained that Gu Achi could get work done by assigning contractors their AAA benefit payments. No work for wages from benefit payments was offered.

Returning to headquarters, the conservationist left the burden of persuasion on the bus driver and local extension man. The latter held a long meeting, during which people hurled questions at these men which they hesitated to ask the headquarters official.... The villagers realized their need of assistance and wanted it. They chose earthen spreader dikes to control some of the worst flooding. (22)

An engineer from the Bureau's Area Office was called in by the agency conservationist to survey a good technical plan for the spreader dikes. But when the village leaders went over the ground they objected.

The conservationist recovered his fumble. He resurveyed according to the wishes of the villagers, who had been ignored by the Area man. He turned planning back to them, restoring the feeling that the project was theirs. "We started to try to go more and more with what the people wanted wherever feasible." Technical advice was not allowed to become planning; over-professionalism was avoided.(22)

The headman and the bus driver led in securing the agreements with the AAA and the contractor for building the dikes. "After surveying and bringing contractor and people together, the administrator left execution of the project to the people." The dikes were completed and the floodwater was spread out to grow hundreds of acres of forage. In the two succeeding years additional agreements were signed by Gu Achi and Ak Chin men for continuing conservation projects (22).

Compared with the Schuk Toak, the Gu Achi case has three general lessons. First, whereas the Schuk Toak people had to be sold the need for adopting the AAA benefit payment project, the Gu Achi people went to the agency needing conservation range work done. Because the need for the work was already there, offering the AAA program as a solution was acceptable to the Gu Achi people. And because the work done satisfactorily helped solve their erosion problem, the Gu Achi people continued making agreements with the AAA for further needed work. But the need of the Schuk Toak people for wages, which need was sold to them by the conservationist, was not satisfactorily met, so they were less willing to continue the projects the following year. Second, there was the factor of how involved the people were in the planning. In Schuk Toak the administrator already had the plan which he sold to the people, albeit rather badly. The people, pressed by the deadline, did not have enough time to discuss, understand, and adopt the plan as their own. In Gu Achi the people planned every step of the way. The administrator presented alternative solutions to their felt needs. Though he erred by letting the engineer from the area office present a plan without consulting the people's wishes and superior knowledge of local terrain, he corrected it by ignoring the engineer's unilateral plan and by laying out the project where the people said it should go. "Furnishing only technical knowledge, he left planning to them." Finally, there was a member of the administrative group living in Gu Achi, but not at Schuk Toak villages. Being accepted as a local member of the village and as a part of the Agency, his position on erosion control was ultimately more significant than that of the headquarters official. Although he was not employed by the conservationist, he convinced the people that the conservationist's recommendation was their best solution, mainly because they could converse with a local village member (*22*).

One result of the AAA and PMA projects was apparent to the soil conservationist in 1952: Papago attitudes were changing. Since no real progress is possible for very long without changed attitudes, the soil conservationist properly called this change "the greatest single accomplishment of the Soil Conservation Division."

Ten years ago few people were aware of their soil problems. Today communities in every district are alarmed over erosion and want to correct it. Ten years ago technicians were met with suspicion and hostility, such words as range conservation and management were taboo to most Papagos and Agency personnel. . . . Through [P.M.A. benefit payments] a great deal of erosion control and waterspreading work was done. This has increased interest in the program.

TABLE 3
Summary of Development Factors

	Requested	Planning	Training	Time to Adapt the Plans	Fits Tribal Social & Political	Fits District Social & Political	Fits Village Social & Political	No Ulterior Benefits	Did Work	Belongs to Papagos
Projects Maintained										
Tribal Herd (post-1966)	x	–	x	–	x	–	–	–	x	x
Gu Achi AAA	x	x	–	x	–	x	x	x	x	x
Gu Achi Stockmen's Association	x	x	–	x	–	x	x	x	x	x
Projects Not Maintained										
Tribal Herd (pre-1966)	0	0	0	–	–	–	–	–	x	0
Schuk Toak AAA	0	x	–	0	–	–	x	0	x	–

The above summary is based on a brief factor analysis of the case studies presented in this chapter. It is the first step to formulating a hypothesis about them.

x = yes; 0 = no; – = unknown or irrelevant.

Requested = project requested by Papagos at some early time.

Planning = Papagos participated in planning, that is, did not simply approve a government-proposed plan.

Training = Papagos trained for unfamiliar roles necessary to maintenance and direction of project.

Time to adapt = Papagos given time to adapt plan to needs, time enough to understand purpose of planning.

Fits . . . social & political = approach and plan of administrators were adapted to social, political and economic life of Papagos, with tribal, district and village distinctions.

Ulterior benefits = project not sold on basis of some other benefit than original one intended.

Did work = Papagos did much of work themselves.

Belongs . . . Papagos = Papagos expressed at some point feeling that project or plans belonged to them, as opposed to government.

If the purpose of the economic development projects is to transfer them from the jurisdiction and responsibility of the administrators to that of the Papagos, these factors are relevant.

Five years ago the need for conservation work had to be pointed out. Today men from communities where no work has been done ask for help. Without P.M.A. (and A.A.A. which preceded it) our progress would have been much slower. (*23*)

Table 3 has been prepared as a first step in formulating a hypothesis about the case studies.

TWO DIFFERENT APPROACHES

Evidently, the Bureau of Indian Affairs had not evaluated the reasons for the success of the AAA range improvement programs, 1946-1949, for in the "Soil Conservation Program Twenty Year Period," 1954, by the Land Operations Department (*24*), Papagos were not to be included in the planning but only in enforcing and carrying out the government-drawn projects. Three major objectives were established: (*a*) "To secure regulation and management which will keep the lands producing a good vegetative cover." (*b*) "To improve suitable lands through water spreading and other practices." (*c*) "To secure economic ranching units through association of smaller ranchers." The Papagos had already clearly expressed a need for the first two objectives and had planned their own projects to achieve them under AAA. The last objective reflects the government's desire to establish a commercial cattle industry. The general "plan of action" was for the Bureau to "establish a framework of regulations" for range management and commercial cattle-ranching that would be adopted by the tribe, districts, and cattle associations, with the hope that responsibilities for such would be transferred to livestock owners. Tribal participation, the report concludes, would involve execution of the above planning. "Much of the cooperation from the tribe will have to be in the form of ordinances, land codes, and enforcement regulations." The problem of this approach to development should have been clear by 1954 if continual evaluation of development projects had been a significant part of government programs on the Papago reservations.

At issue here are two different views of a hypothesis of the proper approach for successful reservation development. One is that the Papago people are still in a back eddy, not in the mainstream of American life. The only way they can hope to enter it is with the immediate help of social and technical experts. Their priorities are backward and will have to be reversed before progress is possible. They are not yet expert enough in American know-how to plan for their own development. Their resources must be developed for them through careful technical planning, and, after they have seen the benefits of such development, they will take it over themselves. The other view of what makes for successful development says: At issue is not whether the Papago people know what is best for them, or whether they will inevitably be integrated into American society, but whether they will have the opportunity to participate in resolving these issues, participate in planning the best ways to carry out these resolutions, and participate in the execution of those plans. Only in this way will they learn how to accept the responsibility for carrying out their own development with the advice of experts when it is needed. And, since they have seen projects technically planned by experts fail or go contrary to their expressed interests, Papagos are likely to resist or not to maintain projects that they do not help plan or do not adapt to their local conditions.

Because the droughts in the 1940s showed the need for permanent stock water facilities, and because the Papago Development Plan of 1949 proposed to support one-third of the population by a cattle industry, there was a substantial effort to build permanent wells and distribute them more widely than before.

The most influential of these programs was completed in 1959. To rehabilitate wells that had not been maintained by the government during and soon after the war, and to drill new ones, $1,000,000 was spent. Much of the overgrazing occurred because the water sources were concentrated along a few washes and near the mountains. The goal was to get a better distribution of cattle over the range, relative to available forage and water. Augmented by charco construction, the program was technically successful.

The desired results were not forthcoming, however, not because the program was poor, but because the Papago tradition of village control rather than district control of grazing had not changed. Each village had its own roundup boss and its own grazing area. The villages would not go along with total district-wide cattle controls, such as grazing regulations. And where the districts have had little influence upon village livestock controls, the tribe has had less. A tribal ordinance limiting the number of horses per family, when enforced only by village and family roundup bosses, was ineffective (15).

According to Whitfield, per Simpson (11), the reservation viewed as one ranch has one of the best water-development ranges in the arid Southwest, excluding the mountain ranches. "The locations of the charcos and wells are just perfect. You can close your eyes and draw in fences, and you've got it, because of the beautiful array of water." But what was apparently a technical success, Simpson believes, contributed significantly to overgrazing again of the range, because the Papago stockmen had not organized to develop controls for the expanding cattle population. As during the period 1900-1920 and in the 1930s, stock water sources were increased in number and spread more widely over the reservation, but the few regulations necessary to prevent overgrazing of the range newly accessible to the cattle were not adequately enforced. The dilemma has still not been settled, and according to Simpson, the range continues to be overgrazed. One reason, he ventured, is that the Bureau lacks the educational staff to complement the stock water development. It needs larger and more effective educational programs to insure that Papagos will understand why and how to limit the number of cattle in their districts, and how to maintain pastures once they are cleared and seeded.

A close analysis of the attitudes of the cattlemen and associations during the 1950s would probably reveal whether range regulations were resisted and not enforced by Papagos because of their view (discussed earlier) that cattle were their "bank reserve," or whether they still viewed cattle as necessary to their subsistence needs and not necessary as an industry. James Simpson believes that the Bureau and/or the Tribe could have required that the stock water wells be maintained by the local people or the cattle associations by refusing to put in more wells unless they agreed. After the wells were drilled, it is improbable that anyone could have or would have forced Papagos into upkeep against their will. According to one view, wells would be maintained only if people were trained and if they viewed them as their own. As it turned out, the Tribe hired one well repairman whose job is to go around the reservation and take care of the pump jacks.

TRANSFERRING RANGE RESPONSIBILITIES

While stock water, range, and pasture development were being carried out, the government, realizing it couldn't and didn't want to maintain its range projects by itself, tried to transfer this responsibility to the districts and to the tribe. These efforts have been only partially successful at best, and the differing views of government officers, students of range management, and Papagos bring out some important factors in bureaucratic development problems.

The most durable effort at fixing range-management responsibilities upon the Papagos was the establishment of grazing districts in 1935; the districts also became the political subdivisions of the tribe. The new superintendent, T. B. Hall,

. . . was told by the Indian Division of the Emergency Conservation Work, that, unless the Papagos adopted in short order an effective plan for range management, the extensive construction programs underway would be curtailed. Under the pressure of depression conditions, the heads of hundreds of Papago families had become economically dependent upon these programs. (25)

Consulting with an Arizona State Teacher's College faculty member who had written on the structure of Papago village groupings, and with a group of Papago-speaking Franciscan fathers with extensive knowledge of the Papagos, Superintendent Hall proposed to a meeting of a provisional tribal council that grazing districts be drawn coterminous with the traditional village-controlled grazing areas. He must have explained the ECWP benefits of accepting his proposal, for the Papagos agreed in principle. Ten days later, the plan was presented to 2,000 village representatives. They met with agency and CCC planners (successors to ECWP) to refine the district boundaries so that roundup camps, cactus forests, wells, waterholes and other intermittent water sources belonging to one village were not fenced off from the area traditionally controlled by another. Compromises had to be made where grazing district boundaries crossed village and dialect lines.

By late winter, 1935, the boundaries were agreed upon, and fencing began that spring. By July, 1936, over 300 miles of fence had divided the main reservation into nine grazing districts, which have remained unmodified to the present. But the tragedy of this achievement, according to Charles Whitfield, was that Hall was unable to get district grazing ordinances and district responsibility for range conservation management adopted along with the boundaries (*15*).

Hall was successful as far as he went, very likely because of the careful way he fitted the conservation demands of the federal government to both traditional Papago land jurisdiction and Papago political organization. And, just as important, he had the authority to offer $4,000,000 worth of wages and development projects, which the Papago needed badly enough that they were persuaded to go along with the districting. The Papagos unwrapped the proffered package, took out what they wanted, and left the government with the rest.

The Agency has failed to sell the Tribe on reservation-wide grazing ordinances as well. As recently as 1959 the tribal council rejected such proposals. The bureau views this intransigence as a result of the successful political opposition of a few cattle "baron" families. These families are politically influential out of proportion to their numbers, as the Bureau sees it, because they employ or otherwise patronize the small Papago cattle owners and through this patronage (in effect) they control "90 per cent of the range." District or reservation-wide grazing ordinances would restrict their freedom, and might reduce or end the $250,000 annual range-management subsidy that benefits them now. The Tribe and the government provide the subsidy through the work of the Agency and the Tribal Herd program. Today there are fourteen families who own 300 and more cattle each (*15*).

Gwyneth Xavier observed the diffuse influence and the feudal nature of the large cattle-herd owner-families back in the 1930s:

The owners of these (large) herds and ranches, employing their relatives, and generally surrounded by their relationship group, take the place of the patriarchal head of the family or leader upon whom all relied for advice and help. They are recognized as leading men in the community and take care of those about them with almost feudal benevolence. Their influence often spreads beyond the family into the whole locality. They introduce a new element into Papago life: prestige and leadership with an economic basis. (*9*)

James Simpson believes that the traditional Papago land-use and subsistence-economy attitudes have had very much to do with Papago opposition to accepting any responsibility for practicing range management. In the face of district and tribal opposition to grazing regulations, the Bureau has repeatedly sought to form district cattle associations that would take responsibility for conserving and improving the range, limiting the number and distribution of cattle on it, and maintaining stock water sources, all with government technical assistance when requested. In the late 1930s Xavier noted the

hostility toward the associations, which the government had attempted to form as early as 1935. Bitterness toward the attempt was so strong that "the Papagos had developed a definite hostility to the idea and the very word 'association' could not be used without arousing an adverse reaction. There was very little to build upon in the way of past organization." (*9*)

Attitudes must have changed after the droughts of the 1940s forced many of the small cattle owners out of operation. The range had to be reseeded and built again. According to the soil conservationist (*23*), stockmen were more conscious of the exhaustibility of their range than during the decade before. As villages and districts completed range improvement projects with PMA funds, people began thinking about reacquiring cattle and rebuilding a herd. The large owners began using more land as their herds increased.

In Baboquivari District the small cattle owners saw this happening and went to the Agency to see what they could do "to stop large owners from taking their range." As the agency had previously helped them organize for AAA and PMA range improvement funds, its advice was likely to be followed. Joe A. Wagoner, the Forest and Range Officer then

. . . explained that I was powerless to help them until they devised a system under which they could take offenders to court. They went back to the people and talked things over, and then asked for a meeting to explain how a permit system would work that could be backed in the courts.

Meetings are still being continued to explain what permits mean. From this they will draw a management plan with all the things they want in it, and provision for protection of tribal resources. We told the people what the Federal law is and what they'd have to do.

There has been good reception to this talking, points being brought up on both sides, and that will continue until we get the plan written and signed. The organization will be backed by the tribal court with appeal to the Federal courts. (*19*)

The Agency men hoped the meetings would lead to the organization of the first district or local cattle associations, but the people were not ready. By 1951 the next best action, a district land code, had been completed. It codified customary land use jurisdictions, which would be enforceable in the tribal court. It was a solution to the original request from the small owners, and was evidence that the agency had not overridden the will of the people by pushing its own solution on them. The land code appeared to reflect customary land-use patterns and was adapted to the tribal political structure.

Land codes and local and district cattle associations were formed throughout the San Xavier and Sells reservations thereafter: San Xavier, 1953; Sif Oidak, 1954; Vaya Chin / KaKa / Ventana, 1957; Gu Achi, 1962; Vamori, 1964. The agency hoped to accomplish through these districts and local associations what it hadn't been able to achieve through the tribal council: a reservation-wide system of grazing codes necessary to the unified development of a productive quality-controlled cattle industry. But these plans were still ahead of the Papagos' priorities, according to Metzler.

None of the livestock associations are very active. Their function seems to be to block any action proposed by the Tribal Council. They tend to keep livestock matters on an individual and a district basis, rather than as a tribal activity. (*14*)

Gu Achi Stockmen's Association

Of all the associations on the reservation, the Gu Achi Stockmen's Association has been rated the most active and successful today. It was formed in the following manner:

A. In district council meetings and to individual Papagos, agency men explained to the people how badly their land was overgrazed and how severe their problem would soon be if they didn't take any action. The agency men recommended the formation of a district-wide cattle association to solve their overgrazing problems.

B. Then the agency men withdrew to let the Papagos think about their overgrazing problems and the proposed cattle association, and make up their minds. While the Gu Achi people were considering the agency proposal, from 1960 to 1962, the agency range specialists kept the problem before them through meetings with individual Papagos. "Initially we got nowhere. There was always some opposition to the association until the district passed it" (*15*).

C. In 1962 the Gu Achi District Council requested technical assistance from the agency to organize a district-wide cattle association.

D. After the request, the agency range specialists told the district they would advise them if they formed a by-laws committee.

They agreed, and each of eight villages selected one person for the District By-laws Committee. We gave top priority on our time to meeting with them and organizing this association so we didn't spread ourselves too thinly to be effective. We provided the transportation for by-laws committee members for each meeting. We went out and got them. At least one of our staff members attended each meeting. (*15*)

Two meetings were held each week for about three months to discuss and establish the organizational structure, purpose, and responsibilities of the association.

E. Later in 1962, the Gu Achi Stockmen's Association and by-laws were adopted by the district and approved by the tribal council and the agency superintendent.

F. From 1962 to the present, the Gu Achi Stockmen's Association has carried out on its own the following organizational and range improvement responsibilities:

(a) Election of officers and directors to the board.

(b) Raising revenue to cover their expenses and build a surplus.

Revenue is collected from (1) Grazing fee per head of livestock. (2) Seven percent of Association cattle sales goes into U.S. Treasury account in our name. The Tribe gets 3% of the cattle sales. (3) We receive 4% interest on our money. (4) Proceeds from the sale of mavericks come to the Association. (*16*)

Association revenue is spent in the following ways. (1) Hay for horses and some cattle during roundups. We've spent $1,000 on hay for this [1968, Spring] roundup. (2) Salt licks. (3) Materials for building corrals and chutes. (4) Sometimes groceries for livestock workers. (5) Miscellaneous to members when they need it. (6) Well and pump repairs. (7) Electricity for pumps and the community building. (8) Gas to run electric generator. (*16*)

(c) Organizing and carrying out two month-long (and longer) district-wide roundups each year.

(d) Arranging for and conducting cattle sales.

(e) Planning, designing, and building range improvement facilities like corrals and chutes.

G. From 1962 to the present the association has requested assistance from the Papago Agency for its major problems, such as providing vehicles to haul heavy building materials or equipment and pumps to replace used ones.

Maintaining the stock wells is up to the Association. The B.I.A. put them in, but we have to fix broken parts. The Bureau helps us out if it's a major problem, but a small problem is up to us. (*16*)

The reasons the Agency's range specialists had been so successful in gaining Papago cooperation in a Gu Achi District range-improvement program, when there had been general Papago resistance to any Agency proposals associated with stock reduction, are discussed under "Schuk Toak and Gu Achi," above. But why had the Gu Achi District then taken on range-improvement responsibilities, which previously had belonged to the Papago Agency?

Though estimates of the numbers of cattle owned by Papagos have varied with the estimators, most experts have agreed that the carrying capacity of the Papago rangelands was exceeded by 1919. It has been mentioned that Papago cattlemen, who did not equate larger herds of cattle with overgrazing and the eventual destruction of their cattle economy, viewed livestock reduction as a visible threat to that economy and to community life. Beef had replaced venison. After the introduction of the Anglo cash economy to the Papagos, especially with wage work during the New Deal, cattle became the "Papago bank." Cattle could be sold for immediate cash. As we have remarked, the large numbers of horses far exceeded those needed to support a cattle economy or provide transportation. When, during the 1940s and 1950s, national and state funds were made available to plan and carry out range-improvement projects (seeding, water-spreading, and water storage), one of the districts receiving funds was Gu Achi.

Five huge earthen water-spreading dikes were constructed in 1947 by private contractors . . . with only survey work being done through Indian Bureau funds. Dikes were thrown across a large arroyo draining about 100 square miles. The staggered series forced flood waters over an area of some 2,000 acres. In 1947 and 1948 all runoff was absorbed, indicating the watershed had been removed from the flood-producing class and rendered practically harmless. A heavy crop of annual grasses grew on the formerly totally unproductive area flooded, and water was held by the dikes for stock grazing on the new forage.

In 1948 three more large dikes were constructed in Gu Achi District. In 1949 sleek, fat cattle grazed on a very heavy crop of grasses seeded over the spreader area. . . . During that year 1,952 structures were built, costing $62,250. (*22*)

The approach of the agency conservationist was an important part of Gu Achi's successful experience in co-operating and working with the government.

Furnishing only technical knowledge, he left the planning to them. . . . The Gu Achi District Papagos planned every step of the way. . . . After surveying and bringing contractor and people together, the administrator left execution of the project to the people. (*21*)

Because planning stayed in the hands of the Papagos, they felt that the project was theirs.

Ultimate political control rests with the Papago village. Land use, grazing regulations, and cattle roundup organization have been under village control. The tribal and district councils have consistently refrained from infringing upon the traditional political autonomy of the villages.

In the traditional organization of the Gu Achi cattle economy, there are five cattle grazing ranges. The villages in each range elected a roundup boss, who directed the roundup operation in his range only. Cattle roundup activities have been village-based for probably a century and more, since Papagos shifted from hunting to herding cattle in the eighteenth and nineteenth centuries. Papago communities have thus had about two centuries to integrate a subsistence cattle economy and organization into their customary way of life, and it has barely changed in half a century of federal direction and assistance.

Interviews with the officers of the Gu Achi Stockmen's Association and the Papago Agency's range-management specialist indicate some of the results of their cooperation.

The officers of the association stated that the association belonged to them from the beginning. In fact, their statements indicate that they participated extensively in the development of the association.

The Agency helped us out. They didn't run things; they just asked us questions, explained things, and left it up to us. It didn't belong to the Agency. It started right out with us. Stephen Listo and his father first had the idea for an association. (*16*)

The development approach of the agency range specialist, which allowed the Papagos nearly two years to discuss their overgrazing program and decide whether to request agency assistance to create a cattle association, must have been appropriate, for not only was their initial hostility resolved, but the officers above believe the proposal for an association was their own, not the Agency's. The formation of the By-laws Committee was based on recognition of the village's authority over the use of rangelands and the organization of cattle roundups. Even in an organization as Anglo and technical as the By-laws Committee, the Papagos participated in the planning of the Association and adapted it to their village-based cattle activities. A representative from each of the district's eight villages attended the three months of meet-

ings. When the officers were asked "How much of the By-laws was your thinking and how much was from the Agency?" they explained their participation this way:

We decided what to put in the By-laws. We looked through the By-laws of the other associations supplied by the Agency. We talked over what we wanted on roundups, and then inserted our ideas. The lawyer of the Tribe wrote some of the words so that it was legal. We met two times a week for two to three months. The Agency helped us so our By-laws wouldn't get mixed up with the Tribe's. (*16*)

The By-laws Committee adapted, with some success, the agency-proposed organization, the Stockmen's Association, to the villages' traditional roundup organizations and local authority. The base of the association is the eight villages with their five traditional range areas for grazing their cattle. The five roundup bosses, traditional, selected by the village in the five ranges, became five of the ten directors on the board. The other five directors are elected at an annual meeting of the whole district: president, vice-president, secretary, treasurer, and foreman. The duties of the association foreman and roundup bosses reflect their customary responsibilities, and have been expanded to include new district-wide responsibilities assumed by the association.

The Foreman shall supervise, coordinate and be responsible for the work of the Roundup Bosses and will report to the Board of Directors as needed or as required. . . . The Roundup Boss shall be responsible for conducting the roundups of their designated range jurisdiction, counting, tallying, and reporting the number of cattle and horses for each brand to the Foreman; the Roundup Boss shall also be responsible for the use of Association pastures, maintenance of corrals, and fences, supervision and collection of any assessments made to members and other duties as assigned by the Board. (*26*)

The traditional roundup organization has been maintained, with resources of the association supporting it. There are usually two District-wide roundups each year, during the long dry periods. During the roundups since the formation of the association, the cowboys have moved from one range to another, camping. The roundup boss of the particular range in which they are working and camping becomes the foreman of all the roundup bosses there. Thus there is a new foreman with each shift of range. The cowboys from each of the five ranges operate on the roundup as separate units, maintaining their traditional village autonomy. Every base camp is divided into five range camps, each with its own food and equipment. The association supplements each range camp's supplies when needed.

The Gu Achi Stockmen's Association was also adapted to the district organization. Every district resident grazing cattle there is a member of the association and may vote at the annual meeting. The association's budget is established at the annual meeting, rather then being the prerogative of the board of directors alone. The association foreman is elected by the total membership and is not one of the five roundup bosses elected by the village. The foreman has the District-wide responsibilities discussed above, which have been traditional for each roundup boss within his own range, but were expanded throughout the district by the association.

The responsibility to check overgrazing on the district rangeland has been only partially accepted by the association. The number of horses a family can own is limited by the association, but it has not yet set limits on the number of cattle a family may have.

We're coming to the point of talking about limiting the number of cattle so they don't overgraze. But we haven't decided anything yet. It's up to the members. We realize it's overgrazed. Those who have more cattle . . .,you know what they will say. You look at the small man and the big man. About four or five families have over 100 head. We've been told 250 head is about the limit, but we've never put up a limit. Two old men have over 400 head. (16)

The Gu Achi Stockmen's Association has apparently not yet agreed that overgrazing is so severe that it has to set limits to the number of cattle a family may own, and its members are reluctant to enforce limits on the largest cattle owners. But at least they clearly recognize the relationship between too many cattle and the overgrazing problem.

The bureau's approach to development in this case, and the Papagos' response, demonstrate both the possibilities and the limitations of changing a subsistence cattle economy into an industry, and of transferring to the Papagos responsibility for the rebuilding and conservation of their range. The government can learn from its successes and failures, if it wishes, for a fragmented record is there to evaluate. The preceding histories of Papago range-management problems, the formation of the Tribal Herd and the Gu Achi Stockmen's Association, and the conflicts between tribal and federal motivations and means in the face of range problems for the last fifty years point out the need for a coherent theory to deal with the transfer of power and responsibility for economic development from government administrators to the local community.

CONCLUSIONS

Three sets of conclusions concerning the development of arid land enclave cattle economies can be drawn here:

a) The development of a Papago cattle industry was stalemated when stock reduction threatened the Papago social system, and when inappropriate techniques were employed to stimulate Papago organization and responsibility for their own economic improvement. In contrast, when recognized techniques for developing community organization were employed, the stalemate was broken and a Papago organization took over.

b) Papagos have not taken responsibility until they could call a development organization and its activities their own. The kinds of community participation that make such efforts theirs are presented as a hypothesis.

c) In order to encourage the development of arid-land

enclave cattle economies that have conditions similar to those of the Papago, to avoid threatening the enclave community and thus thwarting economic development efforts, I recommend the following broad strategy:

1) Recognize an existing community organization that is both widely viewed as belonging to and representative of the enclave group and has expressed an interest in improving the cattle economy (or encourage the community to form such an organization).

2) Make investments of funds, training, equipment, and other resources into the enclave-controlled community organization as they are requested, instead of into a separate outside development agency.

3) Maintain a geographical separation from the community or its policy-making organization sufficient not to be threatening, yet to be available to offer assistance upon request.

4) Negotiate economic-development plans with the enclave community organization under its control and at its request, thus being accountable to that body.

On the Papago reservations, state and national agencies generally agree with "Indian involvement," although it is unclear of what kind or to what degree. The Division of Indian Health, for instance, recently defined its policy in this way:

The Indian Health Program is your program. It is to be carried out in accordance with your wishes and your requirements. The Division of Indian Health is the instrument for providing services which are planned, conducted and evaluated with you as individuals and as organized tribal and community groups. (27)

The agencies agree that Papagos should learn to manage their human and natural resources to begin solving the problems associated with chronic poverty and disease. Observation and conversations with tribal leaders indicate that Papago villagers are skilled at coping with the complexities of organizing for feast days, ceremonies, cattle roundups, adobe-house building, and other aspects of their community life. They have not, however, extended these abilities, nor have they developed many new ones to deal with the nontraditional conditions under which they are now living: those fostered by a cash economy, Western science and technology, and government services within the reservation system. In the face of conditions they do not sufficiently understand, Papagos, however, have let the government provide, resisting when such administration threatened to invade their inner community life. Only when these administratively created policy-making organizations and programs are brought within the jurisdiction of the Papago political organization will authentic self-government and self-development be possible. But what kind of participation in villages, districts, and tribal government will transform a largely administered community into one capable of directing its own affairs? That is the problem for the following discussion.

This hypothesis sets out the kind of community participation necessary to transfer program planning, direction, and implementation from professional administrators and technicians to local people, and the kind of community involvement that encourages the growth of skills to cope with nontraditional demands. The results of projects in several Southwest Indian communities, where the participation of change agents and local community members has been recorded, support the hypothesis (*28–30*). A theoretical explanation of the hypothesis would have to involve a discussion of two or more societies confronting each other through their selected members in a contact situation (*31*). The wider and more intense the conflict between them, the more attention is paid to who possesses the elements of power and status. These may be organizations, positions in organizations, persons, programs in the process of development, facilities, equipment, funds, etc, that belong to "us" as opposed to "them." Such distinctions between elements belonging to one society or the other become more evident to the observer when members of one society believe that the activities of the other society are threatening their most valued institutions. Recognition of who possesses which elements of power must have survival value for the members of the threatened society or group. The decision whether to resist or to cooperate can then be directed toward the appropriate elements. The elements themselves may not be threatening, except as they are controlled and used by members of one or the other society for purposes that threaten. For example, the Papago Community Development program, an element in the Papago-Anglo contact situation, is apparently viewed as less of a threat to Papago village life now than it was originally; the villagers have taken greater control and responsibility for their community-development workers and for their activities than when the program first began. It is essentially run by the communities and used for their own purposes. But the same program, if run by a federal agency, would probably be seen as a direct threat to village community life. This view reflects past experience of the Papagos with too many programs carried out for the purposes of the dominant society.

The hypothesis is this: When members of a community or local group describe a program (project, organization, etc) as their own, as opposed to belonging to an outside group or change agent, they will take responsibility for directing and maintaining it after the change agent has left.

In order for them to come to call a program their own, the members must participate in some combination of the following aspects of the program's development:

a) They have to come to accept the initiator (or change agent) of the development process as a person who can be trusted, who has demonstrated concern for the well-being of the community and has respect for its members and their potential for developmental change. He must possess the knowledge, skills, and potential resources sufficient to help them solve their problems. Once the change agent has been accepted by the community, the following kinds of participation can be initiated.

b) They have to *help identify the problems* that they wish to resolve.

c) They must *participate in planning a program solution* to the problems they helped to identify.

d) They must *believe that there will be benefits for themselves to be derived from the program's success.*

e) They must be allowed the time and opportunity needed to *adapt the planning, organization, project, and evaluation to their own socio-cultural life.*

f) They must *recognize as their own the organization* responsible for carrying out the program, and then participate through it.

g) They have to *(1) accept the major responsibility for organizing and carrying out the program solution* and *(2) do the work themselves or select those who do.*

h) They must *(1) decide whether or not new skills and technical knowledge are necessary to carry out the plans* and *(2) identify and obtain the resources* that can assist them.

i) They have to *recognize the results of their efforts, success or failure, as their own.*

REFERENCES AND NOTES

1. KELLY, W. H.
   1963 The Papago Indians of America, a population and economic study. University of Arizona, Tucson, Bureau of Ethnic Research. 129 p.

2. KELLY, W. A.
   1968 Indian adjustment and the history of Indian affairs. Arizona Law Review 10(3):559–577.

3. HACKENBERG, R. A.
   1964 Aboriginal land use and occupancy of the Papago Indians. U.S. Department of Justice, Lands Division, Contract J–40974. 282 p.

4. SPICER, E. H.
   1962 Cycles of conquest: the impact of Spain, Mexico, and the United States on the Indians of the Southwest, 1533–1960. University of Arizona Press. 609 p.

5. BERGER, E. B.
   1968 Indian mineral interest: a potential for economic advancement. Arizona Law Review 10(3):675–689.

6. VAN WILLIGEN, J.
   1969 Community development: the Papago experience. Unpublished essay, Community Development Office, Sells, Arizona. 22 p.

7. JOSEPH, A., AND OTHERS
   1949   The desert people, a study of the Papago people. University of Chicago Press, Chicago. 287 p.

8. WAGONER, J. J.
   1952   History of the cattle industry in southern Arizona, 1540–1940. University of Arizona Social Science Bulletin 20. 132 p.
   According to this source, the Papago may have known of cattle as early as the sixteenth century, when Coronado brought them into what is now Arizona. By 1586 Spanish ranches ran herds of 33,000 and 42,000.

9. XAVIER, G. H.
   1938   The cattle industry of the southern Papago districts with some information on the reservation cattle industry as a whole. Unpublished manuscript, Arizona State Museum Library, Tucson. 52 p.

10. WOODBURY, R. B., AND N. F. S. WOODBURY
    1962   A study of land use on the Papago Reservation, Arizona. Unpublished manuscript. Arizona State Museum Library, Tucson. 60 p.

11. I am in possession of a tape, dated October 26, 1967, recording an interview with James Simpson in which he discusses his 1967 Master's thesis (unpublished): An economic evaluation of selected range improvement practices on the Papago Indian Reservation (University of Arizona).

12. UNDERHILL, R. M.
    1939   Papago economics. Gila-Sonora file, Arizona State Museum Library, University of Arizona, Tucson. Unpublished manuscript. 29 p.

13. MARK, A. K.
    1960   Description of and variables relating to ecological change in the history of the Papago Indian population. University of Arizona (unpublished Master's thesis). 110 p.

14. METZLER, W. H.
    1960   Economic potentials of the Papago Indians of Arizona. Unpublished monograph prepared for University of Arizona Department of Agricultural Economics, in cooperation with the USDA Agricultural Research Service, Farm Economics Research Division. 129 p.

15. Oral communication from Charles Whitfield, Land Operations Officer, Papago Agency, Bureau of Indian Affairs, Sells, Arizona, May 17, 1968.

16. Oral communication from Gu Achi Stockmen's Association, Sells Papago Reservation, Santa Rosa, Arizona, May 23, 1968.

17. Oral communication from John Artichoker, Superintendent, Papago Agency, Bureau of Indian Affairs, Sells, Arizona, April 9, 1968.

18. DOBYNS, H. F.
    1949   Report on investigations on the Papago Reservation carried out in the summer of 1949 under the direction of Dr. Edward H. Spicer, financed by the University of Arizona, Tucson, Arizona; and Recommendations by staff members and students of Cornell University field class Sociology-Anthropology 642. Cornell University, Department of Sociology-Anthropology. 74 p.

19. DOBYNS, H. F.
    1949   Field notes. Gila-Sonora File, Arizona State Museum Library, University of Arizona, Tucson. Unpublished manuscript cards, unpaged.

20. WHITFIELD, C.
    1964   Tribal herd operations, Bureau of Land Operations, Papago Agency, Bureau of Indian Affairs, Sells, Arizona.

21. Oral communication from M. Joaquin, May 20, 1968.

22. DOBYNS, H. F.
    1965   Experiment in conservation. In E. H. Spicer, (ed), Human problems in technological change, a casebook, p. 209–223. John Wiley and Sons, Inc, Science Editions, New York. 301 p.

23. PAPAGO AGENCY
    1952   Narrative report, 1951–65. Bureau of Indian Affairs, U.S. Department of Interior, Papago Agency, Sells, Arizona.

24. PAPAGO AGENCY
    1954   Soil and conservation program 20 year period. Bureau of Indian Affairs, U.S. Department of Interior, Papago Agency, Sells, Arizona. 42 p.

25. PAPAGO AGENCY
    n.d.   The formation of Papago grazing districts: case study. Bureau of Indian Affairs, U.S. Department of Interior, Papago Agency, Sells, Arizona. 6 p.

26. GU ACHI STOCKMEN'S ASSOCIATION
    1962   The articles of association and by-laws of the Gu Achi Stockmen's Association. Santa Rosa, Sells Papago Reservation, Arizona.

27. U.S. PUBLIC HEALTH SERVICE
    1968   To the first Americans, 2nd annual report on the Indian Health Program. U.S. Public Health Service Publication 1580. 10 p.

28. KENEALLY, H. J.
    1963   Community development with the Gila Bend Papagos. Coordinating Council for Research in Indian Education. Annual Conference April 18–19, 1963, Phoenix, Arizona. [Report]. p. 114–121.

29. KENEALLY, H. J.
    1968   The role of health education in documentation of the cause and effect relationship in program activities. Unpublished manuscript.

30. SPICER, E. H. (ed)
    1965   Human problems in technological change, a casebook. John Wiley and Sons, Inc, Science Editions, New York. 301 p.
    See especially Corn and custom, A. Apodaca, p. 97–111; Sheepmen and technicians, E. H. Spicer, p. 185–207; Experiment in conservation, H. F. Dobyns, p. 209–223. Previous (1952) edition published by Russell Sage Foundation, New York.

31. SPICER, E. H. (ed)
    1961   Perspectives in American Indian culture change. University of Chicago Press, Chicago. 549 p.

Building Design and Planning for Self-sustaining Communities in Remote Localities of Australia's Arid Zone

BALWANT SINGH SAINI

Department of Architecture and Building
University of Melbourne, Melbourne, Australia

ABSTRACT

AGAINST THE BACKGROUND OF DEVELOPMENT of newly discovered resources such as minerals, this chapter examines how people of European origin could meet the challenge of living and working in Australia's arid interior. Population distribution, work force, turnover, and employment patterns indicate the need for an improvement in living conditions so that sufficient inducement can be offered to families to enable them to settle permanently, thus establishing a measure of stability, which is totally lacking at present. Isolated communities require specific physiological, psychological, and sociological needs, and valid approaches to building design and planning must be based on these needs.

Apart from the application of architectural methods to achieve environmental control, various mechanical aids (such as air-circulating fans, evaporative coolers, and systems which use solar energy), as distinct from total air conditioning, are useful and effective in different degrees. Also of particular importance are the external factors such as vegetation, water, and the nature of the surrounding open spaces. If judiciously used, such external factors not only help to reduce heat load but also arrest the flow of sand and dust into buildings in the arid regions. Finally, a case is made for adopting a more compact approach to building design and settlement planning.

BUILDING DESIGN AND PLANNING FOR SELF-SUSTAINING COMMUNITIES IN REMOTE LOCALITIES OF AUSTRALIA'S ARID ZONE

Balwant Singh Saini

UNTIL RECENTLY when signs of growth and change began to appear in the area, Australians showed little interest in settling within their arid regions. Most of their twelve million people live in a few large cities in the temperate area of the southeast coastal strip.

The hot, dry interior is sparsely populated except for scattered pastoral families, a few single-industry (such as mining) towns, and semiurbanized commercial and administrative settlements that depend economically upon the single-industry towns.

The early pioneers who settled in the area established a frontier-type economy, where risks were high and progress was erratic. This pattern was maintained until the end of World War II.

Fresh discoveries of minerals and greater flow of capital have altered this picture dramatically. Some of the spectacular instances of development in Australia's arid zone are already evident. The great copper-silver-lead-zinc mine at Mount Isa in Queensland is being developed at a pace that is reflected in the new town of 20,000, well beyond the limits of the old settlement with its frontier atmosphere. There is gold and copper at Tennant Creek and nickel in Kalgoorlie. The newly discovered rich deposits of iron ore in Western Australia have generated considerable development in the Pilbara region, culminating in a whole new system of railways, deep-water ports, and industrial enterprises. Irrigation proposals on the Ord River have resulted in the construction of a dam and the initiation of many crop growing industries in the Kimberleys. In addition to major enterprises, there are a number of smaller developments associated with fishing, pearling, tourism, and meat industries. All these developments have forced the establishment of new settlements, many of which had to be located in remote and hostile environments.

Human beings form one of the chief resources in a developing region. Apart from building for industry and commerce, the main area of responsibility of an architect lies in satisfying the needs—physical, psychological and sociological—of the people who are instrumental in developing the region and for whose benefit the development is being carried out. In the hot, dry environment of Australia's arid regions, their needs assume special significance. Without a clear understanding of their particular

NOTE: This chapter is an adaptation of a presentation at "Arid Lands in a Changing World," an international conference sponsored by the Committee on Arid Lands, American Association for the Advancement of Science, June 3–13, 1969, University of Arizona, Tucson, Arizona, U.S.A.

requirements, it is impossible to design, build and plan for them.

If we look at population distribution, labor turnover, and employment patterns, we find that, despite air-conditioned houses and excellent amenities in some of the new towns, much of arid Australia fails to provide living conditions and urban facilities normally taken for granted in more developed centers in the temperate areas of the south.

CAUSES OF DISSATISFACTIONS

The main factors affecting living conditions in the region are climate, discomfort, and isolation caused by long distances between population centers. Although climate plays an important part and adds to the difficulties experienced by people in adapting to the hot, dry environment, it seems that physical conditions are not the chief causes of dissatisfaction. Many other considerations accumulate to lower the morale and help set the threshold at which complaints concerning heat and other discomforts associated with the surroundings are used as obvious scapegoats in giving vent to real feelings, which may or may not be causal factors. Hot days and hot nights, flies, and mosquitoes are easy and neutral subjects to blame, and these are often substituted for the real causes of dissatisfaction and discontent in an unhappy community.

A list of these causes would indeed be long, but most seem to be associated with isolation and remoteness from the more urbanized centers of large population on the southern coast. Apart from lack of amenities, such as amusements and sporting facilities, there are problems connected with education of children, expensive holidays, difficulties in obtaining specialist medical opinion and treatment, and a host of day-to-day problems which develop when a small number of persons are constantly forced to live in each other's company.

The cost of living is high, particularly in the older settlements which do not enjoy the benefit of subsidies normally available in one-industry towns. Generally food costs, and costs for most other items for that matter, tend to increase as one moves northward in the continent. For example, according to West Australian economist Dr. Alex Kerr, food prices in Port Hedland are about 30 percent higher than in Perth, in Marble Bar about 40 percent higher, and in Halls Creek about 70 percent higher, though variation in some lines is so diverse as to make general comparisons almost impossible (2, p.

23). The other everyday items are likewise expensive. Newspapers in Derby, for instance, cost twice as much as in Perth, and a bottle of beer in Halls Creek also costs three times as much as in Perth.

Housing costs and rentals follow the same pattern, ranging from 60 percent to 100 percent more than Perth prices, depending upon the inaccessibility of the area. In a recent survey we found that electricity charges were six to eight times higher. To operate a single-unit air conditioner for one night so as to maintain reasonable conditions in one room costs as much as two dollars. Gas and water charges are likewise high.

There are two ways in which this situation can be remedied: increase incomes or decrease costs, either of which would enable families to afford more of the necessary personal comforts. It is difficult to see how costs could be decreased in the foreseeable future, as the reduction in living costs will depend upon the growth of population and manufacturing industries on a scale that cannot yet be foreseen. In the meantime, the cost differential will continue to be made up through the adjustment of wages and other earnings, and by bonuses and offers of subsidies towards rentals and other necessary expenses.

PASTORAL SETTLEMENTS

It has been found that pastoral families are better prepared to accept the challenge of a harsh environment. They, more than others, learn to adapt to their surroundings and change surroundings to suit their needs.

Forming the nuclei of large cattle properties enclosing several hundred square miles, pastoral stations are little more than assemblies of makeshift shelters to house people, animals, and goods.

A typical homestead consists of a number of rooms surrounded by a deep verandah and protected by a galvanized iron roof (see Fig. 1). It is the headquarters of each cattle kingdom, and in the cluster of buildings from which it is constituted many tasks are performed to enable the station to be self-reliant. Its size is directly related to its remoteness, but most stations provide housing for 10 to 20 persons. These include, apart from the manager, a head stockman, bookkeepers, mechanical engineers, a cook, and others (3).

Building for scattered families associated with pastoral industry presents little problem to a designer. Imbued with pioneering spirit, these people are individualists and are primarily resident in the area because they already possess the spirit of adventure, resourcefulness, and adaptability so necessary to survive in a hostile setting. The basic relationship of scale between the pioneers and their surrounding space is very different from that which pertains in more developed areas. Consequently, the fact that they are content with clusters of tin sheds and weird outbuildings must be understood in the light of their associations and more Spartan attitude toward physical comfort.

Courtesy of Australian News and Information Bureau

Fig. 1. Homestead headquarters of the million-acre Noonkanbah Pastoral Station, West Kimberleys, one of the largest sheep properties in Western Australia. Surrounded by green lawns, hibiscus, frangipani, and eucalyptus trees, the structure consists of a number of rooms protected by a deep verandah and a galvanized roof.

SINGLE-INDUSTRY TOWNSHIPS

In contrast to the well-adjusted pastoral communities, building in towns associated with single industries, such as mining, is mainly concerned with people who are specialists and skilled workers. They are a class of people who are already in demand elsewhere. They have to be lured from existing cities by offers of better wages, promises of promotion, better comforts and amenities. The success of most enterprises largely depends upon the happiness and contentment of the people, and unless the settlements can attract and hold men, there is bound to be continuous turnover.

Most employers favor a worker who is married—one who considers his new environment as a particularly fine place to bring up his children—in other words, a man who has learned to identify himself physically with the place. This is hard to achieve in settlements such as mining towns, which, being temporary, mainly attract the single man and therefore lose the stabilizing influence of families.

In my opinion, good, comfortable housing and a pleasant environment can be of considerable help in attracting the right type of new settlers.

To begin with, the new town must be able to compensate for the longstanding associations of previous life which are broken. Psychologically most people resist change. This means that we must pay some attention to their background and institutions. Such understanding could help an architect produce solutions that reduce frustrations, mental stress, and other social problems encountered by new settlers. It means sympathetic planning and layout of buildings, their relationship to open areas, and their proximity and linking to community facilities and services.

In a town dominated by one industry or occupation, life can be dull and monotonous. Sound planning can help people overcome boredom by providing mental and physical stimulation. There must be avenues for self-expression and active participation in healthy outlets. Social activities, for instance, could extend beyond the daily get-together in the canteen bar and offer access to playgrounds, swimming pools, outdoor cafes, community workshops, and picnic facilities, and could include participation in plays and concerts in, say, an open-air theater. Theaters of this type are ideally suited to arid regions with their generally cool, rain-free evenings and pleasant nights.

It is necessary to aim for high architectural and aesthetic standards, because the psychological impact of pleasant surroundings is of considerable help in fostering a spirit of community belonging, civic pride, integration, and personal enjoyment. Buildings should avoid the depressing effects of monotony and drabness. New density patterns may have to be evolved—patterns which encourage human contact and exchange and provide people with conditions in which they can feel territorially on their own ground (4, p. 23).

PHYSICAL DISCOMFORT

The causes of physical discomfort in hot, dry regions are many. First, there is the oppressive climate with its high daytime temperatures and low humidity endured in surroundings devoid of vegetation. The landscape is brown or red, accented by rocky outcrops in some locations. There is high incidence of glare reflected from the light-colored ground.

Then, there are pests — mainly flies. Yet another nuisance is dust and sand. Although not health hazards, both dust and sand are known to lower the morale of the people, particularly the housewife who is forced to clean the house frequently. Clothes have to be washed often, and the domestic appliances and fixtures get clogged with dust, thus requiring frequent repairs and maintenance. This service is expensive and often difficult to obtain in remote areas. The growing dust nuisance is not only limited to its effects on people but also extends to a deterioration of vegetation, buildings, clothing, works of art, and other articles of property.

Air Conditioning

An obvious and efficient, even though expensive, method to obtain physical relief in a building is to air-condition it completely. Assuming that we can overcome the cost of maintenance and repairs in remote locations, the question still remains whether total air-conditioning, which really means turning one's back to nature, is the ultimate answer to living in the hot, dry lands.

Before a decision can be made, we need to consider the psychological effects of living in a closed and confined space for long periods, especially from the point of view of the housewife and the children, who are likely to remain indoors for the best part of the day as well as the night. Air-conditioning is a negative approach which denies the enjoyment of pleasant aspects of the arid climate such as the cool evenings and nights when it should normally be possible to spend a great deal of time outdoors.

A possible solution is to consider partial treatment, such as the use of an air-circulating fan, attic fan, or an evaporative cooler.

Air-circulating fans do not cool the air, but the sensation of coolness is mainly achieved by enhanced evaporation from the skin. These fans are therefore reasonably effective in hot, humid situations, such as during the short periods of the rainy season. For the rest of the summer months, when temperatures are extreme, movement of hot air can be an added source of discomfort, thus reducing the effectiveness of fans in hot, dry areas.

Attic Fans

If a blower fan is installed in a central position near the ceiling of a small building, it can cool the interior space at night by expelling accumulated hot air, thus drawing in cooler air through windows, doors, and other openings. Use of such installations at night assists in providing a cool environment for a considerable period during the next day.

According to E. T. Weston of the Commonwealth Experimental Building Station, Sydney, the results of the Australian tests show that in locations where the diurnal temperature range normally equals or exceeds 20 degrees Fahrenheit, the outside/inside temperature difference in the building using attic fan was about twice that which would occur due to natural ventilation alone (5).

In order to achieve desirable results it is necessary to make the right choice of equipment. W. H. Badgett of Texas Engineering Experiment Station has suggested that the fan should have a capacity large enough to give a complete air change throughout the entire building at least once a minute, or 60 air changes an hour (6). For better results, a fan capable of providing 90 to 120 air changes per hour is recommended.

Attic fans are becoming standard equipment in many American homes, yet they have been used little in Australia's extensive hot, dry areas. In view of their low cost and their ability to move large volumes of air quietly and at low speeds, it is indeed difficult to explain their lack of popularity. Recommended equipment for various areas of Australia is shown in Figure 2.

Evaporative Coolers

In hot, dry areas where there is sufficient supply of water, a blower fan can also be successfully used in conjunction with an evaporative cooler. Evaporative cooling takes advantage of the latent heat of vaporization of water and is perhaps the only known historical method of cooling by actually changing the quality of air.

In many hot, dry regions of North Africa and the Middle East, cooling is achieved by covering all openings with large, thick mats of "hessian" (a loosely woven, coarse fabric), lily pad roots, or mats. These various coverings are continuously soaked in water from a perforated tube or pipe from which water trickles down at a steady rate. Hot, dry outside air is made cool and moist before it reaches the interior. In Australia's central desert, pioneers have widely used this principle in order to cool drinking water and preserve vegetables.

These days, the same principle is applied to unit coolers and larger centralized systems. An electric fan not only accelerates the flow of air, but its motor is used to operate a pump for recirculating the unevaporated water that may collect at the bottom of the soaked felt. According to Frank Wickham of the Commonwealth Department of Works, Melbourne, air-handling capacities ranging from 3,500 to 5,000 cubic feet per minute (c.f.m.), and generally some 20 to 40 air changes per hour are required in domestic and similar situations (7). In arid areas, units normally have humidifying efficiencies of between 60 and 70 percent, with a water consumption of between two and three gallons per hour per 1,000 cubic feet per minute capacity. Company houses cooled by the evaporative method are shown in Figure 3.

In some localities evaporative cooling presents a problem in relation to clean water supply. Bore water has a tendency to block the filter pad by deposition of salts, and the waters drawn from streams or wells lead to blockage of the pad through the growth of algae. River water is improved if it is allowed to settle. It is possible to achieve this by installing two 200- to 300-gallon tanks, one of which could be used while the other is filled and allowed to settle. Such tanks also require sludge cocks for periodic cleaning.

The temperature reduction of air through the coolers varies with the relative humidity and dry-bulb temperature of the outside air and ranges from 10 to 30 degrees Fahrenheit, thus providing a welcome relief from hot, dry conditions.

Some evaporative cooling units now being marketed in Australia are fitted with controls for adjusting the

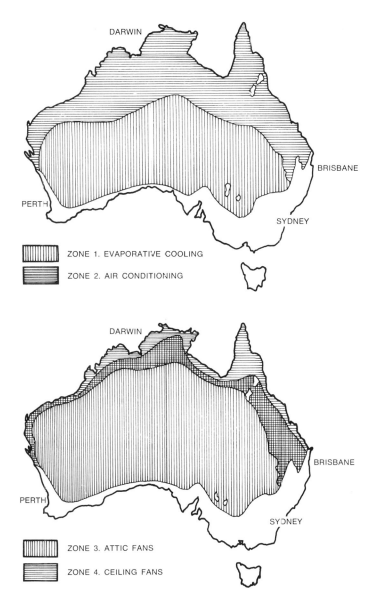

Fig. 2. Air-treatment zone maps (by F. Wickham) indicating large areas of the hot, dry Australian mainland where attic fans and evaporative coolers are effective.

Fig. 3. Company houses (such as these by architects Lund, Hutton, and Newell, at Mount Isa, Northern Queensland) are effectively cooled by centralized systems of evaporative coolers with ducts for uniform distribution of air.

Fig. 4. Completely air-conditioned prefabricated staff houses at Mount Goldsworthy in Western Australia.

Fig. 5. Hospital building at Beulah, a hot, dry, dusty town in the western area of the State of Victoria, Australia, successfully incorporates louvred screens built to give permanent shade to the roof and walls. (Architect: Peter McIntyre)

Fig. 6. Open spaces are reduced to an absolute minimum in the new town of Arad in Israel's Negev desert, due to the high cost of maintenance. In this excellent example of arid-zone town-planning, it is possible for a pedestrian to go from one end of the town to the other in completely shaded areas.

rate of air and water flow. This enables the user to switch off the water supply during periods of high humidity. At certain times, such as during the monsoons when the relative humidity exceeds 80 percent, it is possible to employ the cooler as a straight ventilating system. In such instances, continued air movement alone is sufficient to reduce the effective temperature of the environment.

Window units, though adequate for localized application in domestic structures, possess some limitations, most important of which is the increased noise level created by the fan. If installed in institutional buildings such as schools and hospitals, they tend to interfere with mental effort, sleep, rest, and working efficiency. A possible solution may lie in using either a larger unit, which could be operated at reduced speeds, or in installing a centralized unit set away from the habitable rooms, such as in a plant room, basement, or within the ground floor of a two-story building.

Centralized systems require preplanning in relation to equipment location, and the layout needs to be compact in order to minimize the cost of ducting. Units can also be designed to serve a group of small rooms or a larger space. This has the advantage of reasonable flexibility and permits the use of open and air-conditioned spaces in close proximity to each other. The amount of duct work is much less than that required for one central plant, while the units themselves can be isolated to prevent noise interference in the rooms. An example of this type of installation can be seen in the Montrose Elementary School in Laredo, Texas, where classrooms are grouped in fours. The evaporative coolers, one per classroom, are clustered together on the roof over a central service core.

Evaporative coolers have been successfully used in Australia's hot, dry interior. Their use in most towns is firmly established now. The new mining towns in Western Australia, such as Mount Tom Price and Mount Goldsworthy, though located in the arid zone, have completely air-conditioned houses with units incorporating refrigeration equipment. The fact that equally good results could be achieved by using evaporative coolers, at one-fifth of the cost of the total air-conditioning systems, has been largely ignored. The decision to use air-conditioning units (Fig. 4) in preference to evaporative coolers seems to have been governed by the need to use the publicity value of this luxurious amenity as an aid to attract a work force to such remote areas.

Apart from lower capital costs, evaporative coolers are simple in their construction. The components—motor, pump and fan—can be easily assembled and maintained by relatively inexperienced workmen. This has considerable advantage in remote areas where maintenance services are not easily or cheaply available.

Apart from mechanical means, there are two distinct areas in which an architect and town planner can make a noticeable contribution. There is the structure itself, which can be designed in sympathy with climatic requirements, and there are also external factors, which affect the climate of space between buildings.

Structure

For building design in a hot, dry region, the sun is a far more serious factor to deal with than the breeze. A hot, bright sun overheats all surfaces exposed to it—the roof, walls, exposed terraces, and surrounding ground. These surfaces either reflect light, and therefore heat, into the building, or they transmit heat directly by conduction into the interior. So, for a structure, the first step is to reduce the flow of heat from outside, and this is mainly achieved by creating heavy shade internally and externally (Figs. 5, 6, 7). Add to this some means to reduce glare and stop entry of dust and sand—and one has all the elements that affect design and orientation of buildings.

If the building is much longer than it is wide, then by placing it along the average track of the sun and having reasonably wide overhangs, it is possible to reduce exposure of the walls to the minimum and be fairly certain that at least one main wall will have little or no sun. The important fact is to know how the sun falls in particular localities (8).

Apart from direct sunlight, which can be checked by suitable sunshading devices — louvers, verandahs, pergolas, and the like — heat flows either by conduction or by radiation from the surrounding surfaces.

Heat flow by conduction is inhibited by insulation and by reducing the outside exposed surface area of walls and roofs of a building. Radiation gains are reduced by shading, reflective coatings of surfaces, nonreflective surroundings, and suitable orientation.

The amount of heat transferred through the building fabric under periodic heat-flow cycle not only depends upon the insulation value of the material but also on its heat storage capacity.

Dense or heavyweight materials, such as mud, brick or stone, have a very high heat storage capacity. In hot,

dry regions, such materials have a great advantage over the others (see Figs. 8, 9). They take a long time to absorb most of the heat received during the day before passing it on to the inside surface, thus ensuring a cool interior. This behavior is admirable in such buildings as schools and offices, which are normally occupied only during the daytime. At night, when the outside temperature drops, the interior continues to stay too warm for comfort. To overcome this problem, people in traditional desert settlements, in North Africa and the Middle East for instance, move out to live and sleep outdoors, in courtyards, verandahs, or on rooftops.

This mode of living, however, does not solve the problems of lack of privacy, sudden rains, and duststorms. A possible solution may lie in a house in which bedrooms and other areas reserved for night use only are built up in lightweight structures (see Fig. 10).

Glass

A source of heat gain (and loss) which is often overlooked by many designers is large areas of glass in buildings. Glass has no insulation value, and heat will flow through it whenever there is a temperature gradient from one side to the other. E.G.A. Weiss (9, p.102) maintains that "heat transmission through a single sheet of shaded glass is about sixteen times that through a shaded timber frame wall insulated with 4 inches of mineral wool. Under conditions of tropical sunshine, heat transmission through exposed glass can be as much as eighty times that through a medium colored wall insulated with 4 inches of mineral wool."

In an example calculated for a simple structure in Pretoria, South Africa, S. J. Richards found that during the afternoon of a summer day with unprotected windows in 9-inch brick walls, the ratio of instantaneous heat gains through equivalent glass and wall areas is about 7 to 1 for the north wall, 3 to 1 for the east wall, 11 to 1 for the south wall, and 40 to 1 for the west wall (10). It is obvious that for buildings in hot, dry regions, glassed areas in the walls should be kept to an absolute minimum, compatible with adequate natural lighting. For examples, see Figures 11 and 12.

Roof Shading

Because of its orientation and comparatively large area, the roof can be a major source of heat gain in a building. Intense solar radiation generates considerable heat which, after transmission, raises the temperature of the ceiling surface. Radiation from heated ceilings not only heats other surfaces but adds directly to bodily discomfort. One way to minimize the transmission of heat is to reduce the proportion of the total radiation absorbed by the top surface. Here the color of the roof finish is important. White or light-colored surfaces are better reflectors than dark shades. If the rooms below are used during the daytime, the underside of the top layer could well be a shiny metal, the roof space should be well ventilated, and the ceiling should be well insulated.

Fig. 7. A street in a typical Arab settlement.

Fig. 8. Dense or heavyweight materials such as mud, brick, or stone used in traditional Middle Eastern settlements have a high heat-storage capacity, thus ensuring a cool building interior.

Massive flat roofs are useful only for buildings such as offices and schools, which are seldom used at night. Concrete slabs introduce a time lag into the heat transfer which results in considerable delay before the interior can cool down after sunset. The time lag for stabilization between the upper and lower surface temperatures for an eight-inch-thick concrete slab is six hours. To increase this lag, roofs require additional topping with sand and cement screed and possible insulation at ceiling level. A traditional method commonly used in the Middle

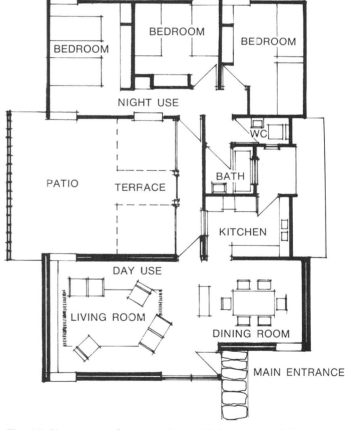

Fig. 10. House plan, incorporating both heavy and light construction. This illustrates the principle that the time lag of temperature increase behind thick walling can be used to ensure cool living space by day, while the light frame can be used to ensure cool sleeping areas by night. (Based on design by Robin McK. Campbell)

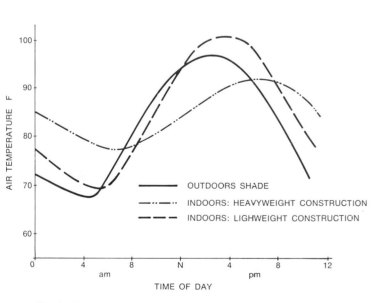

OUTDOORS SHADE
INDOORS: HEAVYWEIGHT CONSTRUCTION
INDOORS: LIGHWEIGHT CONSTRUCTION

Fig. 9. The performance of heavyweight and lightweight structures in hot, dry areas. (From Overseas Building Notes No. 71, Building Research Station, Garston, United Kingdom)

Fig. 11. School building in Chandigarh (by architect J. Malhoutra) keeps glassed window areas to an absolute minimum compatible with adequate natural lighting. In this it follows closely the approach of traditional builders of houses in desert settlements of central India such as those in Udaipur (see Fig. 12).

East is to cover the slab with a layer of earth—a material which has considerable heat-storage capacity. There are further possibilities where grass or some other form of vegetation may be grown on the roof. Apart from cooling the air above the roof, the roots of such a growth preserve a certain amount of moisture which not only helps to keep the temperature down, but also prolongs the life of the concrete by minimizing cracking, which normally occurs under extreme temperature variations.

In areas where the water supply is plentiful, it has been possible to reduce heat transmission by either covering the roof with water or using a series of jet sprays. It has been found that if one inch of water is maintained over a roof, the heat flow is only 35 percent of the amount which would pass through a dry roof of similar construction. With roof sprays, a figure in the region of 30 percent is quoted. Uninsulated concrete roof slabs are likely to be a source of heat gain rather than a heat barrier.

EXTERNAL FACTORS

Apart from the structure itself, indoor (and outdoor) comfort is also substantially influenced by external factors, such as vegetation, water, and the nature of the surrounding open spaces. Vegetation and water not only cool the air but also stimulate the eye. The qualities have always been appreciated by traditional desert communities, which have exploited them as major design elements, both in city planning as well as within individual structures. Residential palaces of a king or an upperclass merchant invariably enclosed a garden cooled by water, which not only delighted the eye on a hot summer's day, but also nourished the plants.

Fig. 12. Houses with small openings in Udaipur, central India. Note the location of windows made suitable for womenfolk who mostly sit and work at floor level.

Water

The pleasant environment in hot, dry situations is made possible by the evaporative process which, as mentioned earlier, increases the relative humidity and decreases the dry-bulb temperature of the surrounding air. The process is greatly assisted by wind movement, whether natural or artificially induced.

Before it is allowed to enter a building, wind can be made to pass over wet barriers, still-water pools (Fig. 13), or water sprays (Fig. 14). But, as J. E. Aronin maintains (*11*, p. 146), the rate of heat loss from the moving air depends upon "the area of water in contact with the air, the relative velocities of the water and air, during contact, the difference between the wet-bulb temperature of the air and the initial temperature of the water, and the time of contact between air and water."

In view of the greater vertical cross-section of air and the vastly increased area of contact between water and air, a spray pond is more effective than a still pond of the same size; the former, however, is likely to consume more water. To achieve the same efficiency in a still pond, it will have to expose a considerably larger surface, as compared to a spray pond. Further improvement could be obtained by carefully zoning the interior so as strategically to place strips of water in unprotected areas around the structure.

Water can be a valuable asset to a designer in hot, dry regions, but its full potential has been rarely exploited.

Vegetation

Vegetation, whether trees, plants, shrubs, or merely grasses, can help reduce the heat load on buildings. Trees need to preserve their leaf temperatures in the same way as human beings—by evaporating moisture. The result is that they have a great cooling effect. Vines and creepers (Fig. 15) act as sun-breakers, shield buildings from dust and sand, and, if located in an open area, also clean the air of petrol fumes and atmospheric impurities. To get the best results, vegetation must be carefully selected and planted in a position where it can be most effective.

Ground surfaces surrounding the buildings in a town need very careful attention. In a hot, dry zone, planting of trees, shrubs, and large stretches of grassed areas materially improve the whole climate of the township. Bare earth, asphalt, concrete and other types of paving have a high thermal absorptive capacity and thus become excessively heated during the day's exposure to solar radiation. As a result, air near the ground is heated beyond the normal. When carried into the building, it helps to increase interior temperatures. External surfaces made up of sand, polished paving, or even water in some instances can reflect radiation directly into or onto a building, thus compounding discomfort by adding radiated heat and glare. In all such cases, vegetation, particularly stretches of grass, are useful elements for reducing the incidence of both heat and glare.

Fig. 13. Still-water pools strategically placed around habitable areas such as this pavilion are an effective means of cooling the air before it enters the building.

Fig. 14. Fountains in a courtyard pool, such as these in the U.S. Embassy building in New Delhi (by architect Edward Stone), not only cool the air but also delight the eye on a hot summer day.

Sand and Dust Control

Vegetation is particularly useful for inhibiting the entry of dust and sand by controlling them at their source (Fig. 16). Plants are nature's soil protectors. They help stabilization and assist in the reclamation of eroded areas. One of the most obvious and simplest methods of achieving these aims is to conserve what vegetation there is, by fencing and keeping the cattle out of the township and its surrounding country.

An excellent example of how this is done can be seen at Broken Hill in the Australian state of New South Wales. This mining town lies in a semiarid region with an average annual rainfall of only 9 inches. The area is highly susceptible to wind erosion; in the past, duststorms turned midday into dark, with dust swirling unpleasantly and burying the sparse vegetation that grew around the city.

In 1936, a local metallurgist, Albert Morris, suggested that the natural vegetation be encouraged in areas behind the shelter of rabbit- and kangaroo-proof fences. The mining company initiated the scheme, which has now been extended to create a pleasant belt of natural vegetation. Thus the town is provided with an excellent screen against duststorms. With a population of over 30,000 people, the city now looks like a giant oasis in a desert country.

The flow of sand and dust presents a complex problem to an architect, as it involves not only planning of settlements as a whole, but also careful attention to design and layout of individual buildings. Dust is finer than

Fig. 15. Vines and creepers such as these on a hotel building in Kunnunnurra in northwestern Australia act as effective sunbreakers and shield buildings from dust and sand.

sand and it travels at higher levels with wind. Although difficult to check, its entry into buildings can be partially reduced by suitable orientation and the use of barriers in the form of overhangs, extension of walls, or simple free-standing buffers. Sand, on the other hand, being heavier, tends to travel near the ground by bouncing, not higher than 4 feet under normal wind speeds of 12 miles an hour. Sand can be controlled at its source by soil stabilization, through the use of such means as fencing and planting of grass, or by spraying with artificial films, such as bitumen, or synthetic products (oil and latex). These are practical measures that cannot be used to avoid the menace of dust with equal success.

LAND SUBDIVISION

The control of heat, dust, and sand by using vegetation, water, and soil stabilization opens up the whole question of the role which open spaces play in modifying the microclimate. The need to design, plan, and maintain open spaces in hot, dry areas, where water is in short supply and vegetation is hard to grow emphasizes the importance of keeping the open spaces to manageable proportions.

Yet the tendency seems to be for designers, particularly in the Australian inland, to follow layouts and subdivisions more suited to temperate, cooler areas.

That such an approach can create very unpleasant environment is obvious when we look at such towns as Alice Springs and Tennant Creek (Fig. 17) in the Northern Territory. Open space, in relation to built-up areas, is far too generous. Roads, even local ones used for light traffic only, are extremely wide. They are unsealed and, with no grassed areas, are a constant source of dust and sand (Fig. 18). Pedestrian traffic is seldom separated from vehicular traffic. When footpaths are provided, there is no attempt to shade them from the sun. Such unimaginative layouts make it difficult to orient houses for sun control.

Too much open space requires large areas of grass, and these need constant maintenance and abundant watering. Even where growth can be encouraged and maintained, the great expanse of dusty background, the roads themselves, and the further extension out into the macroclimatic spaces make it impossible to provide tolerably dust-free and cool spaces within the buildings. With this absence of controlled external space, the four walls of the buildings themselves are the only barriers against hot, dust-laden winds and direct solar radiation (Fig. 19).

In my studies of building in hot, dry lands, I have found that the compact and courtyard type of buildings in traditional desert settlements provide an excellent defense against heat, sand, and dust (Fig. 20). Individ-

Fig. 16. Desert vegetation as an important part of the ground treatment in the new town of Arad in the Negev. In a hot, dry zone, areas planted with trees, shrubs, and large stretches of grass materially improve the whole climate of the township. If judiciously located, these areas reduce the entry of dust and sand by controlling them at their source.

ual dwellings of mud-settlements in North Africa, the Middle East, and northwest India have a tendency to pack together in massive forms and thus reduce their exposed surfaces to the absolute minimum—very much like cactus plant forms (Fig. 21).

The buildings are invariably arranged around simple courtyards, private and communal, which act as cooling wells. By locating outside walls directly on the street boundary, all the habitable rooms face into the inner courtyard which, in turn, serves as a private garden and virtually extends the living area of the house. In summer, a pleasant continuous circulation of moist air is kept up through the use of grass, trees, vines, fountains and waterways (Figs. 22, 23).

To the people of the arid lands, a courtyard not only means enclosing the interior living space, but is also an attempt to frame a part of the sky, to pave a few yards of the desert, and to invest these fragments of nature with man's mark; it is in fact a symbol of shelter. Whatever its emotional origins may be, the courtyard with its infinite possibilities for variations still provides a satisfactory answer to many of the present-day problems of building in hot, dry regions.

The courtyard concept, which has proved so successful in traditional desert communities, is essentially an introvert concept — people turn their backs to the hostile natural surroundings. This attitude is in remarkable contrast to that of the Australian settlers, who, perhaps because of their temperament and antecedents in Europe, where light and heat are scarce, tend to prefer large open spaces around their shelters. This exposure to surroundings, which may have been welcome in temperate regions, tends to create drastic problems of shielding the buildings from the impact of the arid environment. In central Australia, in many cases protection is partly provided by a deep, well-shaded verandah, but the single exposed shelters are still denied the mutual benefits which could be available through more compact layouts, as indicated by traditional settlements.

Compact planning based on the courtyard concept has proved successful in other arid areas of the world, and there is no reason to believe that, if used, it will not be so in an Australian setting. A proposal (Figs. 24, 25) prepared by architect Geoffrey Borrack under my direction during 1963 suggests that the courtyard houses can be designed to meet the needs of typical Australian fami-

Courtesy of Australian News and Information Bureau

Fig. 17. Tennant Creek in Australia's Northern Territory. Rows and rows of small, timber-framed houses with uninsulated roofs sit in a virtual dust bowl on plots far too large.

Courtesy of Australian News and Information Bureau

Fig. 18. Camooweal, in the far northwest of Australia's Queensland. Note open space, unsealed roads, and lack of grass.

Fig. 19. House in a new housing estate in Alice Springs (central Australia), displaying inherent difficulties associated with single-home building in a hot, dry desert setting. The four walls of the house are the only bulwarks against hot, dust-laden winds and direct sunlight.

118

lies in the hot, dry interior of the continent. The principles evolved then, and which are being applied to a new mining township near Kalgoorlie in Western Australia, indicate that the courtyard concept can ensure the same amenities as those found in houses on individual plots of land. The open spaces are judiciously planned; they reduce the total area of the development, thus economizing in the cost of services.

This discussion of external factors indicates that microclimatology, landscape design, and site planning are allied in several ways. The basic objective of course, is to bring good living close to nature without going to extremes that could nullify the benefits to be gained. Achieving the objective is of particular importance in Australia's arid regions, since evidence of a cultivated landscape considerably helps to increase one's sense of security in the middle of the wilderness. To achieve domestic and social security by creating an urban atmosphere, yet with nature close at hand, is a matter of right choice and delicate planning.

Fig. 20. In traditional communities in hot, dry lands, such as Kano, Nigeria (right), and a Pueblo settlement near the Rio Grande (below), mud houses are built close together for collective mass protection against excessive stresses of the environment. These massive and compact forms tend to reduce their exposed surfaces to the absolute minimum — very much like the cactus plant forms (see Fig. 21).

Courtesy U.S. Department of Interior, Washington, D.C.

Fig. 21. Cactus plant form.

REFERENCES AND NOTES

1. AUSTRALIA, COMMONWEALTH BUREAU OF
   CENSUS AND STATISTICS
     1969   Yearbook of the Commonwealth of Australia.
            Commonwealth Government Printer, Can-
            berra.

2. Figures cited from the results of inquiry by West Australian Civil Service Association into living costs north
   of the 26th parallel in:
   KERR, A.
     1967   Australia's north west. University of Western
            Australia Press.

3. HOLMES, M. J.
     1963   Australia's open north. Angus and Robertson,
            Sydney.

4. SHEA, A. D.
     1964   Oil workers' housing and community integra-
            tion in isolated areas. International Labour
            Organisation, Geneva.

5. WESTON, E. T.
     n.d.   Ventilation and cooling of a house with an
            attic exhaust fan. Commonwealth Experimen-
            tal Building Station, Sydney, Special Report 9.

6. BADGETT, W. H.
     1940   The installation and use of attic fans. Texas
            Engineering Experiment Station, Bulletin 52.
            45 p.

7. WICKHAM, F.
     1963   Selection of air movement plant. University of
            Melbourne, Department of Architecture,
            Tropical Building Studies Series 1(4):69–95.

8. PHILLIPS, R. O.
     1948   Sunshine and shade in Australia. Common-
            wealth Experimental Building Station, Sydney,
            Technical Study 23.
   See also, A. and V. Olgyay, Design with climate: Bio-
   climatic approach to architectural regionalism. 1963.
   Princeton University Press, Princeton, New Jersey.

9. SHERIDAN, N. R., AND OTHERS
     1964   Air conditioning. University of Queensland
            Press. 131 p.

10. RICHARDS, S. J.
     1957   Climate control by building design. In Sympo-
            sium on Design for Tropical Living, Proceed-
            ings. South African Council for Scientific and
            Industrial Research, Durban.

11. ARONIN, J. E.
     1952   Climate and architecture. Reinhold Publishing
            Corporation, New York.

Fig. 22. Courtyards and large irrigated gardens such as this in Rajasthan, India, break the monotony and harshness of arid landscape. Outer walls are directly on the street, thus creating a particular style of desert architecture.

Fig. 23. This new neighborhood center in Beersheva in the Negev is based on the courtyard concept and follows closely the pattern set by traditional desert settlements of the Middle East.

120

International Training

The Interamerican Center
for Land and Water Resource Development — A Case Study

BRUCE H. ANDERSON

Director, International Progrcms and Studies
Utah State University, Logan, Utah, U.S.A.

ABSTRACT

THE INTERAMERICAN CENTER FOR LAND AND WATER RESOURCE DEVELOPMENT (CIDIAT) is sponsored by the Program of Technical Cooperation of the Organization of American States and by the government of Venezuela, in cooperation with the Universidad de los Andes. Utah State University provides direction and technical backstopping to the Center. CIDIAT was created to provide selected Latin Americans with a learning experience in harmony with today's development problems. In tracing the historical development of CIDIAT, we indicate the need for it, its sources of support, and its evolving patterns of program formulation and implementation. The total CIDIAT program is evaluated, including the four different types of courses used in the training program. The innovations effected by CIDIAT in national courses were unique contributions. The potentials inherent in CIDIAT can and should constitute a provocative challenge to progressive educational institutions everywhere. A permanent future for CIDIAT is envisioned with national training centers in several South American countries.

INTERNATIONAL TRAINING
The Interamerican Center
for Land and Water Resource Development — A Case Study

Bruce H. Anderson

UNIVERSITY-LEVEL COURSE WORK must be made relevant to current and prospective economic development if academic education is to satisfy the urgent demand for skilled planners, professionals, and administrators in land- and water-resource projects. A creative international educational institution could logically be a unique source of the necessary innovation. In effect, CIDIAT* is a pilot study of the hypothesis as applied to Latin America. CIDIAT was created to provide selected Latin Americans with a learning experience in harmony with today's development problems. At the same time, by keeping current with new technology and social conditions, the Center can offer technical assistance to other educational institutions and development organizations.

Stephen S. Goodspeed states that sincere men have always tried to lessen the insecurity of international life by creating institutions designed to generate some measure of political, economic, and social stability (1, p. vii). The following discussion of CIDIAT illustrates the kind of contribution that an international institution can make to economic security and social stability within and among countries by training personnel concerned with economic development. The discussion will consider the international cooperation required to create CIDIAT, the procedures and techniques used during its formative state, the objectives, problems, accomplishments to date, and the future role of the Center.

THE CREATION OF CIDIAT

CIDIAT, Interamerican Center for Land and Water Resource Development, is Project 213 of the Program of Technical Cooperation of the Organization of American States. The government of Venezuela and the Universidad de los Andes are hosts and cosponsors, and Utah State University is the administrative and technical backstopping agent. The major function of the Center is to train professional personnel for work at various levels in developing land and water resources. All member nations of the Organization of American States can participate in its educative programs.

Training — An Aid to Land and Water Development

The need to intensify and extend training of personnel for work in the many fields and at the various levels of land- and water-resource development is well recognized in South America. The Inter-American Development

Bank, in its 1967 Round Table (2) noted: "One of the most important limitations on Latin American development is the shortage of qualified personnel for program planning, institutional management and project development and execution."

In most Latin American countries, a byproduct of this shortage of qualified leadership has been a low per-capita production of food and fiber. Conservative estimates indicate that by 1980 Latin America would have to produce around 50 percent more carbohydrates and 100 percent more protein foods to meet the minimum nutritional standards recommended by Food and Agricultural Organization (FAO) of the United Nations. It is important to note, moreover, that the policy directives adopted in Punta del Este in 1967 expect Latin American countries not only to care for their own needs but to help to alleviate hunger in other parts of the world.

The analysis of present compound growth rates, as compared to needs (Table 1), emphasizes the enormity of the task of developing and adequately utilizing Latin American land and water resources for food and fiber production. In Latin America, only Mexico has achieved a satisfactory growth rate.

TABLE 1
Present Compound Growth Rates, as Compared to Needs

	Annual Increase (%)	
	Present	Required
Food demand	3.0	4.0
Food production	2.7	4.0
Gross national product	4.0	5.5

Note: Table taken from President's Science Advisory Committee, 3, p. 112.

Much of the technical skill and administrative competence necessary to guide the needed development programs can be made available only through the training and/or the professional improvement of personnel. The methods used will have to include intensified training programs and realistic field experience.

Some Latin American countries, such as Brazil, Colombia, and Venezuela, are starting to develop their vast water and land resources. Others, such as Chile and Peru, have more limited resources, and efficient use is a major problem. While each country has its own unique

*CIDIAT is the abbreviation of the Spanish wording for the Interamerican Center for the Integral Development of Water and Land.

problems, in general, all need professional personnel in most of the major disciplines concerned with land and water development, namely agriculture, engineering, economics, sociology, and law. Competent people are needed to evaluate and inventory the available resources and to participate in the subsequent planning and activation of development projects. Personnel trained in surface and subsurface drainage, flood control, irrigation systems, soil surveys, data collection, cropping patterns under tropical conditions, and especially in the operation and maintenance of irrigation districts, are in very short supply. This listing is far from exhaustive; it simply reflects the magnitude of needs for resource personnel shared by the majority of Latin American countries.

Adequately trained and experienced administrative personnel are also scarce. Most administrators just "happen" into their positions and often lack many of the tools necessary to make proper decisions.

Resource development in Latin America requires an intensive involvement of many agencies of government, and since the political environment is different in every country, the amount of support for development will vary. Even with support, however, coordinating the work of several agencies is difficult. In any given project one agency may be responsible for agricultural production, another for engineering aspects, and still others for markets and credit or colonization. The project leader must be able to integrate all the agencies into a working team. Administrative and public relations skills among technical personnel must be such that reliable data and information are presented to political leaders in a way that aids in making rational decisions. Administrators must know how best to use young professionals returning from abroad with specialized training and graduate degrees.

CIDIAT grew out of the conviction that properly oriented training programs at both national and international levels could materially help the present situation and set the stage for a most productive future.

An International Center

Sound arguments both pro and con can be cited regarding the establishment of international centers. The center concept does allow small nations to participate and benefit from an activity without incurring excessive costs. It also provides a mechanism for trying ideas and innovating without necessitating heavy expenditures and chance-taking by individual countries. A center can serve to develop interest in particular activities that can later be profitably undertaken by local institutions. An international educational institution can, as will be illustrated later, foster training programs on an international basis, while simultaneously benefiting national institutions.

The requirement that one intermittently shift local competent staff to a center's international programs often irritates national agencies, however. It can be argued that more effort should be devoted to strengthening the local institutions rather than to developing international centers.

With specific reference to CIDIAT, its vigorous success tends to validate the need for this training center. The acceptance of the center concept by the member nations of the OAS evidenced a common belief in Goodspeed's idea (*1*, p. *395*):

This is an age when increased knowledge of natural phenomena and command over the sources of energy have placed greatly expanded productivity within the grasp of mankind, and, at the same time, the nations of the modern world, through scientific discovery and technological progress, have gradually been increasing their interdependence.

The CIDIAT approach has capitalized on this interdependence by providing tremendous opportunities for sharing, for helping, and for learning. But the enterprise has not been without problems, and long-term support of the program remains precarious.

DEVELOPMENT OF CIDIAT

Sponsors

The Organization of American States (OAS), a sponsoring international organization, functions through six principal organs. One of these is a council composed of one representative from each state. The council in turn has three parts, one of which (the Inter-American Economic and Social Council) established the Program of Technical Cooperation in 1950.

The Program of Technical Cooperation is a continuing activity of the OAS. Its objectives are (*a*) to promote or establish centers of study in the member states of OAS in the field of advanced training, stressing research and exchange of technical knowledge and experience, all of which are closely related to the country's needs, and (*b*) to strengthen, insofar as possible, national education and research institutions. CIDIAT is one of nine projects sponsored by this program.

The Universidad de los Andes, of Mérida, Venezuela, the second-oldest university in the country, is also a sponsor. The university is comprised of nine schools or *facultades*. The university has about 500 faculty members and 6,000 students.

Utah State University was founded in 1888 as part of the system of land-grant universities established in 1862 in the United States. Its particular research and teaching strengths have traditionally been in disciplines pertinent to natural resources. Since 1950 the university has enrolled a remarkably high ratio of international students as compared to other universities in the United States. These factors combined to make Utah State University a logical choice for involvement in CIDIAT.

History of the Center

The wheels of international diplomacy turn slowly, and many years of persistent effort are usually required to take an idea from initial acceptance to its germination, growth, and maturity. Such was the case with CIDIAT.

The Program of Technical Cooperation accepted the responsibility to promote a training center in land and water development early in 1961. Of the five countries considered for the project, Venezuela displayed the greatest interest, and a program proposal was made to the Ministry of Foreign Affairs in Caracas on October 17, 1961. By December, detailed discussions were in progress with the Universidad de los Andes at Mérida, where Venezuela wished to establish the Center.

In 1962 the OAS contacted Utah State University regarding its interest in providing direction and guidance to the development of the Center. This contact resulted in USU's president visiting the Universidad de los Andes, and later a trip by the rector of the Universidad de los Andes to Logan. Following these visits the OAS submitted a formal proposal to proceed with the Center to its Social and Economic Council at the level of Experts and Ministers. They approved the proposal in November of 1963.

The agreement between the OAS and Utah State University provides the basis for the development and operation of the Center. The stated purpose of the Center is to train, at the graduate level, senior administrators and government planning experts, administrators and planners, and professionals in the field of water- and land-resources development (4).

The agreement defines the responsibility of Utah State University as follows: the University is to provide the technical support, guidance, and administration in the organization and development of the Center. The technical assistance and administration of the project is to be in accordance with the bases and regulations of the program; and the purpose of such assistance is to facilitate the establishment of a permanent educational institution at the University de los Andes in land- and water-resources development.

The agreement between the OAS and the government of Venezuela defines in more specific terms the objectives, functions, and operational details of the project (5).

Specific objectives of the project are: (a) To train, through regular courses and short intensive courses, government administrators and professionals at a high level; to develop their administrative capacity in order to increase the effectiveness of their present work in taking advantage of the indicated resources; to coordinate the operation of existing facilities and better the services of operation and administration of land and water projects. (b) To promote the interchange of technical information and ideas between professionals and administrators in this field. (c) To improve existing institutions concerned with land and water improvement.

The training program is to accomplish its goals through seminars, short courses, international courses, and research related to training.

The Universidad de los Andes is eventually to continue the project as a national center and as a regular part of the University program without financial help from the Program of Technical Cooperation of the OAS.

With contract negotiations completed, Utah State University faced the problem of how to develop a program that would provide the kind of training needed in Latin America. Answers were needed as to what should be taught at the Center, how, and to whom.

Each of the preceding questions, when considered in depth, raises additional questions for consideration. An item of prime importance is the identification of needs and their priorities. Could a general curriculum serve all countries when the development problems vary from country to country? Participants would have different backgrounds and training, thus adding further complexity to the problem. The criteria used in selecting participants would influence subject-matter requirements. Flexibility would have to be maintained in course content, so that changes could be made to meet the needs of the participants.

To provide answers to such questions a series of conferences and seminars was organized at Utah State University by the on-campus project coordinator. The first conferences involved deans and faculty of the University. Later conferences included two groups of specially chosen consultants and a third group of selected leaders from Latin America.

The consultant groups were composed of specialists in agronomy, geography, forestry, engineering, economics, water-resource development, sociology, and private business who had had previous experience in Latin American countries. The two groups met for one week each on different dates to "brainstorm" the questions: What should be taught at the new Center? To whom should it be taught? How should it be taught?

The conclusions and program suggestions generated by the first consultant group were passed on to the second group to aid in their deliberations. A Latin American seminar attended by high-level government and business administrators and educators from 15 South American countries was then organized to review the work of the first groups and to prepare specific recommendations for the Center. These recommendations were used to guide the Center during its development.

It is interesting to note that all consultant groups required time to solve semantics problems. Terms and concepts had to be defined in mutually accepted language before members of each group could communicate well enough to focus on the specific questions under discussion. The language barrier was quite difficult to overcome for the seminar group from Latin America, despite the competence of simultaneous translators.

The seminar group raised many questions about the Center. Why was it located at Mérida, Venezuela, and not in some other country? How could the Center, with a limited number of scholarships, hope to train the numbers of professionals needed by all countries? Where would competent staff be obtained without raiding national organizations? How would students be selected to assure the training of people who would be of use to the country? How could countries individually profit from the Center? How would the policies and programs of the Center coincide with the real needs of each country?

Such questions had to be resolved before the group could deal with specific recommendations for the Center.

The following is a summary of the discussions that provided solutions to most of the problems.

Venezuela had received favorable attention from the OAS as a possible site for several reasons. The government of Venezuela agreed to a substantial monetary contribution, while the Universidad de los Andes agreed to provide facilities and additional support to defray administrative costs for the Center. Venezuela is ideally situated from the standpoint of travel from South and Central American countries and has a wide range of climate conditions. Large expanses of low-lying llanos or plains areas receive heavy rainfall and flood during the wet season and dry out during the dry season. In arid areas nothing grows without irrigation; in mountainous regions of variable rainfall campesinos (farming populace) eke out a meager living on small plots with and without irrigation. The many water projects that are under development in Venezuela affect a wide range of possible educational experiences in planning, implementation, and operation.

Since only 25 scholarships are available for each CIDIAT international course, national training courses seemed essential to train large groups of nationals. A country could request that the Center present a course to prepare technicians and specialists to cope with its national interests. Budget and staff limitations would automatically restrict the number of courses that could be offered each year.

In discussing the selection of students, seminar participants pointed out that many full-time professional students lived a good part of their lives by obtaining scholarships to study abroad. They never stayed in the country long enough to work and contribute to the common welfare. A careful screening of candidates and an endorsement by the organization for whom they work was suggested.

The discussion on staffing the Center brought out the reluctance of development organizations to continually lose their better men to international agencies. The participants realized that CIDIAT had to have a staff, but they resented international groups with adequate budgets that could make it virtually impossible for national agencies to compete in salary negotiations.

The participants stressed practical applications in the course work. A suggestion was made that the seminar participants should select a committee to meet once each year and review the work of the Center. It was not feasible to act upon this suggestion immediately due to budgetary limitations. Finally, emphasis was given to maintaining a close contact with national universities and to involving them whenever possible in national training course programs.

Following the discussions, the seminar group proposed specific recommendations to guide CIDIAT. It was recommended that training be conducted at three levels: (*a*) One- or two-week high-level seminars for persons in policy-making positions.
(*b*) Short courses lasting approximately two months for persons who are at the midmanagement level and who have had eight or more years of working experience.
(*c*) Courses of approximately six months duration for professionals with two to eight years of experience of working with land and water problems.

These three levels of training were considered sufficient to reach the people important in land- and water-development programs. By providing some discussion of development philosophy to all three groups, a basis for communication would be established among individuals from each level within a given organization who attended the CIDIAT courses. The group emphasized the need for maintaining contact with the high-level group as a way to keep abreast of current problems and training needs.

Another recommendation was for national training courses of from one to four weeks, which would be given at a country's request. These courses would provide instruction in a subject area suggested by the country. Through this approach greater numbers of nationals would receive specialized training oriented to national needs.

Perhaps the most arduous task assigned to the seminar participants was that of considering the curriculum. Only a very few of the participants came from educational institutions, and after much deliberation they approved a curriculum much as it had been proposed by the USU consultant groups:

I. Introduction
   A. The theory and basis of economic growth
   B. Social factors affecting use and development of resources
   C. Concepts of interdisciplinary and systems approach to resources planning

II. Resource Data Collection and Evaluation in Terms of Regional Planning Needs
   A. Physical data
   B. Economic data and projections
   C. Legal and institutional data
   D. Socio-anthropological characteristics

III. General Principles of Resource Planning
   A. Consideration of social and political objectives and their translation into economic objectives
   B. Scope of development
   C. Constraints and limitations
   D. Preparation of alternative plans
   E. Economic evaluation with view to optimizing project return
   F. Cost allocation and financing
   G. Priorities and scheduling of development

IV. Logistics of Project Development
   A. Finance
   B. Personnel
   C. Administration
   D. Procurement
   E. Property
   F. Organization

V. Successful Project Operation and Maintenance
   A. Organizational and institutional problems
   B. Extension services and education
   C. Credit and marketing problems

Such a topic outline does not provide enough detail to define the way any particular area is to be taught, thus leaving the focus up to the professor handling that subject matter. The Utah State Planning Committee did later ask some of the consultants to expand the outline to provide specific resource material for use at the Center.

Program Implementation

The Center began operations in May 1965 at the Universidad de los Andes. The first program was held during July of 1965. The Universidad de los Andes provided and is providing temporary facilities including office space, classrooms, and access to simultaneous-translation equipment. From its initiation to the end of December 1970, the Center held four High-Level Seminars, eight International Short Courses, four International Regular Courses, two Regional Courses, and 26 National Courses. To December 1970, 1146 persons had participated in the training programs, representing all countries of the OAS except Barbados, which was only recently admitted to the organization.

How does one evaluate the work of a training center such as CIDIAT? Continuing contact with each alumnus is costly, time-consuming, and difficult. Questionnaires and evaluation forms too often reflect what the participant thinks would like to be heard. Also, the contribu-

tions of individuals to a program do not usually become apparent in short time periods. Notwithstanding these problems, some positive results and examples of progress as a result of CIDIAT's training efforts can be cited.

CIDIAT has been following the general outline of courses recommended at the first High Level Seminar. However, time limitations do not allow full treatment of each subject matter area in the courses.

THE TWO-MONTH SHORT COURSE

In an effort to orient the two-month short courses to the anticipated needs of an average participant, the following format is used:

Orientation. Discussions concerning CIDIAT, how it was formed, its goals and objectives, and how it functions to help train manpower for the member nations of the OAS open each session. Next, a preview is given of what will be covered during the two-month period.

Group Dynamics. During the first part of each course CIDIAT attempts to establish an atmosphere in which mutual trust, acceptance and teamwork would maximize the learning process for each participant. A series of laboratory periods deals with the problems of status and "back-home" situations, then proceeds to modify circumstances that can inhibit teamwork. The techniques used to accomplish the above vary with each course and

Fig. 1. CIDIAT course participants visit rice harvest in Venezuela.

with the competence of the staff available to guide the orientation, but the laboratory approach based on group dynamics and some sensitivity training has produced the best results.

Case-Study Field Trip. Very early in each course, the participants are divided into groups and each group is provided data on an actual project. These data are the basis for group work resulting in a report or feasibility evaluation of the project.

Lectures and discussions help each group define the additional data needed for a feasibility report. A field trip of three to five days then allows participants to check the supplied data, collect more data, and become acquainted with the project through personal contact and discussions with the Venezuelan staff responsible for it.

Lectures and Discussions. The next section of the course includes (*a*) philosophy for land and water development, (*b*) economic considerations at national, regional, and local levels, and mechanics of project formulation, and (*c*) data-collection and analysis techniques for meteorology and hydrology, soils, water-supply and irrigation-water requirements, crops and cropping patterns, urban and industrial water requirements, human resources and institutional problems, and benefit cost-ratio analysis. These classes provide sufficient background so the participants can effectively complete the in-case study assignments during the final two weeks of the course. The academic material is oriented to the case study, to strengthen the participants for their group feasibility report.

The Case Study. The final two weeks of each two-month short course provide time for the subgroups (four groups of approximately six people each) to work on their feasibility reports. The assignment requires considerable team effort and many hours to work with and analyze the data. The encouragement toward initiative and innovation given by the staff has helped the participants recognize and consider the possible alternatives in the development area. During the report-preparation time the staff consults with the groups, as requested, to iron out problems and render assistance.

The final two days of each course are devoted to oral reports by each group to the staff and other groups. The defense of the report provides another opportunity for exchange of ideas and sharpening of concepts and techniques.

The case study selected by the staff varies from course to course to assure diverse experiences. For example, in one course the case study emphasized a project for which few data were available; it presented the problem of how to handle such a situation and the dangers involved in making decisions based on assumptions and scarce data. Another study involved a river basin that had well-advanced urban and industrial areas and new agricultural lands. Another project centered on the problem of colonization in a region where land and water is plentiful but no infrastructure exists.

Table 2 indicates participation by country in the international short courses while Table 3 shows the parti-

Fig. 2. CIDIAT course participants view pineapple production in Venezuela.

cipation by country for each course. These courses have been very favorably received. They fill a great void in the need for professional improvement at the midmanagement level. Many of the participating professionals have been away from school for several years without opportunities to study or learn the latest technology and methodology. They can study at the Center without the second language requirement which is essential if they are to avail themselves of scholarship opportunities in the United States and Europe.

The two-month courses provide insights into the broad problem of development, and they help to alert administrators to the need for considering alternatives. The human social factors have been emphasized in these courses in addition to the technical problems. For, while the world has at hand the tools, the understanding, and the experience to solve most technical problems, solutions to social problems continue to evade us. Thus technical people—engineers, scientists, and administrators—need some training and background to assure their support of programs and research in the socioeconomic area and to bolster their knowledge and acceptance of the advantages of an interdisciplinary approach to project development and management.

THE SIX-MONTH REGULAR COURSE

The six-month course is expected to increase technical skills rather than administrative competence. The participants tend to be younger and less experienced than those in the two-month course. The emphasis on mastery of academic material is greater, and classroom activity is more intensive.

Some elements of the two-month course are carried over. The orientation and group dynamics aspects are essentially the same. A case study is again used to foster a better understanding of all the factors that must be considered in land and water development. Keeping these aspects common to both courses sets the stage for better communication between management and technical personnel.

During a six-month course, the staff can present material on topics such as irrigation, drainage, soils, and crops in relative depth. The academic work offered to the six-month course participants is at the level of a master's program. Participants in the third six-month course (completed in June 1969) who wanted to work toward a master's degree were given credit for the course by the Universidad de los Andes.

Table 2 shows the participation by country in the six-month regular courses, and the participation by course is shown in Table 3.

TABLE 2

Summary of Participation in CIDIAT Courses by Country, from November 1964 to December 1970

Countries	Number of Participants					
	Seminars	Regular Courses	Short Courses	National Courses	Regional Courses	Total
Argentina	4	4	22	67	. .	97
Bolivia	5	6	10	32	. .	53
Brazil	4	6	12	67	. .	89
Chile	5	5	11	19	. .	40
Colombia	6	15	25	111	40	197
Costa Rica	1	. .	3	. .	. .	4
Dominican Republic	1	2	4	. .	. .	7
Equador	2	10	13	54	2	81
El Salvador	3	9	7	32	2	53
Guatemala	2	1	2	. .	1	6
Haiti	. .	1	2	. .	. .	3
Honduras	1	1	6	. .	. .	8
Mexico	2	. .	4	. .	. .	6
Nicaragua	2	2	7	. .	1	12
Panama	3	2	3	39	. .	47
Paraguay	3	1	3	. .	. .	7
Peru	9	14	16	23	. .	62
Puerto Rico	3	. .	. .	. .	2	5
Trinidad and Tobago	. .	. .	. .	16	. .	16
Uruguay	4	3	7	21	. .	35
Venezuela	17	22	25	247	7	318
Totals	77	104	182	728	55	1146

NATIONAL TRAINING PROGRAMS

The national training programs vary each time, since the requests come from different countries and their needs are different. A number of the courses have dealt with irrigation and drainage problems. Another considered the problems of operation and maintenance of an irrigation district. Perhaps the most innovative National Training Program undertaken by CIDIAT involved a course in management and did not include any material of a technical nature. Additional details of some of the above will be given to illustrate specific activities and their results. Table 4 gives statistical data on the national and regional courses.

REGIONAL TRAINING COURSES

Some courses have been handled on a regional basis. Nearby countries have been invited to participate in what was orginally intended as a national course. These courses still emphasize the host country problems, but have more extensive potential applications.

HIGH-LEVEL SEMINAR

The High-Level Seminars have contributed significantly to the success of CIDIAT. They provided a clearing house for ideas, programs, and curricula. Involving seminar participants in discussing these items provided an excellent opportunity for indirect learning. Discussions included development philosophy concepts, management techniques, the process of project formulation, and individual country needs. Participants exchanged ideas and obtained feedback on their approach to development problems. Undoubtedly a close contact with the leaders of development organizations and educational institutions is necessary to CIDIAT's future.

Table 5 indicates the participation by country in the High-Level Seminars.

EXPERIMENTAL COURSE IN EXECUTIVE MANAGERIAL TRAINING

The director of INCORA (Agrarian Land Reform Institute of Colombia) indicated that he did not like to send participants to the CIDIAT courses since he could only send one or two per course, and his needs were substantially greater than this. He also indicated an urgent need for training all of his staff from the "top on down."

CIDIAT sponsored a one-week course in Colombia to help meet some of INCORA's needs (6). The purpose was to provide the executives and administrators of INCORA with new insights into the processes of communication. Men who are competent in their professional disciplines often find themselves engulfed in problems outside of their technical training and experience when they shift to administration. Through the process of dialogue, confrontation, and analyzing the procedures of management, the participants developed their abilities to cooperate and translate the goals and purposes of the organization into desired action.

TABLE 3

Statistical Summary of Participants at CIDIAT International Courses from November 1964 to December 1970

Countries	Regular Courses				Short Courses								Total
	1st	2nd	3rd	4th	1st	2nd	3rd	4th	5th	6th	7th	8th	
Argentina	..	3	1	..	3	4	4	3	2	2	..	4	26
Bolivia	3	2	1	..	2	2	2	1	2	..	..	1	16
Brazil	2	3	1	..	..	..	1	1	2	3	3	2	18
Chile	1	1	2	1	3	1	2	1	..	..	2	2	16
Colombia	4	2	5	4	2	2	1	..	1	..	17	2	40
Costa Rica	..	..	..	..	1	1	1	..	..	..	..	..	3
Dominican Republic	..	..	1	1	2	..	..	..	2	..	..	..	6
Ecuador	4	2	1	3	2	4	..	2	1	1	1	2	23
El Salvador	4	2	1	2	2	..	1	..	2	1	1	..	16
Guatemala	..	..	..	1	1	1	..	..	..	..	..	..	3
Haiti	..	..	..	1	..	..	1	1	..	..	..	..	3
Honduras	..	..	1	..	1	..	..	..	1	1	..	1	7
Mexico	..	..	..	..	2	1	..	1	..	..	..	..	4
Nicaragua	1	1	..	..	2	..	1	..	1	1	..	2	9
Panama	..	1	1	..	1	1	..	..	1	..	..	..	5
Paraguay	..	..	..	1	..	..	1	..	..	1	1	..	4
Peru	6	3	3	2	2	3	4	2	1	3	1	..	30
Uruguay	..	1	..	2	1	..	2	1	1	1	..	1	10
Venezuela	3	5	7	7	3	1	2	1	10	7	..	1	47
Totals	28	26	25	25	30	21	23	15	27	20	27	19	286

TABLE 4

Numbers of Participants in National CIDIAT Courses

Month and Year	Argentina	Bolivia	Brazil	Chile	Colombia	Ecuador	El Salvador	Panama	Peru	Trinidad	Uruguay	Venezuela	Totals
1966													
July 25-Aug 13	..	..	..	..	..	25	..	..	..	..	..	..	25
Nov 14-Dec 2	..	32	..	..	..	..	..	..	..	..	..	..	32
1967													
Mar 21-26	..	..	..	..	31	..	..	..	..	..	..	..	31
April 10-28	..	..	..	..	..	..	..	..	23	..	..	..	23
May 21-27	..	..	..	..	28	..	..	..	..	..	..	..	28
Sept 25-Oct 13	34	..	..	..	..	..	..	..	..	..	..	..	34
1968													
April 29-May 3	..	..	..	..	25	..	..	..	..	..	..	..	25
July 1-30	..	..	..	..	..	..	..	..	..	..	..	29	29
Aug 19-30	33	..	..	..	..	..	..	..	..	..	..	..	33
Aug 19-31	..	..	..	..	..	..	..	..	..	..	..	22	22
Oct 23-Dec 6	..	..	..	..	..	..	..	..	..	..	..	24	24
Nov 4-11	..	..	..	..	..	..	..	..	..	..	..	15	15
Nov 4-22	..	..	..	..	..	..	..	..	..	..	..	30	30
Nov 11-Dec 7	..	..	..	..	33	..	..	..	..	..	..	..	33
1969													
April 12-25	..	..	..	..	..	..	..	..	..	..	..	48	48
May 31-June 6	..	..	..	..	27	..	..	..	..	..	..	..	27
June 9-13	..	..	..	..	..	..	..	..	..	..	16	..	16
June 14-18	..	..	..	..	..	..	..	39	..	..	..	..	39
July 13-18	..	..	..	..	..	..	..	..	..	..	..	50	50
July 28-Aug 1	..	..	..	..	..	..	..	..	..	..	..	29	29
Aug 11-16	..	..	..	..	..	..	..	..	..	..	..	48	48
Aug 4-29	..	..	..	..	..	29	..	..	..	..	..	..	29
Nov 17-Dec 12	..	..	..	..	..	..	..	..	..	..	25	..	25
Nov 17-Dec 12	..	..	..	..	..	..	32	..	..	..	..	..	32
1970													
Oct 5-31	..	..	..	19	..	..	..	..	..	..	..	..	19
Nov 23-Dec 11	..	..	30	..	..	..	..	..	..	..	..	..	30

Techniques employed in the course were designed to build and strengthen an ability for teamwork and to develop more understanding of administrative purposes, while still encouraging individual growth.

The course resulted in the director of INCORA requesting additional courses so that all of his staff could benefit. He also suggested that other organizations in Colombia undergo similar training. Within a year Colombian agencies arranged for and financed 13 additional courses.

Perhaps one of the greatest benefits to CIDIAT from the first INCORA course was the confidence gained by INCORA in CIDIAT's ability to render service and meet their training needs. Shortly after giving the management course, CIDIAT received a request from INCORA for a two-month course in operation and maintenance of irrigation districts.

COURSE IN OPERATION AND MAINTENANCE OF IRRIGATION DISTRICTS

This course, orginally planned as a national course, was broadened into a regional course with El Salvador, Honduras, Nicaragua, and Venezuela participating. The course provided training to project managers and prospective managers in the operation and management of irrigation districts. The Roldanillo Irrigation Project, where the course was held, provided real problems and actual field situations for classroom discussions and laboratory exercises.

The teaching staff included individuals from CIDIAT and INCORA, consultants from other Colombian agencies, and two specialists in operation and maintenance from the United States Bureau of Reclamation. The USBR men provided technical assistance to the project in addition to filling teaching assignments. They were financed by U.S. AID (United States Agency for International Development), Colombia.

One week of sensitivity and management training allowed course participants to look seriously at their own management techniques and operational procedures. The concept of self-analysis and introspection helped participants achieve a new approach to management problems.

The impact of the course was such that the Venezuelans requested a similar program for management and personnel of their irrigation districts. Ecuador also requested a national course in operation and maintenance. CIDIAT complied with both these requests as quickly as possible.

Other benefits from the Roldanillo course included INCORA's realization that the shortage of experienced, trained personnel to cope with the expansion of irrigation in Colombia was critical. Subsequent evaluation of personnel needs indicated that a minimum of 300 persons per year for the next five years must receive training and experience in various aspects of the operation and maintenance of irrigation districts.

INCORA requested CIDIAT to establish a CIDIAT subcenter to help INCORA train its personnel. This has been done, and Colombia now has a subcenter called PADE (Program for Training, Demonstration, and Research). The director of PADE is furnished by CIDIAT, with supporting staff from INCORA and Utah State University with U.S. AID financial support.

The Colombian experience evoked interest in Brazil, and the Under Secretary of the Ministry of Interior of that country traveled to CIDIAT and to Colombia to

TABLE 5

Statistical Summary of Participants by Country at CIDIAT High-Level Seminars

Countries	Logan 1964	Mérida 1965	Logan 1967	Mérida 1970	Total
Argentina	1	1	1	1	4
Bolivia	2	1	2	..	5
Brazil	2	..	2	..	4
Chile	2	1	1	1	5
Colombia	3	1	2	..	6
Costa Rica	..	..	1	..	1
Dominican Republic	1	..	..	..	1
Ecuador	1	..	1	..	2
Guatemala	1	1	..	..	2
Honduras	..	..	1	..	1
Mexico	..	1	1	1	3
Nicaragua	..	2	..	..	2
Panama	1	1	..	..	2
Paraguay	1	1	1	..	3
Peru	3	2	3	1	9
Puerto Rico	1	1	1	..	3
Uruguay	2	..	1	1	4
Venezuela	3	7	6	1	17
El Salvador	1	1	1	..	3
Total	25	21	25	6	77

Fig. 3. CIDIAT participants view an irrigation pump in Venezuela.

discuss the program. Brazil followed through with a request to the OAS to have CIDIAT organize a sub-center of CIDIAT in Brazil. Organizational work is still in process, and if the funds become available, another CIDIAT subcenter will become a reality.

OVERALL PHILOSOPHY OF CIDIAT COURSES

Experience with all CIDIAT training showed that the teaching process has to be oriented to the needs of professionals—of mature, responsible adults. Class schedules therefore provide ample opportunities to discuss and absorb class material. Large groups are subdivided to encourage additional discussion and exchanges of ideas and experiences. The structure of each course is kept flexible to allow changes where necessary to meet the needs of the participants. It is admittedly difficult to provide students with sufficient background in all areas of study related to land and water development in interdisciplinary training programs of short duration. CIDIAT courses are therefore designed to provide at least sufficient material in each discipline so that those not knowledgeable in that area can understand its importance and potential contributions to the development process.

For example, instruction in agronomy emphasizing the relationships between plants, soil, and water, is provided to engineers who usually concern themselves with only the physical features of the project. Stress upon the fact that structures and distribution systems only serve to provide water to farmers for use in growing crops encourages engineers to consider the end use of the developed water and not only the engineering aspects of the facilities to divert, store, and transport it. The economic aspects of water-resource systems are presented to all participants so that alternative uses of money can be evaluated by decision makers and appreciated at all levels. In addition, all courses include a case study that provides the participants with a realistic opportunity to test new concepts and learning relative to project formulation.

Introductory courses in sociology provide engineers, agriculturalists and economists with insights into people-oriented problems that must be solved before a project is successful. Often the social problems present the greatest challenge, especially when the people who will live with a project are never consulted about their part in the development process.

CIDIAT awards 25 scholarships per course for each regular six-month course and for each two-month short course. The scholarship includes travel from place of residence to Mérida and return, housing and a living allowance at Mérida, accident insurance, books, and classroom materials. It is customary for the home institution to continue the full salary for the individual during the course.

Brochures sent to selected agencies outline the next course to be offered and provide sufficient information to assist administrators in selecting applicants. Applications come mainly from government agencies and institutions, although private institutions and agencies receive consideration as well.

A committee comprised of representatives from CIDIAT, the Program of Technical Cooperation, and Utah State University reviews the applications and awards the scholarships. The committee attempts to match participants to the objective of the course in order to obtain as much group homogeneity as possible. This helps minimize unproductive conflicts within the group and fosters a better learning atmosphere.

To test the value of course content and the effectiveness of teaching procedures, the CIDIAT staff adopted a system of student evaluation for national and international programs. Each lecture, or contribution by a staff member is evaluated by the participants. The material is rated with reference to the student's previous knowledge of it and according to its degree of helpfulness to him. The professor is rated on the effectiveness of his presentation. The results of the student course evaluations are used during the staff's planning conferences. Professors' ratings are used to stimulate the development of better methods and techniques in teaching.

The participant evaluations have stressed the advisability of providing competent professors who can teach in Spanish. Simultaneous translation is often inadequate in projecting the desired message for non-Spanish-speaking teachers.

EVALUATIONS AND OBSERVATIONS

The chain of events that led to the sponsoring of CIDIAT by the Program of Technical Cooperation of the OAS and the government of Venezuela indicates a logical and sound approach to international cooperation. The OAS is duty bound to promote economic security for its member nations; it depends upon host countries to provide sites for its programs. Venezuela, by accepting the responsibilities of being host country, including the financial obligations this implies, displayed its willingness to work with and support the OAS. It is possible that a host country could place undue pressures upon such a center to give considerable attention to national problems. Thus far this has not been a problem for CIDIAT. Both the government of Venezuela and the Universidad de los Andes have given CIDIAT every opportunity to develop and carry out its programs.

Utah State University was chosen as the active advisory agent because of its impressive accomplishments and rich background in land and water problems. Although climatic conditions of Utah are not the same as exist in the humid areas of Latin America, the water problems of the arid portions of Latin America resemble those of Utah. For instance, the Río Negro Valley of Argentina, the coastal area of Peru, and the irrigated areas of Chile all have problems similar to those encountered in Utah. In addition, the fundamental principles governing water problems do not change. Nevertheless, judicious selection of staff and consultants was also essential to help negate the lack of absolutely comparable geographic and climatic features between Utah and South America in general.

Undoubtedly another factor in CIDIAT's success was having the concept of the Center generated by the Latin American countries themselves. Although the overall production of food in relation to population growth presents a gloomy picture, any solution imposed from without would be resented. The plan to expedite the training of personnel in management of land and water resources was enhanced by knowledge that money was available to develop sound resource projects.

The early use of South American consultants to advise Utah State assured invaluable insights into the problems of South America and resulted in recommendations that focused the curriculum on real needs. The followup work by selected consultants provided CIDIAT with course material oriented to specific Latin American conditions.

The first High Level Seminar deliberately brought together heads of departments of irrigation, directors of agrarian land reform agencies, university professors, and others knowledgeable about their countries' problems. By continuing contact with these leaders, CIDIAT has gained excellent sources of potential participants and even some staff members. These seminar participants have become legitimizers and firm supporters of CIDIAT's programs.

The patterns of involving the Latin American leaders in planning the training program and of subsequently training members of their staffs were designed to promote a vertical integration of thought and action within each country's land and water agencies. High-level administrators were exposed to the same philosophy of development in much the same manner as individuals at midmanagement and technical levels. This gave a basis for understanding on both sides when changes and innovations were suggested.

In reviewing the program, it is obvious that the desire of the Universidad de los Andes to expand and become involved in international programs was important to CIDIAT's success. The university administrators saw the cooperative program with the OAS as an opportunity to expand its international activities and provide additional services to its own country. They were fully aware that their staff would need training in this new area and that additional facilities would be needed on campus to support the program.

Finally, the project would have faltered and perhaps failed without adequate backstopping and technical guidance during its formative stages. These have been the functions of Utah State University.

Site Selection

Of the five countries approached to host the Center, Venezuela showed the greatest interest and was most willing to provide support. Also, Venezuela is well situated from the standpoint of travel from South America and Central America. The recently completed underwater cable provides excellent telephone communication

Fig. 4. CIDIAT participants from several Latin American countries gain experience with irrigation map, Majaguas project, Venezuela.

between Caracas and the offices of the OAS in Washington, D.C.

Venezuela also has a wide range of climatic conditions as described earlier. The country has many water projects that encompass a wide potential of educational experiences in planning, implementation, and operation. Each training course has included at least one field trip to the developing areas of Venezuela.

Mérida as the site for the Center has advantages and disadvantages. It is approximately 720 km (445 miles) west-southwest of Caracas, 8 degrees north of the equator at an elevation of 1,320 meters (5,300 ft). The climate is pleasant, with an average annual temperature of about 18°C (68°F). Rainfall is approximately 1,520 mm (65 inches) annually. The Universidad de los Andes is the major activity within Mérida. One should not pass over lightly the advantages of locating the center on a major university campus. The operation can thus draw on facilities and staff that would require duplication at most non-university sites. The political activities of the students, however, occasionally disrupt the Universidad de los Andes and cause loss of study time; but this has not yet affected the operation of the Center.

Disadvantages of Mérida include the time required to travel from Mérida to the major project areas of Venezuela. Participants, also, prefer the larger cities where there is more to do on weekends and holidays. Many potential visitors to Mérida bypass the Center since it is out of the way and an extra day or two is required to make the trip.

Evaluation and Recommendations of the Third High-Level Seminar

To provide CIDIAT with feedback and to facilitate re-examination of curricula and procedures adopted, a High Level Seminar was held at Logan, Utah, during September 1967.

Participants in the Seminar were asked to evaluate the CIDIAT program and to recommend possible program changes. The third seminar's recommendations are summarized as follows: (*a*) continue periodic high level seminars; (*b*) orient the two-month course to directors of planning departments, project leaders, or project directors; (*c*) review and modify or eliminate the regular course in favor of more national courses; (*d*) intensify national course training efforts, include subprofessional training, and carry out such courses through local universities; (*e*) participate in research in water management, human social factors, economic factors, water pollution and quality control, and water rights and legislation; (*f*) provide technical assistance in water resource development on an increasing scale to development organizations and to universities concerned with graduate and undergraduate training programs in water resources; (*g*) compile and distribute (and publish if appropriate) data and research results pertinent to land and water development; (*h*) continue CIDIAT's program on a permanent basis and assure finances for the Center.

FUTURE ROLE FOR CIDIAT

CIDIAT's role, while admittedly small in relation to the burgeoning need for training, is sufficiently important to justify continuation and expansion. Under terms of the OAS-USU contract (*4*) and the OAS-Government of Venezuela contract (*5*), CIDIAT will develop a permanent educational institution at the Universidad de los Andes. Approximately four years are estimated necessary for smoothly transferring the responsibility and financing of the Venezuelan National Training Center from CIDIAT to the Universidad de los Andes. During this period (1970 through 1973) the international aspects of the present program at Mérida will necessarily assume lesser importance, even though some international activity will continue, especially in the National Training Program.

A continuation and expansion of CIDIAT as a permanent organization has been proposed. Caracas, Venezuela, is one possible site for the new CIDIAT headquarters. The present commitments of CIDIAT would continue to be honored with such a center, and additional activities would be undertaken to further facilitate the development of land and water resources in Latin America.

A full-time coordinator from the Universidad de los Andes (ULA) has been working with a committee of professors and CIDIAT personnel to develop plans for the program in land- and water-resources at ULA. As plans are approved and implemented, ULA will gradually assume full responsibilities for the program. Selected ULA staff members are involved with presenting CIDIAT courses; others are taking advanced training as preparation for the new university undertaking. A building to house the resources program was slated for construction on the ULA campus during 1969 or 1970. It is expected that the permanent program at ULA will consider strengthening undergraduate training, developing continuing educational programs at various levels, giving graduate training, conducting research programs, and providing some technical assistance to national and regional development organizations of Venezuela. Students from other countries will be welcome to participate in the training programs.

The creation of a national training center in Colombia with CIDIAT assistance hopefully portends a trend. Countries actively involved in land and water development and agrarian land reform programs do not have the personnel to cope with the details involved. As a result, much of the work is contracted to private firms and international agencies. Local technical competence must be developed to take over the operation of such projects and provide continuity to the overall programs. Much of the competence that is lacking can be provided by short course programs and on-the-job training. National centers can satisfy these needs on a continuing basis. Such centers should work in conjunction with local universities or educational centers where possible. They should not try to usurp the basic training offered by local universities. Instead, they should augment the universities in areas of specific need and provide additional opportunities for professional improvement and growth. Each national center should be flexible enough to develop the types of courses needed by that country.

CIDIAT would provide technical guidance and support to the national centers during their formative stages as needed. It would make professors available in special areas on a short-term basis. It would provide assistance to national universities interested in developing and/or improving programs in land and water resources. The ULA (Universidad de los Andes) resource program would be considered as a national center and be entitled to receive continuing support from CIDIAT.

CIDIAT's program of providing national training courses to countries upon request would continue, since not all countries should or would develop national training centers at the same time. CIDIAT would also continue to supply technical competence in the latest technology available. It would lead in demonstrating the applicability of such technology to local problems of resource development.

The 1967 High Level Seminar recommended that CIDIAT involve itself in research, since a serious lack of data and information hampers the solution of development problems in Latin America.

Very little, if any, reliable information is available regarding the maximization of crop production under irrigation. Although drainage problems exist in every country, little is known on how to handle them. Experience, so necessary to designing effective subsurface drainage installations, is practically nil, yet the success or failure of many projects involving irrigation depends on the associated drainage system.

CIDIAT would cooperate with universities and research organizations to identify and initiate research in strategic areas. Many graduates returning to their countries with the doctorate, when left without some guidance and support, fail to develop research competence. Follow-through and backstopping by an international center could provide support and technical guidance to such persons. The proposed cooperative programs would train researchers in addition to generating data and information.

The need for technical assistance in land and water resource development will continue and undoubtedly increase with time. CIDIAT has experienced good success in the past with national courses that provided some technical assistance as well as training classes. If professors or consultants assigned to a national training course can be on the job one to three weeks in advance of the course, they can give technical assistance to the project and use the information obtained to illustrate lectures or laboratory exercises or in problem assignments. Thus both the project and participants in the course reap benefits.

A third benefit inherent in such a program is that the staff members involved become acquainted with field problems and enhance their own abilities to provide service.

Another of the 1967 Seminar's recommendations was to involve CIDIAT in a program of gathering available information pertinent to land and water development and sending it to each country. The need for such a program is self-evident. The future program of CIDIAT would fulfill this recommendation and carry it further.

The emphasis on research would soon produce a continuing flow of data and information. This information would be collected and made available to interested agencies and people. Researchers and extension specialists would be encouraged to publish their work to increase the flow of data and information on Latin American problems.

A serious lack of up-to-date textbooks and teaching materials in the Spanish language exists in the land and water field. CIDIAT would develop textbook materials and publish these for dissemination and use by educational institutions and technical personnel.

A library tie-in would be made with sources in the United States and other countries to keep current with available pertinent information from abroad. This material would be screened and made available in Spanish for use in Latin America.

In summary, a continuing CIDIAT program would require a change in orientation while capitalizing on past experience with, and acquired understanding of, Latin American problems. It would continue to backstop the program at Mérida, Venezuela, but reduce its direct responsibilities as Universidad de los Andes assumed leadership. It would act as a catalyst in other Latin American countries to generate national training centers oriented to national needs. Ideally, the CIDIAT program of national and regional training courses would be continued at an accelerated rate in cooperation with local universities and development agencies. Local universities wishing to strengthen their undergraduate course work or develop competence in graduate training for work in relevant fields could receive technical assistance from CIDIAT. Development agencies could also receive technical assistance, preferably in connection with the sponsoring of a national training course. CIDIAT personnel would instigate and cooperate in research and help to identify problems needing study. In these ways CIDIAT could make an increasing and lasting contribution to the economic, social, and perhaps even the political stability of Latin America.

REFERENCES AND NOTES

1. GOODSPEED, S. S.
   1954   The nature and function of international organization. 2d ed. London, Oxford Press.
2. INTER-AMERICAN DEVELOPMENT BANK
   1967   Agricultural development in Latin America: the next decade. Round Table, Inter-American Development Bank, Washington, D.C., April 1967. 290 p.
3. PRESIDENT'S SCIENCE ADVISORY COMMITTEE
   1967   The world food problem. Vol 1, Report of the panel on the world food supply, The White House, May 1967. Washington, D.C., U.S. Government Printing Office.
4. Agreement between the Secretary General of the Organization of American States on behalf of the Program of Technical Cooperation of the Organization of American States (hereinafter referred to as the PRO-GRAM) and Utah State University (hereinafter referred to as the UNIVERSITY) for the technical support, guidance, and administration of Project 213 of the Program of Technical Cooperation.
5. Agreement between the government of Venezuela and the Secretary General of the Organization of American States for the Establishment in Venezuela of the Interamerican Center for Land and Water Resource Development, of the Program of Technical Cooperation of the OAS.
6. INTERAMERICAN CENTER FOR THE INTEGRAL DEVELOPMENT OF LAND AND WATER RESOURCES
   1967   Innovations in executive training in Latin America. Project 213, Program of Technical Cooperation of the Organization of American States. Directed by Utah State University, Logan, Utah. Mérida, Venezuela.

Organization of Educational Problems in Sparsely Settled Areas of the World

EVERETT D. EDINGTON
Educational Resources Information Center
Clearinghouse on Rural Education and Small Schools
and Department of Educational Administration
New Mexico State University, Las Cruces, New Mexico, U.S.A.

ABSTRACT

THROUGHOUT THE WORLD the rural person is two or three years behind urban dwellers in educational attainment. This situation is especially acute in sparsely populated areas where large distances and few people make it almost impossible to provide adequate educational programs for the inhabitants. Some primary programs are now beginning to appear, but secondary and post-high-school programs are almost nonexistent.

Two of the major problems in remote areas are lack of finances and scarcity of trained teaching staff. Transportation also continues to be a problem.

In the United States the number of small rural schools is rapidly decreasing, primarily due to consolidation of existing school districts. Throughout most of the world rural schools are under the jurisdiction of the Ministry of Education rather than local people.

A number of promising educational programs are emerging in sparsely populated areas of the world. Some of these are financed by Unesco in underdeveloped countries. Others are being conducted in the western United States, Australia, Alaska, and the Arctic regions.

Ministers of education throughout the world are beginning to realize the importance of special educational programs for rural populations in order to upgrade the educational achievements of their total populations. Those aspects of education that seem to be receiving the greatest amount of emphasis in rural areas of the world are teacher education, organization and administration, vocational education, and economic and social development.

ORGANIZATION OF EDUCATIONAL PROBLEMS IN SPARSELY SETTLED AREAS OF THE WORLD

Everett D. Edington

NEED FOR EDUCATION

IN THIS DAY of the knowledge explosion, the major deterrent to an individual's progress is lack of opportunity for an adequate education. Throughout the world the rural person is two, three, or even more years behind his urban cousins in educational achievement. This statement is even more often true for sparsely populated rural areas where the large distances and few people have made it almost impossible to provide economically a program that would enable the people in these areas to compete. Recently government officials and others have begun to realize that it may be far more economical to provide a basic education for these people than to allow them to fall further and further behind. Throughout the world, large numbers of rural children are no longer returning to the land to make their living. Many of them now need to be prepared for other settings in which they will live most of their lives (1).

Developers of new agricultural or industrial areas are concerned with the total development of an area in addition to only the natural resources. Included in both longrange and shortrange planning are programs to meet the educational, medical, and social needs of the individuals involved. In order to attract the types of persons needed, from the laborer to the manager, it is important that adequate educational programs be developed. To encourage people to establish homes in an area, it is becoming increasingly more important that adequate social programs preexist.

Only about one-third of the world's population lives in countries where a complete primary education is provided for children in rural areas. Elsewhere, children living outside of towns either do not go to school at all or else attend schools where teaching does not cover more than two, three, or four years of study (2). This situation is more widespread in the underdeveloped countries in Asia, Africa, and South America.

Availability of Schools

A Unesco study published in 1964 revealed that over half the population in 59 out of 82 countries was rural, and that over 70 percent was rural in at least 32 of the 59 countries. The same study also brought out the fact that most primary students did not leave their home areas to attend school; this means that many children in sparsely populated areas did not have schools to attend (3).

In the United States a small number of "one-room schools" still exist, primarily in thinly populated agricultural areas. Most of these schools are disappearing, however, as consolidation takes place. Buses are used to transport hundreds of thousands of rural children in the United States daily. Many are transported 30 to 40 miles one way to their schools.

Havinghurst (4) found in 1965 that many rural schools in Brazil had only a single room and single class that might contain pupils from four primary grades. A second type of school building had two or three classrooms, with one teacher in charge of each. In this case some classes might be combined, or one class attend in the morning and another in the afternoon. The majority of the primary schools in Brazil are rural. Secondary schools are almost nonexistent in rural Brazil. In order to attend secondary school, most rural children must live in town with friends or relatives.

In 1962 the Committee on International Relations of the National Education Association reported that in Afghanistan fewer than 10 percent of all Afghan boys were enrolled in school, and education of girls outside the home was a recent development. Some of the country's primary schools existed in rural areas, but secondary schools operated only in Kabul and the provincial capitals (5).

The educational situations in Brazil and Afghanistan are similar to those found in many of the less-developed countries of Asia, South America, and Africa. In 1962, for example, 100,000 children of school age in the Sahara did not attend school (6).

Relevance and Cost

Another important factor to consider is that large numbers of rural children will no longer be returning to the land to make their living, and the educational programs in the rural communities should prepare them to make a living in a more urban setting. The school should provide an agricultural type curriculum for those who will remain on the land, a vocational one for those seeking immediate employment after leaving the area, and an academic curriculum for those continuing their education.

The cost of education is becoming an extremely im-

NOTE: This chapter is an adaptation of a presentation at "Arid Lands in a Changing World," an international conference sponsored by the Committee on Arid Lands, American Association for the Advancement of Science, June 3–13, 1969, University of Arizona, Tucson, Arizona, U.S.A.

portant factor in determining which educational opportunities are to be made available. Distance increases the cost for all services that a rural community requires, and education is no exception. Those communities in sparsely settled areas often have the fewest resources and the greatest need for more adequate educational programs.

Other research has shown a much higher level of educational wastage for rural than urban areas. Studies conducted in Burma, Ceylon, and Iran all found this to be true (7). It is generally recognized that living in a rural area often diminishes a person's chances of acquiring a complete education.

SPECIAL EDUCATIONAL PROBLEMS

Due to the nature of sparsely populated areas, a number of educational problems are unique to these areas or are intensified because of the particular situation. One of the most serious difficulties is that of obtaining adequate teachers. A Unesco study (3) found "that the lack of teaching staff and the difficulties of recruitment in backward regions lead to the appointment in rural areas of teachers who have little or no training, or whose qualifications are often inferior to those teachers in towns."

Throughout the world persons trained as teachers prefer to live in the cities and larger towns where facilities are more modern. The majority of teachers come from the populated areas and have no desire to work in the more backward rural areas. Those persons who come from the rural areas also prefer to teach in the urban areas. In Nicaragua in 1966 only 35 percent of the teachers in rural areas were college graduates (8). Most of the rural areas pay lower salaries to teachers than do the urban centers of population.

Proper finance of education, a problem in all areas, is intensified in the sparsely settled regions. Because of the inefficiency due to small numbers of students, it is difficult to provide comprehensive programs in these areas. In 1965 the average cost per student in Alberta, Canada, was above the average cost per student in the urban areas. The cost will probably continue to increase as rural areas attempt to compete with urban areas for teachers and as the demands for vocational and other specialized courses increase (9).

The lower educational level of the rural student is no longer a problem for the rural area only. Both Abramson (10) and Lindstrom (11) point out that rural people migrating to the urban centers are often limited in skills, thereby complicating economic and social adjustments of the migrants. Lindstrom also points out "that systems of support on a wider equalization base must be developed, for rural areas, especially the poor land areas, contribute a large number of youth to urban areas" (11).

Transportation of students has always been a barrier, but a number of unique programs in the experimental stage offer some hope. Better roads, being built in many remote areas, greatly increase the use of buses.

Hobart (12) discovered that children in boarding-school programs undergo drastic changes that are difficult for the parents to understand. He found four possible kinds of changes that a child moving from an Eskimo settlement to a boarding school, and back, might experience. These may be (a) physiological changes in the way of body functions; (b) social-psychological changes in the child's sense of personal security, his attitudes and motives, and his way of relating to other people; (c) changes in his moral conceptions in what he will do and not do; and (d) nonmoral cultural changes — changes in the skills, abilities, and expectations from life. Hobart found also that school was a completely frustrating experience for the students from the more isolated villages. These students were made unfit to live in the land camps that were their homes, but they were not adequately prepared to make a successful transition to wage work in town.

ADMINISTRATIVE ORGANIZATION OF EDUCATIONAL PROGRAMS

The organization of schools in remote areas is far from being the result of a deliberate attempt to adapt and improve the work of the school. Usually it is due to the pressure of circumstances, which tends to render the rural school incomplete and thus deprive children of an adequate education at both the primary and secondary level (2). In some instances such organization may be caused not only by the geographic situation but also by the reluctance on the part of the members of the community to associate with other communities in school consolidation. Improved transportation has made this change possible much faster than many people are willing to accept.

Schools for children from nonurban areas have quite commonly one, two, three, or four classes. In the United States the number of one-teacher schools decreased rapidly from 148,711 in 1930 to 74,832 in 1948 to 15,018 in 1961 (13). Seven plains states of the Midwest had 53 percent of the one-teacher schools in the country in 1961.

A growing tendency is to group pupils from rural areas in central schools. The two most-mentioned ways of doing this are to transport the students by bus or to provide boarding-school establishments. In some cases traveling teachers are being used (3).

The Bureau of Indian Affairs has three programs to provide for the education of the Indian youth on the reservations in the United States. They are (a) boarding schools away from the reservation, (b) bordertown schools that provide education for the Indian youth at public schools while they live in dormitories, and (c) schools on the reservation. A 1965 report showed that the average cost per pupil was $1,409.73 in boarding schools, $1,420.62 in bordertown schools, and $690.01 in reservation schools. The Bureau is building more public schools, especially at the elementary-school level (14). In many cases inadequate roads are creating problems for bus transportation to these reservation schools.

In only a few countries do rural schools come under an administrative department distinct from those dealing with urban schools. In most cases they fall within a department under the Minister of Education. Where educational administration is decentralized as it is in the United States, Scotland, and Switzerland, some educational districts will be wholly rural (2).

Almost all of the rural schools in Jordan are supported by village funds. Formal schooling there is much more advanced for boys than girls.

Australia has no local boards of education, but each state of the commonwealth has a director of education with a professional staff. Australia has an extensive correspondence-course program for some 10,000 students in remote areas. These students are unable to attend regular school. Cook indicated that in the centralized Australian system, education in the sparsely settled arid and semiarid regions probably fares better than it would under a system of predominantly local support (15).

In Israel each kibbutz has its own elementary school up to the sixth grade. Upon graduation from elementary school, the children have completed their compulsory state education requirements. The children can then attend consolidated secondary schools, which may have from 150 to 300 adolescents (16).

A tendency growing through some of the world is to provide more boarding schools for the students from remote areas, even though this practice is beginning to decrease on Indian reservations in the United States (as more schools are being constructed on the reservations themselves). Most of these boarding-school programs are at the high-school level, with emphasis on technical and agricultural education. In the sparsely settled areas of some of the northern European countries, however, the boarding schools have not been satisfactory. In Norway they have proven very expensive. Because the parents, antagonistic toward such schools, are refusing to send their children, the Norwegian government is abolishing boarding programs as much as possible and is establishing additional schools in some of the very small villages. The program aims for a maximum amount of 1¼ hours travel each way for students (17).

In most cases the overall curriculum and syllabi of the various subjects are the same in town and country. The major difference is inclusion of agricultural courses in the rural areas for more than half of the countries in the world (2). In Africa agriculture is now appearing in the curricula of many schools, even at the primary level (18). In Israel, three to four year continuation agricultural classes are often located in agricultural settlements (19).

Cook reported that the method of administration of public schools at the state level in Australia has considerable significance for the sparsely populated and semiarid regions. The general pattern is that the state (within the commonwealth) undertakes to provide the school plant and teachers for an area containing a specified minimum number of pupils. This system often includes the maintenance of living accommodations for the teachers in the more remote places (15).

In North Africa projects for the development of the Sahara soon created a vocational education problem. Vocational training centers for both boys and girls have been set up to meet these needs. Vocational education is one of the most rapidly growing curriculum fields in rural areas of the world. It is a problem in the more sparsely settled areas to have enough students to justify the expense of the costly vocational programs. In an evaluation of the Jimma Agricultural-Technical High School in Ethiopia, the author found there was need to increase practical work training in the vocational schools. This need was especially critical for the graduates who did not continue into higher education. The skills should be closely related to the type of agriculture in the country. This work experience could add productivity that might offset the higher costs of the vocational training (20).

PROMISING PROGRAMS FOR EDUCATION

Innovative programs in rural education are beginning to emerge. These are sponsored largely by Unesco in the underdeveloped countries and by the Elementary and Secondary Act in the United States. Doubtless numerous promising projects are being developed by the various ministers of education throughout the world, but information is difficult to obtain on the majority of these.

One of the oldest and most extensive programs of education in the sparsely populated regions is carried out in Australia. For a number of years in New South Wales, Mobile Instructional Units on wheels served the remote areas. A number of these were vocational and technical training facilities that traveled on railroad cars or trucks. As population increased, many of these units developed into permanent facilities for vocational or technical education (16). Another program pioneered in Australia is that of radio and correspondence courses. The parents are usually responsible for supervising the work, with occasional visits by a traveling teacher.

The State Department of Education in Utah has developed an experimental Mobile Office Education Unit, which is nicknamed MOE. This series of trailers, when joined together, forms a modern office to provide simulated office experiences for the students. The unit travels between three or four different schools and provides experience in a modern up-to-date office that would be impossible for the schools to offer individually (21).

In Omdurman, the Sudanese have built a teacher-training institute designed especially for desert conditions. The design aims at delaying the heating-up process to insure lower temperatures in classrooms during teaching hours. This effect has been achieved largely by roof protection against the sun's radiation through insulation and by special wall thicknesses. A heavy wall facing the prevailing angle of the sun's rays causes the temperature during the day to rise more slowly inside than outside.

During the cold nights, when this same design is used for dormitories, inside temperatures drop more slowly and never reach the extremities of the outside. This entire facility is a laboratory for research on school buildings in arid climates; personnel there, in addition to other duties, analyze space costs for secondary schools in different regions of the world (*18*).

A recent study (1967) recommends the development of a regional high school system for rural Alaska. Each of the regional schools would provide boarding facilities for students. The study recommends six regional boarding high schools for 1975 and eight additional ones by 1980 (*22*). A report from the Alaska State Department of Education pointed out that such a plan may not be feasible, and that more smaller secondary schools are needed closer to the students' homes (*23*).

Alaska is probably doing more to provide education for persons in remote areas than is any other state in the United States. The following is a description of some of the problems involved and programs initiated by Alaska. This information comes from a 1968 report of the Alaska State Department of Education (*24*).

In 1960 only 34 percent of Alaska's native children, ages 14 through 19 years, were enrolled in secondary schools. In 1969 the Department of Education operated village schools across the state, making the rural system the largest school district in area in the United States. "One hundred seventy-three rural day schools are maintained by the State and Bureau of Indian Affairs for elementary pupils. With the exception of the nine largest villages, rural school children must be transported from their villages to metropolitan areas in order to attend high school." In spite of this hardship, surveys show that fewer than one percent of Alaska's native eighth-grade graduates fail to enter high school.

One example of what is happening in Alaska is the Beltz Regional Boarding School at Nome. The school consists of academic and vocational education facilities, a cafeteria, a dormitory for 170 students, and apartments for supervisors and teachers. Such a school must provide for the complete life of the student, which includes recreation and home life as well as the academic. Even strictly academic courses need to be slanted toward practical everyday learning situations. Because this is the last academic preparation the students have for their lives in the north, a great deal of emphasis is given to vocational education. Complete courses combine the academic and the vocational training.

A problem that arises in boarding schools is the frustration and bewilderment caused by the sudden change from the small village to school and dormitory living. To help with this situation at Beltz, the Village Orientation Program attempts to introduce the native student and his family to school personnel and staff while he is yet in his home village. School personnel visit the village during the summer before school starts in order to acquaint the students with the program. They use slides, films, and other visual aids to give the student a more accurate idea of what the boarding school will be like.

Another program developed in Alaska is the Boarding Home Project, under which funds are made available to board students in homes in the larger towns where the schools are located. In the 1968–69 school year over 370 students were accepted in the program, financed by Title I of the Elementary and Secondary Act. After two years of project operation, the state could look on this as a major part of the longrange educational program.

The major problems encountered in this type of program are those stemming from acculturation. The student lives in two different worlds: his native village and his boarding home at school. In order to minimize these problems, a home-school coordinator works with numerous students to help them make the necessary adjustments. He becomes acquainted with the students before they leave the village, finds homes for them in town, and is available through the year to assist them. This program was initiated to serve the overflow of students who could not get into the dormitories at the boarding schools; it has become a regular part of Alaska's educational program.

PLANS FOR THE FUTURE

Emphasis increases on improving education in rural areas worldwide. With improved ground and air transportation, distances are no longer formidable. Major services such as education are much closer to the man who is living away from the center of population.

The goals of the Conference of African States for the Development of Education in Africa were as follows:

a) Universal primary education.

b) Twenty percent of all children leaving primary schools enrolling in secondary schools.

c) University enrollment reaching 300,000 (31,000 enrolled in 1961: 18,000 in Africa and 13,000 abroad).

Fewer than half the 25 million children of school age in middle Africa in 1961 were to complete their primary education. Fewer than three of every hundred were to attend secondary school. Two out of a thousand were to attain some higher education (*18*).

The Conference of Ministers of Education and Ministers Responsible for Economic Planning in Countries in Latin America and the Caribbean in 1966 recommended (*a*) that the differences still existing in the duration of the primary cycle in urban and rural schools, to the determent of the latter, be eliminated, and (*b*) that more widespread use be made of new educational techniques, like television, that can help counteract the scarcity of teachers, help in the work of teaching, bring urban and rural education to equivalent standards, and reduce costs (*8*).

Unesco has reported that Afghanistan, in its third five-year plan (1967 to 1972), provides for 500 new village schools and for transforming 512 one-teacher village schools into two-teacher schools. These are all on the primary level (*25*).

Included in the rural education activities proposed by Unesco for 1969–70 were the following:

a) There will be a "continuation of the experimental project on in-service teacher training at Makerere College (Uganda) with added stress placed on rural development. . . . The project will combine elements of in-service teacher training and the development of new course content, teaching methods, and materials." Member states in Africa will be invited to associate themselves closely with the development of this project by providing support to local study groups.

b) Another experimental project will be based in a predominantly rural country in Latin America, and will foster the use of audiovisual aids, including radio and television, and special prevocational training for boys and girls in the 11–14 age group living in rural areas. This program will last three years.

c) A regional workshop for Arab-speaking member-states will be conducted with the participation of Unesco experts and national specialists to examine the needs for further development of preservice and inservice training of primary school teachers, with particular reference to the improvement of primary education in rural areas.

d) A rural-oriented primary teacher-training institute will be held in Cameroon.

e) Numerous teacher education programs will be developed in Asia, Africa, and South America.

Throughout the world there is an increased emphasis on prevocational and vocational education (6). This area, along with organization and administration, will receive increased emphasis in the next four years. Dr. Abdel El-Koussy, Director of the Regional Centre for Educational Planning and Administration in the Arab Countries, in his report "Trends of Educational Research in Arab Countries," indicated that university research will focus on the current problems of administration and of economic and social development (26). This seems to be the trend throughout the world.

Those involved in agricultural development can ill afford not to have adequate educational programs in operation. It is important that basic educational skills be mastered by all the youth. Those persons choosing agriculture as an occupation need extensive training at whatever level they enter. The lower occupational levels are demanding more technical skills; the more professional levels are demanding higher strata of academic training, usually at the college or university level.

Adult education and extension classes also can be profitable in developing personnel in agricultural communities. Special programs with emphasis on individualized instruction must be developed for those sparsely populated areas where not enough individuals can be brought together for class instruction.

REFERENCES

1. LYONS, R. F. (ed)
   1965   Problems and strategies of educational planning: Lessons from Latin America. Unesco/International Institute for Educational Planning. 117 p.

2. INTERNATIONAL BUREAU OF EDUCATION (GENEVA)
   1958   Facilities for education in rural areas, a comparative study. Unesco, Paris. 241 p.

3. UNESCO
   1964   Access of girls and women to education in rural areas, a comparative study. Unesco, Paris. Educational Studies and Documents 51. 62 p.

4. HAVINGHURST, R. J., AND R. J. MOREIRA
   1965   Society and education in Brazil. University of Pittsburgh Press, Pittsburgh.

5. THOMPSON, E. M.
   1962   Other lands, other peoples; a country-by-country fact book. Rev 2d ed. National Education Association, Committee on International Relations, Washington, D.C.
   A third edition was published in 1964.

6. UNESCO
   1962   The problems of the arid zone, proceedings of the Paris Symposium. Unesco, Paris. Arid Zone Research 18. 481 p.

7. ———
   1967   Bulletin of the Unesco Regional Office for Education in Asia, 1(2).

8. ———
   1966   Conference of Ministers of Education and Ministers responsible for economic planning in countries of Latin America and the Caribbean. Final Report. Unesco, Paris.

9. ———
   1965   Resources for rural development — Census Division, Rural Development Section, Farm Economics Branch, Alberta Department of Agriculture, Alberta, Canada.

10. ABRAMSON, J. A.
    1968   Rural to urban adjustment. Department of Forestry and Rural Development, Ottawa. Agricultural Rehabilitation and Development Agency Research Report RE–4. 160 p.

11. LINDSTROM, D. E.
    1960   Rural social change. Stipes Publishing Company, Champaign, Illinois.

12. HOBART, C. W.
    1968   Some consequences of residential schooling. Journal of Indian Education 7(2):7–17.

13. AMERICAN ASSOCIATION OF SCHOOL ADMINISTRATORS
    1962   School district organization; journey that must not end. American Association of School Administrators/National Education Association, Department of Rural Education. 17 p.

14. U.S. COMMISSIONER OF INDIAN AFFAIRS
    1965   Report to the Senate Appropriations Committee on the Navajo Bordertown Dormitory Program. U.S. Department of the Interior, Bureau of Indian Affairs, Washington, D.C.

15. COOK, W. R.
    1969   An analysis of the adaptation of selected Australian institutional structures to conditions of semiaridity or Cook's tour through the land down under, under the influence of Kraenzel. Paper presented at the Rural Sociological Society Meeting, San Francisco, California.

16. NATIONAL EDUCATIONAL ASSOCIATION
1960 Improvement of rural life; the role of the community school throughout the world. National Education Association, Department of Rural Education, Washington, D.C. 100 p.

17. VESTBY, E.
1969 Rural school problems in Norway. Paper presented at Interskola '69, International Rural Education Meeting, Ørsta, Norway.

18. UNESCO
1966 Africa prospect: progress in education. Unesco, Paris.

19. KLINOV-MALUL, R.
1968 The profitability of investment in education in Israel. The Maurine Falk Institute for Economic Research in Israel, Jerusalem.

20. EDINGTON, E. D., AND I. E. SIEGENTHALER
1965 An evaluation study of the Jimma Agricultural High School in Ethiopia. Research Foundation, Oklahoma State University, Stillwater.

21. STEPHENS, J.
1968 Mobile office education. Paper presented at the meeting of the American Vocational Association, Dallas, Texas.

22. CUMMISKEY, J. K., AND J. D. GARCIA
1967 State of Alaska regional secondary school system: implementation plan; final report to Alaska State Department of Education. Training Corporation of America, Inc, Falls Church, Virginia.

23. ———
1967 Position paper on Training Corporation of American and Secondary Education for Alaska's rural youth. Alaska State Department of Education, Juneau.

24. KADEN, B.
1968 Rural renaissance — new opportunities for young Alaskans. Alaska State Department of Education, Juneau.

25. ———
1968 Bulletin of the Unesco Regional Office for Education in Asia, 3(1).

26. ———
1968 Bulletin of the Unesco Regional Office for Education in Asia, 2(2).

27. UNESCO REGIONAL OFFICE FOR EDUCATION IN ASIA, BANGKOK
1970 Education in rural areas in the Asian region. Bulletin 5(1).

28. SHEFFIELD, J. R. (ED)
1969 Rural education. Michigan State University, African Studies Center, Rural Africana 9.

Early Human Contacts from the Persian Gulf Through Baluchistan and Southern Afghanistan

GEORGE F. DALES

South Asia Regional Studies Department, South Asia Section, University Museum
University of Pennsylvania, Philadelphia, Pennsylvania, U.S.A.

ABSTRACT

THE PERSIAN GULF littoral and islands, Baluchistan, and southern Afghanistan comprise one of the lesser known of the world's arid zones. An archaeological approach to the area offers answers to questions of human settlement, agricultural practices, and trade contacts. The period from about 7000 to 1800 B.C. covers the transition from neolithic to metal-age economics and the rise and fall of the earliest civilizations in Mesopotamia and the Indus valley. Basic to the study are the "relative" and "absolute" chronological frameworks of the zone. The radiocarbon dating method is useful in determining dates and chronologies, although some uncertainties are inherent in the technique.

Prior to 3000 B.C., the major influence on the development of the western half of this zone, minus the Persian Gulf region, was Mesopotamia. As agricultural and related practices improved, the less hospitable ecological regions of southern Iran came under this influence. Meanwhile, the eastern half of the zone — which is very inadequately known — appears to have been developing independently in a sphere of influence which included parts of southern (Soviet) Turkestan.

From 3000 B.C. the entire southern zone, now including the Persian Gulf, became prominent as a transit area for "international" trade and as the source of many of the raw materials essential for the development and support of the Mesopotamian and Indus riverine civilizations. Coastal sea trade, perhaps controlled by the so-called Kulli people of Baluchistan, featured heavily in these contacts. Sumerian and Babylonian economic documents give an idea of the major imports and exports. By 1800 B.C., as the fortunes of the riverine civilizations waned, so did contacts throughout this arid zone.

EARLY HUMAN CONTACTS FROM THE PERSIAN GULF
THROUGH BALUCHISTAN AND SOUTHERN AFGHANISTAN

George F. Dales

THE PREHISTORY AND PROTOHISTORY of human contacts, from approximately 7000 to 1800 B.C., throughout the arid zone stretching from the Persian Gulf to the Indus Valley of West Pakistan are deserving of attention. The time period is of great significance relative to the origins and diffusion of agricultural techniques, plant and animal domestication, and the development of early societies. The era includes the rise and demise of two of the first great civilizations of Asia — the early Mesopotamian (largely, but not wholly, Sumerian) and the Indus (Harappan) civilization of Pakistan and western India. Studies in this zone can provide a testing ground for hypotheses concerning the role of deserts vis-à-vis major centers of urbanized civilization. Are they merely "refuge" areas, or do they serve a more significant function in relation to the affluent riverine civilizations? The establishing of the chronology, ecological situations, and subsistence levels of these early times, adds a necessary temporal dimension to any arid-lands research in the region.

These objectives imply a spectrum of research that includes not only traditional archaeology (bricks, bones, pottery, and gold) but the natural and physical sciences as well. Archaeology is not likely to lose its prime role as the collecting agent for ancient material remains, but the more scientific disciplines must contribute the environmental facts of life if we can ever hope to make meaningful reconstructions of human and societal development. Recent and current field work in western and southern Iran is already making exciting and promising contributions along these lines. But the Persian Gulf area, Iranian Baluchistan, Pakistani Baluchistan, and southern Afghanistan have yet to be liberated from the stagnant object-oriented approach of traditional archaeology.

GEOGRAPHICAL-ECOLOGICAL SETTING

The geographical zone (Fig. 1) under consideration here falls mainly within Meigs's classification (1) of "arid," with cool winters, winter precipitation, and warm or hot summers. The western and southern littorals of the Persian Gulf constitute an exception; those areas have no distinct precipitation season. The zone is part of a phytogeographical region with common characteristics extending from the Atlantic coast of North Africa to the Indus Valley (2). Such classifications are viable and necessary in worldwide perspective, but often not obvious

from the top of a camel! Camelback perspective fractionates the theoretical homogeneity into innumerable ecosystems, each with its distinct environmental characteristics that affect and influence human existence. There is nothing new in this statement, but it must be emphasized here, where the available evidence is still archaeological and deals primarily with restricted geographical locales. One example of the practical significance of this camelback perspective is seen in Raikes' (3) description of five distinct ecological divisions of Pakistani Baluchistan and the differences in human response to them.

Deserts of the World and *Arid Lands in Perspective* (4, 5) provide basic bibliographical information on the physical aspects of the Persian Gulf to Indus zone. To this body can be added the following:

Raikes (6) offers provocative new insights into the ecological picture of the Near East and South Asia from the standpoint of a professional hydrologist. Fisher (7) presents a comprehensive review of the physical and human geography of Iran. The Kirman region of southern Iran has been studied anthropologically by English (8) and archaeologically by Caldwell (9, 10). Barth (11) conducted a human ecological study near Shiraz in southwestern Iran. Hole and his associates (12) used an anthropological approach to the archaeology of northern Khuzistan. Spooner (13) and Italconsult Reports (14) provide excellent observations on the geography, agriculture, hydrology, and sociology of Iranian Baluchistan. Studies on the early history of agriculture in southwestern Iran and southern Mesopotamia are offered by Adams (15), Higgs and Jarman (16), Flannery (17), and Harlan and Zohary (18). The Indian Ocean coastline of West Pakistan has received intensive study by Snead (19–24), as has the same coastline of Iran. The Colombo Plan has produced an invaluable report on the geology of West Pakistan (25). A short geography of the country has been published by K. S. Ahmad (26). LeStrange (27) provides an old but useful introduction to the medieval Islamic geographical accounts, which provide valuable data on Iran, Baluchistan, and Seistan. The many accounts of nineteenth and early twentieth century travelers, political agents, and explorers (28–68) are mines of information on geography and general ecology, as are the volumes of the old Baluchistan District Gazetteer Series (69–71). Additional bibliography of more recent date is grouped as relevant to environmental (72–82) and archaeological (83–92) subjects.

CHRONOLOGY AND DATING

The backbone of history is an agreed chronology. . . . Without an absolute chronology, cultures of different regions cannot be accurately compared, their inter-relationship cannot be assessed; in other words, the vital causative factors of human "progress" cannot be authoritatively reconstructed, and may be widely misunderstood. *(93)*

This statement is especially applicable for the Persian Gulf to Baluchistan zone. Indeed, most of the attention of archaeologists working in the area has been devoted to establishing the chronological relationships of different "Cultures" (usually defined on the basis of their pottery types and styles).

It is only at the western end of this zone, in western and southwestern Iran, that archaeology has matured to the degree that excavation reports can have words like "human ecology" in their titles and be concerned with floral and faunal remains, agricultural practices, and subsistence levels, as well as with the mechanics of dating.

Chronology is basic to this study, however, and must be discussed first. There are basically two types of archaeological and historical dating. One involves "relative" chronology which simply means, "Is Cultural Complex-X earlier or later than Complex-Y, or are they contemporary?" Relative chronology is usually determined by comparing such factors as archaeological objects, art styles, and technological and subsistence levels from site to site, and from culture to culture. So-called absolute chronology ideally tells how many real calendric years ago something occurred. Historical documents, king-lists, coins — virtually any kind of intelligible written record — provide most of the evidence for "absolute" chronology. But for prehistorians and archaeologists working with preliterate or nonliterate societies, outside help is required.

Physics has provided several useful techniques for the determination of the age of archaeological materials. Radiocarbon analysis has been the most widely used — and abused.

Radiocarbon dating is a tool with numerous hidden flaws and limitations, which are often not known to or are not understood by archaeologists and historians who use it. Stuckenrath and Neustupný *(94, 95)* describe some of the inherent problems. More pessimistic are Stuiver and Suess *(96)* who conclude that "it is presently impossible to determine on theoretical ground what the relationship is between a radiocarbon date and the true age of a sample."

Part of this pessimism springs from discrepancies between radiocarbon dates and true dates when radiocarbon tests are made on samples of known age *(96, 97)*. The implication of recent tests is that radiocarbon dates older than 2,000 years ago are too young and that a correction factor will have to be added to the tens of thousands of published dates to bring them in line with true calendric dates. Table 1 (adapted from Michael and Ralph) illustrates the seriousness of the deviations.

Thus, for example, a published date of 2000 B.C. should be corrected to 2350 B.C. I have not added these corrections to the dates published here (Fig. 2), because the whole matter is still in the theoretical stage.

Considering the large numbers of radiocarbon dates available for many parts of the world, the southern Mesopotamia to western India zone is in a state of abject poverty. For the time range of this study there are only 94 logically consistent dates (Iran, 32; Afghanistan, 1; Pakistani Baluchistan, 8; Indus basin, 43; west coast India, 10); of these, only about one-third are from the specific geographical areas covered by this chapter. The remainder, mostly from the Indus Valley and western India, are essential, however, for determining the eastern end of our chronological framework. Figure 2 gives the dates in geographical order from west to east. Several other dates have been published, but they have been excluded here because of their hopeless inconsistencies or known defects in testing procedures.

All dates here are based on a half-life of 5,730 years, which appears to be more nearly correct than the 5,570 years half-life originally used *(94, 96)*. The exact dates and their tolerances and laboratory numbers are listed in the appendix. For additional details, the sample numbers can be easily found in *Radiocarbon (98)*. Huge gaps — both geographical and temporal — are obvious in this radiocarbon framework. New dates from southeastern Iran, all of Pakistani Baluchistan, and Seistan and the Persian Gulf are sorely needed. Especially disappointing was the failure of the tests on the original samples from Mundigak, the only large excavated site in southern Afghanistan. Now, results are awaited on the testing of several new samples, collected during the winter of 1968–69 from the earliest levels (Periods I and II).

Scattered as they are, these dates show an overall pattern that inspires some bit of confidence, especially at the extreme geographical ends of the zone. There is a reasonable consistency between these dates and those

TABLE 1

Proposed Corrections for Radiocarbon Dates

Time period (B.C.) represented by radiocarbon dates	Years to be added to radiocarbon dates to obtain "true" age *
1– 449	+ 50
450– 924	+ 50
925–1324	+ 100
1325–1699	+ 250
1700–2099	+ 350
2100–2499	+ 450
2500–2949	+ 550
2950–3999	+ 600
4000–4499	+ 750

* "True" age of first time period is thus 51 B.C.–499 B.C.

148

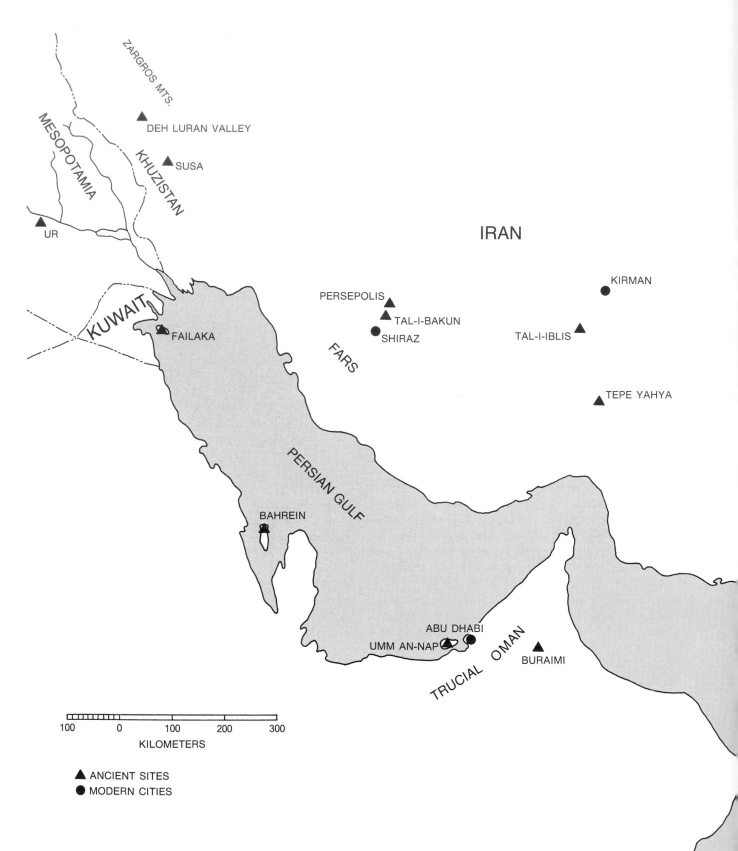

ZARGROS MTS.

MESOPOTAMIA

KHUZISTAN

▲ DEH LURAN VALLEY

▲ SUSA

▲ UR

IRAN

KUWAIT

▲ FAILAKA

PERSEPOLIS ▲
▲ TAL-I-BAKUN
● SHIRAZ

FARS

KIRMAN ●

TAL-I-IBLIS ▲

▲ TEPE YAHYA

PERSIAN GULF

BAHREIN

ABU DHABI
UMM AN-NAP ● ▲ BURAIMI
TRUCIAL OMAN

100   0   100   200   300
KILOMETERS

▲ ANCIENT SITES
● MODERN CITIES

Fig. 1. Ancient and modern cities of Iran, Afghanistan, and Baluchistan.

arrived at by the comparative or relative dating methods, but the addition of the newly proposed "correction factors" would throw the present overall correlations between Mesopotamia and the Baluchistan-Indus region into confusion. The addition of 350 to 450 years to the radiocarbon chronology for the Indus civilization would throw it off by that much in relation to the accepted chronology for early Mesopotamia, which is based on the

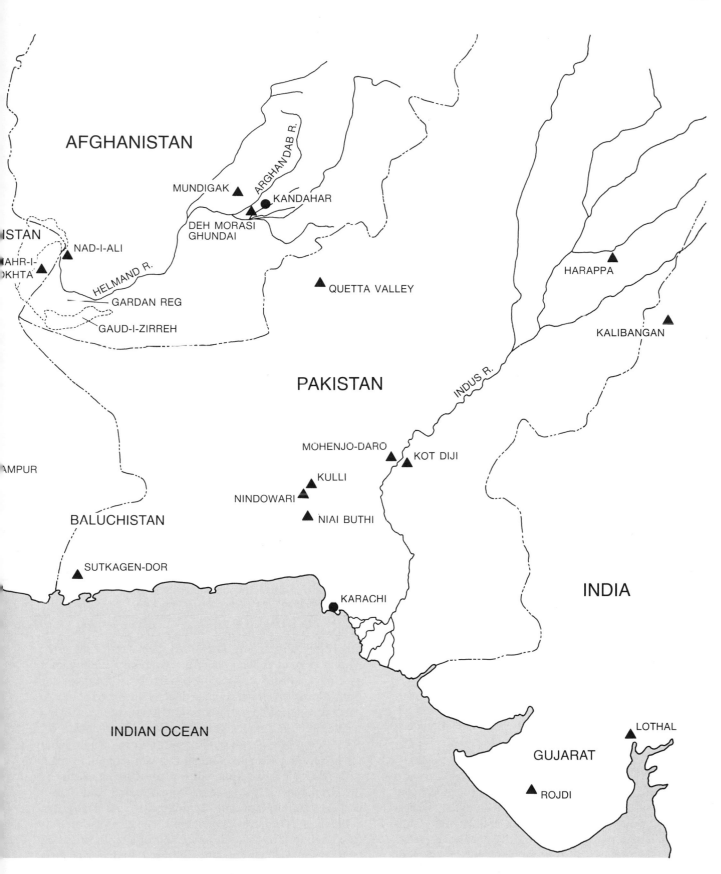

reconstruction of dynasties and reigns from Sumerian and Babylonian written records (99–101).

The details involved in determining the relative cross-dating between Mesopotamia, Iran, Afghanistan, and the Indus region need not concern us here (83–91, 102–109). Figure 3 summarizes the relative chronological framework as I interpret it. Descriptions of the individual sites and cultural phases follow.

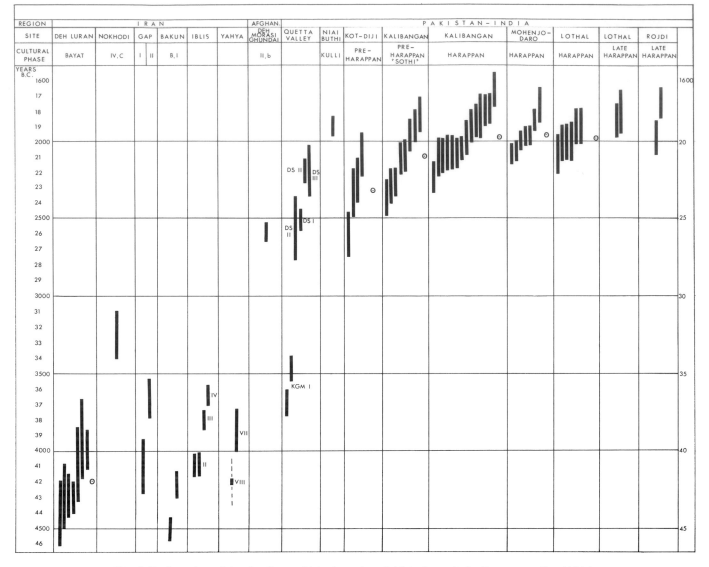

Fig. 2. Radiocarbon dates for the prehistoric and protohistoric periods. Recent studies (*184a*) indicate that pre-Harappan can, in some instances, more reasonably be called Early Harappan.

CULTURAL AND ECOLOGICAL HISTORY OF REGIONS AND SITES

With the help of Figures 2 and 3 we can now survey the available data concerning the early cultural and ecological history of several parts of the Persian Gulf-Baluchistan zone.

Southwestern Lowlands, Khuzistan, Iran

Khuzistan, one of the key environmental regions in the prehistory of the Near East, has yielded the earliest evidence for human settlements in the Persian Gulf-Baluchistan zone. I am excluding the occasional references to discoveries of stone implements of "palaeolithic" date. These are surface finds and are as yet not known from excavations that could confirm their antiquity. Some excellent geographical and environmental descriptions have been published (*12, 74, 75, 77, 78, 110, 111*). Here occur "some of the highest densities of prehistoric archaeological sites in the world."

Geologically, this is an easterly extension of the great Mesopotamian alluvial plain. It provides excellent winter grazing land and is a region admirably situated for the transition from dry-farming (dependent on rainfall) to irrigation agriculture. This is not, however, the region where the *initial* stages in this transition took place (*12, 15*). For those we must look north to the Zagros and other hilly regions surrounding the Mesopotamian plain (*16–18*). What we apparently witness in Khuzistan is the beginning of the gradual spread south and southeastward of agricultural techniques — specifically the cultivation of domesticated plants, the use of domesticated animals, and the development of simple irrigation techniques — which together allowed human societies to develop in regions less hospitable than the fertile regions to the north and northwest. Now this does not imply mass movements of peoples. Neither archaeological nor anthropological evidence forces such a conclusion. It was perhaps more a diffusion of ideas and techniques that were adopted (on a highly selective basis)

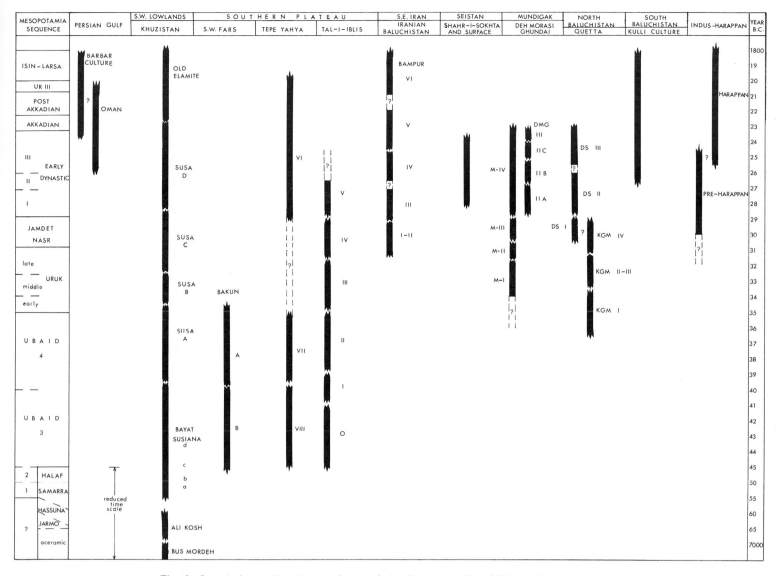

Fig. 3. Correlations of archaeological periods. Recent studies (*184a*) indicate that pre-Harappan can, in some instances, more reasonably be called Early Harappan.

and adapted by the sparse indigenous populations of the various subregions. We could expect a resultant rise in population, subsistence levels, and cultural development.

Archaeological evidence is beginning to document this spread of agricultural techniques and cultural advances from southern Mesopotamia and the northern littoral of the Persian Gulf eastward to Iranian Baluchistan, thus including southern Iran within the general sphere of early Mesopotamian influence. Pakistani Baluchistan and southern Afghanistan appear to have developed quite independently, however, having closer connections with northeastern Iran and Turkestan. As we will see, there is no solid evidence of significant contacts between the southern Mesopotamia / Persian Gulf / southern Iran region and the Pakistani Baluchistan / southern Afghanistan region until at least 3000 B.C. and possibly even as late as 2500 B.C. There is no solid evidence to support the claims so often made that all the basic discoveries and innovations of "civilized" man spread like waves of perfumed benevolence from the Near East to

South Asia. That type of assumption has hobbled South Asian research for far too long.

The recent excavations in the Deh Luran plain of northern Khuzistan (*12*) have uncovered the earliest known human settlements in this region, perhaps as early as 7000 B.C. Previous to these excavations, the archaeological sequence of the region was based largely on the excavations at Susa and a few small sites, plus collections of pottery made during surface surveys (*91, 105, 112, 113*). A flurry of archaeological activity is currently going on in Khuzistan, yearly notices on which appear in *Iran,* the journal of the British Institute of Persian Studies. Again, what is crucial to remember is that, as impressively early as the Khuzistan sequence appears to be, it still represents a later manifestation of cultural and technological developments which started at least a thousand years earlier in the territories to the north and northwest.

Already by the earliest settlements in northern Khuzistan (Bus Mordeh Phase), the inhabitants were practicing agriculturists who were planting emmer wheat and two-row hulled barley, which were *not* native to northern Khuzistan (*114*). They also practiced pastoralism (goats and sheep), hunting, and fishing. The presence of Turkish obsidian and cowrie shells — probably from the Persian Gulf — in this earliest phase indicates some type of long-distance contact. The excavators think that the copious local supply of natural asphalt might have been the basis for trade relations with Turkey.

The next cultural phase, called Ali Kosh, has a consistent set of radiocarbon dates, which range from 6900 to 6200 B.C. The floral remains (*114*) suggest that the cultivation of winter-grown cereals increased while the collection of small-seeded wild legumes decreased from that of the earlier settlement. Clay figurines of goats (?) and human females, plus the presence of formal human burials, reflect something — but what? — about the development of "religious" practices. A tubular bead of native copper helps substantiate evidence from other sites that simple metallurgy was practiced in Iran some 2,000 years before smelting practices were employed (*115–118*). The presence of obsidian, sea-shells, turquoise, and the native copper demonstrates the continuation of long-distance trade relations. So far, only one archaeological site of this phase is reported, and it was apparently a small village with not more than a hundred inhabitants, but this reflects more the paucity of archaeological investigations than the actual intensity of ancient activity.

The Mohammad Jaffar Phase, which follows, is so far represented by only three sites in the Deh Luran Plain, but it displays important innovations. Mud-walled buildings now had stone foundations. Pottery made its first appearance — soft, friable, straw-tempered wares known earliest from other parts of the Near East. Crude clay figurines with female attributes, well known from sites in the Zagros mountains to the north, appeared. Herding continued to intensify, with the statistical balance shifting to a plurality of sheep over goats. Emmer wheat and two-row hulled barley continued as the only cultivated crops, but there was a sharp increase in remains of the perennial legume *Prosopis stephaniana*, which may indicate either a change in the vegetational cover near the site or a change in eating habits.

The Sabz Phase, which follows (equivalent to the older designation "Susiana *a*"), was one involving major developments. The two radiocarbon dates are impossibly inconsistent, but comparison of the pottery with that of other sites in Mesopotamia and western Iran suggests a date of between 5500 and 5100 B.C. It was a period of significant population and territorial expansion. Six villages are known in the Deh Luran plain, each with at least 100 inhabitants, but Adams (*15*) reports 34 sites from other parts of northern Khuzistan. All of these sites are located where the highest annual rainfall occurs. Also, the village settlement pattern shifted from the edges of the seasonally flooded central basin upward to where advantage could be taken of the natural streams, but they still inhabited that ecological range which permitted them to continue rainfall agriculture and at the same time to experiment with simple irrigation techniques. The earliest indication of the local domestication of cattle also appears. Both irrigation and cattle domestication may have been "imported" from the lowland regions to the north and west. The pottery, still handmade, also suggests strong lowlands, southern Mesopotamian connections, while the longrange trade connections with Turkey (obsidian) apparently ceased.

With the following Khazineh and Mehmeh phases (Susiana *b* and *c*), which terminated at approximately 4500 B.C., villages in the Deh Luran Valley had doubled in number and size, but they were in a backwater area in comparison with the general developments in Khuzistan. About 100 kilometers to the southeast, in the plain around the famous archaeological site of Susa, at least 120 sites of this period are reported. Furthermore, at the same time in the southern Tigris-Euphrates valley of Mesopotamia, we see towns and temples springing up during what is termed the Early Ubaid Period. The combination of dry-farming and simple irrigation continues in Khuzistan. A few copper pins are found, but metallurgy was still in its infancy.

The final prehistoric phase in the northern Khuzistan sequence, the Bayat Phase (Susiana *d*) has a radiocarbon time span of 4500 to 3900 B.C. In the Deh Luran valley, the floral remains show a shift toward more halophytic plants, suggesting an increase in salinity in the cultivated areas. This may have been a major factor in the abandonment of the valley. But the rest of Khuzistan was flourishing. Pottery and artifacts point to strong ties with the Ubaid 3 cultural phase of the southern Mesopotamian plain. We now see the diffusion of Mesopotamian influences further south and east into the provinces of Fars and Kirman, along the southern edges of the Iranian plateau, and into the ecologically forbidding region of Iranian Baluchistan.

Southern Iranian Plateau: Fars Province

Geographical references are cited at the beginning of this paper.

Archaeological activity has been concentrated for years in the Persepolis-Shiraz area. Its elevation of approximately 1,500 meters sets it off ecologically from Khuzistan, where the earlier sites are found, but there is little or no information published on the early ecological situation. The emphasis has been almost exclusively on ceramic sequences and chronology. Extensive explorations are currently being made for prehistoric sites. It remains to be seen how much emphasis is being placed on nonceramic evidence. The so-called Bakun B-1 Phase represents the earliest documented human settlements in Fars. The two radiocarbon dates for Bakun B-1 give a maximum range of 4550 to 4150 B.C. Typologically this earliest settlement phase in Fars is equivalent to the Bayat ("Susiana *d*") phase of Khuzistan and the Ubaid-3 phase of southern Mesopotamia. To the east we also find typological equivalents at Tal-i-Iblis and Tepe Yahya south of Kirman. The excavations at Tall-i-

Nokhodi (*119, 120*) and Tall-i-Gap (*121*) both added ceramic and chronological information, but nothing directly related to the reconstruction of the ecological picture. The Tall-i-Nokhodi radiocarbon date is a century or so too young, to judge by the typological comparisons with other sites, but then a single radiocarbon determination is always statistically unreliable.

Southern Iranian Plateau: Kirman Province

Much of the present and current field work in the Kirman province was prompted by the discoveries of that intrepid explorer-archaeologist, Sir Aurel Stein, some 35 years earlier (*90*). The Bardsir valley is the only one running off from the Kirman basin in which extensive pre-Islamic settlement has been reported. The presence of perennial streams and occasional springs there may account for this restricted distribution. Probably it was not until the introduction of the underground water systems called qanats (kariz in Afghanistan) in Parthian times that other parts of the Kirman basin were extensively settled (*8, 92, 122, 123*).

One of Stein's sites, Tal-i-Iblis, in the Bardsir valley some 80 kilometers southwest of Kirman, has been re-excavated, with important results (*9, 10*). The Iblis reports are still preliminary, but the broad outlines are clear. First, the excavators have a semihypothetical phase called Iblis Period 0. No actual settlement has been found, but there are large quantities of the soft, friable, straw-tempered pottery that characterizes the earliest occupations of Bakun, the Fars sites, and the earliest levels at Tepe Yahya to the south of Iblis.

Iblis Period I has yielded the bulk of new information. The settlement consisted of carefully planned multi-roomed mudbrick buildings. Floors were sometimes mat-covered, and roofs were constructed of horizontal poles plastered with mud. A finer, painted, buff-colored ware was introduced; it has strong connections with the early third millennium painted-pottery traditions of Fars and Tepe Yahya in Kirman province.

Crude crucibles were found, and objects of copper. This discovery of the smelting of ores as early as 4000 B.C. is of great importance for the history of pyrometallurgy. Incidentally, the specific study of the Iblis metals is but part of a broader metallurgical research project being carried out in the Persian desert (*115–118*).

Iblis Period I floral remains are sparse. Grain found on a house floor is identified as breadwheat (*Triticum aestivum* L.). Straw temper in one of the coarse-ware pottery sherds is identified as almost certainly emmer wheat. A "mealing bin" with two saddle querns set in a slanting clay bench was found in one of the houses. Apparently the ground meal fell into clay boxes below.

Only part of the animal-bone collection has been studied (*124*). Domesticated breeds include the dog (*Canis familiaris*), goat (*Capra*), sheep (*Ovis aries*), and cattle (*Bos taurus*). Wild breeds include gazelle, aurochs (*Bos primigenius*) (?), onager (*Equus hemonius*), the true horse (*Equus caballus*), and the lion (*Panthera lea*). Clay figurines identified as dogs, sheep, goats, and the aurochs are also found in this period. Bökönyi concludes that this faunal assemblage points to an environment similar to today — desert and semidesert.

With the succeeding Iblis Period II, copper metallurgy was a major activity. Hundreds of earthenware crucibles were found scattered throughout the normal domestic debris. The excavators suggest that smelting was done in the kitchen itself, as part of the regular domestic activities. The volume of remains suggests that copper was being worked in quantities far in excess of local needs and may have been traded. The faunal remains so far analyzed show a preponderance of sheep and goats, followed by domestic cattle.

Iblis Period III is very ill defined, but Iblis Period IV (Aliabad Period) shows important developments, prompted by a pronounced intrusion of Mesopotamian influences. Although most of the pottery continued to be handmade, the potter's wheel was introduced. Copper smelting continues. Sheep and goat bones continue to dominate the faunal remains.

Iblis Periods V and VI are poorly defined, and little is said of them apart from their ceramic assemblages. There follows then a long, as yet unexplained, hiatus in the occupation of this region. Iblis Period VII may be as late as 1100 B.C.

Tepe Yahya

Located about 200 kilometers south of the city of Kirman in the fertile Soghun valley, this is the largest (19 meters high) pre-Islamic site known in southeastern Iran. Excavations began in 1968. Preliminary reports (*125–125b*) indicate the importance of this site for the development of early agriculture, manufacturing, and trade.

Its earliest period, VI (5th millennium B.C.) begins on virgin soil and is called "neolithic." Floral and faunal remains are not yet reported. Five architectural levels in the period consist of small domestic structures of sun-dried thumb impressed bricks. In one room (phase VI D) was found a unique steatite female "idol" lying on a bed of flint implements. Pottery, some painted, is of a coarse chaff-tempered fabric.

Pottery and architecture show continuity into the next period, V (Yahya Period, 3800–3400 B.C.). This period shows an active development in the exploitation and use of local and imported raw materials. Among the manufactured products are chisels, awls, and pins of copper-bronze; microliths of obsidian and flint; beads of carnelian, turquoise, steatite, and ivory; bowls of alabaster and steatite. Stamp seals of clay and stone suggest a concern for establishing identification and ownership. Floral remains are not yet reported. Camel bones are said to be present.

Period IV was the most florescent time in Yahya's history. The first two phases of the period, IV C and IV B (3400–2500 B.C.) are called Proto-Elamite, while Phase IV A (2500–2200 B.C.) is tentatively identified with the mature Elamite period. Evidence for extensive "international" trade is abundant, especially during phases C and B. Parallels in manufactured objects are found from as far west as Syria and as far east as Pakistan. Phase IV C is characterized by massive architecture of sun-dried bricks. Most important was the discovery of six clay tablets inscribed with the earliest script known in southwestern Iran, "Proto-Elamite." Although they cannot be translated in detail, it is certain on analogy with Mesopotamian tablets that they are economic in nature and record commercial or business transactions. With phase IV B the evidence for far-flung trading becomes even more abundant. Especially profuse are bowls, axes, and other objects of steatite. A rich native source is reported nearby. The significance of this trade in elaborately carved steatite is not known, but it seems clear that Yahya, at the eastern end of the Elamite territory, was a major center of manufacturing. Floral and faunal materials are still unreported. Period IV A is still little known, but it is significant that the site was abandoned at the end of the Elamite period, and there is no evidence of occupation for the entire second millennium B.C.

As to the nature of the trade, archaeological parallels suggest that Yahya was involved mainly with *overland* routes through Iranian Baluchistan, southern Afghanistan, and southwestern Iran, with an extension into the Persian Gulf region. There is, on the other hand, little evidence for direct contact with Pakistani Baluchistan and the Indus Valley. The fact, stressed by the excavator, that no evidence of the Kulli people of Pakistani Baluchistan is present at Yahya is not surprising. I have suggested elsewhere (*102, 104*) that the Kulli people were somehow involved in *coastal* trading activities between the Indus region and the Persian Gulf, perhaps as entrepreneurs (?). Yahya is some 40 miles from the coast. Furthermore, Yahya's main period of trading ended by about 2500 B.C. This is centuries before the development of the extensive seafaring activities recorded in the Sumerian and Akkadian written documents of Mesopotamia. As suggested elsewhere in this chapter, the shift from inland to coastal trade is one of the hallmarks of the period of commercial and economic activities between the Near East and South Asia. Yahya was abandoned by this time.

Persian Baluchistan

The southeastern corner of Iran comprises the Province of Baluchistan. The geography, history, and sociology of this little-known region is summarized by Spooner (*13*). The Italconsult Reports of 1959 (*14*) detail recent studies of the agriculture, hydrology, and geology.

Again, it was the explorations of Sir Aurel Stein (*90*) that first revealed the archaeological importance of Persian Baluchistan, especially the Bampur valley. Recently, new excavations were conducted at one of his sites, Bampur (*126–130*). The Bampur excavations focused on chronological problems with the emphasis on ceramics, so again we are left with little or no direct information on the subsistence level of the inhabitants. The painted pottery of Bampur does, however, give some

indirect evidence of the flora and fauna known to the early inhabitants. Pottery of the two earliest periods, I–II, has representations of ibex, deer, snakes, scorpions, and heavy-bodied birds. Floral designs are limited to palm fronds. Period V sees the addition of fish and pipal leaves, whereas Period VI representations seem to be restricted to continuous horizontal bands of horned animals (ibex or goat?).

What is of significance is the discovery at Bampur of archaeological materials that demonstrate contacts with regions as far distant as Kandahar in southeastern Afghanistan, all of southern Pakistani Baluchistan, perhaps Turkestan in southern Russia, and with the southern part of the Persian Gulf. Contacts during Periods III and IV were principally with the north — Seistan, the Upper Helmand, and (Soviet) Turkestan. With Periods V and VI the northern contacts virtually cease, and we see Bampur in the middle of more coastally oriented trading activities, which involved southern Mesopotamia, the Persian Gulf, and the coast of Pakistan.

The Persian Gulf

Today the Persian Gulf's (*4, 75, 111*) age-old commercial and strategic roles have been diminished, partly by the airplane, partly by shifts in traditional trade routes and international power balances. It has lost its prime role as supplier of pearls, dates, fish, salt, gypsum, and, most recently, oil. But 4,000 years ago the Gulf was in the center of the extensive seafaring activities of the Sumerians and Babylonians — activities that might have extended as far as the eastern coast of Africa and the western coast of India.

There is some evidence of "Stone Age" habitation on the Arabian side of the Gulf, but its age and significance are not yet clear. The human story really begins to reveal itself sometime around the middle of the third millenium B.C. Danish archaeologists have been working in the Gulf yearly since 1953. Excavations and explorations have been carried out on Bahrain, in Qatar, Kuwait, and the Trucial Coast (*131–136*).

As yet, little evidence has been published relevant to the ancient environment and ecological relationships, except short notices on the ancient fauna of Abu Dhabi. A few radiocarbon dates have been obtained, but they don't apply to the time range of this paper. The chronology and dating of the early cultures are tied primarily to that of Mesopotamia, Baluchistan, and the Indus Valley, and these connections involve many uncertainties. Stamp seals of a distinctive type, apparently manufactured in the Gulf (*136–139*), have been found at several sites in southern Mesopotamia, at Mohenjo-daro in the Indus Valley, and at Lothal, a Harappan seaport on the west coast of India. In Mesopotamia they have been found in contexts dated to the Old Akkadian, Ur III, and Isin-Larsa periods (approximately 2400–1800 B.C.). The recent discovery of the impression of one of these Persian Gulf seals on a dated cuneiform tablet (*140, 141*) provides the only "absolute" date for the seals and so-called Barbar Culture to which they belong. The cuneiform document is dated to the year 1923 B.C. on the basis of the year of reign of the Babylonian king whose name is given.

The Persian Gulf Barbar Culture was discovered by the Danes during excavation of a site at Barbar on the northern coast of Bahrain. A curious temple, with elaborate stone architecture, was uncovered. It is unique and, just like the stamp seals and the pottery, suggests that the Gulf had its own distinctive and independent culture at the same time it was apparently involved in the international trade and commercial activities of its more powerful neighbors, the Sumerians and Babylonians. The assumption that the Barbarites were engaged in pearl fishing as part of their activities received archaeological confirmation when archaeologists discovered sherds of typical Barbar ribbed pottery mixed in with the huge deposits of oyster shells along the coast of southern Bahrain (*131* for 1958).

An academic exercise of long standing is to try to associate Bahrain and other parts of the Persian Gulf region with geographical names mentioned in ancient Mesopotamian written records, but it is more than just academic in importance. The actual identification of places listed in economic, commercial, and even religious texts would be invaluable for the understanding of the international activities of the period. The literature on the subject is extensive (*99–101, 104, 142–151*).

Magan, Meluhha, and Dilmun are the three places associated by the Sumerian and Babylonian scribes of about 2400 to 1800 B.C. with seafaring and long-distance contacts. It is not necessary to repeat the conflicting arguments concerning the geographical identification of these places, but it can be inferred that one of the names referred to the Persian Gulf region. Dilmun is usually identified with Bahrain, but only on circumstantial evidence. Kramer is virtually alone in advocating the identification of Dilmun with the Indus Valley civilization. Magan and Meluhha have been placed by scholars all the way from Ethiopia to India. I am presently inclined to place Dilmun in the Persian Gulf area, Magan as perhaps southernmost Iran, and Meluhha as the coastal area of Pakistan and perhaps western India.

For reasons detailed in other papers (*104, 152, 153*), I am convinced that a major part of the seafaring had to do with the southern coasts of Iran and Pakistan. Some credence has been lent to this view by the discovery of Baluchistan-related pottery and objects on Umm an-Nar island off the coast of Abu Dhabi and at the Buraimi oasis in Oman (*131, 133*). Pottery and stone vessels from graves and a settlement site are stylistically related to Bampur Periods V–VI and to the so-called Kulli culture of southern Pakistani Baluchistan. This relationship provides a valuable synchronism with regions to the east, but synchronisms within the Gulf itself are tenuous.

The most important relationship to establish is that between the Oman-Kulli material and the Barbar culture of Bahrain, but none of the distinctive incised vessels of Oman-Kulli have yet been found associated with Barbar cultural remains. Conversely, none of the ribbed pottery of Barbar has been found at the Oman sites, nor have any of the characteristic Persian Gulf type seals. The only cross correlation I know of is that pottery of a few types characteristic of the Oman graves and settlement has reportedly been found at Bahrain in levels preceding and contemporary with the Barbar period.

Excavations at the settlement associated with the Umm an-Nar burials have recovered copper fish-hooks and stone net-weights. Stone querns attest to the grinding of grain, but no floral remains have been reported. One preliminary report stated that sheep and goat bones were found in the houses, but a later study (*131,* for 1965, p. 149) failed to identify any in the small collection examined. The majority of the bones are of the sea-cow (*Halicore dugong*). There are also bones of a large whale, of antelope, and of cormorant. Most significant is the presence of camel bones. Relief carvings on stone blocks of the graves clearly depict the dromedary camel. Whether or not the camel was "domesticated" at this early period would be important to know. Other animals depicted on the stones include the goat or oryx and the nonhumped bull. On the painted pottery, however, is the humped bull (*Bos indicus*), hallmark of the Baluchistan Kulli culture. Beads of lapis lazuli are another certain indicator of contact with Baluchistan and ultimately with Afghanistan, its only known South Asian source.

Pakistani Baluchistan

In addition to the references cited at the beginning of the chapter, some others deal specifically with archaeological matters (*3, 85, 88, 89, 152–164*). The relative sequence of the various cultural groups and areas is still far from certain, and the recent attempts to synthesize the data must be treated as working hypotheses and not as conclusive statements (*83, 103, 104, 163, 164*). Not a single site in all of Pakistani Baluchistan has been subjected to large-scale horizontal excavation. Excavations have been limited to small test pits. The internal chronology of the region has been based on ceramic sequences obtained from these small excavations and on typological comparisons of pottery collected on the surface of ancient sites. The radiocarbon dates from the Quetta Valley are either too inconsistent or have such wide tolerances that a new series of samples would be desirable. Virtually no attention by qualified specialists has been focused on the recovery of data relevant to the ancient environment. A lone exception is Raikes (*3, 6, 166*), who has brought his hydrological knowledge to bear on questions of climatic change and ancient water usage. Remains of a gabarband, a water-diverting and/or retarding structure, are shown in Figure 4. The westernmost outpost known (near the present Pakistan-Iran border) of the Harappan civilization of the Indus Valley was at Sutkagen-Dor (Fig. 5).

As mentioned earlier, it does seem that until perhaps as late as the middle of the second millennium B.C., this Pakistani Baluchistan/southern Afghanistan region was experiencing its own development with little or no influence from the southern Iran/Mesopotamian region, but

Fig. 4. Remains of a gabarband near Gwadar on the Makran coast of West Pakistan. Such stone walls have been used in the arid portions of Baluchistan for at least 4,000 years. They retard and divert the scanty rainwater to form artificial terraces for agricultural purposes.

we are seriously hampered by the lack of archaeological confirmation.

Southern Afghanistan

The Helmand river of southern Afghanistan is one of the world's few major streams, such as the Nile, Tigris-Euphrates, and Indus, flowing through extensive desert areas — but it differs markedly from the others in that it does *not* appear to have been the seat of a major early civilization. This curious phenomenon invites investigation. Serious archaeological surveying of the valley proper has barely begun (*167–169*). Attention had previously centered only in the headwater area and the Seistan basin, which straddles the Afghan-Persian border.

In the Kandahar region, at the headwaters of the Helmand, two prehistoric sites have received archaeological attention. The smaller one, Deh Morasi Ghundai (80 by 140 meters in size), was subjected only to trial trenching (*170*), but the discoveries confirmed the presence of third millennium B.C. occupation in southern Afghanistan. The cultural sequence at the site was more fully documented by the excavations at the other Kandahar site, Mundigak. Animal bones from Deh Morasi are few and identified only as sheep or goat (?), bovine, and ungulate. Evidence for ancient plants came from chaff used as temper in a mud brick of Period II, a. It included barley (said to be a 6-row cultivated variety resembling *Hordeum vulgare* var. *afghana*), and a fodder grass (*Aegilops tauschii* syn. *A. squarrosa*). A single radiocarbon date from Period IIb seems reasonable on comparison with the chronology of Mundigak.

Much more extensive were the 1951–58 French excavations (*171*) at Mundigak (Fig. 6), 90 kilometers northwest of Kandahar. The cultural sequence obtained there is the only major one available for all of southern Afghanistan and Pakistani Baluchistan. As noted earlier, the initial radiocarbon tests were unsuccessful. The hope is that tests being run on new samples will date the earliest Periods I and II. Figure 3 shows the relative positions of successive cultural phases to other archaeological sites. Until Mundigak Period IV, there was little or no direct contact with southern Iran or Mesopotamia. Then with Period IV and the apparent disruption of some of the northern trade routes, we see close ties with Bampur, probably via the Helmand route and through Seistan (although there is no archaeological proof of this route).

Considering the extent and duration of the Mundigak excavations, a remarkably small quantity of floral and faunal remains was found. Grains of wheat identified as *Triticum compactum* from Period II are the only significant floral remains reported. The ancient fauna is somewhat better represented (*172*), but the collections are still disappointingly small. A total of 411 bone fragments was collected, said to represent some 106 individual animals and birds. Of these, 103 individual specimens come from Periods I through IV. Table 2 gives the distribution of individual animals by periods.

Of special note are the domestic cattle remains. Except for one bone in a later period (V), these animals are not represented after Period IV. This absence corroborates the evidence of the clay animal figurines where models of cattle are not found in post-Period IV contexts. However, there is the same peculiar inconsistency noted at some of the southern Iran sites, namely, a difference between the types of cattle represented by the figurines and those identified from the actual bone remains. Most of the Mundigak figurines represent the humped zebu (*Bos*

indicus), whereas the skeletal remains are all classified as *Bos taurus,* the common unhumped domestic ox.

The archaeological development at Mundigak is

TABLE 2
Faunal Remains from Mundigak

Animals	Total Individuals	Period			
		I	II	III	IV
Domesticated					
Sheep *(Ovis aries* L.)	22	2	2	12	6
Ass	21	5	7	8	1
Cattle *(Bos taurus)*	12	3	2	3	4
Goat	12	2	2	7	1
Horse	2	2	—	—	—
Dog	1	1	—	—	—
Wild					
Gazelle	19	4	6	8	1
Ibex	9	2	1	5	1
Lynx	2	1	1	—	—
Bird of prey (unidentified)	2	—	2	—	—
Mole	1	—	—	—	1
Totals	103	22	23	43	15

probably housed 2,000 or more persons — garrison troops and merchants — sometime between 2400 and 1900 B.C.

Fig. 5. Sutkagen-Dor, the westernmost Indus Valley (Harappan) civilization outpost in the now-desolate Dasht valley separating coastal Pakistan and Iran. The fortified citadel and lower town

reasonably clear, but there is need of considerable environmental research in this Kandahar region that was so closely related to northern Pakistani Baluchistan and in contact with southern Iran, Seistan, and apparently southern (Soviet) Turkestan.

Seistan

Seistan is comprised mainly of the huge delta depression at the southwestern end of the Helmand valley. It straddles the border of Afghanistan and Iran, a border that has little or no geographical viability. References to publications on this little-known area have been given earlier. It is a baffling and difficult region. Archaeological remains abound in staggering profusion amidst endless scenes of modern desolation and waste. Climate and Mogul invaders have received most of the blame for its present inhospitality, but neither one may prove to have been a major culprit when the story is more fully known.

Archaeological work on the Afghan side of the border has been limited. The French conducted the first serious survey and excavations during the 1930s (*173*). Surveys have since been conducted by Fairservis (*168*), Hammond (*174*), and Trousdale (*169*). Most recently, excavations and surveys were directed in 1968 and 1969 by Dales (*167*).

The French excavations at Sorkh Dagh (Nad-i-Ali) (*175*) held out great promise that the lower part of this high mound — which they did not penetrate — contained occupations of the Chalcolithic and earlier prehistoric periods, but the University of Pennsylvania Museum expedition there in 1968 discovered the disappointing fact that the entire lower half of the mound is a solid mud-brick platform and that there is no evidence of any occupation in the area earlier than Achaemenid Persian times (fifth century B.C. or so).

The Museum's work in 1969 shifted to the desert regions south of the Helmand. Part of Fairservis's earlier route was covered — mainly along the course of the dried up Rud-i-Biyaban from the Helmand to the Iranian border, and south into the desolate Gardan Reg (Fig. 7). There, an astonishing quantity of debris from old copper-smelting operations (Fig. 8) covers the severely eroded surface. Some of the activity can be assigned to the Islamic period, but most of it may date back to Sassanian and Parthian times if not earlier. Fairservis found in the Gardan Reg painted pottery that appears identical to types from Mundigak Period IV and Bampur Period IV. Geomorphological and metallurgical studies are sorely needed and are planned for 1971–72.

Going then further south, we traversed the length of the Gaud-i-Zirreh, the saline depression that once took some of the overflow of the Helmand. A search for early sites at the western end of the depression yielded mainly remains of medieval Islamic occupation (Fig. 9); however, carved alabaster discs and columns were found in abundance on the surface of cemetery areas. These are potentially important because they are identical to alabaster objects found in northeastern Iran during the Hissar IIIc period of approximately 1800 B.C. But no evidence of early contact with either southern Iran, Baluchistan or Mundigak was found on the surface. This lack may occur because of the extreme surface erosion that characterizes the area, and of the heavy sand-dune cover that obscures much of the present surface. An ancient and a modern town in this general part of Afghanistan are shown in Figures 10 and 11.

Fig. 6. Mundigak, near Kandahar, is the only large prehistoric site in southern Afghanistan excavated to date. Seven major periods of occupation, ranging from the third to the first millennia B.C., were excavated by French archaeologists from 1951 through 1958.

Across the border in Iran the picture is infinitely brighter. Stein (*87*) discovered dozens of early sites in the ancient delta areas of the Helmand. One of these, Shahr-i-Sokhta, is presently being excavated by Italian archaeologists, with spectacular results (*176–176a*). Only preliminary reports are available, but it is already clear that the principal occupation correlates with the impressive Period IV of Bampur and with Mundigak Period IV.

Shahr-i-Sokhta is an exceptionally promising site because of its unexpectedly fine preservation. The parched sands of the area have preserved plant and animal materials to a degree almost comparable to Egypt. Large wooden beams and posts have been found *in situ;* they are being studied to determine their species and radiocarbon date. So far only numerous "ox bones" are reported, but much more faunal information is expected. Clay figurines of animals are abundant — more than 200 of them found the first season. They are almost all of humped oxen (*Bos indicus*) plus a few boars.

Most exciting is the discovery of a large stone-working industry at the site. Alabaster vessels of types common throughout Baluchistan and southern Iran were manufactured here. Large quantities of lapis lazuli, both in rough chunks and as finished beads, suggest that Shahr-i-Sokhta was a major partner in the international lapis trade. Inasmuch as the only confirmed source of Near Eastern lapis is Badakhshan in the northeastern tip of Afghanistan, and the major users of the stone were southern Mesopotamia and Egypt, it was an impressive trade indeed.

The Shahr-i-Sokhta workshops — seemingly out in the middle of nowhere — make sense in terms of the shift in supply routes that took place around 3000 B.C. (*100, 102, 177, 178*). Before that date the main routes were across northern Iran and into northern Mesopotamia, but then came a pronounced shift southward in the distribution of lapis. From 3000 B.C. it is found mainly in southern Mesopotamia and Khuzistan where the powerful Elamite capital of Susa may have been the principal import and distribution center. The peak of lapis popularity in southern Mesopotamia was during the Early Dynastic III Period (about 2500–2400 B.C.); it is found in great profusion in the tombs and graves of the so-called Royal Cemetery at Ur. This is the very period of the Shahr-i-Sokhta workshops and of the beginnings of the most extensive contacts between the Persian Gulf, Baluchistan, and southern Afghanistan.

TRADE AND TRADE ITEMS

The precise degree and nature of ancient "international" trade is inadequately reflected in archaeological remains. Only some of the nonperishable, nonconsumable materials survive. Some of the surviving materials, as, for example, obsidian, can be analyzed to determine their geographical source, but this process is still not possible for most stones and metals. Also, the intrinsic rarity of some materials and the limited native sources for them — for example, lapis lazuli — help verify other suggestions of longrange contacts and trade. The ultimate evidence, however, is that of the written word — economic, commercial, and political documents that mention specific products and geographical locations, but such documents are luxuries not available in many archaeological situations. A combination of the discovery of "foreign" items and influences at archaeological sites

Fig. 7. Gardan Reg, in the southern part of Afghan Seistan. Severely eroded and completely abandoned today, it holds abundant remains of ancient copper-working activities and graves of the third millennium B.C.

Fig. 8. One of hundreds of plateaus covered with ancient slag and copper ores in the Gardan Reg region, southern Afghan Seistan.

and references in Sumerian and Babylonian cuneiform documents allows us to make certain statements about early trade contacts between the Persian Gulf and regions to the east. Dilmun, Magan, and Meluhha are the places that featured most prominently in these activities (*99–101, 104, 142–151, 179–181a*).

The maximal period of early "commercial" contact between the Near East and South Asia was between 2500 and 1800 B.C. approximately (during the Early Dynastic III, Akkadian, Ur III, and Isin-Larsa periods of Mesopotamia and the Kulli-Harappan era of Pakistani Baluchistan and the Indus Valley). Actually, the earliest Mesopotamian written references to foreign products arriving in their country suggest tribute rather than commercial trade, but this may be because only the official royal documents have survived. An inscription of King Ur-Nanshe of Lagash (Sumerian Early Dynastic III Period, about 2500 B.C.) states, "The ships of Dilmun brought him [the king] wood as a tribute from foreign lands." King Sargon of Agade (about 2350 B.C.) boasts of gifts and tribute coming to him from the four corners of his realm and of the fact that ships from Dilmun, Magan, and Meluhha docked at his quays.

The bulk of the activity, especially during the Ur III and Larsa periods, was commercial in nature, and most of it involved seatrade through the Persian Gulf. The Ur III Sumerians even had a special title for "seafaring merchants." Not only was the palace interested in these activities, but the temples were deeply involved. One text, for example, tells of merchants getting large quantities of garments, wool, perfumed oil, and leather goods from the temple of Nanna in Ur to be loaded onto boats bound for Dilmun and Magan and exchanged there for copper. Another document lists a rich tithe paid to the temple of the goddess Ningal at Ur out of products brought back from an expedition to Dilmun. This included lapis lazuli, "fish eyes" (pearls?) various types of stone, copper and bronze ingots, gold, silver, antimony (for eye paint?), ivory (both raw and carved into rods and combs), red ochre, and objects of *mesu* wood (identified as Sissoo wood [*Dalbergia sissoo* Roxb.], which is common to Baluchistan) (*182*).

In addition, economic documents list as imports into southern Mesopotamia from Dilmun, Magan, and Meluhha: oils and essences, carnelian beads with etched designs, tortoise shell (?), sea shells, dates, Magan and Dilmun "onions," multicolored birds from Meluhha (peacocks?), wood of various kinds specified for the making of furniture and boats and religious shrines, ready-made wooden products such as chairs and tables

Fig. 9. Approach and closer view of one of hundreds of medieval Islamic structures seen throughout the presently desolate and abandoned region of Afghan Seistan. This particular struc-ture is one of several at the western end of the Gaud-i-Zirreh, on the north side of the dried-up Shela Rud.

inlaid with ivory decorations, and headrests. The items coming through the Persian Gulf into southern Mesopotamia were thus raw materials for building, manufacturing, and decorating, and also metals, jewelry, cosmetics, "exotics," and food staples.

Exports from southern Mesopotamia to Dilmun, Magan, and Meluhha were limited in variety. Byproducts of their cattle and herding industries were the most numerous exports. They included both raw wool and finished fabrics and garments, skins and hides, as well as finished products of leather. Barley, flour, cedar wood (from northern Mesopotamia?), essences, and oils are also listed. Silver frequently figures in the lists; it may have been used as payment for the imports as well as for a raw material.

By the end of the Larsa Period, about 1800 B.C., the names of Magan and Meluhha had dropped out of the economic documents. This date correlates with the end of the Indus civilization and probably the Kulli culture, whose people may have been the entrepreneurs in the east-west trade. The decline in internal stability and fortunes at about 1800 B.C. in both Mesopotamia and the Indus region may account for the cessation of the longrange seafaring activities. The Persian Gulf-Indian Ocean route is a dangerous and costly business. One round trip a year is average for sailing ships, which must rely on the monsoons. The mortality rate, even for modern dhows of more than one-ton weight, averages 1 in 10 (*183-184*). The fact that the name of Dilmun continues to appear in Mesopotamian documents for many centuries after 1800 B.C., and the fact that it was always regarded as an important religious place in Sumerian and Babylonian literature (*149-151*), suggests that Dilmun was more likely located nearby in the Persian Gulf rather than in India, but archaeological confirmation is needed.

CONCLUSIONS AND RESEARCH NEEDS

Prior to about 3000 B.C., the western and eastern halves of the Persian Gulf-Baluchistan arid zone were passing through a Neolithic to Bronze Age development almost totally independent of one another. The western half from about 7000 B.C. experienced the steady intrusion and spread of Mesopotamian culture southeastward into the ecologically difficult areas along the southern edge of the Iranian plateau. Archaeologically this spread correlates with the spread of improved agricultural techniques. New methods of water recovery and distribution, the increasing use of basic irrigation techniques, and the progressive domestication of plants and animals allowed these more difficult areas to be developed and exploited. Archaeological research in this western region is beginning to provide a meaningful picture of these developments and of the prehistoric environmental situation.

There is no evidence as yet that the Persian Gulf region — with the exception of the Iranian littoral — was participating in these developments until after 3000 B.C.

The eastern half of the zone (Baluchistan and southern Afghanistan) was apparently in a different sphere of development and influence that included northeastern Iran and southern (Soviet) Turkestan. Archaeological research has been extremely limited and has concentrated more on chronological than on ecological problems. The role, if any, of Baluchistan in the rise of civilization in the Indus Valley is unknown at the present. A new

Fig. 10. Tarakun, a medieval Islamic fortress-town situated along a presently dried-up bed of the Helmand River, Seistan, Afghanistan. The entire region is now virtually uninhabited

Fig. 11. Juwain, a modern town in northern Afghan Seistan, is undergoing a rapid period of deterioration and abandonment because of the increasing salinity of the surrounding countryside.

study of the third millennium B.C. material from Baluchistan shows that northern Pakistani Baluchistan did play a major role in the rise of the Indus civilization. In fact, what we have been calling pre-Harappan can, in some instances, more reasonably be called Early Harappan (*184a*).

Around 3000 B.C., major events began to influence greatly the fortunes of the southern Iranian desert zone. On the one hand the far northerly land routes vital to such operations as the lapis lazuli trade were discontinued. Apparently the route shifted south, through Seistan, through southern Iran, to the Persian Gulf and Mesopotamia. On the other hand this zone was now flanked by the ascending power and influences of the Near Eastern and South Asian riverine civilizations. There is solid archaeological evidence of extensive contacts ranging from southern Iran to Baluchistan and southern Afghanistan. In addition to the lapis trade, came the rapidly increasing demand for raw materials and supplies to build and support the urban centers of Mesopotamia and the Indus region. The Persian Gulf developed a distinctive culture of its own during the third millenium B.C. at the same time that it was involved in the international seafaring activities of the Mesopotamians — activities that extended probably as far eastward as the coasts of Pakistan and India. These contacts ceased around 1800 B.C. because of major disruptions in the prosperity of both the Mesopotamian and Indus civilizations.

If we are to better understand the role of this Iranian desert region vis-à-vis the affluent riverine civilizations that flanked it, research must be refocused and greatly expanded. There are certain "musts," such as:

1. Distribution maps of natural resources: metal ores, woods, stone (for buildings, arts and crafts), floral and faunal habitat zones, and water resources.

2. Distribution maps of ancient settlements, arranged diachronically.

3. More information on the subsistence potentials of the subregions occupied by the various ancient cultures.

4. More secure chronological frameworks, especially concerning the temporal relationships between the riverine civilizations and the desert cultures. This item is essential for such questions as the effects of the rising prosperity and decline in one area on the others; also for verifying theories of diffusion.

5. More excavations of "fringe" sites and "outposts" to help clarify the nature of relationships between the desert and riverine regions.

6. More information on the present-day ecological and ethnographical scenes.

7. Well-planned interdisciplinary orientation for field work with greater emphasis on nonmaterial aspects of archaeological research.

APPENDIX

Radiocarbon dates are here based on the half-life of 5,730 years and stated in years B.C. Only those dates which are logically consistent are included. The dotted circles in Figure 2 indicate the averages of individual groups of dates. Details of all dates can be found under the laboratory number in *Radiocarbon (98)* unless indicated otherwise. The periods are listed below in the same order as in Figure 2, from left to right.

Ali Kosh Period (Deh Luran, Khuzistan)
From Hole, Flannery, and Neely (12)
Not in Figure 2:

Shell	1246	6712 ± 206	B.C.
Humble	O – 1833	6728 ± 185	
Humble	O – 1816	6728 ± 185	
Humble	O – 1845	6548 ± 180	
	I – 1491	6393 ± 175	
Average		6620	B.C.

Bayat Period (Deh Luran, Khuzistan)

SI – 203	4405 ± 206	B.C.
SI – 204	4292 ± 206	
I – 1502	4292 ± 144	
UCLA – 750A	4302 ± 103	
I – 1503	4086 ± 237	
SI – 205	3921 ± 258	
SI – 156	3993 ± 124	

Not in Figure 2:

I – 1499	4282 ± 144	
Average	4197	B.C.

Nokhodi
BM – 171	3252 ± 152

Gap
GAK – 197	4096 ± 175
GAK – 198	3653 ± 124

Bakun
P – 931	4502 ± 72
P – 438	4220 ± 83

Iblis
P – 925	4091 ± 74
P – 926	4083 ± 75
P – 927	3792 ± 60
P – 928	3645 ± 59

Tepe Yahya
Personal communication	4200
Iran VII (1969): p. 168	3860 ± 140

Not in Figure 2 (125a):

Period VI C/Neolithic
GX – 1728	4800 ± 144

Period VI A/Neolithic
GX – 1509	4244 ± 185
GX – 1737	3729 ± 165

Period V C/Yahya Culture
WSU – 871	3770 ± 144

Period V B/Yahya Culture
WSU – 872	3739 ± 422

Period IV B/Proto-Elamite
GX – 1734	3378 ± 175
WSU – 876	3342 ± 480
GX – 1727	2511 ± 370

Deh Morasi Ghundai, Period II B
P – 1493	2597 ± 55

Quetta Valley
KGM I P – 524	3688 ± 85	
KGM I UW – 61	3468 ± 82	
DS I UW – 59	2510 ± 70	
DS II P – 522	2559 ± 202	
DS II P – 523	2200 ± 76	
DS III UW – 60	2200 ± 160	

Niai Buthi / Kulli Culture
P – 478	1902 ± 64

Kot Diji / Pre-Harappan
P – 196	2605 ± 145	B.C.
P – 179	2335 ± 155	
P – 180	2255 ± 140	
P – 195	2090 ± 140	
Average	2321	B.C.

Kalibangan / Pre-Harappan
TF – 155	2370 ± 115	B.C.
TF – 157	2294 ± 110	
TF – 241	2263 ± 90	
TF – 162	2108 ± 100	
TF – 161	2098 ± 100	
TF – 165	1964 ± 100	
TF – 156	1902 ± 105	
TF – 154	1825 ± 110	
Average	2103	B.C.

Kalibangan / Harappan
TF – 160	2232 ± 100	B.C.
TF – 607	2098 ± 124	
TF – 25	2095 ± 115	
TF – 608	2077 ± 113	
TF – 153	2077 ± 110	
TF – 163	2077 ± 100	
P – 481	2045 ± 75	
TF – 605	1974 ± 108	
TF – 150	1902 ± 103	
TF – 141	1866 ± 110	
TF – 149	1835 ± 140	
TF – 142	1794 ± 100	
TF – 152	1774 ± 85	
TF – 143	1665 ± 110	

Not in Figure 2:

TF – 145	2062 ± 103	
TF – 147	2031 ± 103	
TF – 151	1964 ± 103	
TF – 139	1939 ± 103	
TF – 942	2227 ± 113	
TF – 946	1763 ± 103	
TF – 947	1920 ± 88	
TF – 948	1980 ± 103	
Average	1973	B.C.

Mohenjo-daro / Harappan
P – 1179	2083 ± 66	B.C.
P – 1177	2062 ± 66	
P – 1180	1993 ± 63	
P – 1178A	1967 ± 61	
P – 1176	1966 ± 61	
P – 1182A	1864 ± 65	
TF – 75	1760 ± 115	
Average	1956	B.C.

Lothal/Harappan

TF –	136	2082 ± 130	B.C.
TF –	22	2010 ± 115	
TF –	27	2005 ± 115	
TF –	26	1995 ± 125	
TF –	29	1900 ± 115	
TF –	133	1902 ± 110	
Average		1982	B.C.

Lothal/Late Harappan

TF –	23	1865 ± 110
TF –	19	1810 ± 140

Rojdi/Late Harappan

TF –	200	1974 ± 110
TF –	199	1748 ± 100

Amri, Pakistan/Amri Culture
Not in Figure 2:

TF –	863	2670 ± 113
TF –	864	2900 ± 113

Nindowari, Pakistan/Kulli Culture
Not in Figure 2:

TF –	862	2067 ± 108

REFERENCES AND NOTES

1. MEIGS, P.
   1953 World distribution of arid and semiarid homo-climates. *In* Reviews of research on arid zone hydrology. Unesco, Paris. Arid Zone Programme 1: 203–210.

2. McGINNIES, W. G.
   1968 Appraisal of research on vegetation of desert environments. *In* W. G. McGinnies, B. J. Goldman, and P. Paylore (eds) (*4*), p. 381–474.

3. RAIKES, R. L.
   1968 Archaeological explorations in southern Jhalawan and Las Bela (Pakistan). Origini, University of Rome, 2: 103–171.

4. McGINNIES, W. G., B. J. GOLDMAN, and P. PAYLORE (eds)
   1968 Deserts of the world, an appraisal of research into their physical and biological environments. University of Arizona Press, Tucson. 788 p.

5. McGINNIES, W. G., and B. J. GOLDMAN (eds)
   1969 Arid lands in perspective. American Association for the Advancement of Science, Washington, D.C., and University of Arizona Press, Tucson. 421 p.

6. RAIKES, R. L.
   1967 Water, weather, and prehistory. John Baker Publishers Ltd, London. 208 p.

7. FISHER, W. B. (ed)
   1968 The land of Iran. The Cambridge History of Iran 1. 792 p.

8. ENGLISH, P. W.
   1966 City and village in Iran. University of Wisconsin Press, Madison, Milwaukee, and London. 204 p.

9. CALDWELL, J. R. (ed)
   1967 Investigations at Tal-i-Iblis. Illinois State Museum, Preliminary Report 9. 408 p.

10. ————
    1968 Tal-i-Iblis and the beginning of copper metallurgy at the fifth millennium. Archaeologia Viva 1: 145–150.

11. BARTH, F.
    1965 Nomads of South Persia, the Basseri tribe of the Khamseh confederacy. Universitetsforlaget, Oslo; Humanities Press, N.Y. 159 p. (Also published as Bulletin 8, Universitets Etnografiske Museum, University of Oslo).

12. HOLE, F., K. V. FLANNERY, and J. A. NEELY (eds)
    1969 Prehistory and human ecology of the Deh Luran Plain: an early village sequence from Khuzistan, Iran. University of Michigan, Museum of Anthropology, Memoirs 1. 438 p.

13. SPOONER, B.
    1964 Kŭch U. Balŭch and ichthyophagi. Iran 2: 53–67.

14. ITALCONSULT
    1959 Reports. Rome.

15. ADAMS, R. C.
    1962 Agriculture and urban life in early southwestern Iran. Science 136: 109–122.

16. HIGGS, E. S., AND M. R. JARMAN
    1969 The origins of agriculture: a reconsideration. Antiquity 43 (169): 31–41.

17. FLANNERY, K. V.
    1965 The ecology of early food production in Mesopotamia. Science 147: 1247–1256.

18. HARLAN, J. R., AND D. ZOHARY
    1966 Distribution of wild wheats and barley. Science 153: 1074–1080.

19. SNEAD, R. E.
    1964 Active mud volcanoes of Baluchistan, West Pakistan. Geographical Review 54(4): 546–560.

20. ————
    1967 Recent morphological changes along the coast of West Pakistan. Association of American Geographers, Annals 57(3): 550–565.

21 ————
    1968 Weather patterns in southern West Pakistan. Archiv für Meteorologie, Geophysik und Bioklimatologie B, 16: 316–346.

22. ————
    1969 Physical geography reconnaissance: West Pakistan coastal zone. University of New Mexico Publications in Geography 1. 55 p.

22a.————
    1970 Physical geography of the Makran Coastal Plain of Iran: Final Report, reconnaissance phase. University of New Mexico, Albuquerque, for Office of Naval Research, Contract N00014-66-C-0104. 715 p.

23. ———— and S. A. FRISHMAN
    1968 Origin of sands on the east side of the Las Bela Valley, West Pakistan. Geological Society of America, Bulletin 79: 1671–1676.

24. ———— and MHD. TASNIF
    1966 Vegetation types in the Las Bela region of West Pakistan. Ecology 47(3).

25. CANADA. DEPARTMENT OF MINES AND TECHNICAL SURVEYS, SURVEYS AND MAPPING BRANCH
    1958 Reconnaissance geology of part of West Pakistan. A Colombo Plan Cooperative Project. Surveyed and compiled in 1953–56 by Photographic Survey Corporation, Ltd, Toronto, in cooperation with the Geological Survey of Pakistan. Published for the Government of Pakistan by the Government of Canada, Ottawa. 31 sheets, 29 colored maps. Accompanied by text: Hunting Survey Corporation, Ltd, 1960. Reconnaissance geology of part of West Pakistan. Toronto. 550 p.

26. AHMAD, K. S.
    1964   A geography of Pakistan. Oxford University Press, Karachi. 216 p.

27. LeStrange, G.
    1905   The lands of the eastern caliphate. Frank Cass and Co, London. 536 p.

28. AITCHISON, J. E. T.
    1887   The zoology of the Afghan Delimitation Commission. Linnean Society of London, Transactions 5: 53–142.

29. ———
    1896   Botany of the Afghan Delimitation Commission, 1884–85. Linnean Society of London, Transactions, ser. 2, Vol. 3. 139 p.

30. ANONYMOUS
    1901   Baluchistan desert and part of East Persia. Geological Survey of India, Memoirs 31(2).

31. ———
    1918   Bibliography of Baluchistan natural history. Baluchistan Natural History Society, Journal 1.

32. BELLEW, H. W.
    1874   From the Indus to the Tigris. Trübner & Co, London. 496 p.

33. BLANFORD, W. T.
    1872   Note on geological formations seen along the coasts of Baluchistan and Persia from Karachi to head of Persian Gulf, and some Gulf islands. Geological Survey of India, Records 5(2).

34. BURKILL, I. H.
    1909   A working list of the flowering plants of Baluchistan. Superintendent Government Printing, Calcutta. 136 p.

35. FERRIER, J. P.
    1857   Caravan journeys and wanderings in Persia, Afghanistan, Turkistan, and Beloochistan. 2d ed. J. Murray, London. 534 p.

36. GOLDSMID, F. J.
    1873   Journey from Bandar Abbas to Mash-had by Sistan, with some account of the last named province. Royal Geographical Society, Journal 43: 65–83.

37. ———
    1876   Eastern Persia: an account of the journeys of the Persian boundary commission, 1870–72. Oxford, London. 2 vols.
    Volume 1, Geography; Volume 2, Zoology and Geology.

38. HAIG, M. R.
    1896   Ancient and medieval Makran. Royal Geographical Society, Journal 7: 668–674.

39. HEDIN, S.
    1910   Overland to India. Macmillan and Company, London. 342 p.

40. HOLDICH, T. H.
    1896   Notes on ancient and medieval Makran. Royal Geographical Society, Journal 7(4): 387–405, 677.

41. ———
    1897   The Perso-Baluch boundary. Royal Geographical Society, Journal 9: 416–422.

42. ———
    1901   The Indian borderland, 1880–1900. Methuen and Co, London. 402 p.

43. ———
    1910   The Gates of India. Macmillan and Co, Ltd, London. 555 p.

44. HUGHES, A. W.
    1877   The country of Balochistan, its geography, topography, ethnology and history. G. Bell & Sons, London. 294 p.

45. ———
    1904–   Gabrbands in Baluchistan. Archaeological
    1905   Survey of India, Report.

46. FLOYER, E. A.
    1882   Unexplored Baluchistan: a survey, with observations astronomical, geographical, botanical, etc., of a route through Mekran, Bashkurd, Persia, Kurdistan and Turkey. Griffith and Farran, London. 507 p.

47. HUNTINGTON, E.
    1905   The basin of eastern Persia and Sistan. In R. Pumpelly, Explorations in Turkestan. Carnegie Institution of Washington, Publication 26: 219–317.

48. ———
    1905   The depression of Sistan in eastern Persia. American Geographical Society, Bulletin 37: 271–281.

49. KHAN, M. M.
    1877   Notes on Persian Baluchistan (translation of report to Persian Government). Royal Asiatic Society, Journal 9: 147–154.

50. LACE, J. H., and W. B. HEMSLEY
    1891   A sketch of the vegetation of British Baluchistan, with descriptions of new species. Linnean Society of London, Journal, Botany 28: 288–327.

51. MacGREGOR, C. M.
    1882   Wanderings in Baluchistan. W. H. Allen, London. 315 p.

52. MASSON, C.
    1842   Narrative of various journeys in Balochistan, Afghanistan, and Punjab, including a residence in those countries from 1826 to 1839. London, 4 vols.
    A later printing (1844) is listed as follows:
    Narrative of various journeys in Balochistan, Afghanistan, Panjab, and Kalat, during a residence in those countries from 1826 to 1838. R. Bentley, London. 4 vols.

53. McMAHON, A. H.
    1897   The southern borderlands of Afghanistan. Geographical Journal 9: 393–415.

54. MOCKLER, E.
    1876   On ruins in Makran. Royal Asiatic Society, Journal 9(1).

55. NOETLING, F. W.
    1895–   Fauna of Baluchistan. Geological Survey
    1897   of India, Palaeontologia Indica, Series 16 1(1–3).

56. POTTINGER, H.
    1816   Travels in Beloochistan and Sinde. Longman, Hurst, Rees, Orme, and Brown, London. 423 P.

57. RAVERTY, H. G.
    1880–   Notes on Afghanistan and part of Baluchis-
    1883   tan; geographical, ethnographical and historical. G. E. Eyre & Spottiswoode, London. 3 parts.

58. RAWLINSON, H. C.
    1873   Notes on Sistan. Royal Geographical Society, Journal 43: 272–294.

59. ROSS, E. C.
    1866   A memorandum of notes on Mekran, together with a report on a visit to Koj and route

through Mekran from Gwadur to Kurachee. Bombay Geographical Society, Transactions 18: 36–77.

60. ———
1871 Routes in Baluchistan. London.

61. SAVAGE-LANDOR, A. H.
1903 Across coveted lands; or, A journey from Flushing (Holland) to Calcutta, overland. C. Scribner's Sons, New York. 2 vols.

62. SKRINE, C.
1931 The highlands of Persian Baluchistan. Geographical Journal 78:(321–340).

63. SYKES, P. M.
1902 Ten thousand miles in Persia; or, Eight years in Iran. J. Murray, London. 481 p.

64. TATE, G. P.
1909 The frontiers of Baluchistan. Witherby, London. 261 p.

65. ———
1910– Seistan: A memoir on the history, topography,
1912 ruins, and people of the country. Superintendent Government Printing, Calcutta. 4 parts in 2 vols.

66. TAYLOR, A.
1882 Unexplored Baluchistan. London.
Referred to in several bibliographies. May be same as *46*.

67. VREDENBURG, E.
1901 A geological sketch of the Baluchistan desert and part of eastern Persia. Geological Survey of India, Memoir 31(2): 179–302.

68. YATE, CHARLES E.
1906 Baluchistan. Central Asian Society, London. 39 p.

69. BALUCHISTAN DISTRICT GAZETTEER SERIES
1906 Bolan Pass and Nushki Railway District. 4. Items *69, 70,* and *71* are part of a series of 14 published from 1906 through 1908 in Bombay. C. F. Minchin is listed as editor of the series.

70. ———
1907 Chagai District. 4–A.

71. ———
1907 Quetta-Pishin District. 5.

72. BILLIMORIA, N. M.
1930 Bibliography of publications on Sind and Baluchistan. 2d ed, rev and enl. Author, Karachi. 136 p.

73. FIELD, H.
1959 An anthropological reconnaissance in West Pakistan, 1955, with appendixes on the archaeology and natural history of Baluchistan and Bahawalpur. Harvard University, Peabody Museum, Paper 52. 332 p.

74. HARRISON, J. V.
1942 Some routes in southern Iran. The Geographical Journal 99(3): 113–129.

75. GREAT BRITAIN NAVAL INTELLIGENCE DIVISION
1944 Iraq and the Persian Gulf. *Its* Geographical Handbook Series B. R. 524. 638 p.

76. JANJUA, N. A.
1940 The cultivators of Baluchistan. Indian Farming (New Delhi) 1(12): 596–601.

77. GREAT BRITAIN NAVAL INTELLIGENCE DIVISION
1945 Persia. *Its* Geographical Handbook Series B. R. 525. 638 p.

78. PITHAWALLA, M. B.
1946 Physiographic divisions of the Iran Plateau. University of Bombay, Journal 14(4): 45–51.

79. RAMACHANDRA RAO, Y.
1941 A list of some of the more common plants of the desert areas of Sind, Baluchistan, Rajputana, Kathlawar and southwest Punjab, with their various local names as far as available. Imperial (Indian) Council of Agricultural Research, Miscellaneous Bulletin 43: 1–45.

80. RAY, A. K., and R. C. BHATTACHARYA
1946 A preliminary study of rainfall at Quetta. India Meteorological Department 5(51).

81. SEDLACEK, A. M.
1955 Sande und Gesteine aus der sudlichen Lut and Persisch-Belutschistan. Oesterreichische Akademie der Wissenschaften, Vienna, Sitzungsberichte Abt. 1,164: 607–657.

82. SIAL, N. M., and A. R. GHANI
1953 On the need for a comprehensive multilingual dictionary of Pakistan plant-names; together with a list of existing glossaries and indexes. Pakistan Journal of Forestry 3(1): 14–20.

83. ALLCHIN, B., and R. ALLCHIN
1968 The birth of Indian civilization. Penguin Books, Harmondsworth, Middlesex. 365 p.

84. GHIRSHMAN, R.
1954 Iran. Penguin Books, Harmondsworth, Middlesex. 368 p.

85. HARGREAVES, H.
1929 Excavations in Baluchistan, 1925. Archaeological Survey of India, Memoir 35. 89 p.

86. PORADA, E.
1965 Ancient Iran, the art of pre-Islamic times. (Art of the World, a series of regional histories of the visual arts.) Methuen, London. 279 p.

87. STEIN, SIR AUREL
1928 Innermost Asia. Oxford University Press, Oxford. 3 vols.

88. ———
1929 An archaeological tour in Waziristan and north Baluchistan. Archaeological Survey of India, Memoir 37.

89. ———
1931 An archaeological tour in Gedrosia. Archaeological Survey of India, Memoir 43. 211 p.

90. ———
1937 Archaeological reconnaissances in north-western India and south-eastern Iran. Macmillan and Company, London. 267 p.

91. BERGHE, L. VAN DEN
1959 Archéologie de l'Iran ancien. E. J. Brill, Leiden. 285 p. (Documenta et Monumenta Orientis Antiqui, 6).

92. WULFF, H. E.
1966 The traditional crafts of Persia, their development, technology, and influence on Eastern and Western civilizations. Massachusetts Institute of Technology Press, Cambridge. 404 p.

93. WHEELER, R. E. M.
1956 Archaeology from the earth. Penguin Books, Harmondsworth, Middlesex. 252 p.

94. STUCKENRATH, R., JR.
1965 On the care and feeding of radiocarbon dates. Archaeology 18(4): 277–281.

95. NEUSTUPNY, E.
1968 Absolute chronology of the neolithic and aeneolithic periods in central and southeastern Europe. Slovenska Archeologia 16(1): 19–56.

96. STUIVER, M., and H. E. SUESS
1966   On the relationship between radiocarbon dates and true sample ages. Radiocarbon 8: 534–540.

97. MICHAEL, H. N., and E. K. RALPH
1970   Correction factors applied to Egyptian radiocarbon from the era before Christ. *In* I. U. Olsson (ed), Radiocarbon variations and absolute chronology. Nobel Symposium, 12th, Institute of Physics at Uppsala University, 1969, Proceedings p. 109–120. John Wiley, New York.

98. RADIOCARBON, Vol. 1, 1959 to date. New Haven, Conn. (Title varies: v. 1–2, 1959–60, published as American Journal of Science, Radiocarbon Supplement.)

99. OPPENHEIM, A. L.
1964   Ancient Mesopotamia. University of Chicago Press, Chicago and London. 433 p.

100. MALLOWAN, M. E. L.
1968   The early dynastic period in Mesopotamia. *In* Cambridge ancient history, preliminary fascicles for revised edition, v 1, chap 16. 71 p.

101. GADD, C. J.
1963   The dynasty of Agade and the Gutian invasion. *In* Cambridge ancient history, preliminary fascicles for revised edition, v 1, chap 19. 54 p.

102. DALES, G. F.
1965   A suggested chronology for Afghanistan, Baluchistan and the Indus Valley. *In* R. W. Ehrich (ed), Chronologies in Old World archaeology, p. 257–284. University of Chicago Press, Chicago.

103. ———
1968   A review of the chronology of Afghanistan, Baluchistan and the Indus Valley. American Journal of Archaeology 72:305–307.

104. ———
1968   Of dice and men. American Oriental Society, Journal 88(1):14–23.

105. DYSON, R. H.
1965   Problems in the relative chronology of Iran, 6000–2000 B.C. *In* R. W. Ehrich (ed), Chronologies in Old World archaeology, p. 215–256. University of Chicago Press, Chicago.

106. ———
1968   Annotations and corrections of the relative chronology of Iran, 1968. American Journal of Archaeology 72:308–313.

107. PORADA, E., and D. P. HANSEN
1965   The relative chronology of Mesopotamia 6000–1000 B.C. *In* R. W. Ehrich (ed), Chronologies in Old World archaeology, p. 133–213. University of Chicago Press, Chicago.

108. ———
1968   The relative chronology of Mesopotamia 6000–1000 B.C. American Journal of Archaeology, 72:301–305.

109. WHEELER, R. E. M.
1968   The Indus civilization. 3d ed. (Supplementary vol., Cambridge History of India) Cambridge University Press, Cambridge. 144 p.

110. LAYARD, A. H.
1846   A description of the province of Khuzistan. Royal Geographic Society, London, Journal 16.

111. MEIGS, P.
1966   Geography of coastal deserts. Unesco, Paris. Arid Zone Research 28.

112. LEBRETON, L.
1957   The early periods at Susa, Mesopotamian relations. Iraq 19(2):79–124.

113. McCOWN, D. E.
1942   The comparative stratigraphy of early Iran. University of Chicago, Oriental Institute, Studies in Ancient Oriental Civilization 23. 65 p.

114. HELBAEK, H.
1969   Plant collecting, dry-farming, and irrigation agriculture in prehistoric Deh Luran. *In* F. Hole, K. V. Flannery, and J. A. Neely (eds), (*12*), p. 383–426.

115. DOUGHERTY, R. C., and J. R. CALDWELL
1966   Evidence of early pyrometallurgy in the Kerman range in Iran. Science 153:984–985.

116. LAMBERG-KARLOVSKY, C. C.
1967   Archaeology and metallurgical technology in prehistoric Afghanistan, India and Pakistan. American Anthropologist 69(2): 145–162.

117. ———
1969   Further notes on the shaft-hole pick-axe from Khurab, Makran. Iran 7:163–168.

118. WERTIME, T. A.
1968   A metallurgical expedition through the Persian desert. Science 159(3818):927–935.

119. GOFF, C.
1963   Excavations at Tall-i-Nokhodi. Iran 1:43–70.

120. ———
1964   Excavations at Tall-i-Nokhodi, 1962. Iran 2:41–52.

121. EGAMI, N., and T. SONO
1962   The excavation at Tall-i-Gap, 1959. Tokyo University Iraq-Iran Expedition, Report. v 2.

122. CRESSEY, G. B.
1958   Qanats, karez, and foggaras. Geographical Review 48:27–44.

123. SMITH, A.
1953   Blind white fish in Persia. George Allen and Unwin, London. 231 p.

124. BÖKÖNYI, SANDOR
1967   The prehistoric vertebrate fauna of Tal-i-Iblis. *In* J. R. Caldwell (ed), Preliminary Report 9:309–317. Illinois State Museum, Springfield.

125. LAMBERG-KARLOVSKY, C. C.
1969   Tepe Yahya: Survey of excavations. Iran 7: 197–198.

125a.———
1970   Excavations at Tepe Yahya, Iran: Progress Report I. American School of Prehistoric Research, Peabody Museum, Harvard University, Bulletin 27.

125b.———
1971   The Proto-Elamite settlement at Tepe Yahya. Iran 9.

126. DeCARDI, B.
1967   The Bampur sequence in the third millennium B. C. Antiquity 12:33–41.

127. ———
1968   Bampur, a third millenium site in Persian Baluchistan. Archaeologia Viva 1:151–155.

128. ———
1968   Excavations at Bampur, southeast Iran: A brief report. Iran 6:135–155.

129. ———
    1970   Excavations at Bampur, a Third Millennium Settlement in Persian Baluchistan, 1966. American Museum of Natural History, Anthropological Papers 51(3):233–355.

130. LAMBERG-KARLOVSKY, C. C.
    1970   A re-evaluation of the Bampur, Khurab and Chah Husseini collection in the Peabody Museum. Private manuscript.

131. KUML
    1953   Årbog for Jysk Arkaeologisk Selskab, Moesgard (yearly numbers since 1953).

132. BIBBY, G.
    1958   Excavating a Bahrein citadel of 5000 years ago, and seal links with Ur and Mohenjo-Daro. Illustrated London News 232(6188): 54–55.

133. ———
    1964   A forgotten civilization of Abu Dhabi. British Petroleum Company, London, BP Magazine 13:28–32.

133a.———
    1969   Looking for Dilmun. Alfred A. Knopf, New York.

134. GLOB, P. V.
    1958   The prosperity of Bahrein five thousand years ago: solving the riddle of the 100,000 burial mounds of the island. Illustrated London News 232(6187):14–16.

135. GLOB, P. V., and T. G. BIBBY
    1960   A forgotten civilization of the Persian Gulf. Scientific American 203(4):62–71.

136. ROUSSELL, A.
    1961   A matter of one thousand stamp seals of 4000 years ago. The amazing potential of a Kuwait island tell. Illustrated London News 238(6339):142–143.

137. BIBBY, T. G.
    1958   The "ancient Indian style" seals from Bahrain. Antiquity 32:243–246.

138. GADD, C. J.
    1932   Seals of ancient Indian style found at Ur. British Academy, Proceedings 18:3–22.

139. RAO, S. R.
    1963   A Persian Gulf seal from Lothal. Antiquity 37:96–99.

140. BUCHANAN, B.
    1967   A dated seal impression connecting Babylonia and ancient India. Archaeology 20: 104–107.

141. HALLO, W. W., and B. BUCHANAN
    1965   A "Persian Gulf" seal on an Old Babylonian mercantile agreement. *In* Studies in honor of Benno Landsberger, pp. 191–209. University of Chicago Press, Chicago.

142. OPPENHEIM, A. L.
    1954   The seafaring merchants of Ur. American Oriental Society, Journal 74:6–17.

143. ———
    1967   Essay on overland trade in the first millennium B. C. Journal of Cuneiform Studies 21: 236–254.

144. LEEMANS, W. F.
    1960   Foreign trade in the Old Babylonian period as revealed by texts from southern Mesopotamia. Brill, Leiden. (Studia et Documenta ad Jura Orientis Antiqui Pertinentia, 6. 196 p.)

145. ———
    1968   Old Babylonian letters and economic history. Journal of the Economic and Social History of the Orient 11:171–226.

146. BIROT, M.
    1962   Review of Leeman's "Foreign trade in the Old Babylonian period." Journal of the Economic and Social History of the Orient 5: 91–109.

147. WEITEMEYER, M.
    1964   Notes on a recent study of Old Babylonian trade. Acta Orientalia 28(1–2):205–213.

148. MALLOWAN, M. E. L.
    1965   The mechanics of ancient trade in western Asia. Iran 3:1–8.

149. KRAMER, S. N.
    1963   The Sumerians. University of Chicago Press, Chicago. 299 p.

150. ———
    1963   Dilmun: Quest for paradise. Antiquity 37: 111–115.

151. ———
    1964   The Indus civilization and Dilmun, the Sumerian paradise land. Expedition 6:44–52.

152. DALES, G. F.
    1962   Harappan outposts on the Makran coast. Antiquity 36:86–92.

153. ———
    1962   A search for ancient seaports. Expedition 4:2–10.

154. CASAL, J.-M.
    1966   Nindowari, a chalcolithic site in south Baluchistan. Pakistan Archaeology 3:10–21.

155. DALES, G. F., and R. L. RAIKES
    1968   The Mohenjo-daro floods: A rejoinder. American Anthropologist 70(5):957–961.

156. DeCARDI, B.
    1950   On the borders of Pakistan: Recent exploration. Royal Indian, Pakistan and Ceylon Society, Journal 24(2):52–57.

157. ———
    1959   New wares and fresh problems from Baluchistan. Antiquity 33:15–24.

158. ———
    1965   Excavations and reconnaissance in Kalat, West Pakistan: The prehistoric sequence in the Surab region. Pakistan Archaeology 2: 86–181.

159. FAIRSERVIS, W. A.
    1956   Excavations in the Quetta Valley, West Pakistan. American Museum of Natural History, Anthropological Papers 45(2):169–402.

160. ———
    1959   Archaeological surveys in the Zhob and Loralai districts, West Pakistan. American Museum of Natural History, Anthropological Papers 47(2):277–448.

161. PAKISTAN ARCHAEOLOGY, No. 1, 1964 to date.
    Department of Archaeology, Karachi.

162. PIGGOTT, S.
    1950   Prehistoric India. Penguin Books, Harmondsworth, Middlesex. 289 p.

163. CASAL, J.-M.
    1961   Les débuts de la civilisation de l'Indus a la lumière de fouilles récentes. Académie des Inscriptions et Belles-Lettres, Paris, Comptes-Rendus des Séances de l'année 1960:305–316.

164. FAIRSERVIS, W. A.
1967 The origin, character, and decline of an early civilization. American Museum Novitates 2302. 48 p.

165. MALIK, S. C.
1968 Indian civilization, the formative period. Indian Institute of Advanced Study, Simla. 204 p.

166. Floral and faunal remains from the Quetta Valley test excavations were too sparse to be of much positive use, per Fairservis (*160*, p. 382–385).

167. DALES, G. F.
1968 The South Asia section. Expedition 11(1): 38–45.
Detailed reports on the 1968–69 expeditions are in preparation.

168. FAIRSERVIS, W. A.
1961 Archeological studies in the Seistan basin of southwestern Afghanistan and eastern Iran. American Museum of Natural History, Anthropological Papers 48(1). 128 p.

169. TROUSDALE, W.
1967 Land of the Sistan sands. MidEast 7(7): 7–14.

170. DUPREE, L.
1963 Deh Morasi Ghundai, a chalcolithic site in south-central Afghanistan. American Museum of Natural History, Anthropological Papers 50(2): 59–135.

171. CASAL, J.-M.
1961 Fouilles de Mundigak. Délégation Archéologique Française en Afghanistan, Mémoire 17. Librairie C. Klincksieck, Paris. 2 vols. 260 p. et album de pls.

172. POULAIN, T.
1966 Étude de la fauna. L'École Française d'Extreme-Orient Bulletin 53(1): 119–135.

173. HACKIN, J., J. CARL, and J. MEUNIE (eds)
1959 Diverses recherches archéologiques en Afghanistan (1933–1940). Délégation Archéologique Française en Afghanistan, Mémoire 8. Presses Universitaires de France, Paris. 140 p.

174. HAMMOND, N.
1967 The Afghan Road into Seistan. The Illustrated London News 250(6654): 23–25.

175. GHIRSHMAN, R.
1959 Recherches préhistoriques dans la partie Afghane du Seistan. *In* J. Hackin, J. Carl, and J. Meunie (eds) (*173*), p. 39–48.
Also published as Fouilles de Nad-i-Ali dans le Seistan Afghan. Revue des Arts Asiatiques 13 (1939–42): 10–22.

176. TOSI, M.
1968 Excavations at Shahr-i Sokhta, a Chalcolithic settlement in the Iranian Sistan: Preliminary report on the first campaign, October-December 1967. East and West 18(1–2): 9–66.

176a.———
1969 Excavations at Shahr-i Sokhta: Preliminary report on the second campaign, September-December 1968. East and West 19(3–4): 283–386.

177. HERRMANN, G.
1966 The source, distribution, history and use of lapis lazuli in western Asia from the earliest times to the end of the Seleucid era. Unpublished PhD dissertation, Oxford University.

178. ———
1968 Lapis lazuli: The early phases of its trade. Iraq 30(1): 21–57.

179. DURRANI, F. A.
1964 West Pakistan and Persian Gulf in antiquity. Asiatic Society of Pakistan, Journal 9(1): 1–12.

180. ———
1964 Stone vases as evidence of connection between Mesopotamia and the Indus Valley. Ancient Pakistan 1: 51–96.

181. RAO, S. R.
1965 Shipping and maritime trade of the Indus people. Expedition 7: 30–37.

181a. GELB, I. J.
1970 Makkan and Meluhha in early Mesopotamian sources. Revue d'Assyriologie et d'Archéologie Orientale 64(1): 1–8.

182. GERSHEVITCH, I.
1957 Sissoo at Susa. University of London, School of Oriental and African Studies, Bulletin 19: 317–320.

183. HOURANI, G. F.
1951 Arab seafaring. Princeton University Press, Princeton. 133 p.

184. PRINS, A. H. J.
1966 The Persian Gulf dhows: Two variants in maritime enterprise. Persica 2(1965–66): 1–18.

184a. MUGHAL, MHD. R.
1970 The Early Harappan period in the Greater Indus Valley and northern Baluchistan (c. 3000–2400 B.C.). University of Pennsylvania, Philadelphia, Department of Anthropology, PhD dissertation.

185. I wish to thank Louis Flamm and Barbara Dales for their diligence and care in the preparation of the charts and manuscripts.

Part Two

LAND AND RESOURCE USES

Applications of Remote Sensing to Arid-Lands Problems

ROBERT N. COLWELL
and
DAVID M. CARNEGGIE

School of Forestry and Conservation
University of California,
Berkeley, California, U.S.A.

ABSTRACT

REMOTE SENSING can greatly facilitate the gathering of information needed for preparing inventories of arid lands. In many arid-land regions of the world, there may be no acceptable alternative when one takes into consideration accessibility, costs, time, and the types of information desired. Recently developed film types and remote-sensing devices provide the arid-land manager with greatly increased capabilities for acquiring information about a wide diversity of arid-land problems. The concept of multiband reconnaissance provides a key for maximizing the amount, accuracy, and timeliness of arid-land information. Meaningful classification of arid lands into broad categories can be accomplished from a study of space photographs. Then, more detailed information can be obtained by subsampling each broad category with large-scale aerial photos. For this reason, we see great promise for the use of multistaged sampling techniques that will employ either space photos or very-high-altitude aerial photos in the initial stage, and very-large-scale aerial photos in a later stage. By such means the required information for performing integrated resource analyses of arid lands can best be made, thereby facilitating the wise management of these lands.

APPLICATIONS OF REMOTE SENSING TO ARID-LANDS PROBLEMS

Robert N. Colwell and David M. Carneggie

THE TERM "remote sensing," as employed in this chapter, pertains to the acquiring of information through the use of cameras or other sensing devices, operated from aircraft or spacecraft that are situated at some distance from the area being investigated.

The rationale for applying remote-sensing techniques to the solution of certain arid-land problems is as follows:

a) Because the world's population is rapidly increasing, the need for developing and wisely managing its resources also is increasing rapidly.

b) Vast arid-land areas that are richly endowed with resources that might benefit mankind continue to remain largely undeveloped and unmanaged.

c) An important first step leading to the wise development and management of arid lands is that of obtaining an accurate inventory of their resources.

e) Remote sensing from aircraft and spacecraft shows great promise as the means by which resource inventories of arid lands might best be made.

Consistent with the theme of this book, major emphasis will be given to those remote-sensing techniques that might lead to an increased production of food and fiber on arid lands. It is primarily because of impending shortages of these commodities that mankind soon will be confronted with one of the most serious crises of his existence. There is abundant evidence for the validity of this assertion. For example, the Food and Agricultural Organization of the United Nations recently stated that, to provide a decent level of nutrition to the world's people, the world's production of food will have to be doubled by the year 1980 and trebled by 2000. Other studies have documented the need for a similar rate of increase in the production of fiber. In addition, the supply versus demand problem with respect to these two commodities (food and fiber) has been made even more serious than a mere look at the problem of a population explosion would have indicated, because of another recent development: Within the past decade a greatly increased "awareness of the have-nots," both in the United States and elsewhere, combined with an almost insatiable demand of the "affluents" for an ever-higher standard of living, has resulted in a tremendous increase in the *per capita* demand for food and fiber products at virtually every economic level.

BASIC CONSIDERATIONS
Advantages of Remote Sensing

Remote sensing from aircraft or spacecraft can provide the most suitable means for rapidly collecting data about arid-land resources, especially when the alternative is to make laborious on-the-ground surveys. There are several reasons why this is so: the first of these is clearly implied in the simple statement that "the face of the land looks to the sky." The task of inventorying arid land resources is, first of all, one of delineating boundaries between one resource characteristic and another. When confined to Earth, man often has great difficulty both in recognizing and in delineating these boundaries. This difficulty is attributable mainly to the limited visibility of terrain features that is afforded to him, especially in areas where the topography is heavily dissected. In contrast, a "continuous plotting" of these resource boundaries can be performed on vertical photos taken from aircraft or spacecraft.

A second reason for using aircraft or spacecraft results from the sheer vastness of the arid lands in which earth-resource surveys must be made. It is only from aircraft or spacecraft that the broad synoptic view, so essential for quick and economical delineation of arid-land resource features, can be obtained.

Another advantage is the fact that a vast portion of the total arid lands that are to be inventoried is more frequently cloud-free than humid areas; hence they can be photographed at a single point in time and space and under relatively uniform lighting conditions. This potential advantage is much more difficult to exploit in the more humid areas because of the frequency with which clouds obscure the view. For example, a study recently was made to determine the feasibility of obtaining space photography of an agricultural area in Indiana at various times during the growing season. Because of the high humidity prevailing in that nonarid land, clouds obscure it much of the time. In fact, this study showed that, for a representative 1,000-square-mile area in Indiana to be photographed in a cloud-free condition, only one Earth-orbital pass in 28 would find suitable weather conditions over the area to provide the desired coverage. Even if discrete clouds did not obscure such an area, the high atmospheric humidity would make it far less photogenic than a typical arid land area.

The ability of an aircraft or spacecraft to travel *quickly* from one camera station to another constitutes still another advantage when the objective is to acquire detailed information about a large arid region in a short period of time and with a minimum of discomfort to personnel performing the inventory.

Specifications for Remote Sensing

Despite the many advantages for using remote-sensing techniques in the inventory of arid-land resources, meaningful information is provided to the ultimate user (that is, the land manager) only when the remote sensing is performed to proper specifications. These specifications must consider (*a*) the wavelength bands to be used, (*b*) the spatial resolution obtainable at various flight altitudes in relation to the resource details that must be discerned, and (*c*) the time of day and season of year when these resource details are most photogenic.

With reference to the wavelength bands to be used, one should recognize that there are many remote-sensing devices (for example, single lens and multilens cameras, optical mechanical scanners, thermal and radar scanners) capable of recording wavelengths of reflected and/or emitted radiation from different portions of the electromagnetic spectrum. However, we are here primarily concerned with multiband black-and-white photos and color photos (either conventional color or "false color") secured by aerial cameras as a means for gathering information about arid lands.

In order to appreciate the value of procuring multiband photography (photographs taken, usually simultaneously, in each of several wavelength bands in the visible and near infrared), we first need to answer the following question: why should we expect to be able to inventory Earth resources better on multiband photography than on photography taken in only one wavelength band? The answer to that question, and the rationale behind it, can be expressed as follows: our ability to inventory Earth-resource features on multiband photography rests on the fact that every type of feature encountered on the surface of the Earth tends to reflect and emit radiant energy in distinctive amounts at certain specific wavelengths. Consequently, when remote sensing is done simultaneously in each of several wavelength bands (a process variously known as "multiband sensing," "multispectral sensing," and "multiband spectral reconnaissance"), a multiband "tone signature" or "spectral response pattern" can be determined which theoretically is distinctive for each type of feature. Obviously, if only one band is used, the complete tone signature is not decipherable; consequently, ambiguities may exist regarding the identity of a feature.

Despite the theoretical value of the multiband concept, the photo-interpreter is confronted with a laborious and time-consuming task of extracting useful information from the photos. Specifically, for each feature that is to be identified from its tone signature, the interpreter must observe its gray scale value as imaged on *each* of the multiband series of photographs. Then he must compare this signature with a master set of signatures which has been compiled for a great many types of features. Presumably by this means he will have correctly identified the feature in question when he has found which feature in the master set has the most similar tone signature.

To overcome this troublesome problem, a form of "image enhancement" has been developed that employs the "additive color" process. The purpose of such enhancement is twofold: to increase the total amount of information derivable from the "raw data," and to facilitate the data-extraction process. The image-enhancement process operates in the following manner:

a) A lantern slide (that is, a positive black-and-white transparency) is made of each of the multiband photographs that have been obtained of the area to be interpreted.

b) One of the resulting series of slides is projected onto a screen through a *red* filter, so that shades of gray as seen on the original slide appear as corresponding shades of red on the screen.

c) By means of a second lantern slide projector, another slide from the multiband series is *simultaneously* projected onto the screen through, say, a *green* filter, thereby converting shades of gray on that slide to shades of green on the screen.

d) Similarly, through use of a third projector, a third member of the multiband series is projected through, say, a *blue* filter.

When all the projected images are superimposed so that conjugate images are in common register, each feature that has a unique "tone signature" on the multiband black-and-white photos exhibits a unique "color signature" in the combined color image appearing on the screen. Various color composites can be made in this manner simply by changing the combination of filters. Ideally, the operator seeks those filter combinations which emphasize specific features of interest.

The method that has just been described is an *optical* method of enhancing multiple black-and-white images by the additive color process. The colors exhibited on the composite image are a function of (*a*) image densities on the multiband transparencies and (*b*) the colors of filters through which the transparencies are projected. Other additive color techniques employ an electronic process to produce the enhanced image on a closed-circuit color television screen. While the electronically enhanced images have lower spatial resolution than those produced optically, they have certain offsetting advantages. These include (*a*) the ability of the operator to "level select" for tone or brightness range on one or more images for the purpose of highlighting or encoding spectral densities, and (*b*) the ability to record the electronic signals in analogue or digital form on magnetic tape as the first step in using computer techniques for the automatic identification of features.

With reference to the spatial resolution needed to discern significant resource detail, one should recognize that different remote-sensing devices, when operated at a given level (altitude), inherently yield images that differ considerably in spatial resolution. If, however, we consider a single remote-sensing device, such as an aerial

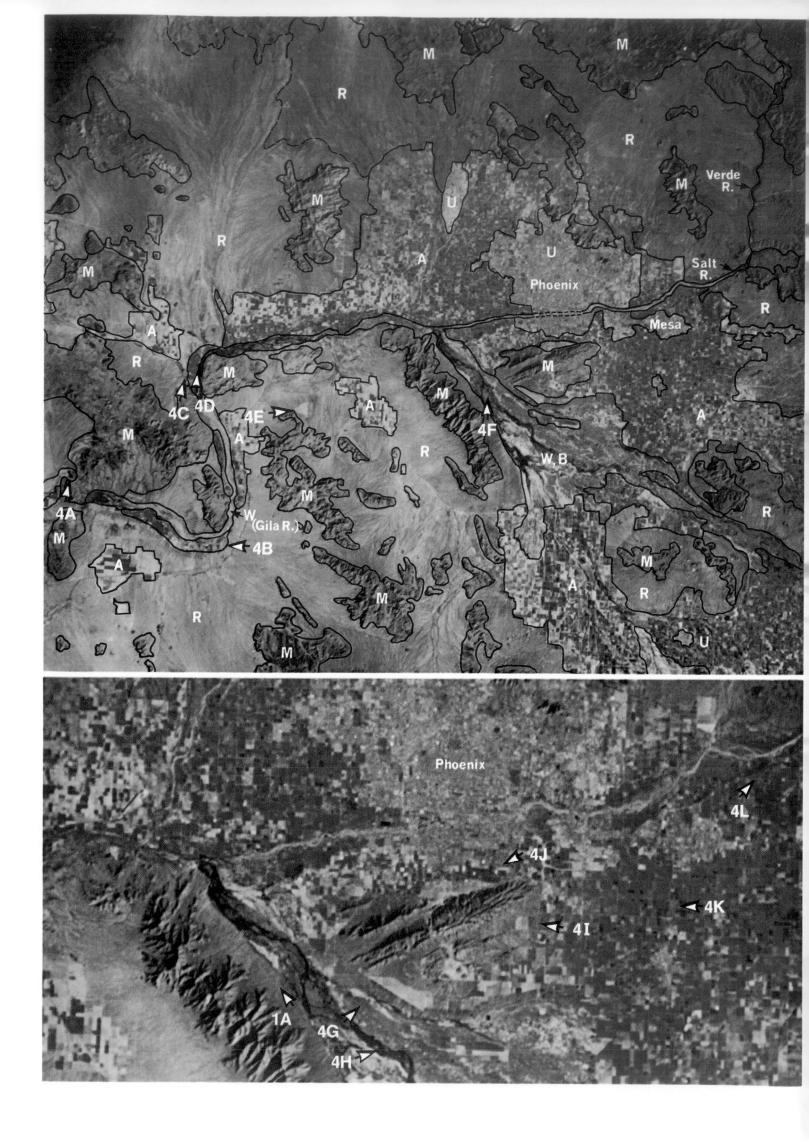

camera that is operated at different levels (altitudes), we obviously would expect to obtain photographs that have higher spatial resolution at lower altitudes. The choice of optimum spatial resolution depends upon the level of detail needed to satisfy certain management objectives and upon the size of features and the tone contrasts between features of interest. If the user is primarily concerned with classifying broad landforms, he may find that the relatively low spatial resolution of space photographs (see Fig. 1) is adequate to perform this task. If, however, the user desires detailed information concerning, for example, the identification and/or quantification of individual plants, he may need to obtain very-large-scale photos of relatively high resolution (such as those seen in Fig. 2).

With reference to the season of the year to obtain images that are readily interpretable, one should recognize that the transient nature or changing appearance that characterizes most features can be exploited as an aid to evaluating such features. Specifically, as vegetation develops, it progresses through distinctive phenological stages. If photographs are obtained at a date when the phenological stage of the vegetation is distinctive, there is a more likely chance that it will be discerned. Oftentimes individual plant species, both natural and cultivated, can be identified because they possess flowers or distinctive leaf coloration. In arid regions, flushes of vegetative growth accompany sporadic rainfall. Photos obtained following this event could be useful for determining the amount and distribution of annuals that germinate and grow in response to the rainfall. Similarly, aerial photos taken at the appropriate time are useful for determining the distribution and availability of water in ephemeral streams or waterholes.

KEY: A — Agriculture
           M — Upland mountains
           R — Range
           U — Urban
           W — Watercourse
           B — Bottomland

Fig. 1. (*Top*): Part of an Infrared Ektachrome taken from an altitude of 145 statute miles on March 12, 1968, by the Apollo 9 astronauts. Natural and cultural resources are being evaluated within this approximately 7,000-square-mile study area near Phoenix, Arizona. High-resolution aerial photos obtained at the same time as the space photos aid in determining the usefulness of the space photos for making resource inventories. The area occupied by the various land-use categories delineated here was determined by an area calculator as: agricultural cropland 20%, rangeland 43%, uplands and mountains 24%, watercourses 8%, and urban 5%. The cropland estimates need to be scaled down by about 10% to account for roads and farm buildings included within the area.

(*Bottom*): Enlargement of a portion of the top space photo. Of particular interest is the extensive agricultural development, where fields as small as 40 acres are readily discernible. Fields containing a dense stand of vegetation (mostly alfalfa and barley) appear red. Note development at extreme lower left, where a semiarid shrub-range area has been converted successfully to productive cropland. The landforms at 1A — upland mountains, upper and lower bajadas, bottomland, and salt flat — are discussed in the text.

Locations 4A through 4L indicate the positions of corresponding low-aerial photos from Figure 4, showing land-use activities in greater detail.

With reference to the time of the day for obtaining interpretable images, one needs to consider how shadow patterns and the amount of available light aid in the process of identifying features of interest. In the case of animal inventories, consideration must be given as well to behavorial patterns as a function of the time of day. For example, many grazing animals graze mainly in the early morning or evening and seek cover beneath trees during the hotter portions of the day. Photos obtained during the middle of the day would no doubt prove of little value for inventorying such animals.

The remainder of this chapter is devoted to a presentation of examples designed to illustrate interpretation techniques by which certain land-resource features can be inventoried on aerial or space photographs.

SPECIFIC EXAMPLES

A major objective of an arid-land manager is that of putting to optimum use each portion of the land that he is charged with managing. Although remote-sensing techniques can be employed at various stages in the management process (inventory, analysis, and operation), emphasis here is on inventory, the most important first step leading to intelligent management.

Inventories Based On Correlations Between Resource Features

In the inventory step an effort is made to determine "how much" of "what" is "where" throughout the area, with special emphasis being placed on landform types, soil conditions, water availability, and vegetation type. In arid regions these four factors tend to be so closely correlated that merely by the inventory of one factor, accurate inferences can be drawn regarding the other three associated factors. As applied to one important arid region, the Sonoran Desert of New Mexico, Arizona, and Mexico, this correlation is well illustrated in Figure 3. The diagram comprising the bottom half of that figure was developed many years ago and published by L. Benson and R. S. Darrow in their excellent book, *Trees and Shrubs of the Southwestern Deserts* (University of Arizona Press, 1954.)

The aerial oblique photo comprising the top half of Figure 3 shows a part of the Sonoran Desert southwest of Phoenix, Arizona, as photographed from an aircraft in March, 1969. Although vegetation can be discerned on the aerial photo, it is obvious that no direct identification of species is possible from a study of it. The same is true with reference to the soil types and moisture regimes present in various parts of the area. Only one of the four previously listed factors, namely the type of *landform* that is present in each part of the area (for example, eroding slope, upper bajada, lower bajada, bottomland, and salt flat) can be clearly identified on this photo. Yet a visit to the area has shown that a very high correlation exists between landform type and each of the other three factors (soils, moisture, and vegetation). Furthermore, the correlation is almost exactly as indicated in the Benson and

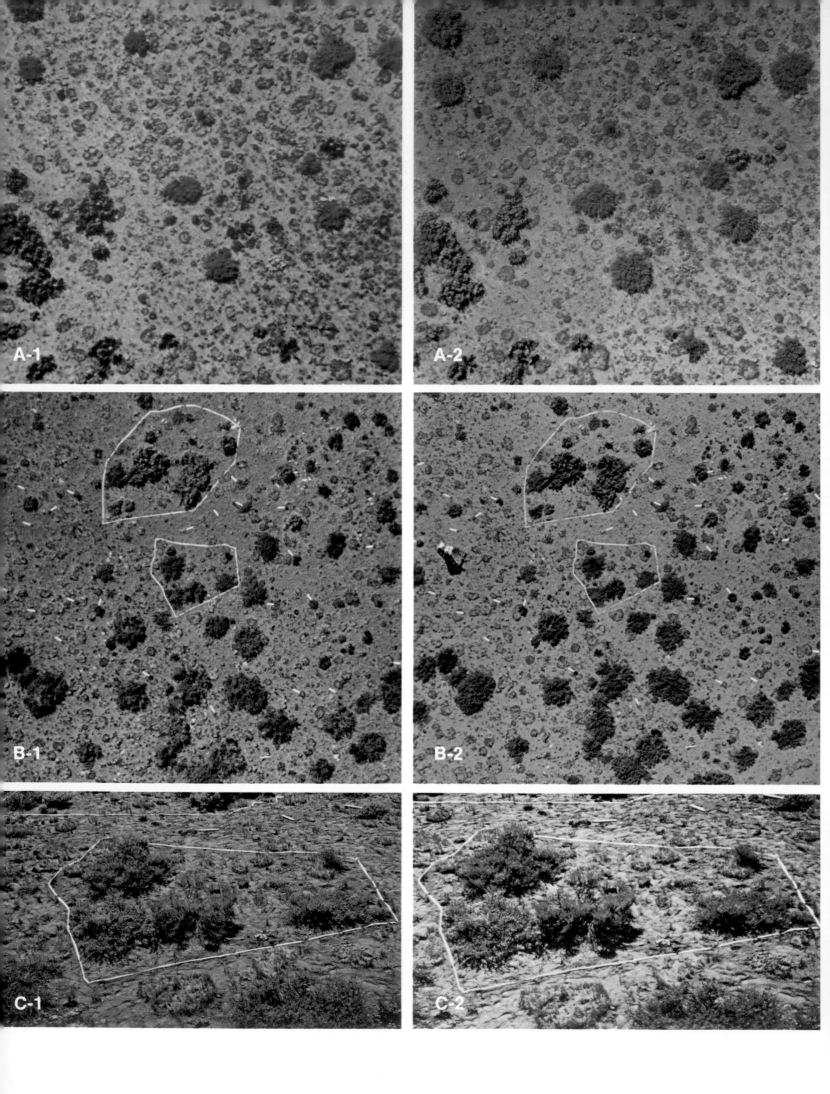

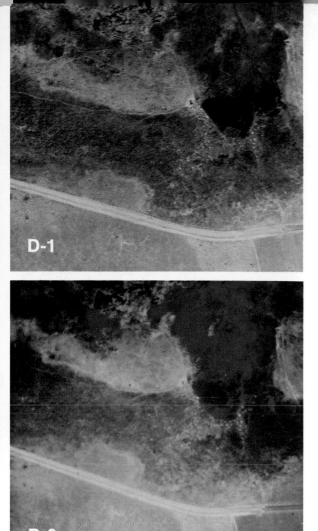

Fig. 2. Large-scale aerial photographs.

A–1 and *A–2,* Infrared Ektachrome photos (scale 1/170), taken July 1967 and July 1969, show native perennial forage species in a range allotment in northeastern California at the fringe of the Great Basin sagebrush type. The conspicuous red shrubs (bitterbrush, *Purshia tridentata),* palatable to cattle and deer, are easily differentiated here from the less palatable pink shrubs (big sagebrush, *Artemisia tridentata).* Note the dramatic increase in shrub size during the two-year interim between *A–1* and *A–2,* when this rest-rotation area was lightly grazed.

B–1 and *B–2* (scale 1/200) were taken, respectively, with color (Anscochrome D-200) and color infrared (Infrared Ektachrome) films to test the photographic interpretability of arid-land vegetation. *B–2* was judged more interpretable in terms of ease of identifying species and determining vigor. Measurements of shrubs in the two irregularly shaped plots were made both on the ground and on the photo; a statistical comparison indicated no significant difference.

C–1 and *C–2* (the latter infrared) were taken from a step-ladder to show additional details in the lower plot of *B–1* and *B–2.*

D–1 (Infrared Ektachrome, scale 1/8000) shows distinctive vegetation-soil types. This photo, taken at the same time as those in Figure 5, and showing conspicuously red lush meadow vegetation near a spring, illustrates the ease of detecting dense vegetation in what might otherwise be classified as semiarid land. *D–2,* same area as *D–1,* is a color composite from the three black-and-whites in Figure 5. The arrow points to a conspicuous vegetation boundary, far better seen here than on *D–1.* This illustrates that subtle but important arid-land features often can be emphasized by use of such image-enhancement techniques.

Darrow diagram. Therefore, in this example and many others, the use of remote-sensing techniques to identify landforms can inferentially provide the resource manager with virtually all of the data that he needs as he seeks to accomplish the inventory phase of his threefold task.

Stratification of a Space Photograph

The question logically arises as to whether this same objective might have been realized through remote-sensing from a spacecraft rather than from an aircraft. The reader is afforded an excellent opportunity to answer that question for himself from a study of the top part of Figure 1. This is part of an Infrared Ektachrome photo taken of the Phoenix, Arizona, area by the Apollo 9 astronauts in March, 1969, from an altitude of 145 statute miles.

The following points seem worthy of emphasis with reference to the analysis of the space photo in Figure 1.

a) The area is highly complex in terms of both the terrain features present and the uses currently being made of the land.

b) Despite that complexity, each portion of the area can be meaningfully characterized through the use of only a few major classification categories (upland, range-land, agricultural cropland, urban land, stream bottom-land, etc).

c) Each of these categories has a distinctive photographic appearance, with the result that category boundaries can be delineated with almost unbelievable ease and accuracy. Thus *uplands,* which are too sparsely vegetated for livestock grazing, exhibit a heavily dissected topography (accentuated by shadows); *rangelands,* which occupy the lower slopes and are suitable for livestock grazing, exhibit much the same color as uplands on Figure 1, but they lack evidence of the heavily dissected topography; *salt flats,* which for most purposes must be regarded as wastelands, are recognized by their topographic site and distinctive white coloration; *urban lands,* which comprise the areas having the highest per-acre value, are gray, and *agricultural lands,* which yield the highest revenue here in terms of earth-resource products, form a coarse patchwork of red and gray, depending on whether the individual fields were vegetated or not at the time of photography.

Preparation of a Photo-Interpretation Key

An essay-type description of features (even one as short as the preceding paragraph) is difficult for the reader to grasp and retain. Consequently, we might do well to consider whether there is some better way than the above to set forth the photo-recognition features of each of the major land-use classes appearing in Figure 1. Even with as few categories as the six which characterize this area, there can be advantages to preparing and using a photo-interpretation "key" that has been designed to facilitate the identification of these categories on Infrared Ektachrome space photography. A dichotomous or two-branched key usually is preferred because, at each step in such a key, the photo interpreter merely decides which

Fig. 3. Oblique aerial photo (*above*) shows (left to right) eroding upland mountain slopes, upper and lower bajada, and bottomland, composed of river sediments, salt flats, and stream channel. This transect of landforms coincides almost exactly with the vegetation and landform-profile diagram (*below*) by Benson and Darrow (*Trees and Shrubs of the Southwestern Deserts*, University of Arizona, March 1954). Many major plant species associated with landforms in the diagram also occur in the photo. The significance of this relationship is apparent in interpreting space photos, as in Figure 1, where vegetative detail is lacking but landforms are conspicuous.

VEGETATIONAL TYPES	YUCCA-AGAVE-SOTOL	PALO VERDE-SAGUARO-CACTUS	CREOSOTE BUSH	MESQUITE	WILLOW-COTTONWOOD	SALTBUSH
	YUCCAS AGAVES OCOTILLO SOTOL TURPENTINE BUSH PRICKLY PEARS FALSE MESQUITE	PALO VERDES SAGUARO OCOTILLO DESERT IRONWOOD PRICKLY PEARS CHOLLAS EPHEDRAS BRITTLE BUSH JOJOBE BUR SAGE	CREOSOTE BUSH BUR SAGE WHITE THORN CAT CLAW	MESQUITE CAT CLAW SALTBUSHES LYCIUMS JUJUBE	WILLOWS COTTONWOOD ARROW WEED BATAMOTE	SALTBUSHES GREASEWOOD PICKLEWEED

BASIN-AND-RANGE PROFILE

LAND FORMS & SOIL CONDITIONS	ERODING MOUNTAIN SLOPE	UPPER BAJADA	LOWER BAJADA	BOTTOM LAND	STREAM CHANNEL	SALT FLAT
	SHALLOW, ROCKY OR GRAVELLY SOIL WITH GOOD DRAINAGE. NO SUBSURFACE WATER.	COARSE-TEXTURED, ROCKY, WELL-DRAINED SOIL. PARTLY UNDER-LAID BY ROCK BENCH. NO SUBSURFACE WATER	SANDY AND FINE-TEXTURED SOIL. OFTEN WITH CALICHE HARDPAN. NO SUBSURFACE WATER.	FINE-TEXTURED SOIL WITH POOR DRAINAGE AND LOW SALT CONTENT. SUBSURFACE WATER AVAILABLE.		SIMILAR TO BOTTOMLAND BUT WITH HIGH SALT CONTENT.

of two possibilities seems most applicable to the photo image he seeks to classify. (The advantage of dichotomy lies in the fact that it usually is much easier to make the correct choice when there are only *two* alternatives than when there are *many*). For each such image, he starts reading descriptions at the top of the key and makes a series of choices depending on which member of a contrasting pair of descriptions best applies to the image he is examining. In this way he eventually should arrive at the correct identification of the feature, provided that the key has been properly prepared for photography of the type that he is interpreting. If the reader were to ignore annotations appearing in Figure 1, and instead were to try to classify each part of the area shown there merely by use of the photo-interpretation key that follows, he would almost certainly be impressed by the ease and accuracy with which the correct identifications are made.

Dichotomous Photo-Interpretation Key for the Identification of Major Land-Use Categories on Infrared Ektachrome Space Photography of the Sonoran Desert Area

1. Man-made pattern discernible, many rectangular blocks of land present ....See 2

1. Man-made pattern not discernible ....See 3

   2. Rectangular blocks large, clearly discernible, forming a coarse pattern; most of the fields either bright red (vegetated) or gray (bare ground) .... *Agricultural croplands*

   2. Rectangular blocks small, some not clearly discernible, forming a fine pattern; area mostly gray but with some small red blocks .... *Urban lands*

3. Mountainous ridges and canyons present; accentuated by shadows .... *Uplands*

3. Mountainous ridges and canyons absent or indistinct .... See *4*

   4. Area consists of either uniformly gray or reddish slopes (depending upon density and vigor of vegetation) adjacent to upland areas *Rangelands*

   4. Area consists of mottled flatlands, not adjacent to upland areas .... See *5*

5. Meander scars present; colors mostly red, gray and black ....*Bottomlands* (Stream channels and washes)

5. Meander scars absent; colors mostly white and gray ....*Salt flats*

For several of the categories described in this key, important subdivisions may be discernible. In such instances, the photo interpreter may find it helpful first to use the above key in order to identify the major land-use category of an area and then to consult an additional dichotomous key that describes the various subdivisions within that category. But in any event, small-scale photo-graphs provide a nearly indispensible base upon which to stratify landscapes into categories that can be classified according to potential productivity and use of the land.

Interpreting Infrared Ektachrome Photographs

Because of the unusual colors with which features are rendered on Infrared Ektachrome film (the type shown in Figure 1, on which the dichotomous key is based) we would do well to describe the unique properties of this film. This is doubly important because image *color* is used to such an extent in identifying earth-resource features on this type of photography. Consequently, unless the photo interpreter knows the spectral responses of the three-layer emulsion comprising that film, he is likely to make many erroneous interpretations of photography taken with it.

Infrared Ektachrome is known as a "subtractive reversal" color film. In such a film the dye reponses, when the film is processed, are inversely proportional to the exposures that were received by the respective emulsion layers. In devising Infrared Ektachrome film, the manufacturers had one major objective in mind — that of causing healthy vegetation to exhibit a strong photographic color contrast with respect to all other features. More specifically, it was decided that a sub-tractive reversal photographic film should be devised on which healthy vegetation would appear bright red while everything else would appear in colors other than red. To accomplish this objective, it was necessary to exploit the fact that healthy vegetation exhibits very high infra-red reflectance and relatively low reflectance of visible light. Thus, when the manufacturers were devising the three-layer emulsion of Infrared Ektachrome film, they linked a cyan dye to the infrared-sensitive layer of the film, and they linked yellow and magenta dyes to the green- and red-sensitive layers, respectively. Although this film has undergone some modification since it was first produced during World War II as "Camouflage Detection Film," the foregoing is an accurate description of its present characteristics. Whatever blue-sensitivity exists in the three layers is rendered inconsequential through use of a "minus-blue" (Wratten 12 or 15) filter over the camera lens at the time of photography.

Consistent with the foregoing, and keeping in mind that the dye responses are inversely proportional to exposures received, the following responses occur in each area on the Infrared Ektachrome film where healthy vegetation is imaged: (*a*) there is a great deal of yellow and magenta dye left in the film after processing, because the film has been only *weakly* exposed to red and green wavelengths to which those dyes, respectively, are linked; and (*b*) there is little or no cyan dye left in the film after processing because the film has been *strongly* exposed by infrared wavelengths to which that dye is linked.

When the processed Infrared Ektachrome film, in transparency form, is viewed over a light table through which white light is shining (that is, light that contains

Fig. 4. Low-altitude oblique aerial photos, showing land-use activities within the semiarid region southwest of Phoenix, Arizona. Positions of these photos are indicated on the space photo of Figure 1 by key numbers 4A to 4L.

A: flood-control dam and catchment basin.

B: new agricultural land on a lower bajada occupied previously by creosote bush (*Larrea divaricata*). Note the irregular drainage networks that characterize the bajadas.

C: large feedlot on a boundary between agricultural land and semiarid shrub land. Relatively large-scale (1/5000) aerial or oblique photos are needed for accurate counts of animals and for assessing livestock-raising potential. Although extensive areas of rangeland can be seen on the space photo (Fig. 1), forage production is relatively low, especially in the semiarid areas.

D: sharp boundary lines between (right to left) mesophytes (mesquite and tamarisk) within the watercourse of the Gila River, agricultural land, and semiarid rangeland. These boundaries are readily discerned on the space photo (Fig. 1). The mesophytes appear black on the space photo, providing a clue to underground water.

E: semiarid rangeland vegetation occupying alluvial fans. The light-toned area (*center*) has had most of the soil scarified and vegetation removed, apparently as a range-seeding experiment designed to improve forage production. In the background is more agricultural development on what was once semiarid shrubland.

F: sharp boundaries between bottomland vegetation, a salt flat, wildland classified as range, and agricultural land. The irregular fields (*center*) within the bottomland vegetation have

lower productivity, due to high salt content, than those fields converted from range land (*upper middle*).

G: large angular field converted for agricultural use, occupying what once was a bajada. The salt flat in the center tends to indicate land not suitable for conversion to cropping.

H: tamarisk-mesquite vegetation along the Gila River channel, and agricultural land within the watercourse. Such fields are marginal in terms of productivity, and usually only salt-tolerant crops can be grown.

I: large cropped area converted from alluvial-fan material that normally supports semi-desert shrub vegetation.

J: citrus-growing area south of Phoenix. Note the sharp boundary between citrus orchards and residential areas to their left.

K: portions of an agricultural area wherein crop type and conditions were accurately determined by on-the-ground observation for all fields in a 16-square-mile area. The unplanted field outlined here in white is conspicuously seen on the space photo of Figure 1, but crop types were not consistently identifiable by the photo interpreter. Sequentially obtained small-scale aerial photography provided considerable improvement in crop identification.

L: citrus orchards (dark tones) occupy land previously designated as semi-desert shrubland. Orchards have a unique rust-brown color in Figure 1. On the aerial oblique photo many orchards appear irregular, suggesting areas marginal for citrus production.

approximately equal amounts of blue, green, and red) the following factors are operative in those parts of the transparency where healthy vegetation is imaged: (*a*) the concentration of yellow dye is so high that *blue* light is almost completely absorbed, (*b*) the concentration of magenta dye is so high that *green* light is almost completely absorbed, and (*c*) the concentration of cyan dye is so low that *red* light is almost completely *transmitted*. The net result of these factors is to cause the healthy vegetation to appear red. Since virtually no other features have this peculiar combination of spectral reflectances, no other features appear red on the transparency.

Multistage Sampling

There are strong advocates for the use of aircraft rather than spacecraft when making arid-land resource surveys. Conversely, there are advocates for the use of spacecraft rather than aircraft when making such surveys. Several of the examples that soon will be presented provide strong support for a third view: that such surveys might best be made by means of a multistage sampling technique that employs spacecraft, aircraft, and ground observers. Each of these data-collecting components, in turn, provides a progressively closer look at a progressively smaller area, but provides progressively more detailed information about that area.

Figure 4 shows black-and-white photographs taken in March 1969 at approximately the same time that the space photo appearing in Figure 1 was taken by the Apollo 9 astronauts. The camera station and camera orientation for each of these photos is indicated on Figure 1. Many of the features shown in the oblique photographs are seen with almost the same clarity on the space photo. (This is true even though the aircraft was at an altitude of only 1,500 feet and traveling at 180 miles per hour while the spacecraft was at an altitude of almost 800,000 feet and traveling at 18,000 miles per hour). Nevertheless, some important details not seen on the space photo are seen on the aerial photo. Furthermore, there are still other details, not discernible even on the aerial photos, that can be obtained only by direct on-the-ground observation.

It follows that the most efficient means of making an inventory of the earth resources in an arid region such as that shown in Figure 1 is through the multistage sampling process. By this process, very small-scale photography taken from either aircraft or spacecraft first is used to delineate significant boundaries throughout the entire area that is to be inventoried. (The use of keys and techniques as previously discussed can be quite helpful at this stage.) Once these "stratification" boundaries have been drawn, further study is made of the small-scale photos in order to select a few representative sample areas within each "stratum" (uplands, rangelands, agricultural lands, etc). For these areas *only,* large-scale vertical or oblique aerial photographs are taken, such as those appearing in Figures 1 and 2, on the basis of which additional information (resource characteristics) about each stratum is obtained.

Finally, a careful study is made of these large-scale aerial photos in order to select a few very small representative sample areas within each stratum for direct on-the-ground observation. In this multistage sampling process, progressively smaller areas are examined in progressively greater detail, and the important details that are thus obtained are, in turn, applied to the progressively larger areas of which they are representative.

Although from its description the multistage sampling process may sound cumbersome, the increased efficiencies resulting from its use are by no means trivial. For example, the process just described recently was tested on space photography that was taken on the Apollo 9 mission, although not in an arid region. In that instance, the test area consisted of the Mississippi bottomlands and surrounding uplands for a 6000-square-mile area between Vicksburg, Mississippi, and Monroe, Louisiana (Langley, Aldrich, and Heller, 1969). The objective was to determine the usefulness of space photography in a multistage sampling system designed to inventory the *timber* resources of this vast area. It was demonstrated that use of the space photography (merely for the purpose of delineating all bottomland hardwood and upland conifer-hardwood stands) increased sampling efficiency in that study by approximately 60 percent.

We are now in the process of studying the effectiveness of these same techniques for inventorying multiple resources in the vast arid lands of Arizona and New Mexico, including the area shown in Figure 1.

The very-large-scale photos in Figure 2 have been analyzed in yet another study to determine how the detailed interpretation and measurement of vegetation parameters and soil conditions on such photos might be incorporated into a multistage sampling scheme in conjunction with smaller-scale imagery. By exploiting the high ground resolution provided by these photos, and color differences between dominant plant species, we have found it possible to identify most of the important *shrub species* to an accuracy of greater than 80 percent. *Herbaceous plants*, including sedges, forbs, and grasses, can be readily discerned; however, many cannot be accurately identified as to individual species. In certain instances plants from two or more species tend to group together to form characteristic shapes which permit the identifying of these species groups. Furthermore, it is possible to distinguish differences in vigor of individual shrub species and evaluate soil surface conditions, including disturbance factors, rockiness, etc.

Acceptable estimates of the density of plant cover (particularly of shrubs) also can be made, either by visual estimation from the photographs or by photogrammetric techniques that include direct measurement with precise scaling devices. In a statistical comparison of shrub intercepts as measured on the photographs and on the ground, no significant difference was found. Such results further substantiate the promise of large-scale photos for quantifying vegetation parameters. We are presently exploring means for automatically determining plant cover of conspicuous plants.

Since our studies have been conducted for three consecutive growing seasons, it has been possible to determine the extent to which large-scale photographs can be used to detect and measure those changes in plant growth and other disturbances that have occurred in the interim period. Obvious changes in plant size due to normal growth can be compared directly on the photos taken at the different dates or by measuring some parameter of the plant such as plant intercept. From these kinds of analyses we can conclude that large-scale aerial photography will be a useful tool for making range condition and trend determinations and for providing detailed information regarding plant species, plant density, plant cover, and other important surface features or conditions. Such information can be of great value to those who seek to manage rangeland resources as intelligently as possible.

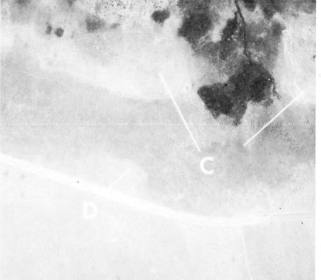

Fig. 5. Multispectral photographs, taken simultaneously with a four-lens spectrozonal camera, show vegetation-soil types in a range in northeastern California. The *upper left* photo was exposed on Infrared Aerographic film through a Wratten 89B (dark red) filter. The *upper right* photo was exposed on Plus-X panchromatic film through a Wratten 25A (red) filter. The *lower right* photo was exposed on Plus-X panchromatic film but through a Wratten 58 (green) filter.

Compare the tones of corresponding vegetation-soil types on the three multiband photos. Observe that healthy meadow vegetation (*A*) is light in tone, caused by high reflectance of near-infrared energy. A marshy area (*B*), created by flow from a spring, appears relatively dark.

On the two panchromatic photos, other range types are differentiated. For example, a low sagebrush type (*C*) and a distinct boundary between two vegetation-soil types (*D*) can be seen. These latter two features are not readily observable on the Infrared Aerographic photo (*upper left*).

A color composite (Fig. *2D-2*) from these three multiband photos simulates an Infrared Ektachrome photo.

BIBLIOGRAPHY

CARNEGGIE, D. M.
1968 Analysis of remote sensing data for range resource management. A report of research performed under the auspices of the Forestry Remote Sensing Laboratory, Berkeley, California, for NASA Office of Space Sciences and Applications, Earth Resources Survey Program. 62 p.

CARNEGGIE, D. M., and D. T. LAUER
1966 Uses of multiband remote sensing in forest and range inventory. Photogrammetria 21:115–141.

CARNEGGIE, D. M., and J. N. REPPERT
1969 Large-scale 70mm aerial color photography — a potential tool for improving range resource inventories. Photogrammetric Engineering 35(3): 249–257.

COLWELL, R. N.
1968 Remote sensing of natural resources. Scientific American 218(1):54–69.

COLWELL, R. N., and Others
1969 An evaluation of earth resources using Apollo 9 photography. Forestry Remote Sensing Laboratory, Berkeley, California, for NASA Office of Space Sciences and Applications, Earth Resources Survey Program.

DRISCOLL, R. S., and J. N. REPPERT
1968 The identification and quantification of plant species, communities, and other resource features in herbland and shrubland environments from large scale aerial photography. Rocky Mountain Forestry & Range Experiment Station, Ft Collins, Colorado, for NASA Office of Space Sciences and Applications, Earth Resources Survey Program.

FRITZ, N. L.
1967 Optimum methods for using infrared-sensitive color films. Photogrammetric Engineering 33(10):1128–1138.

HOWARD, J. A.
1965 Small scale photographs and land resources in Nyamweizland, East Africa. Photogrammetric Engineering 31(2):287–293.

LANGLEY, P. G., R. C. ALDRICH, and R. C. HELLER
1969 Multistage sampling of forest resources by using space photography. In Proceedings of Annual Earth Resources Aircraft Program Review, 2nd.

POULTON, C. E.
1968 Applications and potential of remote sensing in the analysis of range resources. Paper given at two-week course, Remote Sensing of the Environment, sponsored by Engineering and Physical Sciences Extension, University of California, Los Angeles.

POULTON, C. E., B. J. SCHRUMPF, and E. GARCIA-MOYA
1968 The feasibility of inventorying native vegetation and related resources from space photography. Oregon State University, Department of Range Management, Annual Report for NASA Office of Space Sciences and Applications, Earth Resources Survey Program.

WICKENS, G. E.
1966 The practical application of aerial photography for ecological surveys in the savanna regions of Africa. Photogrammetria 21:33–44.

The Gobi Desert of Mongolia

Geographic Description and Prospects for Land Use on the Basis of Soil, Vegetation, Hydrology, and Climate

JERZY LEFELD

Institute of Geological Science
Polish Academy of Science
Warszawa, Poland

ABSTRACT

GEOGRAPHICALLY the Mongolian Gobi Desert may be divided into four subregions. Prospects for possible future land use are based on analysis of soil, vegetation, climate, and hydrology of these arid lands. The most promising areas in this respect are the Great Lakes area in the northwestern part of the Gobi and some areas in the Valley of the Lakes in the north-central part of the Mongolian Gobi where large areas of wasteland can be turned into pastures and arable land. The northern and eastern parts of the desert also offer possibilities for land use, particularly for expansion of good pastures. In all these lands, water must be obtained from below the surface. Groundwater resources in promising areas are much greater than will be needed for any foreseeable use. The "true desert" areas of the Gobi lying south of the Gobi-Altay mountain system (the so-called Trans-Altay Gobi Desert) seem to be unpromising for land use, at least in the near future.

THE GOBI DESERT OF MONGOLIA
Geographic Description and Prospects for Land Use
on the Basis of Soil, Vegetation, Hydrology, and Climate

Jerzy Lefeld

THE NORTHERN PART of the Gobi Desert lies within the boundaries of the Mongolian People's Republic* (Mongolia, sometimes called Outer Mongolia), and the southern part within the boundaries of China (specifically in the part called Inner Mongolia). The Gobi is the northernmost of all Asiatic deserts. Because of a diversified morphology and extremely continental climate, it exhibits various kinds of geographical environments. Regarded as a whole, the Gobi Desert is rather of semidesert character. This is particularly true in the Mongolian part of it. Such situation suggests many possibilities for future land use in this Central Asiatic desert.

The state of knowledge about the Gobi Desert is still far from being satisfactory, however, despite considerable progress in scientific research both in Mongolia and China.

After a period of various expeditions to the Gobi, a systematic research of the Mongolian territory began in the early 1920s. After World War II rapid progress in all fields in the Mongolian People's Republic resulted not only from the great efforts of the Mongolian nation, but also from foreign, chiefly Soviet, aid.

An extensive literature concerning practically all fields of human activity in Mongolia exists at the present time, and the number of publications increases every year. The majority of these works are produced by Soviet research workers and are being printed in the Soviet Union (1). Among these publications, *Trudy Mongolskoi Kommisyi* (Works of the Mongolian Commission) are of particular importance.

The present-day planning of the Mongolian economy is based on scientific grounds. The northwestern and central parts of the Mongolian People's Republic are the most promising, hence planning concerns primarily these territories. Relatively less attention is being paid to the Gobi Desert, but even in these infertile lands considerable progress has been attained.

Nevertheless, much remains to be done. For example, the meteorological survey in Mongolia is starting to develop, but data from the Gobi are scarce and highly unsatisfactory. There is a complete lack of data from the highlands. The limnology of numerous Gobi lakes has been the subject of initial investigations only. Also, the hydrology of many desert areas is insufficiently known

(2). The same can be said about the state of knowledge on morphologic processes and surface geology of many Gobi areas.

The scientific research in the Chinese part of the Gobi has made considerable progress during recent years also, but the results are inaccessible because of the language barrier. Practically all scientific publications in China lack abstracts in European languages. The most recent obtainable information is from Soviet literature (3-5).

My contribution aims to present a general description that will enable the reader to imagine what the Mongolian Gobi is like and to give a rough estimate of the prospects for use of some of the land. The contribution does not, however, pretend to vie with official planning in the Mongolian People's Republic nor give any suggestions in this matter.

CLIMATIC CONDITIONS

The Gobi Desert, being situated in the midst of the Asiatic continent, is characterized by a continental climate carried to the extreme.

The main factors that cause these continental conditions are:

a) Great distance to the oceans. The nearest, the Yellow Sea, is 650 kilometers away.

b) High average altitude of the Gobi Desert, which is from 800 to 1100 meters above sea level.

c) Presence of very high and long mountain ranges that surround the desert on practically all sides.

The combination of all these factors is the chief reason why the Gobi receives very little precipitation.

In addition, the presence of the South-Siberian anticyclone (high-pressure) center in northern Mongolia during the winter months strongly influences climatic conditions in the Gobi Desert. Considerable cooling of the country during winter, caused by the existence of this anticyclone, is the main reason that the permafrost zone in the Great Lakes area extends much further south than in the neighboring regions of Central Asia. The gradual weakening and disappearance of the anticyclone in springtime is reflected by stronger winds of variable directions. During summer, the polar front shifts northward and tropical-continental air masses come to the Gobi Desert. Scanty summer precipitation is caused by the direct contact of two different air masses, namely the tropical-continental and the polar-continental ones.

---

*Those readers wishing to follow in detail the place names used in this article are referred by the author to *Pergamon World Atlas,* Panstwowe Wydawnictvo Naukowe, Warszawa (1968).

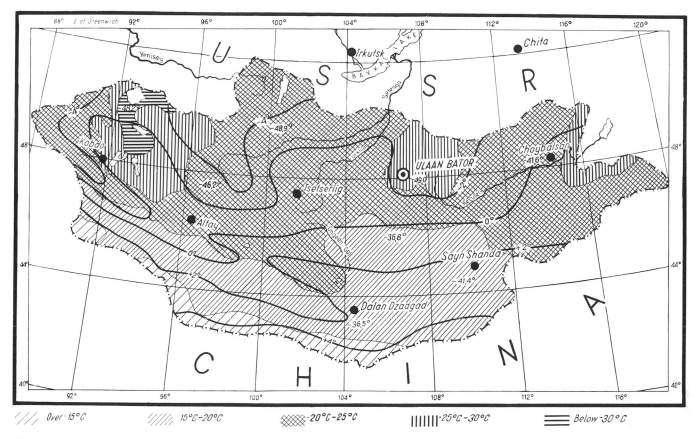

| /// Over -15°C | ///// -15°C-20°C | ⧄⧄⧄ -20°C-25°C | ‖‖‖‖‖ -25°C-30°C | ≡≡≡ Below -30°C |

Fig. 1. Mean January temperatures in Mongolia, in centigrade (from 7). Annual isotherms are shown in thick lines. The minimum known mean winter-long temperature is minus 36.5°C.

The town of Altay, shown on this and subsequent figures, was formerly known as Joson Bulak or Yoson Bulag.

Northerly, northwesterly and westerly winds prevail throughout the year, but some deviations from these directions may also occur according to local morphology. In the Soviet authors' opinion (6), monsoons are of little importance in Mongolia. Even in the easternmost part of the Mongolian Gobi, in the Sayn Shanda area, monsoons are much less frequent than northern and northwestern winds.

In July 1965, in the Nemegt valley (Trans-Altay Gobi) I observed a half-day rainfall that has been connected with a southeasterly wind, probably the "Chinese monsoon." In general, southeasterly winds that bring rain are rare. "Monsoonal" rains during summer months are known from the Alashan desert (5). Scanty summer precipitation in the steppe zone, that stretches roughly from 45° N northwards in central and eastern Mongolia, is connected with cyclones (6). Some rainfall in the Gobi-Altay mountains is caused by vertical movement of air masses, but the adjacent intermontane depressions hardly obtain even minimal quantities of rain, if any.

Exact meteorological data from the Gobi are still insufficient because of the very small number of meteorological stations. It is only 30 years since the meteorological survey began to work in Mongolia.

Mean annual temperatures range from plus 4°C in the southernmost part of the desert to 0° C in the desert steppe zone and in the Gobi-Altay mountains (7). The Great Lakes area shows even lower mean annual

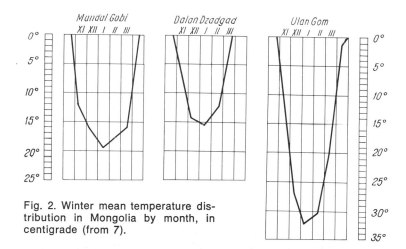

Fig. 2. Winter mean temperature distribution in Mongolia by month, in centigrade (from 7).

temperatures, namely minus 2° C, and in the Ubsa Nur basin, minus 3° C. The latter area is the coldest in Central Asia. The lowest mean monthly temperatures are in January, for example in Kobdo (Jirgalanta) minus 48° C, in Dalan Dzadgad minus 36.5° C and in Sayn Shanda minus 41.4° C (Figs. 1 and 2). Mean January temperature in the southernmost parts of the Mongolian Gobi is minus 15° C, but in the Ubsa Nur basin under minus 30° C. The latter low value is directly connected with the South Siberian anticyclone.

Winter in the Gobi Desert is relatively short when compared to other parts of Mongolia. It lasts from mid November until March. Winds are weaker and many days are windless. The sky is cloudless as a rule. The

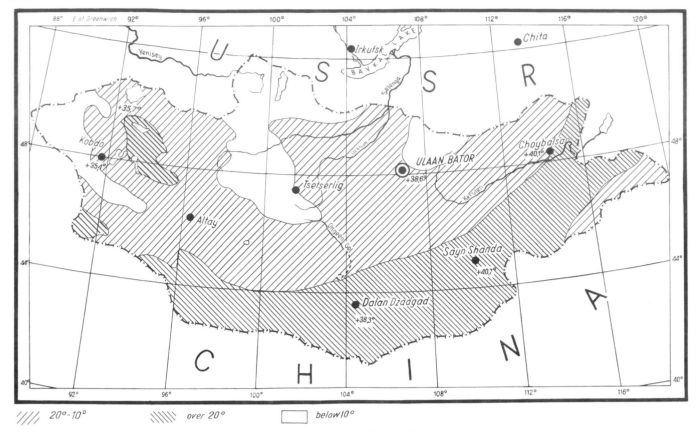

Fig. 3. Mean July temperatures in Mongolia, in centigrade (from 7). The maximum known mean summer-long temperature is 34.9° C.

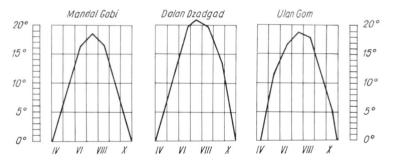

Fig. 4. Summer mean temperature distribution, in centigrade (from 7).

meteorological station at Bayan Dalay, southwest of the Gurvan Saykhon (Gurban Sayhan) mountain range, observed clouds only 10 percent of the days in January. Snowfall in the Gobi is insignificant. It does not even wet the soil and quickly melts in the strong Gobi sun. Snow cover may exist, however, in the highest peaks of the Gobi-Altay mountains. For instance the Ikhe Bogdo mount (4000 meters above sea level) is covered with eternal snow.

Vertical inversions of temperature may be observed in the intermontane depressions, chiefly during winter, but also in spring and autumn when rapid changes of temperature occur (6). This phenomenon is a result of a stagnation of cold air masses that usually happens during windless days.

Prolonged existence of the winter anticyclone is the main reason why the permafrost zone extends so far southward in the Great Lakes area. Its southern limit is

so far unknown. A permafrost zone has been detected in the southeast slopes of the Tannu Ola mountains, along the Tes River, in the northwestern part of the Khangay mountains, some 40 kilometers south of the Mongol Els sand field, and near Bayan Hongor (2). Further east, the already-known southern boundary of the permafrost zone runs some 200 kilometers south of Ulaan Bator toward Choybalsan. Isolated islands of permafrost occur in the Mongol Altay mountains and in Kharhira mountains (southwest of Ulaangom). Generally speaking, the permafrost zone in the Mongolian Gobi is of island-like character. It occurs primarily in narrow valleys and lake depressions, in areas with clayey rocks, in forested slopes poorly exposed to sunlight (slopes of north and northwestern exposure). In such zones soils show negative heat balance. The existence of a permafrost zone is connected with scanty snowfall and very low winter temperatures, and also with the fact that summer is very short in Mongolia. Even during hot summer months, nights are rather cold — another factor favoring the existence of permafrost. According to observations of Dr. Don, a member of the Polish Geological Expedition to Western Mongolia in 1962, permafrost ground was found at depth of 1.5 to 2 meters some 40 kilometers to the southeast of the Mongol Els sand field in the Jirgalanta somon (8, 9) near the Dzabhan river in the desert steppe zone. Jirgalanta somon is situated in an elongated depression that is almost parallel to the course of the Dzabhan river. This newly found site of permafrost is situated much further south than previously found sites (2).

The top of the permafrost layer is usually found at 1 to 2 meters below the surface, and at lower elevations at 3 to 5 meters. Seasonal (winter) freezing of the ground, particularly of loose surface material, may reach 4 meters down (2). The peculiar combination of desert landscape with permafrost ground underneath during summer months is one of the most interesting features of the Mongolian Gobi Desert. The southern part of the Great Lakes area is the southernmost region of permafrost in the Northern Hemisphere.

Mean July temperatures in the Gobi are over 20° C, and in the Ubsa Nur basin from 15° to 20° C (Figs. 3, 4). Maximal summer temperatures are: in Dalan Dzadgad 38.3° C, in Sayn Shanda 40.7° C and in Ulaangom 35.7° C (7). In the Trans-Altay Gobi these temperatures are even higher, but exact data are lacking. In Murzayev's opinion (6) the maximal summer temperatures in the Gobi are lower than in Soviet Middle Asia (10) and in eastern Turkestan. Day-night amplitudes are high. For example, Przewalski noticed at the mouth of Edsin Gol river (Inner Mongolia, Chinese territory) an amplitude of 30° C. At Orok Nur lake, Kozlov's expedition noticed (March 28th to 29th, 1926) an amplitude 42.4° C (from minus 14.2° at night to 28.2° C during the day) (6). These data seem to be exceptional, however. Day-night amplitudes at Murung Els (Inner Mongolia) noted by Horner (11) on June 20, 1932, were 20° C (from 18° to 38° C), but they were usually lower (average temperature amplitude 16° C). Ground maximal temperatures measured at 2 millimeters below the surface (11) at Murung Els were at the same time over 62° C at noon. Night temperatures of ground measured in the same way were about 10° C, but the lowest were only 2° C (11). Horner's meteorological data are very interesting, as they present also humidity, air pressure, and wind velocity.

Relatively low summer temperatures in the Gobi Desert are to some extent caused by the high altitude of the area. Deep bends of mean annual isotherms (Figs. 1, 3) reflect well the considerable influence of the Altay mountain system on the Gobi climate. The Gobi-Altay mountains are characterized by lower temperatures than the adjoining desert depressions and valleys (for example, the Valley of the Lakes).

Summer in the Gobi is short and comprises only June and July. Night frosts may occur even in early August, but daily temperatures are still high. The second half of August belongs definitely to fall, particularly in the Great Lakes area. In Dalan Dzadgad, the first frosts are noted in the second half of September.

Spring is the windiest season in the Gobi. Duststorms are frequent in that season. Wind velocity noted at Dalan Dzadgad in April was 28 meters per second. The Gobi-Altay mountains are free of snow in spring because of rapid rise of temperature and lack of precipitation. Spring is the season of the highest aridity. For example, in Dalan Dzadgad in May relative humidity was only 2 percent; this should be regarded as an extreme value (6). Mean monthly humidity in Sayn Shanda in May

is 35 percent. The air is slightly more humid during summertime: in Sayn Shanda in June, 47 percent; in July, 67 percent; in August, 52 percent. Humidity is lower again during fall, for example in Dalan Dzadgad 35 to 47 percent, in Sayn Shanda 50 to 54 percent. A rise in relative humidity, noted during winter months, is a result of very low temperature, for example, in Dalan Dzadgad 64 to 69 percent, in Sayn Shanda 63 to 72 percent (6). Humidity in the Gobi-Altay mountains is relatively higher, but exact data are lacking. It should be pointed out here that Bactrian camels are resistant to the low temperatures and dry air of the Gobi Desert.

The Gobi Desert receives very small quantities of precipitation. Mean annual precipitation (Figs. 5, 6) in the Trans-Altay Gobi is usually less than 60 millimeters (2). The western part of the Trans-Altay Gobi shows even lower values, which might be expected from its typically desert landscapes (6). At the mouth of the Edsin Gol river (Inner Mongolia) the mean annual rainfall in the years 1928-1929 was only 25 millimeters (6). Fortunately, the remaining parts of the desert show higher values. Uneven distribution of rainfall from year to year is a very characteristic feature of the Gobi climate. Besides that, the diversified morphology of the desert also strongly influences rainfall distribution.

It is a well-known fact that mountain ranges of the Gobi receive more precipitation than the intermontane depressions. Nevertheless, there is an almost complete lack of meteorological data from the highlands. The Ikhe Bogdo and Baga Bogdo mountain ranges obtain about 150 millimeters yearly, but the precipitation in more southerly situated ranges of the Trans-Altay Gobi is probably much lower, as can be judged from their poor vegetation cover. The Mongol Altay and Khangay mountains receive more than 300 millimeters a year (2), but these highland areas cannot be regarded as desert zones. This fact is of importance because the depressions in between these highlands are desert and semidesert in character.

The Valley of the Lakes, most of the Great Lakes area, and the northern and eastern Gobi (except the Sayn Shanda area) receive 100 to 160 millimeters of precipitation yearly. About 70 percent falls during July and August. Usually rains cover only limited areas, which causes to some extent the island-like plant distribution in the Gobi Desert. In the steppe zone this is the chief reason that nomads move from one pasture to another in search of better grass. In the Trans-Altay Gobi it usually rains in torrents yielding up to 25 millimeters in a single rainfall.

One of the most characteristic features of the Gobi climate is the great irregularity of precipitation from year to year. For instance, in Dalan Dzadgad in 1939 precipitation was 179.2 millimeters; in 1941 it was only 74.1 millimeters (6). The Nemegt valley shows typical desert and semidesert landscapes, but during the summer of 1965 it was unusually green and I observed herds of horses on fairly good pastures — unusual in the Trans-Altay Gobi.

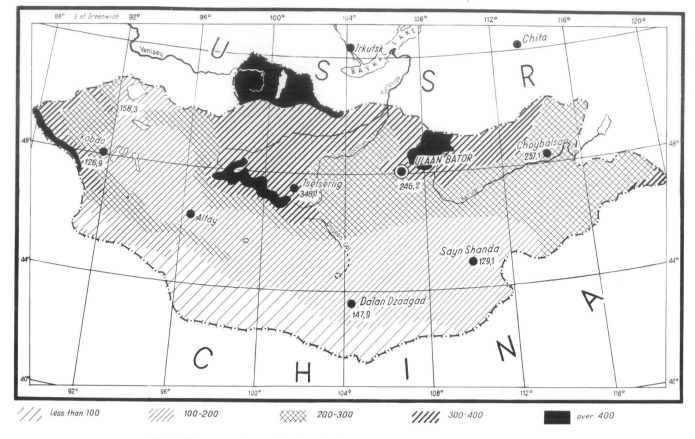

Fig. 5. Mean annual precipitation, in millimeters, at points in Mongolia (from 7).

There are some extremely dry years in the Gobi, but other years seasons may be normal or rather more humid. For example, one day's rainfall may exceed the total annual sum of a dry year. Consequently, the vegetation cover in the Gobi is subject to considerable change from year to year. Irregular precipitation is the chief cause of drought, a menace to Mongolian cattle herds.

The Gobi Desert is an area of ever-blowing winds that are strongest during spring months. The mean wind velocity is 3 to 5 meters per second. Winds are the strongest at noon or an hour later. During springtime, winds reach 15 to 25 meters per second in heavy duststorms, sometimes even 28 meters per second (6). The highest wind velocities noted by Horner at Murung Els (11) were 16 meters per second on June 20, 1932, at 22 hours P.M., but the average was not more than 4 to 5 meters per second. In the opinion of Marinov and Popov (2), the continental character of the Gobi Desert climate is expressed by large amplitudes (both annual and day-night), sharp season changes, considerable aridity of air, low precipitation, and cold, long winters.

REGIONAL DESCRIPTION

The Gobi Desert within the boundaries of the Mongolian People's Republic can be subdivided into several subregions. These are, from west to east:

a) The Great Lakes area, which consists of three major intermontane depressions, two of them with large lakes.

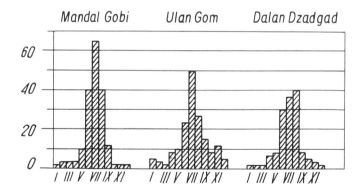

Fig. 6. Distribution of precipitation per month in three Gobi towns, in millimeters.

b) The Trans-Altay Gobi, composed of desert and semidesert areas south of the highland barrier of the Gobi-Altay mountains.

c) The northern Gobi, a desert steppe and semidesert area north of the Gobi-Altay barrier. Also included is the Valley of the Lakes, an elongated depression between the Khangay Mountains and the western part of the Gobi-Altay system.

d) The eastern Gobi, a desert steppe and semidesert area east of the Khangay and Gobi-Altay mountain systems.

Of these four subregions, only the Trans-Altay Gobi may be regarded as a "true desert" area. The three remaining areas are semidesert or, in some parts, of desert steppe character.

The boundaries of the Gobi Desert were defined in detail by Berkey and Morris (*12*). Strict limits of the desert, however, are difficult to establish because of frequently changing climatic conditions that cause considerable changes in the extent of the particular vegetation zones from year to year. Under average climatic conditions, a gradual transition from steppe to semidesert zone may be observed about half way between Mandal Gobi and Dalan Dzadgad. It roughly coincides with the boundary between the steppes with *Stipa Gobica* Roshe, and *Artemisia* to the north, and desert steppes with *Stipa, Allium, Tanacetum* and other halophytic plants to the south (Fig. 7).

In the Central and Eastern Mongolia this transition is very gradual, which is partly due to gentle morphology of the area. In the west, the Great Lakes area is bordered by the Mongol, Altay, and Khangay mountain systems, and by the Tannu Ola mountains in the north.

The Gobi-Altay mountains are bordered by a semidesert zone on their northern side and a semidesert-to-desert zone, the Trans-Altay Gobi, on their south side. This subdivision is clearly shown by climatic differences delineated in Figures 1, 3, and 5.

The Great Lakes Area

The Great Lakes area is the northwesternmost part of the Mongolian territory in which desert conditions are to be found. It comprises three major undrained intermontane depressions; the northernmost, the Ubsa Nur; the middle depression, the Khirgis (Hirgis); and the southernmost, the Sharghin (Shargain) Gobi. All these depressions are surrounded by high mountain barriers: by the Mongol-Altay ranges in the southwest and west and by the Khangay mountains in the east. Central parts

of these depressions show semidesert or even desert character. This is particularly true of the Khirgis depression. Proceeding from the center toward the mountain slopes (that is, to the pediments) one may pass through a semidesert steppe zone and then to the steppes in the higher parts of pediments (slopes).

Desert conditions exist in these northerly depressions chiefly because they are located between high mountain ranges.

The Great Lakes area is a vast country, stretching almost 700 kilometers from north to south. There are considerable differences in geographic character and climatic conditions between its particular subregions. The Ubsa Nur depression is the northernmost desert area in the Eastern Hemisphere. It shows great amplitudes between its winter and summer temperatures. Mean July temperatures are 15° to 18° C, and the highest known summer temperature at Ulaangom is 35.7° C. These fairly high summer temperatures cause strong evaporation. The mean January temperatures are under minus 30° C, and the lowest recorded winter temperature at Ulaangom is minus 48.2° C and at Kobdo minus 48° C.

In the Khirgis depression the mean July temperatures are higher than in the Ubsa Nur basin, over 20° C, but the mean January temperatures are extremely low, under minus 30° C. As is seen from the above figures, the annual temperature amplitudes are extreme in the Khirgis depression — the greatest amplitudes of all desert areas in the Gobi.

The Shargin Gobi depression shows summer temperatures similar to those of the Ubsa Nur, but January temperatures are milder there, from −15° to −22° C.

Mean annual rainfall in the Ubsa Nur and Khirgis depressions is from 100 to 200 millimeters. Most of the

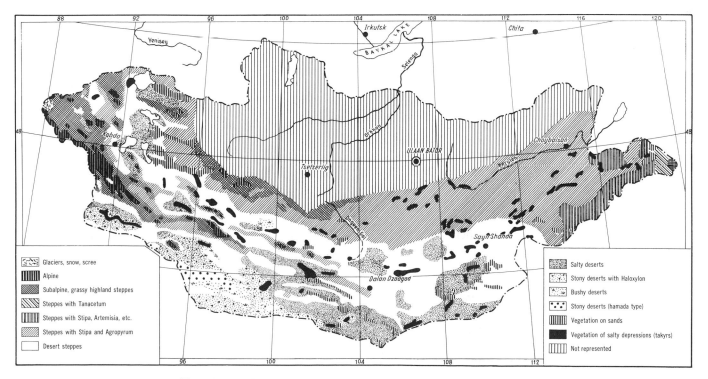

Fig. 7. Vegetation cover in the Mongolian Gobi Desert (after Yunatov).

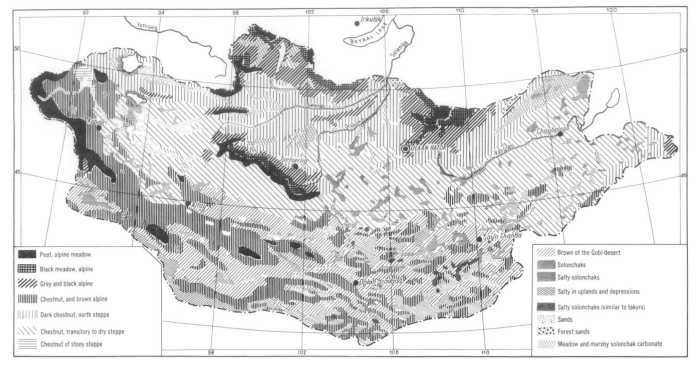

Fig. 8. Soil map of Mongolian People's Republic (after Bespalov).

precipitation falls during the summer months, chiefly in the form of torrential rains. The Shargin Gobi depression shows slightly higher values. However, a large part of the precipitation is poured on the slopes of the Tannu Ola, Mongol-Altay, and Khangay mountain systems, and the depressions obtain small quantities of rainfall (Fig. 5). Extremely low winter temperatures cause considerable cooling of ground, which hampers growth of vegetation and thus favors to some extent desert conditions. General strong salinity of the Great Lakes area (6) is another factor in favor of desert regime.

In central parts of depressions, salty soils predominate (Fig. 8). In meadows there are small areas of salty, carbonaceous, steppe soils. Higher situated areas, chiefly old-rock elevations, show chestnut and brown soils of highland type. Chestnut soils are extensively distributed in the Kobdo area, where they are used in agriculture; also, marginal parts of the Shargin Gobi depression are covered with chestnut soils. The Khirgis depression has the least acreage in this kind of soil and shows predominantly salty soils and sands. Strong, torrential summer rains act destructively on soils in many areas and hardly wet them enough for use by vegetation.

In salty lowlands there occur *Reaumuria soongorica* (Pall.) Maxim. and *Nitraria sibirica* Pall. These plants are also common in the central Gobi (Fig. 7). Near lakes, salty depressions, and springs, *Lasiagrostis* and some reeds occur. *Ephedra, Zygophyllum,* and *Nitraria sphaerocarpa* Maxim. are absent in the Great Lakes area and occur only in the Trans-Altay Gobi. Large areas are covered with desert steppe vegetation including *Stipa, Allium,* and *Tanacetum.* This type of vegetation is common on chestnut and brown soils on the slopes of mountain ranges surrounding the depressions.

The vegetation cover in the Great Lakes area shows typical island-like character, which is an evidence of very arid climate. Continuous cover is almost nowhere seen in this area. The small, halophilous plant *Anabasis brevifolia* C. A. Mey is common in the Khirgis depression (6).

UBSA NUR DEPRESSION

This area shows flat desert and semidesert morphology with larger sand areas in the east and a stony desert floor in the west. In the vicinity of the Ubsa Nur lake is salty, flat terrain with halophilous vegetation; solonchaks (13) are common in local depressions. Pediments — the slopes of the surrounding highlands (the Tannu Ola in the north and the Khan-Khuhey [Han Hohey] mountains in the south) — are dissected by numerous dry sayrs (14). Lower parts of mountain slopes are covered with steppes, on which predominate *Nanophyton erinaceum* (Pall.) Bge., *Stipa,* and *Artemisia.* Higher parts of slopes show subalpine highland grassy steppes.

The Ubsa Nur lake is the largest Mongolian saltwater reservoir. Its area is 3,350 square kilometers; length, 84 kilometers; and width, 79 kilometers. Salinity is 18.7 grams per liter.

TABLE 1

Chemical Composition of Ubsa Nur Water (from 6)

Chemical	Grams per Liter
Na_2SO_4	7.1404
NaCl	6.6972
$MgCl_2$	3.2143
Na_2CO_3	1.1722
K_2CO_3	0.2372
$CaSO_4$	0.0738
Na_2SiO_3	0.0449
Salts of humus acids, etc	0.1619
Total	18.7419

Bathymetry of Ubsa Nur lake is unknown. The lake is surrounded by a desert; arable fields, tilled by Mongols, are infrequent. In salty lowlands occur thick brushwoods of *Lasiagrostis splendens* (Trin.) Kunth. Borders of the lake are covered with sands, marshes, or solonchaks.

KHIRGIS DEPRESSION

Between the Khan-Khuhey and the northern part of the Mongol-Altay mountains is a flat country with occasional elevations of hard but strongly weathered igneous rocks. Large sand fields occupy eastern parts of the Khirgis depression. It is most conspicuous that in this area the sand fields are the largest of the whole Mongolian Gobi Desert. The largest sand field, the Mongol Els (Elesu) (area: 2,724 square kilometers) is situated along the lower course of the Dzabhan river. This field, shown in Figure 9, is connected with the Bara-Nurin-Urdu sand field (area: 1,187 square kilometers). The latter is situated east of Duru lake.

Sand fields in the Great Lakes area are connected with alluvial plains and rivers and are rather shallow. Many smaller sandy areas are scattered all over the area. Sand is being transported eastward by winds from the northwest and west; it covers slopes of the Khangay mountains. Belts of sands border the rivers Dzabhan, Narin Gol and other streams. Permafrost was recently reported underneath the Mongol Els sand field (*8*). Sand temperature of the Mongol Els during summer months is 50° C at noon, and the temperature of the ground underneath is some degrees below zero, which makes an amplitude of about 60° C!

The Khara Usu (Hara Usa) lake is the highest, 1,153 meters above sea level. Its area is 1,760 square kilometers; length, 78 kilometers; width, 26 kilometers. The Kobdo (Hobdo) river flows into the lake, and the Tchono-Kharaikh river flows out of it. Khara Usu is a freshwater lake. The salinity of 0.11 grams per liter (*15*) rises to 0.24 grams per liter in a shallow bay in the southern part of the lake. Maximum depth is unknown. In the center the Khara Usu is about 15 to 20 meters deep (*6*), but in most places it is shallower than 2 meters. In the south and near the Kobdo delta, the banks are marshy and covered by reeds and solonchaks. There is one large island (274 sq km), and several smaller islands.

The Khara (Hara) Nur is another lake in this system. Water comes there from the Khara Usu lake and flows out to the Dzabhan river and south to the Duru lake. Total area of the Khara Nur lake is 530 square kilometers; length, 30 kilometers; and width, 26 kilometers. Salinity is insignificant — only 0.26 grams per liter (*6*); bathymetry is unknown.

The Ayrik (Ayrig) and Khirgis (Hirgis) lakes were

Photo by S. Kozlowski

Fig. 9. A dune zone of the Mongol Els sand field in the Great Lakes area.

once one large lacustrine basin, but at present they are connected only by a strait. Ayrik lake is 20 kilometers long and 13 kilometers wide. Its area is 117 square kilometers. It is shallow (0.9 meters). Its water is fresh, muddy, and troubled. Dzabhan and Khunguy (Hunguy) rivers flow into the lake, depositing silt and sands. Khirgis lake has no outflow; it terminates this chain of lakes. Its area is 1,360 square kilometers; length, 83 kilometers; and width, 31 kilometers. The water level of the lake lies at 1,034 meters above sea level. The water is rather salty.

TABLE 2

Chemical Composition of the Khirgis Lake Water (from 6)

Chemical	Grams per Liter
Na_2SO_4	3.1067
Na_2CO_3	1.7657
$MgCl_2$	1.1376
$NaCl$	0.9038
K_2CO_3	0.2299
$CaSO_4$	0.1681
Na_2SiO_3	0.0140
Salts of humus acids, etc	0.2137
Total	7.5395

Surroundings of the Khirgis lake are covered with scanty, discontinuous, grassy vegetation. *Anabasis brevifolia* C. A. Mey, *Chenopodium,* and *Tanacetum achillaeoides* D. C. are common (*16*). The uninhabited banks are typical desert. On stony desert floor grows *Eurotia ceratoides* (L.) C. A. Mey. Only in a few places do seepages occur, and in such environments marshy plants grow in plenty. In one place known as Khokho-Deresni, small fields of land are cultivated. Old terraces are easily visible on the banks of the lake, thus proving its higher level in ancient, more humid times. The Khan-Khuhey mountains that border the lake on the north are waterless and deserted. There are several smaller lakes of minor importance in the Great Lakes area.

Central parts of depressions in this area show typical desert character. Proceeding from the lowest, central parts of depressions toward the mountain slopes the character changes to semidesert and then to desert steppe. In the valley of Narin Gol stream that flows into the Ubsa Nur lake, there are the Borig Els and Altan Els sand fields, their combined area is over 4,000 square kilometers. The length of these sand fields is 180 kilometers and width about 48 kilometers. Some arable fields exist north of the Borig Els sand field along the banks of the Narin Gol stream. In many places in the Great Lakes area one may see small "islands" of igneous or metamorphic rocks stretching out of the surrounding sands. These are chains of hills almost completely submerged in sand. Barchans and other kinds of sand dunes exist in many places. Between the Duru and Baga-Nur lakes the barchans attain 60 meters in height.

Lower parts of mountain slopes (pediments) are watered by streams flowing down and are the main agricultural terrains in the western part of the Mongolian People's Republic. Some of these streams disappear, and reappear after some distance in form of springs. Good, fertile soils and artificial watering allowed development of many oases of the Ulaangom type in which rye, wheat, barley, and millet are cultivated (*6*). Many streams and rivers of mountainous character as a rule flow down the slopes of the Mongol-Altay and Khangay and other mountain systems to the depressions. The majority of them are of permanent character. Their waters are used for agriculture.

The Dzabhan river shows the character of highland stream in its upper course, but it loses part of its water in the sand fields, chiefly as a result of evaporation in the summertime. Its water is hardly used in agriculture. Khunguy river also flows from the Khangay mountains to the Ayrik lake. Its length is 196 kilometers. Waters of that river are used for agriculture in many places along its course, because its valley is covered with better soils. The largest river of the Mongol-Altay, the Kobdo, is over 500 kilometers long. It has many tributaries that flow down from the Mongol-Altay, Saylugem, Kharhira, Turgen, and Mongkhe-Khairhan mountains. The Kobdo river is navigable. Highest levels are reached in May and June. Several fluctuations in water level are noted during summertime. The largest left tributary of the Kobdo river, the Tsagan Gol, shows highest levels in June-July when melting of the Tabun Bogdo glaciers is at maximum. There are also considerable day and night fluctuations in water level. The night high level is 30 to 40 centimeters higher than day level. Both rivers show rather high streams.

The Buyantu river flows into the Khara Usu lake and on its way passes through the town of Kobdo. Its waters are extensively used for agricultural purposes near the town. Water level is variable, particularly during spring. Waters of the Tsenkher river are also extensively used for agriculture; thus the river does not reach the Khara Usu lake. Other smaller rivers and streams of the Mongol-Altay mountains are poor in water and as a rule do not reach the depressions.

SHARGIN GOBI

This almost-closed intermontane depression is situated between the Mongol-Altay mountains on the west and the Khasaghtu Khairhan mountains on the east. Once the whole depression was occupied by a large lake basin, the only remnant of which is the shallow, solonchak-type Shargin-Tsagan lake. The central part of the Shargin Gobi depression shows desert character. Large parts of the area are covered by saxaul (*Haloxylon*). Numerous sayrs (*14*) cut the mountain slopes (pediments), thus showing that headward erosion is very active at some times there. The lowest point, the lake, is 948 meters above sea level. Large areas of the Khasaghtu and Tayshiri mountains are known as good pastures. Several permanent streams flow down the southern slopes of the Khasaghtu mountains toward the Shargin Gobi; thus, these areas are used for agriculture in a few places, particularly near the town of Altay (Joson Bulak or Yoson

Bulag). In some places fields situated in the lower parts of mountain slopes are watered from springs that appear in this zone.

PROSPECTS FOR LAND USE

Areas selected for soil cultivation (arable land) in the Great Lakes area occur in zones of slightly higher precipitation, but even so the precipitation there is light and irregular (2). Besides that, during May and June, months of sowing, rainfall is insignificant. Higher precipitation, which occurs usually in July and August, is rapid (torrential rains) and results in insufficient soil moisture, particularly in structureless soils. So even during the relatively rainy summer season, plants are under general arid conditions. The situation is even worse when a dry year comes. In such a situation almost the only water supply is from underground (2). Groundwaters are frequently cold (2° to 5° C) and must be warmed in reservoirs to at least 10° C before watering.

Underground water is obtainable almost everywhere, and there are many areas where the water is supplied by surficial waters, from rivers and streams. This is particularly true in the Kobdo area, where numerous streams flow down the slopes of the Mongol-Altay mountains. Groundwaters can be found in weathering zones of basement, chiefly igneous and metamorphic rocks. Such waters are fresh and only slightly mineralized (2). Numerous bore-holes drilled for agricultural purposes have pierced aquifers in Meso-Cenozoic rocks. Such waters occur at depths from 2 to 36 meters, with average discharge of wells about 1 to 5.3 liters per second. Most of these waters have little mineral content. Practically all the drillings were situated in marginal parts of artesian basins. In the central parts of these basins (depressions), waters are highly mineralized (2).

Lands already under cultivation occur in the Kobdo area, in the Ubsa Nur basin, in Shargin Gobi, and near the town of Altay (Joson Bulak). Good pastures exist north of the Mongol Els sand field, along the lower course of Dzabhan river, and also in many highland areas, as for example in the Ikh Bural Ula, Ikhe Bayan Ayragh mountains west of Jibhalanta (old name Uliasutay), and in all other small massifs of the area in question — above 1,200 meters altitude — also in the Khasaghtu and Tayshiri massifs.

Water is supplied from the surface in the area north of Kobdo; the same can be said about the Khan-Khuhey and southern part of the Mongol-Altay mountains. A satisfactory water supply exists in the Beger (Biger) depression, near Joson Bulak (Altay), and around Kobdo. There are also good water resources west of Ulaangom.

As seen from the above review, the hydrological conditions of the Great Lakes area are rather favorable for development of agriculture and better use of pastures, at least in some selected areas. These are primarily the marginal parts of depressions and the slopes, that is, broad pediments of the numerous mountain systems that surround the areas. These foothills are covered with chestnut and brown soils. Such areas are extensive, particularly along the northeastern slopes of the Mongol-Altay mountains and in the Shargin Gobi.

Recent achievements of Mongolian agriculture in this area allow us to foresee much better results in the future.

The Trans-Altay Gobi

The southern part of the Mongolian Gobi Desert, called also the Trans-Altay Gobi, is an area situated behind (south of) the Gobi-Altay mountain system. It is the only region within the Mongolian People's Republic that bears character of a "true desert," at least in some parts. The existence of the Gobi-Altay mountain barrier has a tremendous influence on the climate, vegetation cover, fauna, and landscape.

The Trans-Altay Gobi, unlike many other world deserts, is characterized by diversified and quickly changing landscapes. As in the Great Lakes area, the Trans-Altay Gobi shows also many closed, frequently elongated, undrained intermontane depressions situated between high mountain ranges. The area obtains extremely scanty precipitation, which is usually less than 100 millimeters per year in average. The Gobi-Altay ranges obtain slightly larger quantities, but exact data are unknown because of lack of meteorological stations in the highlands. There occur, however, very dry seasons, and in such cases the desert obtains even lower precipitation. Only definitely xerophytic plants can exist under such unfavorable circumstances.

Summer temperatures in the Trans-Altay Gobi are high. Mean July temperatures are over 20° C and the maximal ones are probably over 40° C, but this has not been proved. Some smaller, closed depressions are known to be the hottest places in the Mongolian People's Republic. The Khaytchi depression, situated some 80 kilometers south of Shargin Gobi and to the northwest of the Aji Bogdo mountains, may serve as an example. Other depressions, usually elongated from east to west, are less hot, as prevalent western and northwestern winds do not allow too extensive heating of ground by insulation.

Winter in the Trans-Altay Gobi is less severe than in the remaining parts of the Mongolian People's Republic. Mean January temperatures are usually above minus 15° C (Fig. 1). Minimal winter temperatures are unknown, but it is probable that heavy frosts are rare there. Slightly more severe winter conditions should be expected in the mountain ranges. Relative humidity data are unknown, but in Dalan Dzadgad in May of 1941 (6) the relative humidity was 2 percent. This should be regarded as an extreme value in the Gobi. Hence the relative humidities in the Trans-Altay Gobi, particularly during spring months, should be even lower. The Trans-Altay Gobi is a windy country.

The soil cover in the Trans-Altay Gobi is poorly developed and as a rule discontinuous. Salty brown soils, frequently saturated with gypsum, prevail there (15), and also very salty white soils occur in solonchaks (13). In salty Gobi depressions south of the Mongolia-China border, a concentric zonation (in plane) of salty soils (solonchaks) was observed (5). Bottoms of smaller, closed depressions show very salty (solonchak-type) soils. Dark brown and black highland-type soils occur in the slopes of the Mongol-Altay and Gobi-Altay mountains (Fig. 8). Large areas in the Trans-Altay Gobi are covered with stony (hamada-type) desert or by sands that are partly fixed by scanty xerophytic vegetation. Chestnut soils are rather seldom found and occur only in the highland parts. Deflation of soil occurs frequently.

The vegetation cover in the Trans-Altay Gobi is distinctly poorer than in the northern parts of the desert. In the southern slopes of the Gobi-Altay mountain ranges a semidesert vegetation with *Stipa gobica* Roshe., *Allium, Tanacetum* and other halophilous grasses occurs (15). Typical desert areas are almost devoid of vegetation except the low-situated parts, as, for example, in the bottoms of local depressions and sayrs. *Ephedra* is a plant most resistant to drought. It is widely distributed in the vast areas of the Trans-Altay Gobi. Takyrs (17) (Fig. 10) and salty flats are frequently covered with small bushes of *Salsola passerina* Bge. and *Anabasis brevifolia* C. A. Mey (6).

Sandy zones with *Nitraria* are very conspicuous among takyrs. This plant grows on sand hummocks in both desert and semidesert zones. Its small, crimson, berry-shaped fruits become ripe in August and are collected and dried by Mongols to be consumed during the difficult winter months. In the opinion of many Soviet authors,

the sand-accumulating role of *Nitraria* is remarkable (6). In fact, sand is fixed by *Nitraria,* but it is blown away from between the hummocks. Such *Nitraria*-overgrown hummocks are 0.5 to 2 meters high. The sandy areas with *Nitraria* are hardly passable for any kind of motor transport. A parasite of *Nitraria,* called by Mongols "goyo," (*Cynomorium soongoricum*) is frequently found on *Nitraria* hummocks (Fig. 11). The roots of this plant are sappy and thus may serve as a rescue for thirsty travelers. Extensive sandy areas are covered with sparse saxaul (*Haloxylon*) vegetation. These bushes (Fig. 12) may attain more than 2 meters in height but are usually smaller. Their presence does not indicate the proximity of groundwater, unfortunately.

Saxaul wood, which is fibrous, is often collected by Mongols for heating during winter. Trees are very seldom found in the Trans-Altay Gobi. Elms (*Ulmus pumila* L.) grow in small groups or as isolated trees in sayr bottoms or other small depressions of terrain with relatively high groundwater level. Elm occurs only in the eastern part of the Gobi Desert, and its extent westward does not pass beyond meridian 100°30′ (16). In the oases of Trans-Altay Gobi, poplars (*Populus diversifolia* Schrenk) form small forests. Generally speaking, an island-like plant distribution is characteristic in the Trans-Altay Gobi (16, 18). The vegetation cover in the highlands is better developed. Subalpine, grassy highland steppes occur (as, for example, in the Aji Bogdo mountains); they are due to more abundant precipita-

Fig. 10. Sparse vegetation on sands in takyr depression, eastern part of the Nemegt basin, Trans-Altay Gobi.

Fig. 11. *Cynomorium soongoricum,* parasite of *Nitraria,* Trans-Altay Gobi. A large example of a parasite, usually 12-20 centimeters tall, is seen growing on roots of *Nitraria* at upper left. The three cigar-shaped objects are the parasites.

Fig. 12. Sands partly covered with sparse vegetation — a poplar tree and saxaul bushes. Tost mountains are in the background. View to the south is the Nemegt basin.

Fig. 13. Southward view of Trans-Altay Gobi, Nemegt basin, showing strongly dissected pediments of the Altan Ula. Desert gravels (dark) are covered with desert varnish. Sand zones and badlands are far distant, Tost mountains on horizon.

tion and general humidity. In other mountain ranges semidesert steppes prevail.

As previously mentioned, many intermontane, undrained depressions occur in the Trans-Altay Gobi. The Nemegt basin may serve as an example. It is situated between the Nemegt (highest peak 2,766 meters above sea level) and Altan Ula mountains in the north, and the Tost (highest peak 2,503 meters a.s.l.) and Noyan mountains in the south. The depression is elongated from west to east and is about 45 kilometers broad. Well-developed pediments — gentle slopes — border both mountain ranges descending to the Nemegt basin (Fig. 13). Two parallel sand belts with occasional barchan dunes and saxaul fields occur in the lowest parts of the basin, running parallel to the main axis of the depression.

A distinct zone of badlands is situated between the sand belts, approximately in the central part of the basin. The badlands are formed by Tertiary clastic rocks such as sandstones and silts. The pediments are covered with stony desert floor that shows shiny, black, desert varnish on all stones lying on the surface. In some places they are strongly dissected by numerous sayrs, in the vertical walls of which clastic Upper Cretaceous rocks are exposed. This formation is one of the largest dinosaur-hunting sites in the Gobi Desert.

The lowest place in the basin is occupied by a solonchak that only occasionally is covered by water to form a saline lake. Salt is being collected there in the Gurvan Tes somon (9). Small springs occur where the Nemegt and Altan Ula massifs contact the surrounding pediments (slopes). Their discharge is so small that only very small herds of camels or sheep can use them. In the central, lowest part of the basin, a few small wells are found, usually in the sayr bottoms. In such cases water is to be found there at depths of 2 to 4 meters. Occasional seepages occur in the badland area. Usually the Nemegt depression is of semidesert-to-desert character, but during the summer of 1965 the grass grew so well that even larger herds of horses grazed in many places.

West of the Nemegt depression, the Gobi Desert is even more arid. The depressions situated to the north and south of the almost insignificant Ongon Ulan range are practically devoid of any surface water. Stony desert stretches from there westward to the Chinese boundary (19). The area between the Aji Bogdo and Atas mountains is probably the wildest in the Gobi. Vegetation cover is extremely scanty there, and large parts of terrain are almost completely devoid of plants. The westernmost part of the Trans-Altay Gobi is, in the opinion of many Soviet authors, an eastern prolongation of the Dzungarian depression (4, 6, 19).

Oases are seldom found in the Trans-Altay Gobi, but they do occur in places where groundwaters come to the surface. The Dzahooy oasis may serve as an example: 96°32′ E, 45° N; it lies about 160 kilometers south of Joson Bulak. The Dzahooy oasis, visited by the Polish-Mongolian Paleontological Expedition in August 1964, is situated in an intermontane depression between the Aji Bogdo mountains and other ranges of the Gobi-Altay mountain system. A small stream flows through the oasis and disappears in marshy depression floor. About one hundred Mongolian yurts (circular tents) are there among poplar trees and willows. Abundant vegetation seems to be vividly green, especially to travelers inured to dull, Gobi landscapes. Near many yurts, small vegetable gardens are cultivated. The Mongols keep camels, goats, and sheep, and horses can also be encountered. In other areas elsewhere in the depression, traces of abandoned land melioration activities exist, which is proof of ancient use of land in this desolate part of the Gobi Desert. The Dzahooy depression is surrounded by a stony desert. Wild Bactrian camels live south of the Dzahooy oasis in the stony desert areas.

Sand fields are of minor importance in the Gobi Desert. Murzayev's (6) evaluation gives 13,500 square kilometers of sand areas in the whole Gobi within the boundaries of the Mongolian territory, which is only 3 percent of the total area. Hongorin Els is a large sand field situated south of the western range of Gurvan Saykhon (Gurban Sayhan) mountains. Its area is about 900 square kilometers. Another large sand field, the Argalant Ulan Shire, lies to the southeast of Sayn Shanda in the eastern Gobi. Much larger sand fields are on both sides of the Edsin Gol river in Inner Mongolia, in Chinese territory (11). The majority of sand fields in the Trans-Altay Gobi are partly fixed by saxaul and other plants.

The hydrology of the Trans-Altay Gobi is very poorly known. There are no rivers or lakes, and in many areas groundwaters do not occur at all (or are unknown yet) or are salty. In the intermontane depressions artesian waters seem to occur, but they have seldom been located (2). Mountain ridges that surround the depressions are built of strongly weathered basement rocks and serve as alimentation areas for artesian waters because of slightly higher precipitation on higher altitudes and condensation of vapor at dew point.

The depressions themselves are filled with clastic Meso-Cenozoic rocks, usually feebly cemented. Artesian water should occur at depth of several dozen meters, and the deeper the aquifer the more mineralized the water (2). Nevertheless, the majority of the southwestern Gobi depressions have never been investigated from the hydrological point of view. It is highly probable that some of them do not show any significant amount of underground water. Springs or wells only 2 to 5 meters deep can be found in the marginal parts of some massifs. In a few places discharge from such springs may reach several liters per second. However, discharge from higher situated springs is considerably smaller. The mountain ridges of the Gobi-Altay are practically devoid of groundwater. In the Bayan Dalay depression southwest of Dalan Dzadgad, groundwater was found at depths of 30 to 44 meters in Meso-Cenozoic clastic rocks (2). Discharge from wells was 0.5 to 2.1 liters per second (2). Below that aquifer, at least two lower-seated aquifers have been found at depths of 65 and 80 meters respectively. The Bayan Dalay depression is of semidesert character.

Prospects for land use in the Trans-Altay Gobi are slight. Good pastures are found only in the slopes of the Aji Bogdo mountains (20); pastures of medium grade are scattered in many parts of the area, chiefly in its eastern and northern parts, but these are rather limited fields. Land can be cultivated only in oases. The situation may change after the hydrology of the Trans-Altay Gobi is better known. Water supply in the Trans-Altay Gobi must always be obtained from underground (20).

The Northern Gobi

North of the Gobi-Altay mountains stretch vast, rather monotonous semidesert and desert-steppe areas, the northern boundary of which cannot be precisely defined because they gradually pass into the steppe zone of central Mongolia. Gentle, wavy terrain is a typical landscape in this region. Just north of the central part of the Gobi-Altay mountains and south of the eastern part of the Khangay mountains is the major elongated depression called the Valley of the Lakes. It was named by a Russian explorer of Central Asia, M. Pievtsov. Eastward, the northern Gobi passes gradually into the semidesert area of the eastern Gobi.

The northern part of the Gobi Desert obtains larger quantities of rainfall than the Trans-Altay Gobi. The mean annual precipitation is 100 to 200 millimeters (Fig. 5). Mean July temperatures are slightly higher than 20° C in the southern part of the zone and 15 to 20° C in the northern part, including the Valley of the Lakes. Maximum summer temperature in Dalan Dzadgad is 38.3° C (Fig. 3) and in Mandal Gobi, 34.9° C. Mean January temperatures are minus 15° to minus 20° C, but the Valley of the Lakes shows values below minus 20° C. Minimum winter temperature in Dalan Dzadgad is minus 36.5° C and in Mandal Gobi, minus 36.8° C. Relative humidity is higher than in the Trans-Altay Gobi (see section on climatic conditions).

Soil cover in the desert steppe is predominantly composed of brown, calcareous soils that contain little decomposed wood (15). In the southern part of the region, salty soils are frequently encountered. In the vicinity of Dalan Dzadgad, brown soils occur again. Large areas show soils covered with loose gravel. Table 3 presents information on chemistry of a typical brown, loamy, slightly detrital soil of the Gobi. At the surface this type of soil is greyish brown with a slightly reddish tinge.

Vegetation cover in the semidesert zone is represented by about 250 to 350 species (16). Extensive areas are

TABLE 3

Chemical Analysis of Grey-brown Soils From the Vicinity of Dalan Dzadgad, After Bespalov (15)

Depth (centimeters)	Humus by Tyurin method, %	CO_2 %	Gypsum *	Dry residue *	Alkali HCO_3*	Cl *	SO_4*
0.5	0.78	0.53	—	0.062	0.050	0.003	—
10-15	0.97	2.27	—	0.063	0.060	0.003	—
25-30	0.81	7.48	—	0.049	0.050	0.002	—
45-50	0.44	5.26	—	0.052	0.050	0.003	—
75-80	—	9.90	—	0.094	0.060	0.016	—
95-100	—	4.86	—	0.219	0.060	0.026	—

* Aqueous extracts in percent of air-dry soil.

covered with sparse vegetation in which *Stipa, Allium, Artemisia,* and *Tanacetum* prevail. Bushy steppes (Fig. 14) with *Caragana pygmaea* (L.) D.C., *C. microphylla* (Pall.) Lam., *C. spinosa,* and *C. bungei* Ldb. occur in the desert steppe zone. Takyrs (*17*) and salty flats are frequently covered with small bushes of *Salsola passerina* Bge. and *Anabasis brevifolia* C. A. Mey. Tamarisks (*Tamarix*) grow near some lakes. In the steppes south of the town of Mandal Gobi, the vegetation becomes sparser and sparser, and near Tsogt Obo one faces a transition to semidesert landscape (depending on the season). According to my observations, the steppe vegetation in this zone is confined to soils developed on basement (igneous or metamorphic) rocks, whereas in the areas of light-colored, clastic, Meso-Cenozoic sedimentary rocks, semidesert or even desert conditions exist.

In desert steppes *Stipa capillata, S. Krylovii* and *S. gobica* Roshev. grow together with *Diplachne squarrosa, Artemisia frigida* Willd., *Kochia prostrata* (L.) Schrad., and *Allium* (*16*). Steppes with *Allium, Carex,* and *Stipa capillata* occur at foothills and in morphological depressions. In the opinion of Soviet authors (*6*), the northern boundary of the semidesert zone coincides with that of occurrence of the kangaroo rat (*Alactaga bullata* Allen), a common rodent in the Gobi. The steppe vegetation cover is subject to changes: during wet years steppes extend far to the south, whereas in dry years the semidesert zone shifts northward. In definitely dry, salty depressions and valleys, *Lasiagrostis splendens* (Trin.) Kunth. is common.

The vast desert steppes of the northern Gobi are the area of maximal population density of camels in Mongolia (Fig. 15). Immense numbers of rodents that can be seen every season in the steppes and desert steppes are a plague to the cattle and threats for agriculture; *Citellus undulatus* and *C. dauricus* are very common in steppes with *Agropyrum* and *Stipa,* particularly in the eastern part of the area. Various hamsters (*Cricetulus dauricus, C. roborovskii, C. barabensis,* and *Phodopus songarus*) are also plentiful in steppes. Various kinds of *Microtus* are most common in the eastern steppes. *Microtus brandti* is probably the most dangerous of all rodents. Its population grows rapidly sometimes, and during such seasons these small rodents change steppes into desert. In many areas the Mongols have to move from one pasture to another because of the destructive activity of these rodents. The population of *Microtus brandti* is subject to change from season to season.

Groundwater can easily be found near the surface in the desert-steppe zone. Usually these are waters from Quaternary (geologically young) aquifers. Deeper-seated Tertiary and Upper Cretaceous aquifers are seldom exploited. The Dalan Dzadgad depression shows favorable hydrological conditions; some wells discharge as much as 20 to 30 liters per second (*2*). In general, plans to use the water supply in the northern Gobi should

Fig. 14. Desert steppe zone west of Dalan Dzadgad — northern Gobi.

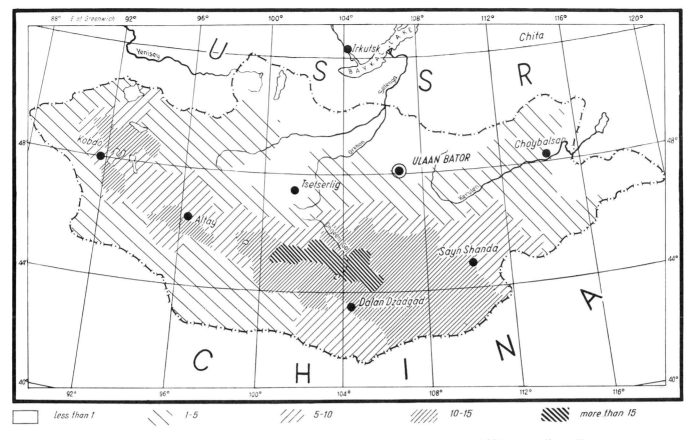

Fig. 15. Bactrian camel distribution in Mongolia — number per 10 square kilometers (from 7).

be based on exploitation of groundwater. Particularly the steppe zone around the town of Mandal Gobi and to the west of the town shows promise in this matter (2). In the future, an expansion of good pastures may be forecasted in the northern Gobi. Large areas of arable land should be cultivated in the future, particularly in some selected areas to the north of Mandal Gobi. Present-day sparse human population hampers progress of this region to a considerable extent.

VALLEY OF THE LAKES

The Valley of the Lakes, situated to the north of the Gurvan Bogdo and Barun Bogdo mountains and to the south of the Khangay mountains, is a chain of elongated depressions that probably once formed a uniform lacustrine basin with outflow eastward. The depression is of tectonic character. In the largest lake basin, and that furthest to the west, is the Boon (Bon) Tsagan Nur lake. Its area is 240 square kilometers; length, 24 kilometers; and width, 16 kilometers. It lies at an altitude of 1,336 meters above sea level. Its water is brackish and its bathymetry is unknown; it looks much like a typical Gobi lake. The old outflow may be seen in the eastern part of the lake.

The second largest lake, the Orok (Orog) Nur, lies north of the highest Gobi-Altay peak, the Ikhe (Yihe) Bogdo. The lake's area is 130 square kilometers; length, 28 kilometers; and width, 8 kilometers — but the narrowest place is only 2 kilometers wide. The altitude is 1,198 meters above sea level. Water level and salinity

are subject to change because of unstable water supply (inflow) from the Tuin Gol river (6). Bathymetry is unknown. It is estimated, however, that the lake is shallow. Orok Nur is covered with ice from the end of October until the middle of April. There are plenty of fish in the lake, and such birds as ducks, geese, gulls, and cormorants (*Phalacrocorax*) are common.

The morphological situation of the last lake, the Ulan Nur, is completely different. It lies in the midst of a flat depression to the east of the eastern termination of the Artsa Bogdo mountains. Ulan Nur is the only freshwater lake in this part of the Gobi. Ongin Gol river flows into it occasionally; as a rule, however, it does not reach the lake but disappears in the desert some 80 kilometers northeast. Near all the lakes of the Valley of the Lakes, old terraces have been investigated by many explorers (6, 12).

The floor of the Valley of the Lakes shows desert or semidesert characteristics, with some sand fields (12) and stony desert areas. Takyrs (17) are seldom found in the western part of the subregion, but they are more frequent in the east, near Ulan Nur lake. Results of deflation processes are clearly observable in the Valley of the Lakes.

Water from Quaternary aquifers is obtainable in marginal parts of these basins, and it is generally good, with only traces of mineralization (2). It is usually used by Mongols for cattle. In central parts of these basins, water is strongly mineralized as a rule and hence is seldom used even for cattle.

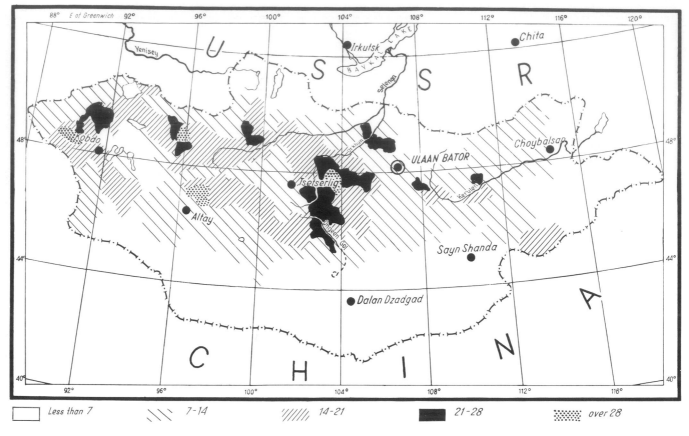

Fig. 16. Sheep distribution in Mongolia — number per square kilometer (from 7).

Good pastures exist on the southern slopes (foothills) of the Khangay mountains at (average) altitudes of 1,600 to 2,400 meters above sea level. This is the reason that a greater population of sheep and goats is noted in these areas (Figs. 16, 17). There are some cultivated lands along the Baydarik (Baydarag Gol) river. The Valley of the Lakes is the second most promising region of the Mongolian Gobi Desert for future land use. Good water resources in its northwestern part and in the slopes of the Khangay mountains would allow further development of agriculture and expansion of good pastures in these areas.

The Eastern Gobi

The uplands of the eastern Gobi Desert comprise an eastern prolongation of the northern Gobi plains. These are monotonous steppes and desert steppes situated at altitudes from 800 to 1,100 meters above sea level. The semidesert character increases and becomes more distinct southward.

The Khentey (Henteyn) mountains (in the northwest) and the Kerulen (Hereleng Gol) river (in the northeast) are the natural northern boundaries of this region. The area is usually flat with some wavy and ridgy uplands as, for instance, the hilly country south of the Kerulen river that is a water divide between the Amur river drainage basin and undrained depressions of the Gobi. Depressions are of little importance but are large as a rule, as, for example, the Eastern Gobi Depression that stretches from Sayn Shanda to the southwest.

The steppes of eastern Gobi Desert receive 100 to 200 millimeters of precipitation yearly (Fig. 5). Only in the far eastern termination of Mongolian territory is precipitation more abundant, from 250 to 350 millimeters yearly. The Sayn Shanda area, situated more southerly, is an exception, being of semidesert character. It receives less than 100 millimeters a year.

Mean July temperatures are 15° to 20° C in the northern steppe zone, south of the Kerulen river, and more than 20° C in the semidesert zone (Sayn Shanda area). Maximal known summer temperature at Sayn Shanda was 40.7° C. Mean January temperatures are minus 15° to minus 20° C (Fig. 1) in the semidesert zone, and minus 20° to minus 25° C in the steppe zone. Minimal winter temperature in Sayn Shanda was minus 41.4° C.

Chestnut soils prevail in the northern part of the eastern Gobi Desert. These soils were formed on the Quaternary alluvial and deluvial clastic deposits (21) that as a rule are not salty. They contain small quantities of humus. Further to the south, brown soils and salty ones dominate (Fig. 8). In general, soil cover in southeastern Gobi is poorly developed and discontinuous. Soil processes are fairly active at the present time in sandy areas of eastern Mongolia (6).

Vegetation cover in the eastern Gobi is similar to that of the northern Gobi. In the steppes south of the Kerulen river grow Stipa, Agropyron cristatum, Diplachne squarrosa, Carex discessa, and C. duriuscula C. A. Mey (6). Artemisia frigida Willd. and Potentilla acaulis are not rare (16). In many areas, Caragana microphylla

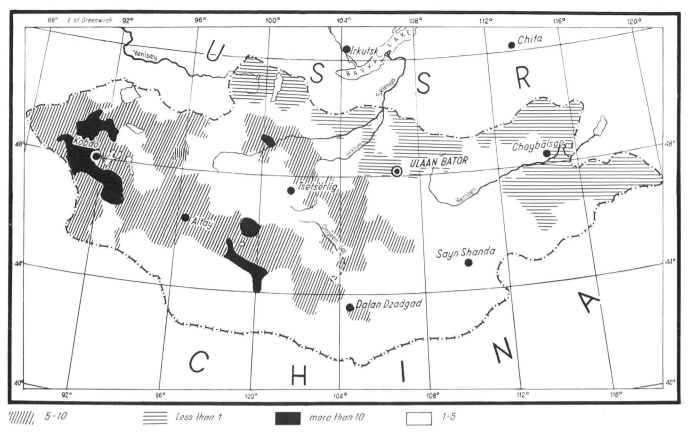

Fig. 17. Goat distribution in Mongolia — number per square kilometer (from 7).

(Pall.) Lam. and/or *C. pygmaea* (L.) D. C. grow together with *Stipa capillata.* Further south these shrubs serve as sand collectors, thus producing phyto-eolian forms of morphology. Sands are frequently stabilized by *Lasiagrostis splendens* (Trin.) Kunth. and *Peganum.* Saxaul (*Haloxylon*) grows in some parts of the Eastern Gobi Depression. Takyrs and solonchaks occur in many parts of this depression.

THE EASTERN GOBI DEPRESSION

This basin stretches to the southwest from the town of Sayn Shanda far behind the Mongolia-China boundary into Inner Mongolia. Its central part is occupied by takyrs and solonchaks that only in some places are covered by scanty saxaul vegetation. Marginal parts of the depression are strongly covered with sand and dissected by numerous sayrs. Elm trees can be found in places of high groundwater level (6). Near the national boundary the depression is practically uninhabited (Fig. 18).

THE DARIGANGA PENEPLANE

This upland is a basalt plateau situated in the eastern part of the Mongolian Gobi, about 300 kilometers east of Sayn Shanda. Its southern part lies in Inner Mongolia. Average altitudes of the Dariganga peneplane are 1,150 to 1,300 meters above sea level (maximal: 1,760 meters a.s.l.). In the Dariganga steppes there grow *Stipa capillata, Potentilla acaulis, Artemisia frigida,* and *Caragana pygmaea.* In the lower parts of slopes (foothills) one may find *Ephedra equisetina* Bge., *Agropyron cristatum,*

and *Artemisia frigida* Willd. Trees, which are seldom found, are represented by elms (*Ulmus pumila* L.). Some areas in the Dariganga peneplane are regarded as good pastures (2). In the upper parts of slopes, soils are hardly developed (6).

The old drainage pattern is still distinct, but the few small streams now carry insignificant quantities of water and this only in their upper courses; thus surface water is of no importance to agriculture. Groundwater reached by drilling in the Dariganga peneplane is highly mineralized (2), in some cases up to 4.76 grams per liter.

Large areas are simply basaltic covers practically devoid of any soil. Extinct volcanic cones are a characteristic of this part of the eastern Gobi.

Two large sand fields, the Ongon and the Moltsog, exist in the Dariganga area. The area of the Ongon sand field is 127 square kilometers (6), and that of the Moltsog is 248 square kilometers. These sand fields are underlain by impervious beds, which results in numerous seepages and small lakes in marginal parts of the sands. Another consequence is that several oases exist among these sand fields.

Artesian waters are known only in the Sayn Shanda and Dzuun Bayan depressions (2). Upper Cretaceous aquifers show considerable discharge, as for example at Sayn Shanda, up to 30 liters per second. Lower Cretaceous waters are highly mineralized (2) and thus useless for people and cattle. In practice only the aquifers of Upper Cretaceous and Tertiary age are suitable for all purposes. In the southwestern part of the Eastern Gobi

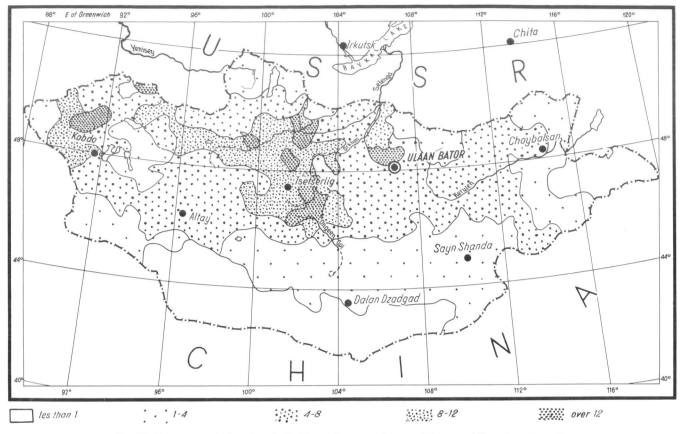

les than 1   . · 1-4   ·.·.·. 4-8   8-12   over 12

Fig. 18. Human population density in Mongolia — number per 10 square kilometers (from 7).

Depression, Tertiary aquifers yield water at average depth of 3 meters. In the Tamtsag depression (extreme east of the country), three aquifers occur in Tertiary clastic rocks (2). In some places, discharge of wells may attain 3 liters per second. The second aquifer shows discharge (in some wells) of up to 9.44 liters per second (2), but the water is highly mineralized. In general, good, fresh, or only slightly mineralized waters are to be found in the marginal parts of artesian basins. In their central parts, and at greater depths, water is highly mineralized. Good aquifers exist in the Ongon and Moltsog sand fields (20). They are able to furnish large quantities of fresh water. The rate of discharge from shallow aquifers strongly depends on precipitation, but that from the Upper Cretaceous (deeper) aquifers is practically independent of precipitation. In general, the eastern Gobi shows poor water resources, the Dariganga peneplane being an exception as it shows medium resources (20).

Prospects for land use in the eastern Gobi of Mongolia are limited to its northern areas which show better (chestnut and brown) soils and more abundant precipitation than other areas of the eastern Gobi. The same can be said about the Dariganga peneplane, in which good pastures may expand in the future. Such an expansion of pastures may be foreseen in many areas of the eastern Gobi, but this needs a water supply from underground.

The extreme eastern termination of the Mongolian Republic shows good soils and better precipitation, but unfortunately this area is very sparsely populated at the present time.

The southern part of the eastern Gobi has practically no prospects for land use, as waters there are highly mineralized as a rule and pastures are of poor quality.

CONCLUSIONS

The above review offers general prospects for future land use in the Mongolian part of the Gobi Desert, being based on a description of geographical environment. The total area of Mongolian wasteland that is potentially arable land is about 3.5 million hectares. The majority of this land, however, lies in central and northern aymaks (9). It is really difficult to estimate how much of this total area is in the Gobi.

The possibilities of cultivation of wastelands exist chiefly in the Great Lakes area and in the Valley of the Lakes. Also, the northern and eastern Gobi areas give some grounds for hope in this matter. Use of the land can be foreseen in the areas covered predominantly with brown and chestnut soils in the steppe and desert-steppe zones, but only in places of sufficient groundwater resources.

It should be kept in mind that immense numbers of rodents in the steppe and desert-steppe zones are a serious threat to possible crops.

Very good insolation practically throughout the year offers possibilities of vegetable plantations in greenhouses. Nevertheless, it is a rather theoretical possibility only, because of the lack of skilled personnel. Even open-field plantations of some vegetables are possible in many areas of sufficient water supply, but a condition sine qua non is the use of fast-growing varieties — these are absolutely necessary because of the short summers. Some vegetables — onions, cabbages, potatoes, cucumbers, etc — are already cultivated in the northern and central aymaks.

There are great possibilities for expanding pastures in the extensive areas of steppes and desert steppes of the Gobi. It is generally estimated that about thrice as many cattle as at present could be kept in the Mongolian pastures. However, the possibilities in the Gobi Desert are less promising.

In the Mongolian Gobi, sheep, camels, and goats are kept in great herds (Figs. 15, 16, 17). Horses are confined to the northern parts of the desert and are rare in the Trans-Altay Gobi.

Cattle in Mongolia (Fig. 19) live in open steppes, even during the winter, without any roofed shelters. Nevertheless, they are well adapted to the severe environment. For example, goats, cows, and horses are hairy, thus, even heavy frosts are harmless to them. However, during hard, snowy winters, many cattle die, as they do in spring, when weak after winter. In early spring, wells are still covered with ice, and cattle die of thirst. To avoid that it is suggested to construct wells deeper than the depth of seasonal frosting of ground (20). Roofless,

stony shelters are built only to protect lambs and kids from severe northerly and westerly winds (Fig. 20). Heavy snow (which is rare), or drought are the main causes of death of cattle in Mongolia.

Animal diseases are no serious problem at the present time. The number of cases decreases southward, and in the Gobi the number is rather insignificant. This situation is a result both of rapid development of veterinary survey and of the fact that cattle population is sparser and sparser southward.

Modern cattle-breeding is being developed in many regions of Mongolia. Mixing is done in stock-raising farms chiefly in the northern and central aymaks. Crosses give good results, but more extensive development of mixed breeds would need large-scale construction of permanent shelters for these newly developed breeds, for they would inevitably be less well adapted to severe climatic conditions. Aside from cross-breeding, there are still great possibilities for improvement of traditional Mongolian cattle-breeding which is developing.

Groundwater resources in the Mongolian Gobi are much greater than will be needed for any foreseeable use. In most of the Gobi Desert, the water supply must be based entirely on groundwater resources. Drilling for water and construction of wells is being done in Mongolia on a large scale, and many new wells appear every year in the desert. Of the Gobi regions described in the preceding sections, only the Trans-Altay Gobi is practically hopeless for land use. Surface-water resources in the

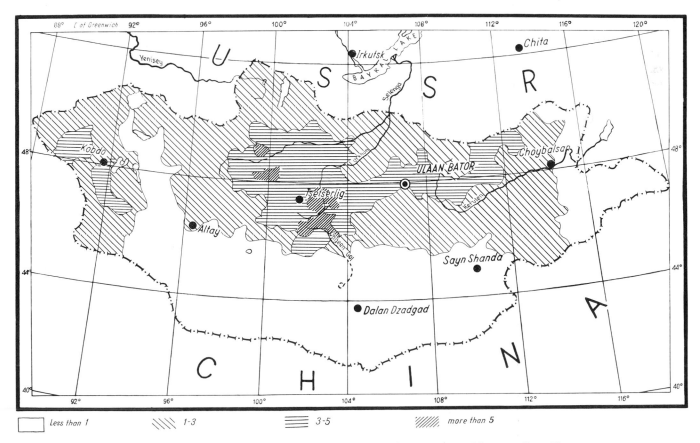

Fig. 19. Cow population density in Mongolia — number per square kilometer (from 7).

Fig. 20. Sheep and goats in front of a stony shelter — northern Gobi, between Tsogt Obo and Dalan Dzadgad.

latter subregion are insignificant, and groundwater resources are unknown (2). Areas situated south of about the forty-fifth parallel show poor water resources (20). The Dalan Dzadgad area is an exception in this respect.

The history of land cultivation is long in the Gobi. Archaeological observations have revealed that in some areas of the Gobi, land was cultivated in prehistoric times. For example, at Bayn Dzak (previous name, Shabarakh Usu), about 90 kilometers to the northwest of Dalan Dzadgad, Neolithic quernstones and sickles have been found recently (8). Sven Hedin's expedition found Neolithic ceramics with casts of corn grains in the Edsin Gol valley (Inner Mongolia).

Agriculture has a limited but old tradition among the Mongolian people. Under Chinese and Manchurian rule, fields in Mongolia were cultivated by foreign, predominantly Chinese, colonists. It is the present aim of the Mongolian government to expand the arable land area and to encourage people to harvest hay.

The Mongolian People's Republic has made considerable progress during last 50 years. This once almost entirely nomadic nation is developing modern agriculture and industry. Mongolian shepherds are gradually chang-

ing their mode of life from typically nomadic to sedentary (22). Nevertheless, the extremely sparse human population in the Gobi Desert is still a most serious problem. The northern and northwestern parts of the desert show only 1 to 4 persons per 10 square kilometers (Fig. 18). The Trans-Altay Gobi is practically uninhabited. In the desert, according to some sources, only the Dariganga and Kobdo (Jirgalanta) areas show from 4 to 8 persons per 10 square kilometers.

Fortunately, the future in this matter looks promising (23). The Mongolian People's Republic shows a higher birthrate than both great neighbors, U.S.S.R. and China. The 1957 birthrate per thousand (from 22) indicates Mongolia 27.0, China 23.2, and U.S.S.R. 8.3. An increased birthrate is conspicuous in Mongolia (22): 1938, 0.5; 1958, 30.6. The total population of the Mongolian People's Republic was in 1966 (22) 1,087,000; in 1967, 1,156,000.

The potential for land use in the Mongolian Gobi is high, but the central and northern (non-Gobi) regions of the country are higher in government priority for expansion of agriculture, and little attention (or funds) is being applied to desert areas.

REFERENCES AND NOTES

1. BALDAYEV, R. L., AND N. N. WASILYEV
1963 Bibliografia Mongol'skoi Narodnoi Respubliki (Bibliography of the Mongolian People's Republic). Knigi i statyi na russkom yazykie (1951–1961). Izdanye Wostochnoy Literatury, Moskva.

2. MARINOV, N. A., AND W. N. POPOV
1963 Gidrogeologia Mongol'skoi Narodnoi Respubliki (Hydrogeology of the Mongolian People's Republic). Gostoptehizdat, Moskva.

3. PETROV, M. P.
1959 The vegetation of the deserts of Central Asia and the regular features in its distribution. International Botanical Congress, 9th, Montreal, 1959, Proceedings 2:298–299. Abstract.

4. PETROV, M. P.
1962 Types de déserts de l'Asie Centrale. Annales de Géographie 71(384):131–155.

5. PETROV, M. P.
1966– Pustyni TSentral'noi Azii. 2 vols. Nauka,
1967 Leningrad. Translated as "The Deserts of Central Asia," by Joint Publications Research Service, Washington, D.C., JPRS 39145 and 42772.

6. MURZAYEV, E. M.
1957 Mongolia. Panstwowe Wydawnictwo Naukowe, Warszawa. (In Polish, translated from Russian).

7. BADAMJAB, D., D. BAZARGUR, J. BUDJAB, AND KH. TSEDENSONDOM
1966 Gazarzuyn Atlas (Geographical Atlas). Bugd Nayramdah Mongol Ard Uls. Ulan Baatar. (in Mongolian).

8. Thanks are due to Dr. Jerzy Don of the Geological Department of the Wroclaw University for his kind information on permafrost zone in the vicinity of the Mongol Els sand field. The author is also grateful to Dr. Janusz Kozlowski of the Historical Department, Jagellonian University, Cracov, for his archaeological information. Thanks are also due to Dr. Stefan Kozlowski of the Geological Institute, Warszawa, for lending a photograph of the Mongol Els sand field for publication.

9. *Somon* — Smaller administrative unit of the Mongolian People's Republic. *Aymak* — Larger administrative unit of the Mongolian People's Republic, comparable to state.

10. In the terminology of Soviet authors, the term "Middle Asia" (Sredniaya Azya) is used for Soviet Asian deserts and the term "Central Asia" (Centralnaya Azya) for other Central Asian deserts.

11. HORNER, N. G.
1957 Some notes and data concerning dunes and sand drift in the Gobi Desert. Reports from the Scientific Expedition to the northwestern provinces of China under the leadership of Dr. Sven Hedin. The Sino-Swedish Expedition, 1927–1935, Publication 40, III (Geology), 5. 40 p.

12. BERKEY, C. P., AND F. K. MORRIS
1927 Geology of Mongolia, a reconnaissance report based on the investigations of the years 1922–1923. American Museum of Natural History, New York. 475 p. (Central Asiatic Expeditions, Natural History of Central Asia, 2).

13. *Solonchak* — A flat salty soil, but also soils with more than 1 percent of salts such as $MgCl_2$, $Na_2SO_4 \cdot 10 H_2O$; $Na_2SO_4 \cdot 3 H_2O$. Comparable to kewir (kavir) or shott (chott).

14. *Sayr* — In Mongolian, dry river bed. Comparable to uadi (wadi) of Arabian deserts.

15. BESPALOV, N. D.
1951 Pochvy Mongol'skoi Narodnoi Respubliki. Akademii Nauk SSSR, Mongol'skaia Kommissiia, Trudy 41. Translated, 1964, as "Soils of Outer Mongolia," by Israel Program for Scientific Translations, Jerusalem. 328 p. Also cited as OTS 64–11073.

16. YUNATOV, A. A.
1950 Osnovnye cherty rastitelnogo pokrova Mongol'skoi Narodnoi Respubliki (General outlines of the vegetation cover in the Mongolian People's Republic). Akademiia Nauk SSSR, Mongol'skaia Kommissiia, Trudy 39.

17. *Takyr* — A clayey, sometimes salty flatland in deserts of Central Asia. Takyrs are formed by clay deposition sedimented from muddy waters. In general: bottoms of undrained depressions. Comparable to playa.

18. GRUBOV, W. I., AND A. A. YUNATOV
1952 Osnovnye osobiennosti flory Mongol'skoi Narodnoi Respubliki (Main features of the flora of the Mongolian People's Republic in connection with its distribution). Botanicheskiy Zhurnal 37(1):45–64.

19. BANNIKOV, A. G., E. M. MURZAYEV, AND A. A. YUNATOV
1945 Ocherk prirody zaaltayskoy Gobi v predelakh Mongol'skoi Narodnoi Respubliki (An outline of the nature of the Trans-Altay Gobi within the Mongolian People's Republic). Vsesoyuznogo Geograficheskogo Obshchestva, Izvestya 77(3):127–144.

20. DEREVIANKO, P. A.
1959 Sielskokhaziajstviennoye vodosnabjenye Mongol'skoi Narodnoi Respubliki (Agricultural water supply of the Mongolian People's Republic). Akademiia Nauk SSSR, Laboratoriia Gidrogeologicheskikh Problem, Trudy 21.

21. GUSENKOV, E. P., AND E. I. PANKOVA
1964 Vodno-fizicheskiye svoistva kashtanovyh pochv vostochno Mongol'skoi ravniny. Pochvovedenye 1964(9):44–51. Translated as "Hydro-physical properties of chestnut soils on the east Mongolian plain," in Soviet Soil Science 1964(9):926–932.

22. DYNOWSKI, W.
1967 Wspolczesna Mongolia (Contemporary Mongolia). Panstwowe Wydawnictwo Naukowe, Warszawa. (English summary).

23. For those readers wishing to investigate further the scientific literature on the Mongolian Gobi, the author suggests *24–35*.

24. GERASIMOV, I. P.
1959 Pustynia Gobi (The Gobi Desert). *In* Ocherki po fizicheskoi geografii zarubiejnyh stran. Geografgiz (Izdatelstvo Geograficheskoy Literatury), Moskva.

25. KIHARA, H.
1940 Biological survey of Inner Mongolia. Tokyo.

26. LAVRENKO, E. M.
    1957 Rastitelnost Gobiyskih pustyn Mongol'skoi Narodnoi Respubliki i yeyo sviaz s sovremyennymi geologicheskimi processami (Vegetation of the Gobi deserts of the Mongolian People's Republic and its connection with the contemporary geological processes). Botanicheskiy Zhurnal 42(9):1361–1382.

27. LITOVCHENKO, G. R.
    1946 Ovtsevotstvo Mongol'skoi Narodnoi Respubliki (Sheep-breeding of the Mongolian People's Republic). Mongol'skoi Narodnoi Respubliki, Ulan Bator, Komitet Nauk, Trudy 3.

28. LUSS, Y. Y.
    1936 Domashnye jivotnye Mongolii (Domestic animals of Mongolia). Akademiia Nauk SSSR, Mongol'skaia Kommissiia, Trudy 22.

29. MAIGNIEN, R.
    1962 Soils des régions désertiques de l'Asie Centrale (Desert soils of Central Asia). Association Française pour l'Etude du Sol 12:553–576.

30. MARINOV, N. A.
    1967 Geologicheskoye isledovanya Mongol'skoi Narodnoi Respubliki (Geological Researches of the Mongolian People's Republic). Moskva.

31. NORLINDH, T.
    1949 The flora of the Mongolian steppe and desert areas. Reports from the Scientific Expedition to the northwestern provinces of China under the leadership of Dr. Sven Hedin. The Sino-Swedish Expedition, 1927–1935, Publication 31, XI (Botany), 4(1). 155 p.

32. SINITSYN, V. M.
    1959 Tsentral'naya Aziya. Geografgiz (Izdatelstvo Geograficheskoy Literatury), Moskva. 456 p. Translated, 1960, as "Central Asia, physical geography handbook," 229 p. JPRS 3420.

33. SINITSYN, V. M.
    1959 Mongolo-Sibirskiy anticyklon i regionalnaya zonalnost eolovyh otlojenyi Centralnoi Azyi (Mongol-Siberian anticyclone and regional zonal distribution of eolian deposits of Central Asia). Akademii Nauk SSSR, Doklady 125(6).

34. SOKOLOVSKY, S. P.
    1960 Vodno-fizicheskiye svoistva kashtanovyh i buryh pochv nekotoryh rayonov Mongol'skoi Narodnoi Respubliki. Pochvovedenye 1960 (10):59–68. Translated as "Hydro-physical properties of chestnut and brown soils in certain districts of the Mongolian People's Republic," in Soviet Soil Science 1960(10):1072–1081.

35. TSEGMID, S.
    1962 Fiziko-geograficheskoye rayonirovanye Mongol'skoi Narodnoi Respubliki (Physico-geographical regionalization of the Mongolian People's Republic). Akademiia Nauk SSSR, Izvestya, ser. Geologicheskaya 5:34–41.

Pastures and Sheep Production
in the Arid Zone of the Soviet Union

NINA T. NECHAYEVA

Desert Institute
Turkmen S.S.R. Academy of Sciences
Ashkhabad, U.S.S.R.

ABSTRACT

DESERT PASTURES are of great importance in animal husbandry of the Soviet Union. Because sheep production is a developed industry, natural vegetation of the arid zone is effectively used, and some unused reserves remain.

By 1980 the sheep population in Kazakhstan, Uzbekistan, and Turkmenistan will reach 55 to 57 million, requiring stabilization of forage reserves. Rational use of pasture resources, developing reserves (particularly for use of Karakul sheep), and increasing productivity are thus of immediate importance in the Soviet Union.

Karakul sheep production in the desert can produce more profit than such highly remunerative forms of agriculture as growing cotton. Some 45 million hectares should be provided with more or better water. Revision and redistribution of pastures on the basis of accurate land evaluation appears needed. Grazing intensity should be controlled and rotation instituted where appropriate. Intelligent use of supplemental feed can help increase Karakul sheep population by 30 percent.

Methods for predicting yields of grazing lands are being perfected. Improvement of pastures can lead to productivity increases of two to five times. Combined use by sheep and camels is efficient on certain lands.

Lenses of fresh water form rapidly after construction of large reservoirs. Artificial watersheds can be created, desalting is under investigation, and the construction of deep and stable wells is no problem. Aqueducts supplying water to pastures are rapidly gaining popularity.

No intensification of desert animal husbandry will pay for itself without rational planned use and preservation of natural grazing lands.

PASTURES AND SHEEP PRODUCTION
IN THE ARID ZONE OF THE SOVIET UNION

Nina T. Nechayeva

NATURAL FORAGE RESOURCES of the desert play a tremendous role in the national economy of the Soviet Union. Because sheep production is a developed industry, the natural vegetation of the arid zone is effectively used, and some unutilized reserves still remain.

As a result of inventory of the natural grazing lands, a compilation was made of characteristics of all pastures according to seasonality and yield capacity. Maps of grazing lands were also compiled.

Study of pastures not being grazed has provided material on seasonal and annual dynamics of yield and nutrients on different types of pastures. Thus a rational system of pasture use has been developed, applying the concepts of pasture rotation and grazing by plots (portions). Later some methods were developed for desert pasture improvement and for cultivation of forage crops — on irrigated lands in small oases, and in the heart of the desert without irrigation.

Together with the research on the desert pastures, organizational and economic work has been carried out: state and collective farms were alloted fixed pastures; wells, dwellings, and other structures were built; large water mains were constructed (for example, the Great Kara-Kum Canal and the Golodnaya [Hunger] Steppe Canal); large reserves of artesian waters were explored in the heart of the desert. Water has brought life into the desert, made it possible to build cultural and economic centers in the heart of the pasture massifs, and led to increase in forage crops in irrigated lands.

In 1969 171 million hectares were used for grazing (124 million in Kazakhstan, 18 million in Uzbekistan, 29 million in Turkmenistan); this hectarage maintains 42.5 million head of sheep (1962 data), including 17 million head of the Karakul sheep. By 1980 the sheep population will reach 55 to 57 million, thus calling for stabilization of forage reserves (1).

In many developing countries of the arid zone, conservation and rational use of pastures is considered the most important national problem. The problems of rational use of pasture resources and procuring reserves for the development of Karakul sheep-raising and increasing its productivity are also rather acute in the U.S.S.R.

NOTE: This Chapter is adapted from a presentation at "Arid Lands in a Changing World," an international conference sponsored by the Committee on Arid Lands, American Association for the Advancement of Science, June 3–13, 1969, at the University of Arizona, Tucson, Arizona, U.S.A.

CHARACTERISTICS AND USE OF PASTURES

Pasture lands occupy 95 percent of agricultural lands in Turkmenistan, 84 percent in Uzbekistan, and 89 percent in South Kazakhstan.

The desert pastures constitute grazing areas for almost year-round use and provide the cheapest forage. But their yield capacity is low (1 to 4 centners per hectare) and sharply fluctuates from year to year. In various types of pastures it may increase twofold and drop down 3 to 5 times below the average, while the extreme values of yield from year to year are even greater.

Nutritiousness of pasture forage sharply fluctuates from season to season. For example, in the Kara-Kum desert, 100 kilograms of forage contain 89 to 75 fodder units in spring, 58 to 33 in autumn, and 40 to 36 in winter (3).

The quantity of forage in pastures is, in summer, two-fifths of the spring quantity; the summer accumulation of nutrients in fodder is only a fifth of the spring accumulation. The content of digestible protein drops down especially sharply: summer content is a twentieth of spring content.

Sheep graze in the pastures all year round with some little extra feed. In the forage balance, procured forage comprises only from 5 to 15 percent in different republics of the Soviet Union and does not compensate for the sharp seasonal and annual fluctuations of yields in natural pastures.

The aforesaid indicates a kind of extensive desert animal husbandry. Intensification, together with mobilization of all available reserves, is the only way to increase the number of sheep and the productivity of sheep-raising.

Desert grazing in sheep production is a highly economical branch of agriculture. In the desert, Karakul sheep-raising successfully competes with such highly remunerative kinds of agricultural activity as growing cotton. For example, in Turkmen S.S.R. in 1963–1966 (3), 1 rouble gained 37 kopecks net profit in cotton growing and 48 kopecks in sheep production (average data).

Intensification of Karakul (astrakhan) sheep-raising will inevitably raise the cost of production, but this can be compensated for by reduction of losses, which are caused by irregular forage supply and mass death in years of low harvest. Besides, the following should be taken into consideration: contemporary requirements for products of animal husbandry are constantly rising in connection with rapid population growth, and any measures increasing production in the field of animal husbandry will be profitable.

RESERVES FOR THE IMPROVEMENT OF KARAKUL SHEEP PRODUCTION

Water Supply of Pastures

One of the reserves for increasing sheep production is the opening up of yet-unwatered pasture massifs for grazing. Toward this end, some 45 million hectares (7 million in Uzbekistan, 6 million in Turkmenistan, 2 million in Tadzhikistan, and over 30 million in Kazakhstan) should be watered or better provided with watering places.

Half of the territory without irrigation contains highly mineralized subsurface waters; therefore, deep wells are of no use in such areas. Other methods of providing pastures with water should be applied: particularly the collection of atmospheric precipitation on existing takyr (claypan) watersheds or artificial concrete plates, the construction of aqueducts from reservoirs with fresh water, and so on.

Estimate of Pasture Resources and Their Rational Use

Grazing lands in the desert are all distributed among sheep-raising farms for permanent use. The distribution played a positive role in the first years of collectivization and organization of state farms. However, during the subsequent 40 years, new methods of estimate of forage quality have been developed. When pasture lands were being distributed in the 1930s, their quality was not defined precisely. As for an estimate in terms of money, it has not yet been done (4). The consequence is erroneous planning of numbers of sheep and of productivity. Often equal demands are made of farms possessing grazing lands of different quality, and monetary estimate of land value causes negligent treatment of pastures and does not favor their protection and maintenance. Revision and redistribution of pastures appear needed on the basis of accurate land valuation and estimate.

For pastures to be in good condition and achieve their normal productivity, determining the intensity of grazing by sheep is of great importance. In order to obtain nutritious forage and conserve grazing areas, livestock should not consume the total annual forage increment but only some portion of it. Experiments carried out for many years toward calculation of grazing standards for shrub-ephemeral sandy pastures of the Kara-Kum in large desert massifs for flocks of sheep (5) showed that if one sheep was supported by 3 hectares of grazing land a year (high standard), 75 percent of edible forage was consumed; by 6 hectares (mean standard), 67 percent; and by 9 hectares (low standard), 54 percent. The most reasonable grazing standard is some 70 percent of forage. It is not wise to allow animals to consume 100 percent of the edible forage; that practice harms both livestock

and pastures. Most scholars from the U.S.A. are of the same opinion.

Alloted pastures should be used by farms on a long-term plan. Pasture rotation is one of its important elements. In the desert, pasture rotation consists of alternation of grazing season during the years without rest. In richer pastures of Kazakhstan, interchanging use of land for pastures and hayfields is very effective (6).

Pasture rotation cannot be applied universally. For example, some pastures with salted watering places should be permanently used in the winter, since salted water cannot be used in any other season. Such use is permissible, since pasture vegetation gets least impoverished in winter.

Pasture rotation is based on large (3000 to 5000 hectares) pastures near wells, determined by a radius of grazing area (with a watering-place in the center, the radius is about 5 kilometers in average). The scheme of pasture rotation is plotted only on a plan of the area; there is no need to mark the boundaries in the field.

Uniform use of lands near wells is realized through subsequent grazing by plots (portions). Daily portions are determined by a radius of 400 to 500 meters from the station of herdsmen. Gradual moves for the stated distance ensure planned all-around grazing. This system has been worked out by Turkmen shepherds as a result of experience accumulated in many centuries. This valuable experience of the people was made known in scientific publications (7).

Pasture rotation providing for a moderate rate of stocking may raise pasture productivity by not less than 15 percent.

Procured Feed

Provision of procured feed is required in conjunction with decline in quantity and nutrient value of grass toward winter and sharp fluctuations of yields from year to year.

Shortage of nutrients in winter pasture forage may be compensated for by hay and forage concentrates, especially in years of low harvest. Standards of procured feed have been calculated as 25 kilograms of forage concentrates and 50 kilograms of coarse fodder per head in the year. In years with rich crops there is plenty of forage in the pastures, and procured forage will not be used — thus creating reserves for compensation of forage shortage in years of low harvest.

Natural hay lands are found only in piedmonts in the desert zone (once in 3 to 4 years) and in river valleys. In piedmonts, dry-farming of drought-resistant forage crops can provide much hay. In piedmonts, natural pastures and dry-farming lands can meet the requirements of Karakul (astrakhan) sheep production in forage. In recent years the water supply for irrigation has improved after creation of canals and artesian holes. The improved water supply (in piedmonts) permits the irrigation and cultivation of high-yield forage crops. The crops can be transported to sandy deserts that have no mowing lands.

V. Manakov's research (*3*) has proved that it is economically reasonable to use feed for 60 days in the desert piedmont pastures in Turkmenia, where a rouble gains 1.27 roubles profit. In places where irrigated lands or those with dry-farming agriculture cannot be used to provide feed, it will be more expensive and less effective, but in any case preservation of livestock is more profitable even if forage is more costly.

Rational organization of feeding supplements for sheep in winter allows us to reduce pasture areas set aside for winter and use them in other seasons when there is more forage in the pastures. It will permit a 30 percent increase in the Karakul sheep population.

Apparently it is expedient to use more feed, up to semistable maintenance of sheep in winter, for sheep farms specializing in other types of wool, and to practice limited feed in raising Karakul sheep, taking into account the great effect of standard and quality of forage upon the formation of an astrakhan.

When animal husbandry based on desert pastures is intensified, predicting forage yields, which allows us to determine the number of sheep that the grass can support, will be of great importance. Taking this into account, methods for predicting yields of the desert grazing lands are being perfected (*8–10*).

Pasture Improvement

Improvement of pastures is an important reserve to help achieve valuable year-round grazing of sheep and increase animal holdings. It should be taken into account that sheep would rather eat grass, and the Karakul turns out to be the best if sheep graze in desert pastures with specific forage composition. Availability of good winter pastures reduces the requirements for procured forage.

Experiments carried out on vast areas (*11* and *12*, among others) have shown that in piedmonts with annual precipitation of 200 to 250 millimeters, radical improvement of pastures (with *Carex pachystilis* and *Poa bulbosa* prevailing) for the interplanting (after strip ploughing) of desert shrubs and undershrubs (genera *Aellenia, Artemisia, Calligonum, Haloxylon, Salsola,* and others) brings very good results.

The productivity of improved ranges can be increased by two to five times, and even more, compared with the controls (natural grazing lands); the improved ranges are fairly drought-resistant and notable for longevity (can be used for not less than 15 to 25 years).

Creation of artificial pastures on 15 percent of the piedmonts, comprising an area of some 10 to 12 million hectares in Uzbekistan, Turkmenistan, and Tadzhikistan, will allow a 20-percent increase in animal holdings. At present, improvement of pastures has started at a large scale in the piedmonts, while experiments are still being conducted in the rest of the territory.

There are still more potentialities for pasture improvement in Kazakhstan. Here longterm pastures and mowing lands composed of drought-resistant perennial grasses or shrubs and undershrubs can be created.

Fig. 1. Natural ranges of the Karakum desert (Haloxylon persicum, Calligonum setosum, Carex physodes).

Fig. 2. Improved perennial ranges in foothill deserts in ploughed up sierosems (Artemisia badhysi).

One of the methods of increasing the output per unit of area is a combined use — grazing two kinds of animals (sheep and camels) one after the other. This method is particularly important on saltwort pastures (*Salsola gemmascens, Anabasis salsa*), where forage is rather monotonous and is not likely to be used by sheep.

CONCLUSIONS AND RECOMMENDATIONS

Development of science and engineering and improvement in the organization of the economy make it possible to solve the problem of watering pastures and stabilization of the forage reserves rather easily.

The aforementioned reserves can be promptly put to productive use. For instance, in the past it took decades for lenses of fresh water to form due to the atmospheric precipitation; now, after construction of large reservoirs, it takes 5 to 6 years. At places where natural watersheds (takyrs) are absent, artificial watersheds would have to be created. Various methods of desalting are being developed, while construction of deep, stable wells is no longer a problem.

Still another way of supplying water to pastures, rapidly gaining popularity, is the aqueduct. Some time ago aqueducts were not used at all in the desert.

A considerable effect, which consists mainly of conservation of yield capacity of natural grazing lands at the highest possible level, is obtained through more full and (what is most important) rational use of pastures.

The problem of improvement of desert pastures in the piedmonts, which occupy vast areas, has been successfully solved owing to practical application of the results of phytocoenological and agroclimatic research. Widespread use of these scientific achievements will make it possible to double pasture productivity.

Grazing supplemented with procured hay and concentrates (which is done after construction of large irrigation canals and discovery of great reserves of artesian waters in the desert) opens up practically unlimited reserves for increasing sheep population and that of other kinds of livestock.

Due to rapid population growth, views and ideas of scholars have to be reconsidered. Nowadays, every means that provides livestock growth and increases its productivity is profitable and worth the expense.

At the same time, the following should be kept in mind: no intensification of desert animal husbandry will pay for itself if proper attention is not paid to rational, planned use of natural grazing lands and their preservation. Modern economics must be used to create all possible conditions for further growth and development of animal husbandry of the arid zone of the Soviet Union.

216

REFERENCES

1. YESAULOV, P. A.
   1963 Animal husbandry and pasture economy in the desert. *In* Natural conditions, animal husbandry and forage reserve of the desert. Academy of Sciences, Turkmen S.S.R., Ashkhabad.

2. MIKHEYEV, G. D.
   1963 Nutrient value of pasture forage in the sandy Kara-Kum Desert. Academy of Sciences, Turkmen S.S.R., Ashkhabad.

3. MANAKOV, V. S.
   1968 Creating of stable forage reserve for sheep-breeding in the Kara-Kum. Problems of desert development, 2:33–40.

4. GAEVSKAYA, L. S.
   1968 About pasture cadastre in the desert zone. Problems of desert development, 2:3–7.

5. NECHAYEVA, N. T.
   1954 Effect of grazing on the Kara Kum pastures as a base for pasture rotation. *In* Deserts of the U.S.S.R. and their development, No. 2.

6. MATVEYEV, V. I., AND A. P. MAKAROV
   1960 Rational use of pastures in Kazakhstan. Ministry of Agriculture, Kazakh S.S.R.

7. NECHAYEVA, N. T., AND I. A. MOSOLOV
   1953 Pasture maintenance of sheep in Turkmenistan. Academy of Sciences, Turkmen S.S.R. Ashkhabad.

8. FEDOSEYEV, A. P., AND N. T. NECHAYEVA
   1962 Some regularities of formation of pasture vegetation of the southeastern Kara-Kum in conjunction with meteorological conditions. Proceedings of the Botanical Institute (7).

9. FEDOSEYEV, A. P.
   1964 Climate and pasture grasses of Kazakhstan. Gidrometeoizdat, Leningrad.

10. GRINGOFF, I. G.
   1967 Pasture plants in the Kyzyl-Kum and weather. Gidrometeoizdat, Leningrad.

11. NECHAYEVA, N. T., AND S. YA. PRIKHOD'KO
   1966 Artificial winter pastures in the piedmont deserts of Middle Asia. Turkmenistan Publishing House, Ashkhabad.

12. SHAMSUTDINOV, Z. SH.
   1963 Improvement of pastures and field production of forage in Karakul-breeding farms of Uzbekistan. *In* Natural conditions, animal husbandry and forage reserve of the desert. Academy of Sciences, Turkmen S.S.R.

Range-Cattle Production Under Dry, Warm Conditions

CARL B. ROUBICEK
and
DONALD E. RAY
Animal Science Department
University of Arizona
Tucson, Arizona, U.S.A.

ABSTRACT

THE DEVELOPMENT of a productive and profitable cattle industry under warm desert range conditions requires an understanding of physiological and genetic mechanisms concerned with heat tolerance, acclimatization, and adaptation. Direct and indirect environmental factors affecting animal productivity include a high heat load, lack of potable water, and seasonal or prolonged feed shortage.

Efficient thermoregulation in range cattle depends primarily upon a suitable protective hair coat and an adequate cutaneous evaporative rate. Endocrine function is directly affected by high environmental temperatures. This results in a decline in metabolic rate and an alteration in hormone production which, in turn, influence virtually all body functions.

The lack of suitable drinking water is a common occurrence on desert ranges. Dehydration results in a corresponding decline in feed intake and also directly affects the maintenance of homeothermy. Saline or alkaline drinking water exaggerates the effects of dehydration.

High environmental temperatures may adversely affect many components of total reproductive efficiency, including puberty, gametogenesis, fertilization rate, and embryonic and neonatal mortality. The total effect of environmental stress on reproductive efficiency can thus be relatively large and may be the most serious effect in terms of total productivity.

Growth performance is generally a reflection of forage availability and is consequently seasonal from weaning to maturity. Energy and protein supplementation may be required by calves during the dry season and for mature cattle during periods of extended drought. Development of adequate potable water throughout the range area is the first essential to good range management.

The prerequisite for a successful breeding program to improve genetic merit is the use of existing adapted native stock as the foundation. New genetic material may then be introduced after appropriate testing. Reproductive efficiency has first priority among selection traits. The response to selection for all traits will be more effective if appropriate action is taken to ameliorate the stress conditions.

RANGE-CATTLE PRODUCTION UNDER DRY, WARM CONDITIONS

Carl B. Roubicek and Donald E. Ray

ADAPTATION TO DRY, WARM CONDITIONS

M OST OF THE improved breeds of domestic animals have been developed in temperate areas, and as a result they are essentially cold-adapted animals. From the point of view of heat tolerance, an environmental temperature of about 25°C does appear to provide a rough critical borderline between temperate and other stocks.

The productivity of temperate stocks tends to decrease with increasing environmental temperatures above 30°C. The maintenance of a high level of production is difficult, due to the high heat load imposed upon the animal. Heat of digestion and normal body functions is being continually produced and must be dissipated into an atmosphere hotter than body temperature. The initial response to this heat stress is a reduction in food intake; this reduction is accompanied by the consequent direct effects on such production factors as body growth and milk production.

The lessened productivity may also result from indirect climatic effects, such as the lack of adequate forage or potable water. The animal on semiarid or desert ranges is often confronted with all possible combinations of direct and indirect climatic effects. The development of a profitable range-livestock industry under these conditions is understandably difficult. A systematic process of development of the industry does require an understanding of the physiological mechanisms concerned with heat tolerance, acclimatization, and adaptation.

Heat tolerance is the animal's ability to escape adverse consequences of the direct operation of hot conditions. To the farmer, a heat-tolerant animal is one whose productivity, be it beef, milk, or bacon, is little affected by a hot climate. To the physiologist, a heat-tolerant animal is one whose homeothermy is maintained by physiological responses to the thermal stress of the environment.

Acclimation refers to the changing responses to a single environmental factor, such as in controlled experiments, or day-to-day responses to the environment such as sweating, respiration, and vasoconstriction.

Acclimatization is a process whereby an organism so modifies its physiological mechanisms that it can more efficiently cope with the stresses of its surroundings. These stresses consist of a complex of environmental factors, as in seasonal changes. The physiological

changes usually, if not always, require time intervals of days or weeks to reach completion. Furthermore, once made, they do not disappear immediately as the stressful situation is removed; a prolonged period is required. Certain alterations may persist for a lifetime, even though the animal has returned to neutral surroundings.

Adaptation, in strict biological usage, is a term limited to phylogenetic changes that have fitted a species, strain, or race of animals for a successful existence within a given range of environmental conditions. Presumably such alterations have been brought about through the operation of the law of natural selection.

Physiological adaptation may be considered as any functional property of an individual that favors continued successful living in an altered environment. Unless an animal is adapted to its environment, physiological stresses impose limitations on its production potential.

General Regulatory Mechanisms

Animals are faced with two distinct environments with which they must be constantly in tune. One is the internal environment conditioned by normal metabolic processes and physical activity, and the second is the array of conditions external to the body. Higher animals have developed a complex control system that enables them to monitor and respond to the multitude of stimuli originating within and outside the body. This system is collectively termed the neuro-endocrine system.

The nervous system is paramount in receiving the various stimuli and initiating reactions to integrate body functions. From this standpoint it can be considered similar to a complex switchboard for receiving information and transmitting instructions in response to the particular type of information received. Some of the instructions received may result in immediate, direct responses within the body, whereas finer adjustments are normally brought about through the interactions of the nervous and endocrine systems over a longer time.

Adaptation to a specific set of climatic conditions involves adjustments in the neuro-endocrine system that enhance survival and allow for productivity (meat, milk, wool). The immediate adjustments are primarily the function of the nervous system, whereas the final changes result from shifts in endocrine balance and their ultimate effect on cellular metabolism. The latter may take a period of several weeks, or even months, before the most desirable integration is achieved. The endocrine system thus plays a critical role in the responses of an animal to his environment and ultimately determines whether productivity may be maintained in a particular environment.

At first inspection it may appear that adaptation to hot, dry climates is primarily one of temperature regulation, but many of the indirect consequences of xeric climates also demand adjustments within the body if survival and productivity are to be achieved. Perhaps the most important of these indirect effects relates to nutrient and water availability. Hot, dry climates are generally characterized by a paucity of forage and potable water, at least during certain seasons, and adjustments to the latter may ultimately determine the actual degree of adaptation attained by a particular organism.

A brief schematic diagram illustrating the interrelationships between the external environment and the regulatory mechanisms within the body is shown in Figure 1. Various feedback mechanisms integrate the multitude of metabolic functions within the body, and these functions, in turn, will determine the physical activity and productive capacity (either positive or negative) possible under the given set of conditions.

As an example, assume that the environmental factor is high temperature. Cutaneous thermoreceptors would sense the heat and transmit the information to the central nervous system. The nervous system would respond by initiating responses in the vascular system (peripheral vasodilation), inducing sweating and increasing respiration. The resulting loss of water would be sensed by certain nuclei in the nervous system and result in a signal to the pituitary gland to release vasopressin, the antidiuretic hormone. Vasopressin would be transported through the circulatory system to the kidney, where water retention would be enhanced. If the heat stress were prolonged, the nervous system might inform the pituitary gland to reduce the production of thyrotrophin, ultimately resulting in a reduction of metabolic rate

through a decrease in thyroxine production. Assuming that the heat stress could be physiologically coped with, resulting feedback information from the endocrine glands and metabolic processes would adjust or "tune" the organism so that homeostasis could be maintained. The animal has thus "adapted" to the environment. If nutrient and water availability is sufficient, actual productive functions may proceed — the ultimate in adaptation.

Although the foregoing is overly simplified, and undoubtedly many details are overlooked, it does indicate in general terms the physiological basis of homeostasis. It is thus evident that an intimate knowledge of the neuro-endocrine system is essential for an understanding of adaptation and offers the basic information for possible methods of exploiting animal production under various climatic conditions.

Thermoregulation

In all normal warm-blooded animals the body temperature remains constant within relatively narrow limits. This ability to maintain a constant temperature despite widely changing external conditions has been of great importance in establishing the superiority of the higher forms of animal life. Homeothermy is attained by maintaining an equilibrium between the heat produced by the animal, together with any heat it may absorb from its environment, and the heat which it emits.

The characteristics of heat tolerance are elusive and vague. They depend on the integrity of various systems such as the respiratory, circulatory, excretory, nervous, endocrine, and enzymatic systems. There may be a number of alternate ways animals may come to body-temperature equilibrium. Also, tolerance to a multiple-stress situation cannot simply be predicted from responses to the individual stresses but must be defined in terms of the complete environment. It is difficult to measure such complex and heavily interacting systems, especially since the coordination of all these systems under thermal stress is different, not only between species, but also between breeds and even individuals within a breed.

The optimum level of efficiency and productivity can be attained only if the animal is functioning at a stable body temperature. The maintenance of homeothermy depends on a dynamic equilibrium between heat production and heat loss under widely varying environmental and nutritional conditions. Some of the major sources of heat gain and heat loss are shown in Figure 2.

Animals, even when nonproductive and nonactive, have a huge "maintenance cost" to supply the energy for such obvious processes as circulation, respiration, excretion, muscle tension, and for many other processes not so obvious. The maintenance energy expense is eventually given off as heat. This heat must be dissipated as soon as it is produced, or the body temperature rises above the normal level. The dissipation of heat in an environment where air temperature is above body temperature taxes the entire animal system.

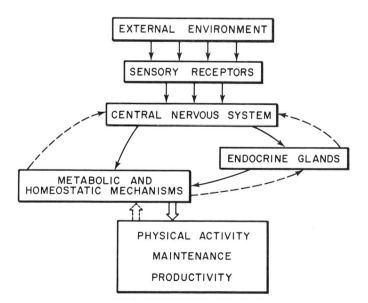

Fig. 1. Major factors involved in homeostasis. Solid arrows indicate direct responses to particular stimuli; broken arrows indicate regulatory feedback mechanisms necessary to integrate the multitude of metabolic functions within the body.

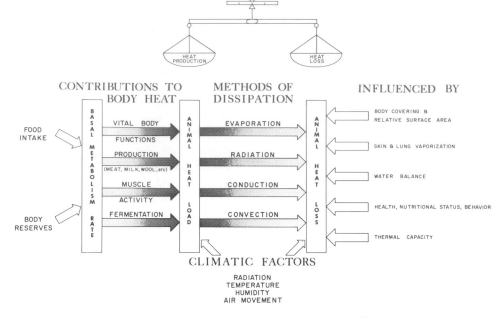

Fig. 2. Sources of heat gain and heat loss in maintenance of homeothermy.

Under standard conditions, metabolic rate is proportional to the three-fourth power of body weight. The metabolic levels of adult homeotherms from mice to cattle average 70 kcal/kg$^{3/4}$ per day, or about 3 kcal/kg$^{3/4}$ per hour (1). This contributes from 35 to 70 percent of the total heat production, depending on the magnitude of the other components. A relatively low rate would favor heat adaptation, but the physiological conditions responsible for a low basal rate may also act to reduce yield or productivity.

Additional heat is produced from the complex chemical changes that follow the ingestion of food. "Specific dynamic action" is used as a general term to include the transformation and storage of absorbed nutrients, glandular secretions, and increased cellular activity associated with nutrient intake.

The animal produces heat when transforming chemical energy of food into production. The additional heat increment depends on the level of production of milk, wool, work, and the functions of reproduction and growth.

Additional heat is produced by muscular activity of the animal, as when walking for food or water. The amount is also affected by animal temperament; the nervous, excitable animal has higher heat production than the more phlegmatic individual.

Heat of fermentation in the ruminant ranges from 0.40 to 1.00 kcal/kg of ingesta per hour, depending on type of ration, breed, etc. This component contributes less than 10 percent of the animal's total heat load.

External sources of heat are extremely important in establishing the total heat load of the animal. Direct solar radiation may exceed 500 kcal/m^2/hr under dry, desert conditions. In addition, as the sun shines upon the ground, the radiant energy is converted to heat and the ground temperature rises. The additional heat load imposed upon the unprotected animal by the temperature

of the ground and surroundings may approach that obtained by direct solar radiation.

Tropical insolation thus imposes a heat load many times larger than that produced internally, and stress of this kind greatly exceeds physiological thermolytic capacity. Accordingly, if lengthy tropical insolation is to be tolerated successfully, a large part of the burden must be encountered physically and externally to the body. The efficiency of a coat as an insulative barrier depends on its type, density, color, and smoothness.

In cattle, both infrared and light rays are effectively reflected by white, yellow, or reddish-brown hair — but not black. Ultraviolet rays are resisted by yellow, reddish-brown, and black skin colors. A white, yellow, or red coat with a dark skin is the ideal combination to render an animal resistant to the intense heat of radiation and short-wave rays. In addition, a smooth, glossy coat is advantageous in warm climates, both as insulation and as a factor affecting efficiency of evaporation. It has also been suggested that a sleek hair coat is correlated with other physiological attributes advantageous to animal growth performance (2).

The heat gain of the animal must be balanced by heat loss. The most important method of heat dissipation is moisture vaporization from the respiratory tract and skin. The loss of heat by evaporation depends on the fact that a certain amount of heat is required to change water to water vapor at the same temperature. The evaporation of one kilogram of water from the body requires about 575 kilocalories, depending on the temperature of the evaporating surface and the temperature and humidity of the surrounding air.

The apocrine glands found on many mammals are a source of cutaneous water loss. The principal function of the glands is to provide a source of water for evaporative cooling for body-temperature control. The apocrine glands are generally distributed over the skin in association with hair follicles. In cattle the total number of follicles is present at birth, and as the surface area of the animal increases during growth, gland density decreases. In mature cattle the number of sweat glands per square centimeter of surface area is about 800 for European breeds and 1,500 for zebu. There is also considerable variation in shape and size of the glands. In Sindhi and Sahiwal cattle, the sweat glands are so large that at normal densities of 1,500 glands per square centimeter the glands touch and form a continuous secretion layer. Sweat-gland shape, defined as the ratio of sweat-gland length to sweat-gland diameter, does serve as an indicator of skin type that could be used for comparing the skin of cattle throughout the world (*3*).

In domestic animals, sweating has been observed in cattle, sheep, horses, donkeys, camels, and some breeds of dogs. With cattle, skin evaporation accounts for 16 to 26 percent of the losses of body heat at air temperatures of 10°C, 40 to 60 percent at 27°C, and over 80 percent at air temperatures above 38°C.

Respiration also acts as a cooling mechanism. Evaporative cooling occurs from the surface of the respiratory passages; the cooled blood drains into the right side of the heart. The amount of cooling obtained by respiration depends upon temperature and vapor pressure of the inspired air, respiratory volume, body temperature, and vapor pressure of expired air. For animals maintained at high environmental temperatures, the beneficial effects resulting from respiratory evaporative cooling are partly offset by increased heat production due to greater muscular exertion required for the increased respiratory activity. Cooling by respiration requires four times more energy than the secretion of sweat. A net increase in cooling effect is accomplished by decreases in thyroid activity (metabolic rate) and consumption of total digestible nutrients.

The body temperature of large animals, such as cattle or horses, lags behind changing ambient temperature because of their large thermal capacity (weight times specific heat). This stored heat can be dissipated into the cool night atmosphere, especially if exposed to the cold sky. Therefore, large animals can withstand very high day temperatures provided the nights are cool. However, the structural features of the blood vessels of cattle tend to assist preservation of heat in a cold environment rather than dissipation of heat in a warm environment (*4*).

Heat can be lost by conduction when the animal is in contact with a cooler body, for example lying on the cool ground. If, however, the ground temperature is warmer than surface temperature, the animal will gain heat.

Convection is important in removing heat from wool and hair tips. In hot weather, when air is as warm as the skin, air movement of 8 to 11 kilometers per hour is adequate for convective purposes. Blowing hot air over a warm, dry body may actually bake rather than cool.

Endocrine Responses

Since the endocrine system is so intimately involved in acclimatization and productive functions, it would seem reasonable that the various endocrine responses to specific and general climatic conditions had been thoroughly studied and their interrelationships determined. Surprisingly, very few studies have specifically emphasized climatic influences on circulating and glandular hormone levels, particularly for large herbivorous mammals. Perhaps the major reason for this deficiency has been the relative difficulty of assaying hormone levels, especially with respect to concentrations in the circulatory system. Recent improvements in assay techniques have alleviated many of these difficulties, however, and it is expected that much of this essential information will become available within the next few years.

Due to this lack of basic knowledge, the following discussion will necessarily rely heavily on subjective interpretation and extrapolation of existing information. The authors assume full responsibility for any possible inaccuracies.

THYROID FUNCTION

Probably more information is available concerning the effect of heat on thyroid function than for any other endocrine gland. Most of this information has arisen from studies conducted with dairy animals in controlled climatic chambers at the University of Missouri.

There is general agreement that exposure of animals to high temperatures results in a decline in thyroxine production. Since thyroxine is intimately concerned with general metabolic level, the resulting decline in metabolism is normally associated with a decrease in productive functions, for example: growth, milk production, and wool production — but it is difficult to separate the direct effects of thyroxine from its interactions with other endocrine glands and the general level of nutrition. It appears, however, that a depression in thyroid activity can occur when feed intake is held constant (*5*). The response is not immediate but occurs only after 60 hours of exposure to high temperature (38°C). The reduced thyroid output may also be associated with impaired fertility (*6*), although work with thyroidectomized animals does not substantiate this conclusion (*7*).

A difference is also noted in thyroid activity between breeds of cattle in that zebu cattle have a lower level of thyroid activity than European breeds. This difference is often cited as a factor in the greater adaptability of zebu cattle to hot conditions. It is also interesting to note that Brahman cattle, when moved from a cool (10°C) to a warm (32-38°C) environment, do not show the depression in thyroxine production that is characteristic of the British breeds (*8*). A difference in thyroid activity has been observed in animals exposed to rising temperatures after being reared at 10° or 27°C, in that the heat-acclimated group (reared at 27°C) exhibited higher thyroid activity than the 10° group (*9*).

The most logical explanation for the decline in thyroxine production is through the influence of heat on the production of the pituitary hormone responsible for thyroid activity, thyroid-stimulating hormone (TSH). Anderson and associates (*10*) have postulated that "warm detectors" in the preoptic-anterior hypothalamic area exhibit an inhibitory effect on the release of TSH from the hypophysis.

ADRENAL GLAND

The importance of the adrenal gland in maintaining homeostasis has long been recognized. Initial responses to many different types of stress are mediated through the adrenal, and thus its reactions are of vital importance when one considers adaptation and productive functions.

Morphologically and physiologically, the adrenal can be divided into two areas — the medulla and the cortex. The two hormones produced by the medulla — epinephrine and norepinephrine — are controlled by neural stimuli. A multitude of steroid hormones are produced by the cortex and are basically concerned with carbohydrate, fat, and protein metabolism (glucocorticoids), and mineral metabolism (mineralocorticoids). The latter are basically controlled by humoral agents. In addition, the cortex also produces varying quantities of the sex steroids (androgens, estrogens, and progestins).

CATECHOLAMINES

The influence of heat stress (hyperthermia) in cattle on production and excretion of epinephrine and norepinephrine has only recently been investigated. Although the findings are not conclusive, it appears that exposure to sufficient heat stress to elevate deep body temperature will result in an increase in production of these medullary hormones (11). Other workers have reported an increase in noradrenaline excretion after exposure to 38°C for 24 hours, although the effect on body temperature was not reported. It would appear that warming of the skin and superficial tissues of cattle in the absence of an increase of deep-body temperature has no effect on the adrenal medulla.

The increases observed in catecholamines are probably responsible for the characteristic changes occurring in cattle subjected to heat stress, for example, increases in concentrations of blood glucose and lactic acid, increases in systemic arterial blood pressure and heart rate, and an increase in the percentage of red blood cells in arterial blood. Since some of the same changes occur with localized heating of the anterior hypothalamus, it is possible that the latter is responsible for monitoring body temperature and signaling for the increased production of catecholamines.

It has recently been determined that epinephrine is a powerful sudomotor agent in cattle, sheep, and goats. Since sweating is a primary means of maintaining temperature homeostasis in the bovine, epinephrine becomes an important consideration in adaptation of animals to hot conditions. Under conditions of mild heat stress, however, the adrenal medulla does not appear to be involved in sweat-gland activity. Even though sweating is controlled by an adrenergic mechanism, it thus appears to be a response acting independently of the adrenal medulla.

CORTICOSTEROIDS

The information concerning corticoid secretion under heat stress is limited and somewhat inconsistent. Considering other responses of the body to hyperthermia and some of the known effects of corticoids, it would seem reasonable that the initial response to heat stress is an increase in activity of the adrenal cortex. Theoretically, this activity would result from an increase in circulating corticotrophin (ACTH) from the pituitary gland. The primary corticoids involved would be the glucocorticoids such as cortisol and corticosterone. Prolonged exposure to heat may eventually result in adaptive adjustments and a consequent general decrease in corticosteroid production.

The situation with respect to aldosterone secretion, the most potent adrenal hormone affecting mineral metabolism, is somewhat different. Aldosterone production is not normally controlled by ACTH, as are other adrenal steroids, but appears to respond to differences in blood volume and electrolyte concentrations. These effects are probably mediated by the renin-angiotensin system. Loss of extracellular fluid, such as occurs during sweating or general dehydration, will therefore result in increased levels of aldosterone. Aldosterone increases the rate of reabsorption of sodium from the kidney tubules, and this reabsorption is associated with sufficient water to maintain isotonicity of the blood. Therefore, considerable changes in the fluid volume may accompany the effects of aldosterone, and it may be the more important mechanism regulating plasma volume under normal conditions.

Recent evidence indicates an alteration in the adrenal production of the female sex hormone progesterone following heat stress. Exposure of ovariectomized cows to high temperature for 24 hours resulted in an immediate increase in blood-progesterone levels (12). The high level following the acute stress is only temporary and is followed by a depressed secretion of the steroid. Blood progesterone levels of lactating dairy cows during the hot summer months were significantly lower than in cows during the spring. The significance of this altered production of progesterone is not known, but it may be related to fertility problems encountered under these summer stress conditions.

PITUITARY FUNCTION

Although the pituitary gland is intimately concerned with regulating the secretion of most of the other endocrine glands, little information is available concerning the influence of heat stress on pituitary function. Evidence has been presented above for the influence of thermal stress on two pituitary hormones, TSH and ACTH. It appears that thyrotrophin secretion is reduced under heat stress, accounting for the general decline in metabolic rate. Although ACTH secretion may be reduced under prolonged stress, it is probable that a transitory increase in this trophic hormone occurs during initial exposure.

Vasopressin secretion is increased following dehydration. This response is probably mediated through osmoreceptors and volume receptors. Classically, vasopressin is known as the antidiuretic hormone because it increases water reabsorption from the kidney in some species. It is thus implicated in water-conserving mechanisms during stresses resulting in dehydration. The exact role of vasopressin in the ruminant is not clear. Some evidence indicates that it effectively reduces urine volume only when the animal is in diuresis, and it may actually have the reverse effect with normal urine production (13).

The effect of heat on growth hormone (GH) production is not known. On the basis of slower growth rates of animals in tropical climates, it would appear that the production of GH is inhibited by heat stress. It is difficult to separate the effects of nutrition on growth in most of these conditions, however, and the reduction in growth may be primarily a function of nutrient intake interacting with other hormones. A recent study with laboratory rats (14) indicates an initial increase in circulating growth hormone following heat stress, with a return to normal levels after one week.

No information is available concerning the influence of heat on the pituitary gonadotrophins (FSH and LH) or on prolactin, but the general decline in reproductive performance and lactation may indicate a decrease in secretion of these hormones in animals exposed to heat stress.

In general, little information is available relating pituitary function to hot, dry conditions. On the basis of animal performance and activities of other endocrine glands, chronic exposure to heat stress probably results in a general hypofunction of the pituitary. Since this gland is so important in determining productivity of animals, it is essential that the interrelationships between climatic conditions and pituitary function be determined.

Water Balance

Heat-regulating mechanisms of sweating and panting depend upon water, which must be supplied by the heat-stressed animal. For animals kept in confinement, ad libitum water can be provided to meet the animal's needs. For range animals in arid areas, water is often scarce or completely lacking, and it thus becomes the limiting factor in efficiency, performance, and utilization of range forages.

Under normal conditions, the body-water content of animals is maintained within very narrow limits. Comparative studies of various species show that the body water, calculated on a fat-free basis, ranges from 70 to 75 percent with a mean value of 73.2 percent. The ruminant may have additional water in the rumen or other parts of the gut equal to 25 percent of the total body water (15).

WATER INTAKE

The maintenance of constant body-water content requires a refined mechanism to regulate water intake. Water enters the body as a fluid, as an integral part of food, and as water of oxidation. The water obtained from food and by oxidation is highly variable; thus the fine adjustment of body water is determined by fluid intake. Under nonstress conditions the quantity of water ingested is metered by the animal during drinking so that intake offsets deficit. Since there are marked individual variations in caloric expenditure, it is obvious that the amount of water required to maintain normal hydration will vary in different individuals, and in the same individual from day to day.

TABLE 1

Water Consumed by Nonlactating Range Cows on Semiarid Range *

Month	Average maximum temp (°C)	Total precipitation for month (mm)	Average daily water consumption (liters)
1	15	15	15.1
2	14	28	8.3
3	15	20	14.7
4	17	3	14.4
5	21	5	21.2
6	29	0	25.7
7	33	53	32.6
8	32	107	30.2
9	28	66	15.9
10	21	0	15.9
11	16	18	22.0
12	11	23	22.7

* Data from Arizona Agricultural Experiment Station Technical Bulletin No. 79 (1938).

The quantity of water intake per unit of dry matter consumed by cattle increases at an accelerating rate with increasing temperatures above 5°C. The number of liters of water consumed per kilogram of dry matter intake more than doubles between 5° and 30°C. The actual water consumption of nonlactating range cows on an Arizona range (Table 1) shows the importance of adequate potable water during the warm summer months. During this period the water consumption of lactating cows would be expected to increase 4 liters for each kilogram of milk produced, resulting in an intake of over 45 liters per day.

An interesting relationship between water metabolism and milk production was noted for range cattle in Australia (16). The water content of calves running together with their dams on the same range was estimated by tritiated water dilution. Turnover of tritiated water gives a measure of milk intake, related to the 0.82 exponent of body weight to compensate for size differences. Measurements were made in the wet season, when the maximum temperature was 32°C. A high rate of water intake and a high content of body solids was found in pure zebu calves, but values were less for zebu-Shorthorn crosses, while banteng, Shorthorn, and buffalo calves had low intakes (Table 2). These data show that in a tropical environment zebu calves with high water turnover have more milk intake and grow faster than the other types.

TABLE 2

Body Composition and Water Turnover of Range Cows and Calves

Type	Adults		Calves	
	Body solids (%)	Water turnover (ml/kg$^{0.82}$/24 hr)	Body solids (%)	Water turnover (ml/kg$^{0.82}$/24 hr)
Zebu	38.6	361	27.3	420
Zebu X Shorthorn	37.1	384	24.5	405
Shorthorn	26.1	484	19.6	353
Banteng	22.7	427	17.5	362
Buffalo	21.4	524	22.1	385

Water-turnover values of adult animals of these different types of cattle were the reciprocal of those of the calves; that is, the highest rate was found in buffalo (*Bubalus bubalis*) and the lowest in zebu.

WATER LOSS

Under warm, arid conditions when water intake may be infrequent, sources of water loss become of critical importance in the maintenance of normal body-fluid balances.

Urine water loss. Urine water loss may be considered under the general categories of "obligatory" and "regulatory" (15). Water is excreted in the urine as a solvent for catabolic products such as urea and for excess minerals such as salt. The higher the levels of minerals and proteins in the diet, the larger the obligatory quantity of water required for excretion. The regulatory water volume serves in the balancing exchange of intake and loss when water ingested exceeds body needs.

Fecal water loss. The high crude-fiber content of the diet of herbivores results in a relatively large fecal water loss. With hay rations, cattle feces average 80 to 85 percent moisture, resulting in a water loss equal to urine output. For example, a beef animal under nonstress conditions with water available ad libitum drinks 20 liters of water daily. Water of oxidation and water in food increases total water intake to 22.5 liters. Under these conditions urine output would be about 9.4 liters, water loss in feces 9.6 liters, and dissipation by evaporation 3.5 liters. Under similar conditions other species, such as sheep, horses, and camels, have a drier feces than do cattle, ranging from 50 to 65 percent moisture. There is a difference in the physiology of the large intestine between European and zebu cattle. The zebu has a lower fecal water content; this may be a part of the greater water economy of zebu cattle.

Cutaneous water loss. The apocrine glands function to provide a source of water for evaporative cooling for body-temperature control. Warming drinking water in the body and excreting it at body temperature dissipates only 3 or 4 percent as much body heat as both warming and vaporizing the same amount of water. The economical use of water by sweating for evaporative cooling thus conserves water and energy compared to respiratory cooling or water loss via urine and feces. This process is an important part of adaptation and water conservation by heat-tolerant animals.

Among cattle, *B. taurus* and *B. indicus* differ significantly in their sweating rate. At ambient temperatures of 45°C and higher, zebu and zebu-cross cattle evaporate over 600 milliliters per square meter of skin per hour. Values for the European breeds range from 140 to 400 milliliters. At ambient temperatures of 20° to 35°C, however, the sweating rate of the zebu is less than 50 percent that of the temperate breeds. This ability of the zebu to conserve body water by restricting cutaneous water loss at moderate temperatures but increasing evaporative cooling at high temperatures is an important factor in establishing the heat-tolerant superiority of the zebu compared to temperate breeds.

Diffusion of water through the skin, apart from the sweat-gland secretion, is referred to as "insensible water loss." Since it is a diffusion process, it is directly affected by the vapor pressure of the skin surface compared with the adjacent air. In domestic animals the air at skin surface is trapped by body hair and is consequently saturated at skin temperature. Changes in ambient air temperature, humidity, and movement thus directly affect the vapor pressure gradient and thereby alter water loss by diffusion.

In cattle, approximately three times as much water is lost by skin diffusion as by respiratory activity. At moderate temperatures, insensible water loss is about 120 milliliters per hour (15). Since the diffusion process of insensible water loss depends upon physical factors, there is very limited physiological control. Under conditions of water restriction, insensible water loss thus becomes critically important as a major source of water loss for the dehydrating animal.

Respiratory water loss. The normal process of breathing results in water loss. The amount of water lost depends upon the respiratory volume and the absolute humidity of the incoming and expired air.

Respiratory water loss in beef cattle ranges from about 23 milliliters per square meter per hour at 27°C to 47 milliliters at 40°C. The respiratory vaporization accounts for 5 percent of total heat dissipation at 25°C to 17 percent at 40°C.

Drooling of saliva is common in some domestic animals. This serves no known useful function and is another source of water loss. Under some stress conditions, cattle may drool 10 to 15 liters of saliva per day. Not only is water lost, but also other saliva ingredients, including minerals.

Dehydration

Under temperate conditions the main requirement for water is to meet metabolic demands. Moderate water restriction may have no appreciable effect on animal performance, since voluntary water intake is usually in excess of body needs. Under warm, xeric conditions the supply of drinking water and water contained in the food may be severely restricted, while at the same time the water demand for thermoregulation is at a maximum. Under such conditions water becomes the limiting factor in animal performance.

With water restriction, urine output is limited to the obligatory portion and fecal moisture content decreases to about 40 percent. Water restriction also results in a voluntary decrease in food consumption. The rate of dehydration determines the decrease in caloric intake for a given species. In the case of temperate cattle breeds, water deprivation results in a decline in food consumption of 50 percent of each preceding day's consumption during the period of water deprivation. Zebu and zebu-cross cattle do not show a loss of appetite during the first 24 to 48 hours of water deprivation, but after two days without water they also have a considerably depressed appetite. Decrease in food consumption decreases fecal output. For cattle, the resulting decline in fecal water loss may be more important than reduction in urine output in conservation of water (*17*).

Dehydration results in changes in the partition of water in the body-fluid compartments. The maintenance of high plasma volume facilitates the circulation and transport of stored heat. This is one of the first functions to suffer during dehydration in nonadapted animals in hot environments. Hereford heifers lost 28 percent of plasma volume after four days of water deprivation (*17*). Thus dehydration is an important factor affecting heat tolerance. The rise in body temperature during heat exposure is proportional to the water deficit, although sweat secretion is active at rather high levels of dehydration (Fig. 3). The actual survival time for domestic animals deprived of water and exposed to solar radiation depends largely upon water reserves in the digestive tract and extracellular compartment. Digestive-tract water reserves may be 80 to 100 liters in mature cattle.

Some herbivores are particularly adapted to warm, dry conditions. Some animals selectively graze plants with high water content. The eland and the oryx do not require water except that obtained from plant food (*18*). The eland, weighing 500 kilograms, increases body temperature 7.3°C during the day, thus storing 3,000 kilocalories. The rise in body temperature as heat is stored reduces the body-to-air temperature gradient and thus reduces the gain of heat. The stored body heat is dissipated by radiation during the cool night. The dissipation of this stored heat by evaporation would require more than 5 liters of water.

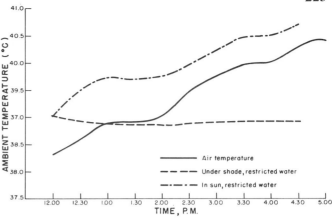

Fig. 3. The effect of solar radiation on body temperature of cattle on restricted water intake.

Ruminants have a distinct advantage in making up a water deficit quickly. The water is stored in the rumen and is then slowly released into the abomasum for absorption. When offered water, dehydrated cattle will drink 15 percent of their weight within 30 minutes. The camel can consume 20 percent of its body weight in 10 minutes. The donkey can drink at the rate of 8 liters per minute and take up 20 percent of body weight in about 2 minutes. The rapid water intake is an obvious advantage for grassland and desert wild animals that must obtain their water from a potentially dangerous waterhole. Domestic cattle also follow the game animal drinking behavior in that cattle rarely drink at night, even when grazing at night.

There do not appear to be permanent changes or any impairment of health due to relatively short periods of dehydration, although the restoration to normal of the various physiological values deranged by dehydration requires more than a day in most instances.

Saline and Alkaline Drinking Water

In many arid regions the available drinking water is often present in small waterholes. The high evaporative rate concentrates the dissolved salts, resulting in an increase in salinity. The utilization of a saline solution requires special effort by the kidney, which must reabsorb all water but the obligatory urine loss. In addition, it must concentrate and excrete the ingested salts. Domestic animals have a limited ability to conserve water by excreting a high osmolal urine. The result of high saline-water intake is a shift in body-water compartments that can be just as harmful to the animal as dehydration resulting from water restriction.

Sheep and cattle can tolerate a 1.5 percent sodium chloride drinking solution during cool weather but are adversely affected by even 1 percent salt in the drinking water during the summer months. A solution of 2 percent sodium chloride is definitely toxic. It causes severe anorexia, weight loss, and anhydremia (*19*). Other solutions also may be detrimental. Water containing up to 7,000 parts per million of sodium sulfate was not harmful to cattle; 10,000 ppm caused scours and weight loss (20).

The effect of water restriction is much more detrimental when the available water is even slightly saline. When the water is saline, drinking induces diuresis; the water consumed is apparently involved in renal osmoregulation rather than in tissue rehydration. Recovery from salt dehydration is much slower than for animals dehydrated by water restriction. When salt-dehydrated animals are offered pure water "free choice" they often collapse and show tetany after drinking. Limiting the water intake for two or three days prevents this occurrence.

EFFECTS OF DRY, WARM CONDITIONS ON RANGE–CATTLE PERFORMANCE
Reproduction

Reproduction in mammals is perhaps the most complex biological phenomenon in nature. It is dependent upon the influence of many discrete factors as well as complex interactions among these factors. Some of the more important influences are hormonal balance, nutrition, climatic factors, and psychological stimuli. Probably no other biological mechanism is so dependent for its successful completion on the general concept of homeostasis as is reproduction.

Under natural conditions, most species have evolved a particular seasonal pattern of reproduction that appears to synchronize birth with the optimal environmental conditions for survival of the newborn. These rhythms appear to be primarily under the control of climatic factors such as photoperiod, temperature, and moisture. The indirect effect of climate on nutrient availability is also a major factor in the regulation of these cycles. As might be expected, these rhythms are much more pronounced in temperate zones than in equatorial areas.

In the process of domestication man has tended to eliminate the seasonality of reproduction in a number of species. This accomplishment occurred through genetic selection within the existing variability of a population and by artificial reduction of the extremes associated with seasonal changes. Even though the "all-or-none" nature of seasonal reproduction may have been eliminated in several species, differences with respect to season may be evident for overall reproductive efficiency and the various components contributing thereto.

The following discussion will be primarily concerned with the influence of high temperature on the various functions concerned with reproduction. Since maximum day length and maximum temperature usually coincide in temperate regions, data from experiments employing some type of climatic control will be relied upon to help separate these influences.

PUBERTY

Puberty was delayed in Brahman and Shorthorn heifers raised at a constant 27°C as compared to a similar group at 10°C. The average age at puberty was 307 and 303 days at 10°C for the Brahman and Shorthorn calves, respectively, whereas at 27°C the comparable values were 463 and 440 days. In relation to mature weight, Brahmans were slower in obtaining sexual maturity than either Shorthorn or Santa Gertrudis heifers (21). There appears to be a closer relationship between body weight and puberty than between age and puberty; therefore, the increased age often observed at puberty may be an indirect effect of the slower growth rate normally encountered under hot conditions.

ESTRUS

Several studies have demonstrated that the duration of estrus is decreased in cattle exposed to hot summer conditions. Gangwar and others (22) reported an average decrease of 6 hours in length of estrus during the summer as compared to the spring in dairy heifers. In addition, the intensity of estrus is also decreased, resulting in difficulty in detection of heat periods. High temperature is also associated with an increase in the frequency of "silent" heat periods, that is, ovulations not accompanied by physical manifestations of estrus.

An anestrous condition may be induced in beef heifers by high temperatures. Experiments by Bond and others (23, 24) demonstrated that heifers moved to a psychrometric chamber maintained at 32°C during the winter months ceased cycling and ovulating after approximately 5 weeks in the chamber, when peak rectal temperatures were recorded. Subsequently shedding of the long hair coat began and rectal temperature and respiration rates decreased to nearly normal levels for adapted animals at 17 weeks. Estrous cycles were subsequently reestablished by the 21st week in the chamber, indicating that the heifers had become acclimated. When summer-conditioned heifers were placed under the same conditions for 14 weeks, only one of six ceased to cycle. If the temperature in the chamber was raised to 38°C, however, sufficient heat stress was imposed to abolish estrous cycles. The estrous cycle can thus be interrupted if heat stress is severe, but the animals may become physiologically adjusted after a period of time and resume normal cycling.

It can be concluded from the foregoing discussion that heat stress can influence the estrous cycle, the length of estrus, and the intensity of estrus in cattle, thus imposing a restriction on reproductive functions in hot climates.

GAMETOGENESIS

Spermatogenesis is adversely affected at normal body core temperatures in most domesticated mammals. It is therefore essential that testes temperature be maintained several degrees cooler than body temperature for normal functioning. This thermoregulatory function is accomplished by the scrotum. If the heat load imposed on the animal is not too extreme, the scrotum is able to maintain a testes temperature favorable for spermatogenesis. As the environmental temperature rises, however, varying degrees of spermatogenic impairment may result from the inability of the scrotum to regulate testes temperature properly.

Bulls and rams normally exhibit summer decline in semen quality and fertility in areas characterized by hot summers. Complete sterility is not usually observed unless the heat stress is extreme. The decline in fertility may become evident within two to three weeks after exposure, again depending on the severity of stress. These detrimental effects may persist for two to three months after normal thermoregulation is achieved. Some individuals may become permanently sterile. It appears that the earlier cell stages involved in the spermatogenic cycle are most susceptible to heat damage.

Definite differences in response to heat stress are noted among individuals and between breeds. In one study, Red Sindhi crossbred males were not as severely affected as Holsteins or Brown Swiss. These differences are probably due to the inherent variation of heat-dissipating mechanisms of the scrotum. Another interesting observation relating to the effect of breed differences on "useful life" was reported by Bonsma (*25*). He stated that Afrikaner bulls in South Africa had a useful life of almost nine years, whereas for Shorthorn and Hereford bulls it was just over three years.

Little information is available concerning the effect of high temperature on gamete production by the female. Previously cited data of Bond and others (*23, 24*) indicates that continuous high temperature (32°C or 38°C) can result in temporary abeyance of ovulation. The incidence of morphologically abnormal ova was increased from 3.7 percent in control ewes to 46.2 percent in ewes exposed to a constant 32°C from time of breeding until slaughter three days later (*26*). It appears that high temperatures may have an adverse effect on gamete production in both sexes.

FERTILIZATION AND EMBRYONIC SURVIVAL

The most critical period in the reproductive cycle of most farm mammals is the interval from mating or insemination through implantation of the developing embryo. It is during this period that fertilization occurs, the zygote passes into the uterus, extra-embryonic membranes are formed, and major differentiation of tissue and organ systems occurs. Until the process of implantation is completed, the developing embryo is essentially "free-living" within the uterus and is dependent on uterine secretions (uterine milk) for nutrition. It is therefore reasonable that most losses would be expected during this period, since so many critical events are occurring and the embryo has no definite attachments to the uterus for nutrient and waste-product transfer. This interval covers approximately 5 to 6 weeks in the cow and is somewhat shorter in the ewe.

Few studies have critically evaluated temperature effects on fertilization rates and subsequent embryo survival through implantation. Most of the studies have been conducted with sheep or laboratory animals. Since the first month or so is generally the most critical period, however, it is logical to assume that any effects due to high ambient temperature will be most important during this interval.

Failure in fertilization may occur if animals are heat stressed shortly before or at the time of breeding. Dutt (*27*) reported a decrease in fertilization rate of ewes exposed to a constant 32°C starting on day-12 of the cycle prior to mating or on the day of mating. Embryonic mortality was also increased if the ewes were heated while the zygote was in the oviduct (prior to three days after breeding), but not after it passed into the uterus. Lambing results indicated that heat stressing ewes prior to eight days postbreeding decreased lambing rates. Embryos in the very early stages of development thus appear to be most susceptible to heat damage.

No comparable data are available for cattle. Stott and Williams (*28*) reported a substantial decline in cows diagnosed pregnant by rectal palpation at 35 to 41 days following service during the hot summer months in Arizona. They attributed this decline in fertility to a low rate of fertilization and a high rate of embryonic mortality. Additional evidence for an effect of temperature on conception rates of cows was presented by Long and others (*29*). Cows with rectal temperatures below 39.7°C at breeding had a higher conception rate (55 percent) than those with rectal temperatures above 39.7°C (24 percent).

Breed differences also exist in fertility response to summer heat stress, probably reflecting inherent differences in adaptability to a hot environment. In Arizona, Guernsey and Holstein cows showed a sharp decline in fertility during the summer, whereas no depression was noted among Jerseys (*30*). Australian studies demonstrated that British cows (Hereford and Shorthorn) produced only 56 percent live calves, whereas zebu-cross cows (Africaner and Brahman) had a calving rate of approximately 75 percent (*31*). British heifers did not express reproductive potential as early as zebu-crosses, the difference being approximately a year.

GESTATION AND PARTURITION

As indicated earlier, the major effects of heat stress are probably expressed in fertilization rates and early embryonic mortality. If the stress is extreme enough, abortions may be induced. For example, cows 4 to 6 months pregnant aborted two days after a 27-hour exposure to 38°C. Adaptation to hot conditions also influences this response, since Australian Illawarra Shorthorns were able to produce normal calves after exposure to 40°C.

Perhaps the most common effect of high temperature on the post-implantation fetus is a reduction in birth weight or the production of "miniature" offspring. British breeds in South Africa produced calves 20 percent lighter in the summer than winter, although no seasonal effect was noted with well-adapted indigenous breeds. Pregnant ewes exposed to high temperatures produced miniature lambs, the reduction in size being proportional to the length of exposure. It has been demonstrated that these lambs are the result of heat stress and are not due to an effect on feed intake of the dam. The miniature lambs were well proportioned and distinct from long-legged, thin lambs produced by underfed ewes. Breed differences are indicated by the fact that Merinos require a higher temperature stress (44.5°C) to reduce birth weights of lambs as compared to an effective temperature of 40.5°C for Romney Marsh ewes.

Difficulties at parturition may also be an indirect effect of temperature stress on the female. Females born and raised under hot conditions may have smaller pelvic canals than normal, resulting in a higher frequency of stillbirths or other types of dystocia.

NEONATAL SURVIVAL

At birth the newborn must begin to make physiological adjustments to fluctuating environmental conditions as compared to the relatively constant temperature and nutrient supply present *in utero*. Many newborn animals are not well adapted to withstand high temperatures early in life. It has been reported that lambs and calves are particularly susceptible to heat stress immediately after birth. Lambs 2 to 7 days old were not able to survive longer than about 2 hours of 38°C temperatures or more than 3 hours of direct exposure to solar radiation.

The reduced milk yields of the dam often associated with heat stress may also contribute to postnatal mortality, since the young is entirely dependent on milk during early postnatal life. High temperatures may thus impose direct and indirect limitations on survival of the newborn and result in a further decline in the overall efficiency of reproduction.

LIBIDO

Sexual activity is normally decreased in animals exposed to hot climatic conditions. In a study in which rams of three breeds were exposed to increasing temperature (up to 43°C), a decline in sexual activity among all breeds was noted (*32*). The effect of temperature on Merino rams was slight as compared to the other breeds, however. Dorset Horns recovered activity after the temperature was lowered, but Border Leicesters did not. Even one or two days of hot weather may lead to a protracted period of lowered sexual activity in some breeds.

PHYSIOLOGICAL MECHANISMS

The foregoing discussion indicates that exposure of animals to high environmental temperatures may adversely affect many components contributing to total reproductive efficiency. Puberty may be delayed, gametogenesis interfered with, fertilization rates reduced, embryonic mortality increased, abortions increased, and higher postnatal losses experienced. A reduction in sex drive may also be observed. The total effect of high temperature on reproductive efficiency can be relatively large, and it may be the most serious effect in terms of total productivity.

The physiological mechanisms through which heat exerts its effects are vaguely understood. Heat could exert a direct thermal effect or it could act through other mechanisms such as nutrient and water availability, appetite, and endocrine balance. Undoubtedly endocrine balance is eventually involved, but little information is available to delineate specific hormonal shifts in response to hyperthermia.

Many workers have ascribed some of the effects of heat to the general decline in thyroid activity and resulting low metabolic rate. Some improvement in fertility of sheep treated with thyroactive compounds has been reported, and it is a common practice to attempt to improve libido in rams with thyroid-stimulating agents. Studies with thyroidectomized cattle indicate that estrus (female) and sex drive (male) are eliminated, although the production of fertile ova and sperm is not affected, since normal calves were born to thyroidectomized parents. Williams and Stott (*7*) studied reproductive performance of thyroidectomized cattle and concluded that the transient hypothyroidism occurring under hot summer conditions could not account for the lowered breeding efficiency.

It is also probable that the adrenal gland is involved in the response of reproductive systems to heat. Experiments with laboratory animals support the contention that excess production of adrenal steroids contributes to embryonic mortality and infertility. Recent studies in Arizona implicate an altered production of progesterone by the adrenal cortex as a factor in lowered fertility (*12*).

The decrease in libido observed in males is probably related to a hypothyroid condition and not to decreases in production of male sex hormones. In fact, there is evidence to indicate that androgen secretion is increased in rams exposed to heat stress.

The mechanism for inducing fetal dwarfing is not clear but may be due to placental defects, reduction in blood supply, a pituitary insufficiency, or an excess of adrenal steroids. Injections of cortisone throughout pregnancy or corticotrophin at intervals did not induce dwarfing, however, and it is therefore unlikely that this condition is due to an overactive adrenal. The most probable explanation is that the fetal pituitary is inhibited in some manner by heat stress but may function normally after birth.

Experiments with rabbits indicate that heat *per se* can have a direct effect on fertilized ova independent of its effect on hormonal function. One-cell fertilized rabbit ova were cultured in vitro for 6 hours at 40°C and subsequently transferred to recipient does. An increase in post-implantation embryonic mortality was noted with eggs treated in this manner, with no effect noted on eggs cultured at 38°C.

Since most body functions are depressed at high temperatures, the most reasonable explanation is that general function of the pituitary gland is depressed. No critical information is available to support this concept, however, and further experimentation will be required to test the validity of this hypothesis.

Growth and Development

With other variables held constant and with an abundant food supply, growth approaches zero near low and high limiting temperatures at which survival is impossible, and the growth rate is greatest at some optimal ambient temperature (*33*). Young animals are more affected at high temperatures than are older cattle. Temperatures as low as 27°C are severe enough to depress growth rate and skeleton growth in some temperate breeds of cattle.

Growth performance of range stock reflects forage availability as well as climatic conditions. Live weight and body measurement data obtained on various breeds and types of range cattle in semiarid areas show that growth in weight is strictly seasonal from weaning to maturity. Body measurements are influenced by season to a lesser extent than body weight. Extended droughts may occur, forcing animals to subsist on mature grasses, browse, and plants ordinarily unpalatable to them. Thus all range stock are subjected to varying periods and degrees of feed restriction. The possible effects of these restrictions on growth and production are of major concern.

A considerable amount of work has been done in various arid parts of the world concerning the effects of feed scarcity (*34–38*). Some general conclusions can be made from these studies.

a) Cattle are able to make complete recovery from short periods of mild restriction, but this recuperative capacity diminishes as either or both the severity or duration of the undernutrition are increased.

b) Older animals are better able to withstand undernutrition and to demonstrate enhanced compensatory growth than young animals.

c) Under conditions of feed scarcity, beef cattle between the ages of 6 and 12 months can be carried at energy levels as low as maintenance, if nutritional needs other than energy are supplied, without lasting harmful effects to growth or milk production. Sexual maturity and reproductive performance may be delayed.

d) Feed restriction of young animals below maintenance levels for any extended period of time does result in smaller mature size.

PROCEDURES FOR IMPROVING RANGE–CATTLE PERFORMANCE
Management

The optimum use of arid ranges by cattle depends on the development of adequate potable water throughout the range area. For range cattle, a distance greater than 1.6 kilometers from water results in a forage-use rate of 25 percent, compared to 50 percent at 0.3 to 0.5 kilometers. At 3 kilometers from water the rate is only 15 percent (*39*). There are seasonal changes in cow distribution on the range during the period of forage growth and maturity. As the forage dries and matures, cows move to other slopes with less-mature forage.

Since range animals do perform better if they drink at least once a day, watering sites should be well dispersed over the entire range area. Site types include earth dams, wells, spring developments, and pipeline distribution points. In some areas, hexadecanal monolayers have been used to retard the large evaporative loss of open waterholes (*40*).

Proper stocking rate is an essential prerequisite to good management practice. It is a too-common tendency to overestimate grazing capacity of semiarid ranges. The unusually favorable year is regarded as normal, and the range is stocked accordingly. This practice leads to overgrazing, range deterioration, and the disastrous losses incurred with prolonged drought. (See the sections by Bauer and Nechayeva in the present volume.)

Since forage growth on desert ranges is seasonal, carotene and protein are available to the grazing animal primarily during the rainy season while the feed is green and growing. Another common characteristic of arid ranges is a deficiency of phosphorus. Many studies in these areas have consistently shown that the addition of a phosphorus supplement results in improved animal performance, most especially reproductive performance (*41–43*). The following general conclusions are based upon the results obtained from many trials designed to determine the value of various types of range supplementation.

Salt and a high-phosphorus mineral supplement should be always available to all range stock. Preferably they should be placed near the watering sites, where they can be most conveniently serviced. Placing salt licks in inconvenient locations within a range results in decreasing salt consumption per animal unit as compared with placing the lick in a convenient location. Locating the lick in a less-frequently used area of a pasture does not cause a significant increase in cows grazing in that area (*44*).

Calves are often weaned during the dry season. An energy-protein supplement should be provided if they are losing weight during the post-weaning period. An attempt should be made to at least maintain animal weight until new food is available on the range. Additional vitamin A may be incorporated in the supplement or may be administered to individual calves by intraruminal injection.

In an 8-year study on a semiarid range in New Mexico with an annual rainfall of 170 mm, providing feed in addition to a mineral supplement did not significantly increase production of mature cows (*45*). It was concluded that mature cows need a feed supplement only in emergency, such as an extended drought, or when suckling calves during the dry season. Even then an energy feed combined with a mineral supplement does as well as a protein feed. In fact a protein supplement, especially when fed as a salt:meal mixture, may serve to complicate further a deteriorating water-balance condition in the animal. The mature animal also builds up a large reserve of vitamin A when green forage is available. Under actual extreme drought conditions, adequate vitamin A reserves were found in liver samples from mature beef cows in a terminal stage of undernutrition (*46*).

The following management procedures should aid in improving reproductive performance:

a) Plan the breeding season for the period of most favorable environmental conditions. A restricted season should be used in most cases.

b) First-calf heifers will require special attention. They should be maintained separately from the cow herd and provided with additional feed if necessary. Early weaning of calves may be required to obtain satisfactory rebreeding performance.

c) Provide vitamin and mineral supplements when required. Vitamin A and phosphorus are usually the most important, as well as a continuous supply of salt.

d) Energy and protein supplementation may be required during extended drought periods. Irrigated pastures are excellent if available, although harvested forage is entirely satisfactory. If natural forage supply is extremely seasonal, try to adjust the management program to have the poor feed conditions coincide with the interval between breeding and calving.

Genetic Improvement

Fitness has been defined as the proportionate contribution of offspring to the next generation (47): it measures the ability of an animal to produce progeny which will themselves survive and produce progeny.

The genetic factors involved in fitness are poorly defined. It is well recognized, however, that natural selection is a principal force, modified by artificial selection and imposed husbandry practices (47, 48). Under conditions of natural selection, the intermediate phenotypes are favored and the population eventually reaches a state of genetic equilibrium. In this situation the gene frequencies are not changing, and therefore the mean values of all metric traits are constant. The fitness of the population is at a maximum for the existing environmental conditions. The force of natural selection is constantly applied to the population to improve fitness. Despite this pressure, the gene frequencies remain constant and population fitness remains stable. This condition indicates that the genetic variance of fitness is primarily nonadditive.

Artificial selection for any metric trait will change the gene frequencies of the original population and thereby upset the genetic equilibrium and reduce mean fitness. If we change any metric trait by artificial selection we must therefore expect a reduction of fitness as a correlated response. Natural selection still operates to resist the change in gene frequencies resulting from artificial selection, and if artificial selection is suspended the population will revert back to its original state or some intermediate plateau of fitness.

Domestication of animals is accompanied by changes in natural environmental conditions, including the use of assortative matings. The improvements that have been accomplished by selection are dependent upon corresponding changes in environmental conditions. Selection has reduced the fitness for life under natural conditions, and only the fact that an artificial environment has been

provided has allowed these improvements to be made. For example, the high-producing dairy cow in a modern dairy herd receives careful individual attention to her nutrition, health, and comfort. A completely artificial environment has been developed to accommodate the selection for maximum milk production. The introduction of these improved European-type stocks in tropical countries has often had disastrous results. It is not inconceivable that, in the course of breeding animals for high levels of production in relatively kindly environments, some of the ability to tolerate a rise in body temperature, relative dehydration, or alkalosis may have been lost (49).

The first step toward establishing and maintaining desirable breeds of livestock suitable to the climate of a region is to study the environmental conditions peculiar to the region and determine the requirements for adaptability and satisfactory performance of livestock under those conditions. The second step is to study the characteristics of the native stock.

General evidence indicates that there are local races of animals highly adapted to particular environments, particularly those of climate (34). Resistance to local parasites or diseases in native stock has been important in establishing superior performance in a given area; for example, the N'Dama cattle appear to have an inherent tolerance to trypanosome infection. The tolerance of the breed appears to be enhanced with exposure to infection. The Criollo cattle, of Iberian derivation, were introduced to the New World 400 years ago by the Spanish. These cattle have become well adapted to tropical and subtropical environments in the Americas and have demonstrated their fitness under natural conditions of extreme drought and high ambient temperatures.

The use of morphological traits or heat-tolerance indexes to aid in selecting for performance and adaptability appears to be of limited value. The low correlations between performance and thermal-stress tolerance measures suggests that producers might well select entirely on the basis of performance. Productivity is the best measure of practical suitability (50).

The first priority in selection traits must be reproductive performance. A high reproductive rate is essential if the enterprise is to be profitable. It also largely determines the genetic progress that can be expected for all economic traits.

Although the direct economic benefits derived from a high reproductive rate are evident, the influence of fertility on progress possible through selection is often neglected. An example illustrating these relationships is presented in Table 3. The basic assumptions are that 20 females and 5 bulls are retained for each 100 cows in the breeding herd. In addition, it is assumed that a large population of animals is involved so that effective variability is near the maximum. The intensity of selection is expressed in standard deviation units for comparative purposes.

Considering only replacement heifers, a reproductive rate equivalent to a 40 percent calf crop weaned would

TABLE 3

Influence of Fertility (Percent Calf Crop Weaned) on Potential Genetic Improvement (Intensity of Selection) *

| | Heifers | | Bulls | | Average Intensity Both Sexes |
Percent Weaned	Percent Saved for Replacement	Intensity of Selection†	Percent Saved for Replacement	Intensity of Selection	
40	100	0.00	25	1.27	0.64
60	67	0.54	17	1.49	1.02
80	50	0.80	12	1.67	1.24
100	40	0.97	10	1.76	1.36

* Assumes that 20 heifers and 5 bulls are saved for replacements for each 100 cows in the breeding herd.
† Standard deviation units.

require that all heifers raised be retained for breeding herd replacements to maintain a static population size. Under these conditions no selection would be practiced, and therefore no improvement in genetic merit would be contributed by the females. As fertility increases, a lower proportion needs to be saved for replacements and correspondingly more improvement is possible. For example, an 80 percent calf crop would permit approximately 48 percent more genetic progress than a 60 percent calf crop.

Since fewer male offspring are required for replacements, the dependency on fertility is not as great. Likewise, the proportionate increments in selection intensity as fertility is improved are not as dramatic. In all cases the greatest potential improvement is possible through selection of males, varying from approximately twice that for females with high reproductive rates to essentially all of the possible improvement in herds with low fertility. The potential for selective improvement from females should never be overlooked, however, since the effective rate will be the average of both sexes (last column of Table 3).

Since most heritability estimates for fertility traits in farm animals are very low, the general consensus is that direct selection for this trait will be relatively ineffective. This position would indicate that most additive variation has been exhausted and any genetic influences remaining would be of a nonadditive nature. Such a situation would appear logical from a theoretical standpoint, since natural selection for reproductive fitness should have been rather intense. Some recent studies (51) with cattle in Florida nevertheless indicate a relatively high heritability (approximately 40 percent) for birth rate. It is possible that the stress conditions normally encountered in Florida may have magnified inherent differences; possibly cattle subjected to climatic stress would respond favorably to selection for reproductive efficiency.

Several crossbreeding experiments have demonstrated that percent calf crop exhibits a significant degree of heterosis. The amount of improvement appears to vary in proportion to natural stress conditions and existing level of fertility, being greatest in more stressful climates where natural fertility is low. The greatest response is normally obtained from a crossbred dam mated to a bull of a different breed. Natural adaptability of the breeds involved is also a major factor determining degree of improvement. Indigenous cattle should form the basis of any crossbreeding program where climatic stress is a factor.

Fertility is the most important trait to consider in any productive enterprise. Both additive and nonadditive genetic influences should be exploited, as should all management procedures economically feasible, to enhance reproduction. The following recommendations should aid in obtaining maximum fertility:

a) Practice direct selection for fertility. The most effective method for the cow herd is to pregnancy-test all animals 2 to 3 months after the breeding season. Cull all open cows. Semen testing and physical examination of bulls should be employed. Although not perfect, this may be the only method of evaluating males in multiple-sire mating pastures. If single-sire pastures are used, conception rate of cows exposed to a given bull is the best measure of male fertility. Observation of the herd during the breeding season is invaluable if feasible.

b) Use some form of crossbreeding to capitalize on heterosis. Crossbred females should form the basis of the breeding herd. Indigenous breeds should be incorporated in the system. Select breeds that will complement each other.

It is not reasonable to assume that growth performance of animals on semiarid ranges will compare to that obtained in high-rainfall areas. Growth and development can be regarded as a stochastic process in continuous time. In certain situations of primary production, certain growth patterns may be more economical, or otherwise more desirable, than others (52). When local environmental conditions impose seasonal periods of food shortage, then a slower-maturing type of animal is most desirable (36). In terms of growth and production the later-maturing zebu unquestionably is thus much better adapted to hot, arid conditions than are breeds evolved in temperate areas. In the temperate zones, European breeds of beef cattle generally have faster growth rates and earlier maturity than the Brahman breeds.

Genetic studies of seasonal growth patterns on the range show that direct selection for growth under these conditions is most effective if based on animal performance following the wet season when range feed is most abundant (53–56).

Replacement animals and those in the breeding herd should be carefully checked for anatomic soundness. It is a waste of effort and may actually be detrimental to the program to include standards of type or conformation as criteria of selection. Lean-meat production and carcass cutout values are practically identical for such diverse types and shapes as eland, wildebeest, zebu, Holstein and Hereford (*57, 58*). The increased fat deposition noted with British beef breeds is due to their earlier maturity, not because of any differences in conformation *per se*.

A successful breeding program to improve genetic production merit of animals on semiarid ranges should use as a foundation existing native stock adapted to the area. If other animals can be obtained from areas with common environmental conditions, a gene pool may be established which would permit selection for improved performance within the adapted stock. If the production potential within the native stock is low, new genetic material may be introduced by top crosses with high-performance temperate breeds. The most appropriate breeds to use can best be determined by a small-scale trial with several breeds.

The response to selection for improved genetic productivity will be more effective if there are simultaneous actions being taken to ameliorate the extreme stress conditions by improved range and water management practices. Present and potential economic returns determine the rate and degree of additional improvement in nutrition and management. Continuing rigid selection will establish the genetic elements needed for optimum productivity.

REFERENCES AND NOTES

1. KLEIBER, M.
   1961   The fire of life: an introduction to animal energetics. Wiley, N.Y. 454 p.

2. SCHLEGER, A. V., AND H. G. TURNER
   1960   Analysis of coat characteristics of cattle. Australian Journal of Agricultural Science 11: 875–885.

3. JENKINSON, D. M., AND T. NAY
   1968   Sweat gland and hair follicle measurements as indicators of skin type in cattle. Australian Journal of Biological Science 21:1001–1011.

4. BIANCA, W.
   1965   Reviews of the progress of dairy science. Section A. Physiology. Cattle in a hot environment. Journal of Dairy Research 32:291–345.

5. YOUSEF, M. K., H. H. KIBLER, AND H. D. JOHNSON
   1967   Thyroid activity and heat production in cattle following sudden ambient temperature changes. Journal of Animal Science 26:142–148.

6. MACFARLANE, W. V.
   1963   Endocrine functions in hot environments. *In* Environmental physiology and psychology in arid conditions, Reviews of research. Unesco, Paris. Arid Zone Research 22:153–222.

7. WILLIAMS, R. J., AND G. H. STOTT
   1966   Reproduction in thyroidectomized and thyro-parathyroidectomized cattle. Journal of Dairy Science 49:1262–1265.

8. BLINCOE, C.
   1958   Environmental physiology and shelter engineering with special reference to domestic animals XLVII. The influence of constant ambient temperature on thyroid activity and iodine metabolism of Shorthorn, Santa Gertrudis and Brahman calves during growth. Missouri Agricultural Experiment Station, Research Bulletin 649.

9. KAMAL, T. H., H. D. JOHNSON, AND A. C. RAGSDALE
   1958   Effect of heat acclimation on physiological and biochemical responses of dairy heifers to various temperatures (35°F–95°F). Journal of Animal Science 17:1227–1228. Abstract.

10. ANDERSON, B., L. EKMAN, C. G. GALE, AND J. W. SUNDSTEN
    1963   Control of thyrotrophic hormone (TSH) secretion by the "heat loss center." Acta Physiologica Scandinavica 59:12–15.

11. ROBERTSHAW, D., AND G. C. WHITTOW
    1966   The effect of hyperthermia and localized heating of the anterior hypothalamus on the sympathoadrenal system of the ox (*Bos taurus*). Journal of Physiology 187:351–360.

12. WIERSMA, F., AND G. H. STOTT
    1969   New concepts in the physiology of heat stress in dairy cattle of interest to engineers. American Society of Agricultural Engineers, Transactions 12:130–132.
    Unpublished information by G. H. Stott, University of Arizona, expands the information given in this source.

13. MACFARLANE, W. V.
    1968   Adaptation of ruminants to tropics and deserts. Chapter 12, *in* E. S. E. Hafez, (ed), Adaptation of domestic animals. Lea & Febiger, Philadelphia.

14. PARKHIE, M. R., AND H. D. JOHNSON
    1969   Growth hormone releasing activity in the hypothalamus of rats subjected to prolonged heat stress. Society for Experimental Biology and Medicine, Proceedings 130:843–847.

15. CHEW, R. M.
    1965   Water metabolism of mammals. *In* W. V. Mayer and R. G. Van Gelder, (eds), Physiological mammalogy, vol II, chap 2. Academic Press, N.Y.

16. MACFARLANE, W. V., B. HOWARD, AND B. D. SIEBERT
    1969   Tritiated water in the measurement of milk intake and tissue growth of ruminants in the field. Nature (London) 121:578–579.

17. WEETH, H. J., D. S. SAWHNEY, AND A. L. LESPERANCE
    1967   Changes in body fluids, excreta and kidney function of cattle deprived of water. Journal of Animal Science 26:418–423.

18. TAYLOR, C. R.
    1969   The eland and the oryx. Scientific American 220:88–95.

19. WEETH, H. J., AND A. L. LESPERANCE
    1965   Renal function of cattle under various water and salt loads. Journal of Animal Science 24:441–447.

20. EMBRY, L. B.
    1959   Salinity and livestock water quality. South Dakota Agricultural Experiment Station, Bulletin 481.

21. DALE, H. E., A. C. RAGSDALE, AND C. S. CHENG
    1959 Environmental physiology and shelter engineering with special reference to domestic animals: LI. Effect of constant environmental temperatures of 50° and 80°F on ovarian activity of Brahman, Santa Gertrudis, and Shorthorn calves with a note on physical activity. Missouri Agricultural Experiment Station, Research Bulletin 704.

22. GANGWAR, P. C., C. BRANTON, AND D. L. EVANS
    1965 Reproductive and physiological responses of Holstein heifers to controlled and natural climatic conditions. Journal of Dairy Science 48:222–227.

23. BOND, J., R. E. McDOWELL, W. A. CURRY, AND E. J. WARWICK
    1960 Reproductive performance of Milking Shorthorn heifers as affected by constant high environmental temperature. Journal of Animal Science 19:1317. Abstract.

24. BOND, J., J. R. WELDY, R. E. McDOWELL, AND E. J. WARWICK
    1961 Responses of summer-conditioned heifers to 90°F. Journal of Animal Science 20:966. Abstract.

25. BONSMA, J. C.
    1949 Breeding cattle for increased adaptability to tropical and subtropical environments. Journal of Agricultural Science 39:204–221.

26. DUTT, R. H.
    1963 Critical period for early embryo mortality in ewes exposed to high ambient temperature. Journal of Animal Science 22:713–719.

27. DUTT, R. H.
    1964 Detrimental effects of high ambient temperature on fertility and early embryo survival in sheep. International Journal of Biometeorology 8:47–56.

28. STOTT, G. H., AND R. J. WILLIAMS
    1962 Causes of low breeding efficiency in dairy cattle associated with seasonal high temperatures. Journal of Dairy Science 45:1369–1375.

29. LONG, C. R., W. A. NIPPER, AND C. K. VINCENT
    1969 Body temperature and estrous control of beef cattle. Journal of Animal Science 28:146. Abstract.

30. STOTT, G. H.
    1961 Female and breed associated with seasonal fertility variation in dairy cattle. Journal of Dairy Science 44:1698–1704.

31. LAMPKIN, G. H., AND J. F. KENNEDY
    1965 Some observations on reproduction, weight change under lactation stress and the mothering ability of British and crossbred-zebu cattle in the tropics. Journal of Agricultural Science 64:407–412.

32. LINDSAY, D. R.
    1969 Sexual activity and semen production of rams at high temperatures. Journal of Reproduction and Fertility 18:1–8.

33. WINCHESTER, C. F.
    1964 Symposium on growth: environment and growth. Journal of Animal Science 23:254–264.

34. McBRIDE, G.
    1958 The environment and animal breeding problems. Animal Breeding Abstracts 26:349–358.

35. SMITH, C. A., AND G. E. HODNETT
    1962 Compensatory growth of cattle on the natural grasslands of Northern Rhodesia. Nature (London) 195:919–920.

36. WILSON, P. N., AND O. F. OSBOURN
    1960 Compensatory growth after undernutrition in mammals and birds. Cambridge Philosophical Society, Biological Reviews 35:324–363.

37. WILTBANK, J. N., W. W. ROWDEN, J. E. INGALLS, K. E. GREGORY, AND R. M. KOCH
    1962 Effect of energy level on reproductive phenomena of mature Hereford cows. Journal of Animal Science 21:219–225.

38. WINCHESTER, C. F., AND N. R. ELLIS
    1956 Delayed growth of beef calves. U.S. Department of Agriculture, Technical Bulletin 1159.

39. VALENTINE, K. A.
    1947 Distance from water as a factor in grazing capacity of rangeland. Journal of Forestry, 45:749–754.

40. MacRITCHIE, F.
    1969 Evaporation retarded by monolayers. Science 163:929–931.

41. BLACK, W. H., L. H. TASH, J. M. JONES, AND R. J. KLEBERG, JR.
    1944 Effects of phosphorus supplements on cattle grazing on range deficient in this mineral. U.S. Department of Agriculture, Technical Bulletin 856.

42. KNOX, J. H., AND W. E. WATKINS
    1958 Supplements for range cows. New Mexico Agricultural Experiment Station, Bulletin 425.

43. REYNOLDS, E. B., J. M. JONES, J. F. FUDGE AND R. J. KLEBERG, JR.
    1953 Methods of supplying phosphorus to range cattle in South Texas. Texas Agricultural Experiment Station, Bulletin 773.

44. WAGNON, K. A.
    1968 Use of different classes of range land by cattle. California Agricultural Experiment Station, Bulletin 838.

45. KNOX, J. H.
    1967 Supplemental feeding of range cattle. New Mexico Agricultural Experiment Station, Memoir 1.

46. GARTNER, R. J. W., AND G. I. ALEXANDER
    1966 Hepatic vitamin A reserves in drought-stricken cattle. Queensland Journal of Agricultural Science 23:93–95.

47. FALCONER, D. S.
    1960 Introduction to quantitative genetics. The Ronald Press Company, N.Y. 365 p.

48. LERNER, I. M.
    1954 Genetic homeostasis. Oliver and Boyd, Edinburgh. 134 p.

49. LEE, D. H. K.
    1959 The status of animal climatology with special reference to hot conditions. Animal Breeding Abstracts 27:1–14.

50. McDOWELL, R. E.
    1968 Climate versus man and his animals. Nature (London) 218:641–645.

51. DEESE, R. E., AND M. KOGER
    1967 Heritability of fertility in Brahman and crossbred cattle. Journal of Animal Science 26:984–987.

52. TALLIS, G. M.
    1968 Selection for an optimum growth curve. Biometrics 24:169–177.

53. ALEXANDER, G. I., D. N. SUTHERLAND, G. P. DAVEY, AND M. A. BURNS
    1960 Studies on factors in beef cattle production in a subtropical environment. Queensland Journal of Agricultural Science 17:123–134.
54. BLACKWELL, R. L., J. H. KNOX, C. E. SHELBY, AND R. T. CLARK
    1962 Genetic analysis of economic characteristics of young Hereford cattle. Journal of Animal Science 21:101–107.
55. PAHNISH, O. F., R. L. ROBERSON, R. L. TAYLOR, J. S. BRINKS, R. T. CLARK, AND C. B. ROUBICEK
    1964 Genetic analyses of economic traits measured in range-raised Herefords at pre-weaning and weaning ages. Journal of Animal Science 23: 562–568.
56. KOGER, M., AND J. A. KNOX
    1951 The correlation between gains made at different periods by cattle. Journal of Animal Science 10:760–767.
57. BUTTERFIELD, R. M., AND R. T. BERG
    1966 Relative growth patterns of commercially important muscle groups of cattle. Research in Veterinary Science 7:389–383.
58. LEDGER, H. P., R. SACHS, AND N. S. SMITH
    1967 Wildlife and food production. World Review of Animal Production 3:13–37.

Shrub Productivity

A Reappraisal of Arid Lands

JOE R. GOODIN

Department of Biology
Texas Tech University
Lubbock, Texas, U.S.A.

and

CYRUS M. McKELL

Department of Range Service
Utah State University
Logan, Utah, U.S.A.

ABSTRACT

FOR MANY REASONS it appears that deep-rooted, perennial shrubs offer a better potential than grasses for improved productivity on harsh sites, where perennial grasses have not been successful. The very fact that a drought-adapted shrub has a massive root system that can use moisture from a greater volume of soil than could a grass plant suggests that long-range arid-environment resource development may be greatly enhanced by concentrating research efforts on desirable forage shrubs.

An effort is being made to assemble information and plant materials of promising species from all arid regions of the world. Surprisingly, many of the better plants for difficult sites are quite high in protein content, and with proper management, breeding, and selection may prove to be very desirable livestock feed.

Increase in arid-land productivity holds out hope for millions of the world's inhabitants who live in arid regions and are otherwise destined to a life of famine or bare existence. Particularly exciting is the possibility of increase in protein for these areas.

A coordinated, international effort in the exchange of information and materials focused on this single theme could mean the difference between starvation and productive lives for millions of people. Time is short, and frustratingly slow progress in arid environments demands that we give more than lip service to one of mankind's cruelest problems.

SHRUB PRODUCTIVITY
A Reappraisal of Arid Lands

Joe R. Goodin and Cyrus M. McKell

THE WORLDWIDE emphasis placed on management and introduction of grasses as forage is understandable and well deserved; however, notable failures of seeded perennial grasses have occurred in many regions where agronomists and foresters would have predicted success. Sometimes this class of problems can be traced to failure of grasses to reproduce following establishment. In many other cases, marginal or erratic precipitation may lead to failure. Even though a single year or few years of favorable rainfall may lead to the conclusion that establishment and adaptation are certain for a given species, a few years of drought may give an entirely different impression.

It appears that under severe moisture limitation deep-rooted, perennial shrubs offer a better potential than does grass for improved productivity on harsh sites. The very fact that a drought-adapted shrub has a massive root system that can use moisture from a greater volume of soil than could a grass plant suggests that long-range resource development may be greatly enhanced by concentrating research efforts on desirable forage shrubs. Even a cursory glance at U. S. Department of Agriculture plant introductions into the United States (and presumably plant introductions for other countries) will convince one that the overwhelming emphasis has been placed on grasses rather than on shrubs or forbs for range improvement. Increasing interest has been shown in shrub introductions, and plant material centers have a number of promising species in their accessions.

Although prickly pear cactus (*Opuntia* spp.) and mesquite (*Prosopis* spp.) indigenous to arid lands have been used for many years as emergency livestock feed in periods of extended drought, almost no attempt has been made to improve or cultivate high-protein shrubs in these less desirable sites.

Shrubs presently serve a useful purpose as a limited source of feed for domestic animals during specific seasons. Some parts of shrubs are more nutritious than others and have been collected or harvested for use as a feed supplement. Shrubs are also used as reserve feed in years of critical feed shortage, and their use as livestock shelter depends upon the plant species present and the severity of the environment.

Because of high shrub productivity, nutritional quality, and adaptability, greater use should be made of shrubs in arid lands. Renewed efforts are justified and needed for the introduction, establishment, and evaluation of shrubs for animal feed. Intensive management of shrubs, including hand- or machine-harvesting of the most nutritious parts, should be considered in light of social and economic necessity and recent advancements in arid-land science and technology.

ARID-LAND PRODUCTIVITY

To obtain some estimate of the potential productivity for arid regions, we must consider productivity in natural ecosystems. The problems encountered in measuring net primary productivity in a complex ecosystem such as a forest (*1*) became considerably greater in an arid environment, although fewer plant species and simpler organization offer a special opportunity for comprehension of bioenergetics (*2*). Massive sampling is required because of the heterogeneity of form, life cycles, and microclimate. Ideally, each individual of each species present should be sampled (*3*) to get a true picture of net primary productivity. Also, sampling should be carried on throughout the year to assess natural abscission of leaves, flowers, and fruits, as well as primary production consumed by animals.

Under desert conditions in which water becomes a dominant factor in growth and development at any given point in time, such a task would prove monumental. Desert annuals, for example, may contribute greatly to net primary productivity, but due to their ephemeral nature, time of use is extremely critical. Even perennial shrubs may lose their leaves in times of severe moisture stress and then replace them quickly as soon as precipitation occurs.

Net primary productivity varies tremendously with form, density, and other factors. Values as low as 300 kilograms per hectare (dry matter) have been recorded for an alpine snowbank (*4*); Whittaker (*5*) reported 10,000 kilograms per hectare for a mixed shrub community and 50,000 kilograms per hectare for *Arundo donax*.

Specific studies of productivity in nonarid regions are well documented (*6*), but few studies have been performed in arid regions. Weaver (*7*) recorded 1,600 kilograms per hectare for a *Bulbilis-Bouteloua* short-grass plain in North America, and Walter (*8*) reported the same figures for an *Aristida* grassland in southwest

NOTE: This chapter is an adaptation of a presentation at "Arid Lands in a Changing World," an international conference sponsored by the Committee on Arid Lands, American Association for the Advancement of Science, June 3–13, 1969, University of Arizona, Tucson, Arizona, U.S.A.

Africa. A "moist southern desert" in the Soviet Union produced 1,200 kilograms per hectare (*9*). In contrast, ephemeral herbs in Nevada produced only 400 kilograms per hectare (*10*).

Chew and Chew (*2*) in 1965 reported the primary productivity of a desert shrub, *Larrea divaricata*. Their detailed study showed a total productivity of approximately 1,000 kilograms per hectare for a relatively young shrub community. They also showed that density varied inversely with the median age of the plants. When first sampled in 1958, 4,460 individuals per hectare were producing an estimated 1,000 kilograms. It was predicted that in 1968 density would be reduced to 2,700 individuals producing 1,400 kilograms per hectare, and that in 1978, 2,075 individuals should be producing 1,500 kilograms per hectare. Near Tucson, Arizona, densities of climax communities of *Larrea* average 1,000 individuals per hectare (*11*).

Productivity estimates for terrestrial communities are difficult to compare due to the variety of measuring and harvesting techniques used. In particular, productivity of below-ground parts is difficult to evaluate. The following measures of productivity have been used: (*a*) absolute productivity of kilograms of dry matter (or other caloric equivalent) per hectare per year; (*b*) relative productivity, measured in kilograms of dry matter of standing weight per hectare per year; and (*c*) net assimilation, measured as kilograms of dry matter per unit of leaves (surface, weight, chlorophyll) per unit of time.

From the standpoint of shrub productivity, we are concerned with estimation of top growth per unit area per unit of time. Partial removal of shoots to encourage regeneration for subsequent productivity is a complicating factor.

In many respects, productivity measurements from forage crops can serve as a standard of comparison for assessing shrub productivity. Loomis and Williams (*12*) reported yields of 40,000 kilograms per hectare per season for forage sorghum (*Sorghum vulgare*) in California in contrast with 96,000 kilograms per hectare per season for Napiergrass (*Pennisetum purpureum*) in Puerto Rico. They calculated a theoretical upper limit of yield at 77,000 kilograms per hectare of dry matter in 100 days, based on available species with given degrees of photosynthetic efficiency.

TABLE 1

Dry Weight Yields (Kilograms per Hectare per Season) of Three Saltbush Species

Species	Yield	
	1968 (irrigated)	1969 (nonirrigated)
Atriplex polycarpa	3,805	7,599
Atriplex lentiformis	6,185	10,169
Atriplex canescens	. .	9,189

Our own recent work in southern California has shown that three *Atriplex* species indigenous to the western United States can be grown under routine agronomic conditions. Observations were made for seedling density, competition, pest control, water requirements, forage yield, harvesting sequence, and forage quality. Cutting *Atriplex* forage as a hay crop was found to be feasible, and yields as high as 10,000 kilograms dry matter per hectare per growing season have been recorded (Table 1). Protein content as high as 19.2 percent was recorded for *Atriplex polycarpa*, whereas carbohydrate levels were extremely low for all species. These results suggest that cultivation of *Atriplex* species as a forage crop has considerable potential in marginal lands subject to prolonged drought or excessive salinity. Similar proposals have been made by Malcolm (*13*) in Australia. Average annual forage production of *Opuntia* in the northern part of Mexico is reported at 5,000 kilograms per hectare dry matter (*14*).

ECOLOGICAL CONSIDERATIONS

Shrubs are ideally suited for growth in arid regions. This suitability is based on a high degree of tolerance to drought, high temperature, high light intensity, and high salinity in the soil. Various physical or morphological characteristics, such as a root system that is either deep or spreading, vertical leaf orientation, gray color, protective leaf surface (often a deciduous habit), high photosynthetic efficiency, and a low respiration rate are among the many features of shrubs that make them highly suited to arid regions. In many cases, the adaptive features of shrubs enable them to persist in desert areas where other plant species may be unsuitable for growth throughout the season. Annual species often make very good use of environmental factors when such factors are favorable. From the standpoint of availability as rangeland forage, however, the lack of consistency associated with annual species presents difficulties in management for short periods of forage production.

Variability within species of shrubs appears very high and provides opportunities for selection to meet specific environmental conditions. Selection within shrub species for particular characters is still in a trial-and-error stage (*15*). Shrubs are probably the least-known group as to their genetic diversity and ecological limits of tolerance. Chatterton and McKell (*16*) showed the high degree of variability within *Atriplex polycarpa* seeds from three populations collected at McKittrick, Bishop, and Lancaster, California, and grown in varying concentrations of sodium chloride. Top growth of seedlings from the Bishop population was four times greater than that of the other populations (Fig. 1), thus indicating a variability that could be used to increase productivity within the species of *Atriplex*. Similarly, it should be possible to select for high-producing strains from within species of many arid-land shrubs.

It is well known that ecological equivalents exist for many plants from similar regions. For example, species of the highly desirable *Atriplex* group may be ecological substitutes for shrub species of considerably lower palatability and nutritive quality in another region. Such eco-

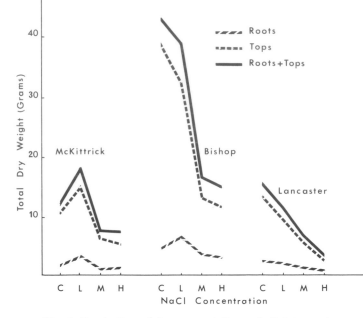

Fig. 1. Production of three populations of *Atriplex polycarpa* grown in four sodium chloride concentrations (50, 13,000, 26,000, and 39,000 parts per million) in solution culture for six weeks.

logical equivalents should be sought out and used to replace the undesirable shrubs as a means of increasing the useful productivity of the arid lands of less-developed countries.

An important aspect of shrub suitability for arid regions is their persistence during less-favorable periods of the year in contrast to other arid-land plants. This feature makes it possible for shrubs to take full advantage of the limited factors necessary for growth under arid conditions. Because of their perennial habit, shrubs are able to use periods of summer rainfall or to grow during winters when temperatures may become favorable for growth during a short period. Because of growth during less-favorable periods of time, the nutritive quality of many shrubs is of a higher order than that of plants that stay dormant during such periods.

Many arid-zone soils are low in fertility, and shrubs are apparently associated with "islands of fertility" in such ecosystems (17). Even though nitrogen fixation can occur from blue-green algae, free-living bacteria, and nodulated plants (both leguminous and nonleguminous), shrubs are important for their longterm influence on the environment. Thus, complete removal of shrubs from a desert ecosystem would be quite detrimental to the recycling of nutrients in the system. Substitution of desirable shrubs for unpalatable shrubs of low nutritional value appears to be imperative.

A careful distinction must be made between rangeland in deteriorated condition and rangeland that still has a high primary productivity. The presence of shrubs in both cases may not indicate the same degree of range value. Criteria for judging range productivity are based on many characteristics of the environment, including a favorable soil condition. Shrub predominance may or may not indicate good range conditions. In a like manner, an abundance of grass may not always assure the best estimate of range productivity. We suggest that the most

rational approach to arid-land productivity depends on an assessment of the capability of the physical environment to support a high degree of production of desirable forage and fodder plant growth. In many cases, it may be necessary to disturb or replace the existing vegetation in order to produce the more desirable animal-fodder species.

Seedling establishment will be a problem in the introduction of many desirable shrubs. Under natural conditions, the establishment of stands of arid-land species takes place very seldom. Only when conditions of soil moisture and temperature over a favorable period of time can be assured will seedling establishment occur under natural conditions. Thus, to achieve establishment of introduced species it will be necessary to modify the seedling environment in order to assure success. Many techniques have been developed to provide a favorable environment and merely need to be adapted for seeding shrubs; however, studies on seed production, seed dormancy, germination, seedbed requirements, and many other subjects will be necessary to provide needed information on establishment of desirable shrubs.

SOIL AND WATER RELATIONS

Approximately 25 percent of Earth's land surface receives less than 300 millimeters of annual precipitation. This area comprises all of the great deserts of the world, but is by no means limited to the desert areas (18). Productivity of these areas is limited primarily by rainfall quantity and distribution; the potential water loss through evaporation and transpiration exceeds precipitation all or part of the year. Thus, absolute precipitation does not in itself define aridity.

Arid-zone soils contain too little water for growth of crop plants without irrigation, except during a brief period after the rainy season. These soils, nevertheless, do support vegetation which may (a) be ephemeral in nature, (b) have a low water requirement, or (c) have root systems that reach water deep in the soil.

Although many arid-zone soils are not high in salinity or alkalinity, these types of soils represent a major segment of water-deficit areas. Thus, problems of salinity and alkalinity are expected in many areas of low rainfall. Some shrubs normally classified as halophytes are adapted to soils of tremendous mineral diversity, yet manage to accumulate ions at a predictable concentration (19).

Except for the most barren deserts, most arid regions support some type of perennial, shrubby vegetation. In many cases, the natural vegetation pattern has been altered drastically, often because of severe animal use. In some areas, valuable forage species have been grazed or harvested to extinction.

Whereas the more desirable perennial grasses usually require a dependable 250 to 300 millimeters of annual precipitation for survival, many perennial forage shrubs can exist on much less moisture. A great emphasis has been placed upon anatomical and morphological modifications that allow a species to become competitive in an arid environment, but little is actually known about the physiology of drought tolerance.

In a typical mesophyte, growing with adequate water, nutrient supply, and normal transpiration rates, only small diurnal water deficits occur. Only slight reductions in growth can be noted when water deficit is restricted to a few hours (*20*). As the soil dries, the plant must cope with a soil moisture deficit in addition to the diurnal lag of absorption behind transpiration. Initially, suppression of metabolism occurs only during the period of maximum water deficit, but this period becomes longer each day. Stomatal closure retards transpiration, increases leaf temperature, and reduces net photosynthesis through a reduction in carbon dioxide and increased respiration. Reduction in the rate of cell enlargement progressively increases as cell turgor falls. As stress becomes more severe, plant water potential becomes dominated by soil water potential and turgor pressure approaches zero. Finally, respiration overtakes photosynthesis and overall growth rates become negative. As protoplasmic dehydration continues, cells and tissues die. If the water supply is renewed before death occurs, the recovery may take a prolonged period of time, and in many cases recovery may never be complete.

Xerophytes are able to survive these extreme water deficits by such adaptations as succulence, in which distended and turgid parenchyma cells accumulate water during periods of precipitation and "ration" the water to metabolism during the dry season. Succulents have low transpiration rates, at least partially because the stomata open at night only and carbon dioxide exchange is accompanied by dark carbon dioxide fixation (*21*). Certain *Opuntia* species, which are extremely palatable high-carbohydrate forage plants, are categorized with this group of succulents.

The nonsucculent, perennial shrubs of arid regions are characterized by rapid elongation of taproots which, in the case of *Prosopis* species, can reach to depths exceeding 35 meters to extract deep sources of moisture. The extensive lateral root system of many species is used to extract a small quantity of moisture from a large soil mass. In such cases, extent of the root system will regulate plant density, and natural thinning will occur. Although many xerophytes transpire water as freely as mesophytes when adequate water is available, they have a control mechanism that allows them to curtail transpiration drastically during periods of drought (*22*). Other factors, such as leaf abscission, reduction in lamina size, sunken stomata, and thickened cuticle, are well known and will not be discussed in detail. Thus, xerophytes differ from mesophytes only in the degree to which they are able to make use of low levels of soil moisture.

In addition to the unusual water relations of xerophytes, they also differ from the normal terrestrial plant in their high temperature tolerance. In *Opuntia*, growth continues even when internal temperatures exceed 56.5°C (*23*). Optimum photosynthesis in *Atriplex vesicaria* occurs at 40°C to 50°C, and continues at least

as high as 55°C (*24*). Seeds of *Carnegia gigantea* have been observed to survive seven days at 82°C (*25*).

Nutrient cycling in desert communities is an important part of the ecosystem. Soil-moisture relations, both long- and short-term, determine the frequency of recycling, almost to the exclusion of other environmental factors.

SALT TOLERANCE

One of the worldwide characteristics of arid lands is an accumulation of salinity or alkalinity because precipitation is not great enough to leach the soluble salts and/ or exchangeable sodium, which accumulate by evaporation, out of the root zone. Often when such lands are irrigated, water quality is so poor that the additional salt added to the soil surface only compounds the problem. Soluble salts produce harmful effects in plants by increasing the salt content of the soil solution and by increasing the exchangeable sodium. Increase in exchangeable sodium occurs when the soluble constituents consist largely of sodium salts, and is more permanent than changes in the salt content of the soil solution, since exchangeable sodium usually persists after the soluble salts are removed (*26*).

Although a specific soil problem may exist due to excess soluble salts (saline soil) or excess exchangeable sodium (alkali soil), both of which reduce productivity, the more common situation in arid regions involves soils that contain an excess of both soluble salts and exchangeable sodium.

A plant survey of saline-alkaline regions of the world reveals many drought-tolerant shrubs which not only exist but seem to thrive under these conditions. Thus, the criteria for selection of desirable shrubs must include a tolerance to salinity and alkalinity. The fact that a particular species is found in a given environment gives some indication of potential in similar environments in other parts of the world. However, the fact that the species is not found in abundance should not be used in assessing its competitive ability, because often the most desirable species have been removed through selective harvesting and grazing.

The remarkable ability of certain browse shrubs to thrive in saline and alkaline soils has been reported for many years, and the suggestion has been made that valuable forage production can be obtained from halophytes grown on saline lands useless for other agricultural purposes. Following Brownell's (*27*) report that sodium is an essential micronutrient for *Atriplex vesicaria,* considerable research has been devoted to experimental studies with desirable forage species of the genus *Atriplex.*

Recently Chatterton and McKell (*16*) reported that *Atriplex polycarpa* continued to grow in culture solutions having osmotic potentials of minus 25 atmospheres (39,000 parts per million of sodium chloride). These values are equal to or higher than total salts from ocean water. Dry weight of plants grown in the same sodium chloride (NaCl) concentration varied among populations, thus indicating considerable variability for salt tolerance. The same species has been grown in culture solutions containing up to 80 ppm boron, and no reduction in growth was noted at 40 ppm (28). In these studies, moisture percentage in the shoots varied only slightly and failed to correlate with increasing salinity.

Mozafar (29) studied the physiology of salt tolerance in *Atriplex halimus* L., a Mediterranean halophyte with exceptional forage potential. He found that low salinity levels, whether induced by Hoagland's solution, NaCl, potassium chloride (KCl), or NaCl + KCl, increased growth and protein content of the plants. Plants receiving equal parts of NaCl and KCl could tolerate a much lower external water potential than plants receiving other treatments. Thus, growth inhibition could be attributed to causes other than osmotic inhibition.

The enzyme system involved in the production of oxalate from glycolate indirectly involves catalase. Mozafar (29) also studied catalase activity in *Atriplex halimus,* and found that increasing salinity strongly reduced catalase activity. As the leaf homogenate aged, catalase activity doubled in a two-hour period, suggesting that the enzyme was bound to some subcellular particle and subsequently released. In control plants, no changes occurred in catalase activity as the homogenate aged.

Plant species grown on saline media may regulate their ion uptake to a certain extent, but in most cases, the increase in salinity causes an increase in ion uptake and the consequent buildup of salts in the plant organs (30). The water-absorbing capacity does not seem to be affected if the plant adjusts osmotically; that is, a constant osmotic potential gradient is maintained between the leaf cells and the growing medium (31). In *Atriplex halimus,* the vesiculated hairs (trichomes), that cover the leaf surface accumulate high concentrations of salts, whereas the salt concentration of whole leaf sap does not increase and remains almost constant over all salinity treatments (32). Thus vesiculated hairs play a significant role in removal of salts from the vascular and parenchymatous tissues of leaves and therefore prevent accumulation of toxic salts in the leaves. Because the primary discriminatory barrier in salt uptake appears to be in the roots, the xylem-sap concentration of sodium and chloride ions may be low, even though accumulation may be taking place in individual cells in the shoots (33).

These kinds of considerations help increase our understanding of how potentially valuable forage shrubs are able to withstand the extreme soil and water conditions imposed in our arid lands.

COMPETITION

The basis for success in competition with other plants is plant efficiency. Efficiency may be seen both in external relations with other plants and within the plant where physiological and biochemical functions can increase the functional capability. In the external category, of greatest importance is the ability of plants to obtain the necessary factors of the environment such as light, water,

nutrients, and satisfactory gas and thermal conditions. Plants with a greater capacity to obtain these elements in sufficient quantity at the same time they are being withdrawn by other plants are generally considered as successful competitors. Oppenheimer (34) reviewed in detail some of the characteristics which enable them to compete successfully in arid regions.

Equally important but less well known as characteristics contributing to successful competition are those physiological and biochemical functions of plants which enable a high degree of efficiency. Black and others (35) divided plants into two groups, efficient and nonefficient plants. They suggested that efficient plants are successful competitors in most ecosystems because they are able to increase growth and vigor as light intensity increases and because their rate of photosynthesis increases and approaches a maximum value two to three times higher than for nonefficient plants. Efficient plants are able to increase growth and vigor when temperatures rise above 15°C to 20°C. Further, growth of efficient plants is not inhibited at normal oxygen concentrations, and there is less inhibition of growth by the oxygen which is normally produced during photosynthesis. In efficient plants, there is less loss of reduced carbon to glycolate oxidation, thus saving energy and substrates for other metabolic processes.

It is rather interesting to find that many of the common weeds are listed in the classification of the efficient plants, and that some of the nonefficient species are important crop plants.

Other characteristics obviously must be considered in order to make an overall classification within the plant kingdom as to competitive ability. However, the classification by Black and his coworkers serves a useful purpose. It is entirely possible that some species of shrubs desirable for their forage value might also fit into the efficient category, but little is known about shrub physiology or biochemistry. One of the serious problems with regard to arid-lands plants is that natural selection has emphasized survival under extreme conditions rather than productivity (36).

Shrubs often afford protection from animal grazing to plants growing up through their crowns. Such protection is one of the ways that perennial grasses escape complete utilization under heavy grazing use. Shrubs also provide a protective cover for many of the ephemeral plants that grow under the crown. Here, soil fertility is higher than in the open spaces between shrubs and there is more protection from the direct rays of the sun, thus providing a reduction of moisture loss. The protection afforded by the shrub canopy, however, is not year long; as soon as moisture stress becomes extreme, the shrub, with its more extensive root system, may then become a high competitor for moisture and eliminate the possibility of survival of the understory species through the duration of the season.

Animal use has a great influence on competition between shrubs and other species. The degree of effect depends upon the preference that animals have for the particular species in question. The balance of competition between shrubs and herbaceous species can thus be shifted through disproportionate use by animals. Continued productivity of a particular group of species will depend upon the regulation of animal use if a mixed stand is desired. Careful animal use must be practiced to maintain a balance of shrub and herbaceous species.

MANAGEMENT

To achieve better use of desirable shrubs for animal feed, improved systems of management will be needed. Improved management will be particularly critical where stands of introduced and improved shrubs have been established. No specific recommendations are given in this paper because social conditions and economic capabilities differ from place to place throughout the arid regions of the world. The approach here is primarily to identify some of the important problems that exist and point out some principles that might be considered in developing a management approach appropriate to the existing conditions.

One of the most obvious conditions of arid lands is a relatively low level of primary productivity. Therefore, a system of use and management must be developed which would not destroy the input-output balance of the ecosystem. The solution appears to be to increase the number of useful shrubs per unit area, thus minimizing the effect of removal through animal grazing. A proper level of use must be established based on the biology of the plant cover and the economic and social requirements.

Shrubs often differ in their period of highest feed quality in comparison with other species. The contrast is even greater when considering various parts of shrubs such as leaves, flowers, and seeds. A management system must take into consideration seasonal differences in feed quality. In some locations, collection by hand may provide the greatest return of usable animal feed such as with mature pods and seeds.

Shrub reaction to animal use must be considered. Reduced root growth occurs in grasses when excess defoliation takes place. Shrubs brought to a weakened condition by overuse undoubtedly show a reduction in root growth in a like manner. Shrub reproduction may also be affected by the degree of use. Nord (*37*) reported that reduced grazing pressures on desert bitterbrush contributed to increased seedling establishment. A system of rest-rotation grazing management, similar to the one described by Hormay and Talbot (*38*) which takes into account the physiological requirements of the plants, may be necessary for maintenance of shrub stands.

Differences in palatability are well known for shrubs and herbaceous forage species. A management system must take into account these differences, which may be related to particular stages of plant development or to fundamental differences among plants normally used by animals.

Concentrating animal use on small areas by herding may overcome objectionable characteristics that some plants present to animals. However, in the process, care must be exercised to assure the persistence of highly palatable species that may be overutilized.

Animal species also vary in their preferences for plants, as Nord (*37*) pointed out in the management of desert bitterbrush. He found that sheep seek out small, tender plants, whereas cattle generally graze the larger and better-established plants.

NUTRITIVE QUALITY

The nutritional value of shrubs as cattle and sheep feed has often been overlooked because of the greater interest in seeding forage grasses on depleted grazing lands. In a study of browse plants of desert ranges in the Great Basin of the Western United States, Cook and others (*39*) stated that shrubs are of a higher quality than grass and generally contribute a greater amount to the grazing animals' diet because of greater abundance. They also stated that browse species are higher in protein, phosphorus, and carotene (Vitamin A). In contrast, grasses are superior to shrubs only in energy-yielding qualities (metabolizable energy). Imperial Agricultural Bureau joint publication no. 10 (*40*) lists the nutritional qualities of 511 species of shrubs and trees throughout the world as animal feed. Data for some of the species are for various growth stages and for various plant parts. Crude protein values range in the area of 20 to 40 percent for buds, flowers, and seeds (Table 2). Protein concentration appears to be lowest in old stems and increases in buds, flowers, fruits, and seeds, respectively. In many areas of the world, it is necessary for man to collect the more nutritious plant parts for his animals, such as the collection of *Prosopis* pods and beans.

Another aspect of shrubs and trees as animal fodder that has considerable promise is the season of use that is afforded by shrubs. Often forage value of grasses is at a low point during the winter season in temperate zones. At this time, many shrubs have their highest feed value levels. The seasonal trend in proteins, carbohydrates, and total digestible nutrients was recently reported for *Atriplex polycarpa* in California (*41*). This report shows that the highest peak for some of the desirable nutritive qualities occurs during the months of January, February, and March (Figs. 2, 3 and 4). At this time the forage value of annual and perennial grasses is generally low, or in some years it may not be of significant magnitude to be of importance to grazing animals. Thus, the availability of browse from fodder shrubs and trees offers opportunities for a year-round program of grazing where otherwise it would not be available.

Palatability is another factor that must be considered with regard to animal use of fodder trees and shrubs. Even though many species may show high nutritive values upon chemical analysis, palatability of some shrubs may be relatively low, particularly in certain plant communities where more desirable or palatable species exist in abundance. In some areas, it may be possible to bring about greater use of shrubby species by including

TABLE 2

Protein and Energy Content of Selected Genera of Palatable Shrubs *

Botanical Name	Sample Description	Source	Crude Protein (% of dry weight)	Nitrogen Free Extract (% of dry weight)
Acacia angustissima	Foliage (May)	Texas	23.6	48.2
Acacia litakunensis	Pods	S. Africa	17.3	49.2
Aesculus californica	Buds, scales, young leaves	California	35.6	. .
Anabasis aphylla	Up to flowering	U.S.S.R.	24.8	48.2
Artemisia cana	Early vegetative	W. Canada	19.3	. .
Artemisia spinescens	. .	Utah	9.3	61.1
Atalaya hemiglauca	Leaves, green	New S. Wales	22.8	35.2
Atriplex nummularia	Young flowering shoots	Queensland	21.9	41.2
Atriplex vesicaria	Leaves, in drought	W. Australia	11.4	46.1
Cassia nicitans	Seed	Alabama	40.9	45.0
Chrysothamnus viscidiflorus	Foliage, winter	Utah	5.9	47.7
Eurotia lanata	Leaf stage	W. Canada	21.9	. .
Kochia brevifolia	Leaves	W. Australia	24.8	33.5
Lucaena glauca	Seed	Rhodesia	21.7	45.5
Moringa oleifera	Leaves, for human use	India	26.6	54.0
Nolina texana	Budding shoots	Texas	24.2	49.0
Opuntia spp	Prickly pear	Texas	4.6	61.0
Prosopis velutina	Seeds	Arizona	37.3	46.5
Purshia tridentata	Leaves, young shoots	Utah	15.4	59.3
Sambucus glauca	Leaves, prefruiting	California	35.9	. .
Sutherlandia microphylla	Average samples	S. Africa	24.3	44.6

* Table extracted from Imperial Agricultural Bureau Joint Publication No. 10, "The use and misuse of shrubs and trees as fodder," published by the Imperial Bureau of Pastures and Field Crops, Aberystwyth. 1947.

different livestock and animal classes such as goats and wildlife. An example of good use of low-quality shrubs is that of a goat-milk cooperative in the northern part of the state of Coahuila, Mexico. Through improved management of the goat herd, including management of the rangeland, the cooperative increased the daily average income threefold in the first four years of operation. Productivity increased, but there was a decrease in the average number of animals per herd, mainly through increased efficiency and greater nutrition for the animals because of the improved use of the shrub-dominated rangeland.

A large number of woody species have been studied in the past with regard to actual or potential use as animal feed. It appears obvious that with the large number of plants available in the world, greater use can be made of the existing shrubs. Some of the more promising genera are *Artemisia, Atriplex, Ceanothus, Ceratonia, Dalbergia, Eurotia, Lucaena, Lupinus, Morginga, Prosopis, Pureria, Quercus, Salix, Sesbania, Sutherlandia,* and *Opuntia.*

Some genera have specific characteristics that offer a deterrent to their use for livestock feed. Toxic substances such as in *Lucaena glauca* exist, but research is presently underway to remove the toxic substance. In *Opuntia,* the main disadvantage is the great imbalance between carbohydrate and protein. Additionally, the presence of spines may be a deterrent for animal use, but plant genetic lines exist which do not have spines. Management may overcome imbalances by either controlled

grazing or through harvesting various parts and providing such parts to livestock in feedlots. Kock (*42*) described feeding experiments in South Africa using *Opuntia* species and old-man saltbush (*Atriplex nummularia*). Animals fed the two species, free choice, used the *Opuntia* to an excessive degree. Separate controlled

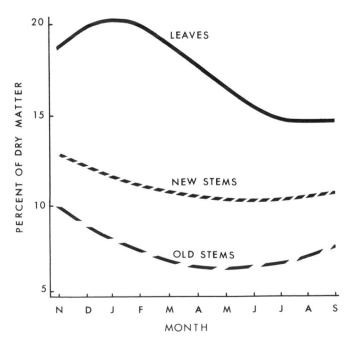

Fig. 2. Seasonal fluctuation in protein content for leaves, new stems, and old stems of *Atriplex polycarpa.*

grazing of the two species for as long as seven years had no deleterious effect on either species.

Some species may be relatively low in palatability, but this may not always be a disadvantage because less palatable species might well serve as a feed reserve for periods of minimum feed availability during years of drought or other stress. With a reasonable amount of research, it is entirely possible that the disadvantages may be overcome through genetic selection and processing to increase the potential of some of these desirable woody species for animal use.

APPLICATIONS—PRESENT AND FUTURE

We can now begin to conceive of some new approaches to increased productivity of arid lands. To be sure, productivity changes come slowly under harsh environmental conditions. Concentrated research efforts require perseverance of the greatest magnitude, and the changes which occur cannot be expected to be as dramatic as those in irrigated agriculture. However, as a resource base, increase in arid-land productivity holds out hope for millions of the world's inhabitants who make arid regions their home and are otherwise destined to a life of famine or bare existence. Particularly exciting is the possibility of increase in protein for these areas.

The possibilities that drought-tolerant shrubs, both succulent and non-succulent, can help fill this food gap for the arid regions is rapidly gaining worldwide attention. Individual researchers have proposed such schemes for many years, but the opportunities for a concerted international effort have never presented themselves. Competent plant scientists exist in most of the world's arid regions, but the moral and financial support has never been great enough to make a major impact.

Current use for the benefit of man is so low that any improvement will be significant. We believe that in some areas the improvement will be sensational. The following concepts may be considered now, or in the future, as a basis for utilization for best advantage:

a) *A study of the possibilities for intensive management of arid-land species for greatly increased productivity.* As early as 1888, the possibilities for *Opuntia* species as a forage were being considered in South Africa, and in 1922 the first formal experiments were conducted (*43*). Sheep would readily consume 5.4 kilograms of cactus per day for more than 250 days without water. Further experimentation showed that rotational grazing between cactus and old-man saltbush (*Atriplex nummularia*) was feasible and good gains were achieved after a two-week adaptation period. Since the high-carbohydrate cactus was considerably more palatable than the saltbush, at least in the beginning mixed stands of the two shrubs would not be feasible.

Our recent studies in southern California lead us to believe that *Atriplex* species might be grown as a high-protein forage on marginal agricultural lands and cultivated similarly to other agronomic forage crops. Such species have considerable potential in marginal lands subject to prolonged drought or excessive salinity. Also, in many arid regions, particularly those with sandy soils, there is reason to believe that brackish or saline waters unsuitable for other crops could be used for irrigation of *Atriplex* and perhaps other species (*44*).

b) *A study of the sociological and political problems which prohibit certain countries from developing their arid-land resources.* Grazing management is probably the single most critical factor in assessing past damage and holding out promise for the future. The general picture is one of gross overstocking with a resultant deterior-

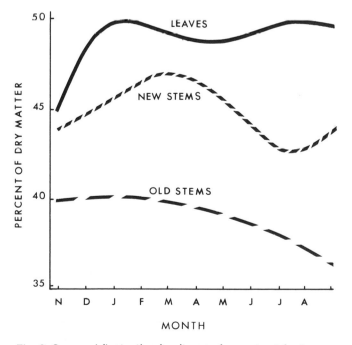

Fig. 3. Seasonal fluctuation in nitrogen-free extract for leaves, new stems, and old stems of *Atriplex polycarpa*.

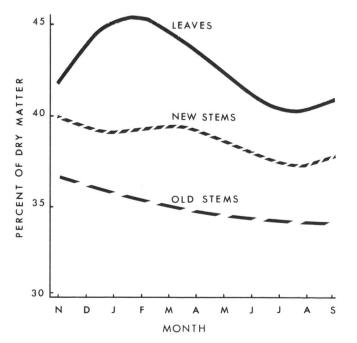

Fig. 4. Seasonal fluctuation in total digestible nutrients for leaves, new stems, and old stems of *Atriplex polycarpa*.

Fig. 5. *Atriplex* species grown for forage under agronomic conditions at the University of California, Riverside.

ation in the vegetative cover and a serious reduction in the capacity of that vegetation to play its primary role, the conservation of soil against water and wind erosion. These shrub and tree associations must continue to be an important source of fodder, at least for low-grade animals producing commodities such as wool, hides, skins, and meat where mere numbers may be more important than quality of animal (40). The speculation has been made that in the great regions of the world where the vegetative cover has been almost or totally eradicated, only a complete cessation of grazing for 10 to 20 years or longer would give it sufficient opportunity to recover. Since it is in these areas that it is most difficult if not impossible to produce any alternative source of fodder from cultivated land, a reduction in numbers of animals grazing the natural vegetation also becomes almost impossible when large human populations depend on them for their food or livelihood.

We contend that a hungry world cannot wait for this return to a so-called climax vegetation, which in some cases would not return in a hundred years, and that the only choice is for creative approaches to the establishment of new vegetation, whether indigenous or introduced species. This ingenuity may take the form of water harvesting via concentration, storage, and spreading techniques (45); chemical fallowing to trap soil moisture

Fig. 6. Adaptation trials of introduced forage species at the University of California, Riverside.

for future establishment (*46*); and impregnation of the soil surface with a sealing material to form a water-repellant coating (*47*).

c) *A clearly defined international program of plant introduction and research cooperation, in which all arid lands would share their material resources* (Figs. 5 and 6). The opportunities for plant introduction are almost unlimited, and in some cases only ecological-physiological studies of native species are required. The detailed studies such as those of Nord (*37*) on the genus *Purshia* and Workman and West (*48*) on *Eurotia lanata* will add substantially to our breadth of knowledge.

We already know that establishment of browse species under these difficult environments is going to require persistence and ingenuity. The kinds of studies needed include (*a*) environmental modification for seedling survival, (*b*) seedling vigor, (*c*) anatomical and morpho-logical investigations, (*d*) rate and extent of root growth, (*e*) the biochemistry and physiology of adaptation to drought, (*f*) reproduction of the species emphasizing regeneration of a given plant (ecotypic variation is immense in many of these species), (*g*) salinity and alkalinity tolerance, (*h*) palatability studies for establishment and longevity, (*i*) modification of plant growth and development with growth regulators for improved quality, use, and ability to compete with less desirable species, (*j*) animal nutrition, and (*k*) predator control.

A coordinated, international effort in the exchange of information and materials focused on this single theme could mean the difference between starvation and productive lives for millions of people. Time is short, and frustratingly slow progress in arid environments demands that we give more than lip service to one of mankind's cruelest problems (*49*).

REFERENCES AND NOTES

1. OVINGTON, J. D.
   1962 Quantitative ecology and the woodland ecosystem concept. Advances in Ecological Research 1:103–192.

2. CHEW, R. M., AND ALICE E. CHEW
   1965 The primary productivity of a desert shrub (*Larrea tridentata*) community. Ecological Monographs 35(4):355–375.

3. ODUM, E. P.
   1963 Ecology. Holt, Rinehart, and Winston, N.Y. 152 p.

4. BILLINGS, W. D., AND L. C. BLISS
   1959 An alpine snowbank environment and its effects on vegetation, plant development, and productivity. Ecology 40:388–397.

5. WHITTAKER, R. H.
   1961 Estimation of net primary production of forest and shrub communities. Ecology 42:177–180.

6. WHITTAKER, R. H.
   1962 Net production relations of shrubs in the Great Smoky Mountains. Ecology 43(3):357–377.

7. WEAVER, J. E.
   1924 Plant production as a measure of environment. Journal of Ecology 12:205–237.

8. WALTER, H.
   1939 Grasland, Savanne, und Busch der arideren Teile Afrikas in ihrer ökologischen Bedingtheit. Jahrbuch für Wissenschaften Botanische 87:750–860.

9. LAVRENKO, E. M., N. V. ANDRUV, AND V. L. LEONTIEF
   1955 Profile of the U.S.S.R. from the tundra to the deserts (translated title). Botanicheski Zhurnal U.S.S.R. 40: 415–419.

10. ODUM, E. P., AND H. T. ODUM
    1959 Fundamentals of ecology, 2nd ed. Saunders, Philadelphia. 546 p.

11. SHREVE, F., AND A. L. HINCKLEY
    1937 Thirty years of change in desert vegetation. Ecology 18(4):463–478.

12. LOOMIS, R. S., AND W. A. WILLIAMS
    1963 Maximum crop productivity: an estimate. Crop Science 3(1):67–72.

13. MALCOLM, C. V.
    1969 Use of halophytes for forage production on saline wastelands. Australian Institute of Agricultural Science, Journal 35(1):38–49.

14. ROJAS, M. P., F. J. MALO, AND D. PALOMA G.
    1966 El nopal forrajero en Nuevo Leon. Agronomia 108:34–39.

15. YERMANOS, D. M., A. KADISH, C. M. McKELL, AND J. R. GOODIN
    1968 Jojoba — a new California crop? California Agriculture 22(10):2–3.

16. CHATTERTON, N. J., AND C. M. McKELL
    1969 *Atriplex polycarpa*: I. Germination and growth as affected by sodium chloride in water cultures. Agronomy Journal 61:448–450.

17. GARCÍA-MOYA, E., AND C. M. McKELL
    1970 The contribution of shrubs to the nitrogen economy of a desert wash plant community. Ecology 51(1):81–88.

18. GOOR, A. Y., AND C. W. BARNEY
    1968 Forest tree planting in arid zones. Ronald Press, N.Y. 409 p.

19. BEADLE, N. C. W., R. D. B. WHALLEY, AND J. B. GIBSON
    1957 Studies in halophytes. II. Analytic data on the mineral constituents of three species of *Atriplex* and their soils in Australia. Ecology 38(2):340–344.

20. SLATYER, R. O.
    1967 Plant-water relationships. Academic Press, London. 366 p.

21. TING, I. P., AND W. M. DUGGER, JR.
    1968 Non-autotrophic carbon dioxide metabolism in cacti. Botanical Gazette 129(1):9–15.

22. DRAR, M.
    1955 A study of the main characteristics of the ecological groups of arid zone vegetation. *In* Plant ecology, proceedings of the Montpellier symposium. Unesco, Paris. Arid Zone Research 5. 124 p.

23. MacDOUGAL, D. T., AND E. B. WORKING
    1921 Another high-temperature record for growth and endurance. Science 54:152–153.

24. WOOD, J. G.
    1932 The physiology of xerophytism in Australian plants. Australian Journal of Experimental Biology and Medical Science 10:89–95.

25. KURTZ, E. B.
    1958 Chemical basis for adaptation in plants. Science 128:1115–1117.

26. RICHARDS, L. A. (ed)
    1954   Diagnosis and improvement of saline soils. U.S. Department of Agriculture Handbook 60. 160 p.
27. BROWNELL, P. F.
    1965   Sodium as an essential micronutrient for a higher plant (*Atriplex vesicaria*). Plant Physiology 40(3):460–468.
28. CHATTERTON, N. J., C. M. McKELL, J. R. GOODIN, AND F. T. BINGHAM
    1969   *Atriplex polycarpa:* II. Germination and growth in water cultures containing high levels of boron. Agronomy Journal 61:451–453.
29. MOZAFAR, A.
    1969   Physiology of salt tolerance in *Atriplex halimus* L.: Ion uptake and distribution, oxalic acid content, and catalase activity. PhD Dissertation, University of California, Riverside.
30. BERNSTEIN, L., AND H. E. HAYWARD
    1958   Physiology of salt tolerance. Annual Review of Plant Physiology 9:25–46.
31. BERNSTEIN, L.
    1965   Osmotic adjustment of plants to saline media. I. Steady state. American Journal of Botany 48:909–916.
32. MOZAFAR, A., AND J. R. GOODIN
    1970   Vesiculated hairs: A mechanism for salt tolerance in *Atriplex halimus* L. Plant Physiology 45:62–65.
33. SCHOLANDER, P. F., H. T. HAMMEL, E. A. HEMMINGSEN, AND W. GAREY
    1962   Salt balance in mangroves. Plant Physiology 37:722–729.
34. OPPENHEIMER, H. R.
    1960   Adaptation to drought: Xerophytism. *In* Plant-water relationships in arid and semiarid conditions, reviews of research. Unesco, Paris. Arid Zone Research 15. 225 p.
35. BLACK, C. C., T. M. CHEN, AND H. R. BROWN
    1969   Biochemical basis for plant competition. Weed Science 17:338–344.
36. MOORE, R. M.
    1960   The management of native vegetation in arid and semiarid regions. *In* Plant-water relationships in arid and semiarid conditions, reviews of research. Unesco, Paris. Arid Zone Research 15. 225 p.
37. NORD, E. C.
    1965   Autecology of bitterbrush in California. Ecological Monographs 35(3):309–334.
38. HORMAY, A. L., AND M. W. TALBOT
    1961   Rest-rotation grazing. A new management for perennial bunchgrass ranges. U.S. Department of Agriculture, Forest Service Production Research Report 51. 43 p.

39. COOK, C. W., L. A. STODDART, AND L. E. HARRIS
    1954   The nutritive value of winter range plants in the Great Basin. Utah Agricultural Experiment Station Bulletin 372. 56 p.
40. IMPERIAL AGRICULTURAL BUREAU
    1947   The use and misuse of shrubs and trees as fodder. Imperial Agricultural Bureau, Aberystwyth, Great Britain, Joint Publication 10. 231 p.
41. TEMBLOR RANGE RESEARCH
    1969   Temblor range research, annual report 1968–1969. University of California, Davis.
42. DE KOCK, G. G.
    1965   The management and utilization of spineless cactus (*Opuntia* spp.). International Grassland Congress, 9th São Paulo, 1965, Proceedings.
43. BONSMA, H. C., AND G. S. MARE
    1942   Cactus and oldman-saltbush as feed for sheep. Union of South Africa, Department of Agriculture and Forestry, Bulletin 236. 72 p.
44. BOYKO, H.
    1966   Salinity and Aridity. Dr. W. Junk Publishers, The Hague. 408 p.
45. LEWIS, D., W. HALL, L. MARTINEZ, AND J. DE LA GARZA
    1969   Analysis and performance of a contour border system for concentration of runoff waters for crop production. Paper given at International Conference, Arid Lands in a Changing World, sponsored by AAAS Committee on Arid Lands, University of Arizona, Tucson, Arizona, U.S.A., June, 1969.
46. EVANS, R. A., R. E. ECKERT, JR., AND B. L. KAY
    1967   Wheatgrass establishment with paraquat and tillage on downy brome ranges. Weeds 15:50–55.
47. HILLEL, D.
    1969   Artificial inducement of runoff as a potential source of water in arid lands. Paper given at International Conference, Arid Lands in a Changing World, University of Arizona, Tucson, Arizona, U.S.A., June, 1969.
48. WORKMAN, J. P., AND N. E. WEST
    1969   Ecotypic variation of *Eurotia lanata* populations in Utah. Botanical Gazette 130(1):26–35.

49. Research was supported in part by a grant from the Dry Lands Research Institute, University of California, Riverside. Dr. McKell's work on this project was accomplished while he was a faculty member at the University of California, Riverside.

Economic Botany of Arid Regions

PETER C. DUISBERG

Natural Resources Division
Inter-American Geodetic Survey
Fort Clayton, Panama Canal Zone

and

JOHN L. HAY

Civil Engineer
Tucson, Arizona, U.S.A.

ABSTRACT

THE ECONOMIC OUTLOOK for desert-plant industries is not promising. Industries derived from wild native plants used for fibers, drugs, rubber, and waxes are declining in volume and offer a poor living to their workers. Industries derived from cultivated desert plants such as sisal and henequen for fiber, gum arabic, wattle for tannins, agaves for alcohol, and cacti for animal feed and human food offer much more promise for subarid and semiarid areas but will have stiff competition from synthetic and substitute products. Possibilities exist of using desert plants as a basis for cultivated crops in nonarid areas or under irrigated conditions. Systematic research on families of plants is yielding potential sources of plant-produced chemical intermediates for industry. The agronomic and technical problems involved in industrialization in each case will be formidable, however.

Although the overall economic outlook for desert-plant industries is somewhat bleak, new unexpected possibilities from research effort may occur from time to time. A recent though still remote possibility is the finding of somewhat promising chemical constituents in the bird-of-paradise plant of potential value in cancer treatment. A partial listing of economic plants of the deserts in the appendix provides information on known uses of these plants.

ECONOMIC BOTANY OF ARID REGIONS

Peter C. Duisberg and John L. Hay

THE BOTANY OF ARID REGIONS has been of fundamental economic importance to the quality of life of every desert-dwelling group from the aborigines of the Americas, Australia, and Africa, through the desert men who were part of the civilizations of Central Asia, the Middle East, and North Africa, to the modern tribesman, nomad, and peasant. For them it played or plays an important role in nearly every aspect of life including food, clothing, shelter, protection, religion, health, and recreation. Although we would judge those peoples as poor, most were in basic harmony with their evironment. This fact has much to do with the independence, dignity, and self-respect that were often noted in desert people.

As "modern" notions of property ownership increasingly began to interfere with the communal use of the whole environment, and profitable irrigable, dry-farming, and grazing lands came under the control of the influential or wealthy, the harmonic pattern began to disintegrate everywhere. Many of the disinherited were forced into the gathering of wild desert plants with commercial uses in order to eke out an existence. This practice kept them living in some of the driest and most rugged areas but without much of the dignity or the satisfaction that they once possessed.

After World War II, the increased opportunities for mobility due to modern transport, road building, and urban growth led these gatherers of economic products from the desert to become increasingly dissatisfied with their lot: low income and insufficient health, educational, and recreational opportunities. Present trends are toward an exodus of workers with a consequent decline and abandonment of traditional desert-plant industries in the more arid regions, even though the alternative attraction for most of these people is only a marginal existence in an urban slum. The process and pattern was clearly established in the United States about a generation ago, when the last industries based on wild desert plants were abandoned for lack of willing labor — even though the same products have continued to be produced and imported in quantity from the adjoining Mexican desert.

It is unlikely, even with unforeseeable breakthroughs in technology, that the production of commercial products from desert plants gathered in the wild will ever again be of significance in the United States. The country in its richness is already finding that the uniqueness and beauty of the desert flora is sufficient economic reason for its preservation as a setting for the tourism, retirement, and specialized industry it can attract.

Most less-developed countries will not have these opportunities for the foreseeable future. In these countries it would be desirable to examine carefully the social and economic situation in the dry areas in an attempt to stabilize it, partially through the use of the economic-botany and desert-plant industries. An attempt would be in order to assess economic benefits for the nation if desert plants were used as resources. Thus the focus of this chapter will be as follows:

1. To discuss illustrative desert-plant industries of the present and the past, not in any detail but with emphasis on assessing their future in the arid and semiarid regions for economic utilization of the land. This subject will include both the industries depending on wild-plant harvest, which are concentrated in the most arid lands, and those in which desert plants have become cultivated crops. The latter are primarily concentrated in semiarid areas or areas of steep topography in humid lands, where they offer an alternative to grazing.

2. To examine the possibilities of arid-zone plants being developed (or improved through breeding or selection) to create new crops that would serve as sources of special industrial chemicals.

3. To consider briefly current events that might possibly have considerable impact in the near future.

4. To present in tabular form an extensive but partial listing of economic plants of the arid region and the known uses of these plants.

STATUS AND OUTLOOK FOR INDUSTRIES BASED ON WILD DESERT PLANTS

The harvest of wild desert plants is primarily centered in the arid areas, which have low carrying capacities for grazing and insufficient rainfall for dry-farming.

Agave lechuguilla for Hard Fibers

The *Agave lechuguilla* is common in northern Mexico and western Texas of the United States in areas averaging 150 to 200 mm of rainfall; it is the source of the hard fiber ixtle. The last attempt at industrialization of this resource failed in the United States about 1950. Mexican production has shown a downward trend but still amounts to some 10,000 tons of fiber annually, which is marketed for export through a national monopoly. The value of the product is about $3,000,000. The harvesting is done by hand, with transport usually by mules. The method of harvesting permits regrowth, and the plant is very abundant; therefore only a small portion of the potential supply is used. The most important external uses are for rope, upholstery tow, and cheap brushes.

The harvesters live at a subsistence level and obtain a meager income by extracting lechuguilla fiber at home at a rate of 6 to 8 kilograms per day. Attempts have been made to develop machines to harvest the wild plant, but these have failed. Decorticating machinery has been developed on the basis of the principles well known for henequen and sisal. The profits have been marginal, however, and they have accounted for only a small part of the production (*1, 2*).

In view of the facts that (*a*) sisal and henequen grow under less arid plantation conditions, (*b*) synthetic fibers are increasingly strong competitors, and (*c*) the harvesters will require increasing financial incentives to remain working isolated under primitive conditions, it is expected that costs will rise and markets continue to contract.

Ephedra sinica and *E. gerardiana* for the Drug Ephedrine

Five thousand years ago the drug ma-huang was produced in China from *Ephedra sinica*. The methods of harvest continue to involve primitive hand-gathering of wild branches, which yield 2 to 5 percent of ephedrine on extraction. Production figures for China are unavailable, but about 1,000 tons of *Ephedra gerardiana* branches are collected annually in Pakistan. Ephedrine was an extremely important drug until recently for treating asthma and afflictions of the ears, nose, and throat. Synthetic vasoconstrictors have already taken part of its market away. Ephedrine has also been synthesized and, if the harvesting and other costs for producing the natural product increase, the synthetic product can be expected to take over the remaining market increasingly (*3*).

The outlook for the drug produced from the wild plant material is not promising.

Parthenium argentatum for Rubber

A guayule rubber boom began in Mexico about 1905. At its height some $100,000,000 had been invested in six extracting plants in Mexico and one in Texas. Too little was known of the ecology of the plant, so it was soon almost exterminated in parts of the area. This did not occur universally only because better-quality rubber from new tropical plantations in Southeast Asia caused a decline in demand and failure of most of the factories by the early 1920s (*4*).

One company survived through a brief revival of the boom in the 1940s, when wartime loss of the rubber plantations of Asia resulted in high prices, subsidized research from the United States government, and a high level of production. Even though guayule rubber had advantages for some special uses, the company was unable to establish a sufficiently strong economic foundation to compete with genetically improved higher-yielding natural *Hevea* rubber trees, so guayule rubber produc-

tion was terminated early in the 1950s and the factory was closed.

The case of guayule is especially discouraging because, after the initial speculatory phase, it represents a serious attempt to form a stable industry backed by much imaginative research. By hybridization, selection, and developing improved cultural practices, it was possible to double the original 7 to 10 percent yield of the native plant, increase disease resistance, increase frost resistance, and decrease the content of resins that render purification of the latex difficult. In the industrial process notable improvements were made in inhibiting oxidation and in eliminating impurities and resins. Probably if this work could have been accomplished thirty years sooner, the guayule rubber industry could have become so well established that it could have progressed sufficiently rapidly in technology to maintain a strong position in the face of its competitors. Even if that had happened, however, it is obvious that to compete effectively the improved plants eventually would have been grown under less arid conditions, probably with limited irrigation. In any event, then, a sound industry based on the original desert climate and a range of 150 to 250 mm of rainfall would have been impossible.

Astragalus gossypenus and Other Species of *Astragalus* for Gum Tragacanth

The production of gum tragacanth from various species of *Astragalus* in Iran requires a very cheap form of labor. In the regions of growth only about 30 plants per hectare occur. These must be cut by hand to expose the cylinder containing the gum, and the 2 to 5 grams which exude must be collected in several trips to the plant. The industry supplements the income of poor peasants, and it yields revenues to Iran of several million dollars annually (*2*). Its hope as an economic plant in its present 125 to 500 mm rainfall area, if living standards improve moderately, would be for it to be cultivated and grown as close together as possible. For this to benefit many present harvesters, some kind of cooperative venture would be involved. In the future the best that might be expected would be for this gum to maintain its present market.

Euphorbia antisyphilitica for Hard Wax

Candelilla wax comprises about 2 percent of the *Euphorbia antisyphilitica* plant and has the relatively high melting point of 67.5°C. It formerly ranked second to wax from the carnauba palms in demand as a hard wax for floor polishes; it was also used in a number of other products from phonograph records to pomades. It grows in northern Mexico and in western Texas. During the shortages of World War II, the plant was harvested without controls in Texas; today laborers in the United States would be unwilling to harvest it at the wages which could be paid even if the supply were abundant. If the plant is harvested correctly, its regeneration occurs within five years. The Mexican government (with some success) has tried to educate the harvesters toward the goal of conservation. The extracting and refining processes involve simple technology (5).

Although the living standards of the candelilla producer were (and are) austere, it seemed that the product was sufficiently unique and desirable to insure stable markets at increasing prices and justify research. The wax content appeared sufficiently associated with aridity and the conditions of natural growth to preclude much possibility that the plant would be cultivated or grown in other areas. There seemed to be ample possibilities for improvement in the efficiency of harvesting and processing and possibilities for developing new uses.

In addition, the Mexican government had a monopoly on world sales, and the area in which candelilla occurred also contained guayule, lechugilla, and other lesser commercial fibers that might be used to supplement the earnings of the candelilla workers during years of lower demand. It appeared that candelilla might play a principal role in stabilizing the economics of the area at a level that would give access to an improved quality of life for the harvesters. Development of synthetics to produce hard self-polishing waxes seems to have destroyed this illusion, however.

The future of candelilla, given the depressing standard of living of the harvesters and the competition with synthetics, is dim. The demand for production may decrease from highs of several million dollars and stabilize at under a million. In such a case it would be hard to justify the social programs and technical research programs needed to develop a stable industry.

Larrea divaricata (Creosotebush) for Antioxidant

One of the most common plants of the Mexican and southwestern United States deserts and the arid lands of western Argentina is creosotebush (Fig. 1). During World War II this plant was found to contain nordihydroguaiaretic acid, which proved to be a powerful antioxidant for fats and oils and was of great value for several years. Once again, however, cheaper synthetic chemical antioxidants succeeded in usurping its market. Meanwhile, interest had been aroused in other byproducts. Studies on the resins indicated good possibilities for paints and varnishes, but they coincided with the period in the paint industry when synthetics began to displace the natural resins, and therefore the resin studies were discontinued. A process for manufacturing a good-quality hardboard was developed but found to be uneconomical unless byproducts could be developed. The de-resined leaf material proved to have a nutritive value for livestock close to that of alfalfa, but the cost of extraction without additional useful byproducts was too high (5). At present, prospects are poor for future development of this plant in spite of its variety of interesting properties and constituents.

Photo by J. R. Hastings

Fig. 1. Creosotebush, *Larrea divaricata.*

Hyocyamus muticus (Egyptian Henbane) as a Drug Plant

Henbane is the only wild plant in the Egyptian deserts of sufficient commercial value to justify its export; it was gathered in such quantities for medicinal drug purposes that it was almost exterminated. Efforts to learn how to cultivate the wild plant profitably have been held back by a declining demand, partly because synthetic and natural substitutes are taking over part of its market (*3*).

General Outlook

The above examples are sufficient to show that the outlook becomes progressively poorer for useful desert plant industries based on harvesting the natural supply and using arid lands which have little other possible economic use. The decline of the production of lechuguilla, guayule, and candelilla between 1950 and 1960, as shown in Table 1, is indicative of the general downward trend and has continued without any prospect for reversal.

Change in Production From Wild Desert Plants, 1950–1960

Product	Tons		Decline (%)
	1950	1960	
Candelilla wax	172,497	142,595	17.5
Guayule rubber	10,857	0	100
Lechuguilla fiber	12,697	10,487	21

Source: Portion of table from Hernandez X. (*10*).

The solution of the United States — to permit limited grazing and to consider the areas valuable as scenery for tourists — can be attractive only to countries like South Africa or Australia. The underdeveloped remainder are faced with the dilemma of having the desert populations become progressively more underprivileged (relative to the general population) as they continue to produce traditional useful plant and animal products, or sit passively by while vast areas between irrigated centers become less occupied than they are today and lose what economic return they still give.

Mexico probably has the best chance to find a solution, and such a solution might be of value to other countries. That nation has an important tradition of desert-plant use and desert-plant industries. It still has considerable to lose as its present desert industries decline. Its general economy and availability of funds are expanding, and it has a good supply of advanced technical personnel. It has made an imaginative start in forming the Fondo Nacional de Fomento Ejidal financed from the sales of candelilla, lechuguilla, ixtle, and guayule (if it revives) to coordinate efforts to improve conditions of the rural populations within its northern deserts. It might be surmised that this is probably already too limited a source of funding, and both government subsidy and financing and a regional approach in which irrigation and other stronger local parts of the regional economy should be considered as sources of additional financing. In addition,

certain public offices might be deliberately dispersed to improve the economic base of the region.

Among other ideas that might broadly be considered under economic botany might be the managed introduction of exotic desert animals capable of making better use of the forage value of the desert shrubs, the subsequent development of sport hunting, and serious consideration of a suggestion by Hernandez X. (*10*) relative to using goats under more careful management. He points out that the destructive reputation of the goat is not so much the fault of the goat as of poor management by man.

STATUS AND OUTLOOK FOR INDUSTRIES BASED ON CULTIVATED PLANTS FROM ARID ZONES

Generally, when an effort is made to industrialize a useful desert plant through cultivation, the plant is grown under semiarid conditions. Under these conditions growth generally is greater, and the local quantity of useful product more, even in cases where the percentage of usable product decreases under less arid conditions. These plants are generally used to extend climatic limits of cultivated-crop agriculture into semiarid and arid areas where ordinary cultivated crops fail because of excessive drought and where grazing gives a lower return than the cultivated desert plant. Through breeding and selection some of these desert plants have been improved. They sometimes are grown even in very humid areas, but in conditions where they have some special advantage. Examples are: on steep topography with thin soils, or on soils whose low moisture-holding capacity leads to temporary drought for ordinary crops in spite of the rainfall.

Agave fourcroydes and *Agave sisalana* for the Hard Fibers Henequen and Sisal

Henequen from *Agave fourcroydes* is primarily produced in the Yucatan, Mexico. Rainfall is variable but averages 900 to 1,000 mm. The climate is classified as semiarid or even subhumid, but actually, because the soils are usually shallow or sandy, conditions are equivalent to more arid environments. More than 100,000 hectares are in production in the Yucatan.

New plantations are started from "hijos de pie," which grow spontaneously from the rhizomes of the existing plants. After 6 to 8 years the outer leaves reach about one meter in length; they are thenceforth cut annually with machetes. A plant produces for 6 to 8 years in fertile soils and 12 to 14 years in poor arid soils before it flowers once and dies. This is a characteristic of all century plants. Thus, it is an economic advantage to grow *Agave* in soils sufficiently poor to postpone the year of flowering.

The average fiber production in the Yucatan plantations is about 1,000 kilograms per hectare annually, or about 7 percent of the leaf weight. The juice and the bagasse remaining are used only for fertilizer. Considerable work has been done to try to develop products from other constituents, including the 0.2 percent of wax. In the case of *Agave sisalana* in Tanzania, the saponin hecogenin, a precursor of cortisone, is produced as a byproduct, even though the concentration in that species is low for the *Agaves*.

Sisal is from *Agave sisalana,* also indigenous to Mexico, but interestingly enough the plant is not grown there to any extent. It is, however, the principal hard-fiber desert plant grown in most other countries.

It is important commercially in Haiti, Israel, Kenya, Mozambique, Cuba, Brazil, Indonesia, Tanzania, and other countries. In the Saint Marc area of Haiti, the rainfall varies from a humid 1,000 to 1,500 mm; the plantations are located on the poorest soils with low water-holding capacity, and short droughts are frequent.

The henequen and sisal industries are the largest and most advanced desert-plant industries (*6*). The United States, for instance, imports a total of 175,000 to 300,000 tons of fiber per year. There is a great competition and consequent fluctuation in price, and the value of imports into the United States has varied between $18,000,000 and $70,000,000 per year. This great fluctuation in price and demand has played havoc with the smaller and higher-cost industries, and many have failed after building expensive plants. Two recent examples occurred in the Azua Peninsula of the Dominican Republic and in western Venezuela, with considerable economic hardship on the employees displaced. Intense competition between the different producing countries leaves industry vulnerable to increasing competition of synthetic fibers.

More than 20 additional arid-zone plants of the genera *Agave, Furcraea, Bromelia,* and *Attalea* are also grown in various countries to produce smaller amounts of hard fibers, generally for their local needs.

Acacia senegal for Gum Arabic

The Sudan produces, in the form of gum arabic, about 85 percent of the world's demand for gums. Most of the remainder is produced in Asia or Africa. Total demand is some 50,000 tons valued at about $30,000,000 (U.S.).

Part of the Sudanese production comes from wild trees from subarid areas, with rainfall generally between 300 and 375 mm, and part from trees cultivated in the Kordofan area, with about 400 mm of rainfall. In both cases the industry provides only parttime work for many of the workers.

The fact that gum-arabic production is the second-most-important industry in the Sudan is a reflection of the underdevelopment of the country. The industry could not survive long without the continued exploitation of very cheap labor (*3*).

Acacia decurrens for Tannins

Plantations of this tree are planted in its native Australia and South Africa, and in semiarid parts of East Africa having 400 to 500 mm of rainfall.

When the plant reaches some 10 meters in height (at the age of about 10 years) it yields 20 to 25 kilograms of bark containing about 35 percent tannin (*7*). Wattle bark remains one of the best and most important tanning agents, but, like all other natural tanning agents, it is increasingly threatened by the growing use of synthetic tanning agents and artificial leathers.

Agaves for Alcohol

In Mexico the alcohol tequila is made from *Agave tequilana* and related species, and the alcohol mezcal from *Agave atrovirens*. In both cases the central portions, or heads, are cooked in ovens to convert the glucosides into sugars. The juices are then extracted, fermented, and distilled. The distilling equipment is generally rudimentary and fiber is produced as a byproduct (*8*).

The production and values of these alcohols are increasing. In the case of mezcal, annual production increased from about 1.90 million liters in 1950 to 2.05 million liters in 1960.

Opuntia Cacti for Food and Forage

No desert plants have been more studied and none are more controversial than the various species of prickly pear *Opuntia*s. Over several centuries, these have spread throughout the world from their American points of origin (*9*).

Peasant peoples in Mexico and many areas of North Africa and the European Mediterranean depend on the tunas, or fruit, as a major source of carbohydrates and vitamins. This commerce within Mexico alone amounts to over 30,000 tons, not including the great quantities used in the subsistence economy. Small commercial plantations have been established in Mexico, Chile, Brazil,

Fig. 2. *Opuntia* cactus with stand of creosotebush (*Larrea divaricata)* in background.

Argentina, the United States, and many other countries; and tunas, jellies, and other products are finding their way into the luxury trade as specialty items. On the other hand, the plant is considered a pest in Australia, South Africa, the United States and elsewhere, and great sums are expended on biological and chemical methods to control its spread through grazing lands.

In the 1960s a comprehensive scientific study was initiated in Mexico aimed at production of high-quality edible fruit, tender pads for vegetables, and forage. Results to date indicate that 8 tons per hectare of fruit is possible at rainfall levels of about 700 mm and is said to be maintainable at least 20 years. In forage species, 200 tons per year of edible pads has been produced per hectare in experimental trials. While the protein content is low, these are impressive results, and the commercial possibilities are similarly impressive (*10*). Unfortunately, comparable work has not been done for the more arid desert regions where the land has no use except grazing at about 100 hectares per animal unit.

Outlook

The outlook for cultivated arid-zone species is considerably better than for the wild species. Volumes of production and prices of products such as sisal, henequen, wattle, and gum arabic — while still greatly dependent on cheap labor — are sufficiently high to permit considerable research on harvesting, processing, and markets. With total world demand for most raw materials increasing due to increased population, these products might, with good leadership, maintain their volumes in the face of competition from synthetics. They show few possibilities for important expansion, however. With the possible exception of prickly-pear forage and fruit, there seem to be no major new possibilities in view for the traditionally useful desert plants.

NEW PLANT POSSIBILITIES

A remote possibility that the bird-of-paradise plant (*Caesalpinia gilliesii*) contains chemicals to control cancer is being investigated by Dr. Jack Cole of the University of Arizona. In a study of over 100 compounds from desert plants, he has found that two plants, bird-of-paradise and "torote blanco" trees (*Bursera microphylla*) have shown the most promise.

The possibility of improving the economies of the arid or semiarid lands through the expansion of present plant-derived industries does not appear very promising. Another aspect of economic botany might offer promise, even though it might not necessarily result in appreciable economic benefits to the arid lands themselves. This opportunity stems from the need for chemicals in processes to make synthetics, which have wrought so much havoc to the established plant industries. In some cases plants may become economical starting materials for synthesis of complicated molecules. One has already been mentioned, the case of *Agave sisalana* as a source of hecogenin used in the preparation of cortisone.

The New Crops Research Branch of the U.S. Department of Agriculture is undertaking an effort to make a systematic study of all genera of plants to detect constituents of possible industrial value. This study, if completed, should provide a fair inventory of the chemical possibilities and be of worldwide interest and value. A few examples of some of the arid- and semiarid-zone plants considered promising will illustrate the possibilities.

Seeds of *Crambe abysinnica* from the semiarid northeast of Ethiopia contain 40 percent oil, of which three-fifths is erucic acid. Erucic acid is used in the production of synthetic rubber. In addition, this acid can be oxidized to form the dibasic brassilic acid and also pelargonic acid, both of which have industrial uses in the synthesis of lubricants, plasticizers, polyesters, insect repellents, and synthetic fibers. Planting tests have indicated possible yields of 2,000 kilograms per hectare in cool semiarid regions.

Lesquerella fendleri, from the arid southwestern United States and northern Mexico, contains oils with high amounts of hydroxylated fatty acids, which have various industrial uses. It is estimated that yields of 1,000 kilograms per hectare of oil might be obtained.

Various species of the genus *Dimorphoteca* have seeds rich in fatty acids with double bonds and hydroxyl groups of potential use in the plastics industry. Ninety-five percent of the fatty acid content in seeds of *Limnanthes* species from California contain compounds with more than 18 carbons in the chain (*11*).

Many arid-zone species of the genera *Agave* and *Yucca* are excellent sources of saponins (*12, 13*). Those with saponins with the hydroxyl group favorably situated in the 3 and 11 positions are especially suitable for some sex hormones and cortisone. There is little possibility that these will be developed for this purpose, however, at least as long as the wild tropical species of Mexican *Dioscorea,* the major commercial source, remains abundant.

The difficulties in developing new crops from these promising plants are formidable. The desert plant *Rumex hymenosepalus,* or canaigre, is a case in point. As early as 1890 the dried roots of the wild plants in the southwestern United States and northern Mexico were the basis of a thriving business for the extraction of tannins for tanning leather. This lasted until the native supply was depleted in the early 1900s.

In 1937, in the face of domestic shortages of tanning materials, the U.S. Department of Agriculture began an intensive effort to develop canaigre as a cultivated crop. This program was continued for some 25 years before the conclusion was reached that the crop could not be a commercial success, and the effort terminated. During this period it was possible to increase the tannin content from 25 to 35 percent, increase disease resistance, solve problems of seed production and harvest, develop machinery and methods for planting, cultivating, harvesting, and storing the roots, and develop varieties for irrigation which would produce 35 to 50 tons per hectare using two-thirds the amount of water needed for ordinary crops. On the industrial side, several excellent methods using combinations of water and solvents were developed, procedures were devised to remove sugars and starches and increase the purity of the extract, and possible by-products were produced. In spite of the solution of so many problems, commercialization was not feasible, partly because of the changes in demand and development of synthetic tanning agents and synthetic leathers during the long research period (14).

Simmondsia chinensis, or jojoba, has received much less attention but has received it continuously over much the same time period as canaigre. It has an unusual chemical characteristic; it contains a liquid wax in place of the fatty esters in most plants. The wax has unique properties, leading to potential uses, not duplicable in other plant products. An example is stability to repeated heating to high temperatures without danger of decomposition. In addition, jojoba could be a source of the 20 and 22 carbon alcohols, eicosol and docosol, for industrial uses. The most difficult problems associated with jojoba have been in connection with learning to cultivate the plant economically. After all the effort it still has far to go (7).

On the other hand, successes have been obtained in commercializing some plants after considerable effort. Two examples will be cited to indicate some of the factors. The fact that both plants, *Cyamopsis tetragonalobo,* or guar, from India, and *Ricinus communis,* or the castor bean, had been cultivated as minor crops for many years in Asia and Europe is of relevance but does not overly diminish the magnitude of the accomplishment. In the case of guar the achievement was to successfully introduce it into the semiarid Southwest of the United States and to develop methods of production, yields, markets, and prices sufficiently high for success. The market development by two companies involved an aggressive program of helping industry to find uses for guar gum and to develop products particularly adapted to each use. Thus 30 different classes of guar gum were produced for industry.

In the case of the castor bean, increased demands for the hydroxylated oil had developed during World War II in the aviation industry. The accomplishments during 15 years involved creating new varieties with higher oil content and nondehiscing characteristics suitable to machine harvesting, developing the machines, preparing chemical sebacic acid and ricinoleic acid from the oil, and developing markets for the oils in the plastics and other industries. Much of this crop is grown in arid lands, but under irrigation. Much of the success is due to the aggressive efforts of one company.

Because many arid-zone plants show unusual physical adaptations to aridity, it is commonly believed that they should be unusually rich in rare and unique chemical constituents. Actually, the most important factor seems the genetic makeup of the plant. Thus plants of the same families or genera from vastly different climatic zones are likely to have more in common chemically than plants of different genetic origin within a zone. Some families are concentrated in specific climatic zones, and thus the special chemical constituents that characterize them are concentrated in these zones. This situation holds true for the families Cactaceae and Euphorbaceae, for instance, in arid zones.

Two genera, *Agave* and *Yucca,* primarily found in arid zones, comprise about 60 percent of the high-saponin plants known. It is likely that desert plants on the whole are relatively richer in essential oil and mucilaginous gums, since both may play a physiological role in adaptation to arid conditions. The systematic study by Wall and others (12, 13) shows that the oil content of high-oil seed plants in arid zones was about 100 percent more than expected, and the protein content of high-protein plants about 80 percent more than expected, based on the average of all plants studied from various climatic zones. There is some economic promise in those figures.

CONCLUSIONS

In the 1950s and 1960s the combination of awakening aspirations on the part of desert peoples and the development of synthetic substitutes for many natural plant products has dimmed the prospects for the field of economic botany in arid regions. Part of the problem, however, has been the backwardness and lack of scientific approach by the industries themselves. Thus some hope lies in applying the same systematic, scientific types of approach used by the drug and chemical companies in developing synthetics. A start has been made in the inventory of the U.S. Department of Agriculture New Crops Program. The promising plants of the inventory might be followed up in a variety of countries and climatic areas. The task of commercializing new crops from even the most promising plants would be difficult, but not impossible. The examples of the castor bean and guar gum are illustrative.

From the standpoint of better land utilization, arid lands with considerable desert shrub and plant vegetation but with little potential for grazing offer dim prospects for existing or potential desert-plant industries. Since in many countries the economic status of the present desert population is poor, regional programs in which desert plants are used might be significant.

Prospects are better to improve cultivated desert plants for use in the semiarid areas and thus to extend further the total area in which dependable cultivation is possible. Improved high-yielding varieties of the prickly-pear cactus being developed in Mexico may have especially important possibilities both for livestock feed and food for man.

In a study of over 100 compounds from desert plants at the University of Arizona, tentative findings of Cole (15) show that two plants, bird-of-paradise (*Caesalpinia gilliesii*) and the torote blanco tree (*Bursera microphylla*), have components that show slight promise for treatment of cancer. This finding has led to continuing U.S. government research grants.

APPENDIX

Select Economic Arid-Lands Plants and Their Uses

(Prepared by John L. Hay, Mary Lu Moore, and William G. McGinnies)

For facilitation of research, a selected list of plants, uses, and citations in *Economic Botany* through April-June 1969 [vol 1–23 (2)] has been compiled from notes prepared by Mr. Hay. The plants listed are not rigidly restricted to those originating in some carefully-defined arid zone; doubtful cases were decided on the basis of usefulness to the reader. Included at times are species from arid-lands genera but marginal in specific geo-graphic distribution, and species not native to the arid zone but grown there (usually under irrigation). Common names for the plants and specific parts of plants in connection with a particular usage are listed here only if given in the journal. Synonymy and nonstandard nomenclature stem from the journal; minor adjustments have been made in spelling and capitalization.

cont. ⟶

Scientific Name	Usage	Citation Data	Scientific Name	Usage	Citation Data
Acacia	Bark — tanning materials	1(1):333 Utilization Abstract	(catclaw)	Food	8(1):3-20
	Food	6(1):23-40		Seed — protein and oil	20(2):127-155 and #326
	Tannins	8(1):3-20	Acacia longifolia	Seed — protein and oil	20(2):127-155 and #329
	Gum arabic; tannins and resins; drilling muds	9(1):93 Utilization Abstract	Acacia nilotica (L.)	Sant or sont; flowers — tea; powdered fruits — incense	22(2):165-177
	. .	18(3):280 Book Review	Acacia pycnantha	Wattle bark; tannin	9(2)108-140
	Alkaloids	20(3):274-278	Acacia senegal	Sudan or Kordofan gum; gum Senegal; gum arabic; water soluble gum in textiles	3(1):3-31
Acacia angustissima	Seed — protein and oil	20(2):127-155 and #320		Gum arabic; vegetable gums and resins	4(2):195
Acacia arabica	Sunt — water-soluble gum in textiles	3(1):3-31		Edible gum; confections	6(1):23-40
	Babul fruits; tannins; pods and bark	8(3):285 Utilization Abstract		Gum arabic	9(2):108-140
	Wood — coffins, paint, adhesives	14(1):84-104		Wood, exudate — coffins, paints, adhesives	14(1):84-104
Acacia aroma	Drink	6(3):252-269		Gum arabic; hair set	20(1):17-30
Acacia baileyana	Seed — protein and oil	20(2):127-155 and #321	Acacia seyal	Talca or tabba; water soluble gum in textiles	3(1):3-31
Acacia campy-lacantha	Edible gum; confections	6(1):23-40		Edible gum; confections	6(1):23-40
			Acacia tortilis	Wood for fuel and construction	8(2):152-163
Acacia cavenia	Flowers — perfume	2(3):334-338 Utilization Abstract	Acantho-cereus pentagonus	Various portions as food	17(4):319-330
Acacia constricta	Seed — protein and oil	20(2):127-155 and #322	Acantho-chiton wrightii	Food	8(1):3-20
Acacia dealbata	Flowers — perfume	2(3):334-338 Utilization Abstract	Agastache spp	Various portions as food	17(4):319-330
Acacia decurrens	Black wattle; green wattle; bark — tannin	2(2):217-218 Utilization Abstract	Agave	Alcoholic beverages — tequila, mescal, pulque	3(2):111-131
	Tannin	4(1):3-36	(century plant)	Food; fibers; alcohol; saponins	8(1):3-20
	Fiber for paper	10(2):176-193		Fiber; beverages — mescal and tequila	9(1):93 Utilization Abstract
Acacia elata	Seed — oil and protein	20(2):127-155 and #324		Precursor of cortisone	9(4):307-375
Acacia farnesiana	Demulcent; emulsifying agent (from stems and branches) — mucilage, infusion	1(1):57-68		Fish stupefaction plant	12(1):95-102
				Medicinal: drugs — steroidal sapogenins from seeds, leaves	15(2):131-132
	Blossoms — perfume	2(3):334-338 Utilization Abstract		Fiber for cordage, hats, mats, basketry, bags, sandals, upholstery, padding	19(1):71-82
	Acacia gum, water soluble gum in textiles, cosmetic and pharmaceutical preparations	3(1):3-31		Maguey; edible fruit	19(4):323-334
				Food	19(4):335-343
	Seed — protein and oil	20(2):127-155 and #325		Sisal, henequin; fiber for textiles, rope, string	22(4):354-358
Acacia greggii	Demulcent; emulsifying agent (from stems and branches) — mucilage, infusion	1(1):57-68		. .	23(2):197-198 Book Review

PLANTS AND THEIR USES — cont.

Scientific Name	Usage	Citation Data
Agave americana (century plant)	Fiber for paper Maguey; paper — writing, paper dolls for witchcraft	10(2):176-193 17(4):361-367
Agave bovicornuta	Fish stupefaction plant	12(1):95-102
Agave deserti	Various portions as food	17(4):319-330
Agave dotrerana	Seed — protein and oil	20 (2):127-155
Agave fourcroydes	Sisal, henequen; fiber Henequen; fiber for paper Fibers; medicinal extracts — sapogenins; pharmaceuticals — steroids, alkaloids; foods and food products; resins Henequen; currency; stencil paper; teabag paper; wall hanging; tear-resistant paper	2(2):158-169 10(2):176-193 13(3):243-260 19(4):394-405
Agave geminiflora	Seed — protein and oil	20(2):127-155 and #89
Agave lechuguilla	Ixtle fibers Mexican fiber; brushes Fiber — brushes, upholstery tow Lechuguilla; leaves — medicine, drugs, steroids Fish stupefaction plants Fibers; medicinal extracts — sapogenins; pharmaceuticals — steroids, alkaloids; foods and food products; resins Yucca; medicinal: drugs — steroidal sapogenins from leaves, seeds Medicinal — drugs, steroidal sapogenins — from leaves, roots, seeds; fiber — brushes	3(2):111-131 4(3):243-252 8(1):3-20 11(1):39 Utilization Abstract 12(1):95-102 13(3):243-260 15(2):131-132 16(4):266-269
Agave lurida	Fibers; medicinal extracts — sapogenins; pharmaceuticals — steroids, alkaloids; foods and food products; resins	13(3):243-260
Agave mescal	Fibers; medicinal extracts — sapogenins; pharmaceuticals — steroids, alkaloids; foods and food products; resins	13(3):243-260
Agave parryi	Various portions as food	17(4):319-330
Agave potrerana	Seed — oil, protein	20(2): 127-155 and #90
Agave schottii	Fish stupefaction plant Soso; extracts from stems and roots	12(1):95-102 14(2):157-159
(yucca)	Yucca; medicinal — drugs, steroidal sapogenins from leaves, seeds	15(2):131-132
Agave sisalana	Sisal; cellulose; fibers; heavy paper Sisal; rope, fiber products, various extracts Sisal; hard fiber; cordage, textiles Sisal; fiber for paper Fibers; medicinal extracts — sapogenins; pharmaceuticals — steroids, alkaloids; foods and food products; resins	3(2):111-131 4(1):83-84 Utilization Abstracts 9(4):376-399 10(2):176-193 13(3):243-260

Scientific Name	Usage	Citation Data
	Sisal; currency; stencil paper; teabag paper; wall hanging; tear-resistant paper	19(4):394-405
Agave strovirens	Maguey; beverage — pulque; poles from stalks, short boards from leaves	17(2):200-210
Agave utahensis	Various portions as food	17(4):319-330
Agave vera-cruz	Fibers; medicinal extracts — sapogenins; pharmaceuticals — steroids, alkaloids; foods and food products; resins	13(3):243-260
Agave zapupe	Fibers; medicinal extracts — sapogenins; pharmaceuticals — steroids, alkaloids; foods and food products; resins	13(3):243-260
Agropyron	Wheat grasses; pollen extracts to treat allergies	5(3):211-254
Aizoon canariense	Hadaq, hudak, samh; seeds for gruel	22(2):165-177
Alhagi camelorum	Gum manna; spice	8(2):152-163
Alhagi maurorum	Gum manna; spice	8(2):152-163
Allium	Food Various portions as food Food seasoning Latex rubber	8(1):3-20 17(4):319-330 20(3):285-301 21(2):115-127
Allium ascalonicum	Garlic; vegetable	8(2):152-163
Allium cepa	Onion; vegetable Onion; bulb — religious functions, embalming Onion; edible root	8(2):152-163 14(1):84-104 20(1):6-16
Allium cf. drummondii	Seed — oil, protein	20(2):127-155 and #78
Allium porrum	Leek; vegetable	8(2):152-163
Aloe	Ornamental plant; leaves — latex for "bitter aloes," cathartic; pulp — astringent, antibiotic; purgative; eyewash; poultice; insect repellent. Used in creams, powders, lotions, tonics, antiseptics	15(4):311-319
Aloe barbadensis	Medicinal drug	7(2):99-129
Aloe barteri	Edible flowers	6(1):23-40
Aloe succotrina	Embalming fluids Exudate; fiber; embalming; wrappings	8(2):152-163 14(1):84-104
Aloe vera	Leaves — medicine, antibiotic	17(1):46-49
Alopecurus arundinaceus	Caryopsis, protein and oil	20(2):127-155 and #17
Alopecurus pratensis	Caryopsis, protein and oil	20(2):127-155 and #18
Amaranthus	Pigweed; food Various portions as food Fiber — cordage, hats, mats, basketry, bags, sandals, upholstery, padding Pigweed; food plant — leafy vegetable	8(1):3-20 17(4):319-330 19(1):71-82 20(1):6-16
Amaranthus graecizans	Tumbleweed; pollen extracts to treat allergies	5(3):211-254
Amaranthus retroflexus	Quelite; extracts from leaves and stems	14(2):157-159
Ammi majus	Seed and pericarp — protein and oil	20(2):127-155 and #592

Scientific Name	Usage	Citation Data
Ammi visnaga	Khella; medicine	7(1):89-92
Ammobroma sonorae	Various portions as food	17(4):319-330
Anabasis aphylla	Alkaloids, insecticide	1(4):437-445
Anacardium occidentale	Cashew, drink	6(3):252-269
Annona cherimola	Cherimoya; edible fruit	1(2):119-136
Annona muricata	Fruit, food	6(3):252-269
	Anonas (masa samba); edible fruit; phytomorphic representation in ceramics, wood, textile, stone, and graphics	16(2):106-115
	Sorsaca; leaves for tea medicinal plant for folk medicine	22(1):87-102
Annona reticulata	Fruit, food	6(3):252-269
Anthemis cotula	Dog fennel, insecticide	8(1):3-20
Aplopappus hartwigi (rayless goldenrod)	Rayless goldenrod; alkaloids, pyridine	3(2):111-131
Aplopappus nuttallii	Roots — tea for coughs	8(1):3-20
Apocynum	Indian hemp; fiber — cordage	8(1):3-20
	Fiber — cordage, hats, mats, basketry, bags, sandals, upholstery, padding	19(1):71-82
Apocynum cannabinum	Tuberous roots — tincture, liniment (stimulant to sensory nerves), depressant	1(1):57-68
	Dogbane, Indian hemp, chewing gum substitute, cardiac stimulant	8(1):3-20
Argemone albiflora	Seed — oil, protein	20(2):127-155 and #178
Argemone intermedia	Seed — oil, protein	20(2):127-155 and #180
Aristolochia	Body ointments	6(3):252-269
Aristolochia maurorum	Seed — oil, protein	20(2):127-155 and #110
Artemisia	Wormwood, insect repellent	8(1):3-20
Artemisia herba-alba	Wormwood; stimulant, tonic, vermifuge	8(2):152-163
Artemisia judaica	Wormwood; stimulant, tonic, vermifuge	8(2):152-163
	Sheeh (wormwood); leaves — tea, respiratory ailments	22(2):165-177
Artemisia maritima	Santonin; floral buds; medicines; drugs; anthelmintic	9(3):224-227
Artemisia tridentata (sagebrush)	Essential oil	3(1):104-106 Utilization Abstract
	Pollen extracts to treat allergies	5(3):211-254
Arundo donax (carrizo)	Arrow shafts	6(3):252-269
	Reed; food	8(1):3-20
	Woodwind instruments; industrial cellulose; ornamental plant; latticeworks; basketry; matting; fishing rods; walking sticks; erosion control; fodder; medicinal	12(4):368-404
	Cane	17(3):200-210
Asclepias (milkweed)	Rubber	8(1):3-20
	Various portions as food	17(4):319-330

Scientific Name	Usage	Citation Data
Asclepias albicans	Wax	13(1):29 Utilization Abstract
Asclepias erosa (milkweed) (hierba lechosa) (desert milkweed)	Rubber	8(1):3-20
	Extracts from stems and leaves	14(2):157-159
	Gums — rubber products	16(1):17-24
Asclepias involucrata	Leaves, shoots, flowers — food and fiber; seeds — floss, oil, wax, plastics, latex	3(3):223-239
Asclepias speciosa (milkweed)	Leaves, shoots, flowers — food and fiber; seeds — floss, oil, wax, plastics, latex	3(3):223-239
Asclepias subulata (desert milkweed)	Gums, rubber products	16(1):17-24
Asclepias syriaca (milkweed)	Food; bast fibers from stems; textiles; seeds — floss, oil, wax, plastics, latex	3(3):223-239
Asclepias tuberosa	Seed — oil and protein	20(2):127-155 and #652
Astragalus	Gum tragacanth; in textile printing	1(4): 402-414
	Gum tragacanth; vegetable gums and resins	4(2):195
	Tragacanth; tannins and resins — drilling muds	9(1):93 Utilization Abstract
	Gum tragacanth; from branches and roots; confections; ice cream; liquors; sizings; medicine; drugs; lotions; jellies; dental creams; waterproofing; textile products	11(1):40-63
	Loco weed; poisonous plant	15(2):119-130
	Various portions as food	17(4):319-330
Astragalus ceramicus	Food	8(1):3-20
Astragalus crassicarpus	Seed — oil and protein	20(2):127-155 and #338
Astragalus gummifer	Gum tragacanth; water-soluble gum; textiles; pharmaceutical preparations	3(1):3-31
	Gum tragacanth; spice	8(2):152-163
	Gum tragacanth; hair set; suspension and thickening	20(1):17-30
Astragalus hamosus	Seed — oil and protein	20(2):127-155 and #339
Astragalus nuttalianus	Seed — oil and protein	20(2):127-155 and #340
Astragalus panduratus	Seed — oil and protein	20(2):127-155 and #341
Astragalus racemosus	Seed — oil and protein	20(2):127-155 and #342
Astragalus tragacantha	Gum tragacanth; spice	8(2):152-163
Atriplex	Pollen extracts to treat allergies	5(3):211-254
	Food	8(1):3-20
	Various portions as food	17(4):319-330
Atriplex canescens (salt bush)	Substitute for baking powder; as ashes	8(1):3-20
Atriplex dimorphostegia	Vegetable (edible leaves)	8(2):152-163
Atriplex halimus	Vegetable (edible leaves)	8(2):152-163
Atriplex rosea	Vegetable (edible leaves)	8(2):152-163

PLANTS AND THEIR USES — cont.

Scientific Name	Usage	Citation Data	Scientific Name	Usage	Citation Data
Atriplex tatarica	Vegetable (edible leaves)	8(2):152-163	Boswellia thurifera (Boswellia serrata)	Frankincense; exudate — embalming	14(1):84-104
Atropa belladonna	Belladonna; pharmaceutical fluid extracts; extracts; tinctures; poisonous drug; alkaloids; anodyne	1(3):306-316	Brahea dulcis	Palm leaf; fiber — cordage, hats, mats, basketry, bags, sandals, upholstery, padding	19(1):71-82
	Belladonna; root, herb; alkaloids and glucosides; medicinal preparations	2(1):58-72	Brassica	Mustard; condiment; seed — oil, fertilizers; cover crop, fodder	13(3):196-204
	Alkaloids; drugs; hyoscyamine hypnotic; alleviates seasickness; hyoscine	3(2):215-216 Utilization Abstract		Seed — oil, protein	20(2):127-155 and #215
	Medicine; drugs; various parts of plants have pain-relieving properties	19(2):99-112		Food seasoning	20(3):285-301
Balanites	Edible gum; confections	6(1):23-40	Brassica hirta	Mustard	2(1):58-72
Balanites aegyptiaca	Desert date (edible fruit); seeds used in bread and soup; edible oil	3(4):436-444	Brassica juncea	Seeds — oil; rubefacient; counterirritant; stimulant; condiment	1(1):57-68
	Desert date; pulp — beverage; fruit — beverage	3(4):436-444		Seed — protein and oil	20(2):127-155 and #206
	Edible seeds	6(1):23-40	Brassica nigra	Seeds — oil; rubefacient; counterirritant; stimulant; condiment	1(1):57-68
(Jericho balsam)	Medicinal oils; antiseptics Higleeg, shaashoat; edible fruit	8(2):152-163 22(2):165-177		Seed — condiment, food flavoring	2(1):58-72
Barbarea vulgaris	Seed — oil, protein	20(2):127-155 and #201		Mustard; condiment	8(2):152-163
Barosma	Buchu leaves; urinary disorders	1(4):402-414		Seed — oil, protein	20(2):127-155 and #213
Barosma betulina	Buchu; medicinal — drug; leaves — aromatic, tonic, beverage, diuretic, antiseptic; ornamental plant	15(4):326-331	Brassica rapa	Seed — oil, protein	20(2):127-155 and #214
			Brodiaea	Various portions as food	17(4):319-330
			Bromelia	Fiber	6(3):252-269
Barosma crenulata	Buchu; medicinal — drug; leaves — aromatic, tonic, beverage, diuretic, antiseptic; ornamental plant	15(4):326-331	Bromelia fastuosa	Fiber	6(3):252-269
			Bromelia serra	Fiber	6(3):252-269
Barosma pulchella	Buchu; medicinal — drug; leaves — aromatic, tonic, beverage, diuretic, antiseptic; ornamental plant	15(4):326-331	Bulnesia	Timber, wood	6(3):252-269
			Bulnesia sarmienti	Resin; glaze for ceramics; varnishes	6(3):252-269
				Guaiac wood; essential oil; cosmetics	6(4):355-378
Barosma serratifolia	Buchu; medicinal — drug; leaves — aromatic, tonic, beverage, diuretic, antiseptic; ornamental plant	15(4):326-331	Bumelia laetevirens	Tempesquitles; food plant — edible fruit	20(1):6-16
			Cadaba farinosa	Edible bark	6(1):23-40
Boronia megastigma	Flowers — perfume	2(3):334-338 Utilization Abstract	Caesalpinia tinctoria	Tara; tannin from pods; pectin from seeds	1(2):119-136
	Essential oil	8(4):316-336	Calendula officinalis	Seed — oil and protein	20(2):127-155 and #814
	Seed — oil and protein	20(2):127-155 and #474	Calochortus	Butterfly lily; food	8(1):3-20
Boswellia	Gum olibanum, candlemaking	1(4):402-414		Seed — protein and oil	20(2):127-155 and #79
	Frankincense; aromatic resin	4(3):203-242	Calochortus nuttallii	Various portions as food	17(4):319-330
	Frankincense; incense	4(4):307-316	Cannabis	Fiber	20(1):106-108
Boswellia bhaw dajiana	Olibanum or frankincense perfume	2(3):334-338 Utilization Abstract	Cannabis sativa	Flowering tops — fluid extract; extract — analgesic, narcotic, sedative, cerebral stimulant	1(1):57-68
Boswellia carteri	Olibanum; frankincense; perfume	2(3):334-338 Utilization Abstract		Hemp; oakum; twine; carpet materials; cordage; coarse cloth	1(3):351-352 Utilization Abstract
	Frankincense; spice for food preservation; incense	8(2):152-163		Hemp; fiber — textiles, cordage, twine, packing, oakum	2(2):158-169
	Frankincense; exudate — embalming	14(1):84-104		Indian hemp, marijuana; fiber; hallucinogen; drugs	4(1):85-92
Boswellia frereana	Olibanum or frankincense perfume; in food products, liquors	2(3):334-338 Utilization Abstract		Hemp; fiber for paper	10(2):176-193
Boswellia papyrifera	Frankincense; spice for food preservation; incense	8(2):152-163		Hemp; fiber	15(2):133-139
	Frankincense; exudate — embalming	14(1):84-104		Medicine — drugs; various parts of plant have pain-relieving properties	19(2):99-112
Boswellia thurifera	Frankincense; spice for food preservation, incense	8(2):152-163		Hemp; tear-resistant paper	19(4):394-405

Scientific Name	Usage	Citation Data
Capparis spinosa	Lassaf, kabar, abaar, or caper; edible flower buds — treatment for colds; capers	22(2):165-177
Capsicum	Chili; food, condiment	7(3):214-227
	Uchu (aji); fruit — condiments; phytomorphic representation in ceramics, wood, textiles, stone, and graphics	16(2):106-115
	Chili; edible vegetable, condiment	19(4):323-334
	Food	19(4):335-343
	Edible vegetable	20(1):106-108
	Food seasoning	20(3):285-301
Capsicum frutescens	Fruit — condiment, chili	1(2):119-136
	Paprika — condiment, food flavoring	2(1):58-72
	Chili; food plant — edible fruit, condiment	20(1):6-16
	Ripe fruit — tincture, irritant in dandruff/baldness preparations	20(1):17-30
Carnegiea	Water soluble gum — adhesives; size and cloth stiffener; pharmaceuticals; cosmetics	3(1):3-31
Carnegiea gigantea	Extracts — fibers, medicinal extracts; sapogenins; pharmaceuticals; steroids; alkaloids; foods & food products; resins	13(3):243-260
	Various portions as food	17(4):319-330
Carthamus tinctorius	Florets — infusion, dye, diaphoretic, laxative	1(1):57-68
	Safflower; seed oil for paints, varnishes; flower heads — dyes	2(1):58-72
	Safflower; seeds — oil; meal — animal feed	3(2):143-149
	Safflower; seed — oil	3(4):427 Utilization Abstract
	Safflower oil; meal; livestock feed	9(2):99-107
	Safflower; red dye from flowers; meal; oil; livestock feed; used in resins, paints, varnishes	9(3):273-299
	Safflower; floret-dye	14(1):84-104
	Safflower; seeds — oil, food products, coatings	17(2):139-145
	Safflower; seeds — linoleic acid, iodine, oleic acid	19(1):53-62
	Safflower; seeds — oil	21(2):156-162
	Safflower; seeds — oil, oleic acid	22(2):195-200
Carum carvi	Caraway; seed — condiment, food flavoring	2(1):58-72
	Various portions as food	17(4):319-330
	Caraway; seeds — oil, perfume	20(1):17-30
Cassia acutifolia	Senna; leaf — embalming	14(1):84-104
Cassia angustifolia	Senna; pods — purgative	1(4):402-414
	Senna; leaf — embalming	14(1):84-104
Cassia alata	Seed — oil and protein	20(2):127-155 and #349
Cassia biflora	Seed — protein and oil	20(2):127-155
Cassia corymbosa	Seed — protein and oil	20(2):127-155
Cassia covesii	Seed — protein and oil	20(2):127-155

Scientific Name	Usage	Citation Data
Cassia durangensia	Seed — protein and oil	20(2):127-155
Cassia hirsuta	Seed — protein and oil	20(2):127-155
Cassia italica or Cassia obovata	Sanna (mekki), senna, samaleika; leaves — purgative tea	22(2):165-177
Cassia leptocarpa	Seed — oil and protein	20(2):127-155 and #357
Cassia roemeriana	Seed — protein and oil	20(2):127-155
Cassia senna	Senna; pods — purgative	1(4):402-414
Catha edulis	Ch'at; leaves — stimulant	14(4):334-335 Book Review
	Miraa; edible leaves and fruit; leaves — tea, stimulant; wood — paper pulp, building materials	21(4):358-362
Celtis	Hackberry; pollen extracts to treat allergies	5(3):211-254
	Various portions as food	17(4):319-330
Celtis australis	Seed and pericarp — oil, protein	20(2):127-155 and #105
Celtis occidentalis	Seed and pericarp — oil, protein	20(2):127-155 and #106
Centaurea cyanus	Entire plant — infusion, mild astringent	1(1):57-68
	Seed and pericarp — oil and protein	20(2):127-155 and #822
Ceratonia siliqua	Locust bean; seeds — flour for mannogalactan gum; vegetable gum; textiles; foods	2(2):223 Utilization Abstract
	Carob, locust bean; seeds — water soluble gum in textiles; fruit — laxative, diuretic, tobacco flavoring, beverages, molasses, syrup; pods — forage crop	3(1):3-31
	Carob; seeds — gum in cosmetics, food products, pharmaceuticals, inks, insecticides; pods — animal feed; ornamental tree	3(4):406 Utilization Abstract
	Carob; pods — livestock forage crop; shade tree; seeds — gum; soil conservation; human food; alcohol; algarroba	5(1):82-96
	Carob pods; vegetable; livestock feed; emergency human consumption	8(2):152-163
	Locust bean; tannins and resins; drilling muds	9(1):93 Utilization Abstract
	Carob; mucilage from seeds, hair set	20(1):17-30
Cercidium floridum	Seed — oil and protein	20(2):127-155 and #367
Cercidium torreyanum	Seed — oil and protein	20(2):127-155 and #368
Cereus	Fruit, food	6(3):252-269
Cereus (Carengiea) gigantea	Saguaro cactus; saguaro fruit; beverages; syrup; alkaloids	3(2):111-131
Cereus grandiflora	Extracts — fibers, medicinal extracts; sapogenins; pharmaceuticals — steroids, alkaloids; foods and food products; resins	13(3):243-260

PLANTS AND THEIR USES — cont.

Scientific Name	Usage	Citation Data
Cereus repandus	Cadushi; stems — food, torches, fuel, fences; trunks — boards	21(2):185-191
Chenopodia- ceae (family)	Goosefoot, chenopod; pollen extracts to treat allergies	5(3):211-254
Chenopodium ambrosioides	Mexican tea, anthelmintic	8(1):3-20
	Leaves — medicinal use for skin disease	21(3):243-272
Chenopodium leptophyllum	Seed and pericarp — oil, protein	20(2):127-155 and #122
Chloris gayana	Caryopsis — oil, protein	20(2):127-155 and #30
Chloris virgata	Caryopsis — oil, protein	20(2):127-155 and #31
Chrysanthe- mum	Pyrethrum	1(2):119-136
Chrysanthe- mum coronarium	Seed and pericarp — oil and protein	20(2):127-155 and # 833
Chrysanthe- mum leucan- themum	Pyrethrum, insecticide Seed and pericarp — oil and protein	1(4):402-414 20(2):127-155 and #834
Chrysoba- lanus icaca	Fruit, food	6(3):252-269
Chryso- thamnus	Rabbitbrush; flowers — yellow dye; inner bark — green dye	8(1):3-20
Chryso- thamnus nauseosus	Rabbitbrush; latex	8(1):3-20
Chryso- thamnus vescidiflorus	Latex	8(1):3-20
Cicer arietinum	Garbanzo; food plant — edible seeds	20(1):6-16
Cichorium	Inulin	20(1):106-108
Cichorium endivia	Endive; edible herbs	8(2):152-163
Cichorium intybus	Rhizomes, roots —infusion; simple bitter; laxative; diuretic	1(1):57-68
	Chicory; coffee substitute; food	8(1):3-20
	Chicory; edible herbs; aphrodisiac	8(2):152-163
Cistus	Labdanum; branches — gum; resin — perfume	2(3):33-338 Utilization Abstract
Cistus salvifolius	Labdanum; spice	8(2):152-163
Cistus villosus	Labdanum; spice	8(2):152-163
Citrullus	Edible vegetables	20(1):106-108
Citrullus colocynthis	Gall (colocynth); cathartic Colocynth; fruit — embalming	8(2):152-163 14(1):84-104
Citrullus vulgaris	Watermelon — edible fruit	3(2):193-212
	Watermelon; edible cooked fruit — used in soups, sauces, and as cooking oil; seeds, fermented, make a food or flavoring; provides water as well as food	3(4):436-444
	Edible seeds	6(1):23-40
	Edible fruit	20(3):285-301
	Watermelon; vegetable; food; beverage; medicine; edible seeds	8(2):152-163
Citrus	Citrus fruits	3(4):436-444
	Edible fruits, seeds	20(1):106-108
	Edible fruit	20(3):285-301

Scientific Name	Usage	Citation Data
Citrus aurantium	Bigaradia, sour orange; blossoms; orange flower — oil, perfumes	2(3):334-338 Utilization Abstract
	Petitgrain; essential oil; cosmetics; soaps	6(4):355-378
	Laraha; medicinal plant for folk medicine; leaf decoction (tea) for various ailments	22(1):87-102
Citrus medica	Etrog, or citron; spice	8(2):152-163
Citrus sinensis	Sweet orange; essential oil	3(1):104-106 Utilization Abstract
	Petitgrain; essential oil; cosmetics; soaps	6(4):355-378
Claviceps	Plant drug; medicine, drugs	20(1):115 Book Review
Cleome	Beeweed; food	8(1):3-20
Cleome serrulata	Stinking clover; black dye Seed — oil, protein	8(1):3-20 20(2):127-155 and #185
Clitoria ternatea(L.)	Seed — protein and oil	20(2):127-155 and #369
Cnidoscolus elasticus	Chilte; latex — gum extender or elastomer with gutta gums to make rubber products	16(2):53-70
Cocculus pendulus	Edible flowers	6(1):23-40
Cocos genus	Palm tree; edible nuts, palm cabbage (food)	6(3):252-269
Cocos botryphora	Miscellaneous purposes	6(3):252-269
Cocos coronata	Miscellaneous purposes Curicuri; wax; polishes, carbon paper	6(3):252-269 7(3):285-286 Utilization Abstract
Cocos nucifera	Cocoanut palm; edible terminal buds and palm cabbages	6(1):23-40
	Oil	20(1):106-108
Colchicum	Plant drug; medicine, drugs	20(1):115 Book Review
Colocynthis vulgaris or Citrullus colocynthis	Handal (ground gourd); edible gourd and seeds; seeds — cathartic tar; gourds — tinder	22(2):165-177
Combretum	Edible tubers and corms	6(1):23-40
Commiphora	Myrrh; aromatic resin	4(3):203-242
Commiphora abyssinica	Myrrh or Balm of Gilead; exudate — embalming	14(1):84-104
Commiphora africana	Bdellium; spice for food preservation; incense	8(2):152-163
Commiphora erythraea	Myrrh or Balm of Gilead; exudate — embalming	14(1):84-104
Commiphora myrrha	Bitter or male myrrh; perfume	2(3):334-338 Utilization Abstract
	Myrrh; spice for food preservation; incense	8(2):152-163
	Myrrh or Balm of Gilead; exudate — embalming	14(1):84-104
Commiphora (Balsamo- dendron) opobalsamum	Myrrh or Balm of Gilead; exudate — embalming	14(1):84-104
Commiphora opobalsamum	Balm of Gilead; medicinal oils; antiseptics	8(2):152-163
Condalia	Seed — oil and protein	20(2):127-155 and #525
Conringia orientalis	Seed — oil and protein	20(2):127-155 and #222
Copernicia australis	Caranday; hard vegetable palm wax	9(1):39-52

Scientific Name	Usage	Citation Data
Cordia boissieri	Anacahuata; extracts from leaves and stems	14(2):157-159
Cordylanthus wrightii	Clubflower; skin bleach	8(1):3-20
Coreopsis basilis	Seed and pericarp — oil and protein	20(2):127-155 and #839
Coreopsis grandiflora	Seed and pericarp — oil and protein	20(2):127-155 and #840
Coriandrum sativum	Fruit, oil; aromatic stimulant; corrective; condiment	1(1):57-68
	Coriander; seed — flavoring	1(4):402-414
	Coriander; seed — condiment, food flavoring	2(1):58-72
	Coriander; essential oil	3(1):104-106 Utilization Abstract
	Coriander; condiment; seeds for flavoring; bread ingredient	8(2):152-163
	Seed and pericarp — oil and protein	20(2):127-155 and #607
Cotoneaster acuminata	Seed — oil and protein	20(2):127-155 and #297
Coursetia microphylla	Lac; sealing agent	8(1):3-20
Crambe abyssinica	Abyssinian kale; seeds — oil, erucic acid for synthetic fibers, polyesters, plasticizers, lubricants, alkyd resins	17(1):23-30
	Seed and pericarp — oil, protein	20(2):127-155 and #224
Crocus sativus	Saffron; stigma — in food products, liquors, perfume	2(3):334-338 Utilization Abstract
	Saffron; condiment; coloring	8(2):152-163
	Saffron; stigma — dye	14(1):84-104
	Saffron crocus; dyestuff from dried stigmas and styles; hair dye	20(1):17-30
	Saffron; dried stigmas of flowers — food, flavoring, coloring, medicinal properties	20(4):377-385
Croton fragilis	Seed — oil and protein	20(2):127-155 and #485
Croton texensis Klotzsch. (skunkweed)	Eyewash; emetic; household insecticide	8(1):3-20
Cucumis melo	Melon; edible cooked fruit (seeds eaten also)	3(4):436-444
	Edible seeds	6(1):23-40
	Muskmelon; vegetable	8(2):152-163
	Cantaloupe; edible fruit	21(4):345-350
Cucurbita	Edible seeds	1(2):239, 240 Utilization Abstract
	Seeds — glutamine	2(2):219 Utilization Abstract
	Gourds, saponins	8(1):3-20
	Wild gourds — oil, protein	9(1):93 Utilization Abstract
	Various portions as food	17(4):319-330
	Cucurbits; food; seeds — oil; gourds — as container; vegetable sponge — sponge products	18(3):279-280 Book Review
	Fiber — cordage, hats, mats, basketry, bags, sandals, upholstery, padding	19(1):71-82
	Squash; edible vegetable	19(4):323-334
	Food	19(4):335-343

Scientific Name	Usage	Citation Data
	Cucurbits; food — fruit, green vegetable, edible seeds	19(4):344-349
	Food	19(4):358-368
	Edible vegetable	20(1):106-108
	Food seasoning	20(3):285-301
	Squash and pumpkins; edible plants	23(1):2-19
Cucurbita digitata	Wild gourds; seeds — food	3(2):111-131
	Cucurbits; seeds — oil	5(1):38-59
	Seed — oil	8(1):3-20
Cucurbita foetidissima	Wild gourds; seeds — food	3(2):111-131
	Seeds — oil	5(1):38-59
	Fetid wild pumpkin; food; seed — oil	8(1):3-20
	Seed — oil and protein	20(2):127-155 and #764
	Gourd; edible fruit; seeds — oil	22(3):297-299
Cucurbita maxima	Edible flowers	6(1):23-40
	Squash; edible vegetable	18(1):92-93 Book Review
	Squash; edible vegetable	22(3):253-266
Cucurbita moschata	Pumpkins; food	7(1):95-96 Book Review
	Zapallo; edible fruits; phytomorphic representation in ceramics, wood, textile, stone, and graphics	16(2):106-115
	Squash; food plant — edible fruit	20(1):6-16
	Winter squash; edible vegetable	22(3):253-266
Cucurbita palmata	Wild gourds; seeds — food	3(2):111-131
	Coyote melon; seed — oil	8(1):3-20
Cuminum cyminum	Cumin; condiment, seeds for flavoring; bread ingredient	8(2):152-163
Cyanopsis tetragonoloba	Guar; vegetable gum from seeds; food; cattle fodder; paper; textile products	2(2):223 Utilization Abstract
	Guar; seeds — water-soluble gum	3(1):3-31
	Guar gum; drugs; cosmetics; paper mfg, mining (flocculent and filter aid); food industry; soil-building properties as plant	11(2):159 Utilization Abstract
	Guar; nitrogen-fixing plant; gum, livestock feed	14(3):241-246
Cycloloma atriplicifolium	Pigweed; food	8(1):3-20
Cydonia oblonga	Quince; seeds — water-soluble gum, demulcent in medicinal and cosmetic preparations	3(1):3-31
	Quince seed; tannins and resins; drilling muds	9(1):93 Utilization Abstract
Cymopterus	Various portions as food	17(4):319-330
Cymopterus bulbosus	Flavoring	8(1):3-20
Cynara cardunculus (L.)	Seed and pericarp — oil and protein	20(2):127-155 and #844
Cynara scolymus	Seed and pericarp — oil and protein	20(2):127-155 and #845

PLANTS AND THEIR USES — cont.

Scientific Name	Usage	Citation Data
Cynodon dactylon	Bermuda grass; pollen extracts to treat allergies	5(3):211-254
	Bermuda grass; lawns; hay; forage; possible use of leaves in blood-pressure reduction	20(1):94-97
Cynomorium coccineum	Cynomorium; vegetable; emergency food	8(2):152-163
Cyperus esculentus	Spanish beverage from tubers	1(3):243-275
	Tiger nut; earth-almond, rush-nut; tubers — candy	3(4):436-444
	Edible tubers and corms	6(1):23-40
Cyperus papyrus	Papyrus; construction paper; boats; mats	8(2):152-163
	Papyrus; fiber for paper	10(2):176-193
	Bulrush; pith of stem — embalming	14(1):84-104
Cyperus rotundus	Edible tubers and corms	6(1):23-40
Dactyloctenium aegyptiacum	Crowfoot grass; edible seeds	7(1):95-96 Book Review
Dasylirion	Various portions as food	17(4):319-330
Datura meteloides	Narcotic	8(1):3-20
	Ololiuque; seed — hallucinogen	14(4):257-262
	Medicine, drug; various parts of plant have pain-relieving properties	19(2):99-112
Datura stramonium	Leaves, flowering tops; extract; fluid extract; tincture; relaxes bronchial muscles; asthma; anodyne	1(1):57-68
	Stramonium; pharmaceutical; fluid extracts; extracts; tinctures; poisonous drug; alkaloids; anodyne	1(3):306-316
	Stramonium; leaves, tops — alkaloids and glucosides; medicinal preparations	2(1):58-72
	Jimson weed; poisonous plant; atropine	8(1):3-20
	Medicine, drug; various parts of plant have pain-relieving properties	19(2):99-112
	Seed — oil and protein	20(2):127-155 and #712
Datura tatula	Leaves, flowering tops; extract; fluid extract; tincture; relaxes bronchial muscles; asthma; anodyne	1(1):57-68
Daucus pucillus	Wild carrot; food	8(1):3-20
Descurainia	Tansy-mustard; food	8(1):3-20
Descurainia sophia	Seed — oil, protein	20(2):127-155 and #228
Dictamnus albus (L.)	Seed — oil and protein	20(2):127-155 and #475
Dorema ammoniacum	Ammoniacum; gum; resin; medicinal	4(4):307-316
Duboisia hopwoodii	Leaves — alkaloid, insecticide	1(4):437-445
	Nicotine	3(2):215-216 Utilization Abstract
	Pituri; alkaloid; scopolamine; drugs	4(1):85-92
	Pituri; alkaloids; drugs; medicinal; nicotine	6(1):3-17
	Medicine, drug — various parts of plant have pain-relieving properties	19(2):99-112

Scientific Name	Usage	Citation Data
Duboisia myoporoides	Alkaloids; drugs — hyoscyamine, hyoscine; hypnotic; alleviates seasickness	3(2):215-216 Utilization Abstract
	Pituri; alkaloids; medicinal; drugs; nicotine	6(1):3-17
	Medicine, drug; various parts of plant have pain-relieving properties	19(2):99-112
Echinocactus	Cactus; various portions as food	17(4):319-330
Echinocactus lewinii (also Mammillaria williamsii)	Mexican peyote; roots — alkaloids, hallucinogen; ceremonial and medicinal use	3(2):111-131
Echinocactus wislizeni	Extracts — fibers; medicinal extracts — sapogenins; pharmaceuticals — steroids, alkaloids; foods and food products; resins	13(3):243-260
Echinocereus	Hedgehog cactus; conserves	8(1):3-20
	Various portions as food	17(4):319-330
Elaeagnus angustifolia	Elaeagnus; fuel and construction	8(2):152-163
Ephedra	Mormon tea, ephedrine beverage, red dye	8(1):3-20
	Plant drug; medicine, drugs	20(1):115 Book Review
Ephedra gerardiana	Ephedra; medicine, drug; ephedrine	11(3):257-262
Ephedra nevadensis	Various portions as food	17(4):319-330
Ephedra torreyana	Mormon tea; diuretic	8(1):3-20
Equisetum laevigatum	Various portions as food	17(4):319-330
Eragrostis tef	Teff; grain	14(4):334-335 Book Review
	Teff; seeds — food, flour, bread	16(2):127-130
	Teff; seeds — source of iron and calcium cereal; hay crop	20(3):268-273
Erigeron canadensis	Horseweed; for diarrhea and dysentery	8(1):3-20
Erigon affinis	Extracts — fibers; medicinal extracts — sapogenins; pharmaceuticals — steroids, alkaloids; foods and food products	13(3):243-260
Eriodicyton californicum	Leaves; fluid extract; aromatic syrup; expectorant; mask bitter tastes	1(1):57-68
Eriogonum	Various portions as food	17(4):319-330
Erodium	Heronbill; tannins	8(1):3-20
Eruca sativa	Mustard; condiment; seed — oil; fertilizers; cover crop; fodder	13(3):196-204
	Seed — protein and oil	20(2):127-155 and #232
Eschscholzia californica	Various portions as food	17(4):319-330
	California poppy; seed — oil, protein	20(2):127-155 and #182
Eucalyptus astringens	Eucalyptus; tannin	8(2):163 Utilization Abstract
Eucalyptus globulus	Eucalyptus; in candies, lotions, ointments, inhalants, gargles, cough preparations, toothpastes, soaps, sprays, cleansers, disinfectants	1(3):350-351 Utilization Abstract

Scientific Name	Usage	Citation Data	Scientific Name	Usage	Citation Data
	Eucalyptus; oil — pharmaceuticals	1(4):402-414	Fumaria officinales	Leaves and juice for infusion — tonic, diuretic, laxative, cholagogue	1(1):57-68
	Eucalyptus; essential oil	3(1):104-106 Utilization Abstract	Furcraea gigantea	Furcraea; fiber	2(2):158-169
	Fiber for paper	10(2):176-193	Gaillardio pinnatifida	Diuretic	8(1):3-20
	Fresh leaves —oil, perfume	20(1):17-30			
Eucarya spicata	Essential oil	8(4):316-336	Garrya fremontii	Leaves for infusion —tonic, antiperiodic	1(1):57-68
Euphorbia	Candelilla wax	9(1):93 Utilization Abstract	Genista canariensis	Canary Island broom; flowers; medicine, drug — minor psychedelic	19(4):383
Euphorbia antisyphilitica	Candelilla wax; waxes; substitutes for carnauba wax	6(1):17	Glycine	Protein	20(1):106-108
	Candelilla; wax; polish	7(3):285-286 Utilization Abstract	Glycyrrhiza glabra	Licorice root; herb — aqueous extract for stabilizer in foam liquid for fire fighting	1(3):275 Utilization Abstract
	Candelilla wax	9(2):99-107		Licorice; extract; chewing gum; candies; tobacco industry; beverage; flavoring; medicinal flavoring; foam; fire extinguishers	12(1):86
	Candelilla waxes; chewing gum; ointments; linoleum polishes; coatings; finishes; candles; inks; crayons; adhesives; cement; carbon paper, phonograph records; plastics; water-proofing agents; lubricant; cosmetics	10(2):134-154			
Euphorbia cerifera (Euphorbia antisyphilitica)	Candelilla wax; polishing materials; chewing gum	8(2):113 Utilization Abstract	Glycyrrhiza lepidota	Licorice; glycyrrhizin	8(1):3-20
			Gossypium barbadense	Fiber — cordage, hats, mats, basketry, bags, upholstery, padding, sandals	19(1):71-82
Ferocactus	Various portions as food	17(4):319-330		Barbados cotton; fiber	22(3):253-266
Ferocactus wislizeni	Barrel cactus; candy	3(2):111-131	Gossypium herbaceum	Cotton; textile; cotton cloth	8(2):152-163
Ferula	Asafetida; medical and aromatic resin	4(3):203-242		Fiber — cordage, hats, mats, basketry, bags, sandals, upholstery, padding	19(1):71-82
	Galbanum; gum; medicinal; asafetida	4(4):307-310	Gossypium hirsutum	Cotton; fiber for paper	10(2):176-193
Ferula assafoetida	Asafetida; gum resin; essential oil	3(1):71-83		Fiber — cordage, hats, mats, basketry, bags, upholstery, sandals, padding	19(1):71-82
Ferula galbaniflua	Galbanum; resins and gum resins for perfume	2(3):334-338 Utilization Abstract		Seeds — oil	19(4):323-334 20(3):285-301
	Galbanum; spice for food preservation; incense	8(2):152-163		Cotton; fiber	22(3):253-266
	Seed and pericarp — oil and protein	20(2):127-155 and #614	Gourliaea decorticans	Drink — beer	6(3):252-269
Ferula gummosa	Galbanum; resins and gum resins for perfume	2(3):334-338	Grindelia hirsutula	Leaves and flowering tops — fluid extract; stimulating expectorant in bronchitis	1(1):57-68
Ferula rubricaulis	Galbanum; resins and gum resins for perfume	2(3):334-338 Utilization Abstract	Grindelia robusta	Leaves and flowering tops — fluid extract; stimulating expectorant in bronchitis	1(1):57-68
Ferula schair	Galbanum; resins and gum resins for perfume	2(3):334-338 Utilization Abstract	Grindelia squarrosa	Leaves and flowering tops — fluid extract; stimulating expectorant in bronchitis	1(1):57-68
Ficus carica	Fig; shade tree; leaves; basketry; juice; medicinal poultice; food plant	8(2):152-163		Gumweed; asthma treatment; antispasmodic, stomachic	8(1):3-20
	Fig; edible fruit	19(2):124-135	Haplophyton cimicidum	Cockroach plant; leaves — alkaloids; insecticide	1(4):437-445
Ficus sycomorus	Fig; construction; shade tree; leaves; basketry; juice; medicinal poultice; food plant	8(2):152-163	Harrisia	Various portions as food	17(4):319-330
	Egyptian fig; wood for coffins	14(1):84-104	Helenium autumnale	Bitterweed; poisonous plant	15(2):119-130
	Fig; edible fruit	19(2):124-135	Helianthus (sunflower)	. .	7(1):95-96 Book Review
	Sycamore; edible fruit	22(2):178-190			
Flourensia cernua	Tarbush; for indigestion	8(1):3-20	Helianthus annuus	Flowers, seeds, leaves; tincture; bitter tonic; astringent	1(1):57-68
Foeniculum vulgare	Leaves; oil; stimulant; carminative; galactagogue; condiment	1(1):57-68		Seeds — food and edible oil	1(1):114
Fouquieria splendens	Ocotillo; relief of fatigue, swelling; wax	8(1):3-20		Seeds — oil, food, purple and black dye; leaves — yellow dye	8(1):3-20
Franseria deltoidea	Rabbit bush; pollen extracts to treat allergies	5(3):211-254			
Fremontia californica	Outer and inner bark for poultice; demulcent	1(1):57-68		Edible fruit, nuts and vegetables	19(4):323-334

PLANTS AND THEIR USES — cont.

Scientific Name	Usage	Citation Data
Helianthus annuus	Seed — oil for hair oil and shampoos	20(1):17-30
	Seed and pericarp — oil and protein	20(2):127-155 and #890
	Food plant	21(3):199-214
Helianthus ciliaris (DC.)	Seed — oil and protein	20(2):127-155 and #891
Helianthus tuberosus	Inulin	20(1):106-108
	Food plants	21(3):199-214
Hoffman-seggia densiflora	Hog potato; food	8(1):3-20
Hordeum agriocrithon	Barley; food — hay, feed, straw, cover crop, malt	7(1):3-26
Hordeum deficiens	Barley; food — hay, feed, straw, cover crop, malt	7(1):3-26
Hordeum distichon	Barley; food — hay, feed, straw, cover crop, malt	7(1):3-26
	Barley; food plant	8(2):152-163
	Barley	12(2):192-204
Hordeum hexastichon	Barley; food plant	8(2):152-163
Hordeum irregulare	Barley; food — hay, feed, straw, cover crop, malt	7(1):3-26
Hordeum jubatum	Squirrel-tail grass; poisonous plant	15(2):119-130
Hordeum sativum	Barley; fiber for paper	10(2):176-193
	Barley; printing and writing papers; corrugating board; boxboard and structure boards	19(4):394-405
Hordeum vulgare	Barley; cereal	3(4):436-444
	Barley; food — hay, feed, straw, cover crop, malt	7(1):3-26
	Barley; enzyme production; malt	8(2):99-113
	Barley; food plant	8(2):152-163
	Barley	12(2):192-204
Hydrophyllum occidentale	Squaw lettuce; food	8(1):3-20
Hylocereus undatus	Food plant — edible fruit	20(1):6-16
Hymenopappus lugens	Toothache remedy; emetic	8(1):3-20
Hyoscyamus muticus	Alkaloids — atropine, scopolamine	1(4):402-414
	Henbane; alkaloids and glucosides, medicinal preparations	2(1):58-72
	Sakaraan; leaves — tea, tobacco	22(2):165-177
Hyoscyamus niger	Henbane; leaves; pharmaceutical; poisonous drug; alkaloids; sedative; anodyne; fluid extracts; extracts; tinctures	1(3):306-316
	Henbane; alkaloids and glucosides, medicinal preparations	2(1):58-72
	Black henbane	15(2):133-139
	Medicine, drug; various parts of plant have pain-relieving properties	19(2):99-112
Hyoscyamus reticulatus (L.)	Seed — oil and protein	20(2):127-155 and #713
Hypericum perforatum	Extracts; fibers; medicinal extracts — sapogenins; pharmaceuticals — steroids, alkaloids; foods and food products; resins	13(3):243-260
Hyphaene thebaica	Gingerbread palm; food — terminal buds, palm cabbages, edible fruit	6(1):23-40

Scientific Name	Usage	Citation Data
Hyphaene thebaica (Cucifera thebaica)	Doam, dom palm or gingerbread tree; edible fruit, beverage	22(2):165-177
Hyssopus officinalis	Hyssop; perfume; aromatic flavoring and spirits	4(1):3-36
Iberis amara	Seed — oil, protein	20(2):127-155 and #234
Idria columnaris (elephant tree)	. .	19(4):427-428 Book Review
Jatropha spathulata	Sangre de drago; extracts from	14(2):157-159
Juniperus mexicana (mountain cedar)	Oil — leather dressings, perfumes, polishes	1(2):136 Utilization Abstract
	Cedarwood; essential oil	3(1):104-106 Utilization Abstract
	Cedarwood; wood; oil — in greases, inks, polishes, insecticides, leather products, and aromatic in perfumes and soaps	3(2):217-218 Utilization Abstract
Juniperus oxycedrus	Juniper oil; hair and scalp preparations for dandruff and baldness	20(1):17-30
Krameria parviflora	Range ratany; roots — dye; preparation for sore eyes	8(1):3-20
Larrea divaricata	Creosote; anti-oxidant; resins; cattle feed	6(3):270 Utilization Abstract
Larrea divaricata (creosote bush)	Preservative for fats and oils, resin, varnish	8(1):3-20
	Antioxidant — nordihydroguaiaretic acid	9(1):93 Utilization Abstract
	Seed — protein and oil	20(2):127-155 and #472
Larrea tridentata	Creosote bush; resins	3(2):111-131
	Wax	13(1):29 Utilization Abstract
Lathyrus silvestris	Plat pea; seeds — amino acids	17(2):107-109
Launaea capitata (L. glomerata)	El huwwaia (dandelion); edible leaves	22(2):165-177
Lavandula	Seed — oil and protein	20(2):127-155 and #687
Lavandula latifolia	Spike, spike lavender; leaves — essential oil for scenting soaps, cosmetics	1(3):350-351 Utilization Abstract
	Spike; essential oil	3(1):71-83
Lecanora esculenta	"Manna"; emergency food; vegetable	8(2):152-163
Lemaireocereus griseus	Datu; edible stems and fruit; living fences	21(2):185-191
Lemaireocereus thurberi	Cactus; various portions as food	17(4):319-330
Lepidium draba	Seed — oil, protein	20(2):127-155 and #239
Lepidium lasiocarpum	Seed — oil, protein	20(2):127-155 and #240
Leptadenia lancifolia	Edible flowers	6(1):23-40
Lesquerella angustifolia	Seed — oil, protein	20(2):127-155 and #244
Lesquerella argyraea	Lesquerella; seeds — oil	16(2):95-100
	Seed — oil, protein	20(2):127-155 and #245
Lesquerella densipila	Seed — oil, protein	20(2):127-155 and #247
Lesquerella engelmannii	Seed — protein and oil	20(2):127-155
Lesquerella fendleri	Lesquerella; seeds — oil	16(2):95-100
	Lesquerella; seeds — oil	16(3):206-211

Scientific Name	Usage	Citation Data
	Seed — oil, protein	20(2):127-155 and #251
Lobivia	Hahuaccellai; food (tunas)	16(2):106-115
Lophophora williamsii	Peyote; mezcal; roots — alkaloids, hallucinogen, ceremonial, medicinal use	3(2):111-131
	Peyotl; poison; medicine; drugs	4(1):85-92
	Peyote; religious rites; curative properties; extract — alkaloids, antibiotic	14(3):247-249
	Peyote; hallucinogen	14(4):257-262
	Peyote; drug — hallucinogen	19(4):429-330 Book Review
Lupinus albus	Lupine; seeds — asparagine, chemical nitrogen	2(2):219 Utilization Abstract
Lupinus termis	Lupine seed; alkaloids	21(4):367-370
Lycium pallidum	Rabbit thorn, food	8(1):3-20
Lygeum spartium	Esparto grass; printing and writing papers	19(4):394-405
Majorana hortensis	Sweet marjoram; leaves and flowers — oil, perfume	20(1):17-30
Mandragora officinarum	Mandrake; emetic; purgative; narcotic; aphrodisiac	8(2):152-163
	Mandrake; medicine, drugs; various parts of plant have pain-relieving properties	19(2):99-112
Martynia	Various portions as food	17(4):319-330
Martynia parviflora Wooten	Devil's claw; fiber, basket weaving, seed, oil	8(1):3-20
Martynia parviflora	Devil's claw; edible oil, protein	9(1):93 Utilization Abstract
Matricaria chamomilla	Chamomile; essential oil	3(1):71-83
Medicago sativa	Alfalfa; forage crop; hay; meal; livestock and poultry feed	3(2):170-183
Melia azedarach	Canelo; extracts from flowers	14(2):157-159
Melilotus indicus	Seed — oil and protein	20(2):127-155 and #426
Melilotus officinalis	Seed — oil and protein	20(2):127-155 and #427
Mentzelia albicaulis	Various portions as food	17(4):319-330
Mentzelia aspera	Dal pega; leaves — decoction; beverage; medicinal plant for folk medicine	22(1):87-102
Mimosa dysocarpa	Fish stupefaction plant	12(1):95-102
Mimusops globosa	Gum for fastening points and feathers to arrows	6(3):252-269
Monolepis nuttalliana	Poverty weed; food	8(1):3-20
Myrtus communis	Myrtle; fuel and construction	8(2):152-163
Nerium oleander	Seed — oil and protein	20(2):127-155 and #648
	Ornamental plant	20(3):285-301
Nicotiana attenuata	Tobacco	8(1):3-20
Nicotiana glauca	Tree tobacco; insecticide	1(4):437-445
	Tree tobacco; insecticide	8(1):3-20
Nicotiana trigonophylla	Tobacco	8(1):3-20

Scientific Name	Usage	Citation Data
Nigella sativa	Nutmeg; flower; condiment	8(2):152-163
Nitraria retusa (Nitraria tridentata)	Ghardaq; leaves — tea, medicinal beverage	22(2):165-177
Nolina durangensis	Seed — oil, protein	20(2):127-155 and #82
Nolina microcarpa	Beargrass; fiber; brooms	8(1):3-20
	Beargrass; broom straw	9(1):93 Utilization Abstract
Nolina texana	Sacahuiste; cellulose; fibers — heavy paper, brooms, basket materials, sisal	3(2):111-131
	Medicinal, drugs — steroidal sapogenins from seeds, leaves	15(2):131-132
Notholacna sinuata	Helechillo; extracts from leaves and stems	14(2):157-159
Olea	Oil	20(1):106-108
Olea europaea	Olive; olives and oil	2(4):341-362
	Olive; oil; fuel; medicinal; food plant	8(2):152-163
	Olive; edible fruit; oil	9(3):228-232
	Olive; ripe fruit — oil, hair oil, shampoos	20(1):17-30
	Olive; medicinal oil; fruit; religious and ceremonial rites	20(3):223-243
Opopanax chironium	Opopanax; perfume	2(3):334-338 Utilization Abstract
Opuntia	Water soluble gum — adhesives, size and cloth stiffener, pharmaceuticals, cosmetics	3(1):3-31
	Seeds; vegetable oil; gum; mucilage	3(2):111-131
	Prickly pear; essential oil, perfume; cattle feed; poultice, rheumatism treatment	8(1):3-20
	Cactus apples; edible fruit, jelly; vegetable — pods (stems)	13(1):66 Utilization Abstract
	Hahuaccellai (tunas)	16(2):106-115
	Tuna; edible fruit	17(3):200-210
	Various portions as food	17(4):319-330
	Fiber — cordage, hats, mats, basketry, bags, sandals, upholstery, padding	19(1):71-82
	Edible fruit	19(4):323-334
	Food	19(4):335-343
	Nopal cruzeta; food plant — leafy vegetable	20(1):6-16
	Food seasoning	20(3):285-301
Opuntia engelmannii (prickly pear)	Edible fruit; stock food; alcoholic beverage; gum — mucilage, laxative	3(2):111-131
Opuntia ficus-indica	Cellulose — fiber	3(2):111-131
	Extracts; fibers; medicinal extracts — sapogenins; pharmaceuticals — steroids, alkaloids; foods and food products; resins	13(3):243-260
Opuntia fulgida	Cholla; gum	3(2):111-131
	Cholla; gum; size; stiffener; contains L-arabinose, D-galactose, and glucosonic acid	8(1):3-20

PLANTS AND THEIR USES — cont.

Scientific Name	Usage	Citation Data	Scientific Name	Usage	Citation Data
Opuntia inermis	Extracts; fibers; medicinal extracts — sapogenins; pharmaceuticals — steroids, alkaloids; foods and food products; resins	13(3):243-260	Pedilanthus pavonis	Candelilla wax; waxes; substitutes for carnauba wax	6(1):17
Opuntia vulgaris	Latex; gum — rubber substitute	3(2):111-131		Jumete; wax; polishing materials; chewing gum	8(2):113 Utilization Abstract
	Extracts; fibers; medicinal extracts — sapogenins; pharmaceuticals — steroids, alkaloids; foods and food products; resins	13(3):243-260	Peganum harmala	Seeds; alkaloids; anthelmintic; narcotic; diuretic; febrifuge	21(3):284
Orchis	Edible tubers and corms	6(1):23-40	Pelargonium graveolens	geranium; essential oils; cosmetics	6(4):355-378
Origanum majorana	Marjoram; condiment, food flavoring	2(1):58-72	Peniocereus greggii	Various portions as food	17(4):319-330
Orobanche	Various portions as food	17(4):319-330	Pennisetum	Bulrush millet; cereal; seeds — beverage	3(4):436-444
Oryzopsis hymenoides	Various portions as food	17(4):319-330	Pennisetum	. .	20(1):106-108
Oryzopsis asperifolia	Various portions as food	17(4):319-330	Pennisetum ciliare	Caryopsis; protein and oil	20(2):127-155 and #40, #45
Pachycereus marginatus (Mexican cactus)	Alkaloids	3(2):111-131	Penstemon fendleri	Seed — oil and protein	20(2):127-155 and #728
Pachycormus discolor	Copalquin or torote blanco	19(4):427-428	Perezia nana	Pipitzohoic acid; alkalinity indicator	8(1):3-20
Panicum hirticaule (panic grass)	Edible seeds	7(1):95-96 Book Review	Perezia wrightii	Pipitzohoic acid; alkalinity indicator	8(1):3-20
Panicum sonorum (panic grass)	Edible seeds	7(1):95-96 Book Review	Pergularia tomentosa	Edible stems and shoots Ghalqa; root — treatment for piles	6(1):23-40 22(2):165-177
Panicum turgidum	Edible seeds Thommaam, shoosh; grass — animal traps	6(1):23-40 22(2):165-177	Petalostemon purpureum	Seed — oil and protein	20(2):127-155 and #440
Papaver rhoeas	Seed — oil, protein	20(2):127-155 and #183	Phaseolus acutifolius	Teparies; food	7(1):95-96 Book Review
Papaver somniferum	Poppy; seed — alkaloids and glucosides; medicinal preparations	2(1):58-72		Tepary beans; food	8(1):3-20
	Poppy; latex, rubber	2(2):198-216	Phoenix dactylifera	Date palm; pollen extracts to treat allergies	5(3):211-254
	Medicine, drugs; various parts of plant have pain-relieving properties	19(2):99-112		Date palm; dates — syrup, liquor, alcohol, vinegar; bud — salad; firewood logs, thatching, rope, basketry	5(3):274-301
Parkinsonia aculeata	Seed — oil and protein	20(2):127-155 and #435		Date; food plant; ornamental leaves; thatching; dusters; basketry; brooms; fiber; rope; beverage; syrup	8(2):152-163
Parthenium	Rubber plants; rubber products	18(2):192		Date palm; fruit	11(2):174-177
Parthenium argentatum	Guayule; latex, rubber	2(2):198-216		Dates; syrups	12(1):41 Utilization Abstract
	Guayule; rubber	3(2):111-131		Date palm; exudate — embalming	14(1):84-104
	Guayule; rubber	5(3):255-273		Tanning; fruit — in beer; apical part — stuffing, cleaning pots	18(4):329-341
	Guayule; rubber	5(4):311-337			
	Guayule; rubber	7(1):75 Utilization Abstract		Date — edible fruit, date honey, date beer; leaves — religious rites, thatching; bast fiber; trunks — beams	21(4):320-340
	Guayule; latex	8(1):3-20			
	Guayule; rubber	9(1):93 Utilization Abstract		Nakhl (date palm); beverage; spines — animal traps	22(2):165-177
	Guayule; rubber	9(2):99-107	Phoenix reclinata	Wild date palm; beverage	3(4):436-444
	Guayule; latex, gums, rubber products	16(1):17-24		Dwarf date palm; terminal buds — food; palm cabbages — edible	6(1):23-40
	Laticiferous	20(1):106-108	Phoradendron californicum (mistletoe)	Food	8(1):3-20
	Latex; rubber	21(2):115-127			
Parthenium incanum	Mariola; latex	8(1):3-20	Photinia salicifolia	Various portions as food	17(4):319-330
Pectis papposa	Essential oil; flavoring Chinchweed; essential oil — food and beverage flavoring	8(1):3-20 3(4):407-412	Phragmites communis	Various portions as food	17(4):319-330
Pedilanthus	Candelilla wax	9(1):93 Utilization Abstract	Pistacia	Turkish mastic; varnish	4(3):203-242

Scientific Name	Usage	Citation Data
Pistacia lentiscus	Mastic; resins and gum; resins for perfume	2(3):334-338 Utilization Abstract
	Lentisk; spice	8(2):152-163
	Gum-mastic; exudate — embalming	14(1):84-104
	Chios mastic; resin; bark; perfume fixer	20(1):17-30
Pistacia terebinthus	Terebinth; fuel and construction	8(2):152-163
Pistacia vera	Pistachio nuts; edible nuts	8(2):152-163
	Pistachio; edible nut; resin	11(4):281-321
	Pistachio; edible nuts	14(2):129-144
Pisum arvense	Field pea; food; cattle feed	12(2):192-204
Plantago	Indian wheat; mucilage	8(1):3-20
Plantago fastigiata	Indian wheat; substitute for psyllium seed	8(1):3-20
Plantago psyllium	Seeds; mucilaginous gum; textile sizing; printing; paper manufacturing; mild laxative	3(1):3-31
	Psyllium; seed — mucilage, hair set	20(1):17-30
Plantago purshii	Indian wheat; substitute for psyllium seed	8(1):3-20
Pluchea sericea	Marsh fleabane; treatment for sore eyes	8(1):3-20
Poa bulbosa	Bulbils — oil, protein	20(2):127-155 and #48
Poliomintha incana	Seasoning; food	8(1):3-20
Populus fremontii	Fremont cottonwood, aspen; twigs; baskets, antiscorbutic	8(1):3-20
Portulaca oleracea	Purslane; food	8(1):3-20
	Purslane; food plant (used as food for pigs)	21(3):243-272
Prosopis	Mesquite gum; water soluble gum — adhesives, size and cloth stiffener, pharmaceuticals, cosmetics	3(1):3-31
	Mesquite; gum — sugars, wax, alcohol	3(2):111-131
	Beer	6(3):252-269
	Various portions as food	17(4):319-330
Prosopis alba	Drink	6(3):252-269
Prosopis algarrobilla	Mesquite; dyes	3(2):111-131
Prosopis chilensis	Algarrabo; living fence	20(4):407-415
Prosopis glandulosa	Mesquite; pollen extracts to treat allergies	5(3):211-254
Prosopis juliflora	Mesquite; seeds — food, gum; fuel; fence posts; lumber; tannin; alcohol	3(2):111-131
	Mesquite; edible seeds and pods	7(1):95-96 Book Review
	Mesquite; gum similar to gum arabic; food; fiber; baskets	8(1):3-20
	Wax	13(1):29 Utilization Abstract
	Extracts; fiber; medicinal extracts — sapogenins; pharmaceuticals — steroids, alkaloids; foods and food products; resins	13(3):243-260
	Seed — oil and protein	20(2):127-155 and #445

Scientific Name	Usage	Citation Data
Prosopis limensis	Algarrobo (thacco); representation in ceramics of Peru	16(2):106-115
Prosopis odorata	Screwbean; edible pods and seeds	7(1):95-96 Book Review
	Screwbean; food; wound treatment	8(1):3-20
Prosopis ruscifolia	Extracts; fibers; medicinal extracts — sapogenins; pharmaceuticals — steroids, alkaloids; foods and food products; resins	13(3):243-260
Prosopis tamarugo	Seed — oil and protein	20(2):127-155 and #446
Punica granatum	Pomegranate; food plant; fruits; beverages; sherbets; dyes, medicine, tanning	8(2):152-163
	Pomegranate; food plant (edible fruit)	20(1):6-16
	Pomegranate; edible fruit; beverage	21(3):215-229
Purshia tridentata	Fish stupefaction plant	12(1):95-102
Quercus ilex	Oak; fuel and construction	8(2):152-163
Reseda odorata	Seed — oil and protein	20(2):127-155 and #285
Rhamnus palaestina	Hedges; fuel and construction	8(2):152-163
Rhus oxycantha (Rhus oxycanthoidee)	Ereen; wood pulp to cure skins	22(2):165-177
Rhus trilobata	Squawbush; berries — mordant in dyeing; fiber; baskets	8(1):3-20
	Seed — oil and protein	20(2):127-155 and #510
Ricinus	Oil	20(1):106-108
Ricinus communis	Bean (seed) — castor oil; purgative, lubricant	1(1):57-68
	Castorbean; seeds — oils in inks, plastics, soaps, ointments; leaves — insecticide; stems — heavy paper products	2(3):273-283
	Castorbean; seeds; oil; castor oil; medicine; lubricating; paints; varnishes	8(1):3-20
	Castorbean; shade	8(2):152-163
	Castorbean; fiber for paper	10(2):176-193
	Castorbean; poisonous plant	15(2):119-130
	Castorbean; seeds — medicine, hemagglutins, blood-typing reagents	18(1):27-33
	Seed — oil	19(1):3-15
	Castorbean; seed — oil in hair spray, shampoos, hair oil and brilliantines	20(1):17-30
	Seed, with or minus seed coat — oil and protein	20(2):127-155 and #499, #500, #501, #502
	Castor-oil plant	21(3):243-272
Rivea corymbosa	Hallucinogen	14(4):257-262
Roccella tinctoria	Lichen; blue dye	8(2):152-163

PLANTS AND THEIR USES — cont.

Scientific Name	Usage	Citation Data	Scientific Name	Usage	Citation Data
Rumex hymeno-sepalus	Canaigre; roots — tanning materials	1(3):333 Utilization Abstract	Sesamum	Oil	20(1):106-108
	Canaigre; roots — tannin	2(1):58-72	Sesamum alatum	Edible seeds	6(1):23-40
	Canaigre; roots — tannin	3(2):111-131	Setaria	. .	20(1):106-108
	Canaigre; tannin	4(1):18-19	Setaria italica	Foxtail millet; edible seeds	7(1):95-96 Book Review
	Canaigre; tannin	5(4):367-377			
	Canaigre; roots — mustard-colored dye; food; tannin	8(1):3-20	Setaria macrostachya	Seeds; oil	19(4):323-334
	Canaigre; tannin (roots)	8(3):286 Utilization Abstract	Sideritis	Leaves — Greek mountain tea	1(3):243-275
	Canaigre; tannin	9(1):93 Utilization Abstract	Sideritis montana	Seed and pericarp — oil and protein	20(2):127-155 and #706
Ruta	Rue oil; vegetable oil; flavoring; perfumes; soaps	1(1):25 Utilization Abstract	Simmondsia californica (jojoba)	Seeds — oil, liquid wax; edible fruits; beverage; livestock feed	1(4):401 Utilization Abstract
Ruta chalepensis	Rue; condiment	8(2):152-163		Seeds — liquid wax, medicinal properties, waxes, polishes, lubricant oil; meal — livestock feed	3(2):131 Utilization Abstract
Ruta graveolens	Rutin; medicinal, drug	6(1):68 Utilization Abstract			
	Rue; condiment	8(2):152-163		Oil from beans; feed; lubricant (wax)	5(1):38-59
Salicornia fruticosa	Saltwort and jointed-glasswort; soaps	8(2):152-163		Liquid wax, seed oil	6(1):41-47
Salsola kali	Saltwort and jointed-glasswort; soaps	8(2):152-163		Wax	21(1):69-80
Salsola pestifer	Russian thistle; pollen extracts to treat allergies	5(3):211-254	Simmondsia chinensis	Seeds — food, cattle forage, waxes	3(2):111-131
Salvadora persica	Edible leaves	6(1):23-40		Seeds — vegetable wax	8(2):165 Utilization Abstract
	Araak; twigs; for toothbrushes	22(2):165-177		Wax	9(1):93 Utilization Abstract
Salvia carduacea Benth.	Seed and pericarp — oil and protein	20(2):127-155 and #695		Seeds; polishes; wax	9(2):99-107
				Edible seeds, oils, waxes	12(3):261-295
Salvia officinalis	Sage; condiment; food flavoring	2(1):58-72		Browse plant; wax; oils; livestock feed; alcohols; acids; in lubricants, cosmetics, food preparations; medicinal properties	12(3):296-304
	Sage; leaves — oil for dandruff and baldness treatment	20(1):17-30			
Salvia sclarea	Clary sage, musky sage; upper stems — perfume	2(3):334-338 Utilization Abstract		Wax	13(1):29 Utilization Abstract
Salvia syriaca	Seed and pericarp — oil and protein	20(2):127-155 and #701		Various portions as food	17(4):319-330
				Seed — oil and protein	20(2):127-155 and #503
Sansevieria deserti	Sansevieria; fiber — cordage, rope, twines	14(3):175-179	Solanum elaeagni-folium (bullnettle)	Enzyme similar to papain	8(1):3-20
Sapindus drummondi	Soapberry; soap	8(1):3-20		Seed — oil and protein	20(2):127-155 and #719
Sapium biloculare	Sapium, yerba de la fleche; latex; poison for arrows, fish, warm-blooded animals	10(4):362-366	Solanum fendleri	Wild potato; food	8(1):3-20
	Fish stupefaction plant	12(1):95-102	Solanum jamesii	Wild potato; food	8(1):3-20
Sarcobatus vermiculatus	Greasewood; acid — antioxidant for food products	3(2):111-131	Sophora secundiflora	Seed minus seed coat — oil and protein	20(2):127-155 and #450
	Extracts; fibers; medicinal extracts — sapogenins; pharmaceuticals — steroids, alkaloids; foods and food products; resins	13(3):243-260	Sorghum almum	Caryopsis —oil, protein	20(2):127-155 and #50
			Sorghum vulgare	Sorghum; edible grain; flour; stalks — fuel, basketry, furniture, mats, fences, brooms, silage, forage, livestock feed; sirup; oil; starch; alcohol; wax	1(4):355-371
	Various portions as food	17(4):319-330			
Schinopsis lorentzii	Quebracho; tanning materials	1(3):333 Utilization Abstract			
	Quebracho; wood, tannin	2(2):217-218 Utilization Abstract		Sorghum; food (grain, bread, cereal), animal fodder; used in beer making	3(3):265-288
	Quebracho; tannin	4(1):3-36			
	Quebracho wood; tannin	9(2):108-140		Sorghum (dhura); food plant; seeds eaten, ground into flour	8(2):152-163
Schinus	Ornamental tree	1(2):119-136			
Scorzonera hispanica (L.)	Seed and pericarp — oil and protein	20(2):127-155 and #954, #955		Durra; grain; sorghum; cereal crop; cattle feed from stocks and leaves; roofing; shelters; fuel	12(2):192-204
Scorzonera tau-saghyz	Latex; rubber	2(2):198-216			
	Laticiferous	20(1):106-108		Sorghum; poisonous plant	15(2):119-130
Selloa glutinosa	Diarrhea treatment	8(1):3-20		Durra; grain; sorghum	18(2):149-157

Scientific Name	Usage	Citation Data
Spartium junceum	Spanish broom; fiber; textiles; cardboard; conveyor belts	12(1):107 Utilization Abstract
	Seed — oil and protein	20(2):127-155 and #452
Sphaeralcea	Globe mallow; mucilage; chewing gum substitute; treatment for bowel disorders	8(1):3-20
Stanleya pinnata	Desert plume; food	8(1):3-20
Stevia ribaudiana	Extracts; fibers; medicinal extracts — sapogenins; pharmaceuticals — steroids, alkaloids; foods and food products; resins	13(3):243-260
Stevia salicifolia	Fish stupefaction plant	12(1):95-102
Stipa tenacissima	Esparto; wax	7(3):285-286 Utilization Abstract
	Esparto grass; fiber for paper	10(2):176-193
	Esparto grass; printing and writing papers	19(4):394-405
Stipa vaseyi	Porcupine grass; fiber; brooms; brushes	8(1):3-20
Suaeda	Seepweed	8(1):3-20
Tamarindus	Edible flowers	6(1):23-40
Tamarindus indica	Tamarind; edible fruit, seeds	3(4):436-444
Tamarix	Tamarisk; wood — coffins	14(1):84-104
Tamarix gallica	Exudation — laxative infusion	1(1):57-68
	Seed — oil and protein	20(2):127-155 and #566
Tamarix mannifera	Gum manna; spice	8(2):152-163
Thelesperma ambiguum	Seed and pericarp — protein and oil	20(2):127-155
Themeda triandra Forsk.	Caryopsis — oil, protein	20(2):127-155 and #53
Thymus capitatus (Coridothymus capitatus)	Origanum; scenting soaps; flavoring foods; disinfectant in oral and other pharmaceuticals	1(3):350-351 Utilization Abstract
Thymus cephalotes	Marjoram; food seasoning	1(3):350-351 Utilization Abstract
Thymus mastichina	Marjoram; food seasoning	1(3):350-351 Utilization Abstract
Thymus vulgaris	Thyme; leaves — oil, perfume	20(1):17-30
	Thyme; germicidal antiseptic; disinfectant in pharmaceuticals and oral preparations; scenting soaps; food flavoring	1(3):350-351 Utilization Abstract
Thymus zygis	Thyme; germicidal antiseptic; disinfectant in pharmaceuticals and oral preparations; scenting soaps; food flavoring	1(3):350-351 Utilization Abstract
Tribulus terrestris	Edible leaves	6(1):23-40
	Gokhru oil	9(2):99-107
	Extracts; fibers; medicinal extracts — sapogenins; pharmaceuticals — steroids, alkaloids; foods and food products; resins	13(3):243-260

Scientific Name	Usage	Citation Data
Trichocereus	Alkaloid; malic acid	3(2):111-131
	Hahuaccellai (tunas)	16(2):106-115
Trichocereus candicans	Alkaloid; anhaline	3(2):111-131
Trichocereus pachanoi	Mescaline cactus (San Pedro); medicinal plant; potion used as folk remedy by healers to cure various ailments	22(2):191-194
Trigonella foenum-graecum	Fenugreek; seeds; extracts; perfume	2(3):334-338 Utilization Abstract
	Leek; vegetable	8(2):152-163
	Fenugreek; food; medicine	12(2):192-204
	Seed — oil and protein	20(2):127-155 and #454
Triticum dicoccoides	Wheat; food plant; flour; bread	8(2):152-163
Triticum monoccum	Wheat; food plant; flour; bread	8(2):152-163
Typha	Fiber — cordage, hats, mats, basketry, bags, sandals, upholstery, padding	19(1):71-82
Typha australis	Pith; food; edible flowers	6(1):23-40
Typha latifolia (cat tail)	Pith; food; edible flowers	19(4):323-334
Verbesina encelioides	Crown-beard; boils and skin disease	8(1):3-20
Vicia leavenworthii	Seed — oil and protein	20(2):127-155 and #458
Vicia villosa	Seed — oil and protein	20(2):127-155 and #462
Washingtonia filifera	Various portions as food	17(4):319-330
Withania somnifera	Ashwagandha; medicinal drug (roots) — sedative, tonic, alkaloids, antispasmodic; leaves — antibiotic, local application	15(3):256-263
Xanthium commune (cockleburr)	Oil for paints, varnishes; diarrhea treatment	8(1):3-20
Yucca	Medicine-drug, cortisone	7(1):75
	Fiber — cordage, hats, mats, basketry, bags, sandals, upholstery, padding	19(1):71-82
Yucca arizonica	Medicinal, drug — steroidal sapogenins from seeds, leaves	15(2):131-132
Yucca baccata (bayonet yucca)	Fiber — paper pulp	3(2):111-131
Yucca baccata (soapweed; spanish bayonet)	Fiber; baskets; mats; sandals; rope; perfume	8(1):3-20
	Medicinal drug — steroidal sapogenins from seeds, leaves	15(2):131-132
Yucca brevifolia (joshua tree)	Medicinal drugs — steroidal sapogenins from seeds, capsules, floral parts, wood	15(1):79-86
	Medicinal drug — steroidal sapogenins from seeds, leaves	15(2):131-132
Yucca elata	Fiber — heavy paper	3(2):111-131

PLANTS AND THEIR USES — cont.

Scientific Name	Usage	Citation Data	Scientific Name	Usage	Citation Data
Yucca elata (soapweed; spanish bayonet)	Fiber; baskets; mats; sandals; rope	8(1):3-20	Yucca mohavensis	Medicinal drug — steroidal sapogenins from seeds, leaves	15(2):131-132
Yucca elata (bear grass)	Fiber for paper	10(2):176-193	Yucca schottii (yucca)	Medicinal drug — steroidal sapogenins from seeds, leaves	15(2):131-132
	Medicinal drug — steroidal sapogenins from seeds, leaves	15(2):131-132		Seed — oil, protein	20(2):127-155 and #88
	Seed — oil, protein	20(2):127-155 and #86	Yucca whipplei	Medicinal drug — steroidal sapogenins from seeds, leaves	15(2):131-132
Yucca glauca (soapweed or bear grass)	Fiber — heavy paper	3(2):111-131	Zizyphus mistol	Beer	6(3):252-269
(soapweed; spanish bayonet)	Fiber; baskets; mats; sandals; rope	8(1):3-20	Zizyphus spina-christi	Edible flowers	6(1):23-40
			Zygophyllum fabago	Seed — oil and protein	20(2):127-155 and #473

REFERENCES AND NOTES

1. Peña, M. T. de la
   1943   Ixtle de lechuguilla. P. 167–171, in Chihuahua económico, tomo II. State of Chihuahua, Mexico. 443 p.

2. Duisberg, P. C.
   1963   Utilización industrial de las plantas del desierto. Unesco Latin American Arid Lands Conference, Buenos Aires, Proceedings 135–169.

3. Unesco
   1960   Medicinal plants of the arid zones. Unesco, Paris. Arid Zone Research 13. 96 p.

4. Lloyd, F. E.
   1911   Guayule (*Parthenium argentatum* Gray), a rubber-plant of the Chihuahuan desert. Carnegie Institution of Washington, Publication 139. 213 p.

5. Duisberg, P. C.
   1953   Chemical components of useful or potentially useful desert plants of North America and the industries derived from them. Desert Research, Proceedings International Symposium, Jerusalem, 1952, sponsored by Research Council of Israel and Unesco. Research Council of Israel, Special Publication 2:281–294.

6. Mesa, A. M., and R. Villanueva V.
   1948   La producción de fibras duras en Mexico. Banco de Mexico, S.A., Monografías Industriales. 572 p.

7. Kidder, M. C., and W. H. Finney
   1948   Commercial development of acacia wattle trees in California as a source of tannin and wood pulp. Conference on the Cultivation of Drug and Associated Economic Plants in California, 1947, Proceedings 3:153–165.

8. Rzedowski, J.
   1957   Vegetación de las partes áridas de los Estados de San Luis Potosí y Zacatecas. Sociedad Mexicana de Historia Natural, Revista 18:49–101.

9. Hare, R. F., and D. Griffiths
   1907   The tuna as a food for man. New Mexico Agricultural Experiment Station, Bulletin 64, 88 p.

10. Hernandez X., E.
    1970   Mexican experience — problems and potentials of arid lands in North America. *In* Arid lands in Transition. American Association for the Advancement of Science, Washington, D.C., Publication 90:317–343.
    Book article adapted from paper given at International Conference, Tucson, Arizona, June, 1969.

11. Earle, F. R., and Q. Jones
    1962   Analysis of seed samples from 113 plant families. Economic Botany 16:221–250.

12. Wall, M. E., and others
    1954– Steroidal sapogenins. Survey of plants for
    1961   steroidal sapogenins and other constituents. American Pharmaceutical Association, Journal, Scientific Edition 43(1):1–7, (8):503–505; 44(7):438–440; 46(11):653–684; 48(12):695–722; 50(12)1001–1034.

13. ————
    1954– Steroidal sapogenins. Supplementary table of
    1955   data. U.S. Department of Agriculture, Agricultural Research Service Circular AIC–363, AIC–367; and ARS–73–74.

14. Gilbert, N. W., and D. S. Black
    1959   Canaigre, potential domestic source of tannin. U.S. Department of Agriculture, Production research report 28. 32 p.

15. Reported by Marilyn Drago in *The Arizona Daily Star*, Section B, page 1, September 19, 1969.

Soil Management: Humid Versus Arid Areas

THOMAS C. TUCKER
and
WALLACE H. FULLER

Department of Agricultural Chemistry and Soils
University of Arizona
Tucson, Arizona, U.S.A.

ABSTRACT

SOIL MANAGEMENT is concerned with decision making. Presumably, these decisions will maximize output with efficient use of production inputs in a given environment. The first step in effective soil management is to consider the basic properties of each soil and its capabilities for production of specific plants.

Problem areas or general soil problems must be identified and plans made for improvement or correction. In humid areas, soil acidity is a common problem that can be corrected by addition of lime, whereas in arid regions accumulations of excess salts and specific toxic ions often require leaching and/or other reclamation procedures for their correction. The quality of irrigation water as well as frequency and amounts to be applied also are important considerations.

All physical manipulations of soil should be planned for maintenance or improvement of soil structure, which controls water and air relationships. The need for plant nutrients must be ascertained and fertilizer programs adjusted to soil properties and cropping practices. All practices must be compatible with a permanent system of crop production — conserving, maintaining, or improving the land and water use for good soil management.

SOIL MANAGEMENT: HUMID VERSUS ARID AREAS

Thomas C. Tucker and Wallace H. Fuller

SOIL AND WATER MANAGEMENT under arid and semi-arid climatic conditions is rivaled in importance only by water availability as a factor in maintaining permanent food and fiber production and the consequent development of a stabilized society. On a worldwide basis, historic relics of agricultural lands that once flourished and since have become barren due to improper management are almost equal in acreage to the irrigated land now in production. The key to the maintenance of a stabilized society in arid and subhumid habitats is proper soil and water management; these elements, by their very nature, must go hand in hand. Soil and water are so interdependent that one cannot be considered without the other. Soil management is aimed at maintaining suitable water infiltration, percolation, aeration, salt balance, drainage control of salts accumulation within the root zone, and plant-water interrelationships for economic plant production.

GENERAL MANAGEMENT ASPECTS

Soil management is defined as the sum total of all tillage operations, cropping practices, fertilizer, lime, and other treatments conducted on or applied to a soil for the production of plants (Glossary of Soil Science Terms. Soil Science Society of America Proceedings 29:330–351, 1965). This definition is not adequate under arid climatic conditions. Tillage and soil-treatment operations must conform to sound water-management practices also, or else the productivity of the land will deteriorate. The lack of an understanding of this interrelationship and employment of humid-climate management practices has been responsible for the condition of vast acreages of arid land that deteriorated, became unproductive, and finally were abandoned during the early part of the twentieth century in the United States.

Soil management becomes a series of decisions for the farmer, grower, manager, and even the home gardener. Good soil management depends upon wise decisions relative to the sum total of all crop-production factors that can be controlled by man. Practices that may be prudent for one situation may be unsuitable for a different soil and/or in a different climate. The wisdom of certain management practices will therefore depend upon the characteristics of the soil in a given climate. The dominant characteristics of a soil should be recognized, since the soil properties will greatly influence management practices necessary for maximum production. Properties of humid and arid soils that have a particular bearing on management are discussed in this chapter.

Soils

Soils are dynamic. They are constantly changing, though slowly by comparison with the life of man. Soils are formed from rocks and minerals by the forces of weathering and are changed with time (Fig. 1). These forces include (*a*) physical actions: wind, water, gravity, and temperature changes that cause rock disintegration; (*b*) chemical reactions that cause changes in composition; and (*c*) biological actions that magnify physical and chemical forces.

Soil-forming processes in arid regions are slow compared with those in most humid climates, and as a result the soil may closely resemble the parent rocks and minerals in its composition, except for fineness of individual particle size. Under natural conditions plant growth is sparse and organic-matter accumulation is low. As a result, soils are young or immature in terms of profile development. Often little distinction can be made visually between the surface soil and the soil at various depths. When differences occur, most often they are textural or particle-size stratifications resulting from deposition of different soil materials by water or wind during another period in geologic time. Man-made differences of this type also result from leveling operations of irrigated land.

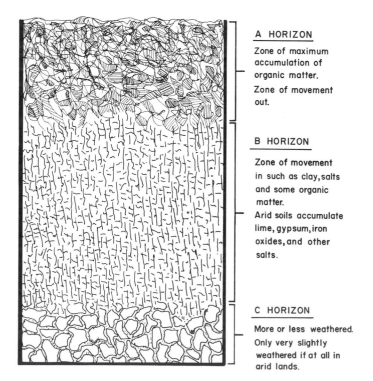

A HORIZON
Zone of maximum accumulation of organic matter.
Zone of movement out.

B HORIZON
Zone of movement in such as clay, salts and some organic matter.
Arid soils accumulate lime, gypsum, iron oxides, and other salts.

C HORIZON
More or less weathered. Only very slightly weathered if at all in arid lands.

Fig. 1. Soil profile showing the separate horizons or layers.

In contrast, in humid regions, soil formation has proceeded at an accelerated rate due to higher rainfall and greater chemical action. Likewise, with more moisture biological activity is more intense, accelerating physical and chemical weathering. An excess of rainwater over the capacity of the soil to hold this water results in movement downward through the soil profile — the vertical cross section of the soil from the surface to the underlying parent material. This downward movement of water, called leaching, carries soluble cations and anions (salts) below the zone of rooting of most plants. In relatively mature or old soils, the small-sized clay and organic particles are moved downward, accumulate at some depth below the surface, and form layers or "pans" which restrict further the downward movement of water and often of plant roots.

Many of the soils in arid climates are remnants of soils developed when the land was in a humid cycle (Fig. 2). Soils with an argillic or clay-mineral-accumulation layer are particularly characteristic of humid climates. Buried, truncated, and degraded remnants of well-developed soils are common in desert regions. Thick lime accumulations (caliche), gypsum, salt deposits, fossil materials, and pollen all indicate that many desert regions were inundated by lakes or oceans in a humid cycle. Soils of arid climates are usually alkaline in reaction and contain free carbonates (lime, caliche) and soluble salts; whereas humid soils generally are acid in reaction and devoid of free carbonates and excess soluble salts.

The physical nature of desert surfaces varies widely from smooth plains to rough bouldered pediments. Because vegetation is sparse, wind and water erosion active, and salt accumulations common, surfaces of arid lands offer prominent features, according to Dregne (1), Fuller (2), and others (3-6). Some of these features that influence the capability of the land for food and fiber production are clay depressions or flats, salt flats, sand dunes (ergs), playas, alluvial plains, desert pavement, stone pavement, gobi, gibber, arroyo outwash, bouldered pediments, and slopes.

Certain subsurfaces also influence food and fiber production in arid lands. Some of these (2, 6) are *Argillic:* (relic of a humid era), subsurface silica clay mineral layer that may or may not impede downward water movement; *Cambic:* altered layer of prismatic soil structure, light in color, low in organic matter and usually limy; *Natric:* enriched with sodium, which has an unfavorable influence on soil structure; *Salic:* enriched with salts; *Calcic:* very limy, often containing an indurated lime layer (caliche); *Gypsic:* enriched with gypsum ($CaSO_4 \cdot 2 H_2O$); *Duripan:* hard compact layer cemented by silica and lime as an accessory.

Water

The most productive soils in arid lands are those that are irrigated; some of the most unproductive soils are irrigated soils that have been mismanaged.

Because large quantities of irrigation water are required to produce food and fiber in arid-region soils on

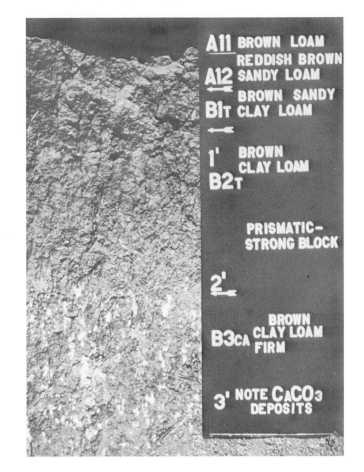

Fig. 2. Mohave soil profile showing clay accumulation in the B horizon and lime in the lower B horizon.

a competitive economic basis, water quality is vitally important. For example, Colorado River water in the All-American Canal contains a little over one ton of salt per acre-foot (2,240 kilograms per hectare). If 4 acre-feet of water (122 centimeters depth) are used to grow an acre crop of alfalfa during the year, 4 tons or 8,000 pounds of salt would have been added per acre of land, or 8,960 kilograms per hectare. This amount can be tolerated by certain crops, provided that some extra water is used for leaching to move the salts below the root zone (see Fig. 3).

Waters vary considerably in quality from those almost free of salts to those brinier than the sea. Subsurface waters are more variable than surface (7). Some waters contain salts toxic to plants even in small amounts; the toxic salts most commonly encountered are boron, lithium, and sodium. Well waters also can contain high amounts of chlorides and nitrates, sufficient to affect crop production adversely (7). Waters high in salt may be used effectively for leaching and on salt-tolerant plants provided suitable management practices are employed (8-10).

Waters of humid regions contain little salt compared to those of arid lands.

Waters have been rated according to their salinity hazard from low to very high (10). The presence of sodium, above certain levels and in relation to the con-

Fig. 3. Winter lettuce beds and canal ditch ridges showing white salt accumulation in central Arizona.

tent of calcium plus magnesium, adversely affects water quality — according to the U.S. Department of Agriculture diagram for classifying water (10).

SOIL ORGANIC MATTER

Organic matter has long been used with confidence as an indicator of the level of soil fertility. Dark soils are still considered to be more fertile than are light soils in some climates. Soil organic matter has been given special attention because of its disproportionately favorable effects on the chemical, physical, and microbiological characteristics of the soil as compared with the mineral matter. These characteristics also are key factors in food and fiber production.

On the other hand, desert soils, which for the most part are notoriously low in organic matter, have proved to be among the most productive soils in the world with proper management. This fact seems to emphasize the importance of continual replenishment or "turnover" of organic matter rather than abundance. Another factor that seems to come into play is that sands are more prominent than silts and clays in desert soils (compared to humid-climate soils); organic matter is more essential to good soil structure in silts and clays than in sands.

Native desert soils range from 0.1 to 1.0 percent organic matter for the most part (11), whereas soils of humid climates of the Midwest and East (U.S.A.) range from 2 to 5 percent, except for peats and mucks.

Forest soils also have a higher content of organic matter than arid-land soils. The distribution of organic matter in desert soils is highly variable because of the dominance of wind action. Organic residues are likely to collect around the base of an isolated desert shrub, in a hollow, or in a microdepression.

There is little difference in the chemical composition of organic matter as related to climate except at the temperature extremes. For example, both the carbon/nitrogen (C/N) and carbon/phosphorus (C/P) ratios are smaller in warm desert soils than in cold humid soils.

Three classes of compounds make up the bulk of soil organic matter (11, 12). These are (a) substances produced by the alteration of lignin of plants, (b) compounds related to carbohydrates (13, 14) (including bacterial gums, slimes, and molds), and (c) material probably derived from proteins (the nitrogen carriers). Other substances, such as organic phosphorus (see Fig. 4), growth regulators (growth-inhibiting and growth-promoting), and organic acids are also present, but in small amounts.

Soil organic matter may also be categorized on the basis of its functions in soil (11).

Plant-Nutrient Source

Organic matter supplies necessary plant macronutrients (such as nitrogen, phosphorus, and potassium) and micronutrients (such as iron, manganese, copper, and boron). Calcium and magnesium also are present.

ORGANIC PHOSPHORUS-PPM

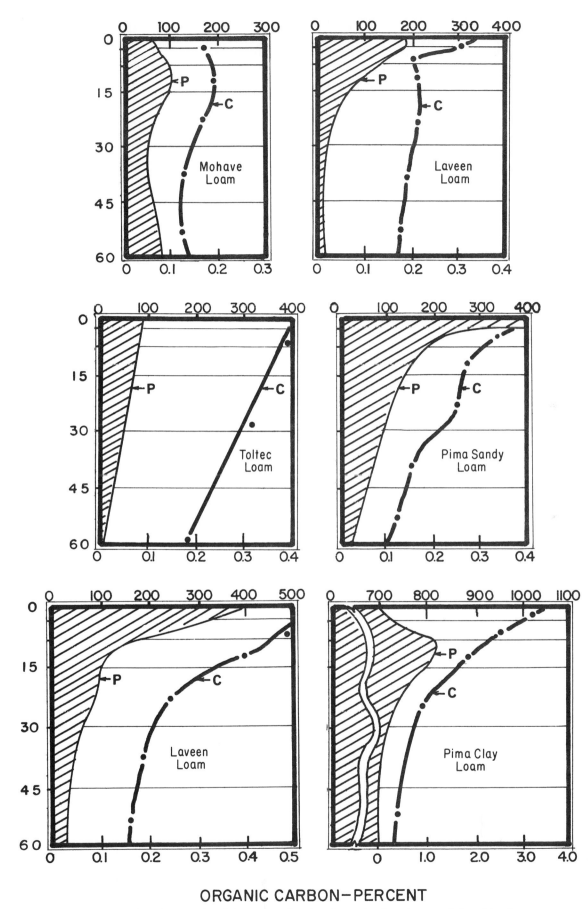

DEPTH IN CENTIMETERS

ORGANIC CARBON-PERCENT

Fig. 4. Examples of the carbon/nitrogen and carbon/organic phosphorus relationships in some virgin Sonoran Desert soils.

Native soil nitrogen is almost wholly in the organic form. The ratio of carbon to nitrogen (C/N) in desert soils ranges from 8:1 to 10:1; in soils of humid climates it is wider or greater, being nearer 10–12 to 1 for the most part. Organic C/P ratio is 19–38 to 1 in desert soils of southwestern United States (*11*) and greater in soils from other climates (*15*).

Food Sources for Organisms

Most soil organisms require organic matter to live. The microorganisms decompose plant and animal residues deposited on and in the soil. Although soil microorganisms are very small, they are present in such large numbers that many pounds of nitrogen per acre are tied up in their cells. Organic residues as well as organic matter also are a source of life for larger organisms such as earthworms, mites, and microinsects. Thus the abundance of the soil organisms is directly proportional to the abundance of the food source, soil organic matter.

Benefits to Physical Properties

Soil organic matter greatly influences the physical properties of the soil. It improves the soil structure and water and air movement, assists in the control of soil compaction, and affects the temperature. Soils with an abundance of organic matter are considered to be in "good tilth."

The characteristic moisture and temperature relationships in arid environments limit the accumulation of organic matter under virgin conditions. The paucity of moisture permits meager vegetation, whereas the warm desert temperatures for extended periods of time permit decomposition to progress very rapidly when moisture will permit.

Under irrigated conditions arid-land soils demonstrate (*a*) a very rapid rate of decomposition of plant and animal residues, (*b*) an increase in organic matter over virgin conditions where legumes and other soil-building crops are used and where large amounts of crop residues, such as straw, are plowed down, and (*c*) the fact that a low content of organic matter does not particularly inhibit maximum crop production, provided, of course, that a small amount is maintained by continued addition of organic residues.

Maintenance

Organic-matter maintenance is critical to favorable food and fiber production. Plants use only small amounts of organic matter directly; thus the principal benefit derived by its maintenance is through its favorable influence on the physical condition of the soil. There are four means of replenishing organic matter in soil: (*a*) mature crop residues, (*b*) animal manures, (*c*) green manures, and (*d*) municipal wastes and composts (*16*).

Because of hauling costs, the additions of animal manures and municipal wastes and composts to agricultural soils usually are not profitable except on certain poor lands and isolated spots. Homeowners will pay more than farmers for manures; thus they often represent a better sales outlet. Manures from pen-fed cattle

Fig. 5. Field experiments showing fertilizer effects on plant growth and color.

under arid climates are often excessively high in undesirable salts (*17*). This is not necessarily the case for manures from feeding pens in humid climates, where rain is sufficient to leach away the salts. The maintenance and renewal of organic matter in arid lands is dependent almost wholly on the *return of crop residues,* since even green-manuring at the present time is uneconomical where land values are high and water is limited. Return of crop residues under arid conditions usually requires an accompanying amount of nitrogen fertilizer to prevent nitrogen hunger in the early growth of the succeeding crop. The wisest management practice is to use the high-analysis inorganic fertilizers in combination with the nutrient-poor manures, composts, and crop residues where organic matter is seriously deficient.

Burning of crop residues, such as straw, is a soil-depleting practice and cannot be defended.

PLANT NUTRIENTS

Soil fertility is concerned with the supply of essential plant nutrients. Some of the most important decisions in soil management are associated with the application of fertilizers to supply these nutrients (see Fig. 5).

As far back as 2500 B.C., writings refer to the fertility of the land. According to Greek mythology, animal manure was used on the land some nine centuries before the birth of Christ. The value of leguminous green-manure crops and wood ashes was recognized by ancient agrarians. The early Indians placed fish under each hill of corn to improve the crop. In spite of these early favorable practices, a number of quaint ideas were proposed around the beginning of the seventeenth century. For example, the classical experiment of van Helmont led him to conclude that water was the sole nourishment of plants. Just as erroneous was Jethro Tull's belief that plants ingest soil directly into the plants' "circulatory system." These and other ideas were not dispelled until the nineteenth century. The work of Liebig probably had the most profound effect on the thinking during the last half of the nineteenth century. It is significant to note that a clear understanding of plant nutrition was not possible until the development of basic sciences such as chemistry and physics.

According to present knowledge, plants require 16 elements for the completion of their vegetative and reproductive life phases (*18*).

Macronutrients

Plant nutrients that are accumulated in relatively large amounts are called macronutrients. Carbon (C), hydrogen (H), and oxygen (O) are supplied by air and water; therefore they are not considered in fertilizer use. Nitrogen (N), phosphorus (P), and potassium (K) are the most common macronutrients supplied by commercial fertilizers. These are the nutrients most commonly deficient in acid soils of the humid regions. However, in arid soils N is usually deficient, P to some extent, and K is usually abundant.

Calcium (Ca), magnesium (Mg), and sulfur (S) are secondary macronutrients. Deficiencies of Ca and Mg are more common in highly weathered soils under humid conditions. When these soils receive ground limestone to neutralize acidity, ample Ca and often Mg are supplied. Sulfur deficiencies have been reported in rural areas of humid regions as well as under dry-land conditions throughout the world. Sulfur deficiencies have become more acute with the advent of high-analysis fertilizers. Fertilizers such as ammonium sulfate and single superphosphate contain S as an ancillary element to N or P and supply more than adequate sulfur. Under irrigation, S deficiencies of arid lands are unusual, since most sources of irrigation water as well as most soils contain S as sulfate adequate for plant growth.

Micronutrients

In recent years more attention has been given to the micronutrients. Some fertilizers are formulated to contain one or more of these known to be essential to plants. Recognized essential micronutrients are: iron (Fe), zinc (Zn), manganese (Mn), copper (Cu), boron (B), molybdenum (Mo), and chlorine (Cl). The heavy-metals micronutrients are more available to plants in acid-humid region soils than in arid soils. However, molybdenum has a higher degree of plant availability in soils alkaline in reaction.

Principles of Fertilizer Use

Management decisions relative to fertilizers are concerned primarily with the following: (*a*) the nutrient or nutrients needed, (*b*) amount of nutrient to apply, (*c*) chemical form of nutrients, (*d*) placement or position in the soil, and (*e*) timing of fertilizer application.

AMOUNT AND KIND OF NUTRIENT NEEDED

Ideally a chemical soil test can be used to measure the level of available soil nutrient. For this value to be meaningful, values for the test must be correlated with some function of yield over the range from acute deficiency to complete sufficiency. Unfortunately, the empirical tests that have been developed do not meet the ideal clearly. Nevertheless, land-grant universities have established soil, water, and plant testing laboratories as a service function for improving food and fiber production. These programs have met with a reasonable degree of success and have proved useful for broad recommendations of amount and kind of nutrients needed. Sophisticated field experiments using radioactive isotopes positively identify the uptake and translocation of nutrients (see Fig. 6).

CHEMICAL FORMS

Nitrogen fertilizers contain urea, ammonium, or nitrate. Most soil scientists agree that sources are equally effective when applied at the same rate of the element N. However, exceptions do occur; for example, nitrate may leach below the root zone very rapidly in a coarse-textured sandy soil, whereas ammonia is less likely to

Fig. 6. Radioautograph of clover and bean leaves from plants grown in soil having received radiophosphorus fertilizer in nutrient study.

leach. The most widely used sources of P are the ammonium phosphates and superphosphate. Sources are available that contain P in forms that are less soluble in water. Important sources of K are the soluble chloride, sulfate, and nitrate salts. Most often the decision regarding fertilizer forms is based on the cost per pound of nutrient and ease of handling and application.

PLACEMENT IN THE SOIL

Placement methods can be considered in two broad categories — *band* and *broadcast*. Band placement usually involves the injection of fertilizer in a continuous band to the side of the plants and below the seed position so that maximum contact is made with the root (see Fig. 7). When such placement occurs at planting, it is often called "starter fertilizer," and when placement is made in this manner during the growing season it is called "side-dressing." Broadcast placement covers a number of practices whereby fertilizer is applied more or less uniformly on the surface of the soil and is incorporated into the soil by discing, plowing, or cultivating. Application of soluble fertilizers in irrigation water would be in effect a broadcast placement.

TIME OF APPLICATION

Timing of fertilizer application has to be considered for each crop and in view of established cultural practices. Generally fertilizers that supply P or K must be applied at planting or early in the season while N is applied in two or more split applications.

Plant Growth/Nutrient Relationship

The general relationship between nutrient supply and plant growth or yield is given by the Mitscherlich equation which reduces to

$$\text{Log}\,(A - y) = \log A - cx,$$

where A is the maximum yield with all nutrients adequate, x is the quantity of nutrient present, y is the yield with x amount of nutrient, and c is a constant depending on the nutrient, plant, climate, and other factors. Figure 8 depicts this relationship. The appropriateness of the Mitscherlich concept depends on the mobility of the nutrient in the soil as has been reviewed by Bray (*19, 20*). It follows from this equation that the yield (y) approaches the maximum yield (A) as the nutrient supply increases from a deficient to an adequate level. The absolute value of A depends on the soil, water supply, plant potential, climate, plant population and planting pattern, insect and disease problems, and the presence of toxic substances. These factors are interrelated and influence management decisions about the use of fertilizers.

Fertilizers are added to the soil to supplement the soil supply of nutrients. Ideally an amount is added that will permit the yield to approach the maximum. Part of the nutrient requirement for maximum yield from a given plant will be supplied by the soil. The difference between the total requirement and that supplied by the soil can be provided by fertilizers and is then termed "the fertilizer requirement."

Nitrogen in nitrate form is relatively mobile in the soil, moving with the flow of soil water and to some extent diffusing in soil water. Since it is mobile, nitrate-N can be used more completely in the soil rooting zone in a single season than any other nutrient (*20-22*). Also due to its mobility, nitrate-N can be lost from the root zone by leaching when excessive amounts of water pass through the soil. Thus, timing of nitrogen application goes hand in hand with water-management practices. Herein lies one of the greatest differences in management of fertilizer use between arid- and humid-region soils. On irrigated arid soils, water application can be controlled more completely than under the extreme conditions that prevail under humid conditions. Often two or more applications of N fertilizers are made to improve efficiency or correct suspected deficiencies.

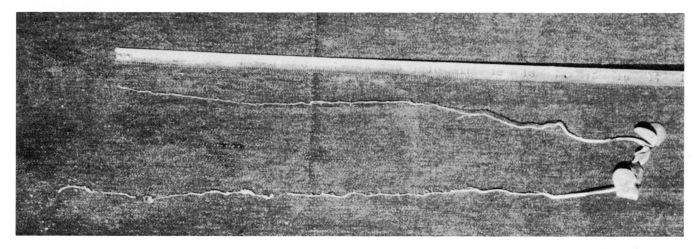

Fig. 7. Deep root penetration (16-17 inches) of a 16-day-old cotton seedling plant.

Nitrogen in the soil also occurs in organic and ammonium forms, which are relatively immobile. Plant residues and soil organic matter are decayed by soil microorganisms with the ultimate release of N in ammonium form. Specific aerobic organisms transform N from the ammonium to the nitrate form. Consequently, under aerobic conditions most N that plants absorb is in nitrate form. Nitrogen fixed from the air by symbiotic N-fixers in association with host legume plants eventually is converted to nitrate (23), also.

Total phosphorus is usually higher in arid soils than in humid soils. However, availability to plants is frequently a problem due to the low solubility of complex calcium phosphates associated with high calcium-carbonate accumulations. Low availability of P in extremely acid soils is also a problem, which is caused by formation of insoluble iron and aluminum phosphates.

Management of acid soils includes the application of ground limestone to neutralize acidity; among the favorable effects of this procedure is increased solubility of iron and aluminum phosphates. Phosphorus is a relatively immobile nutrient in soils, moving very little from point of fertilizer placement or location of native phosphate (23). As a result placement of fertilizers must assure that plant roots will contact the fertilizer for maximum utilization. The time of phosphate application should either be before the seeds are planted or early in the season when the plants are small.

Highly leached acid soils are commonly much lower in K than soils of the arid regions. Plant-available potassium is found in exchangeable (attached to soil particles, notably clay) and water-soluble forms. Water-soluble K is usually quite low under humid conditions as contrasted with arid soils. In addition to the large amounts of plant-available K in arid soils, tremendous K reserves are present in primary minerals such as the micas and feldspars. Potassium in the soil is relatively immobile, although K moves over greater distances than P. The presence of soluble salts increases the mobility of K and other cations as a result of successive exchange between the exchangeable and the water-soluble cations. Potassium fertilizers are generally applied about planting time or injected into the root zone early in the growing season.

Soil, Water, and Plant Testing

LABORATORY TESTS

For a *chemical soil test* to be successful, the extracting solution must remove the chemical forms of a nutrient that plants can absorb, without dissolving those forms plants cannot absorb. Test values must be correlated with crop response to added nutrients (24). If these requirements are satisfied, soil tests can serve an important function in management decisions regarding fertilizer use. By determining the soil-nutrient level, judicious use of fertilizers is enhanced — crop yields are improved and less waste of fertilizer is likely. Soil tests for P are more successful on acid soils of the humid region than on calcareous arid soils. Nitrate-N tests provide some helpful information on arid soils but are of little value on humid-region soils. Tests for K are reasonably satisfactory for both humid and arid soils. Soil tests for the secondary macronutrients can be made although they are not routinely utilized. Micronutrient soil tests have been investigated and are used to a limited extent, principally on acid soils.

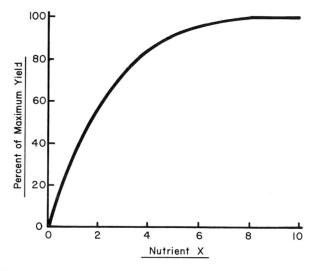

Fig. 8. Idealized relationship between yield and soil-nutrient level (Mitscherlich).

Many *irrigation waters* contain plant nutrients in sufficient amounts to contribute to the nutrition of a crop. Therefore, in addition to the usual analyses concerned with salinity and Na accumulation, analysis for nitrate-N will provide information that can alter the N fertilization program. City sewage effluent represents a potential supply of water for irrigation and contains appreciable amounts of N, P, and K. Often the effluent will contain more nutrients than required by the crop. By analysis and careful planning, the effluent can be mixed with other sources of irrigation water, adding to the total supply of water as well as providing adequate inexpensive nutrition for crops.

Plant analysis is a technique that can be used to evaluate the nutritional status of the plant at the time of sampling. The tests usually are made on a specified plant tissue and usually on plant extracts containing the inorganic form of a nutrient. Interpretation of the values depends on a knowledge of levels of adequacy for each nutrient in a certain plant part at different stages of growth. With many nutrients, a certain minimum level must be maintained in inorganic form, or the metabolic functions will be affected. When only one nutrient is being tested for, its adequacy is perhaps based on that single value without a complete knowledge of other nutrients or external factors. For example, an adequate test value for nitrate-N for a plant very deficient in P would indicate sufficient N under existing conditions. If the P deficiency were removed, the N level might no longer be adequate. Likewise a plant under water stress could become N deficient after water is added and growth rate is accelerated. A tissue analysis can help detect approaching deficiencies before symptoms appear. In this way it makes possible fertilizer application in time to prevent a nutritional deficiency. Tissue analysis techniques have been developed for N, P, and K in a number of plants of economic importance. Information is available regarding critical levels of certain micronutrients in only a few plant species.

FIELD TESTS

When a satisfactory soil test is not available, field experiments (see Fig. 9), field trials or test strips can be used to determine the need for a nutrient and the approximate amount to apply. Research workers attempt to quantify the relationship between a function of yield, soil nutrient level, and fertilizer requirement.

Crop Rotation

The practice of changing crops periodically rather than growing the same crop continuously on the same land dates to ancient times. Probably the most important benefit of crop rotations prior to the development of N fertilizers was the N added to the soil by bacteria in association with legume plants. Much of this fixed N was utilized by nonlegume crops during the following season. With an almost unlimited supply of inexpensive commercial N fertilizers available, a rotation with legumes for the N gain is not justified.

In humid regions on sloping to steep land, sod crops are necessary for erosion control. Whether or not a row crop can be safely grown depends on the slope of the land and danger of erosion. Control of erosion by water is not the same problem on dead-level irrigated arid land; however, cropping practices can aid in the control of erosion by wind.

Crop rotations probably are best justified when the cropping sequence will aid in control of disease or insect problems. Growing a crop with an extensive fibrous root system often promotes a more favorable soil physical condition.

SOIL CONDITIONING
Concept

Soil conditioning is a concept developed in the management aspects of arid lands. The natural tendency for those who move from a humid to an arid climate and who are concerned with planting the soil is to want to put something in the soil or do something with the soil because it looks so barren and because lime, often confused with alkali, is prominent. For most arid-land soils, addition of soil conditioners is not necessary. Water and good soil management are all that are needed for virgin soils except in isolated spots or where drainage is poor. Man's mismanagement creates most of the problems that require correction in arid lands. Soil conditioning and reclamation may be necessary to correct these problems.

The term *soil conditioner* is used to identify a host of organic and inorganic materials, not fertilizers, which are added to soils to influence the biological, chemical, and physical condition, primarily for the purpose of making the soil a more suitable medium for plant growth (*25*). Some materials such as manures have both fertilizing and soil-conditioning properties that cannot be separated

Fig. 9. The effect of fertilization on the growth and yield of wheat: left plot unfertilized and right fertilized.

(*17*). Soil conditioning may also refer to physical manipulation, such as tilling, draining, or mixing of the soil to improve its value for plant growth (*9*).

Functions

Soil conditioning functions most often by influencing the physical and/or chemical properties, in turn favorably improving the structure of the soil. Soil structure controls air and water retention and movement. The productivity of a soil is critically dependent on adequate air and water interrelationships and the way these interact with plants. Soil structure thus is the key to the rate salts accumulate or move in the soil. Soil structure appears to be more transitory in arid- than humid-land soils. Thus more attention must be given to maintaining a proper salt and particularly calcium balance. Some of the ways conditioners affect the soil may be enumerated as follows:

a) cementing particles together, manifested in the number and size of soil particles a conditioner is capable of holding together in desirable water-stable aggregates,

b) direct influence on soil-moisture movement, retention, and availability,

c) influence on air space and temperature,

d) influence on plant-nutrient availability,

e) control of undesirable salt accumulation,

f) influence on microbial activity in the decay of organic residues.

Classes

Soil conditioning and conditioners have been classified into groups by Fuller (*26*) according to their general function in soils.

INORGANIC CHEMICALS

Inorganic chemicals include sulfur, sulfuric acid, gypsum, polysulfides, iron sulfate, and others. Of this group gypsum ($CaSO_4 \cdot 2H_2O$) is most abundantly used today (*27*). It is applied to arid-land soils to supply calcium ions to counteract excess sodium and not for nutritional purpose of its composition. Sulfur is rarely deficient for plants in desert soils because of the abundance of sulfate in the soils and water. Gypsum also prevents sluggish water movement in dispersed soil, which may result when excess salts are leached away by overirrigation with low-salt water or by high-intensity rainstorms. The calcium of gypsum favors flocculation (aggregation of particles) of the soil colloids, which permits more pore space for water and air movement. Not all soils are favorably influenced by the use of gypsum (*27*). Its value lies mostly in reclamation.

Sulfur (S) oxidizes in the soil by action of the soil microorganisms to eventually produce sulfuric acid (H_2SO_4). The acid in turn unites with the lime, making gypsum. Sulfur oxidizes very slowly by microbial action unless it is highly subdivided and dispersed thoroughly in the soil.

Sulfur dioxide (SO_2) forms sulfurous acid with water; that acid is then further biologically oxidized to sulfuric acid.

Polysulfides are complex mixtures of sulfides, sulfates, thiosulfates, and molecular sulfur, usually marketed in the calcium form. Elemental sulfur precipitates in the colloidal form when liquid polysulfides are placed in contact with water.

Pyrites and pyrrhotite appear on the market as a mixture that is almost exclusively pyrites originating from mine waste. Pyrites are so insoluble and so slow to oxidize they have not been shown to be effective in soil conditioning.

Experimental results using iron sulfate (ferrous and ferric sulfate, $FeSO_4 \cdot 7H_2O$) as a soil conditioner have not been encouraging. Iron sulfate reacts rapidly with the soil to form the insoluble iron oxides, which are generally abundantly present in the soil already. Gypsum also forms as a result of the reaction between the sulfate and soil lime.

INORGANIC DELUENTS

Inert deluents are materials that alter properties of the soil of temperature, texture, structure, and/or pore space as dilutants, through their physical makeup. Most are unreactive or very slowly reactive in the soils. Some common examples are sand, gravel, ground rock, exploded silicates, coal ash, coke, carbon black, petroleum mulch (*28*), vermiculite. These substances may or may not condition soils, depending on (*a*) the nature of the soil and the problem, (*b*) the amount added, that is, dilution factor, and (*c*) the method of application.

NATURAL ORGANICS

Many of the natural organic materials are woody. Some examples are peat, sawdust, bark, straw, leaves, grass clippings, and less woody materials such as manures, composts, sludges, and fish meals. Manures are the best known and most widely used. The quality and composition of manures vary widely. Because of the low organic-matter content of arid-land soils, home gardens and landscaping respond well to additions of manures if they are not too high in salt (*17*). Composts and certain other organic materials have been used by home gardeners and nurserymen with success (*16, 29*).

SYNTHETIC ORGANICS

The synthetic organic group includes synthetic organic compounds, such as polyelectrolytes (hydrolyzed polyacrylonitrile) and wetting agents as represented by detergents and surfactants. These compounds vary considerably in their effectiveness. There is a renewed interest for surfactant agents to improve wetting in soils where dehydration is a serious problem. Effective use of these materials requires a thorough knowledge, not only of the soil and its history, but of the physical chemistry of wetting.

BIOLOGICAL MATERIALS

Inoculants, microbial cultures, enzymes, vitamins, and trace growth-regulating substances are represented here. Since desert soils have been exposed to the rigors of high temperatures and desiccation, there is a feeling among those who come from humid climates that the soil is either deficient in soil organisms or sterile. Some suggest that the right kind of organisms may not be present.

These concepts are not accurate. Desert soils are abundantly supplied with the necessary microflora (2, 30-34). Any attempt to add cultures of organisms, vitamins, enzymes, or trace growth-regulators results in their rapid disappearance (12, 34-37). Legume inoculation, as under any climatic conditions, however, can result in better inoculation of the proper host plant where that certain legume or cross-inoculation plant-group has not been grown before (38). Enzyme and/or vitamin preparations have not proved beneficial for plant growth when added to soils.

PHYSICAL CONDITIONING

Tillage operations that are used to improve the physical condition of the soil are in the physical soil-conditioning class. They include ripping, chiseling, deep plowing, or vertical mulching to break up horizontal layering of stratified textures, compacted layers, or hard spots (9). Tilling will be discussed more fully in the following section.

Soil conditioning is a very complex operation. It is not well understood by nontechnical people and has not been well learned by people of humid lands, since it is not as necessary to their food and fiber production as in arid lands. The lack of understanding of soil conditioning has been costly to food and fiber production in arid lands. Almost as much land has been ruined as is now in crop production, because of a lack of understanding of proper soil, water, and conditioning management.

PHYSICAL MANIPULATION AND PRACTICES

Practical agriculture for food and fiber production began with physical manipulation of the soil, such as stirring or probing with a stick to plant seeds. Manipulation of the soil has since developed to the extent of almost complete mechanization of land operations. Indeed, landscapes of all dimensions are mechanized with carts, cars, trucks, sand buggies, and even hill climbing in remote areas by four-wheel-drive vehicles. Traffic over the soil has created problems, and more are to come. Working the soil to get a crop tilled and harvested causes serious problems of tillage pans (hard layers just below the plow), compaction at various depths, and puddling when harvesting vegetables in soil that is too wet. The unfortunate end result is restricted water and air infiltration, hard spots, poor root penetration, and finally crop failures (see Fig. 10).

Soils of arid lands, more often than not, are worked either too wet or too dry. This is not as serious in humid lands, since moisture is almost always present and nature provides ample opportunity to till and work the soil at or near an optimum condition.

Arid lands also must be manipulated to protect against serious loss by wind erosion. Desert soils are dry for such long periods of the year that special wind-control measures are necessary (2, 3, 5, 6). The seriousness of water erosion is compounded in arid lands because of the scant vegetation and the torrential characteristics of the rain-

fall. Notwithstanding the greater annual accumulated precipitation of humid climates, the great intensity of rainfall in arid lands washes, cuts, and dissects the land beyond the worst expected in most humid areas. Erosion control of soil and water is essential to survival in arid lands all over the world.

The necessity of maintaining practices for the control of salts as well as reclaiming salt-affected soils adds to the burden of the agricultural industry in arid lands. A thorough knowledge of soil- and water-management principles and practices is necessary to survive, in an economic food and fiber production program.

Water must also be managed to prevent salt accumulation within the root zone and yet prevent drainage problems that could quickly salt up the land, reducing or eliminating production.

Soil Compaction

Soil compaction may take place as a result of almost any kind of traffic over the land. Arid-land soils, in addition, may become compacted by the action of irrigation water. Pore space is compressed as the soil particles disperse upon saturation, particles reorient to a tighter configuration, the soil becomes dense with globular pore space more prominent than vesicular. This is called "induced" compaction. "Geologic" compaction is found most prominently in alluvial materials (valley floors) as a result of water or glaciers having deposited materials in layers of different density. "Genetic" compaction is found in developed soils and occurs during the period of formation. Correction and prevention of further compaction involves the use of tillage practices such as deep plowing, knifing, chiseling, tilling at different depths, reducing traffic, and working the soil at proper moisture

Fig. 10. Yield of long-staple cotton, in a clay loam soil, is affected adversely by tillage operations that compact soil.

content (*9, 25*). Organic matter applications and crop rotation practices have been used effectively in controlling compaction.

Minimum Tillage

Soil compaction restricts root growth and reduces crop yields (*25*). Sluggish air and water movement can be minimized by adopting certain tillage practices according to Harris, Erie, and Fuller (*9*). This arid-land research often has been called "minimum or rough" tillage and includes (*a*) plowing to a depth below the compacted layers at a moisture content at which the soil turns up cloddy, (*b*) irrigating to "melt down" the clods, and (*c*) tilling only sufficiently to prepare a suitable seedbed for the crop to be grown.

Usually it is most advantageous to irrigate deeply the first time to remove salts from the root zone and store water deep for deep root penetration.

Control of Salts

Salts must be controlled in soil if a permanent food and fiber production is to be attained on arid lands. Salts are derived from (*a*) the water applied and (*b*) the soil itself. All waters contain some salt. The quantity and quality varies with the source from almost pure rainwater to brackish well or surface water. To prevent accumulation of salts, enough water must be added to leach the salt down below the root zone.

Fuller (*7*) enumerates some of the causes of salt accumulation as (*a*) frequent and shallow irrigation, (*b*) poor soil drainage, (*c*) high water table, (*d*) insufficient leaching of compact, dense, slowly permeable soil, and (*e*) continued use of low-quality (sodium, salty, etc) water.

Arid-land soils cannot be managed as indifferently as those of humid lands. Upward movement of salt and deposition on the surface by capillary rise has defeated many agriculturalists. Zones of salt may also accumulate below the surface where textural changes occur and at lowest point of water penetration. These layers require breaking up and leaching.

PLANT

The kind of plant to be grown also requires special attention in arid-land food and fiber production. Plants differ in their ability to tolerate salts (*7, 10, 39*). Growers must know (*a*) the salt-tolerance level of different plants, (*b*) seed placement to avoid excess salts killing seedlings (see Fig. 11), (*c*) germination and stand-establishment tolerance, and (*d*) the effects of high osmotic tension of salts in solution bathing the roots.

SOIL

Following are some examples of soil-management principles and practices used for the control of salts in arid-land soils:

a) Avoid working the soil excessively — employ the minimum tillage principles.

b) Avoid compact layers in soils and eliminate them by ripping, knifing, or deep plowing.

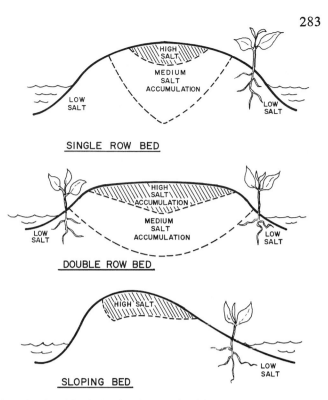

Fig. 11. Seed-bed shaping is practiced in arid-land agriculture to reduce the possible damaging effect of excess salt accumulation due to capillary movement.

c) Eliminate alluvial soil stratified by different textures of sand and clay by deep plowing and mixing.

d) Shape beds for row-crops and plant to avoid ridge areas of high salt accumulation.

WATER

Examples of water-management principles used for the control of salts are listed:

a) Carefully apply water to avoid overwatering. Excessively wet soils adversely affect absorption of plant nutrients and nutrient availability.

b) Avoid overtaxing drainage systems. Improper drainage causes salts to accumulate.

c) Keep salts moving down below the root zone.

d) Intermittent leaching with less water can be as effective as continuous leaching. A little water will carry a lot of salt which is concentrated at or near the wetting front during irrigation.

e) Salty water can be as effective as good water in leaching salts below the root zone.

Reclamation

Although salt accumulation occurs naturally in soil, more land is salted because of man's mismanagement than because of natural processes. Reclamation of salt-affected soil involves a number of practical management procedures. They have been listed as follows (*40*):

DRAINAGE

Good drainage must be provided before reclamation is begun. The water table must be lowered to prevent salts rising by capillary action and accumulating near and on the surface.

LEACHING

Salts must be transported by water below the root zone. All soils of arid lands require a definite minimum leaching percentage depending on the dissolved solids (salts) in the irrigation water.

CHEMICALS

Gypsum is most commonly used where sodium is a problem. Other materials and means are discussed under soil conditions and soil conditioning.

TILLAGE

Because salts tend to collect in compact and very fine-textured soils, various practices are necessary to improve the rate of water penetration.

STRUCTURE

Any means to improve soil structure improves water movement and reclamation. Soil structure controls water relations in soils.

CROP MANAGEMENT

Some crops are more effective in speeding up reclamation than others. Deep-rooted crops are considered the most effective.

Soil Conservation

Soil conservation in arid lands is involved in wholly different practices than in humid climates. Water conservation is as important as soil conservation. Rains fall at such a rapid rate that land-washing, gullying, and arroyo outwash take place at a fantastic rate. After the rains, the soils dry and a strong wind picks up his dirty partners (see Fig. 12) — dust, sand, and debris — and hurls them great distances. Many types of erosion are characteristic only of desert soils. Piping is one of these. Soils move under hydrostatic pressure underground in tubular fashion (26, 41). Before conservation practices can become effective, arid-land conditions must be known, soils studied, and the erosion characteristics understood.

Sand movement is particularly intense in arid regions where trees, grass, and other vegetation is sparse, allowing wind to blow unabated (1-3). Sand covers up productive land and crops as dunes form and reform in restless fashion. Wind fences (also known in humid climates as snow fences) often must be erected about residential areas as well as around food- and fiber-producing land, highways, and water holes to help protect them from inundation by sand.

Water-control measures to protect food- and fiber-producing land from being eroded and dissected by gullies are programmed by series of dams and water-spreading dikes. Excess flow of water and movement of soil and rock debris is, however, still a common occurrence under even the most elaborately controlled practices. Straightening of river channels and eradication of phreatophytes are also soil- and water-conservation measures recommended for arid lands. Because soils of arid, as compared with humid, lands are so readily dispersed by water and because of the sparsely scattered vegetation, erosion caused by a given amount of rain in desert areas exceeds that caused by equal rain in humid climates.

In addition to the literature cited above, bibliographical information about general references (33, 42-58) has been provided for those readers who might find such information of use.

Fig. 12. Serious erosion caused by extremely heavy rains on highly dispersible arid-land soils.

CONCLUDING STATEMENT

In arid as well as humid climatic regions, effective soil management begins with a knowledge and consideration of all factors involved. Of primary consideration is the inherent capability of the basic resource — soil. Although many specific practices differ between humid and arid regions, the major feature of arid-land management is the interdependence of water and soil management. This statement should not be interpreted to mean that water management is not important in humid areas. Certainly, the importance of soil drainage, surface-water control (usually in terms of excesses), and supplemental irrigation must not be minimized. In arid regions throughout the world the role of soil science in land management should be expanded in evaluating water quality, improving the use of water, reducing waterlogging, controlling salinity and alkalinity, reclaiming saline and alkaline areas, preparing and tilling land, and insuring adequate nutrition of crops for maximum food and fiber production in all instances.

REFERENCES

1. DREGNE, H. E.
   1968 Appraisal of research on surface materials of desert environments. *In* W. G. McGinnies, B. J. Goldman, Patricia Paylore (eds), Deserts of the world: An appraisal of research into their physical and biological environments, p. 287–377. The University of Arizona Press, Tucson. 788 p.

2. FULLER, W. H.
   1969 Geologic and geomorphic aspects of deserts. *In* G. W. Brown Jr (ed), Desert biology, vol 2, chapter 3. Academic Press, New York.

3. ACADEMIA SINICA, SAND CONTROL GROUP (ED)
   1962 Chih-sha Yen-chiu. Peiping. 203 p. (Translated 1963 by Joint Publications Research Service as JPRS 19,993. 508 p. Also cited as OTS 63–31183).

4. DOWNES, R. G.
   1956 Conservation problems on solodic soils in the state of Victoria (Australia). Journal of Soil and Water Conservation 11:228–232.

5. LOBOVA, E. V.
   1960 Pochvy pustynnoi zony SSSR. Akademiia Nauk SSSR, Pochvennyi Institut im V. V. Dokuchaeva. 362 p. (Translated 1967 by Israel Program for Scientific Translations, Jerusalem, as "Soils of the desert zone of the USSR." 405 p. Also cited as TT 67–51279.)

6. U.S. DEPARTMENT OF AGRICULTURE, SOIL CONSERVATION SERVICE, SOIL SURVEY STAFF
   1960 Soil classification comprehensive system, 7th approximation. 265 p.

7. FULLER, W. H.
   1965 Water, soil and crop management principles for the control of salts. Arizona Agricultural Experiment Station and Cooperative Extension Service, Bulletin A–43. 22 p.

8. SMITH, H. V., G. E. DRAPER, AND W. H. FULLER
   1964 The quality of Arizona irrigation waters. Arizona Agricultural Experiment Station, Report 223. 96 p.

9. HARRIS, K., J. ERIE, AND W. H. FULLER
   1965 Minimum tillage in the Southwest. Arizona Agricultural Experiment Station and Cooperative Extension Service, Bulletin A–39. 16 p.

10. U.S. DEPARTMENT OF AGRICULTURE, AGRICULTURAL RESEARCH SERVICE, SALINITY LABORATORY STAFF
    1954 Diagnosis and improvement of saline and alkali soils. L. A. Richards (ed). Agriculture Handbook 60. 160 p.

11. FULLER, W. H.
    1969 Soil organic matter. Arizona Agricultural Experiment Station and Cooperative Extension Service, Bulletin A–40. 17 p.

12. ALEXANDER, M.
    1961 Introduction to soil microbiology. John Wiley and Sons, Inc., New York. 472 p.

13. FULLER, W. H.
    1946 Evidence of the origin of uronides in the soil. Soil Science Society of America, Proceedings 2:280–283.

14. MARTIN, J. P.
    1946 Microorganisms and soil aggregation. 2: Influence of bacterial polysaccharides on soil structure. Soil Science 61:157–166.

15. NORMAN, A. G.
    1943 Organic matter in Iowa soils. Iowa Agricultural Experiment Station, Bulletin P57:827–848.

16. FULLER, W. H.
    1966 Pelleted compost now turns former waste into useful products. Progressive Agriculture in Arizona 18(3):17.

17. ABBOTT, J. L.
    1968 Use animal manure effectively! Arizona Agricultural Experiment Station and Cooperative Extension Service, Bulletin A–55:1–11.

18. TISDALE, S. L., AND W. L. NELSON
    1966 Soil fertility and fertilizers. 2d Ed. The Macmillan Company, New York. 694 p.

19. BRAY, R. H.
    1954 A nutrient mobility concept of soil-plant relationships. Soil Science 78:9–22.

20. BRAY, R. H.
    1963 Confirmation of the nutrient mobility concept of soil-plant relationships. Soil Science 95:124–130.

21. TUCKER, T. C., AND B. B. TUCKER
    1968 Nitrogen nutrition. *In* F. C. Elliot, M. Hoover, and W. K. Porter, Jr (eds), Advances in production and utilization of quality cotton: principles and practices, p. 183–211. The Iowa State University Press, Ames. 532 p.

22. GARDNER, B. R., AND T. C. TUCKER
    1967 Nitrogen effects on cotton. II: Soil and petiole analyses. Soil Science Society of America, Proceedings 31:785–791.

23. FULLER, W. H., AND H. E. RAY
    1965 Basic concepts of nitrogen, phosphorus and potassium in calcareous soils. Arizona Agricultural Experiment Station and Cooperative Extension Service, Bulletin A–42. 28 p.

24. BRAY, R. H.
    1948 Requirements of a successful soil test. Soil Science 66:83–89.

25. FULLER, W. H.
    1958 Soil compaction. Arizona Agricultural Experiment Station, Report 168. 7 p.

26. FULLER, W. H.
    1970 Desert soils of Arizona. Unpublished manuscript.

27. FULLER, W. H., AND H. E. RAY
    1963 Gypsum and sulfur-bearing amendments for Arizona soils. Arizona Agricultural Experiment Station and Cooperative Extension Service, Bulletin A–27. 12 p.

28. FROST, K. R.
    1967 Strip tillage and petroleum mulch. Progressive Agriculture in Arizona 19(5):14–16.

29. FULLER, W. H., E. W. CARPENTER, AND M. F. L'ANNUNZIATA
    1968 Evaluation of municipal waste compost for greenhouse potting purposes. Compost Science 8:22–26.

30. CAMERON, R. E.
    1961 Algae of the Sonoran Desert in Arizona. University of Arizona (Ph.D. Dissertation). 118 p.

31. CAMERON, R. E., AND C. B. BLANK
    1965 A. Soil studies — microflora of desert regions. VIII: Distribution and abundance of desert microflora. California Institute of Technology, Pasadena, Jet Propulsion Laboratory, Space Program Summary 37–34, vol. 4:193–202.

32. AL-DOORY, Y., M. K. TOLBA, AND H. AL-ANI
    1959 On the fungal flora of Iraqi soils. II: Central Iraq. Mycologia 51:429–439.

33. FULLER, W. H., R. E. CAMERON, AND N. RAICA, JR.
    1960 Fixation of nitrogen in desert soils by algae. International Congress of Soil Science, 7th, Madison, Transactions 2:617–624.

34. GREAVES, J. E., AND L. W. JONES
    1941 The survival of microorganisms in alkali soils. Soil Science 52:359–364.

35. LOCKHEAD, A. G.
    1957 Qualitative studies of soil microorganisms. IV: Capabilities of the predominant bacterial flora for synthesis of various growth factors. Soil Science 84:395–403.

36. LOCKHEAD, A. G., AND M. O. BURTON
    1957 Qualitative studies of soil microorganisms. XIV: Specific vitamin requirements of the predominant bacterial flora. Canadian Journal of Microbiology 3:35–42.

37. MARTIN, J. P., AND J. O. ERVIN
    1956 Soil organisms — fact and fiction. University of California Citrus Experiment Station, Riverside.

38. MARTIN, W. P., L. N. GOODING, L. P. HAMILTON, L. BENSON, AND J. E. FLETCHER
    1948 Observations on the nodulation of leguminous plants of the Southwest. U.S. Department of Agriculture, Soil Conservation Service, Region 6, Bulletin 107, Plant Study Series 4:1–10.

39. BERNSTEIN, L., AND H. E. HAYWARD
    1958 Physiology of salt tolerance. American Review of Plant Physiology 9:25–46.

40. FULLER, W. H.
    1962 Reclamation of saline and alkali soils. Plant Food Review 19:7–9.

41. FLETCHER, J. E., K. HARRIS, H. B. PETERSON, AND V. N. CHANDLER
    1954 Piping. American Geophysical Union, Transactions 35:258–263.

42. BEAR, F. E.
    1965 Soils in relation to crop growth. Reinhold Publishing Corporation, N.Y. 297 p.

43. BREAZEALE, J. F., AND H. V. SMITH
    1930 Caliche in Arizona. Arizona Agricultural Experiment Station, Bulletin 131:419–441.

44. COOK, R. L.
    1962 Soil management for conservation and production. John Wiley, N.Y. 527 p.

45. FULLER, W. H.
    1969 Phosphorus and phosphorus cycle. In R .W. Fairbridge (ed), Encyclopedia of earth sciences. IV: Encyclopedia of geochemistry. Reinhold Book Company, New York.

46. GOTASS, H. B.
    1956 Composting. Sanitary disposal and reclamation of organic wastes. World Health Organization, Geneva.

47. GRAY, T. R. G., AND D. PARKINSON
    1968 The ecology of soil bacteria, an international symposium. University of Toronto Press, Toronto.

48. GUSTAFSON, A. F.
    1941 Soils and soil management. McGraw-Hill Book Company, Inc, New York. 424 p.

49. HUNT, C. B.
    1960 The Death Valley salt pan, study of evaporites. Geological Survey, Professional Paper 400B: 456–458.

50. KELLOGG, C. E.
    1941 The soils that support us. The Macmillan Company, New York. 370 p.

51. KLINGEBIEL, A. A., AND P. H. MONTGOMERY
    1961 Land-capability classification. U.S. Department of Agriculture, Soil Conservation Service, Handbook 210:1–21.

52. KONONOVA, M. M.
    1961 Soil organic matter, its nature, its role in soil formation and in soil fertility. Translated from the Russian by T. Z. Nowakowski and G. A. Greenwood. Pergamon Press, Oxford and New York.

53. MARTIN, J. P.
    1945 Microorganisms and soil aggregation 1: Origin and nature of some aggregating substances. Soil Science 59:163–174.

54. MAYLAND, H. F., T. H. McINTOSH, AND W. H. FULLER
    1966 Fixation of isotopic nitrogen on a semiarid soil by algal crust organisms. Soil Science Society of America, Proceedings 30:56–60.

55. SHIELDS, L. M., C. MITCHELL, AND F. DROUET
    1957 Alga- and lichen-stabilized surface crusts of soil nitrogen sources. American Journal of Botany 44:489–498.

56. THIMANN, K. V.
    1963 The life of bacteria. 2d ed. Macmillan, New York.

57. THORNE, D. W., AND H. B. PETERSON
    1954 Irrigated soils, their fertilizer and management. The Blakiston Company, Inc., New York.

58. WAKSMAN, S. A.
    1952 Soil microbiology. John Wiley and Sons, Inc., New York.

Part Three

WATER AND AGRICULTURE

Exploitation of Groundwater for Agricultural Production in Arid Zones

DAVID J. BURDON

Land and Water Development Division
Food and Agriculture Organization of the United Nations
Rome, Italy

ABSTRACT

THE PRINCIPAL USES of groundwater for agricultural production in desertic and semi-desertic zones may be grouped into four major categories: irrigation of desert oases, irrigation of riverain alluviums, stock water on pastures, and water for transdesert movement. Many challenges present themselves in the change from traditional to modern concepts of groundwater investigation, development, use, and conservation. Numerous technical problems interact with the social, economic, and personal factors involved in changing a subsistence, self-contained society into one of interdependence and integration.

Irrigation with groundwater is not very different from irrigation with surface waters in the desert oases and in the valleys of major rivers flowing into and through the deserts and subdeserts, although important differences do exist. Understanding of the origin (past and present) of groundwater and of its aquifers is important in the determination of the amount of water available for use. Methods of developing and extracting groundwater in arid areas depend on such problems as high evaporation, lack of drainage, and aeolian sand encroachment. The problem of salinity is the main technical problem facing irrigation with groundwater in the arid and semiarid regions. Other problems relate to recharge and conservation, since groundwater is being drawn from storage and water levels and pressures will fall.

Groundwater is used to supply drinking water for herdsmen and beasts grazing the pastures of the desert, as they make the traditional transhumance journeys to utilize the vegetation of steppe and savannah to maximum advantage, and for all types of travel and movement across the desert. Quality as well as amount of water frequently is limiting. Controlled use of watering points is helpful in managing grazing and preventing overgrazing; the problem of the maintenance of desert watering points is not yet fully solved.

Projections made by the Food and Agriculture Organization of the United Nations (FAO) in its Indicative World Plan for Agriculture (IWP) suggest major irrigation developments, involving surface and groundwater, in fifty-two developing countries of Asia, the Near East, Africa, and Latin America. These projects also cover future use of groundwater to increase animal production on the rangelands of the world. Future action to facilitate effective groundwater utilization for agriculture must be comprehensive, and must consider water as but one of many inputs required to increase production.

EXPLOITATION OF GROUNDWATER FOR AGRICULTURAL PRODUCTION IN ARID ZONES

David J. Burdon

THIS study is intended to present information on present uses and possible future uses of groundwater for agricultural production, mainly through irrigation and stock watering, in the arid regions of the world (1). However, the work is not restricted to fully arid regions, but extends into the steppe, savannah, and dry pampas, where rainfall provides vegetation for pastures; it deals mainly with the deserts and steppe-savannahs of the Old World, regions with which the author is personally acquainted.

The extent and limits of deserts are defined with difficulty, since they grade into surrounding or more elevated areas of higher precipitation. As used here, the terms *desert, steppe,* and *savannah,* as well as *arid* and *semiarid,* refer mainly to the natural ecology and the potential for agriculture and grazing of regions. *Fully arid* and *arid* may be considered as referring to desert regions, emphasizing the dryness aspect; *semiarid* also lays stress on lack of moisture and embraces steppe and savannah. Thus, *desert* (fully arid or arid) refers to seasonally warm to hot regions whose rainfall is insufficient to produce pastures, averaging less than 100 millimeters yearly, with no precipitation at all in some years. Rivers from afar may flow through arid lands, and while such regions remain arid, they are not deserts; the Nile Valley in Egypt is a very striking example of such conditions. *Steppe* (semiarid) refers to regions where the seasonal cold precipitation is sufficient for pastures, scattered trees, and some crops, but where the average precipitation is less than 400 millimeters yearly. *Savannah* (semiarid) refers to regions where rainfall occurs mainly in the hot weather, where it is sufficient for pastures, for scattered trees, and for some crops, but where the total precipitation does not average more than 500 millimeters.

The comparative sizes of the arid lands of the world, as well as cultivated areas, both rainfed and irrigated, are given in Table 1. The six million square kilometers of arid land in this tabulation refer to arid lands not suited to surface water (river) development (2); according to other classifications, arid and semiarid lands account for as much as 13 percent of the total land surface. Table 1 also presents data on the waters of the world, indicating the importance of the underground as a storage reservoir of non-oceanic waters. These sets of figures are very general, and are given only to place this presentation in its proper perspective.

GROUNDWATER USE IN THE ARID LANDS

Groundwater is the principal water resource of the arid lands, and to a great extent of the semiarid lands. Other waters, mainly rare and irregular precipitation that sometimes gives rise to flash floods, support some vegetation — particularly in areas where runoff concentrates sparse precipitation — and may fill seasonal waterholes; however, these other sources of water and pasture can best be utilized in conjunction with groundwater obtained from wells in the past and now also from boreholes. Dew and fog are of interest, but have little practical significance. Great rivers such as the Nile, the Tigris-Euphrates, the Oxus, or the São Francisco, or long irrigation canals as in California, bring large quantities of surface water into the arid regions of the world. There is groundwater in the alluvium of their valleys, which was and is developed for irrigation.

Because the emphasis herein is on groundwater use in the deserts and steppe-savannahs of the Old World, many of the references and examples are to the Sahara, the Arabian and Syrian desert, and to the Indus-Ganges basins. These are the regions where our concepts and structure of civilization took form, based to no small extent on the use of water for irrigation. It is here that the change from traditional to modern uses presents the greatest challenge to groundwater development and use.

The different uses of groundwater within the arid and semiarid zones include (a) irrigation in the true deserts and thus the formation of oases; (b) irrigation in the valleys of great rivers flowing into and through the arid and semiarid zones; (c) watering stock and herdsmen on the pastures of the semiarid regions on the periphery of deserts; and (d) use for travel routes across desert and steppe, with emphasis on the transhumance movement of animals. At the same time, modern large-scale development and use of such groundwater makes it necessary to consider how much groundwater exists, how much is currently recharged, and where the groundwater originated in time and space. To present in simple form the various uses and origins of groundwater considered in this presentation, Table 2 has been prepared.

The percentages given in Table 2 are estimated only to indicate the relative importance of past and current groundwater infiltration in supplying water for the four

TABLE 1

Generalized Summary of Land Uses and Water Volumes of the Earth

Land Areas, in Square Kilometers		*Water Volumes, in Cubic Kilometers*	
World land surface	144,600,000	Total world waters	1,360,000,000
Forest use	40,000,000	Oceanic waters	1,320,000,000
Meadows — pastures use	26,000,000	Atmospheric waters	13,000
Cultivated (rainfed)	13,000,000	Continental waters	37,800,000
Cultivated (irrigated)	1,600,000		
Total use	80,600,000	Icecaps and glaciers	29,200,000
		Groundwater (to 4,000 meters)	8,350,000
Waste	44,500,000	Freshwater lakes	125,000
Unsuitable for Surface Water		Saline lakes and seas	104,000
Development		Soil moisture	67,000
Ice-cap and cold	13,500,000	Rivers (average instantaneous	
Arid	6,000,000	volume)	1,250
Total unused	64,000,000		

NOTE: The cultivated area is some 10% (14.6 million square kilometers) and the irrigated area is some 1.1% of the Earth's land surface. Arid lands cover about 4.1% of the land surface. To irrigate 1.6 million square kilometers, some 3,000 cubic kilometers of water have to be applied to the land (*2-4*).

different uses listed. Thus it is thought that more than half of the quantities used or planned for irrigation in desert oases comes from past infiltration, from great volumes of groundwater stored during the pluvials. For the smaller amounts required for stock watering, transhumance, and desert travel, only some 30 percent may be drawn from past storage; shallow wells and boreholes in wadi alluvium can supply such water from presently existing local infiltration in many areas of the steppe and savannah.

Historical Background

The history of groundwater development and use has been treated, directly and as background material, by many writers, historians, and archaeologists. It is not intended to cover such fields here. It does seem desirable to indicate that the historical background to man's exploitation of groundwater for food production purposes in the arid and semiarid zones can be considered in terms of ingenious adaptations of natural discharge phenomena to supply man's needs and desires. This background is to be found in almost all the older records and has been inferred by the archaeologists and historians. Recently, it has been subjected to new analyses (*5*, p.

TABLE 2

Indications of Origin of Groundwaters as Developed and Used in Deserts and Subdesertic Regions of the Sahara, Arabia, and India-Pakistan

	Origin of Groundwater (%)			
	Past		Present	
Use of Groundwater	Local	Distant	Local	Distant
Irrigation; desert oases	5	50	Nil	45
Irrigation; riverain alluvium	5	Nil	95	Nil
Stock water for pastures	5	25	45	25
Transdesert travel, etc.	5	25	45	25

NOTE: Rainfall, surface runoff, and river flow make varying contributions to the uses of groundwater listed in this table (except for the irrigation of desert oases).

208) by bringing hydrological knowledge to bear on achaeological data. These data are most abundant in the regions of North Africa, Asia Minor, Mesopotamia, and the Indus-Ganges valleys. Some idea of the care with which water was husbanded and used in the Negev and Sinai deserts over 3,000 years ago may be gathered from such a recent article as "Ancient technology and modern science applied to desert agriculture" (*6*).

The oases of the true deserts are located where deep groundwater found natural discharge points. Aeolian forces eroded down on favorable geological structures until an aquifer, often confined, was cut, and seepages or springs developed. Moisture in the form of swamps and lakes resulted and arrested aeolian erosion; Kharga-Dakhla, in the Western Desert of Egypt, is such an oasis. Faulting has also opened up deeper confined aquifers and led to springs; the great Hufhuf Oasis in eastern Saudi Arabia, with a discharge of some 14 cubic meters per second, is such an oasis. In all cases, the oases occur in closed depressions, and the problem of drainage was always present and seldom solved; waterlogging and salinity are characteristic features of natural oases.

The development of river alluviums for agriculture has been through wells, and appears to have been an invention and not an adaptation of a natural phenomenon. However, by extending this technique outside the river valleys, man opened up the way to more effective travel across the full deserts and to more reliable, safer use of the steppe and savannah for grazing. Initially, desert crossing would be based on water in pools and riverbeds after rain, plus water from springs in the oases. Wells in wadi alluvium were followed by deeper wells in solid rock, until such major transdesert routes as the "Queen's Highway" (Queen Zubaydah, about A.D. 800) from Baghdad to Medina could be based on wells over 200 meters deep and spaced at regular intervals.

Wells are also sunk to use the deeper waters of the oases. In so doing, the problem of constructing wells into artesian aquifers was encountered and solved. In such oases as Touggort and Ouargla, in Algeria (7), wells were sunk and maintained by divers who dug and worked under water. The exertion and peril involved is a measure of the value of water in the arid zone, a measure which we — with rotary drill, casing, turbine pump, and motor — are apt to forget.

The devices used by man to lift water from wells and depressions and to bring it under control for efficient use are extremely numerous. The most important devices have been listed and described by Molenaar (8), but each area has had its own adaptations of more standard devices. In addition to human and animal power, moving water was also employed to lift some water where possible, but for extraction from wells, human and animal power were employed almost exclusively. To obtain water for stock, such power is still of major importance in the Old World. The use of wind-power and solar energy to lift groundwater on the steppe and savannah is comparatively new, though windmills were used elsewhere to pump and grind grain.

The famous fogarra (kharez, khanat, chain-of-wells) is an adaptation of spring discharge, and is essentially an infiltration gallery. It must have originated in regions of small seasonal rainfall (say 300 to 500 millimeters), where there is sufficient topographical difference to allow the gallery to be constructed with a discharge slope less than that of the water table but greater than that of the land surface. Zones of weathered chalk, as the Senonian of the Syrian steppe, are favorable to such development, though the longest known khanat (60 kilometers) is found in Persia. The fogarra has been adapted for use in full desert and elsewhere, but it is essentially a man-made spring, originating in zones of natural springs.

Challenge of Changes

The techniques and methods of modern groundwater investigation, development, use, and conservation present many problems and challenges in themselves, even in the development of virgin lands and new aquifers. Such problems and challenges are compounded when an existing groundwater utilization system has to be improved, expanded, and rendered more efficient; the traditional users and owners are then immediately involved, and these are people who have survived due to careful adherence to procedures and practices that have stood the test of time.

Such groundwater development may be undertaken by governmental organizations, with greater or lesser contribution by private enterprise, mainly by the oases-dwellers. An outline of the present groundwater development procedure in Saudi Arabia is given by Burdon and Otkun (9). The main challenge of change is presented to the inhabitants of the oases and to the herdsmen, and to the traditional infrastructure which they require and support. But such groundwater development presents challenges to government in the need to create suitable laws, institutions, and organizations, to carry out the basic resource surveys, to obtain finance and con-

tractors to develop, and then to see that the development projects are established on a sound economic and continuing basis, with adequate maintenance.

Some of the main challenges of change introduced by modern groundwater development and use for food production in arid and semiarid regions may be listed as follows: (a) too much influence of the past, so that water is developed in or close to older installations, whereas modern methods give a wider range of development locations; (b) inability to take full advantage of modern concepts (such as well construction, efficient pumps and maintenance) due to difficulty in applying technical knowledge to the development of water, and teaching the cultivators how best to use this water; (c) danger of overextraction of groundwater, upsetting old balances, causing waterlogging and salinity; where confined aquifers are cut, interaquifer leakage and uncontrolled surface flow are common; (d) failure to plan ahead for falling pressures and water level, for increased salinity, and for the new economy that must be established; (e) introduction of dependency on external supplies (as equipment and fuel) and maintenance, whereas before water was produced and used within a self-contained subsistence economy and technical tradition; and (f) overgrazing of pastures by stock, and related overuse of natural resources based on increased availability of water for man and beast. It is not possible to expand here on any of these items, or to give a longer list. However, the main outlines of the challenge can readily be seen, and it is clearly not only a technical challenge but also a social and economic one.

Alternative Uses

Since groundwater is used in conjunction with surface water in river alluvium irrigation schemes, and since groundwater is essential for drinking on pastures and transhumance travel routes, serious consideration to alternative uses can be given only to groundwater used for irrigation in the oases of the full desert.

Groundwater could be used for purposes other than irrigation in many desert areas. Considering the high consumption of water for irrigation, the limited amount might in some cases be better used for industry, or as an essential contribution to a tourist and recreational trade or in other lower consumptive uses. Likewise, the return per cubic meter of water invested in food production is low, even where a ready and wealthy market exists for the products. For Tucson, Arizona, Wilson (10) has given useful figures: income per cubic meter used for agriculture was $0.171, for mining $3.042, for manufacturing $2.610, and for government $8.00. Figures such as these suggest that before one decides to use groundwater for irrigation in desert oases, alternative uses should be considered.

In regions where settlements already exist and where the development objectives are to improve the lot of the inhabitants or to expand the region, it is almost always necessary as a first step to expand agriculture and increase food production. In turn, this may lead to simple and then more complex processing of the agricultural production, and so in time lead to the introduction and development of industry and tourism.

PRESENT USE OF GROUNDWATER FOR IRRIGATION IN ARID ZONES

This section of the study deals with irrigation uses of groundwater in the true deserts, leading to the formation of oases, and with the use of groundwater in connection with surface water where major rivers flow into and through arid and semiarid regions.

Location and Amount of Groundwater

Within an arid zone, groundwater in amounts suitable for irrigation has originated by infiltration from past precipitation, from current precipitation outside the arid region, or from local infiltration from major rivers flowing into the arid zone from regions of higher precipitation. Rarely, floods from storms within the arid zone itself may contribute considerable recharge to gravels in wadis and so to deeper aquifers, but such water is more important for stock pastures and watering than for irrigated agriculture. In coastal deserts, some groundwater is to be found close to the sea; it is replenished by rare floods and is useful for ports.

Groundwater originating outside the arid regions is found in the deeper aquifers, not infrequently under confined or artesian conditions. This applies to the great Australian artesian basin, and to the groundwaters of the epicontinental (Nubian facies) sandstones of the Sahara, Libyan, and Nubian desert of North Africa and the Arabian Desert (as in *11*). Some of this water is old, and provisional ages of as great as 24,000 years have been assigned to it by radiocarbon dating. Such old water may also have originated outside the present arid region where it now occurs; in past, pluvial time, there may have been much precipitation on present arid regions, but the geological conditions did not always favor deep infiltration. Loss of older groundwater by natural discharge would have provided a mechanism for replacement of older by younger groundwater.

Where deserts are underlain by crystalline rocks, as the pre-Cambrian of the Afro-Arabian Shield, in much of Australia, and in the northeast of Brazil, groundwater is not found in quantities sufficient for irrigation, but runoff from such impermeable regions may lead to considerable recharge to bordering aquifers.

Groundwater replenished from current river flows is found in the alluvial valleys of all the great rivers (the Nile, Tigris-Euphrates, Indus-Ganges, Oxus, the São Francisco, for example) that flow into the semiarid and arid zones. This water is found at shallow depths and is generally of recent origin. There is active replenish from, and sometimes discharge to, the main river. Where such rivers flow over rock outcrops, there may be current recharge to deep or artesian aquifers. Development may stop or increase such recharge; the High Dam at Aswan is expected to increase infiltration to the sandstone aquifers by extending lakes Nasser and Nubia over outcrops of these rocks.

The global approach to hydrology has been emphasized by the International Hydrological Decade. It has meant that work has been done on the hydrology, including infiltration rates, in remote regions; these are often the recharge regions for the deeper aquifers of the great deserts. Likewise, the emphasis on the basin concept has applied not only to surface drainage but to the more difficult concept of groundwater basins. In so doing, much has been learned about the origin, amount, and composition of groundwater.

Development and Extraction of Groundwater

The development and extraction of groundwater for irrigation in the arid and semiarid zones was initially, and until very recently, based on ingenious improvements of natural-discharge phenomena, as outlined in Historical Background, above.

The investigation of groundwater for irrigation in arid zones follows the classical methods and need not be expanded here. In general, factors other than purely hydrological (existing oases, soils, drainage, communications, people) determine where groundwater should be developed for irrigation.

In some cases preliminary geophysical investigations may be justified, mainly seismic to determine structures and electrical resistivity for physical aspects of water; airborne geophysical work, such as magnetometer measurements to determine depth to basement in Shield areas has been useful. Use of remote sensors for temperature and humidity do not as yet appear to have been employed on any large scale.

Where groundwater lies in deep aquifers, exploratory and development drilling by rotary methods are usually employed; in some areas, air-flush can be employed with advantage until water is struck. In artesian basins, very deep exploratory boreholes are justifiable, and deep production wells are not uncommon; in Saudi Arabia, exploratory boreholes have been drilled to over 2,000 meters, while production boreholes of 700 to 1,000 meters are now producing irrigation water in the Sahara of Egypt and Libya.

Flowing artesian boreholes give cause for great initial rejoicings, but uncontrolled discharge can lead to rapid loss of pressure and decreasing flow, while surplus water can create waterlogged areas (swamps). Serious losses, and mixing of waters of different chemical qualities, can occur between aquifers in badly constructed boreholes. Since many of these boreholes are drilled within existing oases that developed in closed depressions, waterlogging from free-flowing boreholes is a common problem; the chain of oases of the Qasim Region in central Saudi Arabia provides a good example of this problem — which is being overcome by effective control as well as remedial measures.

In general, extraction is by deep-well turbine pumps, with diesel (and sometimes electrical) motors. Costs of water are hard to determine and are extremely variable;

there is often an approach which considers water in the desert so precious as to be above price! A recent careful study of costs of groundwater on the High Plains area of Texas (12) indicated variations from 0.15 to 5.90 U.S. cents per cubic meter, with an average of 1.26 cents per cubic meter; rates of production averaged 100 cubic meters per hour, and cost of pumping per cubic meter per foot lift ranged from 0.0035 cents to 0.225 cents. With such variations within a well-developed area and with ready access to maintenance facilities, supply of fuel, lubricants, and so on, it can be imagined that cost of irrigation water for irrigation schemes in the deserts will indeed be high and variable.

In the river alluvium regions, where extraction on a large scale is feasible, and standard production tubewells can be built with power supplied from an extensive electrical grid, the position is very different. Thus in India alone some 5.3 million wells, tubewells, and waterpoints extract annually 74,000 million cubic meters (20 percent of estimated annual recharge) for irrigation, mainly in the Indo-Gangetic plains (13, p. 42). The rate of increase is some 41,000 tubewells per year (14), while electrical power is being extended to existing wells at the rate of 14,000 per year. Dhir (15) has given extensive background data to this present expansion of tubewell irrigation.

Irrigation with Groundwater

In general terms, irrigation with groundwater is not very different from irrigation with surface waters. Factors such as plant requirements, rates, and amount of application, remain somewhat unchanged; much of the information in *Surface Water Development in Arid Regions* (16) applies to groundwater, as do standard FAO publications on irrigation, such as those by Marsh (17) and Booher (18), and basic work such as that of Blaney (19).

But there are important differences. With pumped groundwater, and with proper completion of artesian boreholes, the cultivator has direct control of the amount of water supplied and the times of supply. This will tend to lead to the use of minimum quantities in economizing direct costs for water production. In itself, such economy is good and may be combined with the three-story (palms, fruit trees, and vegetable-forage) system to use the water to the utmost. Since such irrigation is taking place in areas of high evapotranspiration however, too little water may be supplied to satisfy crop requirements (as listed in 20). Certainly too little water is generally supplied to provide sufficient drainage water to carry off as much salts as are introduced with the irrigation water; salts tend to accumulate in the soil. The possibilities of subirrigation to reduce evaporation and salt concentration are excellent in areas where skill and capital are available. However in many areas such systems are entirely too complicated and costly to find favor with the inhabitants.

The problem of windblown sand, as encroaching dunes and as sediments arrested in and on any moist surface, is a general problem in most oases. Irrigation canals, drainage canals, and the evaporating pans are all moist, and once aeolian sand and dust reaches them it is arrested and settles to block the canals and drains. Such a problem may be less acute where groundwater is used, for the canals are shorter and more readily maintained as the property of an individual owner-user; the problem of blocked drains remains.

In many areas, sprinkler irrigation is used, mainly to conserve groundwater. Pressures are obtained by the same pump that extracts the groundwater; pressures may be from an overhead tank of small dimensions. Sprinkler irrigation cannot be used with poor-quality water, as evaporation from leaf surfaces will cause burns and other damages.

With groundwater, each cultivator, cooperative, or scheme has its water resources, water reservoirs, and supply mechanism under its own control. The cost of development and production is composed of many self-contained units of expenditure based on a borehole production unit. This facilitates investment and production.

Other Inputs for Agricultural Production

More than sand and water are required for efficient agricultural production. Although here we are dealing with the groundwater input to agriculture under arid-zone conditions, it is necessary to name some of the other essentials.

On the resources level, irrigation should be on the most suitable soils; this covers texture, natural fertility, form, slope, extent, and form of areas, and the all-important drainage and water-retaining capabilities. The input of fertilizers must be considered; these will have to be imported with high transport costs, though in North Africa and the Middle East phosphate deposits do occur near several of the oases. In general, there will have to be an attempt to build up fertility within the oasis and reduce importation of fertilizers. Good seeds and plants must be supplied, with the full backing of government research and extension services.

On the investment and marketing side, funds must be made available to carry out the necessary studies, decide on the economic feasibility of the project, and, if positive, arrange for investment. In turn, a marketing study and organization will be required, for products must be exported and sold to pay for development and operations. Unlike traditional water development, modern development must be linked to external inputs paid for by sale of products.

Here only a mention can be made of the legal and institutional aspects of the work. Who owns the water and has rights to its use? Existing users of water may be affected by falling water levels. Primitive producers (as of forage for caravans) may be driven out of business by large-scale production with deep groundwater. The allocation of water to the best lands may involve a change in land-tenure patterns. Cooperatives must be established to deal with new inputs (water, fertilizers, new varieties and species) and outputs (storage, transport, markets).

Salinity and Other Problems

Salinity, including alkalinity in many oases, is the most serious problem facing irrigation with groundwater in the arid zone. This is due to (*a*) initial high salinities and temperatures, (*b*) high evaporation, (*c*) lack of drainage, (*d*) recirculation of groundwater, and (*e*) operation in closed basins.

Basically, the problem of salinity is the same as in other areas, and the lines laid down by such guides as the U.S. Department of Agriculture (*21*, chapter 4) and more recently by the National Technical Advisory Committee (*22*, table IV-14) are followed. Trace elements may also cause difficulties (as high fluorine in gypsum aquifers); they are less important in regions of transhumance than in the oases.

Groundwater is generally more saline than surface water, since it comes in close contact over long periods with the minerals of the aquifers, many of which are soluble. Arid zones of today were often arid zones in the geological past, so evaporate sediments are not uncommon and they are readily soluble. Likewise, flushing of the aquifers of connate and other waters is less effective due to low groundwater put-through. High pressures and temperatures increase solubility of some minerals, and gases in solution also affect solubility of other ions.

Once the groundwater is produced at the surface, it is exposed to rapid evaporation due to high temperatures and extreme dryness of the atmosphere and environment. The initial high temperatures mean that, even at night, evaporation is high. Use of special techniques, such as covered tanks and subsurface irrigation conduits, could reduce evaporation; in fact they are seldom used. So, the initial salinity of the groundwater is increased by evaporation even before it reaches the plant. In the course of actual irrigation, evapotranspiration losses are continued; and limited applications of water often mean that the salinity of the fluid in actual contact with the roots may become very high.

Salinity buildup in the soils occurs because the irrigation water supplied is seldom enough to carry away as much salt as is introduced with the irrigation water. To maintain a salt balance, incoming salt should equal outgoing salt. Since the amount of drainage water is small, and since drainage systems are absent or inefficient in many regions, there is a buildup of salt in the root zones and in the cultivated areas. Attempts to wash the salt down often result in a rising water table (there being no drainage outlet), so that the salt is brought up into the zone of capillary evaporation, and salt crusts increase.

Vertical drainage is not uncommon in the sandy soils and closed-basin conditions of soil morphology of oasis irrigation. Thus, water which, due to evaporation and some soil leaching, has increased its salinity, is fed back to the upper aquifer. From this it is likely to be reextracted through pumping from wells or shallow boreholes. Thus, the groundwater applied becomes progressively more saline, and it can be clearly seen that a system is established for circulating water continually enriched in salts.

Finally, the fact that oases tend to occur in closed depressions means that while water can escape by evapotranspiration, the salts must remain behind, except for some limited aeolian removal. In the best cases, the evaporation pans and so the residual areas for salts (saltpans, salinas, sebkhas) lie far from the irrigated area. But they must occur at the base level for each depression and basin of closed drainage, which is also the level of cheapest groundwater extraction; so in some areas they encroach onto the irrigated area, which is thereby destroyed and has to be moved elsewhere. The more that irrigation water is extracted and used to combat high salt content in the root zone, the greater the amount of water that reaches the evaporation pans and the larger they become. Man has, as usual, upset the water, salt, and vegetation balances in his own interests; but if he goes too far, he obtains only diminishing returns from the transformation.

Similar problems occur in the wide areas irrigated from tubewells in the river alluviums of the arid and semiarid zones, but here the problem is less acute. The groundwater is often newly recharged from the rivers and is of good composition. The amount is larger, lifts are lower, and water can be applied. But the essential fact is that the operation is not taking place within a basin of closed drainage, and return groundwater or irrigation flow to river or canal moves much of the salt out of the area. Of course there are dangers; high water tables will bring water into the zone of evaporation, and salinization and waterlogging are not unknown. Such dangers tend to be much more frequent with surface-water irrigation schemes, and, in fact, introduction of tubewell irrigation is often effective in dealing with the many problems induced by a high water table.

This matter has attracted much attention by FAO, in particular in the Tigris-Euphrates and Indus valleys; attention is drawn to (*23*) *Seminar on Waterlogging in Relation to Irrigation and Salinity Problems* held in Lahore in 1964, and to (*24*) the joint FAO/Unesco publication, *International Source-book on Irrigation and Drainage of Arid Lands in Relation to Salinity and Alkalinity* (1967).

Saline groundwaters can, of course, be demineralized, and indeed are so treated for limited drinking supplies. Solar stills operate well under desert conditions, while the electrodialysis and reverse-osmosis methods are suitable for improving groundwaters with medium to low salinity. However, all such treatments are still far too costly to apply to irrigation water, except in certain unusual circumstances. Kuwait is seriously involved in work on the use of desalinated seawater for agricultural uses, possibly mixed with brackish groundwater which can be produced in large quantities from the Dammam Formation and from the Kuwait Group aquifers.

Where salinity is absent, it is often replaced by another danger, due to the low mineralization of the groundwater. Hot groundwater, often with dissolved gases, and with few ions (say less than 400 parts per million) in solution is extremely corrosive. It will attack all standard types of casings, screens, pumps, and ancillary equipment; cathodic protection may be helpful. In certain areas, such equipment must be made (at great expense) of noncorrosive material or coated with plastic or other corrosion-resisting covering.

Recharge and Conservation

In using groundwater for the irrigation of expanded areas in preexisting oases, it is almost certain that some groundwater is being drawn from storage. If it be accepted that prior to development a balance existed between current recharge and current discharge under natural conditions, then the additional water from new boreholes comes initially from storage. The release of pressure head and the increased drawdowns will, however, draw more water into the production area, and in due course a new balance should be reestablished. Since the irrigated areas of oases in a desert zone may represent only 1 percent of the total area, it is possible to draw from storage under an area very much bigger than the irrigated area.

Experience shows, however, that extraction is usually composed of current recharge plus withdrawal from storage and that water levels fall until arrested not by technical or legal control but by economic factors. However, steps should be taken (as by dating the water, maintaining records of extraction and levels) to determine the rate and danger of overdepletion. Legal steps should be taken to control and limit overextraction. But the final control will, in normal cases, be economic; when the cost of using the water to produce food, fiber, manufactured goods, or services to tourists exceeds the value of the product, then groundwater extraction will decline, water levels will recover, and a water balance and an economic balance will be restored, sometimes only at an unwelcomed cost in terms of human suffering.

In the wide spaces of the deserts, with recharge zones located in distant hills in a neighboring country, and with slow rates of groundwater movement, effective induced recharge to aquifers is not possible. In areas where there is some seasonal precipitation, and so limited runoff concentrated into a few floods, underground storage of water presents many advantages as compared to surface storage.

The proper manipulation of groundwater reservoirs, with the help of analog and digital models, is a wide and expanding technique that has application to the conservation and proper use of the groundwaters of the arid zone, but it is a subject that need not be elaborated here.

PRESENT USE OF GROUNDWATER FOR STOCK-WATERING IN ARID ZONES

This section of the study deals with the use of groundwater for stock and herdsmen utilizing the grazing resources of the semiarid regions on the periphery of the deserts, and for travel routes across the deserts, with emphasis on the transhumance movement of animals. Efficient use and conservation of these pastures and their

management in conjunction with irrigated fodder can greatly increase the production of meat, milk, wool, hair, and other animal products from these arid and semiarid regions of the world.

Location and Development of Water for Stock-watering

There is a certain amount of freedom in siting boreholes and wells for stock and movement, since their use is not restricted to certain soils, slopes, or habitations. The hydrogeologist can seek the best point from a technical aspect. But the bore may be the first in the area and the only one for many kilometers; all such boreholes are exploratory in nature. In general, geophysical work is not justified, since whatever they find will need a borehole for full interpretation; if the geologically-sited borehole proves "dry" (humid), then a geophysical survey, to identify a different lithological sequence area, may be justified.

The borehole is also seeking water in minimum quantity, sufficient to water man and beast for a limited time. Reliable yields of as little as 1 liter per second can justify equipping such boreholes on the steppe and savannah. If the water rises so that pumping lifts are small (or nil), so much the better. If hydrogeological conditions are favorable, and abundant groundwater of good composition is encountered, there may be possibilities of creating a small oasis.

The development of the groundwater is closely related to the way in which the water will be extracted. The simpler the method, the better. So, it is often advisable to enlarge the borehole to an open well, so that extraction can be by animal power and maintenance is free of external skills and material. In better-developed areas, hand pumps or windmill pumps can be used (25) and boreholes cased and screened so as to improve the hygiene of the installation. Where conditions are good, an effective operation-supply-maintenance organization established, and a budget ensured, then quite elaborate pump-houses, pumps, and power can be installed, with the water piped to well-distributed drinking troughs built well away from the water-supply point.

Quality of Water for Stock-watering

Some information on the amount of water required for stock and the limits of tolerated salts is given in Table 3. These figures are very general and vary with the seasons and the temperature, the amount of moisture in the pastures, travel distance, shade availability, and whether or not the animal is in milk. The practice of transhumance, in which there is seasonal movement to crop the scanty vegetative production of the desert, steppe, and savannah at the times most suitable to man and beast, has inherent dangers; these include long journeys at certain seasons when the amount and quality of the water intake will depart far from standard requirements. Certain salts have damaging effects, such as magnesium sulphate, while others, such as calcium and magnesium carbonate, are much less harmful. Very poor quality water may be tolerated for short periods but proves harmful over long periods; migration is helpful in this respect. Sykes (27) gives details of water requirements in relation to ambient temperatures and other factors for more humid areas, such as the United States.

TABLE 3

Normal Water Requirements and Safe Limits to Salt Content (Mainly NaCl) of Drinking Water for Animals in Arid and Semiarid Regions

Animal	Water Required (liters/day)	Limits of Salts (ppm of total dissolved solids)
Beef cattle	26–45	10,000
Dairy cattle	38–60	7,200
Horses	30–45	6,500
Swine	11–19	4,300
Sheep and goats	4–15	12,000
Camels (Syria steppe)	20	>12,000

NOTE· These data are based on Australian experience (22, tables IV-8 & IV-10), except for the camel, for which data are based on Syrian experience (26).

The bacteriological and other health aspects of the water are not elaborated here. Almost always, groundwater itself presents no health hazard but is liable to contamination between leaving the aquifer and the point of consumption.

Control of Grazing

While new watering points are usually designed to open up fresh areas to pasture, to allow animals to remain longer on good pastures, and to facilitate movements to pasture and marketing zones, it must be admitted that in many areas they have led to overgrazing followed by deterioration of rangelands, and have had the opposite effect from what was intended. The greatest damage can be caused by a few water points scattered well apart (say 40 to 50 kilometers between points); overgrazing occurs all around such points. Where water points are more numerous and are well distributed (say at an average of 10 kilometers, but varying for bovines, sheep, or camels), the grazing is spread more evenly over the whole region; if there is too much pressure on vegetation, however, the whole will be overgrazed.

New water points may also allow animals to remain in areas where hitherto there were only seasonal pools that dried up before grazing could damage the grass seeds. Here an additional month of grazing may do untold damage.

It is considered that one of the best tools for effective range management is control of the new (and old) water points. Water should be cut off at certain times each year so that the animals have to be moved off the surrounding range after optimum use has been achieved but before damaging overgrazing has commenced.

Problems of Maintenance

It must be emphasized that the task of keeping such watering points in efficient operation is much more diffi-

cult than their establishment; from the beginning, plans must be made to keep them in use. Much more could be written on the subject of maintenance, but it is a problem which is only too well known, if often overlooked in practice.

FUTURE OF GROUNDWATER USE IN ARID-ZONE AGRICULTURE

This section of the study indicates future trends in the use of groundwater for irrigation and for animal production in the arid and semiarid regions of the world; these trends are based on the studies carried out by the Food and Agriculture Organization (FAO) under the Indicative World Plan for Agriculture. There is much to be done. As was said recently, "We are still in the hunting and gathering stage of water use."

Future of Irrigated Agriculture

It is not possible to separate the problem of exploiting groundwater for irrigation in the deserts and semiarid zones from the general expansion of irrigation in the drier parts of the world. The importance of proper control of moisure in increasing agricultural production is now generally realized, and bringing land under irrigation will contribute enormously to supplying the food required by an expanding world population with higher social and economic objectives. In the Indicative World Plan prepared by the Food and Agriculture Organization of the United Nations, an analysis has been made of the

TABLE 4

Data on Selected Areas (in Square Kilometers) Under Cultivation in 1962, with Projections to 1985

Regions	Rainfed		Irrigated	
	1962	1985	1962	1985
Asia	2,236,000	2,683,000	494,000	1,028,000
Near East	349,000	404,600	160,400	189,500
Africa, South of Sahara	642,000	(no est.)	11,400	(no est.)
Latin America	948,000	(no est.)	62,000	98,000
Totals	4,175,000	. .	727,800	. .

NOTE: These projections for four developing regions of the world cover 52 countries with some 1,100 million inhabitants. [Data are extracted from publications by the Indicative World Plan for Agricultural Development (IWP) of FAO.]

Asia: 8 countries (670 million people) — Ceylon, China (Taiwan), India, Korea, Malaysia (West), Pakistan, Philippines, and Thailand. Harvested areas.

Near East: 10 countries (103 million people) — Afghanistan, Iran, Iraq, Jordan, Lebanon, Saudi Arabia, South Yemen, Sudan, Syria, and UAR (Egypt). Total areas.

Africa, South of Sahara: 24 countries (170 million people) — Central Africa: D.R. Congo (Kinshasa), Cameroon, Central African Republic, Chad, Congo (Brazzaville), Gabon; East Africa: Ethiopia, Kenya, Madagascar, Malawi, Tanzania, Uganda, Zambia; West Africa: Nigeria, Gambia, Mali, Mauretania, Niger, Senegal, Upper Volta, Dahomey, Ghana, Ivory Coast, and Togo.

Latin America: 10 countries (152 million) — Argentina, Bolivia, Brazil, Chile, Colombia, Ecuador, Paraguay, Peru, Uruguay, and Venezuela.

probable areas to be brought under irrigation in the coming 15 years to 1985. From the preliminary reports, Table 4 of this study summarizes the present and future position of irrigation for 52 developing countries, many of which have territories within the arid and semiarid zones of the world.

It is not easy to say how much of the increased areas under irrigation in 1985 will be from surface and how much from subsurface sources of water. In the Near East region it will be noticed that the overall increase in area will be small; this is due to the fact that the surface-water resources are already well developed, and the amount to be obtained from underground is not as yet fully known. In this connection, FAO (28, p. 273) considers that "in many parts of the world groundwater represents the greatest unexploited potential for increasing water supplies . . . it is now possible to obtain a much better and cheaper evaluation of potential groundwater supplies." So it must be assumed that future exploitation of groundwater for agriculture will be on an expanding scale. In part, this will be in the valleys of the rivers flowing through the arid and subarid lands, but it will also come from beneath the deserts. There large groundwater areas can be developed at selected sites, and water resources can be drawn through the aquifer system extending under a wide area to concentrate the water and to irrigate a restricted but valuable area of oases.

Future Use of Groundwater on Steppe and Savannah

The used of the grazing resources of the steppe and savannah for animal production is as yet in the subsistence stage throughout most, if not all, of the desert and semiarid zones of the Old World. In the New World the use of ranges to produce meat for market is general but is based on extensive rather than intensive use of land and water resources.

The need for increased protein production has been stressed by FAO as one of its five areas of concentration, and the Indicative World Plan has analyzed the problems and possibilities of achieving such increased production in the years to 1985. Some of the implications for the countries of the Near East and South America have been summarized in a recent publication (29). In the Near East, emphasis has been placed on increased milk production, but Nestel (29) notes, "Water shortage or misuse is an important factor which frequently inhibits agricultural production," and he stresses the advantages of growing fodder crops (for supplementary feeding) on irrigated lands. In Latin America, there still remain rangelands that are not fully used; expansion appears likely to be based on groundwater development, as in a project now under way in the Grand Chaco of Paraguay.

Full development and utilization of the savannah, pampas, and steppe regions of the world call for a multi-disciplinary approach, in which the investigation, development, and use of groundwater will play an important part; control of watering points may prove to be a most important method of controlling grazing and preventing overgrazing.

Action for Groundwater Exploitation And Management

The exploitation and management of groundwater in the arid and semiarid regions is and will be a difficult task, calling for government support to a greater or lesser degree. Government must establish the national policy for such development and see that the legal, institutional, and organizational bases are established to allow the work on groundwater to go ahead, with support from private initiative at appropriate levels.

It is clear that the normal sequence advocated and used by FAO for resource surveys has direct application to the future action on groundwater (30, p. 85-105). Resource surveys for groundwater should be followed by appraisal of the results, and a study of the way in which the resource can best be used and conserved for the benefit of the people and the country. This work is followed by development planning and by the selection of groundwater projects for execution; such development often commences with pilot schemes, followed in due course by full-scale development. In the case of groundwater for irrigation, pilot schemes can be used as full-scale test-pumping of the aquifer and will enable the results (as of aquifer coefficients, changes in salinity, etc) to be tested on a large scale before full development takes place. The need for related agro-climatological investigation and research must be stressed; the lines on which such work has been undertaken jointly by FAO/UNESCO/WMO are described in the publication Agroclimatology in Semi-Arid and Arid Zones of the Near East (31).

The tasks of groundwater utilization, maintenance, and conservation call for a continuing organization to collect the necessary data (as level changes, effects of recharge, chemical composition, increase of salinity), to record and analyze it, and to be able to suggest necessary modifications. A mechanism for the enforcement of such modifications (after approval by the competent authorities) must be set up and must be functional. Likewise, the organization must see that all public installations for groundwater are maintained, and that neither private nor public installation (in particular boreholes) are in bad condition and cause waste or deterioration of groundwater resources.

Economics of Future Use of Groundwater for Arid-Zone Agriculture

In considering future use of groundwater for arid-zone agriculture, it is wise to consider briefly the economics of such use, possibilities of alternative uses, and the other sources from which water may be obtained.

With regard to the use of groundwater for watering stock, for transhumance, and for general travel in the deserts and semidesert areas of the world, groundwater appears to have no real competition, either as to supply or as to cost.

For irrigation in the valleys of the major rivers flowing through the arid and semiarid regions, groundwater is in direct competition with surface water. Since, however, both surface and underground waters have a common source in these valleys, increased use of surface water will diminish groundwater replenishment, while increased use of groundwater will facilitate increased recharge to underground. The two uses are therefore complementary, and the amount of each utilized will depend on technical, economic, and social factors. There should be no real difficulty in establishing a satisfactory balance between tubewell irrigation and irrigation with surface waters along the great rivers flowing through the arid and semiarid zones of the world.

In the case of irrigation of oases in the full desert, where the only source of supply at present is groundwater, thought must be given to its possible uses for purposes other than agriculture. Tourism, recreation, and industry all have their possibilities. Again, desalination may turn waters that are now too mineralized for use into potential resources — provided, of course, that such uses can stand the cost of desalination.

REFERENCES AND NOTES

Since it would be quite impossible to achieve a list of references which would even approach completeness, it seemed best to indicate the wideness of the subject by selecting papers that could lead a reader to different types of publications, as those under the United Nations, Unesco, and FAO; those by the U.S. Department of Agriculture; and various books, proceedings, and miscellaneous papers that deal with, or touch on, the exploitation of groundwater for agricultural purposes in arid zones.

1. The thanks of the author are due to the Food and Agriculture Organization of the United Nations for permission to publish this paper, which was prepared in August, 1969.

2. UNITED NATIONS, DEPARTMENT OF ECONOMIC AND SOCIAL AFFAIRS
   1958   Integrated river basin development; report by a panel of experts. New York, 60 p. (UN Document E/3066).

3. FOOD AND AGRICULTURE ORGANIZATION OF THE UNITED NATIONS
   1961   Agricultural production yearbook. FAO, Rome.

4. NACE, R. L.
   1967   Water resources, a global problem with local roots. Environmental Science and Technology 1:550–560.

5. RAIKES, R. L.
   1967   Water, weather and prehistory. John Baker, London. 208 p.

6. SHANAN, L., M. EVENARI, AND N. H. TADMOR
   1969   Ancient technology and modern science applied to desert agriculture. Endeavour 28: 68–72.

7. BURDON, D. J.
   1953   Some aspects of the development of the water resources of Algeria. Water and Water Engineering 57:406–416, 467–471.

8. MOLENAR, A.
   1956   Water lifting devices for irrigation. FAO, Rome, 60 p.

9. BURDON, D. J., AND G. OTKUN
   1968   Hydrogeological control of development in Saudi Arabia. International Geological Congress, 23rd, Prague, 1968, Report 12:145–153.

10. WILSON, A. W.
    1963   Tucson: A problem in uses of water. *In* C. Hodge and P. Duisberg (eds), Aridity and man. American Association for the Advancement of Science, Washington, Publication 74:483–489.

11. AMBROGGI, R. P.
    1966   Water under the Sahara. Scientific American 214:21–29.

12. ULICH, W. L., AND A. W. SECHRIST
    1968   Irrigation costs and efficiency research. The Cross Section 15(5):1– .
    Quoted in Johnson's Drillers Journal, January–February 1969.

13. MANSINGHAL ASSOCIATES
    1968   The location of information sources regarding water resources in India, a project report prepared for U.S. Aid, New Delhi.

14. WILLIAMS, D. A.
    1968   Water utilization in India — Situations, developments, and recommendations. The Ford Foundation, New Delhi.

15. DHIR, R. D.
    1953   Hydrological research in the arid and semi-arid regions of India and Pakistan. *In* Arid zone hydrology. Unesco, Paris. Arid Zone Programme 1:96–127.

16. DE VAJDA, A.
    1952   Some aspects of surface water development in arid regions. Food and Agricultural Organization Agricultural Development Paper 21. 45 p.

17. MARSH, A. W.
    1967   Applied irrigation research. FAO, Rome. 60 p.

18. BOOHER, L. J.
    1967   Surface irrigation. FAO, Rome. 165 p.

19. BLANEY, H. F.
    1955   Water and our crops. *In* Water, the yearbook of agriculture, p. 341–345. U.S. Department of Agriculture, Washington, D.C.

20. THORNTHWAITE, C. W., AND J. R. MATHER
    1955   The water budget and its use in irrigation. *In* Water, the yearbook of agriculture, p. 346–358. U.S. Department of Agriculture, Washington, D.C.

21. U.S. SALINITY LABORATORY
    1954   Diagnosis and improvement of saline and alkali soils. L. A. Richards, ed. U.S. Department of Agriculture, Agriculture Handbook 60. 160 p.

22. NATIONAL TECHNICAL ADVISORY COMMITTEE
    1968   Water quality criteria, Report to the Secretary of the Interior. Federal Water Pollution Control Administration, Washington, D.C. 234 p.

23. FOOD AND AGRICULTURE ORGANIZATION OF THE UNITED NATIONS
    1965   Report on the seminar on waterlogging in relation to irrigation and salinity problems, Lahore, 16–28 November 1964. FAO, Rome. EPTA Report 1932. 191 p.

24. FOOD AND AGRICULTURE ORGANIZATION OF THE
    UNITED NATIONS/UNESCO
    1967   International source-book on irrigation and
            drainage of arid lands in relation to salinity
            and alkalinity. 667 p.

25. GOLDING, E. W.
    1961   Windmills for water lifting and the generation
            of electricity on the farm. FAO, Informal
            Working Bulletin 17. 104 p.

26. BURDON, D. J.
    1961   Groundwater development and conservation
            in Syria. FAO, Rome. EPTA Report 1270.
            86 p.

27. SYKES, J. F.
    1955   Animals and fowl and water. *In* Water, the
            yearbook of agriculture, p. 14–18. U. S. De-
            partment of Agriculture, Washington D.C.

28. FOOD AND AGRICULTURE ORGANIZATION OF THE
    UNITED NATIONS
    1965   The state of food and agriculture, 1965. FAO,
            Rome. 273 p.

29. NESTEL, B. L.
    1967   The indicative world plan for agriculture.
            World Review of Animal Production 3:34–39.

30. FOOD AND AGRICULTURE ORGANIZATION OF THE
    UNITED NATIONS
    1968   The role and importance of land and water
            resources surveys and appraisals. *In* Seminar
            on land and water use in the Near East. FAO/
            UNDP/TA Report 2425.

31. FOOD AND AGRICULTURE ORGANIZATION OF THE
    UNITED NATIONS
    1962   A study of agroclimatology in semi-arid and
            arid zones in the Near East. FAO/UNESCO/
            WMO, 23061/2.

Water Conservation for Food and Fiber Production in Arid Lands

LLOYD E. MYERS

U.S. Water Conservation Laboratory
Soil and Water Conservation Research Division
Agricultural Research Service
U.S. Department of Agriculture
Phoenix, Arizona, U.S.A.

ABSTRACT

WATER SUPPLIES IN ARID REGIONS are rarely available in amounts sufficient to meet demands for producing food and fiber. There are two ways to solve this problem. One is to increase the water supply, and the other is to reduce the amount of water needed. Major emphasis has been placed on obtaining more water through river diversions, groundwater pumping, and futuristic proposals for artificially inducing precipitation and desalting seawater. These practices and proposals have serious limitations in future application and will not solve all water-supply problems. Consideration should also be given to less-well-known sources of water supply such as water harvesting, horizontal wells, and runoff farming. At the same time, opportunities for reducing water requirements must not be ignored. These opportunities include the adoption of more efficient irrigation methods, better irrigation scheduling, reduction of seepage and evaporation losses, and the development of plants that produce more food and fiber with less water. Agricultural water-supply problems in arid regions can be solved by recognizing and utilizing the many possibilities for simultaneously increasing supply and reducing demand.

WATER CONSERVATION FOR FOOD AND FIBER PRODUCTION IN ARID LANDS

Lloyd E. Myers

MAN HAS LIVED and has thrived in arid regions by altering the natural disposition of water. Most of this alteration has involved the collection or diversion of water for the irrigation of cultivated crops. The amount of water available varies from place to place and from year to year. Ancient man usually adjusted his needs in accordance with the available supply. Modern man, on the other hand, has not been willing to restrict his water use and instead devotes his efforts toward increasing the supply to meet his rapidly expanding desires. He has not been completely successful, and in most arid areas of the world the demand still exceeds the supply.

Proposals for matching water supply and demand should not be bound by convention, and both sides of the coin should be considered. Streamflow diversions and groundwater pumping are the most commonly used water sources, but they are not the only available sources. Current work on desalination and precipitation inducement is imaginative and futuristic. There are also the ancient and essentially forgotten practices such as water harvesting and runoff farming. These alternative water sources, ancient as well as futuristic, should be realistically evaluated and utilized as progress in their development permits. Similarly, many opportunities exist for reducing the amount of water required for irrigation. This can be done by adopting more efficient irrigation methods, better scheduling of water application, reduction of losses to seepage and evaporation, and the development of crops that use water more efficiently. Some of these things can be done now, some will be possible in the near future, and some will need many years of research and development before they are feasible. Although the technology may be complex and difficult, the basic principles and ideas are simple. These principles and ideas should be of interest to anyone concerned with the production of food and fiber in arid lands.

WATER SOURCES

Streamflow and Groundwater

The most common sources of irrigation water in arid regions have been rivers that originated in more humid areas and then flowed through the arid lands on their way to the sea. The first systems irrigated lands immediately adjacent to the rivers by direct diversion. As technology increased, water was conveyed to lands not necessarily adjacent to the rivers. This resulted in the possibility of irrigating more land than the originally diverted river could support. Some recent schemes have proposed transferring water from one state or country to another and transporting it for distances exceeding 5,000 kilometers (1). Although some of these schemes are technically possible, their economic and political feasibility is in doubt. States and countries of origin do not wish to give up their water, and tremendous costs require financial support of the projects by people outside the area to be benefitted.

Groundwater pumping has been an important source of irrigation water since the development of efficient pumps and engines to power them (2). Despite some wishful thinking to the contrary, there is always a limit to the amount of water that can be safely pumped from any natural aquifer, particularly where replenishment by precipitation occurs at a slow rate. In some arid areas, pumping is now being allowed to exceed the rate of replenishment. This means that the water is being mined and that the water supply will ultimately be exhausted. Although there are some arguments in favor of groundwater mining, long-range plans should be based on limiting pumping to the safe, sustainable yield. Groundwater mining can solve an existing water-supply problem only at the cost of creating a much more difficult problem for the future.

Precipitation Inducement

Experiments have shown that precipitation is sometimes increased and sometimes decreased by seeding clouds with very small particles of silver iodide (3). The results of cloud-seeding are not always predictable, because our knowledge of the physical processes governing precipitation from clouds is still imperfect. Basic research on cloud physics is slowly unraveling the mysteries. At the same time, large-scale field studies of cloud-seeding have been providing information on the results obtained under a wide variety of topographic and climatic situations. Research and experience indicate that the best opportunities for increasing precipitation are in areas where cold wet air masses are swept upward over mountain ranges (4); but these are areas where precipitation is already relatively high. Prospects for increasing precipitation directly on lower-lying arid lands do not seem promising. This means that arid lands that can benefit from precipitation inducement will probably be those obtaining water from streamflow originating in areas of high precipitation.

Desalination

Low-cost desalting of seawater would unquestionably be a tremendous boon to arid lands bordering oceans, seas, or salty lakes. Many proposals have been made, widely advertised, and intensively promoted for building huge desalting plants to produce water for agriculture (5). Although new and improved desalting methods have been developed, no developments can yet promise truly low-cost desalted seawater. Current promises of cost reduction are based on the assumption that the cost of the product is reduced as the size of the machine is increased. A practical limit exists to cost reduction achieved by this means. Also there are problems in disposing of huge quantities of hot brine, pumping and conveying desalted water to the point of use, and storing it until needed for irrigation. The most optimistic proponents of desalting schemes agree that the water will be too expensive for use in irrigation as it is practiced today. They state that this problem can be overcome by improved irrigation technology. It has been suggested that a better plan for immediate action would be to use such technology to improve production with the water that is already available (6).

Horizontal Wells

The qanats of Iran, invented about 3,000 years ago by Persian engineers, represent an ingenious solution to the problem of obtaining water in an arid region (7). They consist of underground tunnels or large horizontal wells that intercept aquifers, usually in the upper portions of an alluvial fan, and convey the water by gravity to the ground surface at lower elevations. Qanats, called by other names, have been built in other countries, including Pakistan, Iraq, Syria, Spain, and Sicily. In Iran about 22,000 qanats, with 270,000 kilometers of underground conduits, discharge 75 cubic meters of water per second, which is about 75 percent of all the water used in that country.

Qanats are expensive and dangerous to build by the primitive hand-tunnelling methods used in the past, but they do have certain advantages. They deliver water without pumping costs, and they can develop water where pumped wells are not feasible. The principle is sound, and horizontal wells built by modern construction methods are often practical. These do not have to be large-diameter tunnels but can consist of relatively small-diameter pipe installed by a horizontal drilling machine.

Dew and Fog

The possibility of condensing water from the atmosphere by some simple scheme has intrigued a number of writers, and it has been proposed that ancient civilizations accomplished this for agricultural purposes. Jumikis writes of "dew mounds" in the Negev desert and the "aerial wells" of Theodosia, which were piles of rocks that supposedly cooled during the night and then condensed the early morning dew (8). He describes several experimental aerial wells, which were all failures. The supposed dew mounds of the Negev have now been shown to be the result of soil-smoothing operations to increase rainfall runoff (9). Although dew will condense on piles of rock, this phenomenon does not appear to be of agricultural significance.

Fog occurs in some arid areas adjacent to oceans, and at one location in northern Chile experimental nets of nylon thread collected about 22 gallons of water per year per square yard of net. Other proposals include large condensers cooled with pumped seawater. Although these devices can collect water from the atmosphere, the cost will be too high for agricultural use.

Water Harvesting

Some of the precipitation lost to nonbeneficial evapotranspiration can be harvested by treating the soil to increase runoff (10). The potential of this method can be illustrated by pointing out that 1 millimeter of rain equals 10,000 liters of water per hectare. Water harvesting is practiced in many ways: by building contour ditches to collect hillside runoff, by treating soil to reduce its permeability, by covering soil with impermeable sheeting, and by building pavements or roof-type structures. Pavements and ground covers are too expensive for agricultural use, but soil modification could be practical. Simply clearing and smoothing soil surfaces can double the rainfall runoff. An elaborate system of smoothed-soil catchments is used in Western Australia to obtain water for small towns, livestock, and rural domestic use (11). These are called "roaded catchments," for the soil is graded into a series of parallel ridges resembling roadbeds. Performance has been good, and several thousand hectares of these catchments have been installed. When large areas of low-cost land are available, areas of cleared and smoothed bare soil may be the most economical method of water harvesting.

A promising low-cost method of water harvesting involves spraying soils with chemicals that cause them to become hydrophobic or "water repellent" (12). A wide variety of chemicals can be used. Sodium methyl silanolate, sprayed on sandy loam soil, has caused from 60 to 70 percent runoff over a 4-year period of rainfall averaging 225 millimeters. The treatment has produced water for less than 10 cents per cubic meter. Even lower-cost treatments consist of applying metallic salts and soluble stearates, at rates of about 100 kilograms of each material per hectare, to form insoluble, hydrophobic stearates in the soil pores. Although much more research remains to be done, treatments with hydrophobic chemicals appear to be feasible at the present time on sandy soils that do not swell and shrink upon wetting and drying.

Runoff Farming

Runoff farming was developed almost 4,000 years ago to permit crop production on lands receiving as little as 100 millimeters average annual rainfall (9). Extensive investigations in the Negev desert of Israel have shown that ancient farmers cleared hillsides to increase runoff and built contour ditches to collect and convey the water to lower-lying fields. The irrigated fields were laid out as large terraces surrounded by stone walls in the bottoms of the wadis or valleys. Application systems were designed to apply water to the upper field first; excess water, if available, was applied to the lower-lying fields in turn. These ingenious systems allowed the development of agricultural civilizations in desert regions which today support only a small human population and produce little in the way of food and fiber crops. The ancient runoff farms were destroyed by warfare and by political upheavals that resulted in mismanagement and neglect. Current investigations have shown that the principles of runoff farming are sound and can be used to grow high-value crops such as artichokes, asparagus, flower bulbs, and some fruits and nuts. Barley and other grain crops are well adapted to runoff farming. Crop failures will occur but can be offset by income from good years. Modern runoff farms, utilizing improved technology and adapted crop varieties, could be of tremendous benefit to many desert regions.

Wastewater

Reuse of wastewater is obviously desirable and is widely practiced. Wastewater has not ordinarily been subjected to any purification treatment prior to use for irrigation. As a result, wastewater such as raw sewage effluent could be safely used only for irrigating pasture or other crops not directly consumed by humans. Increasing demands for water have emphasized the desirability of treating polluted wastewater so that it can be used, without restriction, for the irrigation of high-value crops. Studies now under way show that the remarkable chemical and biological properties of natural soils offer great promise for large-scale, low-cost wastewater renovation (13). One system involves intermittent flooding of recharge basins, which allows the polluted water to percolate downward through aerated soil to the water table. It then flows laterally through saturated aquifers to pumped wells which withdraw it for reuse. The resulting sequence of aerobic and anaerobic processes can remove, except for some soluble salts, essentially all undesirable constituents including pathogens, phosphorus, and nitrogen. The complete cost of this treatment, including pumping for recovery, can be as little as 0.5 cent per cubic meter.

Agricultural wastewater occurs as surface runoff from irrigated fields, spills from water delivery systems, and the water collected in drainage systems installed to lower the water table. The first two sources of wastewater disappear as intensive, efficient water management develops, leaving saline groundwater drainage as the only source of agricultural wastewater. This water is now sometimes usable on salt-tolerant crops. Looking to the future, however, the amount and quality of drainage water will decrease as seepage and percolation losses from irrigated lands are reduced. Long-range plans for utilizing agricultural wastewater are dependent upon the development of low-cost desalination techniques. It is apparent that the best way to reuse agricultural wastewater is to stop all preventable waste before it occurs.

IRRIGATION EFFICIENCY

Irrigation Methods

Irrigation is the artificial application of water to soil to maintain soil moisture conditions necessary for plant growth (14). A successful irrigation operation must provide water at the place, at the time, and in the quantity required for optimum plant growth. The selection of an irrigation method for a particular field is dependent upon many factors including land topography, depth and texture of soil, climate, water quality and availability, crops to be grown, availability of money for equipment purchase, and the quantity and quality of available labor. The method and system chosen should be one which results in optimum crop production or cash income per unit of the most limited resource available, whether it be labor, capital, arable land, or water. In arid lands the limiting resource is usually water.

Irrigation methods can be generally divided into four categories: surface drip or trickle, subsurface, surface flow or flooding, and spray or sprinkler. The first method trickles or drips water from small-diameter plastic tubing laid on the ground surface in the plant rows. There are also subsurface application systems where the plastic pipe is buried in the crop root zone. Both methods are claimed to obtain increased crop yield by maintaining optimum soil moisture conditions, and to reduce water requirements by reducing the area of wetted surface soil. Both methods are expensive (costing from $1,000 to $2,500 per hectare for vegetable crops), require the use of filtered water, and require removal of the pipe during tillage operations. Should these systems prove successful they may be useful for the production of specialized, high-value crops. Subsurface irrigation is sometimes practiced where natural conditions are favorable, including permeable soils, an impermeable barrier at shallow depth, and essentially salt-free water. These conditions rarely occur in arid regions. The choice of irrigation methods ordinarily available to the irrigator lies between surface flow or flooding and sprinkler irrigation.

Some surface irrigation methods, such as flooding directly from stream or river diversions, provide essentially no control over the water after it leaves the stream or ditch. Water distribution depends upon the natural slope of the ground surface and is inefficient. The only advantage of this primitive method is the low cost.

Some systems restrict lateral water movement with furrows or border ridges but depend upon ground-surface slope for longitudinal distribution. Rate of movement down the slope depends upon the degree of slope, the rate at which water soaks into the soil, retardance of flow by soil roughness or vegetation, and the rate at which water is applied to the upper end of the furrow or border. Flow retardance and infiltration rates change during the growing season. Uniform replenishment of soil moisture throughout the irrigated field is very difficult to achieve. Good irrigation efficiency can be obtained, however, by deliberately applying water at high rates, allowing the excess water to flow out of the lower ends of the furrows or borders, collecting the runoff, and pumping the water back into the application system.

Level-surface systems control water movement by completely confining the applied water within essentially level furrows or basins. Good level-irrigation systems rapidly apply large volumes of water to furrows or basins limited in size so that uniform water distribution is obtained before any appreciable infiltration occurs. These systems permit applying predetermined quantities of water at predetermined rates, reduce dependence upon operator skill for good water distribution, and allow high irrigation efficiency.

Sprinkler or spray systems distribute water through placement of the application devices throughout the area to be irrigated. Land leveling or smoothing is not ordinarily required. Application patterns for individual sprinklers are not uniform, particularly when spray is blown by wind, and overlapping of spray patterns is necessary. Application rates in excess of infiltration rates will cause surface runoff and uneven distribution on soil surfaces that are not level. The irrigator must, therefore, exercise some judgment in adjusting application rates to compensate for seasonal variations in infiltration rates when sprinkling slowly permeable soils. A wide variety of sprinkler equipment is available, and design criteria have been developed to permit the use of sprinklers under most conditions. Installation of a sprinkler system does not guarantee good irrigation efficiency, but properly designed and operated systems are highly efficient.

The proper choice of an irrigation method is often not difficult, despite the apparently numerous factors to be considered. Sloping-surface systems with pumpback, level-surface systems, and sprinkler systems are all capable of irrigation efficiency exceeding 80 percent if properly designed and managed. This means that 80 percent of the water applied is retained in the root-zone soil for use by the crop. All three systems can be automated, although this is of concern only where labor costs are high.

The choice of system is usually governed by dominant factors that are readily apparent. Sloping-surface systems will be preferred where shallow soils or steep slopes make land leveling not feasible, where infiltration rates are moderate to low, and where wind is a problem for sprinklers. Level systems are preferable where soils and topography permit leveling, where infiltration rates are moderate to low, and wind is a problem. Sprinklers are most advantageous where infiltration rates are high, soil surfaces cannot be leveled, and where frequent, light applications of water are required. When conditions do not clearly favor one method, the choice should depend on an economic analysis of relative installation and operating costs (*15*). The choice should not be made on the false assumption that any one particular method is always the most efficient.

Irrigation Scheduling

Efficient irrigation is not possible without proper scheduling of timing and amounts of irrigation. Crops should be irrigated whenever the soil water content is reduced to some amount below which desirable plant growth is retarded. Experienced irrigators sometimes use plant stress symptoms, such as a change in leaf color, as an indication of soil becoming too dry. The quantity applied should not exceed the amount the soil in the root zone can retain against the forces of gravity, plus a small amount for leaching salt from the soil. Irrigation scheduling is thus dependent upon the water-holding characteristics of the soil and the rate at which water is removed from it. Pertinent soil characteristics can easily be determined by any competent agricultural soils technician. Determining the rate of soil moisture depletion is not a simple matter. As a result, irrigation scheduling has been based primarily on experience and has not usually been conducive to high irrigation efficiency.

Under efficient irrigation practices, soil moisture depletion is caused by evapotranspiration (evaporation from the soil surface plus transpiration from the crop plants). The evapotranspiration rate is governed by climatic factors such as solar radiation, air temperature, and wind velocity, plus crop factors such as plant height, structure of leaves, area of plant cover, and the stage of plant growth. Early work to define the relative influence of climatic and plant factors on evapotranspiration was quite naturally empirical.

The first predictive equation to be widely used for irrigation scheduling was developed by Blaney and Criddle in 1945 (*16*). Their equation used mean monthly temperature, monthly percentage of total annual daytime hours, and empirical crop coefficients determined by field studies. Subsequently, a relatively large number of scientists and engineers have attempted to formulate better predictive methods (*17*). Good progress in this direction has been achieved, particularly in analyzing the climatic and aerodynamic factors. It has not yet been possible to develop a universally applicable physical or mathematical model to characterize the plant factors.

A number of investigators are now working on simplified procedures for characterizing crop coefficients, to be used with energy balance and aerodynamic equations, for calculating evapotranspiration rates and irrigation schedules. The Agricultural Research Service and the Soil Conservation Service, U. S. Department of Agriculture, with the Bureau of Reclamation, U. S. Department of Interior, are preparing a handbook to present these procedures. Publication is scheduled for 1972. In the meantime, planners and irrigators can use one of the many empirical equations available, or can use actual field measurements of evapotranspiration (*18*). Judgment must be used in selecting equations or data that were developed or obtained under climatic and crop conditions similar to those existing where predictive information is needed.

Percolation

Water percolating from the soil surface to below the crop root zone leaches harmful salt from the soil. It also removes plant nutrients, contributes to drainage problems, and is a direct water loss to the farmer. When inefficient irrigation methods and application schedules are used, percolation losses can exceed 50 percent of the water applied. The problem is particularly severe in sandy soils and has caused millions of hectares of these soils to be considered as unsuitable for irrigation.

Percolation losses can be minimized by careful management of efficient irrigation systems. Improvements in sprinkler systems, in level flooding systems, and in methods for calculating irrigation schedules are aiding in solving the problem on soils of moderate or low permeability. Another interesting development promises to solve the problem on sandy soils with high permeability. This is the development of techniques for constructing subsurface membranes or barriers which restrict the downward movement of water. Equipment and procedures have been essentially perfected for installing sprayed asphalt membranes at a depth of 60 centimeters in sands (19). Costs of the membrane are estimated to be 150 to 350 dollars per hectare in the United States. Carefully instrumented tests on sandy soil near Yuma, Arizona, showed that carrots grown over the asphalt membrane required less than half the amount of irrigation water needed by carrots without the subsurface barrier. This is a development of major significance, because definite possibilities exist for reducing the cost. Application of subsurface membranes to sands can be of great importance to arid regions. Except for their high percolation rates, sands have desirable characteristics for irrigated soils, particularly when salty irrigation water is a problem. Development of low-cost subsurface membranes can permit the abandonment of salty clay soils in favor of easily leached sandy soils.

Leaching

All irrigation water contains soluble salts, which will accumulate in the soil unless removed or leached by water percolating downward through the soil. Salt accumulation is frequently a serious problem in arid regions where evaporation rates are high and leaching by rainfall is inadequate or nonexistent. Irrigation water in excess of plant requirements must be applied to accomplish mandatory leaching. The amount of water needed for this has long been a matter for debate. Leaching requirements were initially established on the basis of experience and personal opinions. Later work developed equations for estimating leaching requirements if the salt concentration of the irrigation water and the maximum allowable concentration of salt in the soil-water were known (20). It was thought that leaching should be done by ponding water continuously on the soil surface until the required amount of water had percolated through the

soil. Field studies of leaching were conducted in this way. Both the predictive equations and the field studies were based on the assumption that all of the salty soil solution could freely and rapidly mix with the water applied for leaching.

Recent work has shown that much of the soluble salt in a soil does not rapidly mix with the leaching water. Most soils contain many microscopic pores which hold water but do not freely transmit it. Salt movement from these pores depends largely upon diffusion, which is a slow process. Studies by Biggar and Nielsen have shown that the amount of water required for leaching can be minimized by periodically applying small quantities of water (21). Ten centimeters of intermittent winter rainfall removed the same amount of salt from one plot of soil as did 183 centimeters of continuously ponded water. In addition to using less water, intermittent water application has two other advantages over ponding. The soil is not completely saturated, water movement is slower, and salt removal is more efficient. Second, deterioration of soil structure caused by lengthy periods of ponding is avoided.

We are faced with the fact that we cannot always accurately predict the amount of water required for leaching. Fortunately, it appears that the quantity required may be considerably less than has previously been thought necessary. It also appears that irrigation systems permitting frequent application of small quantities of water offer the best opportunity of reducing leaching requirements.

WATER LOSSES

Seepage

Enormous quantities of water are lost by seepage from channels and reservoirs. Some of the water lost is recovered, but most of it is not. The magnitude of the problem can be illustrated by the situation in arid regions of the United States, where seepage losses from artificial irrigation channels average more than 80 million cubic meters per day and losses from natural watershed channels exceed 115 million cubic meters per day. In addition, about half the livestock-watering reservoirs are unreliable or inoperative because of seepage losses.

Seepage can be reduced by lining the leaky structures or by sealing the soil (22, 23). Concrete is usually the preferred lining material for canals and reservoirs because of its relatively high strength and resistance to mechanical damage. It is subject to cracking, and sealing the cracks is necessary if good seepage reduction is to be obtained. Low-cost linings of polyethylene, vinyl, or sprayed asphalt are useful where they can be covered with soil or gravel. Soil covers are usually undesirable because they cannot be installed on steep side slopes, and weed growth in the cover is a problem. Reinforced artificial rubber sheeting is resistant to weathering and can be installed without a protective cover. This material is relatively expensive and can be damaged by animal hooves or, if not properly installed, by wind. Recent studies have shown that strong, relatively low-cost linings can be constructed by laying unwoven fiberglass matting over the soil surface and then spraying it with asphalt emulsion. The portion of the lining exposed above the water line may have to be resprayed every 5 to 10 years with a protective coating such as asphalt-clay roofing emulsion.

Materials are available that can be added to the water in a canal or reservoir to reduce seepage by plugging soil pores. These include dispersions of various resins, waxes, and asphalt. Successful treatment generally requires that treated soils do not crack upon drying, pretreatment seepage rates must exceed 30 centimeters per day, weed growth is eliminated, mechanical damage to the seal is prevented, and water is kept in the structure continuously. It is also possible to reduce seepage in soils containing 15 percent clay, or more, by dispersing or swelling the clay. Seepage through many soils containing montmorillonite clay can be essentially stopped with sodium carbonate at a cost of $250 per hectare or less. Retreatment every 3 to 5 years may be necessary.

Seepage control for water salvage has not received adequate attention. Much of the existing seepage loss could be saved at a cost of less than one cent per cubic meter. These costs are lower than the costs of some other water sources now receiving serious consideration.

Evaporation from Water Surfaces

Evaporation from irrigation reservoirs can exceed 200 centimeters per year. Not only is water lost, but the salinity of the remaining water is increased. Interest in the problem was aroused when it was discovered that certain materials, such as cetyl alcohol, have long, slender molecules that align themselves side by side on a water surface to form a film that is one molecule thick. A considerable amount of research and publicity has been devoted to the possibility of using such films to reduce evaporation from water surfaces (24). These monomolecular films form a somewhat leaky physical barrier to evaporation but do not reduce the amount of solar energy absorbed by the water. Absorption of this energy, without the cooling effect of evaporation, causes an increase in water temperature. Any disruption of the film results in above normal evaporation from the heated water. Effective evaporation reduction requires maintaining an essentially continuous, undisrupted film. This has been difficult to do when winds exceed 25 kilometers per hour. New materials and methods of application may solve this problem (25).

Another approach toward evaporation reduction involves the use of floating granules, blocks, or covers of plastics, elastomers, foamed concrete, and other materials. Some of these materials reduce water temperatures by reflecting incoming solar radiation. They can be compatible with other water uses such as swimming, boating, and fishing, and in some cases improve water quality by reducing undesirable plant growth.

There is, as of this writing, no satisfactory method for reducing evaporation from irrigation reservoirs. We can only hope that future research will develop suitable methods and materials.

Evaporation from Soil Surfaces

There are some instances where evaporation from soil surfaces is desirable, as in hot-weather cooling of lettuce beds. Ordinarily, however, such evaporation is a waste of water and can result in damage by causing movement of salt to the soil surface. Evaporation can be reduced by covers or mulches (26). Plastic mulches, usually of black polyethylene film, form an impermeable barrier to water in either liquid or vapor form. Yields of corn per unit of water evaporated and transpired were doubled in one experiment where the soil was covered with plastic film, indicating that up to half the water used by unmulched corn may be lost through evaporation from the ground surface. Besides conserving water, plastic mulches can improve germination, seedling emergence, and early growth rate of some crops, and can improve the quality of some fruits by keeping them off the soil. Small slits or holes can allow desirable penetration of rainfall through the plastic into the soil. Plastic film mulches also have certain disadvantages in that they are expensive, may be difficult to apply and remove, and do not always improve crop growth.

The movement of water by vapor diffusion through a dry, porous material is a relatively slow process. Accordingly, placing a layer of dry, porous material over the soil surface can substantially reduce evaporation. Mulches of plant residues are not particularly satisfactory because they are difficult to hold in place, decay when anchored in the soil, and absorb water from the soil. Gravel mulches can be effective in reducing evaporation, are not moved by wind, and allow rainfall penetration. But they may be expensive to install, can interfere with tillage operations, and need some sort of soil-sieving machinery to periodically clean and replace them. Some attempts have been made to use the surface soil itself as a mulch by treating it with hydrophobic chemicals to keep it dry. This work has not yet been successful because of high treatment costs plus the fact that the hydrophobic soil prevents infiltration of rainfall. The problems encountered in the use of gravel and hydrophobic soil mulches are not insurmountable, but they have not yet been solved.

Evaporation from soil in high-water-table areas can be reduced by using drains or pumped wells to lower the water table. Although this is usually done to prevent salt accumulation in the surface soil, it also saves water. Still another approach lies in previously mentioned irrigation systems which wet the soil only in the crop row and are claimed to reduce evaporation losses because most of the soil surface is always dry.

There is no simple, inexpensive way to reduce evaporation from soil surfaces. A number of opportunities exist, however, through the use of mulches, water-application methods, and water-table control. Except for water-table control, these methods can presently be justified only for the production of high-value crops. This is not necessarily an unfortunate situation. The production of high-value crops is ordinarily the best use of limited water supplies in arid regions.

CROP ADAPTATION AND SELECTION

Most crops now irrigated in arid regions have been imported from more humid regions, and most of them are not particularly efficient in their use of water. Some may require over 2,000 kilograms of water to produce 1 kilogram of useable dry matter. Some work has been done on the selection of forage plants adapted to arid rangeland conditions. Almost no work has been done on the selection and breeding of crops that will efficiently use limited supplies of irrigation water. The numerous possibilities should be thoroughly explored.

Older civilizations made extensive use of wild desert plants (27). In the southwestern United States, for example, the Pimas harvested stems, leaves, flowers, bulbs, seeds, and fruits from about 65 different varieties of wild plants. We might think that desert plants, adapted to drought conditions, would be the best candidates for highly efficient irrigated crops. Unfortunately, most of these desert plants have adapted to drought by means such as tough leathery leaves, thick bark, resinous coatings, dropping leaves during stress, and dormancy during periods of drought. Except for the succulents, they do not grow during periods of drought, and when water is available they may not use it efficiently. Even worse, the survival of desert plants is usually more dependent upon resistance to foraging animals than resistance to drought. Accordingly, the larger plants have developed defensive mechanisms such as thorns and bitter flavors. Some improvement can be made, as in the development of spineless cactus (*Opuntia* spp). It should be noted, however, that early agriculturists long ago domesticated the most promising drought-resistant plants, including barley, melons, sorghum, and beans. The process of plant selection to increase yields has resulted in the loss of the drought-resistant characteristics of the original plant ancestors. We are faced with the probability that desert plants that have not already been domesticated are generally poor candidates for the process. Although native desert plants should not be ignored, the best candidates for highly useful plants with low water requirements seem to be the plants we have already domesticated.

Plant breeders have not yet paid much attention to developing plants that use water more efficiently. They have made remarkable progress in developing high-yielding grains, sorghum with uniform height to facilitate harvesting, insect-resistant alfalfa, and tough-skinned tomatoes to reduce shipping damage. They could also develop plants with lower water requirements. Transpi-
ration of plants can be reduced in many ways. Agave and pineapple, for example, can store carbon dioxide that they take in through open stomates at night, and can close their stomates during the day while the stored carbon dioxide is used for photosynthesis (28). Water loss is much less than for plants that cannot as efficiently store carbon dioxide and must open their stomates during the day to take in carbon dioxide. Selection or breeding of plants to increase carbon-dioxide storage could reduce water loss. Ability to withstand higher leaf temperatures resulting from reduced evaporation could, if necessary, be achieved by research on the chemistry of plant metabolism.

An alternative to stomatal closure lies in the number and placement of stomates. Studies have shown that navel orange trees, with most of their stomates on the undersides of the leaves, transpire at about half the rate of cotton having stomates on both sides of the leaves. Leaf structure is also important. Safflower, with a tall, open leaf structure ideally suited to absorbing heat from the air, transpires water at a rate 1.5 times that of sugar beets with a lower, more compact leaf structure. Although water-use efficiency can and should be increased by research on the possibilities just mentioned, perhaps the most immediate opportunity lies in increasing photosynthetic efficiency. An outstanding example of this lies in the recent development of short-strawed wheat varieties with double or triple the yield of older wheat varieties. These increased yields should be achieved with no appreciable increase in water use, which means that water-use efficiency has also been increased two or three times.

Increased water-use efficiency can also be achieved by other means, such as using plants having a short growth cycle in relatively cool weather, or the ability to thrive on low-quality water. Lettuce is not a drought-resistant plant, but it has a low water requirement because it has a short growing season during relatively cool weather. Water requirement of lettuce in southwestern United States is only 22 centimeters as compared to 110 centimeters for cotton. Other plants may have a relatively high transpiration rate but can use salty water that is not suitable for most crops. Bermudagrass hay, grown on salty soil with salty water, makes a premium livestock feed when pelletized under high pressure to break down the fibrous tissue. This procedure has resulted in the production of 900 kilograms of beef per acre from land that was originally thought to be of low productivity. The above examples illustrate the fact that there are many ways in which plants can be selected and managed to use water more efficiently. Although we need to develop more efficient plants, we can also use more ingenuity and common sense in selecting and managing the plants we already have.

REFERENCES

1. EASTER, K. W.
   1968 Interbasin water transfers — economic issues and impacts. American Water Resources Conference, 4th, Urbana, Illinois, 1968, Proceedings 6:191–200.

2. TOLMAN, C. F.
   1937 Ground water. McGraw-Hill Book Company. New York. 593 p.

3. SCHLEUSENER, R. A.
   1967 Evolution of uses of cloud-seeding technology. American Society of Civil Engineers, Irrigation and Drainage Division, Journal 93:187–197.

4. HURLEY, PATRICK A.
   1968 Augmenting Colorado River by weather modification. American Society of Civil Engineers, Irrigation and Drainage Division, Journal 94:363–380.

5. JOINT UNITED STATES/MEXICO/INTERNATIONAL ATOMIC ENERGY STUDY TEAM
   1968 Nuclear power and water desalting plants for southwest United States and northwest Mexico. Clearinghouse for Federal Scientific and Technical Information (CFSTI), Springfield, Virginia, TID–24767.

6. CLAWSON, M., H. H. LANDSBERG, AND L. T. ALEXANDER
   1969 Desalted seawater for agriculture: is it economic? Science 164:1141–1148.

7. WULFF, H. D.
   1968 The qanats of Iran. Scientific American 218:94–101.

8. JUMIKIS, A. R.
   1965 Aerial wells: secondary sources of water. Soil Science 100:83–95.

9. EVENARI, M., L. SHANAN, N. TADMOR, AND Y. AHARONI
   1961 Ancient agriculture in the Negev. Science 133:979–996.

10. MYERS, L. E.
    1967 New water supplies from precipitation harvesting. International Conference on Water for Peace, Washington, D.C., 1967, Proceedings 2:631–638.

11. PUBLIC WORKS DEPARTMENT OF WESTERN AUSTRALIA
    1956 Roaded catchments for farm water supplies. Department of Agriculture of Western Australia, Perth, Bulletin 2393.

12. MYERS, L. E., AND G. W. FRASIER
    1969 Creating hydrophobic soil for water harvesting. American Society of Civil Engineers, Irrigation and Drainage Division, Journal 95:43–54.

13. BOUWER, H.
    1968 Returning wastes to the land, a new role for agriculture. Journal of Soil and Water Conservation 23:164–168.

14. ISRAELSEN, O. W., AND V. E. HANSEN
    1962 Irrigation principles and practices, 3rd ed. John Wiley and Sons, Inc, New York. 447 p.

15. KELLER, J.
    1965 Effect of irrigation method on water conservation. American Society of Civil Engineers, Irrigation and Drainage Division, Journal 91:61–72.

16. BLANEY, H. F., AND W. O. CRIDDLE
    1950 Determining water requirements in irrigated areas from climatological and irrigation data. U.S. Department of Agriculture, Soil Conservation Service, TP–96. 48 p.

17. JENSEN, M. E. (ed)
    1966 Evapotranspiration and its role in water resources management. American Society of Agricultural Engineers Conference, St. Joseph, Michigan, 1966, Proceedings. 66 p.

18. ERIE, L. J., O. F. FRENCH, AND K. HARRIS
    1965 Consumptive use of water by crops in Arizona. Arizona Agricultural Experiment Station, Technical Bulletin 169.

19. HANSEN, C. M., AND A. E. ERICKSON
    1968 The use of asphalt to increase water holding capacity of droughty soils. Symposium on New Uses for Asphalt, Division of Petroleum Chemistry, American Chemical Society, Atlantic City, New Jersey, Preprints 13:C164–169.

20. WILCOX, L. V.
    1963 Salt balance and leaching requirement in irrigated lands. U.S. Department of Agriculture, Technical Bulletin 1290. 22 p.

21. BIGGAR, J. W., AND D. R. NIELSEN
    1962 Improved leaching practices save water, reduce drainage problems. California Agriculture, March, p. 5.

22. MYERS, L. E. (ed)
    1963 Seepage Symposium, Phoenix, Arizona, Proceedings. U.S. Department of Agriculture, Agricultural Research Service, ARS 41–90. 180 p.

23. ——
    1968 Second Seepage Symposium, Phoenix, Arizona, Proceedings. U.S. Department of Agriculture, Agricultural Research Service, ARS 41–147. 150 p.

24. UNESCO
    1962 Symposium on water evaporation control, Poona, India, Proceedings. New Delhi, India. 330 p.

25. FRASIER, G. W., AND L. E. MYERS
    1968 Stable alkanol dispersion to reduce evaporation. American Society of Civil Engineers, Irrigation and Drainage Division, Journal 94:79–89.

26. HANKS, R. J.
    1969 Soil water evaporation control. Seminar, Modifying the Soil and Water Environment for Approaching the Agricultural Potential of the Great Plains. Kansas State University, Manhattan, Proceedings 1:65–69.

27. SHANTZ, H. L.
    1956 History and problems of arid lands development. *In* G. F. White (ed), The future of arid lands. American Association for the Advancement of Science, Publication 43:3–25.

28. EHRLER, W. L.
    1967 Daytime stomatal closure in *Agave americana* as related to enhanced water-use efficiency. Seminar on Physiological Systems in Semi-arid Environments, Proceedings, Albuquerque, New Mexico, p. 239–247.

Runoff Agriculture in the Negev Desert of Israel

MICHAEL EVENARI

Department of Botany, Hebrew University, Jerusalem, Israel

LESLIE SHANAN

Engineering Consultant, Tel Aviv, Israel

and

NAPHTALI H. TADMOR

Department of Botany, Hebrew University, Jerusalem, Israel

ABSTRACT

IN ANCIENT TIMES the Negev foothills and the central Negev highlands were intensively cultivated, although they were then a desert as they are today. The ancient farmers collected the runoff from watersheds 20 to 30 times larger than the cultivated area. In 1958 and 1959 two ancient farms were reconstructed and planted to a variety of fruit trees, pasture plants, field crops, and vegetables. On the reconstructed farms, runoff, floods, and rainfall were measured and their relationship was analyzed. It was shown that even with not more than 80 to 100 millimeters of yearly rainfall, enough runoff water can be collected to ensure good growth and satisfying yields of most of the cultivated plants, the majority of which proved to be quite drought-resistant.

Water use of various trees and pasture plants was also studied using the neutron-moderation method. It was shown that the plants studied performed well with a relatively low water use. It is concluded that runoff agriculture without additional irrigation is feasible in vast desert areas — with climatic, edaphic, and hydrological conditions similar to Israel's Negev — where irrigation water is either not available or too expensive.

RUNOFF AGRICULTURE IN THE NEGEV DESERT OF ISRAEL

Michael Evenari, Leslie Shanan, and Naphtali H. Tadmor

THE NEGEV DESERT, which occupies about 60 percent (one million hectares) of the total area of Israel, is shaped like an irregular triangle. Its base line stretches from Gaza to Ein Gedi on the Dead Sea, and its two sides run from Gaza and from Ein Gedi to Elat on the Red Sea. It links the Sinai Peninsula in the west and the deserts of Arabia in the east. It is part of the immense desert belt stretching from the Sahara to Saudi Arabia. Physiographically, the Negev can be divided into various regions (Fig. 1). We are here concerned only with two of

Fig. 1. The Negev, showing (I) foothills and (II) central highlands. The dashed-and-dotted line indicates the boundary between the Negev and Jordan and Sinai. The triple dots indicate the ruins of ancient cities (together with Massada).

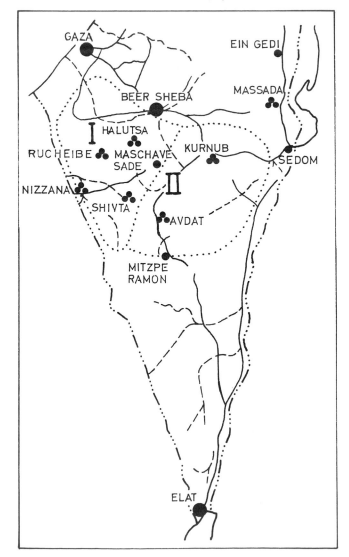

them: the foothills and the central highlands. Geomorphologically, the foothills consist mainly of Eocene limestone hills separating wide rolling plains with the elevation ranging from 200 to 450 meters above sea level. The hillsides carry a very shallow, gravelly, highly saline soil with an immature profile. Its surface is covered by smaller and larger stones forming a typical desert pavement. The soil of the plains and depressions, however, consists of loess, which may reach a depth of 3 meters. This loess is a fine, windblown soil which contains 70 percent silt and clay. Loess is very fertile when enough water is available.

The highlands contain a series of parallel anticlines with an elevation of 450 to 1000 meters above sea level. They are composed of Cenomanian-Turanian limestones and cherts. The soil conditions of hillsides and plains between the high ridges are as described for the foothills.

The rain in both subregions falls during the winter months — November to March. It averages 80 to 100 millimeters (3-4 inches) annually. As typical for all deserts, the seasonal variations of rainfall are extreme (1, 2). The following example illustrates this nicely: During the 1962/63 season our farm at Avdat received 25.6 millimeters of rain, and the following 1963/64 season, 152.7 millimeters! The mean yearly temperature for Shivta in the foothills region is around 20°C, for Avdat in the highlands 18°C. The two regions differ climatically also in another respect. The temperatures of the foothill region rarely fall below freezing point, whereas in the highlands the temperatures in the valleys reach minus 4 to minus 5°C, and an average of 40 to 50 nights have temperature minima below 0°C.

ANCIENT DESERT AGRICULTURE

The ruins of six large ancient cities are situated in the foothill and highlands subregions of the Negev (Fig. 1) together with innumerable remains of extensive ancient agriculture dating back to the Israelite period (about 950 to 700 B.C.) and the Nabataean and the Roman-Byzantine periods (about 300 B.C. to 630 A.D.) (3–5). The area, which today has no perennial streams and no underground water supply, was once intensively cultivated and supported a thriving civilization. Could it be that at those

NOTE: This chapter is adapted from a presentation at "Arid Lands in a Changing World," an international conference sponsored by the Committee on Arid Lands, American Association for the Advancement of Science, June 3–13, 1969, at the University of Arizona, Tucson, Arizona, U.S.A.

Fig. 2. Oblique air photograph of an area near Shivta. Many ancient runoff farms are visible, each with its terraced cultivated area surrounded by a wall and its catchment basin. The faint lines leading from the catchments to the various farms are runoff channels that led the floodwater to the cultivated area. The second farm from the bottom has a large farmhouse next to its fields.

times the Negev was not a desert as it is today, and that it had a rainfall of 400 to 500 millimeters per year, sufficient for "normal" agriculture? The arguments against a climatic change are strong.

a) The Bible describes the Negev as a desert without water (*see,* for example, Genesis 21:14–19).

b) If there had been a more humid climate in ancient times, there would have been no need to develop the ingenious ancient agriculture based on maximum water conservation which we will describe later on.

c) The most convincing argument is that working in our reconstructed farms with the same relationship of size of catchment area to size of cultivated fields as is typical for the ancient farms, we could always collect enough runoff water to sustain agricultural crops (see Agricultural Results, below).

We do not deny the existence of definite variations in the average annual rainfall. We only point out that at least since about 1000 B.C. there has occurred no major climatic change in the Negev, which then was already a desert more or less as it is today.

If there was no climatic change, how could the ancient farmers have cultivated the land under a 100-millimeter rainfall regime without any source of additional water for irrigation? It took us many years of field work to answer this question (*6-9*).

Our investigation proved, first of all, that all ancient agriculture in the Negev foothills and highlands was based on the utilization of surface runoff from small and large watersheds; hence we call the agricultural type "runoff farming." The ancient farmers used various methods for this purpose. We will describe only the most common and successful one, the "runoff farms" that received their water from relatively small watersheds. Each farm consisted of two parts (Fig. 2): the farm proper (that is, the cultivated area) and the catchment basin. Each cultivated area was situated in a narrow valley bottom on loess soil 2-3 meters deep. It was terraced by low stone walls. The farm's catchment basin (20 to 30 hectares in size) was on the surrounding slopes. When a rain occurred heavy enough to cause runoff, the runoff caused a flood (Fig. 3), and the floodwater collected in channels that led it to the various terraces of the farm proper. The terrace walls kept part of the water standing on the field, where it slowly soaked into the ground. The surplus went through drop structures in the terrace walls to the next lower terrace. The water harvest from the catchments averaged 150 to 200 cubic meters per hectare per year. Since the ratio of cultivated land to catchment area in all farm units was more or less the same (1:20 to 1:30) one hectare of cultivated land collected runoff from 20 to 30 hectares of hillside catchment. This means that each hectare of cultivated land received on an average about 3,000 to 6,000 cubic meters of runoff water per year. These high water yields were possible because of certain characteristics of the loess soil discussed below. One to five floods could be expected annually, producing enough runoff water to deep-wet the loess soil of the cultivated farm area. This water enabled the ancient farmers to grow successfully wheat, barley, legumes, almonds, and grapes, as reported in documents of the time found in Nizzane (10).

RECONSTRUCTED FARMS

In 1958 and 1959 we reconstructed two farms, near the ancient cities of Shivta and Avdat (1, 11). We superimposed on the ancient system all the instruments needed (flood and rain gauges, water meters, and so on) and a network of pipes which guaranteed an equal distribution of the floodwater inside each experimental unit. The aims of the reconstruction were: (a) to test experimentally whether our theories about the working of the ancient desert agriculture were correct, (b) to collect hydrological data typical for the area and to analyze the relationship between rainfall and floods (a vital point, since floods are the only water source for runoff agriculture), (c) to find out whether desert runoff agriculture is economically feasible today, and (d) to determine water use and drought-resistance of cultivated plants under conditions of runoff farming.

Flood Analysis

When rain starts falling, it first hits the vegetation, which prevents some rainwater from reaching the soil (interception storage). In deserts like the Negev this factor is negligible because of the scantiness of the vegetation. The first raindrops reaching the ground infiltrate the soil. Whenever the rate of rainfall is greater than the

Fig. 3. A flood at the Avdat farm. In the background is a floodgate with an automatic gauge.

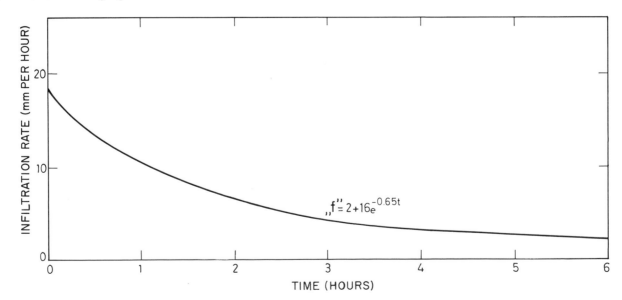

Fig. 4. Typical infiltration rate curve for Avdat watersheds. e: base of the naparian logarithm; t: time in hours.

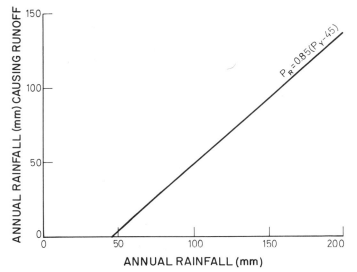

Fig. 5. Relationship between annual rainfall (P_y) and annual rainfall causing runoff (P_R).

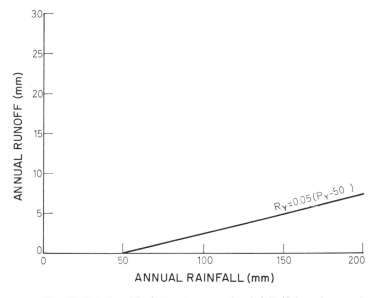

Fig. 6. Relationship between annual rainfall (P_y) and annual runoff (R_y) from a large catchment (350 hectares).

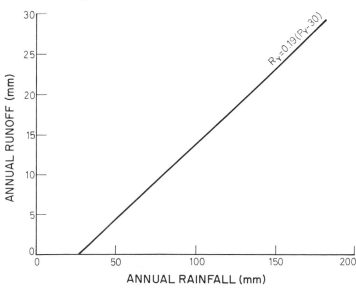
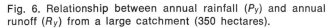
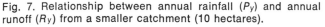

Fig. 7. Relationship between annual rainfall (P_y) and annual runoff (R_y) from a smaller catchment (10 hectares).

infiltration rate, part of the rainwater will fill the surface depressions (depression storage). Our measurements have shown that in our region this factor amounts to very little. When the depression storage has been filled, runoff starts. Therefore the amount of runoff under our desert conditions is mainly determined by the rate of rainfall on the one hand and the infiltration rate on the other.

The infiltration rate depends much on physicochemical qualities of the soil. After being wetted, loess, because of its composition, forms on its surface a very thin crust nearly impermeable to water (7). This quality explains why its initially high infiltration rate drops rapidly during the first half hour of wetting and reaches a steady rate of 2.5-3.5 millimeters per hour (Fig. 4). It is interesting that the infiltration rates we measured in various catchment areas in Avdat were more-or-less similar, even though these catchments differ in their soils and physiographical structure.

Though the rainfall rate is an exceedingly variable quantity, the average annual rainfall rate can be calcu-

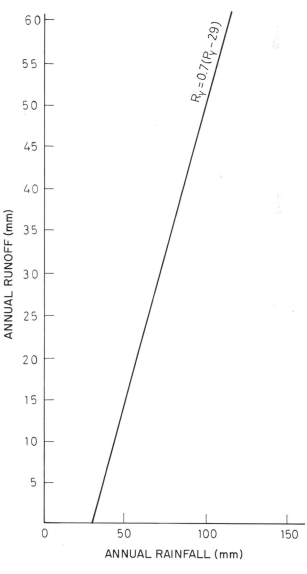

Fig. 8. Relationship between annual rainfall (P_y) and annual runoff (R_y) from a microcatchment.

Fig. 9. Runoff plots at Avdat. To the right, one stone-covered control plot; to its left, a bare plot; further left, a plot with stones raked together into mounds. The runoff from each plot is collected in the barrels visible in the foreground.

TABLE 1

Total Annual Runoff (in Millimeters) Measured on Twenty Runoff Plots in Avdat During the Periods 1965/66 and 1966/67

Treatment	Slope (%)	1965/66 runoff (mm) (90.7 mm rainfall)	1966/67 runoff (mm) (69.3 mm rainfall)	Runoff increase (%) of treatment over control plots (2-year average)
Control	10.0	23.64	9.46	. .
	13.5	21.23	8.03	. .
	17.5	17.88	7.41	. .
	20.0	13.61	4.67	. .
Mounds	10.0	26.02	11.85	13
	13.5	26.22	11.35	29
	17.5	20.70	8.77	17
	20.0	17.60	8.26	43
Bare	10.0	30.26	13.82	32
	13.5	21.83	9.53	8
	17.5	21.02	9.19	20
	20.0	16.91	8.15	32
Mounds, wet rolled	10.0	31.46	14.86	40
	13.5	27.59	12.24	36
	17.5	20.75	9.56	21
	20.0	20.08	9.23	61
Bare, wet rolled	10.0	31.46	15.29	41
	13.5	24.54	10.37	23
	17.5	25.27	11.67	45
	20.0	17.22	9.17	36

lated from existing rainfall records. The rainfall depth-time equation has the approximated form of $R = 5\sqrt{t}$, where R is the depth of rain in millimeters and t the period of the rainstorm in hours. This means that a half-hour rain will produce about 3.5 millimeters, an hour rain 5 millimeters, a 10-hour rain 16 millimeters, and so on. This is naturally only an approximation, and there are rainfalls of one hour duration which produce more than 50 millimeters, but they are comparatively rare phenomena.

The next step in our analysis is to look at the relationship between rainfall causing runoff and annual rainfall, and between annual rainfall and annual runoff, from our various catchment areas in Avdat. These calculations are based on our observations during the seven-year period 1960/61 - 1966/67 made on 8 catchment areas of various sizes and 20 runoff plots.

Figure 5 indicates the amount of rain that will cause runoff out of a given annual rainfall. It shows that about 30 millimeters of an annual rainfall of 80 millimeters will cause runoff, 50 millimeters of an annual rainfall of 100 millimeters, about 90 millimeters of an annual rainfall of 150 millimeters, and so on. Figure 5 does not tell us how much actual runoff we will get in each case. Figures 6 through 8 show actual runoff for three catchments of different size. The largest catchment (350 hectares) produces only about 2.5 millimeters of runoff with an annual rainfall of 100 millimeters, a smaller catchment (10 hectares) produces about 13 millimeters for the same rainfall, and a very small catchment of 0.1 hectares (a "microcatchment") produces about 50 millimeters runoff with the same rainfall.

The import of the figures is that the smaller the catchment the larger the percentage of rainwater which appears as runoff. Or, in other words, the smaller the catchment the larger the amount of runoff per unit surface. Naturally this does *not* mean that the smaller catchments produce the most runoff in absolute terms. There is an additional advantage of the microcatchments. Rains which are ineffective, that is, do not cause runoff on the large catchments, are effective on the microcatchments. During the 1967/68 rainy season, for example, we had one large flood on the 350-hectare catchment area and 11 on the microcatchments.

It is obvious that topography and nature of the surface of a catchment affect the runoff. We studied this experimentally in 20 small runoff plots of equal size (80 square meters) (Fig. 9). We arranged the plots in 4 replicates, each with 5 different surface treatments. Each replicate had a different slope ranging from 10 to 20 percent. The different treatments were: (*a*) stones of the desert pavement removed and soil bare, (*b*) the same, but soil once gone over with a roller when first wet ("rolled"), (*c*) stones removed but piled into heaps on the runoff plot, (*d*) the same, but bare soil between stone heaps rolled, and (*e*) untreated control.

The results of two typical rainy seasons given in Table 1 show that the steeper the slope the smaller the amount of runoff. This is true for all treatments, including the control. The other important point is that in all treatments in which the stones were partly or completely removed, the runoff yields are increased in comparison with the controls. The highest runoffs were obtained on the rolled plots. This obvious effect of the stone-clearing has an interesting implication. It had already been found by Palmer (*8*), the first investigator to detect and study the ancient agricultural systems in the Negev, that the catchments surrounding the ancient fields were covered by innumerable stone mounds and strips (Fig. 10), and

Fig. 10. Oblique air photograph of stone mounds and strips on the catchment basin of an ancient farm.

all the later investigators speculated about the function of these strange man-made structures (3) and agreed only on one point: they must have had something to do with agriculture. Our runoff experiment has now shown that the structures apparently were made by the ancient farmers in order to increase the water harvest from their catchment areas. The device is simple, ingenious, and efficient.

It is interesting to note that the yearly average of soil washed down from the stone-cleared slopes by the runoff waters amounts to not more than 0.1 - 0.2 millimeters. This means that if we take the ratio of cultivated land to catchment area as 1:20, the loess in the bottomlands receives yearly only an additional 2-4 millimeters of soil from the stone-cleared slopes.

All our experience concerning the relationship between runoff and rainfall is schematically summed up in a nomogram (Fig. 11). It shows that with an annual rainfall of 100 millimeters, for example, one can expect, from a 100-hectare catchment area with a 5 to 10 percent slope and an untreated surface, about 70 cubic meters of runoff per hectare. The same area when cleared of stones will produce about 130 cubic meters per hectare. A microcatchment, however, will produce under the same conditions about 160 cubic meters and 210 cubic meters per hectare respectively. If we assume that the ratio size of cultivated fields to size of catchment area is 1:30 (as mentioned above), under the conditions stated above a field of one hectare will receive 2,100 cubic meters (equivalent to 210 millimeters of rain), 3,900 cubic meters (=390 millimeters), 4,800 cubic meters (=480 millimeters), and 6,300 cubic meters (=630 millimeters) respectively.

Agricultural Results: Fruit Trees

After the reconstruction of the farms, fruit trees were planted there in 1960 and 1961 (Fig. 12). They were olives, carobs, figs, pomegranates, peaches, apricots, almonds, apples, grapevines, cherries, and loganberries. Various varieties of each tree species were tried out, and different combinations of root stock and scion (1). The trees received a small amount of water at planting time and from thereon lived exclusively on runoff water. The most successful were almonds, apricots, grapes, and figs. A promising species is the pistachio; it is too early to pass final judgment because the pistachios, which grow very slowly, were planted only in 1963. All the trees grow well (Fig. 12), and the yields of the most successful species are satisfying (Table 2). Most of the trees proved to be astonishingly drought-resistant, if we take this term in its most simple meaning — the ability to survive prolonged periods of drought without damage. (Plant physiologists have tried to define "drought-resistant" in terms of physiological mechanisms making plants resistant to water stress. We are unable to do this, since in our case we lack the necessary information.)

TABLE 2

Yields of Various Fruit Trees Over a Two-year Period (tons per hectare)

Fruit	Yield
Peaches	8-12
Apricots	5- 8
Grapes	12-15
Figs	6- 8
Almonds (dry shelled)	0.43-0.93

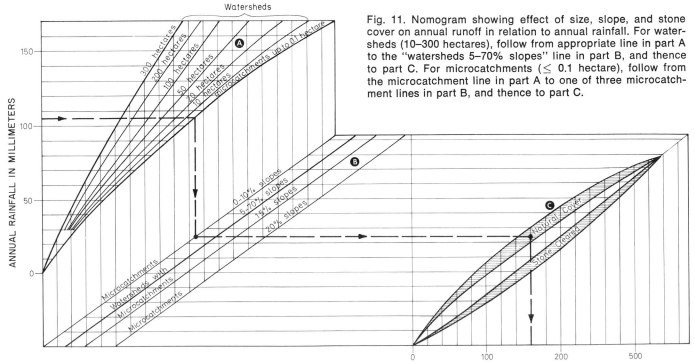

Fig. 11. Nomogram showing effect of size, slope, and stone cover on annual runoff in relation to annual rainfall. For watersheds (10–300 hectares), follow from appropriate line in part A to the "watersheds 5–70% slopes" line in part B, and thence to part C. For microcatchments (≤ 0.1 hectare), follow from the microcatchment line in part A to one of three microcatchment lines in part B, and thence to part C.

During the rainy season 1962/63, the Avdat farm received only 27.7 millimeters of rain, and only two small floods occurred. Because the last small flood of 1962/63 came in February 1963 and the first one of the 1963/64 season occurred in December 1963, the trees did not receive a drop of water for 9 months and survived well with little available water.

TABLE 3

Moisture Depletion (millimeters) of the Soil Volume in Which Apricot and Peach Trees Are Rooted

Period	Apricots	Peaches
Dec 26/63-Dec 3/64	306.2	242.7
Jan 2/65-Sept 22/65	288.1	308.0
Oct 15/65-Oct 10/66	223.9	262.2
Nov 22/66-Aug 3/67	133.8	142.0

In 1963 we began to measure the water use of apricot and peach trees by the neutron-moderation method (*1, 14*). Table 3 shows the moisture depletion of the whole soil volume in which the trees are rooted. In order to know what the real water use of the trees is, we have to know how much water evaporates from the soil and how much is lost by internal drainage. Evaporation from the soil is 8 to 11 millimeters per year. This is an astonishingly low figure for an area with a very high potential evaporation; it is explained by the formation of the crust immediately after the loess is wetted by the floods. The internal drainage could not be measured exactly. However, the main point here is that the fruit trees can grow and yield well with comparatively little water. Since irrigated orchards in the vicinity of our farms are watered with about 1,000 millimeters of water, it becomes obvious that our figures point the way to a much more rational water use and the possibility of decreasing water wastage.

Agricultural Results: Pasture Plants

In 1961 we started in Avdat an experiment to test 127 different species and cultivars of perennial and annual pasture plants for drought resistance, yields, and water use under conditions of runoff agriculture (*1, 15*). Table 4 shows as an example the performance of the best species for two periods. The yields in 1964/65 were high, as this was a relatively good year with 140.7 millimeters of rain and six floods. In 1965/66, in spite of 84.3 milli-

Fig. 12. Almond trees of the Avdat orchard. Some access tubes for the neutron probe are visible.

meters of rain and one good flood, the pasture plants received little water, because the pipe distributing the flood water to the pasture plant plots was accidentally blocked. The yields therefore were low, but the main point is that after 14 months without any appreciable amount of water the plants did not die; they survived until the next flood, showing a high degree of drought-resistance. The most efficient water user was the annual *Avena sterilis,* which produced 2.6 and 2.9 kilograms of dry matter for each cubic meter of water used.

After we knew what the most promising pasture plants were, we used them in an experiment with simulated floods given at different times of the year with different frequencies and with different amounts of water (*11*). We conducted this experiment because the effect on yield and performance of a natural flood occurring in October is very different from one occurring in April. The same is true for different numbers of floods per season and for floods of different depths. We also tested the effect of a

TABLE 4

Yields, Water Use, and Water Requirements of One Annual (*Avena*) and Four Perennial Pasture Plants in Avdat

Species	1964/65 season			1965/66 season		
	Yield (dry weight, kg/ha)	Water use (m³/ha)	Water requirement (kg of dry matter/ m³ water)	Yield (dry weight, kg/ha)	Water use (m³/ha)	Water requirement (kg of dry matter/ m³ water)
Agropyrum elongatum	7,800	5,660	1.40	870	1,650	0.53
Medicago sativa *	8,080	6,000	1.35	1,630	1,500	1.10
Oryzopsis miliacea	8,400	5,500	1.50	1,560	1,570	1.00
Phalaris tuberosa	8,240	5,800	1.42	660	1,570	0.42
Avena sterilis	10,660	4,000	2.65	9,650	3,320	2.90

* *Medicago* also produced 590 kilograms per hectare of seed in addition to herbage.

TABLE 5

Harvesting Yields of Alfalfa *(Medicago sativa)* in 1967/68 Under Various Simulated Flood Regimes With Triple, Double, and Single Harvesting Programs

Flood regime (cubic meters per hectare)	Fresh weight (tons per hectare)			Dry weight (tons per hectare)		
	Triple	Double	Single	Triple	Double	Single
5,520	26.6	22.8	16.4	6.4	6.0	4.1
4,400	24.9	21.0	10.7	5.5	5.3	3.3
3,900	23.3	17.8	9.8	5.2	3.9	2.7
2,680	16.7	11.3	5.4	4.6	3.4	1.6
1,940	12.5	4.5	5.0	3.5	1.7	..
1,440	3.5	4.4	5.0	1.7	1.3	..

single, double, and triple harvest on yields. Table 5 shows some of the results for alfalfa *(Medicago sativa)*. A triple harvest gives the highest yields. This is most pronounced for the lowest flood regimes. The yields again prove (as was already shown in Table 4 for natural floods) that runoff agriculture produces satisfying yields even with as little water as 3,000 to 4,000 cubic meters per hectare. Alfalfa irrigated with 13,000 cubic meters of water (usual in Israel) produces considerably higher yields, but in the light of our experience the question arises as to whether this practice is worthwhile, taking into account the scarcity and high price of water in Israel. This point is accentuated when we calculate the optimal water use, that is, the highest amount of dry matter produced per one cubic meter of water applied. Alfalfa (triple harvest) reaches this point (1.8 kilograms of dry matter per 1 cubic meter of water) at an annual water use of about 2,000 cubic meters per hectare (Fig. 13). With higher amounts of water the efficiency of water use decreases, although the yields increase (Fig. 13), but this increase is very low.

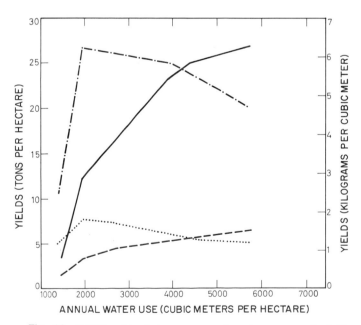

Fig. 13. Relationship between annual water use and yield of alfalfa (triple harvesting) under various simulated flood regimes. The solid line indicates fresh weight (tons/hectare); broken line, dry weight (tons/hectare); dash-dot line, fresh weight (kg/m³ water used); dotted line, dry weight (kg/m³ water used).

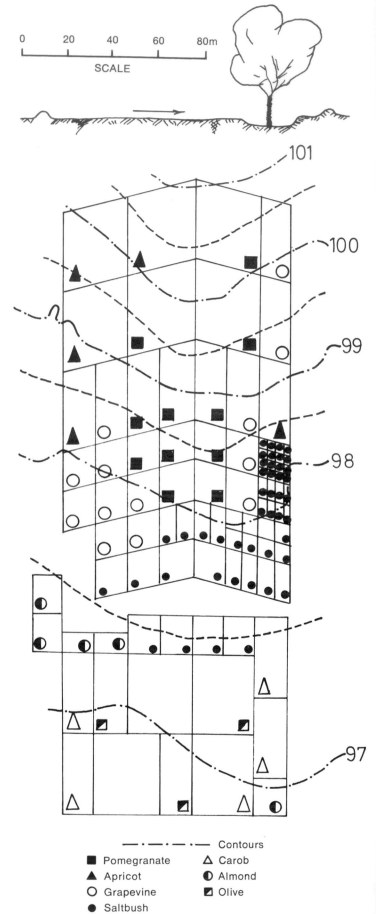

Fig. 14. Layout and typical cross section of microcatchments.

Agricultural Results: Field Crops and Vegetables

Barley, wheat, peas (seeds), sunflowers (seeds), and onions (seeds) were the most successful field crops we raised over the years (Table 6). The best vegetable was asparagus, which seems to be well adapted to the conditions of runoff farming.

TABLE 6
Yields of Various Field Crops
(tons per hectare)

Crop	Yield
Barley	1.2 -5
Wheat	1.1 -4.5
Peas	5 -6
Sunflowers	2.2 -5.5
Onions (seeds)	0.52-0.65

MICROCATCHMENTS (NEGARIN)

After we had found that the amount of runoff per unit surface increased as size of catchment area decreased, we decided to plant trees, each in its own small catchment, instead of having a whole farm with a comparatively large catchment. We chose as the experimental area a barren loessial plain of high salinity (0.9 to 1.2 percent total soluble salts). An area of 1.8 hectares of this plain with a natural slope of 1.5 percent was artificially divided by small border checks into 92 microcatchments (or negarin, after a Hebrew word) ranging in size from 16 to 1,000 square meters. At the lowest point of each microcatchment, a square basin was dug to collect the runoff water and to plant the trees (Fig. 14). In 44 of the microcatchments, ranging in size from 250 to 1,000 square meters, fruit trees were planted. In

Fig. 15. Negarin with a three-year-old pomegranate tree in the foreground.

48 of the negarin (16 to 250 square meters in size), seedlings of the salt bush (*Atriplex halimus*) were installed. The various plants were placed in different-sized microcatchments because we had to find the catchment size optimal for each species.

An important result of our experiment was that after one or two rainy seasons the salt of the planting basins of all the microcatchments was almost completely leached out. So far the best trees are pomegranates (Fig. 15), almonds, olives, and grapevines, but it is too early to arrive at a final judgment, as the trees are still too young. The optimal catchment size for pomegranates, olives, and almonds is apparently 250 square meters. The optimal catchment size for the saltbush is 32 square meters (Fig. 16). Such a saltbush plantation produces a yearly average of 650 kilograms per hectare fresh weight and 400 kilograms per hectare dry weight.

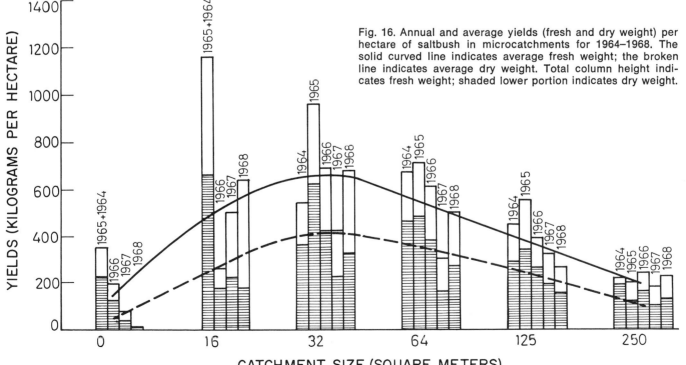

Fig. 16. Annual and average yields (fresh and dry weight) per hectare of saltbush in microcatchments for 1964–1968. The solid curved line indicates average fresh weight; the broken line indicates average dry weight. Total column height indicates fresh weight; shaded lower portion indicates dry weight.

APPLICABILITY OF RESULTS

Our experiments have shown that runoff agriculture as practiced by the ancient farmers in the Negev highlands functions properly today as it did 2,000 to 3,000 years ago (*17*). It enables man to turn wasteland into an agriculturally productive area without any additional irrigation. We think that it is also economically feasible. The microcatchments are the most promising method. We take the saltbush as an example. Using modern machinery, the cost of construction of microcatchments is between $5 and $20 per hectare, depending on the microcatchment size. If planted to saltbush, one hectare will produce 160-170 Scandinavian feed units, which is equivalent to 30 kilograms of protein per hectare. Since the area in its natural state produces only 5 to 10 feed units per hectare, this means a fifteen- to thirtyfold increase of productivity.

Since in Israel today a feed unit costs 2 to 3 cents, it is certain that the barren desert loessial plains can be turned into well-producing rangelands. It is equally certain that the same system can be adapted to other range plants such as alfalfa and *Oryzopsis miliacea,* and that

with these plants the rentability could be increased. One can envisage a grass-legume-saltbush pasture as the best range under our conditions. It will also be possible to increase the efficiency of the microcatchment system by soil treatments which will increase the water yields and by other means. Fruit-tree plantations, especially of almonds and probably pistachios, are also of potential practical value, but in this respect our data are not yet complete enough to be sure.

The labor needed for the maintenance of the large catchments is negligible. Our experience shows that 5 man-days were enough in 10 years to maintain 40-hectare watersheds in proper order. The labor needed to maintain the microcatchments is even less. If nomads should use the runoff agriculture system they could set these up and visit them only once or twice a year.

The practical possibilities of extensive runoff agriculture are not limited to the Negev of Israel. Similar rainfall, climatic, and soil conditions are found outside of Israel in large desert areas of the world, where runoff agriculture appears most promising as the best and cheapest way of land use.

REFERENCES AND NOTES

1. All basic meteorological, hydrological, and agricultural data are summarized in four progress reports to the Rockefeller Foundation, and in a new Harvard University Press book:
    EVENARI, M., L. SHANAN, AND N. H. TADMOR
    1963– Runoff-farming in the Negev desert of Israel
    1968 I–IV. National and University Institute of Agriculture, Rehovot; Hebrew University, Jerusalem, Department of Botany.
    1971 The Negev: The challenge of a desert. Harvard University Press, Cambridge.
2. SHANAN, L., M. EVENARI, AND N. H. TADMOR
    1967 Rainfall patterns in the central Negev desert. Israel Exploration Journal 17:163–184.
3. EVENARI, M., Y. AHARONI, L. SHANAN, AND N. H. TADMOR
    1958 The ancient desert agriculture of the Negev. III: Early beginnings. Israel Exploration Journal 8:231–268.
4. AHARONI, Y., M. EVENARI, L. SHANAN, AND N. H. TADMOR
    1960 The ancient desert agriculture of the Negev. V: An Israelite agricultural settlement at Ramat Matred. Israel Exploration Journal 10: 23–36, 97–111.
5. EVENARI, M., L. SHANAN, N. H. TADMOR, AND Y. AHARONI
    1961 Ancient agriculture in the Negev. Science 133: 979–996.
6. TADMOR, N. H., M. EVENARI, L. SHANAN, AND D. HILLEL
    1957 The ancient desert agriculture. I: Gravel mounds and gravel strips near Shivta. Ktavim 8:127–151.
7. SHANAN, L., N. H. TADMOR, AND M. EVENARI
    1958 The ancient desert agriculture of the Negev. II: Utilization of runoff from small watersheds in the Abde (Avdat) region. Ktavim 9(1–2): 107–128.
8. TADMOR, N. H., L. SHANAN, AND M. EVENARI
    1960 The ancient desert agriculture of the Negev.

    VI: The ratio of catchment area to cultivated area. Ktavim 10:193–221.
9. SHANAN, L., N. H. TADMOR, AND M. EVENARI
    1961 The ancient desert agriculture of the Negev. VII: Exploitation of runoff from large watersheds. Israel Journal of Agricultural Research 11:9–31.
10. KRAEMER, C. J. JR.
    1958 Excavations at Nessana. III: Non-literary papyri. Princeton University Press, Princeton, New Jersey. 355 p.
11. EVENARI, M., L. SHANAN, AND N. H. TADMOR
    1968 "Runoff farming" in the desert. I: Experimental layout. Agronomy Journal 60:29–32.
12. HILLEL, D.
    1959 Studies on loessial crusts. Israel Agricultural Research Station, Bulletin 63. 87 p. In Hebrew with English summary.
13. PALMER, E. H.
    1871 The desert of Exodus (2 vols.). Deighton, Bell, and Co, Cambridge.
14. COHEN, O. P., M. EVENARI, L. SHANAN, AND N. H. TADMOR
    1968 "Runoff farming" in the desert. II: Moisture use by young apricot and peach trees. Agronomy Journal 60(1):33–38.
15. TADMOR, N. H., O. P. COHEN, L. SHANAN, AND M. EVENARI
    1966 Moisture use of pasture plants in desert environment. International Grassland Congress, 10th, Helsinki, 1966, Proceedings: 899–906.
16. TADMOR, N. H., L. SHANAN, AND M. EVENARI
    1969 Consumptive water use of range plants under desert conditions. Hebrew University, Jerusalem, Research Report. 130 p.
17. The very generous help of the Ford Foundation (archaeological field survey), and of the Rockefeller Foundation, the Edmond and James de Rothschild Memorial Group, and Israel's Prime Minister's Office (restoration of farms and agricultural research) enabled us to carry out our work.

Artificial Inducement of Runoff as a
Potential Source of Water in Arid Lands

DANIEL HILLEL
Hebrew University
Jerusalem, Israel

ABSTRACT

RUNOFF INDUCEMENT, also called water harvesting, is the practice of artificially sealing land surfaces to decrease infiltration and induce the rain to trickle downslope as surface runoff. Runoff inducement can be an important potential source of water, heretofore generally unrecognized. Perhaps the simplest method of runoff inducement is to cover the surface with an impervious apron of such materials as plastic, metal, or concrete. Possibly a more economical approach is to cause the soil itself to shed, rather than absorb, the rain. The author and his colleagues have studied the interrelations of rainfall, infiltration and runoff as affected by various soil treatments.

The results point to the following methods for treating arid-zone soils to collect runoff: (*a*) eradication of vegetation and removal of surface stones, to reduce interception of rain and obstruction of overland flow, and to permit the formation of a continuous surface crust; (*b*) smoothing the land surface, to obliterate surface depressions and prevent the retention of water in puddles; (*c*) compaction of the soil top layer to reduce its permeability, which can be done by means of a smooth roller at optimal soil moisture content; (*d*) dispersion of soil colloids to induce self-crusting, by means of sprayable solutions of sodium salts; (*e*) impregnation of the surface with a sealing and binding material, among which the most promising materials are sprayable petroleum solutions that can form a water-repellent and stable coating. For treatments *c* and *d* to succeed, the soil must contain sufficient clay to be compactible and dispersible, but not so much as to exhibit marked shrinkage and cracking.

These methods are listed in the order of increasing complexity and cost. Which of the methods, whether alone or in combination, will be most suitable in any particular area will depend upon local physical conditions and cost factors.

With these and other methods described in this chapter, it is possible to increase runoff several-fold. In a desert area with a seasonal rainfall of 250 millimeters (10 inches), yields as high as 200,000 cubic meters of water may be obtainable per season per square kilometer of treated area.

ARTIFICIAL INDUCEMENT OF RUNOFF AS A POTENTIAL SOURCE OF WATER IN ARID LANDS

Daniel Hillel

IN MANY REGIONS, large tracts of land remain unused owing to insufficient or unstable rainfall, poor soils (shallow, stony, or saline), or irregular topography. The possibility of controlling and increasing the amount of surface runoff obtainable from such lands can be of great importance, particularly in arid and semiarid zones, where the runoff thus obtained can augment the meager water supply for crops, livestock, industrial and urban reservoirs, and groundwater-recharge projects.

Where rainfall is insufficient but the soil is otherwise arable, it may be possible to utilize the land in a system of "runoff farming," in which alternate "runoff strips" are treated (surface-sealed and stabilized) so as to contribute their share of the rainfall as runoff to adjacent "runon strips" in which crops can be grown.

The importance of runoff inducement is possibly greater than the mere increase in total runoff yield which it may produce. Effective inducement of runoff can also lower the runoff threshold, that is, the minimal rainstorm (in terms of size and intensity) needed to start runoff. This decrease of the threshold may correspondingly increase the probability of obtaining adequate runoff a sufficient number of times during the season (even from small rains) and thus effectively decrease the incidence of drought.

Runoff utilization is in fact a very old art, practiced in the Negev of Israel as well as in other parts of the Near East by the ancient Israelites, Nabataeans, Romans and Byzantines (1-4). Archaeological evidence of the widespread use of runoff abounds, particularly in the Negev Highlands region, where 80 to 90 percent of the area consists of bare, rocky hills. Rain falling on these hills is, for the most part, wasted. Runoff yield under natural conditions consists of only a small percentage — generally no more than 5 percent — of the annual rainfall (5). We use the term "yield" in the agricultural sense, since the desert farmer of ancient times actually harvested his supply of water from the slopes, depending almost entirely on runoff to fill his cisterns and irrigate his terraced fields (which were invariably located in bottomlands and wadi beds). But the Negev runoff-farmer did more than merely gather natural runoff — he actually attempted to induce more of it, by clearing the surface

gravel and thus exposing, and possibly smoothing, the finer soil beneath in order to facilitate crust formation (6,7).

Even at best, however, the ancients were able to collect no more than an average of 10 percent of the annual rainfall, while the remaining 90 percent or so soaked into the ground. In desert areas the latter amount can be considered a complete loss, as it soon evaporates either directly or through the leaves of useless vegetation. The ancients therefore needed a runoff-to-runon area ratio of between 20:1 and 30:1 in order to supply their fields with enough water (8).

Modern technology and chemistry hold the promise of more effective runoff inducement, as indicated by the pioneering work of the U. S. Water Conservation Laboratory (9–13). By means of mechanical treatments (stone-clearing, smoothing, and compaction) as well as by a variety of chemical treatments to seal and stabilize the surface, it is possible to increase runoff several-fold, thus permitting significant reduction of the runoff-to-runon area ratios in runoff farming. Furthermore, even where runoff farming is impractical or unnecessary, runoff inducement (water harvesting) can serve as a vital source of water for other uses. In some areas, especially in high-altitude plateaus far inland from the sea, even allowing for the cost of land preparation and surface treatment, water harvesting may eventually prove to be more economical than desalination and conveyance of seawater, often held to be the ultimate solution to the problem of water supply in arid regions. Water harvesting is already more economical than desalination in some special locations, such as Gibraltar and Hawaii (L. E. Myers, personal communication).

This brief and necessarily sketchy account of some general aspects and problems of runoff inducement is based largely on the experience of the author and his colleagues working in Israel.

INFILTRATION AND RUNOFF

Infiltration is the term applied to the process of water entry into the soil, generally through the soil surface in downward flow. *Runoff*, or *overland flow*, is the term applied to the water that fails to infiltrate but trickles off the surface.

The infiltration of water into the soil begins with the onset of rain. For any given soil in a given condition there is a maximal intake rate which can be termed *soil infiltrability* or *infiltration capacity*. This rate is relatively

NOTE: This chapter is an adaptation of a presentation at "Arid Lands in a Changing World," an international conference sponsored by the Committee on Arid Lands, American Association for the Advancement of Science, June 3–13, 1969, held at the University of Arizona, Tucson, Arizona, U.S.A.

high at first, especially in an initially dry soil, but decreases as the soil wets and generally approaches a steady rate sometimes called the *final infiltration capacity* (*14*).

As long as the rate of water supply to the soil surface, determined by *rainfall intensity,* is less than soil infiltrability, the rain is absorbed as fast as it falls, and no runoff can form. When rainfall intensity exceeds soil infiltrability, the excess of the former over the latter (which we shall call the *excess rainfall*) begins to accumulate over the surface. In most cases the soil surface is fairly rough and includes numerous depressions, or pockets, in which this excess water is temporarily stored in the form of puddles. The volume of these puddles per unit area of soil surface is called the *surface storage* capacity. Water stored in puddles can eventually infiltrate into the soil, either after the cessation of the rain or during respite periods when rain intensity falls below soil infiltrability. Where the ground surface is nearly level and/or very rough (that is, with deep pockets giving a high surface storage capacity) there may quite possibly be no runoff even though the rain intensity may greatly exceed the soil infiltrability.

On the other hand, where the surface is relatively smooth and steep the small surface storage capacity fills up soon after the onset of excess rainfall; and, should the rainfall persist at an intensity exceeding infiltrability a while longer, the puddles must begin to overflow and runoff is thus produced.

Obviously, the amount of runoff depends upon rainfall characteristics (the number, frequency, size, and intensity of rainstorms) on the one hand, and on soil characteristics (surface storage capacity and infiltrability) on the other. Since we cannot easily control the rain, the way to increase the probability and amount of runoff is to decrease surface storage and infiltration rate.

Surface storage can be reduced by smoothing the land surface. However, effective control of soil properties influencing infiltration must depend upon a fundamental understanding of the infiltration process. Since practical treatments are necessarily limited to the top layers of the soil, it is of particular interest to establish how the properties of these layers, relative to the properties of the entire profile, influence infiltration.

Although the presence of a surface crust has long been known to impede the infiltrability of the soil, until recently no direct methods were available to predict the magnitude of this effect from knowledge of the basic hydraulic properties of the crust and underlying soil. Some progress in the theory of infiltration into crusted soils has been achieved recently (*15-17*). The theory now available permits calculation of relative infiltrability for various conditions of the surface and subsurface. It indicates quantitatively how infiltrability decreases with increasing *hydraulic resistance* of the crust (this resistance being directly proportional to the thickness of the crust and inversely proportional to its hydraulic conductivity) and with increasing coarseness of the subcrust soil. This theory was verified experimentally. Typical infiltration versus time curves for uncrusted and for various

ously crusted soil columns are shown in Figure 1. It is seen that the denser and/or thicker the crust, the lower the infiltrability, and the more quickly the soil attains its final constant infiltration capacity.

These findings can be applied where it is desirable to decrease soil infiltrability artificially. Under such conditions, the approach cited can aid in determining the desirable crust and subcrust soil properties that can be induced to best advantage. In principle, the ideal soil profile for runoff inducement should consist of a smooth, stable, and dense crust of low hydraulic conductivity, underlaid by a coarse-structured, porous zone of low unsaturated conductivity.

Once formed, runoff begins to accelerate as it flows downslope. As the velocity of the water increases, so does its erosive power. Rapidly flowing water can detach and transport soil particles and scour the soil surface. It can cut right through a surface crust and dig into the soft soil beneath. This process of erosion muddies the water and results in the deposition of sediment wherever the runoff water is eventually collected. For this reason, it is not enough for the purposes of runoff inducement to *smooth* and *seal* the surface. It is also necessary to *stabilize* the surface against water action and to route the water (by controlling the steepness and length of the overland flow path) in such a way as to minimize destructive erosion (*18*).

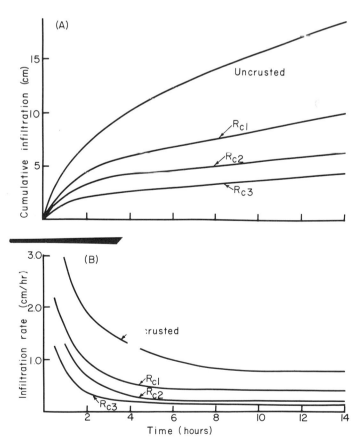

Fig. 1. Time-dependence of cumulative infiltration (*A*) and of infiltration rate (*B*) for uncrusted and crusted columns of Negev Loess. Crust resistance values R_{c1}, R_{c2}, R_{c3} are 3.2, 9.1, and 17 days, respectively.

SURFACE TREATMENTS

In general, it is possible to classify runoff-inducing surface treatments as follows (*19*): (*a*) *mechanical treatments,* such as smoothing and compacting the surface; (*b*) *colloidal dispersion treatments,* to cause slaking of soil aggregates and to induce self-crusting of the soil; (*c*) *hydrophobic treatments,* to reduce the wettability of the soil surface by applying water-repellent surface active agents; (*d*) *surface-binding treatments,* designed to permeate and seal the surface pores with an adherent material so as to cement the initially loose soil into a firm matrix; (*e*) *surface-covering treatments,* in which a thin film of an impervious and stable material is placed over the soil; and (*f*) various combinations of these treatment types.

Perhaps the simplest method of runoff inducement is to cover the surface with an impervious apron of such materials as plastic, metal or concrete. An approach that is possibly more economical is to cause the soil itself to shed, rather than absorb, the rain.

In the search for efficient treatments for runoff inducement, it is necessary to have a standard screening procedure for the preliminary evaluation of soil-sealing methods and materials. Such a procedure has been developed, consisting of microwatersheds which can be packed with soil, given various treatments, and tested under natural rain or under a rain simulator, and monitored for the continuous determination of rain-infiltration-runoff-erosion relationships (*19*).

Preliminary tests of soil surface treatments have revealed three possible effects:

a) The *hydrophobic effect,* in which an initially dry soil surface is made to resist wetting. This effect may be only temporary, however, as once the surface pores do become saturated (under the beating action of raindrops or under the positive pressure of ponded water) the hydrophobic effect no longer comes into play. For this reason the use of the term *waterproofing* may be misleading. Hydrophobic materials can render the soil water-repellent but not waterproof. Since this property has only a temporary effect, it is obvious that a hydrophobic treatment alone is not sufficient for effective runoff inducement.

b) The *structure-stabilizing effect.* Low application rates of certain chemical treatments were found to stabilize soil structure, thus preventing soil slaking and self-crusting, and actually increasing the infiltration rate over that of the untreated soil. While the stabilizing effect is necessary to prevent surface erosion, it will not by itself induce runoff.

c) The *surface-sealing effect.* This effect occurs only when the surface pores are actually blocked by the treatment material and the hydraulic conductivity of the surface layer is reduced. This effect will not last if the material is not sufficiently adherent to the soil. For maximum efficiency, therefore, a treatment should have both stabilizing and sealing effects, and it should be applied in sufficient quantity for the sealing effect to be fully developed.

Laboratory tests in microwatersheds of loessial soils from the Negev showed promising results with sprayable treatments such as sodic dispersants and fuel-oil-based solutions of petroleum derivates. The best results were achieved with a combination of these treatments (*18*).

In treatment of the soil with sodic dispersants, the application quantity should be sufficient to form a slaked layer several centimeters deep. Treatment of the very top of the soil is not enough, as too thin a crust will quickly crack, disintegrate, and erode. In general the effect of dispersants will be greatest in soils with a high content of clay, though it can still be pronounced in arid-zone soils of relatively low clay content. Since, however, the dispersion of soil colloids and slaking of soil aggregates not only decreases permeability but also increases soil erodibility, it must be coupled with a surface-binding and -sealing treatment.

Of the various materials that can bind and seal the soil surface, petroleum solutions appear to be among the most feasible as they have the advantages of universal availability; relatively low cost; ease of handling, storage, and application; water repellency; and ability to bind loose soil (provided it is not too clayey). Various additives may improve the weathering resistance and stability of the treated surface. The application and dilution rates can be adjusted so as to influence penetration depth and adherence to the soil. In Israel, the petroleum materials described performed best in the sandy loam and silt loam of the Negev but did not perform nearly so well in the clay of the Judean Hills. One disadvantage of petroleum materials is that contamination of the runoff water can be a problem.

For the proper evaluation of soil treatments in the field, consideration must be given to the rainfall pattern of the particular experimental season and location. Different rainfall patterns obviously have different potentials for producing runoff. A method was therefore developed for evaluating and characterizing the runoff-producing power of individual rainstorms and of whole seasons. The method is based on an analysis of the rain intensity distribution and gives an estimate of the amount of water obtainable from runoff areas of given (hypothetical) average infiltration rates. These calculations, which were programmed for a computer, can allow the comparison of different seasons and locations (*18*).

Analysis of rainfall intensities versus quantities shows that the intensity distribution is highly skewed and strongly weighted toward the lower intensities. The peak prevalence in the Negev desert apparently occurs at an intensity of 1 to 2 millimeters per hour, and the median intensity at about 6 millimeters per hour. This means that about half the seasonal rainfall occurs at intensities smaller than 6 millimeters per hour, a rate which the soil in its natural state can generally absorb readily. Seasons of lower rainfall totals also appear to have relatively less rain of high intensity. Thus, in the wettest of the seasons monitored by the author and his colleagues in the Negev, 32 percent of the rain occurred at intensities exceeding 10 millimeters per hour, whereas in the driest season only 22 percent of the rain occurred at such intensities. This further emphasizes that the lower the seasonal rainfall the greater the probable importance of runoff inducement methods, without which the chance of obtaining appreciable runoff may be nil.

Rain intensities as well as total amounts are generally higher in humid than in arid regions. The prediction of potential runoff probabilities in any region is very difficult, as the total runoff yield and its seasonal distribution depend upon the seasonal pattern of rainstorms and their intensities. In general, the more arid the climate the greater the relative variability of the rainfall pattern. For these reasons, as well as for reasons of soil heterogeneity, specific experimental studies of runoff-inducement methods must be conducted in each separate region where runoff utilization is proposed.

The results of field runoff-plot tests in Israel have shown great differences among treatments. Whereas cultivation and especially aggregation of the surface reduced the runoff yield to relatively low and often negligible quantities, such treatments as compaction, sodic dispersants, and the application of sealing and hydrophobic materials can increase runoff several-fold, particularly during relatively dry seasons. The gross comparison of seasonal runoff totals disregards the differential behavior of the various treatments during large versus small storms, early versus late-season storms, high-intensity versus low-intensity storms, and so on. One of the simplest ways to consider the effects of the season's various storms is to calculate the regression of per-storm runoff on storm size. The assumption of a linear relationship gives reasonable results where the soil infiltration rate is relatively low, but the data generally cannot be expected to fit a linear regression where infiltration rates are above 10 millimeters per hour.

The runoff-to-rainfall regression line generally intercepts the rainfall axis. This intercept, which indicates the minimal rain generally needed to form runoff, has been termed "threshold rain." Its value is directly related to the infiltration rate of the soil and is thus lowest for the most effective treatments. Some of the treatments tried in the Negev gave runoff from practically every rain, while the less-effective treatments gave runoff only from the few large storms and none from the numerous small storms of the season.

A proper evaluation of the efficiency of runoff-inducement treatments should be based on knowledge of each treatment's durability as well as of its initial effect. A treatment durability trial in the northern Negev showed that the sodic, silicone, and fuel-oil treatments retained most of their effect over three consecutive seasons without reapplication. In practice, however, a light reapplication will probably be required periodically for long-term maintenance of runoff areas (*18*).

The results of small-scale trials cannot simply be extrapolated to predict the quantitative behavior of treatments when applied on a large scale. Owing to such factors as areal nonuniformity and the longer path of overland flow, large-scale applications of runoff-inducement treatments cannot be expected to yield runoff rates as high as those obtained in the small and meticulously treated experimental plots, or to last as long, or to erode as little.

RUNOFF-INDUCEMENT METHODS

The results of studies by the author and his colleagues (*18*) point to a number of practical methods for treating arid-zone soils to collect runoff. These methods, mentioned above, can be summarized as follows: (*a*) eradication of vegetation and removal of surface stones, to reduce interception of rain and obstruction of overland flow, and to permit the formation of a continuous surface crust; (*b*) smoothing the land surface, to obliterate surface depressions and prevent the retention of water in puddles; (*c*) compaction of the soil top layer to reduce its permeability, which can be done by means of a smooth roller at optimal soil moisture content; (*d*) dispersion of soil colloids to induce self-crusting, by means of sprayable solutions of sodium salts; (*e*) impregnation of the surface with a sealing and binding material, among which the most promising materials are sprayable petroleum solutions that can form a water-repellent and stable coating. For treatments (*c*) and (*d*) to succeed, the soil must contain sufficient clay to be compactible and dispersible, but not so much as to exhibit marked shrinkage and cracking.

These methods are listed in the order of increasing complexity and cost. Which of the methods, whether alone or in combination, will be most suitable in any particular area will depend upon local physical conditions and cost factors.

With these and similar methods, it is possible to increase runoff several-fold. In a desert area with a seasonal rainfall of 250 millimeters (10 inches), yields as high as 200,000 or more cubic meters of water may be obtainable per season per square kilometer of treated area (Fig. 2).

RUNOFF FARMING

Runoff farming depends on the collection of runoff water and its storage in the soil for the subsequent use of crops. Since plants must respire constantly, while the supply of irrigation by runoff is only occasional (and rather infrequent), it is essential that the soil have a sufficient water-storage capacity within the rooting depth to sustain the crop between irrigations. It is also essential that the crop chosen be adapted to the special circumstances of runoff utilization, not only from the physical and physiological standpoints but from the economic standpoint as well. That is to say, it must produce returns commensurate with the investment. A very important aspect of the problem, therefore, is the selection of the crop to be grown under runoff farming. We shall therefore address ourselves primarily to this aspect. Limitations of space preclude a more comprehensive review of any other engineering and agronomic aspects of runoff farming.

In general, runoff farming techniques can be applied to the production of pasture (grasses or shrubs), field crops (annual or perennial), or trees. The following are some of the considerations involved:

a) Depth of soil water storage and utilization. For greater efficiency, the soil moisture reservoir should be as deep as possible, provided crop roots are capable of tapping it. Annual crops are likely to be shallow-rooted and must start their root development anew seasonally. Perennial crops, once established, have ready root systems capable of utilizing soil moisture more efficiently. What is needed is a perennial crop requiring only low-cost annual maintenance operations, of sufficient hardiness to withstand drought seasons (with a minimum amount of supplementary irrigation, or even without any) and having the ability to recover and produce a high return during favorable seasons. Certain fruit trees are likely to be more deep-rooted, drought resistant, and economically productive than most other species of perennial crops.

b) Danger of trampling. To be economical, surface sealing or coating treatments must necessarily be applied rather thinly (that is, just thickly enough to bind and seal or cover the soil so as to withstand weathering, but probably not enough to resist destruction by repeated trampling). Rupturing, pitting, and loosening of a surface crust by sharp-hooved animals accelerate deterioration and erosion. Where animals are to be led onto the runon grazing areas, they can be prevented from trampling over the treated catchment areas only by means of expensive fencing. This objection to runoff inducement

for pasture production may not apply to the supply of drinking water for livestock (*20, 21*) if the treated runoff-producing areas can be enclosed or otherwise protected economically.

c) Rainfall variation. The inter- and intra-seasonal variation of rainfall is so great and unpredictable as to make annual cropping highly hazardous. In a mediterranean-type climate (rainy winter, dry summer) the hazard is greatest for winter crops, which must be planted before the farmer can know anything about the season's runoff potential or the system's actual water supply. Even for summer crops, which are planted at the end of the rainy season when the soil moisture reserve is already measurable, the season-to-season variation and drought frequency may still preclude stable cropping. Furthermore, the requirement of repeated cultivations for annual crops may make the runoff area itself entirely too susceptible to erosion.

Though a complete runoff-farming system should include certain reserve areas that could be irrigated and planted to annual crops in case of an overflow from the primary cropping area (that is, during the wetter seasons), tree crops should probably be given first priority in planning this primary area. Second priority should perhaps be given to other drought-resistant perennial crops, and third priority to annual summer crops (especially deep-rooted ones). A schematic representation of a runoff orchard is shown in Figure 3.

ECONOMIC FEASIBILITY

The ultimate problem is that of economic feasibility, and it is a problem for which there can be no simple or single answer. The economics of a runoff-inducement system will depend on the local factors that will determine

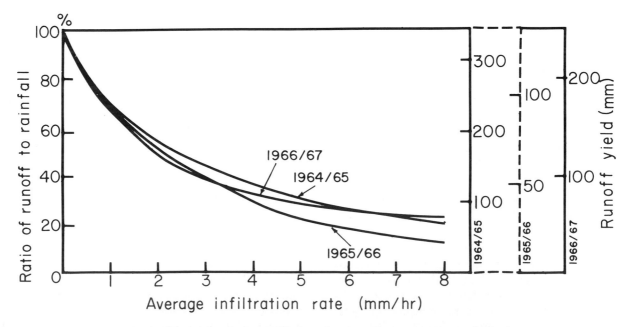

Fig. 2. Seasonal runoff potential (runoff ratio and expected water yield) as function of hypothetical average infiltration rate (Φ–value), for the three experimental seasons in Gilat (Northern Negev, Israel).

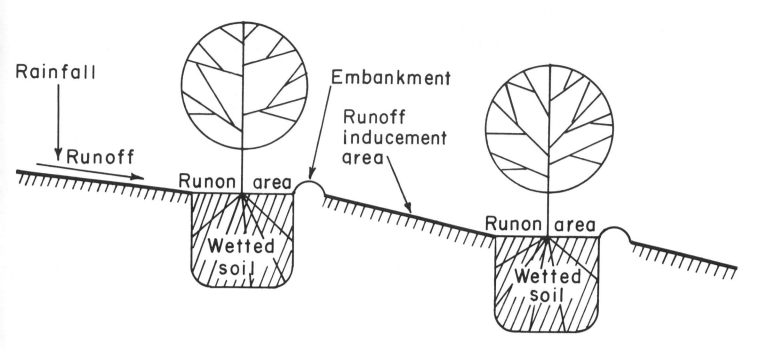

Fig. 3. Profile of runoff-irrigated orchard (schematic).

the longterm cost of a unit volume of water at the site in relation to the price of the products that it can produce. This cost must be compared with the cost of water from alternative sources. In order to make a proper judgment, one would have to take into consideration the fact that runoff water is not available on demand or in a completely dependable manner (except on the basis of statistical probabilities), as it is subject to the vagaries of rainfall. Hence the runoff collected must be stored, and the considerable costs of storage facilities must be included in the overall costs of any project considered. Since all these factors differ from one location to another, and certainly from one country to another, no universal prescription can be given.

In principle, runoff inducement and utilization will be most feasible in regions where alternative supplies are most expensive, yet where rainfall is sufficiently dependable and predictable to permit designing an efficient system. Such regions might include remote and high-altitude sites in deserts, or islands lacking fresh surface or ground waters, where the requirements of settlement and industry justify the costs of utilizing runoff water.

NEED FOR FURTHER RESEARCH

In studies carried out to date, basic methods, theoretical as well as experimental, were developed for the evaluation of runoff-inducement techniques. Using these methods, tests were conducted in the laboratory and in the field which showed that, in the case of certain types of arid-zone soils at least, sodic dispersants as well as several chemical sealants and hydrophobic materials gave excellent results. The progress achieved offers prospects for the large-scale economic application of these techniques for supplying water in arid regions.

Additional laboratory screening trials are needed to evaluate and compare surface sealing, waterproofing, and stabilizing methods. Adherent materials (such as asphaltic solutions) should be tried in combination with sodic dispersants. Plastic materials should be tested as they become available. A detailed study is needed of application methods and amounts, and of the effects of dilution rates and various additives designed to improve weathering resistance. A study is also needed of initial soil preparation, vegetation control, stone clearing, land planing, and surface compaction. Runoff-inducement experiments should also be combined with research into methods of runoff storage and utilization, for agricultural as well as other uses.

The research should also be extended to study some of the more fundamental physical and hydrological aspects of the problem. The heretofore separate sciences of surface hydrology and soil physics should thus be combined in a unified context to achieve a more exact evaluation of runoff-inducement methods under hypothetical and actual conditions of soil, climate, and technology (*22, 23*).

IN CONCLUSION

Although considerable data have been collected and analyzed so far, and the results appear to be highly promising, it should be obvious that many problems yet remain before runoff inducement can become generally feasible or widely applicable, and that research along these lines must continue. Better knowledge of the surface hydrology of arid regions will help to spur greater efforts toward more efficient development and management of the precious water resources in the dry lands where so many millions of mankind now barely subsist.

REFERENCES AND NOTES

1. EVENARI, M., AND D. KOLLER
   1956   Masters of the desert. Scientific American 44:
   39–45.

2. EVENARI, M., Y. AHARONI, L. SHANAN, AND N. TADMOR
   1958   The ancient agriculture of the Negev. III:
   Early beginnings. Israel Exploration Journal
   8:231–268.

3. EVENARI, M., L. SHANAN, N. TADMOR, AND
   Y. AHARONI
   1961   Ancient agriculture in the Negev. Science
   133:979–996.

4. LOWDERMILK, W. C.
   1953   Floods in the desert. *In* Desert Research, Proceedings International Symposium held in
   Jerusalem, May 7–14, 1952, sponsored by the
   Research Council of Israel and Unesco. Research Council of Israel. Special Publication
   2:365–377.

5. HILLEL, D., AND N. TADMOR
   1962   Water regime and vegetation in the Negev
   highlands of Israel. Ecology 43:33–41.

6. HILLEL, D.
   1960   Crust formation in loessial soils. International
   Congress of Soil Science, 7th, Madison, Wisconsin, 1960, Transactions 1:330–339.

7. TADMOR N., AND OTHERS
   1957   The ancient desert agriculture of the Negev.
   Israel Journal of Agricultural Research 8:127–
   151.

8. SHANAN, L., N. TADMOR, AND M. EVENARI
   1958   The ancient desert agriculture of the Negev.
   II: Utilization of runoff from small watersheds
   in the Abde region. Israel Journal of Agricultural Research 9:107–128.

9. MYERS, L. E.
   1961   Waterproofing soil to collect precipitation.
   Journal of Soil and Water Conservation 16:
   281–282.

10. MYERS, L. E.
    1967   Recent advances in water harvesting. Journal
    of Soil and Water Conservation 22:95–97.

11. MYERS, L. E., AND OTHERS
    1967   Sprayed asphalt pavements for water harvest-
    ing. American Society of Civil Engineers,
    Irrigation and Drainage Division, Journal
    93(IR3):79–97.

12. MYERS, L. E.
    1967   New water supplies from precipitation harvest-
    ing. International Conference on Water for
    Peace Proceedings, Washington, D.C. Paper
    No. 391.

13. MYERS, L. E., AND G. W. FRASIER
    1969   Creating hydrophobic soil for water harvest-
    ing. American Society of Civil Engineers,
    Irrigation and Drainage Division, Journal
    95(IR1):43–54.

14. HILLEL, D.
    1971   Soil and water, physical principles and pro-
    cesses. Academic Press, New York.

15. HILLEL, D.
    1964   Infiltration and rainfall runoff as affected by
    surface crusts. International Congress of Soil
    Science, 8th, Bucharest, 1964, Transactions
    2:53–62.

16. HILLEL, D., AND W. R. GARDNER
    1969   Steady infiltration into crust-capped profiles.
    Soil Science 108:137–142.

17. HILLEL, D., AND W. R. GARDNER
    1970   Transient infiltration into crust-capped pro-
    files. Soil Science 109:69–76.

18. HILLEL, D., AND OTHERS
    1967   Runoff inducement in arid lands. Final report
    submitted to U.S. Department of Agriculture
    on P.L. 480 Project No. A10–SEC–36. 142 p.

19. HILLEL, D., AND OTHERS
    1969   Laboratory tests of sprayable materials for
    runoff inducement in a loessial soil. Israel
    Journal of Agricultural Research 19:3–9.

20. CLUFF, C. B.
    1967   Water harvesting plan for livestock or home.
    Progressive Agriculture in Arizona 19:3.

21. LAURITZEN, C. W., AND A. A. THAYER
    1967   Rain traps for intercepting and storing water
    for livestock. U.S. Department of Agriculture
    Information Bulletin 307.

22. HILLEL, D., AND E. RAWITZ
    1968   A preliminary field study of surface treatments
    for runoff inducement in the Negev of Israel.
    International Congress of Soil Science, 9th,
    Adelaide, 1968, Transactions 1:303–311.

23. This chapter is a contribution from the Corn Belt
    Branch, Soil and Water Conservation Research Divi-
    sion, Agricultural Research Service, U.S. Department
    of Agriculture, Madison, Wisconsin, in cooperation with
    the Wisconsin Agricultural Experiment Station.

Physiological Basis for Plant Growth Inhibition Due to Salinity

JAMES W. O'LEARY

Environmental Research Laboratory
University of Arizona
Tucson, Arizona, U.S.A.

ABSTRACT

INFORMATION ON PLANT GROWTH INHIBITION due to salinity, collected from several sources, has been integrated into a unified theory to explain how salinity inhibits plant growth. The resistance of roots to waterflow increases in plants grown in saline solutions. This results in development of water stress in leaves when plants are grown in an environment with a high evaporative demand. Hormone delivery from roots to leaves is reduced in plants grown in saline solutions. Possibly this causes a decrease in cell-wall extensibility in leaf cells, and this factor becomes of importance when plants are grown in an environment with a low evaporative demand so that water stress does not develop. Suggestions for overcoming growth inhibition due to salinity include growing plants in enclosures to maintain high humidity, and foliar applications of plant-growth hormones.

PHYSIOLOGICAL BASIS FOR PLANT GROWTH
INHIBITION DUE TO SALINITY

James W. O'Leary

BEFORE THE EXTENSIVE AREAS of arid land in proximity to sources of saline or brackish water can be utilized for agricultural purposes, the physiological bases for growth inhibition under saline conditions must be identified, and suitable techniques must be developed to compensate for these physiological disorders. While considerable research has been conducted on the responses of plants to salinity, only a small percentage of the effort expended has been directed toward a delineation of the specific metabolic changes that are responsible for the observed effects. Furthermore, no unified theory has been advanced to explain the interrelated effects that typically result from an increase in salinity in the root environment.

Such a theory is a prerequisite before suitable techniques can be devised to attempt to overcome the harmful effects of salinity. What I shall attempt to do, thus, is to integrate experimental data of our own with that from other laboratories and try to present the general picture that emerges from such an approach. This systems analysis approach is intended to provide a working hypothesis to guide research toward developing suitable techniques that will compensate for the growth inhibition resulting from increased salinity in the root environment.

When a plant is subjected to increasing salinity in the root environment, even if the increase is a gradual one, growth is reduced. Since the increased salt concentration of the solution around the roots raises the osmotic pressure (OP) of the solution (and lowers the water potential), it was thought that this created a "physiological drought." However, if we look at halophytes, or plants that normally grow in areas with extremely low soil water potentials, we find that they have high osmotic pressures in their leaf sap. This high plant OP probably represents an adaptation to the environment.

Thus it should not be too surprising to find that when glycophytes are subjected to gradually increasing salt concentration in the root environment, they also adapt. That is, the plant osmotic pressure increases as the osmotic pressure of the solution around the roots increases (1–4). The osmotic gradient for water absorption thereby apparently is maintained (5, 6), and physiological drought should not result. However, salinity

toxicity symptoms often resemble drought symptoms (3, 7), and growth inhibition has been found to be proportional, in cases, to the osmotic pressure of the growth solution (3). This seems to create a paradox, the apparent development of drought symptoms in plants that have maintained the necessary driving gradient for water absorption. We also have grown plants in saline solutions and found that water stress does develop in plants that have osmotically adjusted to the saline environment (8).

OSMOTICALLY ADJUSTED PLANTS AND WATER STRESS

This, then, raises the paradoxical question of how water stress or physiological drought can occur in osmotically adjusted plants. Recently, Oertli (9) has pointed out that the evidence for osmotic adjustment in leaves is subject to severe criticism. That is, the osmotic pressure of expressed leaf sap has been taken as indicative of the osmotic pressure of vacuolar sap, when in fact the expressed leaf sap is a mixture of cell sap and solution from the protoplasts' surroundings, such as cell walls. If the solution reaching the leaves has a high concentration of salts, and the leaf cells themselves are unable to accumulate salts within their vacuoles in sufficient quantity to result in true osmotic adjustment, then the leaf cells actually may experience true physiological drought. As Oertli points out, expressed leaf sap is a mixture of the solution within and without the vacuoles. Therefore no definitive statements can be made concerning the true environment of leaf cells based on evidence collected to date.

Even if leaf cells do adjust sufficiently, however, to maintain the water potential gradient from external solution to leaves necessary for continued water movement to the leaves, there still is another factor that would result in the leaf cells experiencing physiological drought. This is an increased resistance in the pathway from external solution to the leaf cells. In general, the flux (q) of water through the pathway from external solution to the leaf cells can be described by the following simple equation:

$$q = \frac{\triangle \psi}{\Sigma R}$$

where $\triangle \psi$ is the water potential gradient from external solution to the leaves, and ΣR is the sum of the resistances in the pathway. In all previous work on salinity and osmotic adjustment, ΣR has been neglected com-

NOTE: This chapter is an adaptation of a presentation at "Arid Lands in a Changing World," an international conference sponsored by the Committee on Arid Lands, American Association for the Advancement of Science, June 3–13, 1969, at the University of Arizona, Tucson, Arizona, U.S.A.

332

pletely or assumed to be a constant. It is easy to see, however, that an increase in the resistance term would lead to a lowered rate of water flux through the pathway, even if the driving gradient is maintained. Where in the pathway from external solution to leaf cells would one expect to find significant change in resistance? Since the greatest resistance to water movement between the external solution and the leaf cells is found in that portion of the pathway across the intervening tissue between the root epidermis and xylem under normal conditions (*10*), appreciable changes in the resistance term of the above equation would be expected to result from changes in permeability of the roots to waterflow.

When permeability of roots to waterflow was measured by a pressure-chamber method (*11–13*), I found that the permeability decreased considerably in osmotically adjusted plants (*8*). This can be seen easily from Figure 1. For such change in root resistance to be of physiological significance, the changes must be great enough to cause reduced delivery of water to leaves sufficient to result in loss of turgor. Furthermore, the loss in turgor should be great enough to reduce stomatal aperture. It was important, therefore, to measure the resistance of leaves to vapor loss. Such measurements indicated increased leaf resistance in those plants (Fig. 2).

It thus appears that plants growing in saline solutions can be subjected to a form of physiological drought even though osmotic adjustment may occur, if the resistance of some component of the pathway increases significantly. When such a situation occurs, the reduced water delivery to the tops can cause partial stomatal closure which will reduce the net uptake of carbon dioxide and result in lowered photosynthesis.

Since the turgor of the cells in the leaf is a function of the balance between water delivery to the leaves and water loss from the leaves, the reduction in turgor could

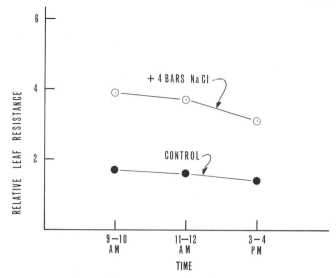

Fig. 2. Leaf resistance to vapor diffusion in bean plants growing in nutrient solution with and without sodium chloride. Chart is a modification of Figure 2 of O'Leary (*8*, p. 4). Reprinted with permission of *Israel Journal of Botany*.

be prevented by reducing the rate of water loss from the leaves. The plant does this naturally by closing the stomates. The effect of this is reduced carbon-dioxide uptake, however. On the other hand, the rate of water loss from the leaf could be reduced without causing stomatal closure by lowering the evaporative demand of the air, that is, by raising the relative humidity of the air. On the practical side, this means that moderately saline water could be used for irrigation if the plants are grown in an atmosphere of high relative humidity. Small-scale experiments have shown that growing plants in atmospheres of high relative humidity does partially reduce the growth inhibition due to salinity (*14, 15*), and techniques for providing these high humidity conditions in areas close to sources of brackish or saline water have already been developed (*16*).

DIRECT EFFECT OF SALINITY ON GROWTH

High humidity, however, does not completely eliminate salinity-induced growth inhibition. There still is some direct effect of salinity on growth. What, then, is the cause of growth inhibition under saline conditions in environments of low evaporative demand where water stress does not develop? Under these conditions (and this includes growth-chamber experiments, for the most part), rather than water-deficient leaves, one finds leaves that seem to have more than enough water. That is, a typical response to salinity is an increased succulence of leaves. Why then do leaves expand less and become more succulent? Let's consider the characteristics of leaves from osmotically adjusted plants and see if this question can be answered.

The total leaf area of salt-treated plants is reduced considerably. The thickness of the leaves also is different. Yet one must be cautious in describing the anatomi-

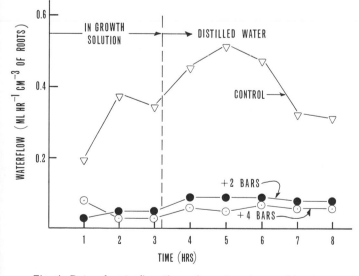

Fig. 1. Rate of waterflow through root systems of bean plants grown in nutrient solution with and without sodium chloride added. Excised root systems were placed in a pressure chamber and water forced through them at a pressure of 2 bars. Chart is a modification of Figure 1 of O'Leary (*8*, p. 4). Reprinted with permission of *Israel Journal of Botany*.

334   JAMES W. O'LEARY

cal features of the leaves since it makes a difference
whether the leaf is one that was present and expanding
during the period of osmotic adjustment, such as the
first trifoliate leaf, or one that has been initiated subsequently. We find that the leaves expanding during the
osmotic adjustment are not as thick on the plant undergoing osmotic adjustment as they are on the control
plant. If we compare leaves that have been initiated subsequent to the period of osmotic adjustment, however,
we find that the leaves from osmotically adjusted plants
are thicker than the corresponding leaves from control
plants. It is somewhat surprising to find that the increased
thickness is not due to increased numbers of cells or
increased cell size, but rather to increased intercellular
space. A summary of the characteristics of the leaves is
presented in Table 1. All indicators of leaf size — area,
dry weight, and fresh weight — show the reduction due
to salinity. The data presented here, particularly the
amount of water per unit of leaf area, all indicate the
typical response to salinity, that is, increased succulence
of leaves.

TABLE 1
Leaf Characteristics of Bean Plants

| | Added NaCl | |
	0 bars	4 bars
Leaf Area (LA)	28.6 dm^2	5.1 dm^2
Dry Weight (DW)	11.6 g	3.6 g
Fresh Weight (FW)	64 g	16 g
DW/LA	0.4	0.7
FW/LA	2.2	3.1
gH$_2$0/LA	1.8	2.4
Leaf Thickness	0.12 mm	0.19 mm

(Leaves initiated after osmotic adjustment)

This, then, raises the interesting question of why leaf
expansion is less in osmotically adjusted plants. If leaf
expansion is a turgor pressure driven process, then there
are basically two ways that leaf expansion could be
inhibited: (*1*) not enough water absorption by the leaf
cells, so the turgor pressure is insufficient to cause cell
expansion, and (*2*) decreased extensibility of the cell
walls, so that even if enough water is absorbed, the cell
wall resists stretching. The increased succulence of the
leaves leads one to suspect that the first possibility is not
too likely, even though it usually is the explanation
offered. So let's look at the second possibility. If the cell
walls are less elastic or plastic in the osmotically adjusted
plants, then the turgor pressure should be higher in these
plants due to the resistance offered by the walls.

We have measured water potential and osmotic pressure in leaves from plants subjected to different levels of
added sodium chloride, and we can calculate the turgor
pressure in the leaves. The data presented in Table 2
show that the turgor pressure increases with increasing
salinity as predicted by the discussion above. Table 3
shows turgor pressures calculated from data of Boyer
(*17*), and the same relationship is evident. What, then,
is the cause of the decreased extensibility of the cell
walls?

TABLE 2
Osmotic Quantities (bars) of Bean Leaves

NaCl Treatment	Osmotic Pressure	Water Potential	Turgor Pressure
0 bars	6.7	—6	0.7
2 bars	7.4	—6	1.4
4 bars	9.4	—7	2.4

TABLE 3
Turgor Pressure in Osmotically Adjusted Cotton Leaves
(all values in bars)

Osmotic Pressure of Nutrient Solution*	Osmotic Pressure of Plant*	Water Potential of Plant*	Turgor Pressure of Plant
0.5	10	— 6	4
3.5	14	— 9	5
6.5	18	—11	7
8.5	21	—13	8

*Taken from Boyer (*17*)

From Table 1 we find that the dry weight per unit leaf
area in osmotically adjusted plants is almost double what
it is in control plants. Since the number of cells per unit
leaf area is not greater in osmotically adjusted plants,
the dry weight per cell must be greater. Does this mean
the cell walls are thicker? It might, but we can't say for
sure at this point. Probably it is more than coincidence,
furthermore, that the higher turgor pressure (and more
rigid cell walls) observed here also is characteristic of
mature or senescing cells. Also, there is increasing evidence to show that the symptoms of water and salt stress
often resemble the symptoms of senescence in leaves.

CYTOKININS

The presence of cytokinin-like substances has been
detected in roots (*18, 19*) and in the bleeding stream
from truncated root systems (*18, 20*), and they appear
to be continuously synthesized in roots (*18, 19*). Since
cytokinins seem to control activities in the leaves, particularly those associated with development of senescence (*20*), it seems likely that a normal feature of a
plant's physiology involves synthesis of cytokinins or
similar substances in the roots and subsequent transport
to the leaves. Furthermore, any perturbation of this
activity would be expected to result in a disruption of
hormone balance in leaves and possibly even accelerate
the onset of senescence.

It has been shown by Poljakoff-Mayber and coworkers
that one of the effects of increased salinity in the root
environment is a decrease in malic dehydrogenase activity in roots (*21*). If this means a decrease in Krebs cycle
activity in general, then it is not unlikely that processes
such as purine synthesis, which depend on the Krebs cycle
for building blocks, would likewise be inhibited by increased salinity in the root environment. It already has
been shown by Vaadia and coworkers that less cytokinin
can be recovered from the bleeding stream of root systems
subjected to water stress (*22*). And it has been found

that the amount of cytokinins reaching the shoots under normal conditions drops off dramatically when the plants reach their final size, that is, just about when leaf senescence starts (*23*). Based on this reasoning, it is hypothesized that a significant effect of increasing salinity in the root environment is a decreased delivery of cytokinins to the leaves.

How does one test such a hypothesis? Unfortunately, the extracting and quantitative analysis for cytokinins (which are the hormones of primary concern) is extremely difficult, and most approaches have to be indirect. Some indirect evidence for lower endogenous cytokinin levels in leaves from stressed plants already has been obtained (*24, 25*). Thus, the evidence rapidly is accumulating to support the idea that a primary effect of salinity in the root environment is to decrease delivery of cytokinins and/or other hormones to the leaves. This hypothesis and the evidence that supports it have recently been eloquently stated by Itai and coworkers (*26*). What needs to be done now is to put all of the pieces together. As stated at the outset, I will try to integrate all of these relatively independent pieces of information and present the unified overall theory that emerges. It is recognized that there will be some points that are much better supported by data than others, but this does not weaken the overall theory nor does it provide sufficient cause to avoid attempting such a postulation.

UNIFIED THEORY

The general scheme is presented in Figure 3. As the osmotic pressure of the solution around the roots is increased, alterations in root metabolism occur, and this is reflected by the increased resistance of the roots to waterflow and the decreased hormone delivery to the leaves. This decreased hormone delivery, thus, disrupts the hormone balance in the leaves, and one of the effects of this (among others, of course) is reflected by the increased rigidity of the cell walls. During this time the plant is accumulating solutes, and the plant osmotic pressure is increasing accordingly (osmotic adjustment). This apparently maintains the osmotic gradient necessary for water absorption.

Now if the plant is growing in an atmosphere with a high evaporative demand, the increased root resistance restricts the flow of water, and in spite of the maintenance of the driving gradient, water delivery to the leaves is insufficient to keep up with water loss from the leaves, and water stress develops (this interaction is indicated by A in Fig. 3). This is reflected by a decrease in leaf turgor. However, if the plant is growing in an atmosphere with a low evaporative demand so that the increased root resistance is not of significance and water delivery to leaves is sufficient to keep up with water loss from the leaves, then the increased rigidity of leaf cell walls becomes of significance (this interaction is indicated by B in Fig. 3). Thus, in spite of sufficient water delivery to the cells in the leaf, the expansion is restricted, and again the leaf expansion is reduced. This is reflected by an increase in leaf turgor. Of course by now the increased osmotic pressure of the leaf cells has direct effects on photosynthesis and hormone metabolism (indicated by dashed arrows in Fig. 3), and these all contribute to the reduction of growth typically observed in osmotically adjusted plants.

REMEDIAL MEASURES

This theory raises many questions and indicates many aspects of the physiology of osmotically adjusted plants that need to be investigated further. However, what is more important, it indicates remedial measures that should be attempted to overcome salinity-induced growth inhibition. If the theory presented here holds, then one should be able to offset some of the growth inhibition by growing plants in atmospheres with high relative humidity. This already has been tried with success on a small scale, as already mentioned. Furthermore, one should be

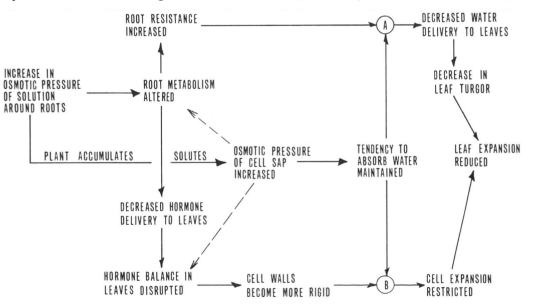

Fig. 3. Unified theory to explain growth reduction due to salinity in environments of high or low evaporative demand.

able to offset some of the growth inhibition by applying the proper hormones to the leaves directly, in the proper concentrations, at the proper times. We have tried this also on a small scale with success. The details of this work have been presented for publication elsewhere, but in general we found that spraying plants with benzyladenine or gibberellin offered promise.

It will not be an easy task to determine which hormones should be applied when, but it seems that this is a task well worth the effort. All evidence indicates that the chances of success are good. In fact, the combination of high humidity and hormone spraying may even result in yields from plants irrigated with brackish water that will be comparable to those from plants grown with non-brackish water. But even if yields are not as good, if it makes it possible for people to inhabit areas that now are uninhabitable, this is equally important. In my opinion, it is more important.

REFERENCES

1. EATON, F. M.
   1927   The water requirement and cell-sap concentration of Australian saltbush and wheat as related to the salinity of the soil. American Journal of Botany 14:212–226.
2. ———
   1942   Toxicity and accumulation of chloride and sulfate salts in plants. Journal of Agricultural Research 64:357–399.
3. BERNSTEIN, L.
   1961   Osmotic adjustment of plants to saline media. I: Steady state. American Journal of Botany 48:909–918.
4. SLATYER, R. O.
   1961   Effect of several osmotic substrates on the water relationships of tomato. Australian Journal of Biological Science 14:519–540.
5. BERNSTEIN, L.
   1963   Osmotic adjustment of plants to saline media. II: Dynamic phase. American Journal of Botany 50:360–370.
6. KRAMER, P. J., O. BIDDULPH, AND F. S. NAKAYAMA
   1967   Water absorption, conduction and transpiration. In Irrigation of agricultural lands, R. M. Hagan, H. R. Haise, and T. W. Edminster (eds), p. 320–336. American Society of Agronomy, Madison, Wisconsin.
7. BERNSTEIN, L., AND H. E. HAYWARD
   1958   Physiology of salt tolerance. Annual Review of Plant Physiology 9:25–46.
8. O'LEARY, J. W.
   1969   The effect of salinity on permeability of roots to water. Israel Journal of Botany 18:1–9.
9. OERTLI, J. J.
   1966   Effect of external salt concentrations on water relations in plants. II: Effect of the osmotic differential between external medium and xylem on water relations in the entire plant. Soil Science 102:258–263.
10. SLATYER, R. O.
    1967   Plant water relationships. Academic Press, New York. 366 p.
11. O'LEARY, J. W.
    1965   Root-pressure exudation in woody plants. Botanical Gazette 126:108–115.
12. MEES, G. C., AND P. E. WEATHERLEY
    1957   The mechanism of water absorption by roots. I: Preliminary studies on the effects of hydrostatic pressure gradients. Royal Society, London, Proceedings ser B 147(928):367–380.
13. ———
    1957   The mechanism of water absorption by roots. II: The role of hydrostatic pressure gradients across the cortex. Royal Society, London, Proceedings ser B 147(928):381–391.

14. NIEMAN, R. H., AND L. L. POULSEN
    1967   Interactive effects of salinity and atmospheric humidity on the growth of bean and cotton plants. Botanical Gazette 128:69–73.
15. RILEY, J. J., AND J. W. O'LEARY
    1968   Effect of humidity on plant growth inhibition induced by brackish water. Arizona Academy of Science, Journal 5 (Proc suppl): 13. Abstract.
16. HODGES, C. N.
    1969   A desert seacoast project and its future. In Arid lands in perspective, W. G. McGinnies and B. J. Goldman (eds), p. 121–126. American Association for the Advancement of Science, Washington, D.C., and University of Arizona Press, Tucson.
17. BOYER, J. S.
    1965   Effects of osmotic water stress on metabolic rates of cotton plants with open stomata. Plant Physiology 40:229–234.
18. KENDE, H.
    1965   Kinetin-like factors in the root exudate of sunflowers. National Academy of Sciences, Proceedings 53:1302–1307.
19. SETH, A., AND P. K. WAREING
    1965   Isolation of a kinin-like root factor in Phaseolus vulgaris. Life Science 4:2275–2280.
20. LETHAM, D. S.
    1967   Chemistry and physiology of kinetin-like compounds. Annual Review of Plant Physiology 18:349–364.
21. PORATH, E., AND A. POLJAHOFF-MAYBER
    1965   Effect of salinity on metabolic pathways in pea root tips. Israel Journal of Botany 13:115–121.
22. ITAI, C., AND Y. VAADIA
    1965   Kinetin-like activity in root exudate of water-stressed sunflower plants. Physiologia Plantarum 18:941–944.
23. SITTON, D., C. ITAI, AND H. KENDE
    1967   Decreased cytokinin production in the roots as a factor in shoot senescence. Planta 73:296–300.
24. BEN-ZIONI, A., C. ITAI, AND Y. VAADIA
    1967   Water and salt stresses, kinetin, and protein synthesis in tobacco leaves. Plant Physiology 42:361–365.
25. O'LEARY, J. W., AND J. T. PRISCO
    1970   Response of osmotically stressed plants to growth regulators. Advancing Frontiers of Plant Science 25:129–139.
26. ITAI, C., A. RICHMOND, AND Y. VAADIA
    1968   The role of root cytokinins during water and salinity stress. Israel Journal of Botany 17:187–195.

The Protoplasmic Basis for Drought-Resistance

A QUANTITATIVE APPROACH FOR MEASURING PROTOPLASMIC PROPERTIES

EDUARD J. STADELMANN
Department of Horticultural Science
University of Minnesota
St. Paul, Minnesota, U.S.A.

ABSTRACT

DROUGHT-RESISTANCE OF PLANTS is based on water-stress avoidance and/or water-stress tolerance. Water-stress avoidance (usually found in higher plants) is realized, for example, by reduction of cuticular transpiration, higher suction force of the root, or a deeper root system. Water-stress tolerance concerns the protoplasm of the cells and indicates the ability of the protoplasm to survive conditions of suboptimal water supply.

Water is present in the cell protoplasm at different degrees of mobility (extremes: structurally bound water and free water), depending upon the binding force and the molecules (for example, proteins, nucleic acids) involved. Water shortage in the protoplasm of cells with a large central vacuole eventually causes increase of cell sap concentration and thus an increase of the water-retention power of the cell. By this mechanism these (mostly higher) plants are able to compensate for water shortage between certain limits (homoiohydric plants). Cells of lower plants are small and often without vacuoles. These cells have no regulatory mechanism and the plants dry out under water-stress conditions (poikilohydric plants) but revive after remoistening. The protoplasm of homoiohydric plants may become water-stress hardened by drought, and an increase in cell sap concentration will persist after removal of water stress.

Protoplasmic water stress should be measured as the difference between optimal hydration and actual hydration of the protoplasm. A measure for the hydration of the protoplasm is the osmotic potential of the cell sap. Water-stress tolerance of the protoplasm may be calculated as the quotient of the osmotic potential of the cell sap at which the cell can survive divided by the osmotic potential of the cell sap for water saturation of the cell. The basis for water-stress tolerance seems to be the ability of the protoplasm to rearrange the spacing of the reactive sites of its macromolecules, so that proper functioning for cell metabolic reactions and maintenance of the protoplasmic structure is assured under low water content.

Protoplasmic qualities accessible for experimental investigation, and which are closely related to protoplasm structure and architecture, are protoplasmic permeability and viscosity. Permeability determination is based on observation of the deplasmolysis time course in a hypertonic solution of a permeating solute (solute permeability) or of a nonpermeating solution (water permeability). Protoplasmic viscosity can be determined by observation of the changes in plasmolysis form of the plasmolyzed protoplast after it has completely separated from the cell walls. Numerical values indicating relative viscosity result from measurement of the rate of the surface change of the protoplast.

There is an urgent need for further work on the protoplasmic aspects of drought-resistance because of applications to human nutrition under conditions of world population growth. Another problem requiring attention is the training of research scientists in specialized techniques for research on protoplasmic drought-tolerance.

337

THE PROTOPLASMIC BASIS FOR DROUGHT-RESISTANCE
A Quantitative Approach for Measuring Protoplasmic Properties

Eduard J. Stadelmann

ABOUT ONE-HALF OF THE LAND SURFACE of the earth is arid or semiarid (*1*, p. 362), and, other than temperature, water is the most important determining factor of vegetation. With increasing population and its need for food and shelter, the interest in more efficient use of the arid areas for plant production will continue to increase rapidly. Many quite varied approaches are being attempted to further plant growth and production in arid areas. Practical and immediate problems are being undertaken with great concentration, while basic, more fundamental aspects are generally studied to a lesser degree, since the importance of understanding basic principles is still not yet widely enough recognized.

One of these basic factors for any plant growth in arid regions is the drought-resistance of plants. This term describes the totality of factors that constitute a plant's ability to withstand water shortage. Drought-resistance varies widely from species to species. Most higher plants, especially those species of primary economic importance, require a minimum amount of available water above that found in large areas of the arid zones. However, by development of new varieties of economically important plant species with a higher degree of drought-resistance, the growth area for those species may gain a much-needed extension.

Drought-resistance is the result of special anatomical, morphological, physiological, and biochemical features of the plant. These features may result in the avoidance of water stress or in water-stress tolerance. That the basis of drought-tolerance lies in the protoplasm was recognized more than 40 years ago (*2*, p. 385; *3*, p. 395). However, just as all plant parts ultimately are formed by the plant protoplasm, also the factors which determine drought avoidance have their ultimate cause in the protoplasm of the plant.

In many higher plants, drought-resistance is based mainly on water-stress avoidance, as for example in the succulent plants (water spenders and water savers, *3a*). The protoplasm may possess no drought-tolerance at all. However, generally the vegetative organs of higher plants all exhibit some degree of drought-tolerance.

Features that increase drought-resistance by water-stress avoidance may be (*a*) any qualities that decrease water loss, as for instance (*4*, p. 11): reduction of cuticular transpiration, faster stomatal closure, leaf modifications resulting in smaller leaf surface, deep insertion of stomata; or (*b*) factors that increase water uptake from the soil, as for instance: higher suction potential of the roots, extension of the root system through larger volumes of soil, deeper root systems.

The water-stress tolerance of the protoplasm is defined as its ability to survive adverse conditions of suboptimal water supply. Great differences exist in the degrees of water-stress tolerance between various plants, varieties, and species, as well as between different cell types of the same plant.

An increased water-stress tolerance of the protoplasm often will lead to considerable differences in functioning and structure of the plant (cf *5*, pp. 162 ff). Its morphological and anatomical characteristics may be changed and physiological functions such as respiration may be altered. A higher cell sap concentration, smaller cell size, and changes in protoplasmic permeability and viscosity are found when compared with similar cells of a drought-sensitive plant of the same variety.

Plants exposed to severe drought beyond their limits of drought tolerance become damaged and ultimately die. The mechanism of drought injuries and the changes in the plant and its protoplasm after drought stress is applied have been discussed frequently (cf *5–8*) and are outside the scope of this contribution.

PROTOPLASM AND WATER

Water is needed in the plant for four important functions (*9*, p. 609 ff):

As a component of the protoplasm. In the apical meristem, often 50–80 percent of fresh weight is water. Woody tissue is about 50 percent water.

As a reagent. Water is an important participant in the chemical reactions that take place inside the living cell. In photosynthesis and in hydrolytic reactions, water is directly involved.

As a solvent for solid (salts, sugars), liquid, and gaseous (carbon dioxide, oxygen) substances. The material transport inside the plant always takes place in an aqueous solution. For this function, that portion of the water content of a plant present in the cell walls is important also.

As a factor for mechanical stability of nonlignified herbaceous organs or parts of organs. The maintenance of the cell turgor gives such plant parts their mechanical

NOTE: This chapter is an adaptation of a presentation at "Arid Lands in a Changing World," an international conference sponsored by the Committee on Arid Lands, American Association for the Advancement of Science, June 3–13, 1969, at the University of Arizona, Tucson, Arizona, U.S.A.

strength, and the turgor can develop only when sufficient water is in the vacuole.

The first three qualities of water concern the integrity and normal function of the protoplasm, which are of crucial importance, since the protoplasm is the most significant part of the plant cell. There occur all the processes aimed at maintaining the living state of the plant and at synthesizing new organic matter for growth and propagation. These plant products often are used for human needs.

The role of water in the protoplasm recently has been more and more investigated (cf *10*); however, it is far from being completely understood. Water is present in the cell at different degrees of mobility. At one extreme there is the free water, for example, in the vacuole: in most cells the vacuolar sap obeys closely osmotic laws; water within the vacuole is thus proven to be completely free of permanent molecular bonding. The other extreme is the structurally bound water (constituent water), for example, that is present in crystals. A large portion of the protoplasm dry mass consists of macromolecules. Water is held by these macromolecules at a low degree of mobility (imbibition water; cf *11*, p. 5) with variable strength resulting from dipole attraction of water to proteins or hydrogen bonds. The function of this water may be to maintain the necessary spatial relationship between reactive sites that carry out specific reactions, primarily enzyme reactions. Water also is incorporated into the membrane systems of the protoplasm (*12, 12a*).

The "binding power" of a particular molecular site for a water molecule varies at the different sites and may also decrease as distance between the H_2O molecule and this site increases. Thus, when water is removed from the cell by any external force (for example, osmotic forces, mechanical pressure) the amount of imbibition water removed depends upon the strength of the "water-removing" force, that is, the method applied. This relation between water-withdrawing power of the external force and the amount of water remaining in the cell must be considered for determination of bound water.

The degree of protoplasmic swelling can be extremely high: under the pathological conditions of cap plasmolysis (cf *13*, p. 205) the amount of imbibition water may be more than 50 to 100 times that of the normal water content; however, protoplasmic structures are not permanently disturbed, since swelling is reversible. Stocker (*14*, p. 702) considers bound water to be of great importance for drought-tolerance.

WATER STRESS AND WATER-STRESS TOLERANCE

Water shortage in the plant from decreased supply or increased water loss leads to a lowering of protoplasmic imbibition, which in turn has far-reaching consequences: the change in the spatial relationship of the macromolecules involved causes alteration of metabolic activity. When hydration decreases below a certain minimum value, first growth becomes slow; when the protoplasm is more severely dehydrated, growth stops and no further increase in dry weight takes place. A latent state (latent point; cf *1*, p. 363) is entered. Under still more severe drought, the plant decreases in dry weight through intensive use of storage materials and proteins for respiration, since stomatal closure prevents assimilation and causes starvation.

It is generally assumed that an optimum degree of protoplasmic hydration exists. However, it may vary for different functions (for example, assimilation, growth, flowering, fruiting). For instance, it is well known that leaf development is favored in many plants when water supply is abundant and that flowers develop only under certain conditions of specifically limited water supply (cf *15*).

A certain minimum degree of protoplasmic hydration is required to avoid irreversible damage to the protoplasm. This minimum may differ for different cell types.

As for many external factors, it is not certain if a maximum degree of protoplasmic hydration exists, since the uptake of water, even when unlimited externally, is limited by wall pressure in the water-saturated state of the cell. Therefore it might never be possible to reach the maximum value for hydration, beyond which functioning of the protoplasm would be impeded.

Information about the optimum degrees of protoplasmic imbibition would be of utmost interest for horticulture and agriculture, since maximum growth, flowering, fruiting, and so on, could then be obtained with minimum water expense. However, the intracellular status of water has been little studied and information about water-stress effects on individual growth processes is scarce (cf *16*; for a recent review of results on water stress on plant or organ growth, see *15*).

When water-stress conditions occur, the mesophytic higher plant reacts not only passively by stomatal closure to reduce transpiration, but also actively by increased capacity of the plant organism to retain water. Increased water-retention power is achieved by the active transport of osmotically effective solutes from the protoplasm into the vacuole; thus their concentration in the vacuolar sap is increased and the external water stress is counteracted (cf *5*, p. 168). The increased concentration of the cell sap causes decrease of hydration of the protoplasm, since the water-holding forces of the protoplasm are always in equilibrium with the water-holding forces of the cell sap vacuole.

To some extent, higher plants compensate for water stress and protoplasmic dehydration with this regulatory mechanism of cell sap concentration, if environmental conditions are mildly severe and do not fall below the survival limit of the protoplasm. Plants that respond to changes in external water-stress conditions by such a regulatory mechanism are called *homoiohydric plants* (Greek: "equally moistured") (*17*, p. 9; *18*, p. 695). The homoiohydric plants have only limited drought-resistance and are more-or-less heavily damaged if the water supply drops below a critical limit. Homoiohydry may be considered a drought-avoidance mechanism.

However many lower plants and a few higher plants react in a different way to water stress. The hydration of their protoplasm varies with the degree of atmospheric humidity (*poikilohydric plants;* "with varying moisture," *17*, p. 9). The most familiar example of this kind of plant is the lichen, which can survive complete drought over extended periods of time in a state of anabiosis (latent life). Besides lichens, only some ferns and a few flowering plant species are known to belong to the poikilohydric plants (for example: *Ceterach officinarum, Myrothamnus flabellifolia* [Myrothamnaceae, Rosales], *Ramondia nathaliae, Haberlea* [Gesneriaceae], *Trilepis pilosa* [Cyperaceae]).

Poikilohydric plants exhibit the highest degree of drought resistance, if survival is used as a criterion. Lack of a vacuole in the cell of some of these poikilohydric plants, small cell size, little volume change during dehydration or solidifying cell saps obviously prevent disorganization of the submicroscopic structure of the protoplasm during desiccation.

The decrease in protoplasmic hydration and the increase in cell sap concentration are reversible only when the water-stress condition is present for a short time. A more pronounced (several days) decrease of protoplasmic hydration causes drastic and permanent alterations: this process usually is described as *drought hardening.* The nature of these protoplasmic alterations is not known; however, they manifest themselves by changes in many physiological functions and morphological structures. These changes generally adapt the plant to the drought condition so that it is able to grow and reproduce under these changes in environment, although often to a lesser degree. Examples of physiological functions that change throughout the entire plant are photosynthesis and respiration. In *Beta vulgaris* ssp. *vulgaris,* var. *altissima rapa* (sugar beet), photosynthesis decreases shortly after application of water stress. However, it later recovers and becomes even higher than before application of the water stress. Respiration responds in the opposite direction (cf *14*, p. 708; *16*).

This recovery of physiological functions indicates an increase in drought-resistance of the plant initiated by the application of water stress. The causes for this increase are changes in the protoplasm structure and activity, which also involve an increase in water-stress tolerance of the protoplasm. The most visible manifestation of the protoplasmic changes is the alteration in its morphogenetic activity: the protoplasmic changes in the meristematic cells lead to the development of new organs more xeromorphic than those developed before the increased drought-tolerance.

MEASUREMENT OF PROTOPLASMIC WATER STRESS AND WATER-STRESS TOLERANCE

The importance of water supply for plant growth was already recognized in ancient times. Hales (*19*) made the first experimental determinations, although quantitative measurements of the water relations were started relatively recently (cf *20*, p. 5ff).

The fundamental equation for the water relation of the cell is (*21*, p. 344; *22*, p. 145):

$$\psi = \psi_s + \psi_p \tag{1}$$

(water potential＝osmotic potential＋pressure potential).

This relationship was first recognized by Ursprung and Blum (*23*, p. 530) and described by the equation:

$$S_z = S_i - P$$

(suction potential of the cell = suction potential of the cell content − wall pressure).

A similar relationship was postulated by Thoday (*24*, p. 110), who did not refer to the work of Ursprung and Blum. Later articles still occasionally mention Thoday as the author who first established this relationship (cf *25*, p. 73). However, priority must be given to Ursprung and Blum, whose earlier work is often overlooked.

Meyer (*26*, p. 535) introduced the same equation with different symbols which were widely adopted by American plant physiologists:

$$DPD = OP - WP$$

(diffusion pressure deficit = osmotic pressure − wall pressure).

In comparing the three equations above, the following relations can be established:

> Water potential (ψ) = − suction potential of the cell ($-S_z$) = − diffusion pressure deficit ($-DPD$). Osmotic potential (ψ_s) = − suction potential of the cell content ($-S_i$) = − osmotic pressure ($-OP$). Pressure potential (ψ_p) = wall pressure (P) = wall pressure (WP).

During the intervening decades, an ever-increasing number of new methods were developed to measure water status, water stress, and water content of whole plants, organs, and cells. A vast number of publications describing water relations and water conditions in plants resulted. However, it is surprising to note that most of this work is concerned exclusively with the extra-protoplasmatic aspects of the water relations: with water uptake, water potentials, and water loss. Therefore, without denying the importance of these aspects, it seems to be appropriate to emphasize the importance of the water within the protoplasm, since it determines the synthetic activity, survival, and production of organic matter of the whole organism. Of course, it is to a certain extent understandable that preference is given to those areas of water relations presently under intensive investigation. These aspects can be studied much more easily by using mechanical instrumentation in experimental methods, while measuring the water potential or osmotic potential of individual cells is laborious, involves delicate manipulations, and requires great experience of the investigator to achieve reliable results.

Fortunately, the importance of the intracellular status of water is becoming more and more recognized at present, so that greater emphasis on this line of work can be expected in the future (cf *16,* p. 4ff; *15*).

Water stress in plants is presently evaluated by external criteria, for example, pattern of soil water supply, or nutrient solution, or change in transpiration. For comparative work with the same plant species, a number of indexes for the water content or the internal water status in the plant have been used, for example, relative water content, water potential of nutrient solutions.

The results of these works often give important information for a specific problem on water stress for a particular plant species. Seldom, however, does a similar amount of external water stress cause an identical water stress in plants of different species (cf "Osmotic spectra," *27,* p. 291; *28,* p. 313; *15*).

Measurement of water stress (*28a*) must be based on the determination of the degree of protoplasmic hydration (*28b*), and the measure of protoplasmic water stress should be the *difference between optimal hydration and actual hydration of the protoplasm* (*15*).

It is not possible to determine the protoplasmic hydration directly. However, protoplasmic water-holding forces, which cause the hydration of the protoplasm, will be always in equilibrium with the water-retention power of the cell sap, which originates from osmosis. Thus, the determination of the osmotic potential of the cell sap is an adequate measure for the hydration of the protoplasm.

The osmotic potential of the cell sap can be measured by different methods. The two most common methods will be briefly described. A more detailed discussion which mentions possible sources of error of the two methods has been given elsewhere (*13,* p. 152ff).

a) Determination of the osmotic potential ($\psi_s = -OP$) *of the cell sap for an individual cell in its normal state.* For detailed analysis of water stress and water-stress gradients in a plant, it may be necessary to determine the osmotic potential of the cell sap for individual cells. This can be done for normal-sized, differentiated plant cells by plasmolytic methods. In doing so, one first determines the osmotic ground value O_g. The osmotic ground value is the osmotic value for the "ground state" of the cell, when the cell wall is completely relaxed and without turgor — that is, it is the osmotic value at the state of incipient plasmolysis (*29*).

For the *method of incipient plasmolysis* a series of (for example, glucose) solutions are prepared with concentration increasing by steps, and a piece of tissue is immersed in each solution. After sufficient time and at intervals of 10–20 minutes, the tissue strips are checked under the microscope and the plasmolysis frequency is determined by counting of the cells or by estimation. The concentration in which 50 percent of the cells are found to be plasmolyzed is the concentration O_g. The value for O_g may be interpolated from the two results nearest 50 percent (see formula 2). An example for the determination of O_g is given in Table 1 (from *30,* p. 1232) for a leaf of *Elodea* sp. The middle third of the lower leaf side was tested, but cells of the "vein" were not included. Plasmolyticum was saccharose. Entire and undamaged (except at the cut surface) leaves were immersed in the saccharose solutions, a different leaf in each solution. Leaves of three consecutive whorls were used.

For interpolation the equation

$$\frac{O_g - C_1}{C_2 - C_1} = \frac{50 - P_1}{P_2 - P_1} \tag{2}$$

is used, where P_1 is the percent frequency of plasmolyzed cells in the concentration C_1, P_2 is the percent frequency of plasmolyzed cells in the concentration C_2, C_1 is the highest concentration of the series where the frequency of the plasmolyzed cells is less than 50 percent, and C_2 is the lowest concentration of the series where the frequency of the plasmolyzed cells is higher than 50 percent.

As an example for formula 2, using the results given in Table 1: the value of O_g may be calculated for the time $t = 30$ minutes. At this time:

$C_1 = 0.34$ M (*moles of saccharose per liter solution*),

$C_2 = 0.36$ M, $P_1 = \frac{8}{56} \cdot 100\%$, and $P_2 = \frac{62}{62} \cdot 100\%$;

thus $O_g = 0.348$ M.

TABLE 1

NUMBER OF TOTAL AND PLASMOLYZED CELLS AT DIFFERENT TIMES IN THE CONCENTRATION SERIES*

Time elapsed after transfer to the test solution (min)	Concentration (moles of saccharose per liter solution)													
	0.28		0.30		0.32		0.34		0.36		0.38		0.40	
	Number of cells plasmolyzed (PI) and total number of cells evaluated (T)													
	PI	T	PI	T	PI	T	PI	T	PI	T	PI	T	PI	T
5	0	48	0	52	2	47	2	56	12	58	32	61	45	56
15	0	75	0	74	3	61	4	60	23	64	61	66	68	68
30	0	68	1	71	4	58	8	56	62	62	68	68	70	70
50	0	65	2	78	5	74	8	80	All cells plasmolyzed					
70	0	63	2	81	5	72	11	73						
100	0	66	2	80	5	79	12	65						

* From Ursprung (*30,* p. 1232, table 16d).

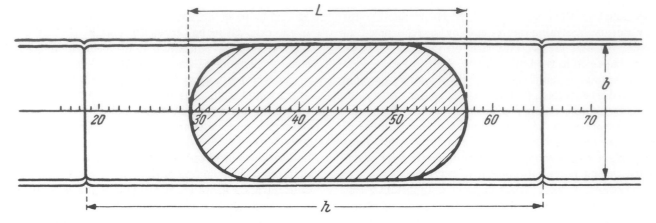

Fig. 1. Cylindrical cell with plasmolyzed protoplast in the state of final plasmolysis (smallest possible protoplast surface for a given volume of the vacuole). L = length of protoplast, b = inner cell diameter (cell width), h = inner cell length. A micrometer scale is superimposed on the cell to demonstrate how L is measured. (The diagram shows L = 28 units.)

The *plasmometric* method (*31*) is based upon the geometry of a cylindrical cell. At least the cell walls should be parallel in their longest dimension and the cell should have approximately a cylindrical cross section. In the ideal state of final plasmolysis, the protoplast will be a cylinder with two hemispheric ends (Fig. 1).

The protoplast volume, V_p, can be calculated as the sum of the volume of a cylinder plus the volume of two hemispheres:

$$V_p = \frac{\pi b^2}{4} \cdot (L - \frac{b}{3}), \qquad (3)$$

where L is the length of the protoplast (Fig. 1), and b the inner diameter of the cell. L and b may be measured in micrometric units and need not be calculated as μ or any other absolute unit at this point.

When, as can be assumed for most cells, the quantity of osmotically effective molecules and ions remains constant inside the vacuole during the time of the experiment, the reciprocity law for the concentration and volume of the vacuole can be established:

$$V_1 \cdot C_1 = V_2 \cdot C_2 = V_n \cdot C_n, \qquad (4)$$

where V represents the volume of the vacuole and C the corresponding concentration of the cell sap for this volume.

The concentration of the osmotic ground value O_g is derived from the external concentration C and the volume V_p of the protoplast in osmotic equilibrium with C:

$$O_g = C \cdot \frac{V_p}{V_z}, \qquad (5)$$

where V_z is the volume of the vacuole at turgor pressure = 0. Since the volume of the protoplasm in such an adult cell is very small, the volume of the vacuole can then be taken as the volume of the cell lumen. V_z is calculated with the formula of a cylinder:

$$V_z = \frac{\pi b^2}{4} \cdot h, \qquad (6)$$

where h is the inner length of the cell and has to be measured in the same units as L and b (for example, in micrometric units).

Equations (3), (4), and (5) combined result in:

$$O_g = C \cdot \frac{L - \dfrac{b}{3}}{h} \qquad (7)$$

The great advantage of the plasmometric method of determining O_g is that only one concentration of the solution is used rather than many different concentrations. A disadvantage is that this method usually can be used only with cylindrical or nearly cylindrical cells.* As an example for calculation, the values of an epidermal cell of *Zebrina pendula* are used (*13*, p. 161–162). Solution: 0.5 m glucose (molal concentration). $h = 90.0 - 47.0 = 43.0$ micrometric units (MU); $L = 76.8 - 58.0 = 18.8$ MU; $b = 54.0 - 40.0 = 14.0$ MU. With these values, formula (7) gives $O_g = 0.164$ m (molal concentration).

To obtain the osmotic pressure of the cell sap, one must consider the change in the cell sap concentration, which is caused by the change in the cell volume between the normal state and the turgorless state when plasmolyzed. The osmotic value for the normal state O_n is also calculated from the reciprocity law (formula 4):

$$O_n = O_g \cdot \frac{V_g}{V_n}. \qquad (8)$$

The determination of V_n may be difficult if the cell expansion is different in the direction of the cell depth from its expansion in width, since the change in the cell cross section cannot then be determined with sufficient accuracy. (For further details cf *13*, p. 167.)

When the value for the concentration O_n is calculated, the osmotic pressure (osmotic potential) of the cell sap can be found from tables for the osmotic pressures for the solution used in the plasmolysis experiment (*33*).

b) *Determination of the osmotic potential of a tissue or plant organ.* The expressed sap method is one of the most widely used methods, especially in ecological

---

* Recently Härtel (*32*) developed a procedure applying the plasmometric method also for irregularly shaped cells or cells with concave plasmolysis form by using an integrating eyepiece.

studies, to determine the osmotic potential. Samples of plant parts or tissues are selected for sap extraction. The sap is expressed after killing the material by a mechanical or hydraulic press, and the concentration of osmotically active substances usually is determined by freezing-point depression. The details of the basically very simple procedure and the precautions are discussed elsewhere (*17*, p. 26ff; *28*, p. 332ff; *30*, p. 1309ff; *34*, p. 119ff; *35; 36*, p. 221ff).

Results from plasmolytic measurements and from values derived for the osmotic potential of the expressed cell sap occasionally have shown discrepancies (cf *37*, p. 139ff; *13*, p. 169ff), but evidence of good agreement between values found by both methods is also available (*28*, p. 335ff).

The water potential of the cell was often thought to be an indicator of the degree of water stress of the protoplasm. With a few exceptions the water potential and protoplasmic hydration (measured as osmotic potential of the cell sap) change in the same direction. The water potential decreases during water stress faster than the osmotic potential and thus is a more sensitive indicator of changes in water conditions than the latter. However, since the degree of hydration of the protoplasm has to be measured, the osmotic potential (and its difference between the optimum state and the actual state) has to be determined, because it is the more direct measure.

Water-stress tolerance, as a protoplasmic quality, can only be measured at a cellular level, where interference with drought-avoidance mechanisms is eliminated. There is, at present, no generally accepted method of quantitative measurement of water-stress tolerance. The osmotic potential of the cells (cf *38*, p. 699ff) or the critical water content (cf *4*, p. 59) is no absolute measure for water-stress tolerance. The determination of water-stress tolerance must be based on protoplasmic hydration. The degree of protoplasmic hydration when the cell is in the state of water saturation should be compared with that at the lowest degree of water content the protoplasm can survive (measured conveniently as the osmotic potential of the cell sap at the 50 percent cell survival level).

Levitt (*5*, p. 158) proposed that the quotient of these two values (determined as cell sap concentration) be used as a measurement for water-stress tolerance:

$$D_d = \frac{C_{dk}}{C_o},$$ (9)

where D_d = measure for water-stress tolerance, C_{dk} = cell sap concentration at the drought killing point (50 percent cell survival), and C_o = cell sap concentration of the water-saturated cell.

The quotient of the osmotic potentials may also be used to characterize the water-stress tolerance:

$$D_d = \frac{\psi_{sdk}}{\psi_{so}}.$$ (10)

While determination of ψ_{so} generally does not pose any problem, measurement of ψ_{sdk} may become impos-

sible when cells survive in humidities so low that the vacuole disappears completely (for example, in epidermal cells of leaves of *Brassica oleracea*, var. *capitata* [red cabbage], the vacuoles disappear at a relative humidity of about 80 percent, which corresponds to an osmotic potential of about −300 atm [*39*, p. 742]). In this case an equivalent osmotic potential corresponding to the relative humidity must be introduced in the above formula (cf Tables I–III, of *17; 11*, p. 6).

Survival of vacuolated cells depends not only upon the protoplasmic drought-resistance ("desiccation resistance," *40*, p. 324), but also upon mechanical injury of the protoplasm from volume changes, consistency of the vacuole, and the rigidity of the cell wall during the drying and remoistening process. Therefore, water-stress tolerance of the entire cell (cellular water-stress tolerance) may be quite different from that of the protoplasm (*41*, p. (94)).

The equivalent values for ψ_{sdk} often vary considerably for the same cell material, depending upon the speed of desiccation and remoistening. A gradual change in external humidity, instead of an abrupt one, significantly increases the survival rate of the cells and thus results in a lower equivalent value for ψ_{sdk} (*39*). For determination of D_d, in experiments with different desiccation and remoistening speeds, obviously, the lowest (equivalent) value found for ψ_{sdk} should be taken. It is difficult to assess whether the different speed for drying and remoistening causes an alteration of the protoplasmic drought-resistance also, or if the mechanical damage alone is decreased or avoided.

In plants having small cells with small vacuoles or none, or with easily pliable cell walls, the mechanical forces on the protoplasm will be insignificant (see *41*, p. (96)) and the resulting equivalent ψ_{sdk} value should be independent of the speed of desiccation and remoistening.

Protoplasmic water-stress tolerance was studied especially in species of the Jungermanniales (Hepaticae), and considerable differences were found between different species and between different parts of the same plant (*41*, p. (99ff), *42*, p. 37).

Protoplasmic water-stress tolerance might be quite high and some protoplasm might be able to survive complete dryness (*38*, p. 697), when mechanical damage during drying and remoistening is carefully avoided. However, protoplasmic water-stress tolerance varies with different plants and increases after precedent water-stress exposure (*38*, p. 621; *43*).

The experimental procedures for determination of water-stress tolerance are quite simple: sections of the tissue or entire plant parts or organs are exposed in a closed glass vessel of suitable size to an atmosphere with a defined degree of humidity (established by solutions of NaCl, H_2SO_4, etc, of corresponding vapor pressure (see *42*, p. 7).

For drying, the material is transferred through a series of those vessels with stepwise decreased humidity, and vice versa for remoistening. Sufficient exposure time in each of these humidities and small intervals between the steps of relative humidity are essential to avoid development of the intracellular mechanical forces.

The common tests for viability (usually plasmolysis) are applied to determine the humidity at which 50 percent of the cells survive.

ANALYSIS OF WATER-STRESS TOLERANCE

Water-stress tolerance as observed in protoplasm of a given plant is the result of a multitude of individual factors in the protoplasm, that is, the ground plasm, the organelles, and the protoplasmic membranes. The individual factors are little known in detail, do not lend themselves easily to experimental analysis, and require discussion based largely upon indirect conclusions; hence the more detailed theories of water-stress tolerance generally show lack of direct experimental evidence.

The basis of water-stress tolerance must be sought at the molecular level. When water shortage in the protoplasm increases, the proper distribution and spacing of the macromolecules (especially enzyme proteins) becomes more and more disturbed and the availability of water as solvent and reactant decreases. Therefore, water-stress tolerance of the protoplasm is based on its ability to function under such adverse conditions and to rearrange the molecular reactive loci of the enzymes and other macromolecules in such a way that (a) metabolism can proceed with less protoplasmic water content, and (b) the protoplasmic structure is maintained. This rearrangement at the molecular level depends upon the architecture of the protoplasm of the individual cell, the rate of water-stress increase, and the duration of water stress. It can be expected that under longer-lasting, non-damaging water-shortage conditions, the progress of the molecular rearrangement in the protoplasm may reach such a degree that it will no longer be reversible, even when normal water supply is reestablished later: the protoplasm retains its increased water-stress tolerance; it has become hardened for water stress.

The importance of structural changes at the submicroscopic level was early recognized (for literature see 14, p. 701). Levitt (44) recently advanced a more specific explanation about the effect of water-stress conditions on the protein molecules (45; see also 46, p. 225ff). Water-stress injury involves an unfolding of proteins and protein denaturation by formation of intermolecular S-S-bonds from SH-groups (Sulfhydryl-hypothesis). Water-stress tolerance therefore would imply the presence of proteins that avoid formation of S-S-bonds at low protoplasmic water content and are thereby resistant to such denaturation.

The fundamental change in the protoplasmic architecture when water shortage begins causes a number of alterations in cell function. Changes that are merely consequences of the water stress can be distinguished from those changes brought about by the increase in protoplasmic water-stress tolerance: after removal of water-stress the former will discontinue, while the latter will persist. The permanent increase in osmotic cell sap concentration after an increase of water-stress tolerance is an example.

Since water-stress tolerance is closely related to protoplasmic structure and arrangement, study of qualities of the living protoplasm will be useful to investigate water-stress tolerance.

The two protoplasmic qualities fairly accessible for experimental investigation of living intact protoplasm are permeability and viscosity. Both qualities change under water-stress conditions, although the reports of changes for viscosity are controversial (14, p. 699; 4, p. 64). Permeability for polar substances (for example, urea) was found to increase after water stress. From experiments with frost hardening (47, p. 291; for literature see 48, p. 7ff), it can be expected that this permeability increase is associated with the increase in water-stress tolerance of the protoplasm (cf 5, p. 172).

The evaluation of protoplasmic permeability is a powerful tool for close investigation and analysis of the protoplasmic factors of water-stress tolerance. Recently, methods for quantitative measurement of permeability became available (calculation of the permeability constant as shown in the next section of the present chapter; cf 13, p. 189ff), so permeability and permeability changes can be expressed numerically.

Plasmalemma and tonoplast are the main sites for the permeation resistance, and determination of permeability will indicate primarily the changes in these membranes. These alterations may concern the lipid portion of the membrane or its proteins. Sulfhydryl-disulfide conversions in membrane protein are already known for erythrocytes (cf 49) and may be of importance for water-stress hardening also. Since membrane structure is discussed in the present literature in great detail, selection of specific permeators and of appropriate additional treatments and conditions may yield a more advanced analysis of permeability differences between water-stress-tolerant and -sensitive cells. A more detailed insight into the membrane changes involved may result. Structural alterations of the plasma membrane are closely related to the activity of the mesoplasm, and thus changes in the main part of the protoplasm may be inferred.

Viscosity reflects the water status mainly of the mesoplasm. It will be affected by any change in water content and also by shifting of water between free water and bound water. A reevaluation of viscosity changes with regard to water-stress hardening seems to be especially promising, as more new approaches open possibilities of quantitative measurement for protoplasmic viscosity (see next section).

Of course, other approaches for analysis of water-stress tolerance have to be applied, especially cell-morphology studies (microscopical and electron-microscopical). Cell fractionation and biochemical analysis are very promising in contributing important information, which, when combined, one hopes, will provide a closer understanding of (a) the processes occurring in the protoplasm during water-stress-tolerance increase, and (b) the nature of water-stress tolerance.

DETERMINATION OF CELL PERMEABILITY AND VISCOSITY

While permeability and viscosity of the protoplasm were recognized quite early in the progress of experimental cytology, only recently methods became available which allow measurement, although only under specific conditions, of both qualities quantitatively, so that numerical values can be obtained.

For the determination of the passive permeability of the protoplasm layer of an individual cell, the absolute permeability constant is calculated (cf *13*, p. 189). This represents the amount of substance passing through the protoplasmic layer from the external solution into the vacuole per surface area and time unit, when the driving force for this permeation process is given a unit value. When the driving force is a concentration difference, as in this instance, its unit value is 1 mole per cubic centimeter (a value generally much too high to be realized in an experiment). In other words the concentration of the permeating substance is 1 mole per cubic centimeter higher on one side of the membrane than at the other side of the membrane. The absolute permeability constant is derived from a formula closely related to Fick's diffusion equation, but the formula differs from it by the lack of a term for the length of the pathway (membrane thickness).

The basic equation for the permeability constant is:

$$\frac{dn}{dt} = K \cdot A \cdot (C - k), \qquad (11)$$

where n = the amount of the penetrating substance, K = the permeability constant, A = surface area of the protoplasm layer, C = external concentration of the permeating substance, k = actual internal concentration of the permeating substance in the cell vacuole, and t = time. When the amount (n) is indicated in moles (C and k in moles per cubic centimeter), and the CGS system is used, the dimension for K is centimeters per second.

This equation can be used to calculate the permeability constant, K_s, of the plasma layer with respect to a given substance (for example, urea, glycerol) for an individual cell of cylindrical shape.

To measure the permeability of plant cells by the plasmometric method, a tissue consisting of cylindrical cells is selected. Cuttings of this tissue of sufficient thickness to retain one or two undamaged cell layers are first immersed in a hypertonic solution of a nonpermeating substance, for example, a sugar solution. Plasmolysis then proceeds to its final stage, at which the protoplast has the shape of a cylinder with two hemispheres at its end (Fig. 1). The length of the protoplast (L_0) and the inner width of the cell (b) are determined at this point. Next, the sugar solution is replaced by the solution of the permeating substance (permeator) with the same osmotic potential as the sugar solution (equimolar concentrations of these substances may have approximately the same osmotic potential; however, at higher concentrations the differences are greater and may already be considerable at the concentrations applied in these experiments). Since the permeator (and water, in an equivalent amount to maintain concentration equilibrium) will gradually penetrate the protoplasm layer into the vacuole, the protoplast will expand.

When the permeability constant for the permeator is less than 1/500th of the permeability constant for water, as is the case for most substances used in these experiments, the experimental procedure can be simplified by omitting the transfer of the cutting into the nonpermeating plasmolyzing solution. In this case the cutting is directly transferred into the hypertonic solution of the permeator. The cells will become plasmolyzed, since the penetration of the permeator is much slower than the movement of water out of the vacuole. The value for L_0 is in this experiment obtained by extrapolation of the protoplast length-time diagram to the time of the transfer. According to theory (*50*, p. 387ff), this dilatation of the protoplast takes place proportional with time. Thus, a time plot of the protoplast length will give a straight line (see Fig. 2). This is a valuable test for cell damage, since damaged cells show nonlinear dilatation and hence altered permeability. As the protoplast expands, the ends of the protoplast eventually touch the transverse cell wall. Finally the entire cell lumen will be again filled by the protoplast. For permeability measurements, only that segment of the dilatation during which the protoplast ends are hemispheres and do not touch the transversal cell walls can be used.

Equation (11) can be integrated for the dynamics of the dilatation phase, yielding the equation:

$$K_s = \frac{b}{4} \cdot \frac{L_2 - L_1 - \frac{b}{3} \cdot \ln \frac{L_2}{L_1}}{(L_0 - \frac{b}{3}) \cdot (t_2 - t_1)}, \qquad (12)$$

where

K_s = The absolute permeability constant for a solute in centimeters per second,

b = cell diameter in centimeters,

L_0 = the protoplast length in the state of complete plasmolysis in a sugar solution of the same concentration (water activity) as the solution of the permeating substance to be tested, in centimeters,

L_1 and L_2 = the lengths of the protoplast at the times t_1 and t_2, in centimeters,

t_1 and t_2 = the times of the measurements, both times being in the phase of deplasmolysis of the experiment, in seconds.

For the convenience of the observation and for avoidance of excess handling and damage to the tissue-cutting examined, the cutting should be mounted in a perfusion chamber and photomicrography used to record the experiment; thus the protoplast-length changes can be measured on the film strips (*51, 52*). In this way the permeability constants can be determined for a larger number of cells, and hence a more valuable mean value for K_s can be calculated.

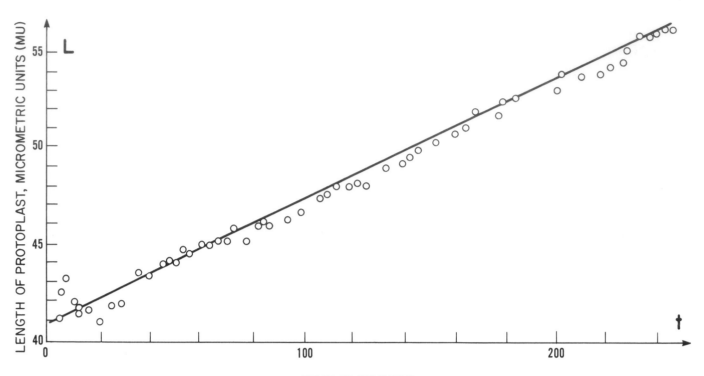

TIME IN MINUTES

Fig. 2. Example for the dilatation of the plasmolyzed protoplast of a cylindrical cell in a permeating hypertonic solution (1 mole urea per liter). 1 MU = 2.78 μ; h = 75.5 MU; b = 23.0 MU. Material: *Zebrina pendula;* parenchyma cell of the stem. The section was directly transferred into the solution of the permeator without precedent plasmolysis in a sugar solution. The time of the transfer was 0 minutes. From the time course of the deplasmolysis for t = 0, the value of L_0 = 40.9 MU results (50: Fig. 1a, p. 379).

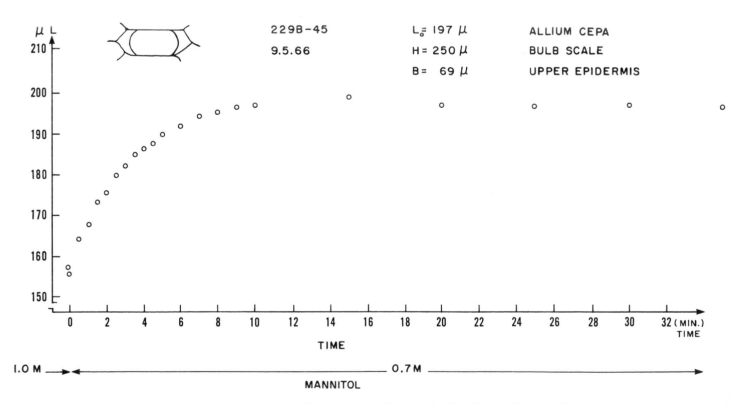

Fig. 3. Protoplast length-time diagram of an upper epidermal cell of the bulb scale of *Allium cepa.* H = inner cell length; B = inner cell width; L_0 the final length of the protoplast after reaching the equilibrium in the final plasmolyticum. Note that the final equilibrium is reached about 14 minutes after the change in the external concentration. The experiment was performed in a perfusion chamber and recorded with photomicrographic time-lapse pictures. The protoplast expanded from about 157 μ to 197 μ — that is, the expansion range is 40 μ. L_1 for formula 13 should therefore be 157 + 50% $\times$ 40 = 177 μ, and L_2 = 157 + 75% $\times$ 40 = 187 μ. The corresponding values for t resulting from the graph are: t_1 = 2.1 minutes and t_2 = 4.3 minutes. (The middles of the optimum ranges for L_1 and L_2 are 50% and 75%.)

An experiment with a piece of epidermis of the upper midrib of a leaf of *Sonchus laciniatus* can be used as an example for calculation of K_s. The tissue was not first plasmolyzed in sugar, but the epidermis piece was directly transferred into a 1.0 M urea solution. The protoplast length-time diagram (cf. *13*, p. 190, Fig. 5b) shows the following values: the protoplast length for the time of the transfer of the section into the permeator solution was $L_0 = 11.7$ micrometric units (MU). At $t_1 = 19$ minutes during expansion of the protoplast, its length L_1 was 12.4 MU. For the later time t_2, the protoplast length L_2 was 14.1 MU. The inner cell width was 5.5 MU. The calibration of the eye-piece micrometer gave 1 MU=2.78 μ. When these values are introduced into formula (12), K_s becomes $2.7 \cdot 10^{-8}$ centimeters per second. (For intermediary steps in this calculation, see *13*, p. 192.)

Water Permeability

Water permeability of the protoplasm layer can be determined in a similar way. In this case, two solutions of a nonpermeating substance (for example, a sugar) are used. The first, more highly concentrated solution is applied to obtain a plasmolyzed protoplast like that described above, with the form of a cylinder and two hemispheric ends. When this final state of plasmolysis is reached, a second solution with lower concentration of the same substance is applied to the section, and partial deplasmolysis is observed and recorded. Since this deplasmolysis is often quite fast, photographic recordings of its progress are strongly recommended.

For this experimental procedure, an equation similar to formula (11) can be used. Integration for the conditions here leads to the final equation (cf *13*, p. 193):

$$K_{wo} = 32.0 \cdot \frac{b}{C \cdot L_0 \cdot (t_2 - t_1)} \cdot A \quad (13)$$

$$A = (L_0 - \frac{b}{3}) \cdot \log\frac{L_1 - L_0}{L_2 - L_0} + \frac{b}{3} \cdot \log\frac{L_1}{L_2},$$

where K_{wo} = the absolute permeability constant, which can be compared with the absolute permeability constant for solutes in centimeters per second,

b = the inner cell width in centimeters,

C = the concentration of the nonpermeating solute (for example, mannitol) used as the deplasmolyzing concentration in the experiment, in moles per liter,

L_1 and L_2 = the length of the protoplast at the times t_1 and t_2, in centimeters,

t_1 and t_2 = the times of the measurements of L_1 and L_2 in seconds,

L_0 = the length of the protoplast after final equilibrium is reached in the deplasmolyzing solution.

Figure 3 shows an example of the dilatation of a protoplast in an epidermal cell of the bulb scale of *Allium cepa*. The dilatation is shown after the cell had been transferred from a 1.0 M to a 0.7 M solution of a nonpermeating solute (mannitol).

From a mathematical analysis of the equation (13) and examination of the experimentally found time course, it can be shown for *Allium* epidermal cells (*53*) that the L_1 value should be chosen in the range of 40 percent to 60 percent — that is, about the middle part of the total expansion range observed in the experiment. The L_2 value should be taken between 65 percent and 85 percent of this range.

An experiment with epidermal cells of the stem of *Lamium purpureum* can be used to illustrate the calculation of K_{wo}. The concentration of the nonpermeating plasmolyticum (glucose) was 1.0 moles per liter for the first and 0.6 moles per liter for the second solution. The inner cell width was $b = 34.0$ μ; $L_1 = 40.6$ μ, $L_2 = 53.2$ μ; L_0 (the protoplast length at the final equilibrium) $= 58.8$ μ; $t_1 = 43$ minutes 55 seconds; $t_2 = 44$ minutes 56 seconds. These values introduced in formula (13) give $K = 11.6 \cdot 10^{-4}$ centimeters per second. (For intermediary steps of this calculation see *13*, p. 194ff.) Computer programs to perform the calculations can be easily written, so that the time for computation of the constants becomes negligible.

Measurement of Protoplasmic Viscosity

From the various methods described for measuring protoplasmic viscosity (cf *54*, p. 345ff), only a few can be used more generally for the differentiated cell of a higher plant with a large central vacuole. Pekarek's method, which measures the Brownian motion, gives absolute values for viscosity; however, a sufficiently large volume of protoplasm must be present at one place in the cell so that Brownian motion is not restricted by the limits of the plasma body. This, of course, is seldom realized in fully differentiated cells of higher plants, which usually have only a thin plasma layer between vacuole and cell wall.

The plasmolysis time and form method is more suited for cells with large central vacuoles; however, until now only relative values could be obtained for viscosity by this method. Schaefer (*55*) improved the evaluation and obtained a numerical (but still relative) value for the protoplasmic viscosity: the rounding coefficient α. The cells must be plasmolyzed to such a degree that the protoplast is completely separated on all surfaces from the cell wall, since this contact (wall attachment) considerably alters the time course of the volume and surface changes of the protoplast.

For calculating the value of α, the ratio (r) of the surface of the protoplast (A) at a given time to its minimum surface (A_{min}) was determined:

$$r = \frac{A}{A_{min}} \cdot$$

These calculations were made for a series of different protoplast surfaces occurring during plasmolysis of a cell. With these values, a table was established (Table 2), which gives the value of r for the different plasmolysis forms. The surface area and the volume of more complicated forms of the protoplast (concave, angular,

TABLE 2

SURFACE RATIO *r* FOR DIFFERENT PLASMOLYSIS FORMS

	Plasmolysis form		Surface ratio $r = \dfrac{A}{A_{\min}}$		
Stage	Diagrammatic presentation	Designation	Mean	Minimum*	Maximum*
1		extreme concave III ("cramped" plasmolysis)	3.5	3.0	4.0
2		extreme concave II	2.5	2.0	3.0
3		extreme concave I	1.8	1.6	2.0
4		concave III	1.55	1.5	1.6
5		concave II	1.45	1.4	1.5
6		concave I	1.35	1.3	1.4
7		angular	1.25	1.2	1.3
8		irregular convex	1.15	1.1	1.2
9		convex	1.0	1.0	1.05

*The maximum and minimum values correspond to those plasmolysis forms which are at the limit between each consecutive pair of the forms illustrated above. After Schaefer (55, p. 426, table I).

convex), as they are needed for the calculation of *r*, were calculated by composing these forms from simpler stereometric bodies.

During plasmolysis, of course, the surface area of the protoplast tends to decrease; thus the value of *r* becomes smaller with time, and reaches its minimum value $r = 1$ when the minimum surface area $A_{\min}$ is reached.

Since it can be assumed that the decrease in protoplast surface takes place exponentially with time, $\ln(r)$ will be inversely proportional with time and the proportionality coefficient α can be used as measure for protoplasmic viscosity. Since the final value of $r = 1$, $\ln(r) = 0$, so that the formula for α becomes:

$$\alpha = \frac{\ln r_o}{t_e - t_o}.$$

The rounding coefficient α has the unit reciprocal minutes. The maximum surface ratio r_o, at the time t_o, is taken from Table 2. To make measurement as accurate as possible, the value of r_o should be determined as soon as possible after the protoplast has separated completely from the cell walls.

An example for calculation of the rounding coefficient (after 55, p. 430) follows.

Material: *Lemna minor,* root, second subepidermal cell layer, 2.2 millimeters above root apex. After the

protoplast was completely separated from the cell wall, the plasmolysis form was similar to stage 6 (concave I, Table 2), so that $r = 1.35$. It took 25 minutes from the time of this measurement until the final convex plasmolysis form was reached. With these data

$$\alpha = \frac{\ln 1.35}{25} = 0.012 \text{ min}^{-1}.$$

An appreciable advantage of Schaefer's method is its applicability to any cell shape as long as complete separation of the protoplast from the cell wall can be obtained. However, it may be difficult to recognize separation of the protoplast from the upper or lower cell wall. In addition, dehydration of the protoplasm by plasmolysis may also modify the viscosity of the protoplasm.

PRELIMINARY OUTLINE FOR WORK ON PROTOPLASMIC ASPECTS OF DROUGHT-RESISTANCE

The complexity of the problems discussed in the previous chapters suggests a list of many areas of investigation of protoplasmic drought-resistance. Of course, such a listing does not establish a definite plan or work program. In fact, the approach may have to be modified as work progresses. Certainly other ways to attack the protoplasmic aspects of drought-resistance also may be used to obtain as complete information on drought-tolerance as possible with the presently available methods and based upon our current knowledge of cell structures and processes.

The need for detailed knowledge about drought conditions and protoplasmic research on drought-resistance results from the plant protoplasm being the site of all synthesis of organic matter. The urgency for such work is obvious, since many of these organic materials are important in human nutrition and industry. The expansion of their production is compulsory for a fast-growing world population.

Large gaps exist in our knowledge of water relations and drought-resistance, since previous work deals with a few individual aspects only and investigates different plant species. Thus, the resulting picture is quite fragmentary.

For investigation of water relations, the first step might be the acquisition of a coherent body of quantitative data for such qualities as water potential, osmotic potential, and water content. These quantities must be determined for the different organs (or parts of organs or tissues) of a plant under normal and water-stress conditions during the whole lifespan of the plant. This work, of course, is very laborious, and it is most unlikely to be mechanized, so that such comprehensive investigation may be done at first on plants of one species. To choose this species is a difficult task. One approach would be to use one of the few species that have already been investigated in some respect in earlier studies on drought and drought-resistance, so that some basic data would be verified and the results expanded. Also, species with a significant sensitivity to water stress as well as those with high degrees of drought-resistance are promising research material. These plants should have, if possible, some economic importance also. Later, other plant species with different degrees of specialization for drought conditions may be investigated.

Next, the optimum of protoplasmic hydration for important tissues or organs (for example, leaves) of the chosen plant species should be determined. Probably two optima will be observed, one for vegetative development and the other for fruiting or seed development. It can be expected that the value of optimal protoplasmic hydration is not constant but varies during the different phases of plant growth, so that a timetable for the protoplasmic hydration has to be established for optimal plant development. It is self-evident that other external factors important for plant development should also be controlled and always maintained above a limiting threshold.

Knowledge of the dynamics of the water relations within a plant during water stress will give for the first time a quantitative insight as to how the different plant organs react to water stress, and interrelationships between different tissues will be indicated.

The data may also be used to relate actual hydration of the protoplasm to frequently used indexes for water conditions, such as soil water content, tensiometer values, relative turgidity, and others. Such a relation may be of great practical value when known for economically important plants. When the soil water content can always be adjusted so that the optimal hydration of the protoplasm results, plants will develop best, while excess watering and subsequent waste of water are avoided.

In the work on drought-resistance and drought-tolerance, two equally important lines should be followed:

a) A causal analysis of protoplasmic drought-tolerance involving the structural and functional factors at the subcellular level should be made. Here, advanced understanding of the basic mechanisms for drought-tolerance or drought-sensitivity of the protoplasm is the final goal. Work in permeability and viscosity (and the

interpretation of their changes for selected test substances and conditions), combined with biochemical analysis, is a most promising area for this research work. While this work is of a basic type, it will nevertheless be of fundamental relevance for the much-needed applied work in drought-resistance of agriculturally important plants.

b) Methods should be developed for a rapid determination of drought-tolerance with regard to their application in plant genetics and plant breeding for selection of varieties with increased drought-tolerance. Here the basic aim is to facilitate plant-breeding programs for increase of drought-resistance. Usually the time needed for testing of drought-resistance is very long, since test-plot planting and harvest evaluation may be involved and often require a quite advanced stage in plant development. By using appropriate types of tissues for testing, it should be possible to determine the protoplasmatic or cellular component of drought-resistance at quite an early stage of plant development, and thus considerably accelerate the screening procedures. Here the development of methods for the measurement of the actual drought-resistance (as defined in formula 9 or 10) is the goal, rather than the analysis of drought-resistance itself.

Another aspect of research work on drought-resistance, equally important for other areas of plant physiology, is the adequate training of research scientists in the specialized techniques needed for research on protoplasmic drought-tolerance. Such research may not involve instrumentation as much as extensive knowledge and experience in working with living cells and the ability to interpret microscopical observations correctly.

While biochemical and electron-microscopical training is relatively easily available in many laboratories in the United States, the opportunities for the acquisition of knowledge in handling and experimenting with the living protoplasm are less frequent. Thus, study programs in protoplasmic research are a basic requirement of successful work on drought-resistance at the cellular level (*58*).

REFERENCES AND NOTES

1. STOCKER, O.
   1947   Probleme der pflanzlichen Dürreresistenz. Naturwissenschaften 34:362–371.

2. MAKSIMOV, N. A.
   1926   Fiziologicheskie osnov'i zasuhoustojunvosti rastenij (The Physiological basis of drought resistance). Trudy po Prikladnoi Botanike, Genetike i Selektsii, Prilozhenie 26. 436 p. *See* (*3*), following, for English translation.

3. MAKSIMOV, N. A.
   1929   The plant in relation to water, a study of the physiological basis of drought resistance. G. Allen and Unwin, London. 451 p. *See* (*2*), above, for original Russian version.

3a. LEVITT, J.
   In press   Responses of plants to environmental stresses. Academic Press, New York.

4. LEVITT, J.
   1958   Frost, drought, and heat resistance. *In* L. V. Heilbrunn and F. Weber (eds), Protoplasmatologia VIII/6. Springer, Wien. 87 p.

5. LEVITT, J.
   1956   The hardiness of plants. Academic Press, New York. 278 p.

6. ILJIN, W. S.
   1933   Über Absterben der Pflanzengewebe durch Austrocknung und über ihre Bewahrung vor dem Trockentode. Protoplasma 19:414–442.

7. NIR, I., S. KLEIN, AND A. POLJAKOFF-MAYBER
   1969   Effect of moisture stress on submicroscopic structure of maize roots. Australian Journal of Biological Sciences 22:17–33.

8. KLEIN, S., AND B. M. POLLOCK
   1968   Cell fine structure of developing lima bean seeds related to seed desiccation. American Journal of Botany 55:658–672.

9. Kramer, J.
   1959   Transpiration and the water economy of plants. *In* F. C. Steward (ed), Plant physiology, a treatise. Vol 2, p. 607–723. Academic Press, New York.

10. Dick, D. A. T.
    1966   Cell water. Butterworths, London. 155 p.

11. Biebl, R.
    1962   Protoplasmatische Ökologie der Pflanzen. *In* L. V. Heilbrunn and F. Weber (eds), Protoplasmatologia XII/1. Springer, Wien. 344 p.

12. Hechter, O.
    1965   Role of water structure in the molecular organization of cell membranes. Federation Proceedings 24(2:3):S91–102.

12a. Schultz, R. D., and S. K. Asunmaa
    1970   Ordered water and the ultrastructure of the cellular membrane. Recent Progress in Surface Science 3:291–332.

13. Stadelmann, Ed.
    1966   Evaluation of turgidity, plasmolysis, and deplasmolysis of plant cells. *In* D. M. Prescott (ed), Methods in cell physiology. Vol 2, p. 143–216. Academic Press, New York.

14. Stocker, O.
    1956   Die Dürreresistenz. *In* W. Ruhland (ed), Encyclopedia of plant physiology, Vol. 3, p. 696–741. Springer, Berlin.

15. Walter, H., and Ed. Stadelmann
    In press   A new approach to the water relations of desert plants. *In* G. W. Brown, Jr. (ed), Desert Biology. Vol 2. Academic Press, New York.

16. Gates, C. T.
    1964   The effect of water stress on plant growth. Australian Institute of Agricultural Science, Journal 30:3–22.

17. Walter, H.
    1931   Die Hydratur der Pflanze und ihre physiologisch-ökologische Bedeutung. G. Fischer, Jena. 174 p.

18. Walter, H., and Ed. Stadelmann
    1968   The psysiological prerequisites for the transition of autotrophic plants from water to terrestrial life. Bioscience 18:694–701.

19. Hales, S.
    1727   Vegetable staticks, or, an account of some statical experiments on the sap in vegetables. W. and J. Innys, London. 376 p.

20. Stocker, O.
    1956   Einführung. *In* W. Ruhland (ed), Encyclopedia of plant physiology. Vol 3, p. 1–9. Springer, Berlin.

21. Taylor, S. A., and R. O. Slayter
    1961   Proposals for a unified terminology in studies of plant-soil-water relationships. Unesco, Paris. Arid Zone Research 16:339–349.

22. Slatyer, R. O.
    1967   Plant-water relationships. Academic Press, London. 366 p.

23. Ursprung, A., and G. Blum
    1916   Zur Methode der Saugkraftmessung. Deutsche Botanische Gesellschaft, Berichte 34:525–539.

24. Thoday, D.
    1918   On turgescence and the absorption of water by the cells of plants. New Phytologist 17:108–113.

25. Crafts, A. S., H. B. Currier, and C. R. Stocking
    1949   Water in the physiology of plants. Chronica Botanica, Waltham, Massachusetts. 240 p.

26. Meyer, B. S.
    1938   The water relations of plant cells. Botanical Review 4:531–547.

27. Walter, H.
    1949   Einführung in die Phytologie. 1st ed. III (1): Grundlagen der Pflanzenverbreitung: Standortslehre (analytisch-ökologische Geobotanik). E. Ulmer, Stuttgart. 332 p.

28. Barrs, H. D.
    1968   Determination of water deficits in plant tissues. *In* T. T. Kozlowski (ed), Water deficits and plant growth. Vol I, p. 235–368. Academic Press, New York.

28a. Levitt (*3a*) more accurately calls the unfavorable external condition "stress" and distinguishes it from the "strain" the plant or the plant part is experiencing by this condition. This clear distinction is not introduced in this chapter, and the term "stress" is used for both meanings.

28b. The term "hydration" is used here instead of the less familiar "hydrature" (*18*, p. 697). It rather describes the chemical activity of the water in the protoplasm and is not meant to merely indicate the amount of water per gram dry weight.

29. Plasmolysis is the separation of the living protoplasm layer (envelope) from the cell wall. The separation is caused by the action of a more concentrated external water-withdrawing solution. Plasmolysis is contingent upon presence of the differential permeable protoplasm layer which lines the interior of the cell wall and surrounds the large central vacuole. The vacuole contains an aqueous solution of a multitude of substances, many of which are osmotically active (for example, sugars or salts). Differential permeability (free passage of the solvent only, with little or no net passage of the solute through the membrane) is a quality of the protoplasmic membranes. The protoplasmic layer is limited adjacent the cell wall by the plasmalemma and adjacent the vacuole by the tonoplast. The protoplasm between these two membranes is called the mesoplasm. It consists essentially of ground plasm, organelles, nonliving inclusions (for example, oil droplets) and the endoplasmic reticulum. The vacuole and its enveloping protoplasm are called the protoplast. When an adult cell is transferred in a hypertonic solution (hypertonic: concentration of the external solution is greater than the concentration of the osmotically active material in the vacuole), the water will leave the vacuole until the vacuolar concentration (osmotic potential) equals the concentration (osmotic potential) of the external solution. Therefore, the vacuole shrinks and the surrounding protoplasm layer separates from the cell wall.

Incipient plasmolysis is that degree of separation of the protoplasm layer first recognizable with the microscope. Thus it corresponds to the state of the cell at which the volume of the vacuole (with the surrounding protoplasm envelope) is most nearly equal to the lumen of the relaxed cell.

Osmotic value is defined as that concentration of a nonpermeating nonelectrolyte (for example, sucrose, glucose) which produces the same osmotic potential (osmotic pressure) as the cell content (*56*, p. 88; *57*, p. 290). Walter (*17*, p. 12) introduced the term osmotic value to designate the osmotic potential of the cell content.

The osmotic ground value O_g can be found directly by evaluation of incipient plasmolysis or by measurement of cells of a tissue strip immersed in a sufficiently high hypertonic solution. This latter method is possible for cylindrical cells only (plasmometric method).

30. URSPRUNG, A.
    1939 Die Messung der osmotischen Zustandsgrössen pflanzlicher Zellen und Gewebe. Die Messung des Widerstandes, den das Substrat (Boden, Lösung, Luft) dem Wasserentzug durch die Pflanze entgegensetzt. *In* E. Abderhalden (ed), Handbuch der biologischen Arbeitsmethoden 11(4:2): p. 1109–1572. Urban und Schwarzenberg, Berlin.

31. HÖFLER, K.
    1917 Die plasmolytisch-volumetrische Methode und ihre Anwendbarkeit zur Messung des osmotischen Wertes lebender Pflanzenzellen. Deutsche Botanische Gesellschaft, Berichte 35: 706–726.

32. HÄRTEL, O.
    1963 Über die Möglichkeit der Anwendung der plasmometrischen Methode Höflers auf nichtzylindrische Zellen. Protoplasma 57: 354–370.

33. The citations for such tables for the most frequently used plasmolytica are given in:
    STADELMANN, ED.
    1966 Osmotic quantities: Protozoa and lower plants. *In* P. L. Altman and D. S. Dittmer (eds), Environmental biology, p. 541–551. Federation of American Societies for Experimental Biology, Bethesda, Maryland.

34. WALTER, H., AND H. KREEB
    1970 Die Hydratation und Hydratur des Protoplasmas von Pflanzen in ihrer öko-physiologischen Bedeutung. *In* L. V. Heilbrunn and F. Weber (eds), Protoplasmatologia, II/C/6. Springer, Wien. 306 p.

35. MEYER, B. S.
    1929 Some critical comments on the methods employed in the expression of leaf saps. Plant Physiology 4: 103–112.

36. WALTER, H.
    1960 Einführung in die Phytologie. 2d ed., III(1): Grundlagen der Pflanzenverbreitung: Standortslehre (analytisch-ökologische Geobotanik). E. Ulmer, Stuttgart. 566 p.

37. BENNET-CLARK, T. A.
    1959 Water relations of cells. *In* F. C. Steward (ed), Plant physiology, a treatise, Vol 2, p. 105–191. Academic Press, New York.

38. ABEL, W. O.
    1956 Die Austrocknungsresistenz der Laubmoose. Österreichische Akademie der Wissenschaften, Mathematisch-naturwissenschaftliche Klasse, Sitzungsberichte Abt. I, 165:619–707.

39. ILJIN, W. S.
    1935 Lebensfähigkeit der Pflanzenzellen in trockenem Zustand. Planta 24:742–754.

40. ILJIN, W. S.
    1931 Austrocknungsresistenz des Farnes *Notochlaena marantae* R. Br. Protoplasma 13:322-330.

41. HÖFLER, K.
    1942 Über die Austrocknungsfähigkeit des Protoplasmas. Deutsche Botanische Gesellschaft, Berichte 60:(94)–(107).

42. BIEBL, R.
    1962 Protoplasmatische Ökologie der Pflanzen. *In* L. V. Heilbrunn and F. Weber (eds), Protoplasmatologia XII/1. Springer, Wien. 344 p.

43. HÖFLER, K.
    1950 Über Trockenhärtung des Protoplasmas. Deutsche Botanische Gesellschaft, Berichte 63: 3–10.

44. LEVITT, J.
    1962 A sulphydryl-disulfide hypothesis of frost injury and resistance in plants. Journal for Theoretical Biology 3:355–391.

45. GAFF, D. F.
    1966 The sulphydryl-disulfide hypothesis in relation to desiccation injury of cabbage leaves. Australian Journal of Biological Sciences 19: 291–299.

46. PARKER, J.
    1968 Drought-resistance mechanisms. *In* T. T. Kozlowski (ed), Water deficit and plant growth. Vol 1, p. 195–234. Academic Press, New York.

47. LEVITT, J., AND G. W. SCARTH
    1936 Frost-hardening studies with living cells. Canadian Journal of Research, Section C 14: 286–305.

48. LEE, O. Y.
    1967 The influence of decenyl succinic acid on permeability of plant cells to water and its possible effect on frost hardiness. University of Minnesota, St. Paul (M.S. thesis). 103 p.

49. SUTHERLAND, R. M., AND A. PIHL
    1968 Repair of radiation damage to erythrocyte membranes: The reduction of radiation-induced disulfide groups. Radiation Research 34:300–314.

50. STADELMANN, ED.
    1952 Zur Messung der Stoffpermeabilität pflanzlicher Protoplasten. II. Österreichische Akademie der Wissenschaften, Mathematisch-naturwissenschaftliche Klasse, Sitzungsberichte Abteilung I, 161:375–408.

51. STADELMANN, ED.
    1959 Eine einfache Durchströmungskammer (mit historischem Überblick über Durchflusskammern). Zeitschrift für Wissenschaftliche Mikroskopie 64:286–298.

52. STADELMANN, ED.
    1962 Eine mikrophotographische Zeitrafferanlage mit Auswertevorrichtung. Zeitschrift für Wissenschaftliche Mikroskopie 65:172–185.

53. PEDELISKI, V., AND ED. STADELMANN
    n.d. Computation of the water permeability constant K_{wo} and its relationship to the choice of the value of the dependent variable. Manuscript in preparation.

54. SEIFRIZ, W.
    1955 The physical chemistry of cytoplasm. *In* W. Ruhland (ed), Encyclopedia of Plant Physiology. Vol 1, p. 340–382. Springer, Berlin.

55. SCHAEFER, G.
    1955   Ein Versuch zur quantitativen Auswertung
            der Plasmolyseform- und -zeitmethode. Proto-
            plasma 44:422–436.

56. URSPRUNG, A., AND G. BLUM
    1916   Über die Verteilung des osmotischen Wertes
            der Pflanzen. Deutsche Botanische Gessell-
            schaft, Berichte 34:88–104.

57. HÖFLER, K.
    1920   Ein Schema für die osmotische Leistung der
            Pflanzenzelle. Deutsche Botanische Gesell-
            schaft, Berichte 38:288–298.
58. This work was supported by the Minnesota Agricul-
    tural Experiment Station and constitutes Scientific
    Journal Series, Paper No. 7071. The author wishes to
    thank Mrs. Virginia Pedeliski, University of Minnesota,
    for revision of English and style.

Part Four
ECOLOGY OF ARID REGIONS

Sand-Stabilization Methods in Arid Lands

Protection of Agricultural and Settlement Areas

MIKHAIL P. PETROV

Department of Geography
Leningrad State University
Leningrad, U.S.S.R.

ABSTRACT

VEGETATIVE AND PHYSICOCHEMICAL MEANS are used to stabilize sands in the main sandy deserts. Mobile sands, often the result of man's activities, are sometimes seriously damaging to the economy of an area. Worldwide experience in stabilization is extensive. Vegetation can stabilize sands in semiarid and arid regions, but not in extremely arid zones. Vegetative methods may be preventive or reclamative. Vegetation-establishment projects are efficient only when there is sufficient knowledge about local sand types and of the activities of the human population. Three major types of sandy areas and eight major groups of vegetation sites are described for the world's deserts. The richest assortment of plant species can be established in the semiarid areas, where tree-shrub species achieve the best results. A more restricted number of species can be established in the arid areas if protected for 2 to 5 years by fences or different covers: shrubs, low shrubs, and perennial and annual grasses are used. In extremely arid areas, physicochemical reclamation is possible using mechanical protections, bitumen, clay, rubber, polymers, and other covers. In semiarid areas fixation may take place in 5 to 10 years; in arid areas, tens of years may be required.

Plant species may be introduced from other areas to stabilize sands, or their relatives may be sought among native plants. Sand-fixing plants are sown or started from seedlings or cuttings. The processes are mechanized as much as is presently practicable. Physicochemical stabilization is still expensive, but if the costs can be reduced such methods will be used over larger areas.

Research and development work required: physicogeographical study in the immediate regions requiring stabilization, perfection of methods for establishing vegetation (including mechanization), development of measures for rational use and restoration of destroyed natural vegetation, creation of oases, improvement of physicochemical methods, and investigation of geomorphological processes in the sandy desert.

SAND-STABILIZATION METHODS IN ARID LANDS
Protection of Agricultural and Settlement Areas

Mikhail P. Petrov

IMPORTANCE OF THE PROBLEM

As MEASURES TO RECLAIM ARID LANDS and make use of their natural resources intensify, the unfavorable effects of man's activities upon these natural landscapes become more pronounced. This situation has led in many deserts to formation of large areas of loose sand not fixed by vegetation. Such regions sometimes seriously damage the economy.

In older times, formation of loose sandy dune fields was connected with the irrational practice of denuding the sand of vegetation, which was used for cattle graze or fuel. Consequently the dunes would rise near wells or at the margins of oases. Later, after railroads and highways had been built, dunes appeared near railway stations. During recent years a considerable acreage of vegetative growth, which had stabilized sands in arid regions, has been destroyed during the construction of oil and gas mains, industrial enterprises, and nearby populated places.

Besides, in many deserts and semidesert regions, where exploratory drilling and exploitation of oil and gas deposits are going on, shrub vegetation has been cut down over large areas, or destroyed in connection with the transportation of heavy drilling equipment.

In the process of reclamation of new lands for irrigated farming, sand islands are sometimes encountered amidst these lands. Newly reclaimed lands at times lie in immediate proximity to fields of mobile sands.

As a result, prevention of sand from drifting onto fields and settlements, either in sandstorms or as encroachment of solid dunes (Fig. 1), demands great efforts everywhere in the arid areas. Man has been confronted with this unfavorable natural feature since time immemorial. Very pronounced sand-movement processes already were taking place at the margins of ancient farming oases, where the unplanned effects of man upon the natural desert were most damaging. Moreover, we are still today confronted with the same problem, irrational use of soils and vegetation, even in moister areas.

Thus, the problem of protection of sandy areas from irrational use is most important and calls for intensification of work to establish vegetation on drifting sands and of other methods of combating sand deflation and dune

Fig. 1. Shifting sands approach the southern border of Tunghuan oasis, Central Asia.

356

f) Mobile sands of varying thickness lying on loose-sandy lacustrine and alluvial deposits (in all Asian deserts) with shallow, fresh underground waters available to plant roots.

h) Mobile sands of varying thickness lying on land formerly cultivated and irrigated but now abandoned — in river valleys and deltas of the whole Asia desert zone. Favorable soil conditions; shallow underground waters (often fresh) available to plant roots.

Each group of described types includes several subtypes of sands, depending on the thickness of sand dunes:

a) small barkhan dunes, up to 1 meter high

b) medium barkhan dunes, up to 3 meters high

c) high barkhan dunes, above 3 meters

This typology of sands can acquire specificity under various climatic conditions. It is necessary, for instance, to distinguish specific types of sands in the extremely arid, arid, and semiarid regions. There may be different moisture and salinity conditions, and afforestation and fixation projects should be modified in conformity with new conditions.

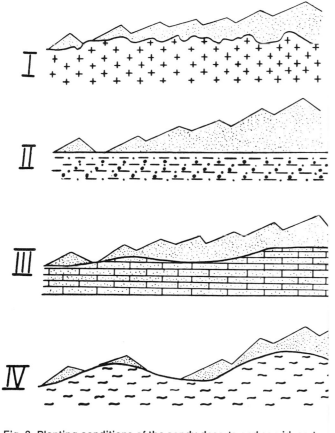

Fig. 3. Planting conditions of the sandy deserts and semideserts of Central Asia. In types *I-V*, the groundwater is located at a great depth and is inaccessible to plant roots.
I. Sands of varying thickness lying on water-impermeable solid metamorphic and sedimentary rocks of low mountain ranges.
II. Sands of varying thickness lying on Cretaceous sandy-pebble plains.
III. Sands of varying thickness lying on Cretaceous and Jurassic sandstones.
IV. Sands of varying thickness lying on loess and clay deposits.

movement. This combat has to be carried on under different, sometimes very severe, conditions.

Nature used to win the combat of man and sands — in older times when the strategy was not planned. Man had to shift for new places, leaving fields and homes covered with sand. Such cases have been described many times in the literature for deserts of North Africa, Middle and Central Asia, and South and North America. (Soviet geographers use "Central Asia" to refer to territory east of the Tien Shan mountains and the Pamirs, and "Middle Asia" to refer to Soviet territory only.)

At present, the fight against drifting sands and their deflation is based on detailed study of the processes of movement and accumulation of sands, the ecology of the vegetation of sandy deserts, features of tree-growing conditions in deserts, and certain other factors.

The principal regularities of sand movement and of the formation of sandy topography have been revealed from studies of sand accumulation and deflation. Relationships between sand accumulation and relief-formation have been established, some protective measures against sand deflation and drifts have been developed, an assortment of plant species for stabilization of sands has been found, and cultural practices for growing the plants on shifting sands have been developed.

Worldwide experience in the afforestation and stabilization of drifting sands in arid areas is rather extensive. These measures have been applied in all countries containing sandy deserts and semideserts. The bibliography section of this chapter provides a brief review of the scale and geographical dispersion of such work. Analysis of the bibliography assures us that specialists in the field

have developed practicable methods of establishing vegetation and stabilizing mobile sands in semi-arid and arid regions of all natural zones where the sands contain enough moisture (see Fig. 2). As for extremely arid deserts, methods of combating mobile sands have not yet been properly developed.

Since vegetative means of stabilizing sand in extremely arid regions are impossible, new methods have been found by soil physicists and chemists. They suggest fixation covers for the purpose; for example, clay, pebble, cement, oil, bitumen, rubber suspensions, polymers (physicochemical reclamation).

METHODS OF STABILIZING MOBILE SANDS

The chief principle for combating sand movement successfully over vast territories is to plan the revegetation carefully. In all cases, vegetation establishment on sands should be preceded by the creation of detailed project plans, to be made after analyzing the results of studies of vegetation in the same sand-dune fields.

Projects for establishing vegetation on mobile sands fall into two categories: (*a*) prophylactic or preventive measures, which are aimed at limiting any increase in the amount of land covered with mobile sands and at excluding the possibility of the development of new fields of active sand dunes, and (*b*) vigorous vegetation-establishment measures aimed at reclaiming mobile sands and turning them into fodder or forest acreage.

Prophylactic or preventive measures include regulations to ensure more rational use of the vegetation

Photo by K. Pashkovsky

Fig. 2. Plantings of *Haloxylon aphyllum* on desert sands near the Zeravshan river, Middle Asia.

growing on sands. For example, the utiliza
tremely loose (broken) sands is officially re
many cases cattle-grazing is prohibited and
fuel of tree-shrub and herbaceous vegetation
sands is forbidden in sandy regions bordering
lands or industrial enterprises and settlement

When vegetation regenerates on protectec
use is controlled by reclamative afforestatior

Active measures to control mobile sands in
tation establishment and covering of sand su
fixating agents (physicochemical reclamation

Measures to prevent sand deflation and co
movement in deserts and semideserts use th
approaches:

a) Reclamation of sands near oases aim
afforestation, and protection of various obje
irrigated land, canals, populated places, and
encroaching sand dunes.

b) Landscape gardening near wells, railw
and other settlements in the desert.

c) Protection of railroads and highways fr
sand dunes.

d) Protection of industrial enterprises, pij
from encroaching sand dunes.

e) Protection of irrigation systems from m

Man is confronted with deflation and with
in the process of combating moving sands. D
be controlled by way of fixation of deflat
through vegetation establishment and use
covers. Combating mobile dunes is far me
cated. Sometimes all measures possessed by tl
tion agencies should be applied.

Typology of Conditions for Vegetation Establishment in Sandy Areas

Vegetation-establishment projects are ef
when there is sufficient knowledge of the effec
ent types of sands on growth and sufficient i
on the economic activity of the population
sandy areas. Hence it is clear that all affore
sand-fixation projects should begin with
establishment surveys. Planting conditions
sands comprise and combine all environmenta
that influence growth and development of
plants.

It is necessary to sum up all available data
conditions and on the economic utilization of s
should be used as basic information to c
vegetation-establishment projects. The follo
lems are to be considered the most essential:

a) The typology of growing sites on terr
jected for shelter-belt growing, afforestation, a
of sands.

b) The working out of corresponding
reclamative afforestation projects in confo
observed growing sites.

c) The evaluation of the reclamative ei
trees, shrubs, and grasses recommended for
belt growing, for fixation, and for afforestatic

In subtropical and tropical zones with mild winters and abundant precipitation, intercontinental semiarid regions (*S*) cover large areas in Australia, South, Central, and North America, South and North Africa, and South (India) and Southwestern (Middle East) Asia. In these regions the richest assortment of plant species can be used for afforestation and fixation of mobile sands. The best results here have been achieved by tree-shrub species.

Establishment of vegetation on these sands quickly provides positive results. Annual plantings show a high survival rate. At places the sands can be used for growing such crops as grapes, dates, and alfalfa.

In arid subtropical and tropical zones (*A*) within the same continents, where conditions for vegetation are more severe, reclamation of mobile sands faces many obstacles. Young plants here should be protected from deflation, burial, and cutting by sand. Different types of protections are necessary for the purpose. Sown and planted species should be protected for the first 2 to 5 years. After this period they will be strong enough to withstand unfavorable effects of drifting substratum.

The assortment of sand-fixing plants for establishment in arid deserts is restricted. Trees will not grow in such conditions. Principal life forms are represented by shrubs, low shrubs, perennials, and annual grasses. Due to repeated droughts, planting is not always successful, but vegetation establishment on sands is possible.

In extremely arid regions of subtropical and tropical zones (*E*), that is, the central Sahara, the southern part of the Arabian Peninsula, the Atacama, and the lower reaches of the Colorado River, establishment of vegetation on the sands is completely impossible. Here, due to negligible precipitation and high summer air and soil

TABLE 1
The Main Sand-Stabilizing Plants of the Deserts and Semideserts of the World

Middle Asia and Kazakhstan *	Central Asia *	India and Pakistan	Northern Africa and Middle East
Trees		*Trees*	
Populus nigra	Populus simonii	Prosopis spicigera	Pinus pinaster
Elaeagnus angustifolia	Elaeagnus moorcroftii	P. juliflora	
		P. glandulosa	P. halepensis
		Acacia senegal	
		A. arabica	Eucalyptus gomphocephala †
		Salvadora oleoides	
		Zizyphus nummularia	E. camaldulensis †
		Balanites roxburghii	
		Gymnospora montana	
		Casuarina equisetifolia †	Tamarix articulata
		Calophyllum inophyllum	
		Tamarix articulata	Casuarina equisetifolia
		T. aphylla	
Shrubs			
Haloxylon aphyllum	Haloxylon ammodendron		
H. persicum	Tamarix ramosissima	*Shrubs*	
Tamarix ramosissima	T. laxa	Tamarix ramosissima	Tamarix ramosissima
T. laxa	T. chinensis	Capparis aphylla	T. pseudopallasii
Salix caspica	Salix flavida	Calligonum polygonoides	Retama ratam
S. rubra	S. mongolica	Ricinus communis	Acacia cyanophylla
Calligonum arborescens	S. cheilophila	Leptadenia spartium	A. cyclops
C. caput-medusae	Calligonum mongolicum	Clerodendron inerme	Nitraria retusa
C. aphyllum	C. roborobskii		Calligonum comosum
C. turkestanicum	C. caput-medusae †		C. arich
Hedysarum scoparium	Hedysarum scoparium		C. azel
Salsola richteri	H. mongolicum		Ricinus communis
S. paletzkiana	Caragana microphylla		
Nitraria schoberi	C. korshinskii		
N. sibirica	Nitraria schoberi		
	N. sibirica		
Undershrubs, Semishrubs		*Undershrubs, Semishrubs*	
Artemisia arenaria	Artemisia sphaerocephala	––––	Artemisia monosperma
	A. ordosica		
	A. halodendron		
Perennial Plants		*Perennial Plants*	
Aristida karelini		Cenchrus ciliaris	Ammophila arenaria
A. pennata	Psammochloa villosa	C. biflorus	Aristida obtusa
Elymus giganteus		Arundo donax	A. pennata
		Agropyron elongatum	A. pungens
Annual Plants		Ehrhartia brevifolia	Arundo donax
Agriophyllum arenarium	Agriophyllum arenarium	Saccharum spontaneum	Agropyron junceum
A. latifolium	Pugionium cornutum	S. munja	Lotus creticus
Horaninovia ulicina	Stilpnolepis centiflora	Citrullus colocynthis	Imperata cylindrica
Salsola praecox	Salsola praecox	Panicum antidotale	Pennisetum dichotomum
S. turcomanica		Elionurus hirsutus	Polygonum equisetiforme
Bibliography		*Annual Plants*	
M. Petrov 1950	M. Petrov 1967	Kochia indica	Salsola praecox
		Bibliography	
		M. Khalil 1960	Karschon 1953
			M. Motte 1963
		M. Sharif 1960	J. Messines 1952
			D. E. Tsuriell 1961
		M. Petrov 1963	C. Sulzlee 1963
			D. Sesmaisons 1963

temperatures, the most extreme conditions for vegetation are encountered. Scanty local vegetation inhabits only temporarily-watered riverbeds or depressions with shallow subsurface waters, often saline.

In such conditions, reclamation by vegetative means is replaced by physicochemical reclamation — various mechanical protections, bitumen, clay, rubber, polymers, and other covers.

In the mobile sand fields of the temperate zone, conditions are less favorable for vegetation than in subtropical and tropical belts, due to the severe winters. Sand fields of this type are found in the deserts of Middle Asia, Central Asia, and at places in the north of Nevada in North America.

Low winter air and soil temperatures (up to minus 40°) limit utilization of psammophilous plants of subtropical and tropical belts for fixation of mobile sands. However, experience indicates that deserts of the tem-

perate zone possess enough local shrubs and herbaceous plants for fixation and growth on mobile sands (see Table 1).

The assortment of sand-fixing plant species and the type of reclamation used varies, depending on the aridity of the area. Thus, in the semiarid temperate belt (*S*) (precipitation over 250 millimeters), tree stands can be cultivated on mobile sands. Sand of this type can be used for pastures if necessary. Such pastures of *Artemisia arenaria, A. sphaerocephala,* and *A. halodendron,* with an admixture of *Agropyrum desertorum,* have been grown in the north of Asia.

The rate at which sands can be covered with vegetation in semiarid regions is rather high. Fields of drifting barkhan dunes can be turned into fixed ones in 5–10 years, depending on all relevant conditions.

In arid deserts of the temperate zone (*A*), where the water regime is more severe, vegetation-establishment on

TABLE 1 (cont.)

Southern Africa	Australia
Trees	
Eucalyptus gomphocephala †	Eucalyptus gomphocephala
Casuarina equisetifolia †	Casuarina equisetifolia
Pinus pinaster †	Banksia integrifolia
Shrubs	
Acacia cyanophylla	
A. cyclops	
A. saligna	
Myoporum insulare	Acacia cyanophylla
Leptospermum laevigatum	
Myrica cordifolia	A. sophorae
Metalasia muricata	
Stoebe vulgaris	Atriplex vesicaria
Brachylaena discolor	
Osteospermum species	
Acanthosicyos horrida	
Perennial Plants	
Arthraerua leubnitziae	Spinifex hirsutus
Zygophyllum stapfii	Pennisetum clandestinum
Aristida kalahariensis	Senecio crassiflorus
A. brevifolia	Carpobrotus aequilatorus
Ammophila arenaria †	Arctotis stoechadifolia
Ehrharta giganthea	Ammophila arenaria †
Annual Plants	
Digitaria sanguinalis	Canavalia maritima
Bibliography	
C. Laver 1936	
J. Loubser 1959	J. Sless 1958
J. van der Westhyuzen 1957	

North America and Mexico	South America A, Atacama; P, Patagonia
Trees	
Prosopis juliflora	Prosopis argentina (P)
	P. strombulifera (P)
	P. tamarugo (A)
	P. chilensis (A)
Shrubs	
Atriplex canescens	Atriplex canescens (A) †
Poliomintha incana	A. nummularia (A) †
	A. semibaccata (A) †
	Plectocarpa tetracantha (P)
Undershrubs, Semishrubs	
Artemisia filifolia	Kochia brevifolia (A) †
Perennial Plants	
Agropyron smithi	
Ammophila arenaria †	
A. brevilegulata	
Andropogon scoparius	Panicum urvilleanum (P)
A. gerardi	
Bouteloua curtipendula	Pennisetum chinense (A)
B. gracilis	
Calamovilfa gigantea	Hyalis argentea (P)
C. longifolia	
Elymus giganteus †	
E. canadensis	
Panicum virgatum	
P. havardii	
Redfieldia flexuosa	
Sporobolus flexuosus	
Bibliography	
C. Whitfield and R. Brown 1948	
McGinnies and others 1968	McGinnies and others 1968
M. Gonzalez (personal communication) 1969	A. J. Prego (personal communication) 1970

---

* Soviet geographers use "Central Asia" to refer to territory east of the Tien Shan mountains and the Pamirs; "Middle Asia" is used to refer to Soviet territory only.
† Introduced from other deserts of the world.

mobile sands is possible also. However, the assortment of plant species is more restricted here. Besides, the rate of coverage by vegetation will be far slower. In such conditions tens of years instead of several years will be required for the complete fixation of sands.

The management of sand-fixation work will also be more complicated in arid deserts of the temperate zone. In semiarid sandy areas, plantings can survive without special measures against deflation and sand burial of young plants. In arid deserts, young plants should be protected from sand with row mechanical protections or special covers. The assortment of sand-fixing plants is represented by local shrubs, annuals, and perennial grasses. They are heat-tolerant and drought-resistant, and atmospheric precipitation is enough for their growth.

In extremely arid deserts of the temperate belt (Central Asia) as well as in tropical and subtropical deserts, vegetation establishment on sands is impossible, due to extreme desiccation of sandy grounds, and physico-chemical measures should be used.

Within every climatic belt, several subtypes of sands can be distinguished: intercontinental, coastal, and oasis sands. Practices of sand fixation and afforestation on coastal plains and intercontinental deserts differ very much.

Coastal sands are characterized by favorable conditions for establishment of vegetation. They are better moistened than other sands because of shallow subsurface waters as well as condensation of water vapor. Air humidity is higher and its temperature lower on the coastal plains than in the intracontinental deserts. Salinity of surface horizons of coastal sands is usually higher, but this factor is not critical for the development of a number of salt-tolerant species. In conformity with the saline conditions, sand-fixing plants grown on coastal sands often possess ecological features of halophytes and are known as psammohalophytes. They include *Spinifex hirsutus* in Australia, some *Tamarix* species, the psammophilous perennial grass *Ammophila arenaria* in the Mediterranean area, and others (see Table 1).

Sands bordering on oases or forming dune groups inside them benefit from the most favorable conditions for vegetation establishment. These consist of the occurrence of subsurface waters close to the surface and the periodical moistening of sands when irrigation waters are discharged outside oases. In such conditions, prospects for afforestation and fixation of mobile sands are more favorable. They can be used after certain reclamative measures not only for growing of tree stands but for cultivation of crops (for example; cereals, fruits, melons, and grapes).

The most severe conditions for establishment of vegetation are observed in intercontinental sands of arid and extremely arid deserts, and their fixation faces many obstacles.

Assortment of Plant Species

The assortment of plant species used for the afforestation and fixation of sands includes mostly local endemic plants growing on dunes and some plants introduced from other deserts. The choice is rich enough for all reclamative vegetation-establishment projects.

In different natural zones of the world, the assortment of sand-fixing plants varies. It includes trees, shrubs, sub-shrubs, and perennial and annual grasses belonging to different botanico-geographical provinces and families (Graminaea, Chenopodiaceae, Leguminosae, Compositae, etc) (see Table 1).

The ecology of all these plants is peculiar to psammophilous plants. They possess some adaptive features allowing them to grow and reproduce on the drifting substratum. They easily withstand deflation of root systems and sand accumulation on above-ground plant parts. They can form adventitious roots on branches covered with sand during one season and new shoots on deflated roots. Besides, they are drought-resistant, heat-tolerant, and frost- and salt-tolerant (to chloride and sulphate salinization) in the temperate zone.

Similar ecological properties of psammophilous plants in different botanico-geographical regions make them very useful for introduction into new, often very remote geographical regions to stabilize sand dunes. Thus workers interested in establishing vegetation on sand fields are provided with great opportunities, which can be used in two principal ways:

a) Introduction of species tested during the fixation and afforestation of sands in other deserts with similar site conditions. For instance, *Casuarina equisetifolia* from Australia has been introduced into shelter belts on coastal sands in India, South Africa, the Mediterranean coast, and the Pacific coast of North America. On the other hand, psammophilous grass from the Mediterranean coast (*Ammophila arenaria*) grows very well on Australian sands.

b) Search in the local flora for plants systematically related to species with well-known sand-fixing properties. Thus for instance, if an *Artemisia* species is being used for sand fixation in the deserts of northern Africa, probably some other *Artemisia* species from the deserts of Asia can be successfully used for the purpose.

Cultural Practices for Establishing Vegetation on Sands in Arid Areas

Cultural practices on mobile sands have some peculiarities. Sands are loose, which prevents the usual preparation of soil for sowing and planting, all attention being concentrated upon construction of mechanical shelters aimed at protection of planted sites from deflation and sand accumulation.

Plantings are protected with various mechanical shelters (row, linear checkerboard, pole, mat-like), which break down the wind, stop deflation, and stabilize the mobile sands (see Fig. 4).

In some countries (for example, Morocco) expensive shelters made of desks are being constructed (G. Souleres, 1963). It should be mentioned that from the point of view of their aerodynamic efficiency, these shelters are not justified. Observations have shown that the efficiency of rows of wooden panels is almost equalled by that of cheaper shelters made of reed or other grasses.

Mechanical protections are used to stop immediately the movement of sands, to check the surface from blowing, and to hold in place the seeds of wild plants. In some cases mechanical protections are used for managing the relief and sand movement — to level barkhan topography, remove sand from areas on which it has drifted, or to accumulate sand where necessary.

Recently, to protect psammophytes and other plants from deflation, bitumen and clay films have been used, as well as hydrophilous polymers, which fix the sand's surface layer and prevent deflation and dune formation by the wind. These films do not affect the physical properties of sand. Precipitation easily penetrates them, and the water regime of sands covered with such a film does not change. It is not an obstacle for seedlings. They easily break it without lowering its overall effectiveness. Sand-fixing plants are started from seedlings grown in nurseries or from cuttings prepared in advance in natural stands on sands. From 3000 to 5000 hills are planted

per hectare. The space between plants in the row is 1 meter; the interrow space is 2 to 3 meters. When mechanical protection is given to row planting, the plants are set 15 to 20 centimeters to the lee side of the protection.

Seeds of sand-fixing plants are sown either as the principal operation or as an additional planting in existing stands. Only in sandy areas where conditions are most favorable is seeding alone successfully practiced.

Establishment of vegetation on sands is mechanized as much as possible. Extensive areas are sown with sand-fixing plants from airplanes and trucks. Cultivation, sowing, and digging in nurseries are also mechanized.

The most favorable opportunities for mechanization of vegetation-establishment methods exist along railway lines. A special device on a plaform was constructed by the Tashkent Institute of Railway Engineers (U.S.S.R.). It simultaneously sows seeds of sand-fixing plants, applies fertilizers, and prepares and spreads bitumen suspension for fixation of the sand surface.

Experiences with aerial sowing of seeds of sand-fixing plants in different desert conditions have shown the best

Fig. 4. Mechanical checkerboard protections at a railroad in the Tengeri sandy desert near Djunway, Central Asia.

results on partially overgrown sands with shallow subsurface waters (buried cultivated lands).

In the heart of the desert where conditions are more severe, especially on bare barkhans, efficiency of aerial sowing is lower because the wind blows seeds into depressions and because sand blows away or onto seedlings. To raise the efficiency of aerial sowing, preliminary preparation of sowing sites should be carried on by establishing row or checkerboard mechanical protections along the interbarkhan depressions.

The installation of mechanical protections is the most labor-consuming process of all measures for establishing vegetation on sands. Nevertheless, the wide utilization of such devices is necessary, especially for protection of plantings and for planting of sand-fixing species.

Physicochemical Stabilization of Sands

During the last decades, use of mineral resources — oil, gas, ferrous and nonferrous ores — has increased greatly. As a result of this process, large industrial enterprises and populated places have appeared in the deserts. Oil works in Kuwait in the Arabian Peninsula and industrial enterprises in the deserts of Middle and Central Asia are vivid examples.

Some of these industrial facilities are located in barkhan sands with extremely unfavorable conditions for vegetation establishment (high salinity of sands or precipitation below 50 millimeters). Attempts to establish vegetative cover on such sands have not yet brought any positive results.

In conjunction with this problem, more interest has been shown during the last years in fixation of mobile sands with various "artificial" covers. These measures follow several approaches:

a) Utilization of local materials (gravel, road metal, clay, town refuse, vegetation residues) for fixation of sands at places where it is economically advantageous.

b) Bitumenization of sands for the purpose of creating a durable cover possessing high resistibility to the wind action (see Fig. 5).

c) Creation of various films of water-soluble polymers and other chemical substances, which can raise the resistibility of sands to wind erosion. Examples are polyacryloamide, nerozin, divinylstyrol latex ARM-15, and others. Such research has been carried out in several countries.

A hydrophilous polymer of the K-4 series, developed in Tashkent (Middle Asia, U.S.S.R.), can fix the surface of mobile sands (K. Akhmedov and others, 1969). A British firm suggested utilization of the synthetic latex of rubber for this purpose. In steppe regions of Hungary solacrol is used for sand fixation and soil reclamation.

The laboratory and field experiments have shown that sometimes these covers ensure stability of sand fixation in regard to deflation and mechanical damage. Formation of a crust on the sand surface by means of polymer suspensions and other fixators consists of formation of a sort of bridge made of fixators and fastening the sand grains together.

For the fixators to be efficient, they should possess sufficient mechanical durability, water impermeability,

Photo I. Ivlev

Fig. 5. Fixation with bitumen cover of sands near a road. Middle Asia.

and resistibility to atmospheric influence (wind action especially) and to salts (chlorides and sulphates).

Thus a new trend in controlling moving sand dunes and deflation of sands has appeared: physicochemical stabilization. But methods of construction of mechanical covers are still too expensive, and their application is limited to small areas. Therefore they have not yet been widely used, because the technology of producing fixators has not yet been properly developed, because special machines and devices for covering sands with fixators are absent, and because costs of fixators as well as their application costs are high (see Fig. 5).

Introduction of new, cheap fixators and mastery of the application technology will make it possible to reduce the cost of protection from encroaching sand dunes. Thus expensive mechanical protections will gradually disappear from the list of sand-stabilization practices. There is hope that fixators will be more widely used in the immediate future to stabilize moving sand dunes in extremely arid deserts.

Another method related to engineering reclamation is that of sand drift across the protected objects without accumulation (Znamensky, 1968). The wind blows the sand across the protected object (highway) or along its side (separate houses and other structures). A. Znamensky suggested a special profiling of highways in Western Turkmenia to help sand drift across the roads without it accumulating on them.

THE IMMEDIATE SCIENTIFIC PROBLEMS

Considering the present scale of industrial and agricultural activity in the sandy deserts, the data available to reclamation engineers are too few for solution of the problems put forward by the practical work. A large complex of new research has to be completed, its results serving as a foundation for the forthcoming projects to protect farmlands and populated places against encroaching sand dunes and deflation.

The research work is aimed at the following:

a) Complex physico-geographical study of the sandy deserts in the immediate regions of the reclamation aimed at compilation of special detailed maps of conditions for plants and measures for establishment of vegetation.

In conjunction with the aforesaid classification, a typology of sands should be perfected and types of vegetation-establishment conditions determined for each type (water regime, thickness and mobility of sands, content of nutrients, etc) together with the assortment of plant species and establishment methods. Besides, maps and schemes should be compiled showing distribution of dune forms, their origin and morphology, location of contemporary deflation centers, ways and intensities of sand transportation and accumulation, and position of sand dunes in certain objects of economic value.

The data obtained will be the basis for compilation of a scheme of deflation state of a sandy desert and principal regularities of sand movement and accumulation.

Thus while roads, industrial enterprises, and populated places are planned, it will become possible to ensure the correct distribution of economic objects and work out a general scheme of protective measures.

b) Development of perfected measures for establishing vegetation on sands. Research should follow several approaches: selection of fast-growing sand-fixing plants, improved cultural practices, and mechanization.

Introduction of new plants from other deserts of the world should be carried on at the experimental stations in different countries for the purpose of enriching the assortment of sand-fixing plant species in the desert, together with selection of the most promising species and development of agrotechnic methods for their cultivation in the new desert conditions. The selection of new species, most resistant to unfavorable life conditions, remains one of the principal tasks.

The negligible success achieved in mechanization of sand-dune-stabilization procedures is an important gap. Very little has been done so far. Most of the work is being done by hand, which makes it very expensive.

The most easily solved part of the problem is to ensure mechanization in nurseries and in seeding sand-fixing species on mobile sands (aerial, from trucks, railway platforms, etc). However, the nature of the work demands machines for putting up mechanical protections and forest-planting machines able to operate in highly dissected barkhan relief.

c) Development of measures aimed at rational use and restoration of destroyed natural desert vegetation. This research is directly connected with the problem of establishing vegetation and preventing deflation of mobile sands. It is known that centers of deflation originate when natural desert vegetation is extensively used for grazing or procuring fuel. Such use may lead to formation of fields of mobile sand dunes.

To prevent this consequence, strict regulation of all types of utilization of natural desert vegetation is indispensable. Besides, research should continue into the processes of restoring natural vegetation and the methods of accelerating the processes in keeping with the concrete ecological conditions of each desert.

d) Creation of oases in the desert near wells and new populated places using subsurface waters and waters of local runoff. This research will make it possible to reduce the unfavorable effect of hot climate and duststorms on the desert inhabitants. Simultaneously landscape gardening around populated places will help protect them from encroaching sand dunes. For this purpose, methods of landscape gardening of populated places in the heart of the desert — near industrial enterprises and near wells in cattle-breeding areas — should be perfected. The assortment of shrubs and trees growing on sands with improved water regime has to be extended, and the most rational cultural practices of growing plants irrigated with saline water must be specified.

e) Development of the most rational methods of stabilizing drifting sands by means of physicochemical reclamation and selection of fixators for this purpose is an important section of future research. This is of special significance for the extremely arid deserts where vegetative stabilization of mobile sands is impossible. The assistance of physicists and chemists specializing in polymers is indispensable to solve this exceedingly important problem.

f) A number of problems of afforestation and fixation of mobile sands to protect farmlands and populated places from encroaching dunes and deflation are closely

connected with some purely theoretical problems. The most important is research on geomorphological processes in the sandy desert. Of special interest is the study of aerodynamics of relief-formation on drifting sands, in the laboratory as well as in the field.

The results of this work will make it possible to provide more valid practical recommendations on control of sand dunes and deflation.

BIBLIOGRAPHY

U.S.S.R.

AKHMEDOV, K. S., E. A. ARKHIPOV, AND T. NURYEV
1969   Zakreplenie podvizhnykh pescov polymerom K-4 (Sand fixation by polymer K-4). Akademiia Nauk Uzbekskoi SSR, Doklady 4: 20–21.

BEZHANBEK, E. A., AND F. K. KOCHERGA
1951   Zakreplenie i oblecenie peskov Uzbekistana (Sand fixation and afforestation in Uzbekistan). Goslesbumisdat, Moscow-Leningrad. 64 p.

GAEL, A. G.
1952   Oblesenie bugristykh peskov zasyshlivykh oblastei (Afforestation of hummocky sands in droughty regions). Geografgiz, Moscow. 218 p.

IVANOV, A. E., AND M. M. DRUCHENKO
1969   Komplexnoe osvoenie peskov (Complex development of sands). 2d ed. Isdatelstvo Lesnaya Promyshlennost, Moscow. 302 p.

KAUL, R. E., ed.
1970   Afforestation in arid zones. Junk, The Hague. 435 p. (Monographiae, Biologicae, 20)

KHODZKHAYEV, A. A.
1947   Borba s peschanymi zanosami na zheleznykh dorogakh (The protection of railways from sand drifts). Transzheldorizdat, Moscow. 104 p.

LEONT'EV, A. A.
1962   Peschaniye pustini Srednei Azii i ikh lesomeliorativnoye osvoyeniye (Sandy deserts of Middle Asia and their reclamative afforestation). Gosudarstevennoe Izdatel'stvo Uzbekskoi SSR, Tashkent. 159 p.

NECHAEVA, N. T., AND OTHERS
1959   Opyt uluchsheniya pustynnykh pastbisch v Turkmenistane (Experience in improving desert pastures in Turkmenistan). Ed. Minselkhoz, Ashkhabad, Turkmenistan SSR. 246 p.

ORLOV, M. A.
1940   Peski Astrakhanskoi polupustini, methody ikh ukrepleniya i khozyaistvennogo ispolzvaniya (Sands of the Astrakhan semidesert, methods of their fixation and economic utilization). Goslostekhizdat. 136 p.

PALETZKY, V. A.
1956   Izbrannye trudy po lesorazvedeniyu i ghidrologhii (Selected works on afforestation and hydrology). Akademiia Nauk Uzbekskoi SSR, Tashkent. 144 p.

PETROV, M. P.
1950   Podvizhnye peski pustin Soyuza SSR i borba s nimi (Drifting sands of the USSR deserts and their control). Geografgiz, Moscow. 454 p.

PETROV, M. P. (cont.)
1952   Agrolesomelioratsiya peskov v pustinyakh i polupustinayakh Soyuza SSR (Reclamative afforestation of sands in the USSR deserts and semideserts; a bibliography of Russian publications, 1768–1950). Akademiia Nauk Turkmenskoi SSR, Ashkhabad. 208 p.

1957   La phyto-amélioration des déserts de sable en URSS. Annales de Géographie 66(357): 397–410.

1967   Pustyni Tsentral'noi Asii. Nauka, Leningrad. 286 p. Issued in translation as Joint Publications Research Service (Washington) 42,772, "The Deserts of Central Asia," Vol 2. (Vol 1 available as JPRS 39,145)

For information on the afforestation of moving sands in Central Asia, see p. 255–272 (p. 367–393 in translated version).

PODPRYADOV, N. A.
1958   Borba s peschnanymi zanosami na zhelezhykh dorogakh (The protection of railways from sand drifts). Transzheldorizdat, Moscow. 136 p.

YAKUBOV, T. F.
1951   Opyt obleseniya i zakrepleniya peskov Severnogo Prikaspiya (Experience in afforestation and fixation of sands in North Caspian region). Akademiia Nauk SSR, Moscow. 100 p.

1955   Peschnaye pustyni i polupustyni Severnogo Prikaspiya (Sandy deserts and semideserts of the North Caspian area). Akademiia Nauk SSR, Moscow-Leningrad. 532 p.

ZAKHAROV, A. G., AND B. REVUT
1964   Zakreplenie podvizhnykh peskov s pomoshchyu bitumnoi emulsii (Sand fixation by bitumen emulsion). pp. 449–464. Sbornik: Pustyni SSSR i ikh osvoenie. Akademiia Nauk SSR, Moscow-Leningrad.

ZNAMENSKYI, A. I.
1968   Experimentalnye issledovania protsessov vetrovoi erosii i voprosy zashchity ot peschanykh zanosov (Experimental investigation of wind erosion processes and questions of protection from sand invasion). Sbornik: Materialy issledovanya v pomoshch proectirovanyu i stroitelstvu Karakumskogo kanala. Akademiia Nauk Turkmenskoi SSR, 3:13–31.

Central Asia

GAO SHAN-U
1964   Sand fixation and afforestation in Juilin (translated title). Scientia Silvae 9(2):114–133.

LI MIN-GHAN AND OTHERS
1960   About the protection of railway Baotou-Lanjou from sand drift (translated title). Line-zsi-kan 3:1–47. Peking.

Li Min-ghan, Van Kan-fou, Lu Shou, and
Chou Tsi-uan
  1965 Sand dune stabilization by vegetation in Shapotou region (translated title). Chzhisha Jantsu 7:109–119.

Petrov, M. P.
  1967 Pustyni Tsentral'noi Asii. Nauka, Leningrad. 286 p. Issued in translation as Joint Publications Research Service (Washington) 42,772, "The deserts of Central Asia," Vol 2. (Vol 1 available as JPRS 39,145)

For information on the afforestation of moving sands in Central Asia, see p. 255–272 (p. 367–393 in translated version).

India and Pakistan

Anonymous
  1956 Soil conservation in the Rajasthan desert. Work of the Desert Afforestation Research Station, Jodhpur. Delhi. 22 p.

Anonymous
  1959 Sand dunes in West Pakistan. Unasylva 13(2): 102–103.

Bhimaya, C. P., and others
  1964 Species suitable for afforestation of different arid habitats of Rajasthan. Annals of Arid Zone 2(2):162–168.

Das, E. S.
  1958 Problem of wind erosion in the Punjab and experience gained so far to meet it. Indian Forester 84(9):566–567.

Kaul, R. N.
  1965 An approach to provenance trial in relation to tree introduction in arid lands. Annals of Arid Zone 4(2):164–171.

Khalil, M. A. K.
  1960 The role of vegetation in checking erosion in arid and semi-arid tracts. Symposium on Soil Erosion and Its Control in Arid and Semi-Arid Zones, Karachi, 1957, Proceedings.

Muhammad Khan
  1960 Some afforestation problems and research needs in relation to erosion control in arid and semi-arid parts of West Pakistan. Symposium on Soil Erosion and Its Control in Arid and Semi-arid Zones, Karachi, 1957, Proceedings 223–232.

Petrov, M. P.
  1963 Obleseniye i zakrepleniye pustyni Thar v Indii (Afforestation and fixation of sands in Thar Desert in India). Repetekskoi Peschanoi Stantsii, Ashkhabad, Opit raboti.

Prakash, M., and M. Chowdhary
  1957 Reclamation of sand dunes in Rajasthan. Indian Forester 83(8):492–496.

Sharif, M.
  1960 Mastung sand dunes — their formation and control. Symposium on Soil Erosion and Its Control in Arid and Semi-arid Zones, Karachi, 1957, Proceedings 285–307.

Middle East

Goor, A. J.
  1947 Sand dune fixation in Palestine. Israel, Department of Forests, Annual report (Append. 1). 8 p.

Karschon, R.
  1953 Forestry in Israel. Unasylva 7(4):165–167.

Karschon, R.
  1955 Techniques de reboisement in Israël. Journal Forestier Suisse 106(4):215–221.

Mayerson, S.
  1961 Fixation and afforestation of coastal sand dunes (translated title). La-Yaaran 11(1):17–19. In Hebrew, English summary p. 41.

Tsuriell, D. E.
  1961 Fixation of shifting sand dunes. La-Yaaran 11(2):12–14.
  1966 Sand dune stabilisation in Israel. Israel, Ministry of Agriculture, Tel-Aviv. 19 p.

North Africa

Ben Aissa, J.
  1968 Fixation et reboisement des dunes littorales en Tunisie (Coastal sand dune fixation and reforestation in Tunisia). FAO World Symposium on Man-Made Forests and their Industrial Importance, Canberra, 1967, Special Invited Papers 2:1087–1097.

Boudy, P.
  1948– Les dunes de Mogador et d'Agadir. Economie
  1951 forestière nordafricaine. Larose, Paris (1, 2); Moncho, Rabat (3). 3 vols.

  See especially vol 1, p. 273 ff; 2, p. 471 ff; 3, p. 211 ff.

Maheut, J., and Y. Dommergues
  1961 La fixation par le reboisement des dunes de la presqu'île du Cap Vert. Bois et Forêts des Tropiques. 24 p.

Messaudi, M. B.
  1961 Notes sommaires sur la conservation des sols en Libye (en particulier sur la fixation des dunes et l'érosion éolienne). Colloque sur la conservation et la restauration des sols, Teheran, 1960. Imprimerie Dellée, Coutances.

Messines, J.
  1952 Sand-dune fixation and afforestation in Libya. Unasylva 6(2):50–58.

Motte, M.
  1963 Fixation et reboisement des dunes maritimes en Tunisie et plus spécialement dans le région de Bizerte. Revue Forestière Française 1963 (5):449–466.

Sesmaisons, D.
  1963 La fixation des sables dans le sud Tunisien. Revue Forestière Française 1963(5):431–448.

Souléres, G.
  1963 Protection contre le sable par palplanches. Revue Forestière Française 1963 (5):419–430.

Sulzlee, C.
  1963 Les dunes d'Essaouira [Mogador]. Revue Forestière Française 1963(5):401–418.

South Africa

Kett, J. D. M.
  1927 Afforestation and conservation in South West Africa. Windhoek. 66 p.

Laver, C. G.
  1936 Reclamation of drift sands. Farming in South Africa 11(119):53–57.

Loubser, J. H.
  1959 Fight against shifting dunes in the Kalahari sandveld. Farming in South Africa 34(2): 22–23.

Westhuyzen, J. J. N. van der
  1957 Combating sand dunes at Port Edward. Farming in South Africa 33(7):37–39.

Australia

CONDON, R. W., AND D. A. BARR
  1968 Guidelines for real estate development near coastal dunes and beach areas. New South Wales Soil Conservation Service, Journal 24(4):237–245.

HORE, H. L.
  1940 Sand drift and control measures. Victoria Department of Agriculture, Journal 38(5).

SLESS, J. B.
  1958 Planting guide for coastal sand drift stabilisation. New South Wales Soil Conservation Service, Journal 14(3):184–190.
  1957– Coastal sand drift, pts 1-2. New South Wales
  1958 Soil Conservation Service, Journal 13(3):146–158; 14(1).

North America

MCGINNIES, W. G., B. J. GOLDMAN, AND P. PAYLORE, (EDS)
  1968 Deserts of the world, an appraisal of research into their physical and biological environments. University of Arizona Press, Tucson. 788 p.

WHITFIELD, C. J., AND R. L. BROWN
  1948 Grasses that fix sand dunes. U.S. Department of Agriculture, Yearbook of Agriculture, p. 70–74.

WHITFIELD, C. J., AND J. A. PERRIN
  1939 Sand dune reclamation in the Southern Great Plains. U.S. Department of Agriculture, Farmers' Bulletin 1825. 13 p.

WHITFIELD, C. J., AND F. NEWPORT
  1938 The reclamation of a sand dune area. Soil Conservation 3(7):191–193.

South America

FERRANDO, J.C.
  1965 Praderización de médanos y dunas con el empleo de productos derivados del petróleo. IDIA, Suplemento 19:45–47.

MCGINNIES, W. G., B. J. GOLDMAN, AND P. PAYLORE (EDS)
  1968 Deserts of the world, an appraisal of research into their physical and biological environments. University of Arizona Press, Tucson. 788 p.

PREGO, A. J., AND J. E. CALCAGNO
  1960 Técnica rápida de fijación de médanos mediante siembras protegidas. IDIA, Suplemento 1:208–210.

PREGO, A. J., F. RIAL ALBERTI, AND F. J. PROHASKA
  1964 Forestación de médanos en la región pampeana semiárida (Primera contribución). IDIA, Suplemento Forestal 12:73–82.

PREGO, A. J., AND OTHERS
  1966 Estabilización de médanos mediante forestación en la región pampeana semiárida. IDIA, Suplemento Forestal 2:75–92; INTA, Instituto de Suelos y Agrotécnia, Publicación 100.

Stabilization of Sand Dunes in the Semiarid Argentine Pampas

ANTONIO J. PREGO

ROBERTO A. RUGGIERO

FLORENTINO RIAL ALBERTI

Centro Nacional de Investigaciones Agropecuarias
Instituto Nacional de Tecnología Agropecuaria (INTA)
Castelar, Buenos Aires, Argentina

and

FEDERICO J. PROHASKA*

Department of Geography
University of Wisconsin — Milwaukee
Milwaukee, Wisconsin, U.S.A.

ABSTRACT

THE MOBILE SAND DUNES of the Argentine Pampas are a spectacular product of wind erosion. They destroy soils, disrupt transportation, damage crops, and are costly in other ways. Organized research on stabilizing the dunes of this economically valuable semiarid area began in 1947. In 1950 the first experimental area was set aside for dune stabilization by means of pasture establishment, and in 1959 experiments started with direct afforestation of active dunes.

In the fall seeds of the following plants are used successfully to stabilize dunes by pasture establishment: *Secale cereale* (which can stabilize the dune alone), *Agropyrum elongatum*, and *Bromus brevis*. In the spring the task is accomplished with *Zea mays, Panicum mileaceum, Sorghum caffrorum, S. sudanense, S. saccharatum, S. almun*, and *Eragrostis curvula*. The mixture of *Panicum mileaceum* (a sorghum) and *E. curvula* is sufficient most of the time to guarantee stabilization. Varieties and cultivars of the following trees have been used for afforestation: *Ulmus pumila, Eucalyptus viminalis, Populus deltoides, P. nigra, P. canadensis,* and *Salix babylonica* X *Salix alba*.

U. pumila and *E. viminalis* should be planted on dunes previously stabilized with pasture in order to obtain good stands. *E. viminalis* gives a fine response in the eastern half of the region. *P. deltoides* does not thrive under conditions of the semiarid zone. *P. nigra* gives a mediocre response. *P. canadensis* cultivars 154 and 214 have shown good adaptation to the environment, particularly cultivar 214. *P. canadensis* cultivar 214 has an overall establishment rate of 75–90 percent.

In the eastern portion of the semiarid zone, plantings with the *Salix* hybrids gave excellent rooting and growth in areas with annual rainfall between 600 and 700 millimeters. Results were similar to those obtained with poplars regarding rooting, and better for growth. In the western sector (between 500 and 600 millimeters of rainfall), where the poplars continue to show good responses, the hybrid willows are not thriving. Pasture establishment on dunes is more complicated than afforestation, but in 3 to 6 months it stabilizes the dune. Afforestation is simple, but it takes 2 to 3 years to halt the shifting of the sand completely.

* Deceased, December 22, 1970.

STABILIZATION OF SAND DUNES IN THE SEMIARID
ARGENTINE PAMPAS

Antonio J. Prego, Roberto A. Ruggiero, Florentino Rial Alberti,
and Federico J. Prohaska

ACCORDING TO moisture availability for crop production, the Argentine Republic, which occupies the southeasternmost portion of the South American continent, contains three large zones: humid, semiarid, and arid. Approximately 75 percent of the area of Argentina, comprising the central and western portions of the country, suffers from problems of aridity, which intensify from east to west (1).

To a greater or lesser degree, all the arid and semiarid environments undergo wind erosion, but it is in the western sector of the Argentine Pampas, known as the semiarid Pampas, where erosion created by wind shows its most typical manifestations: formation of sand dunes. It is well to make clear here that in Argentina the sand accumulations along the sea coast are called *dunas,* and the same type of sand formations located within the continental area and produced by destruction of soil, are termed *médanos.*

The semiarid Pampas extend over an area of approximately 22,000,000 hectares, and nearly the whole region undergoes an active process of soil erosion caused by wind. The soils belong mainly (and in order of importance) to the chestnut, brown, and regosolic great soil groups (2).

The above-mentioned region can be delineated precisely because the erosion caused by wind results in the formation of sand dunes — which does not happen in the semiarid Chaco (dry Chaco), which occupies the north-central section of the country — owing to the larger grain size in the Pampas.

IMPORTANCE OF THE PROBLEM

The semiarid Pampas, despite the enduring condition of semiaridity, have great socioeconomic importance, part of which is reflected in the following data: about 5 million hectares sown with grain, more than 5 million hectares in cultivated pasture, and more than 10 million head of livestock, counting cattle and sheep (5 sheep equal one cow) as the two major species raised. The school, health, highway, and rail systems are excellent, and the inhabitants willingly accept the technology which agricultural extension recommends, overcoming critical

NOTE: This chapter is an adaptation of a presentation at "Arid Lands in a Changing World," an international conference sponsored by the Committee on Arid Lands, American Association for the Advancement of Science, June 3–13, 1969, at the University of Arizona, Tucson, Arizona, U.S.A.

periods of edaphic destruction that extended from occupancy at the beginning of the century to the last severe drought in 1962. The following figures from the general evaluation made from 1945 through 1948 by the Instituto de Suelos y Agrotecnia (3) indicate the magnitude of the destructive process during the first half of the twentieth century:

Predominantly slight erosion ...... 4,500,000 hectares

Predominantly moderate erosion 7,000,000 hectares

Predominantly severe erosion .... 4,000,000 hectares

Predominantly extreme erosion .. 500,000 hectares

The sand dunes, a typical characteristic of this phenomenon in the semiarid Pampas, cause the following destruction evidenced in the area: (*a*) destruction of land located in leeward areas, (*b*) burying of fences and watering places for animals, (*c*) disruption of roads, railroad tracks, and telephone and telegraph lines, (*d*) invasion of homes and rural communities, and (*e*) occasional disturbances of manufacturing functions and life in rural communities owing to suspended particles.

There is no doubt that the continuing destruction of new and fertile land is the most serious threat that the dunes pose when they move onto neighboring areas, starting irreversible processes since the blowing out of the fine integral elements of the edaphic complex causes a degradation that is permanent because it is impossible to return the soils to their original condition.

The dunes are permanent foci for new destruction. The only method of avoiding this is to achieve their fixation or stabilization (interchangeable terms) as a means of halting the movement of grains and thus neutralizing effects on adjacent lands. This aspect determines the high priority of specific research directed toward finding techniques of stabilization in order to protect the basic natural heritage of humanity: land.

PREVIOUS WORK

The problems of inland (médanos) and coastal (dunas) dunes have held the attention of students and agricultural technologists in Argentina since the preceding century, as can be verified from references 4 to 9.

The province of Buenos Aires (10) devoted much attention to the problem and succeeded in developing a special method of pasture establishment and afforestation of the coastal sand formations, a method that has been used for some time.

In 1947 organized research on inland dunes began, and in 1948 Tallarico (*11*) published the first results. In the vicinity of Catriló (*12*) in 1950, the first experimental area in Argentina devoted to investigations of pasture establishment was started.

Experiments began in the Pampas in 1959. These efforts were aimed at substantiating the possibility of direct afforestation of active sand dunes in the area of Trenque Lauquen, taking cognizance of the method of deep planting described by Ragonese (*13, 14*). Since then the field tests have continued without interruption in order to establish said possibility.

In 1960 the first contribution (*15*) appeared; it was followed by a series of partial reports that reflected the continuity of effort and slow progress in the study of sand-dune dynamics, in the establishment on dunes of pastures protected initially by stubble or asphaltic mulch, and in afforestation of dunes.

This chapter attempts to describe synthetically the environmental characteristics, the research accomplished, and the results achieved in 20 years of continuous investigation in the area (*16*).

ENVIRONMENTAL CHARACTERISTICS OF THE AREA OF INVESTIGATION

The area of investigation, mapped in Figure 1, comprises the central portion of the semiarid Pampas zone and extends from 35° to 37° S and from 62° to 65° W. A brief description of characteristics of climate, soil, vegetation, and land use will aid the reader in understanding the results and their possible applicability to analagous regions.

Climate

Tests for sand-dune stabilization were conducted in a wide area in the provinces of Buenos Aires and La Pampa. Table 1 indicates the locations of the investigations, their geographic coordinates, the annual averages of rainfall for two periods (1913/37 and 1921/50), and annual totals in 1965, a decisive year for afforestation with poplars.

RAINFALL REGIME

In the semiarid Pampas, the annual rainfall average decreases in a northeast-southwest direction, from approximately 750 millimeters (at Paso and América) to about 450 (General Acha). In this area, provided that dry-farming methods are utilized, the average rainfall amount guarantees normal development of crops. Thus, in years below this average, problems of drought arise in one form or another.

The great variation in summer rains makes for noticeable differences in yearly totals. The deviations in both directions may assume values larger by far than are seen in Table 1, with a tendency to maintain them for several years. For example, the rainfall occurring in 1965 was much below normal in those locations where afforestation experiments were made on dunes that year (Trenque Lauquen, Roosevelt, Vértiz, and Caleufú). In this regard, Weber (*17*) and Prohaska (*18, 19*) conducted investigations specifically related to characteristics of precipitation as a factor in drought and erosion.

An analysis of the rainfall regime of this area indicates that nearly 75 percent of the yearly rain falls in the summer season, from October to March, while the remaining 25 percent occurs in winter. This division into two periods is a peculiarity of the continental regime of the subtropical zone in South America (*20*).

TABLE 1

Sites of Dune-stabilization Experiments and Annual Rainfall

| Site | Geographic Coordinates | | Rainfall Averages | | Year |
			1913/37 (mm)	1921/50 (mm)	1965 (mm)
Paso	35° 51′ S	62° 17′ W	745	720	587
Casbas	36° 46′ S	62° 30′ W	737	719	468
San Fermín	36° 48′ S	62° 37′ W	735	. .	408
Trenque Lauquen	35° 58′ S	62° 44′ W	770	689	519
América (F. Olavarría)	35° 28′ S	62° 58′ W	788	726	725
Pellegrini	36° 17′ S	63° 10′ W	676	. .	627
Maza	36° 49′ S	63° 20′ W	739	674	500
Agustoni (Roosevelt)	35° 46′ S	63° 23′ W	. .	631	460
Catriló (Lonquimay)	36° 25′ S	63° 26′ W	659	636	489
Intendente Alvear	35° 14′ S	63° 33′ W	. .	664	651
Quemú-Quemú	36° 04′ S	63° 34′ W	. .	614	532
Macachín	37° 08′ S	63° 41′ W	627	601	623
General Pico	35° 39′ S	63° 44′ W	694	625	456
Speluzzi (Vértiz)	35° 30′ S	63° 47′ W	687	644	431
General Acha	37° 22′ S	64° 35′ W	476	456	417
Caleufú (La Maruja)	35° 35′ S	64° 33′ W	. .	576	381
Victorica (C. Quemado)	36° 14′ S	65° 26′ W	559	512	318

NOTE: The sites in parentheses lack individual data; the general conditions are represented by the accompanying station.

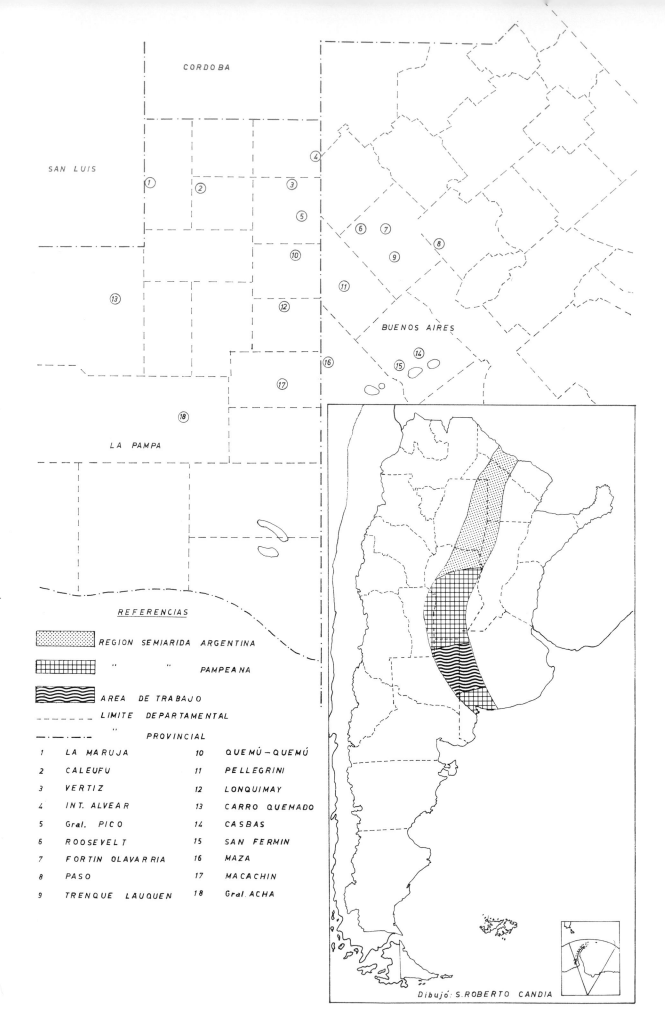

CORDOBA

SAN LUIS

BUENOS AIRES

LA PAMPA

<u>REFERENCIAS</u>

REGION SEMIARIDA ARGENTINA

" " PAMPEANA

AREA DE TRABAJO

LIMITE DEPARTAMENTAL

" PROVINCIAL

1	LA MARUJA	10	QUEMÚ-QUEMÚ
2	CALEUFU	11	PELLEGRINI
3	VERTIZ	12	LONQUIMAY
4	INT. ALVEAR	13	CARRO QUEMADO
5	Gral. PICO	14	CASBAS
6	ROOSEVELT	15	SAN FERMIN
7	FORTIN OLAVARRIA	16	MAZA
8	PASO	17	MACACHIN
9	TRENQUE LAUQUEN	18	Gral. ACHA

Dibujó: S. ROBERTO CANDIA

Fig. 1. Area of investigation: location of dune-stabilization experiments by afforestation. Experiments were made in the region keyed "area de trabajo."

TEMPERATURE REGIME

The temperature regime corresponds to the situation of the area in the subtropics combined with a strong oceanic influence, owing to the small width of the continent in these latitudes. In the Pampas region the monthly average temperatures are 5° C to 10° C in winter and 20° C to 25° C in summer, but every year temperatures between minus 10° C and plus 40° C may occur.

WIND REGIME

In the Argentine Pampas the atmospheric circulation depends primarily on the position and intensity of the Atlantic high-pressure system. Weak north and northeast winds prevail throughout the year with speeds between 2 and 3 meters per second. Only in late winter and spring (August through November) does the mean wind speed increase to 3 to 4 meters per second. This is the result of the increase of the pressure gradient inland, due to the differential heating of continent and ocean in springtime. The diurnal variation of wind speed shows a well-expressed maximum between the late morning hours and afternoon and mostly a lull during the night.

The blowing of sand from the dunes depends not so much on the mean wind velocity as on the frequency with which the wind speed rises above certain threshold velocities. They depend upon the granulometric constitution of the dune and are normally above 3 to 5 meters per second (*21*).

WATER BALANCE

Strong insolation, high saturation deficit, and regular winds are the reason for the high evaporative power of the air during the entire year (*22*). Using monthly temperature and precipitation measurements, the water balance was computed by the Thornthwaite and Mather method (*23*) for several Argentine locations where stabilization of sand dunes by afforestation was tried; see Table 2.

It was not possible to compute the balance for all sites, owing to lack of meteorological data, but it is thought that the values obtained for General Acha, General Pico, Victorica, Trenque Lauquen, Macachín and Quemú-Quemú, by their geographic locations, may be similar to the remaining sites (see Fig. 1). Thus, for example, the data obtained for Trenque Lauquen may be applied without serious error to Paso and Fortín Olavarría; those for Quemú-Quemú to Roosevelt and Pellegrini; and those for General Pico to Vértiz and Intendente Alvear.

The negative yearly water balance in the whole semiarid Pampas, particularly in the westernmost part (Victorica, General Acha), is a consequence of intense evaporation, owing to the temperature and rainfall regimes (Table 2). This is clearly seen when the annual total of potential evapotranspiration is compared with actual evapotranspiration (*24*). The resultant moisture deficit indicates that in this area between 100 and 300 millimeters of water are lacking annually.

In general, in the semiarid Pampas, temperature and humidity undergo great daily and yearly fluctuations; the scarcity of precipitation during the winter makes its effects felt at the beginning of spring — the wind velocity then reaches its maximum. The water balance is consistently negative. The exposure of the crests of the dunes to the prevailing winds, added to the aforementioned climatic conditions, explains why this area is affected so greatly by wind erosion — especially in spring, when the most intensive erosion phenomena are observed.

Soils

The soils in which the major part of the experiments were located belong to three great soil groups: chestnut, brown, and regosols. These are more or less susceptible to wind erosion. Naturally the investigation sites are, properly speaking, typical regosolic accumulations, azonal formations scattered throughout the whole area, concentrating in regions of greatest susceptibility. In general, prevalent textures are sandy-loam and loamy-sand; the sand content increases with depth (*25*).

Vegetation

In order to describe the natural vegetation of the area, we will use the nomenclature applied by two distinguished botanists who have studied it. Ragonese (*26*) calls it "Pampas grasslands" and Cabrera (*27*) "western Pampas district of the Pampas province" consisting of a generally grassy steppe and edaphic communities, among which the "psammophytic steppes," growing on the regosols, are notable. To the west a considerable part of the region forms a province called woods and dunes of San Luis by Ragonese; Cabrera calls this western zone "the Caldén district (*Prosopis calden*) of the spine province."

Land Use

The natural condition of the land permits its use for livestock and agricultural production, with the former predominating. It is an excellent area for fattening cattle, and on it are produced Argentine meats known throughout the whole world. Nevertheless, lack of technical agricultural skills mainly, plus the small plots of land (minifundia), created a dwindling cereal exploitation which degraded the soils and produced the erosion processes that build sand dunes.

The institution of a livestock-agriculture rotation program consisting of (*a*) long cycles of perennial mixed grasslands and short periods of grain production, (*b*) subsurface tillage and plantings under stubble mulch, and (*c*) the practice of seasonal fallowing in all instances constitutes a protection against wind erosion and assures beneficial returns (*28*).

The Sand Dunes

The sand accumulations called dunes, *médanos*, or sand banks are formed as a consequence of the disintegration of soil aggregates and their subsequent movement; the finest grains, smaller than 0.1 millimeter, are carried by the wind and moved in suspension, getting scattered outside the area. Those larger than 0.1 millimeter move by saltation and creeping and may form dunes. Table 3 presents laboratory analysis of samples

TABLE 2

Water Balance (in millimeters) of Some Argentine Stations Where Stabilization of Active Dunes Was Performed

GENERAL ACHA

	J	F	M	A	M	J	J	A	S	O	N	D	Year
Precipitation 1913/1937	46	62	60	33	26	10	12	18	27	71	54	57	476
Potential Evapotranspiration	144	108	88	51	28	14	14	22	38	64	95	130	796
Actual Evapotranspiration	48	63	60	33	26	10	12	18	27	64	56	59	476
Deficit	96	45	28	18	2	4	2	4	11	0	39	71	320
Excess	0	0	0	0	0	0	0	0	0	0	0	0	0

MACACHÍN

	J	F	M	A	M	J	J	A	S	O	N	D	Year
Precipitation 1913/1937	64	76	81	44	31	15	16	29	39	83	69	80	627
Potential Evapotranspiration	140	108	87	52	29	16	16	24	38	65	95	128	798
Actual Evapotranspiration	73	79	82	45	29	15	16	24	38	65	73	88	627
Deficit	67	29	5	7	0	1	0	0	0	0	22	40	171
Excess	0	0	0	0	0	0	0	0	0	0	0	0	0

TRENQUE LAUQUEN

	J	F	M	A	M	J	J	A	S	O	N	D	Year
Precipitation 1913/1937	72	94	92	71	36	22	33	31	52	80	87	100	770
Potential Evapotranspiration	141	109	89	54	32	16	17	24	40	65	93	129	809
Actual Evapotranspiration	117	102	89	54	32	16	17	24	40	65	92	122	770
Deficit	24	7	0	0	0	0	0	0	0	0	1	7	39
Excess	0	0	0	0	0	0	0	0	0	0	0	0	0

VICTORICA

	J	F	M	A	M	J	J	A	S	O	N	D	Year
Precipitation 1913/1937	72	72	72	30	25	17	10	18	28	76	64	75	559
Potential Evapotranspiration	145	110	90	53	28	15	15	24	40	67	98	130	815
Actual Evapotranspiration	75	73	72	31	25	15	10	18	29	67	66	78	559
Deficit	70	37	18	22	3	0	5	6	11	0	32	52	256
Excess	0	0	0	0	0	0	0	0	0	0	0	0	0

GENERAL PICO

	J	F	M	A	M	J	J	A	S	O	N	D	Year
Precipitation 1951/1960	82	79	91	59	22	32	29	13	44	76	84	71	682
Potential Evapotranspiration	145	117	89	48	29	15	15	23	39	65	96	118	799
Actual Evapotranspiration	103	89	89	48	24	15	15	17	39	65	89	89	682
Deficit	42	28	0	0	5	0	0	6	0	0	7	29	117
Excess	0	0	0	0	0	0	0	0	0	0	0	0	0

QUEMÚ-QUEMÚ

	J	F	M	A	M	J	J	A	S	O	N	D	Year
Precipitation 1951/1960	78	78	102	52	20	34	27	13	41	86	60	55	646
Potential Evapotranspiration	145	110	89	44	29	15	13	23	39	65	96	115	783
Actual Evapotranspiration	99	86	89	44	23	15	13	17	39	65	77	79	646
Deficit	46	24	0	0	6	0	0	6	0	0	19	36	137
Excess	0	0	0	0	0	0	0	0	0	0	0	0	0

TABLE 3

Topography and General Soil Characteristics of Dune-Stabilization Experiments by Afforestation

Sample No.	Owner and Location	Topographic location	Depth in cm	Plant growth	Physical analysis (% of dry soil)			Physico-chemical Analysis		Chemical analysis (mg per 100 g dry soil)				Sands (measured % of particles)				
					Moisture	Calcareous in CO_3 Ca	Organic material	pH Chemical reaction	Salinity and reaction to phenophthalein	Nitrogen assimilated by ratio C:N 10:1	Phosphorus assimilated	Potassium assimilated	Calcium assimilated	Coarse 0.5–1 mm	Medium 0.25–0.5 mm	Fine 0.1–0.25 mm	Very fine 0.05–0.1 mm	Total % of particles
1	Trenque Lauquen (Rodríguez Bros.)	1. depression	0–40	poor	0.5	0.0	0.03	7.2	trace negative	0.002	0.4	24	120	0.6	5.0	55.0	32.0	92.6
2	"	"	40–80	"	0.5	0.0	0.03	7.3	"	0.002	0.4	24	160	0.5	5.0	52.0	35.0	92.5
3	"	"	80–120	"	0.6	0.0	0.03	7.9	"	0.002	0.4	24	120	0.6	9.0	57.0	29.0	95.6
4	"	2. crest	0–40	good	0.6	0.0	0.07	7.2	"	0.004	0.4	16	120	0.2	9.0	77.0	13.0	99.2
5	"	"	40–80	"	0.6	0.0	0.12	6.8	"	0.007	0.4	24	120	1.0	10.0	74.0	12.0	97.0
6	"	"	80–120	"	0.7	0.0	0.15	6.6	"	0.009	0.4	24	120	0.6	11.0	71.0	15.0	97.6
7	"	depression	0–40	—	0.7	0.7	0.07	7.7	"	0.004	0.8	16	120	1.0	10.5	53.0	31.0	95.5
8	"	3. slope	0–40	—	0.8	—	0.08	7.1	"	0.005	—	—	—	0.2	6.0	69.0	18.0	93.2
9	"	crest	0–40	—	0.5	0.0	0.07	7.2	"	0.004	0.4	16	120	0.0	3.0	84.5	12.0	99.5
10	Roosevelt (Ferrero Bros.)	depression	0–40	—	0.6	—	0.04	7.3	—	0.003	—	—	—	6.0	4.0	44.0	40.0	94.0
11	"	slope	0–40	—	0.5	—	0.07	7.1	—	0.004	—	—	—	2.0	5.0	63.0	28.0	99.0
12	"	crest	0–40	—	0.4	—	0.09	7.3	—	0.005	—	—	—	0.0	4.0	79.0	16.0	99.0
13	Int. Alvear (A. Urbina)	depression	0–40	poor	0.7	0.0	0.04	7.9	trace medium	0.003	1.6	16	160	0.5	2.0	30.0	40.0	72.5
14	"	"	40–80	"	1.4	1.7	0.07	8.7	trace strong	0.004	1.6	16	>800	2.0	4.0	33.0	37.0	76.0
15	"	"	80–95	"	1.8	6.8	0.09	8.7	trace negative	0.005	3.2	24	>800	6.0	10.0	35.0	33.0	84.0
16	"	slope	0–10	excellent	0.5	0.0	0.04	8.6	"	0.002	0.4	24	160	2.0	11.0	48.0	35.0	96.0
17	"	4. hollow	40–80	"	0.6	0.0	0.04	8.3	"	0.002	0.4	24	160	1.0	4.0	53.0	39.0	97.0
18	"	"	80–120	"	0.6	0.0	0.04	7.8	"	0.002	0.4	24	160	5.0	11.0	48.0	33.0	97.0
19	"	crest	0–40	—	—	0.0	0.03	7.2	"	0.002	0.4	32	160	0.0	6.0	73.0	20.0	99.0
20	"	slope	0–40	—	0.5	0.0	0.08	7.9	"	0.005	0.4	24	160	2.0	7.5	57.0	30.0	96.5

obtained from dunes near Trenque Lauquen, Intendente Alvear, and Roosevelt.

The size of the dunes is quite variable: from 50 centimeters to 15-20 meters in height and from less than one hectare to hundreds of hectares. Within these limits, fully active dune areas exist in the semiarid Pampas. Nevertheless, the active eroded areas generally cover between 2 and 10 hectares (29).

The form is quite variable as well. The topography of the dune areas is typically heterogeneous; transversal ridges of several meters in height alternate with depressions of varying depths (ranging from 2 to 5 meters) and with intermediate, nearly level stretches, called "playas." The orientation is a function of the prevailing north winds; if the dunes are crescent-shaped, like barchan or transverse dunes, the longitudinal axis is located from east to west. For this reason, the windward, north-oriented face of the dunes has a gentle slope, and the leeward or slip face is steep. Active dune areas have an ovoid shape, with the main axis located from north to south (29).

Special studies have been carried out on the internal and external dynamics of the dunes (30). The dune mass itself does not advance, but sand from active ridges invades noneroded lands, kills or covers their vegetation, and begins anew on them the erosive process of accumulation and excavation, forming new dunes. In fact, dunes do not shift, but spread, since sand in motion communicates the "disease" and unleashes the process, particularly on leeward areas (toward the south in the Pampas).

Research has been done on internal dynamics, especially for dunes involved in the afforestation process. In order to determine the size of vertical movements of the

Fig. 2. A typical ridge progressing toward the south (leeward), showing the steep face.

sand, the same cuttings planted for stabilization in a 5 by 5 meter checkerboard have been used as fixed points. Each year measurements are made in relation to a mark or ring painted in red on the trunk of the small tree, thus ascertaining how much the sand has risen or fallen.

The numerical data gathered were classified or grouped into three categories for study: (a) variations of less than 5 centimeters, which were considered to represent insignificant activity, (b) variations of more than 5 centimeters, (c) variations greater than 30 centimeters. All values could be positive or negative, depending upon whether sands accumulated or were excavated.

TABLE 4

Percentages of Surface Affected by Positive and Negative Changes in Dune Height

LOCATION	Paso	Trenque Lauquen				Roosevelt			Int. Alvear	Vértiz	La Maruja	
YEAR	63-65	64-65	65-66	64-66	65-68	63-65	64-65	64-66	64-65	64-65	64-65	64-66
Variation of Soil Level												
Accumulation of more than 5 cm	27	21	50	42	38	20	33	42	26	31	53	56
Accumulation of more than 30 cm	10	9	22	25	16	6	19	39	21	11	24	32
Excavation of more than 5 cm	50	63	31	35	34	58	53	39	56	44	23	24
Excavation of more than 30 cm	5	14	3	13	12	17	26	20	26	5	12	16
No difference	22	21	19	22	24	20	10	20	18	24	24	20

NOTE: When there is more than one column for the same year for a site, each column corresponds to a different test within the same sandy area. Percentage figures for variations of more than 5 centimeters include variations of more than 30 centimeters, which are also given separately.

Table 4 presents the percentages of surface in the measurement area that rose or fell according to the aforementioned classification. Mobility of the dunes is reflected in the fact that approximately 80 percent of the surface has experienced a vertical change of more than 5 centimeters per year. The 20 percent of the area without variation is very uniform throughout all the observation sites, save in one instance. The visual observation that accumulations are higher than excavations is confirmed by the data.

In general, the data reveal the impossibility of carrying out normal procedures of pasture-establishment without some system of temporary immobilization of the dune surface. Regarding afforestation, the great size of the shifting movements indicates the grave danger of injury and loss (by unearthing or burying) if common species are planted and exposed to damage by sand in motion.

Water Table

In the area of Trenque Lauquen for several years (1960-1965), a system of phreatometers measured variability in depth of water table, which is generally between 5 and 10 meters deep in the area. In the area of investigation, an average depth of about 6-7 meters was found. Beneath the afforestation experiments, the measurements indicated a depth of approximately 7 meters, with annual fluctuations that, depending upon the rain variability, could reach as much as 50 centimeters per year. The water analyses established that in the main there were bicarbonates, chlorides and sulphates, sodics, calcics, and magnesics, with a pH (hydrogen ion concentration) fluctuating around 8. The annual variation of groundwater temperature was around 18°C, the annual mean air temperature being 16°C (*31*).

MATERIALS AND METHODS

Species Used

The herbaceous species used in pasture establishment and the trees used in afforestation are as follows.

Pasture establishment on dunes (with a mixture of forage for purposes of stabilization) can be accomplished best either in fall or in spring in the Pampas. In fall the following species were tried: rye (*Secale cereale* L.), tall wheatgrass (*Agropyrum elongatum* [Host] Beauv.), and Pampean wild barley (*Bromus brevis* Nees.). For spring sowings, maize (*Zea mays* L.), millet (*Panicum miliaceum* L.), grain-bearing sorghum (*Sorghum caffrorum* [Retz.] P. Beauv.), Sudan grass (*Sorghum sudanense* [Piper] Stapf), sweet sorghum (*Sorghum saccharatum* [L.] Moench), perennial or black sorghum (*Sorghum almun* Parodi), and weeping lovegrass (*Eragrostis curvula* [Schrader] Nees.) were tried. In all cases, commercial seed was used in the field tests.

In field experiments of dune afforestation, in succession or simultaneously, the following forest species were tried:

Ulmus pumila L. var. *arborea* Litv.
Eucalyptus viminalis Labill.

Populus deltoides Marsh. cultivar I:63/51, cultivar 64/51 and cultivar 72/51.
Populus nigra L. var. *italica* (Münch.) Koehne.
Populus canadensis Moench cultivar I-154 ("Arnaldo Mussolini").
Populus canadensis Moench cultivar I-214.
This hybrid is a female cultivar created by a spontaneous crossbreeding in Italy, with early foliation, late dropping, and rapid growth; it is resistant to diseases. It is the most widely cultivated poplar in Argentina. Ninety percent of the dune afforestation studies were done with this hybrid.
Salix babylonica × *Salix alba* cultivar A-131/25 male.
This is a hybrid created by INTA at the Centro Nacional de Investigaciones Agropecuarias in Castelar; it is a tree with good wood, cylindrical trunk, and small, suberect, reddish-chestnut-colored branches.
Salix babylonica × *Salix alba* cultivar A-131/27 female. This is also a hybrid created by INTA in Castelar. It is very robust, with excellent wood, a straight, cylindrical trunk, little ramification, and small, greenish-colored, suberect branches.

For *Populus* and *Salix* species, in all but one instance (cultivar 154, in which rooted plants were also used) cuttings were used exclusively. These cuttings consisted of terminal shoots, called *guías* in Argentina, from 2 to 5 meters long, preferably from 1 to a maximum of 2 years in age. Small 2-year-old plants of eucalyptus were utilized on sand dunes previously stabilized with pasture. Small 2-year-old *Ulmus* plants were used (one from a bed and one from a nursery), or large *Ulmus* cuttings 5 to 6 meters long and 5 to 7 centimeters in diameter.

Experimental Technique

The experimental research for pasture establishment consisted of two basic aspects: preparation of the ground and planting of protected seeds.

The work of lowering and smoothing the crests was done with animal traction and with a caterpillar or track-type tractor, or a 60-HP Fiat, drawing various types of equipment: bulldozers, trenchers, graders, and heavy beams or rails. Wind-intensifiers were tried also (piled-up sacks, transverse planks, and wind channels opened with shovels).

The experimental seedings began in Catriló (La Pampa) in 1950 and consisted of the establishment of strips of different species, with repetitions. This initial experimental area in Argentina, which marked the beginning of research that has continued uninterruptedly until the present, was examined by H. H. Bennett in November 1950 in his first visit to Argentina, and merited his approval (*32*).

Pasture-establishment trials of an experimental and demonstration type have been carried out throughout the semiarid region during the past 20 years; the conclusions reached after these trials are based upon voluminous technical information.

The experiments with direct afforestation on fully active sand dunes began in Trenque Lauquen in 1959 in the central demonstration area of pasture establishment. They have been conducted without interruption for the past 10 years.

For planting the cuttings, three tools for preparing the ground have been tried: (*a*) a small, heavy, solid, sharp-pointed stake 2 meters long, (*b*) a manual (soil) auger, like those used to extract soil samples, 5 centimeters in diameter, and (*c*) a spiral drill driven by a two-horse-power motor.

The experimental procedure comprises two types of experiments: on depth of planting and on distance between plants. In all the tests conducted with *Populus canadensis* this procedure was followed.

In the test of planting depth, the uniform distance between plants was 5 by 5 meters; the cuttings were set in a square, and three depths were tried: 100 centimeters, 150 centimeters, 200 centimeters. There were two types of experiments: large and small. The large one contained plots of 21 plants, set in 3 rows of 7 trees each in a Latin square and distributed in 3 replications. In the small test each plot had a row with 9 plants; the distribution was at random in 3 blocks or repetitions.

In the test on distance, three spacings were tried: 5 x 5 meters, 6 x 6 meters, and 7 x 7 meters. The depth was uniform at 150 centimeters. The plots also had 21 plants in 3 rows and were randomly distributed in 3 blocks or replications.

Three basic observations were made: rate of budding or striking, diameter at a height of 1.30 meters, change in altitude of the sand surface.

Before proceeding with any kind of planting, it is necessary to fence the sandy area so that livestock or other animals cannot intrude and cause damage. In areas with more favorable water balance, simple fencing can bring about the slow but certain establishment of spontaneous vegetation until it stabilizes the whole dune. In general it does not happen in this manner, but isolation almost always insures a growth of hardy, spontaneous vegetation on the periphery of the problem area. The vegetal ring then stops outward mobility of sand and the dune, still active in its interior, remains within its environs and no longer damages neighboring areas.

DISCUSSION

Smoothing the Sand Ridges

If sowing of pasture is to be done mechanically, it is first necessary to level the rugged relief of the sand ridges and mounds to such a degree that the seeding machines may travel over them. This smoothing also facilitates striking of the cultivated vegetation (*11*).

The dunes may have steep slipfaces of several meters. The objective of the undertaking is just to smooth out the ridges — not to level the terrain, which would be extremely costly. The smoothing measures tried on the Argentine sand dunes have been diverse.

Fig. 3. A scoop drawn by two horses and guided by a crew of two men opening a wind channel near steep sandy depression.

Use of wind-intensifiers proved that placement of boards in a vertical position perpendicular to the prevailing winds and alignment of sacks filled with sand on top of the dune crests, although they create a decline in height, are measures that are laborious and costly in upkeep — thus their use cannot be recommended. On the other hand, the opening of wind channels by means of two expertly operated passes by an ox-hauled scoop is feasible. One scoop drawn by two horses and guided by a crew of two men can open up 100 channels in a day, 2 meters apart; in other words, 200 lineal meters of crest line per day can be treated this way. The wind, increased in this form, then tends to scatter the sand; if necessary, the channels can be opened again to allow even greater smoothing.

The 60-horsepower tractor allows working with a scoop in front, or with a medium-sized grader with blades, or drawing a combination of 2 or 3 beams or rails mounted parallel to each other, a meter apart and 3 to 4 meters in length. Rails are very useful on medium-height dunes where the tractor goes ahead dragging the rails that shove great masses of sand, help smooth the ridges, and push down mounds formed around isolated clumps of vegetation. Graders can also be used, but rails have an extraordinary advantage in their extreme primitiveness, since they have no moving parts to deteriorate with the rough and heavy work that occurs with the graders. A good alternative is to combine use of the grader or bulldozer with the rails. Nevertheless, in the great majority of cases, rails alone will smooth the ridges and mounds adequately to expedite sowing and rooting. In addition, if only a small 30- to 40-horsepower tractor is available, the work can still be done, only more slowly, drawing one or perhaps two rails. This tractor is quite useful also for drawing the seeder and discs, and for transporting the straw. In very high dunes, however, the track-type tractor cannot descend the leeward side of

Fig. 4. Smoothing the sand surface with tractor-drawn rails.

the ridge. In this case, the bulldozer must be used, which pushes the sand from the top of the ridge by means of its forward blade, without itself descending. It then backs up and repeats the operation until it has reduced the angle of descent. Then it is safe for the tractor and grader or rails to descend without danger.

Pasture Establishment on Dunes

Once the objective of smoothing the surface has been attained, the sowing ought to be done immediately, because otherwise the wind activity creates ridges again. To prevent this problem, it is advisable to continue smoothing the sowed surface daily.

The most favorable season is fall (March, April) because of the decrease in wind speed and increase in rain and atmospheric humidity. Sowing can be done with a grain seeder, but it is better to use a forage seeder equipped with two chutes — one for large- and medium-sized seed, the other for very fine seed. Sowing of a mixture consisting of 60 kilograms rye (*Secale cereale*), 5 kilograms tall wheatgrass (*Agropyrum elongatum*), and 5 kilograms Pampean wild barley (*Bromus brevis*) is advised. In fact, the excellent response of the rye allows it — even when sown by itself — to stabilize the treated area, even when sown with other seeds. Because seeded areas are not grazed or harvested during the first years, the rye ripens and lodges its seed, so that the following year there is again a spontaneous growth. This repetition lasts several years, but gradually the rye is replaced by spontaneous vegetation that stabilizes the dune permanently.

In spring the mixtures should include a series of crops that differ in time to maturity, aggressiveness, and volume; these gradually cover the dune until they immobilize it completely. It is advisable to use a mixture of approximately 5 kilograms per hectare of each of these species:

Zea mays, Panicum miliaceum, Sorghum caffrorum, S. sudanense and *S. almun;* the mixture should include 1 kilogram of *Eragrostis curvula*. The species develop in the order mentioned; naturally, one can achieve good results with only three of the species (including a perennial forage crop [*S. almun* or *E. curvula*] and applying the same amount of seed per hectare) but the greater number of components gives more assurance of pasture stabilization.

Given the extreme mobility of sand, immediately after seeding the surface should be protected with a thin mulch of small-grain straw in the spring, and of sorghum or maize in the fall. The tests without cover on active dunes always failed because the seedlings were damaged by the action of the sand and were buried or laid bare. Furthermore, the mulch should be continuous. To save materials, mulching in strips was tried, but blowing sand from the uncovered sections piled up on the protected ones, ruining the whole sowing.

Another requirement is the use of a simple disc harrow, which will pack down the mulch distributed on the seeded crop. The straw remains anchored and the wind cannot pull it off, assuring protection of the little plants during the first month until they themselves decrease the surface wind speed to below critical velocity.

One should note that the high amounts of seed sown per hectare are necessary because there are great losses of seed and plants: seed is planted too deep and plants are damaged by wind action.

Smoothing, sowing, covering, and discing require about 10 working days of 8 hours daily per hectare. Complete mechanization of mulching the seeded dune with stubbles could mean reduction in work days of from 30 to 40 percent.

When the technique described is used, the fully active sand dune is stabilized and pasture is established in 3 to 6 months, according to the weather.

EQUIPMENT

It is necessary to point out here that pasture can be established on schedules that vary, depending upon the equipment available.

Travel within the dunes requires a track-type tractor, since those on wheels cannot move about there except on the playas. Notwithstanding, according to the experiments conducted, given more time, all the tasks necessary for pasture establishment can be done by animal traction and manual labor. Equipment needed for this method is: (*a*) a grader with about a 1.50 meter blade; (*b*) one or two rails or I-beams 3 meters long (rails can sometimes substitute for the grader, but their operation together is more efficient); (*c*) an ox-hauled scoop 50 to 60 centimeters wide; (*d*) a grain or forage seeder; (*e*) a mowing machine to cut the straw and stalks or stubbles; (*f*) a small, mechanized rake to gather the straw into windrows; (*g*) pitchforks for piling up and scattering the straw (when straw, stalks, or stubbles are devoid of thorny weeds, distribution on the dune can be done quickly and to advantage by hand); (*h*) a loading sled

Fig. 5. Undertaking pasture establishment with forage seeder and track-type tractor.

Fig. 6. A disc-type harrow used to pack the straw down.

or cart to carry the straw; and (*i*) a simple harrow of 6 to 10 discs to pack the straw down. This equipment can be obtained in any developing agricultural area.

The equipment just described can be replaced in various degrees by mechanized traction and more sophisticated implements, depending upon the circumstances and feasibilities. In industrialized countries with a high degree of agricultural mechanization, all the operations can be performed with machinery, and, as a result, the work time could be reduced to 30 or 40 hours per hectare. This is highly desirable when it can be achieved, because speed is a very favorable factor in conserving effort and assuring success. Provision of mechanized equipment for stabilizing dunes, manned by trained personnel, would be the solution.

In Argentina the situation is intermediate, since all facets have been mechanized except that of loading and distributing straw on the surface of the dune — which requires about 50 percent of the approximately 80 hours per hectare which the process of pasture establishment requires.

INORGANIC MULCH

Even if the mulching of the seeded dunes with a layer of straw does assure pasture establishment, there is not always such material available, especially in critical seasons. For this reason a series of experiments was performed with a binding substance derived from bentonite to make a crust or thin compacted covering over the surface of the recently seeded dune for the purpose of temporary stabilization (30 to 60 days) until the vegetation guarantees permanent fixation. The experiments did not permit verifying with certainty the efficiency of bentonite as a binding material (*33*).

Usage of asphalt to achieve temporary consolidation is very well known. In Argentina state agencies and a private petroleum company have successfully performed experiments in this regard (*34*). At the present time, particularly by utilizing economical equipment, protection by organic mulch is proving somewhat less expensive than the aforementioned asphalt mulch. Nevertheless, asphalt binding materials are an important resource to consider whenever planning dune stabilization.

Afforestation of Sand Dunes

The first attempts at afforestation began in 1959. In the following year, organized experiments described above in **Experimental Technique** began, and they continued without interruption — hundreds of test plots having been set up in the past 10 years. In all instances the tests have been performed in farmers' fields in locations representative of the area. Thus many of the experiments were damaged, particularly by animals (especially cattle) that could enter where fencing was deficient. The mass of numerical data available is very large; therefore a selection was made. For the further discussion, the areas chosen were those best protected against animals and distributed from east to west throughout the whole semiarid Pampas, so that the results give a cross section of reactions of the trees to their natural environment.

The results, summarized in Tables 5 through 9, will be commented upon. Sprouting of the cuttings will be discussed first, then their growth, considering in each case the various species.

First of all, it is essential to investigate the capacity of a species to establish itself (referred to in Argentina as sprouting — *brotación* — or *prendimiento* or *arraigo*. We need to know what percentages of plants remain alive one or two years from planting. Then it is necessary to evaluate growth, for which purpose measurements were taken of the diameter at a height of 1.30 meters.

PLANTING EQUIPMENT

Three different types of equipment were tested for digging holes in order to insert cuttings at various depths.

A heavy, solid bar that dug holes by pressure was tried in two sizes. It proved practicable only for depths less than 1 meter; thus it was not useful for experimental purposes, since planting of cuttings should be deeper.

Also used was a portable motor, operated by two men, at 1½ horsepower (two cycles), powered by a naphtha-

TABLE 5

Monthly Rainfall and Depth to the Water Table at Trenque Lauquen
(Buenos Aires Province) — 1960 to 1963

Month	1960 P*	1960 GW†	1961 P*	1961 GW†	1962 P*	1962 GW†	1963 P*	1963 GW†
January	223.3	—	60.0	692	39.4	735	54.5	789
February	44.0	—	96.4	(700)	19.8	743	18.4	796
March	72.9	660	47.0	(706)	19.8	753	170.4	796
April	4.6	668	25.3	712	66.2	760	118.9	793
May	13.1	666	50.8	(715)	25.6	762	36.2	790
June	41.9	664	10.6	718	5.0	767	27.7	(784)
July	48.6	671	8.8	722	6.2	(769)	85.2	778
August	1.9	670	50.6	(726)	64.5	770	27.2	762
September	53.3	(674)**	28.8	730	23.0	775	59.2	761
October	104.0	678	141.6	(730)	42.7	778	118.0	(757)
November	21.2	685	35.6	730	155.5	782	125.6	742
December	63.8	(689)	110.1	736	48.2	781	154.4	705
TOTAL	692.6		665.6		515.9		995.7	
AVERAGE		(670)		(720)		(765)		(770)

* P: Monthly precipitation in millimeters; data obtained from the National Meteorological Service.
† GW: Depth of the water table in centimeters; measurements taken once per month, by the authors.
** Values in parentheses are interpolated.

oil mixture. The motor operates a meter-long helicoid drill, which can be of different diameters. This arrangement proved useful, since it dug the hole in 20 to 30 seconds. For greater depths the boring was continued with hand drills. Improvement of these mechanized helicoid drills so that they can operate at greater depths will allow planting to speed up noticeably.

In the majority of instances, a soil auger was used for planting. It is operated by hand and consists of 80-centimeter sections that permit lengthening the end so that the equipment will excavate holes up to several meters in depth. Considering the size of the cuttings to be planted, it is suitable to use a soil auger with a 6- to

8-centimeter diameter, like those used to extract soil samples. Given the nature of the sand, the holes open up rapidly (less than 5 minutes for each) so that for a planting distance of 5 x 5 meters and a depth of approximately 1.50 meters, this hand method requires about 4 work days per hectare.

ULMUS PUMILA VARIETY ARBOREA

Two experiments were carried out. Planting of elm cuttings (shoots 5 to 7 centimeters in diameter and 5 to 6 meters in length), done at the suggestion of a nursery, failed at all depths and in all places. Initial striking was lower than 5 percent and surviving specimens were dying.

Fig. 7. Helicoid drill, operated by portable motor powered by a naptha-oil mixture.

Fig. 8. Soil auger shown with cultivar I-214 during winter season.

Conversely, small plants about 90 centimeters high (one year in a bed and one in a nursery), raised in the same area and planted on active dunes only in the vicinity of Trenque Lauquen, had a high percentage of striking. Subsequent movement of the sand brought about death of 50 percent of the plants by burial or unearthing (baring the roots). Hares (*Lepus europaeus*) also inflicted great damage. In view of these consequences, the experiments were not continued. Anyone attempting to plant these elms on active dunes should use taller plants (three years old), organize protection against hares, and, what is most important, look for areas of the dune with minimum mobility (easily ascertainable by experts). It should be added that the surviving elms made good progress.

EUCALYPTUS VIMINALIS

Owing to the type of plants and rhythm of growth, direct afforestation on active dunes was not attempted. The plantings were made on a massive scale by the owner

TABLE 6

Sites of Experiments With Hybrid Poplar *Populus canadensis* Cultivar 214

(Figures in body of table are percentages of cuttings sprouting or establishing themselves.)

Location	Planting Year	Fully Active Sand Dunes									Stabilized Dune		
		Experiments with Depth (meters)			Experiments with Spacing (meters)			Experiments with Depth (small scale) (meters)			Experiments with Depth (small scale) (meters)		
		1	1.5	2	5	6	7	1	1.5	2	1	1.5	2
Paso	1962	88	80	96	..	..	..	..	..	..	..	..	..
	1963	98	97	100	98	98	92	..	..	..	..	..	..
Casbas	1963	..	..	..	..	..	..	100	88	92	..	..	..
San Fermín	1963	..	..	..	..	..	..	92	96	100	..	..	..
Trenque Lauquen	1962	.96	84	92	..	..	..	..	..	..	..	..	..
	1963	87	92	98	100	100	100	..	..	..	60	65	85
	1963	98	95	82	..	..	..	..	..	..	..	..	..
	1964	98	89	84	..	..	..	..	..	..	48	40	48
	1966	64	70	81	..	..	..	..	..	..	30	22	38
	1967	74	96	93	..	..	..	82	97	96	..	..	..
F. Olavarría	1963	..	..	..	..	..	..	..	..	..	74	71	71
Pellegrini	1963	..	..	..	..	..	..	98	100	100	..	..	..
Maza	1963	..	..	..	..	..	..	92	84	84	..	..	..
Roosevelt	1963	95	87	90	92	98	94	..	..	..	82	74	82
	1964	..	..	..	..	..	..	89	96	93	..	..	..
	1966	..	..	..	..	..	..	44	78	93	..	..	..
	1967	..	..	..	..	..	..	30	74	93	..	..	..
Lonquimay	1963	..	..	..	94	98	95	92	96	92	..	..	..
Intendente Alvear	1963	74	87	86	68	68	60	..	..	..	..	..	..
	1965	..	..	..	..	..	..	50	50	70	..	..	..
Quemú-Quemú	1963	90	96	85	82	92	100	..	..	..	..	..	..
Macachín	1963	..	..	..	..	..	..	76	84	80	..	..	..
General Pico	1963	..	..	..	..	..	..	..	..	..	92	92	96
Vértiz	1964	74	66	60	95	87	97	..	..	..	..	..	..
	1966	44	78	93	76	62	65	..	..	..	..	..	..
	1967	30	74	93	..	..	..	..	..	..	..	..	..
Caleufú	1963	..	..	..	..	..	..	68	64	80	..	..	..
	1964	94	95	97	89	89	94	..	..	..	..	..	..
	1966	59	72	74	65	75	84	..	..	..	..	..	..
General Acha	1963	..	..	..	..	..	..	56	64	84	..	..	..
Carro Quemado	1963	..	..	..	..	..	..	84	96	98	..	..	..
La Maruja	1964	..	..	..	..	..	..	78	97	100	..	..	..
	1966	..	..	..	..	..	..	78	59	63	..	..	..
Average		79	84	82	76	77	88	75	83	89	64	61	70

of the farm where the experimental area of Trenque Lauquen is located, on a dune 6 hectares in size stabilized with rye 4 months previously, in two areas 500 meters apart, in two consecutive years, applying only an initial irrigation of some 10 liters per plant. In the first year the planting was completely successful and more than 90 percent of the individual plants took root; in the past nine years they have achieved a remarkable growth, forming 6 hectares of beautiful forest. Nevertheless, one circumstance must be pointed out. Specimens planted on

crests and playas developed in similar fashion to those located in noneroded fields; on the other hand, those planted in the depressions excavated by the wind displayed a much more limited growth. This effect is constantly observed in all the plantings done in the area, with all the species used.

Planting in the following year, which was drier, had the reverse effect: only about 10 percent of the plants took root. These grew with as good results as the planting of the previous year. The eucalyptus is a most interesting species for continuing to explore afforestation of active dunes after pasture establishment in semiarid zones.

POPULUS DELTOIDES

During two consecutive years, three types of hybrid deltoides (I:63/51; I:64/51; I:72/51) consisting of cuttings 4 to 6 meters long were tried in the vicinities of Paso, Trenque Lauquen, Roosevelt, and Intendente Alvear — that is, from the subhumid to the typically semiarid area. Rooting of the cuttings fluctuated between 2 and 10 percent on the average. The best results were obtained with the I:72, but in no instance was 25 percent establishment exceeded. It then was agreed not to continue the experiments with *P. deltoides* (*35*).

POPULUS NIGRA CULTIVAR ITALICA

Two experiments were performed (1961 and 1962) at Trenque Lauquen, using both rootless cuttings 2 meters high and plants with roots. A large portion of the former dried up, and something more than 50 percent of those with roots took hold. Possibly the drier than average spring and the use of cuttings that were too small were involved in the poor performance. Nevertheless, in the same year the hybrids established themselves well. Growth was good, but the number of branches — an important aspect — was inferior to the hybrids. The tests were not continued with this species, as it was realized that *P. canadensis* was far superior.

POPULUS TREMULA X POPULUS ALBA

Experiments were not made with this hybrid. The only specimen planted — a cutting 6 centimeters long —

TABLE 7

Experiments With Hybrid Poplar *Populus canadensis* Cultivar 214, Planted in Fully Active Sand Dunes

(Figures in body of table give increase of diameter of trunk, in millimeters, measured at a height of 1.30 meters, between the year of planting and winter 1968.)

Location	Year of Planting	Depths Tested (meters)			Spacings Tested (meters)		
		1	1.5	2	5	6	7
Paso	1962	73	127	140	..	..	..
	1963	39	35	31	102	56	63
Trenque Lauquen	1962	46	61	62	..	..	..
	1963	..	..	..	46	54	55
	1964	95	103	103	..	..	..
	1966	10	14	13	..	..	..
	1967	21	20	21	..	..	..
Roosevelt	1963	65	64	74	107	113	126
	1964	67	58	57	..	..	..
Pellegrini	1963	87	90	96	..	..	..
Lonquimay	1963	92	95	94	..	..	..
Quemú-Quemú	1963	46	51	57	88	94	88
Vértiz	1964	69	66	61	78	83	102
	1966	..	..	..	22	23	24
	1967	9	12	8	..	..	..

NOTE: The 1965 planting failed in all areas because poplar cuttings that were too small were used in a drought year. In the willow, with medium-sized cuttings, there was no failure.

TABLE 8

Experiments With Hybrid Willows, *Salix* 131/25 and *Salix* 131/27, Planted in Fully Active Dunes

(Figures in body of table are percentages of cuttings sprouting or establishing themselves.)

Location	Year of Planting	Salix 131/25			Salix 131/27		
		Depth Tested (meters)					
		1	1.5	2	1	1.5	2
Trenque Lauquen	1964	52	81	92	..	..	..
	1965	62	90	86	71	90	86
	1966	19	81	96	78	87	100
	1967	81	100	100	41	67	80
	1968	74	89	98	..	..	..
Roosevelt	1965	78	72	100	56	85	78
	1966	40	40	67	93	100	100
	1967	67	85	89	78	96	100
	1968	..	..	..	78	70	91
Vértiz	1965	33	60	52	..	..	..
	1966	7	7	30	22	74	96
	1968	..	..	..	15	30	41
Average		51	71	81	59	78	86

TABLE 9

Experiments With Hybrid Willow *Salix* 131/25, Planted In Fully Active Sand Dunes

(Figures in body of table are growth of trunk diameter, in millimeters, during 1968/1969 biotic cycle.)

Location	Year of Planting	Depths Tested (meters)		
		1	1.5	2
Trenque Lauquen	1964	25	25	27
	1965	6	12	14
	1966	5	16	8
	1967	3	6	4
	1968	1	4	4
Roosevelt	1965	20	22	25
	1967	15	14	15
Average		11	14	14

gave an excellent response; therefore the observations were included to suggest a line of investigation.

Five years after having been planted in the center of the active sand dune at Trenque Lauquen, the individual exhibited the following appearance: (*a*) height: 8-9 meters, (*b*) diameter at height 1.30 meters: 140 millimeters, (*c*) forty "offspring" or shoots within a radius of 10 meters, (*d*) some thirty "offspring," the most vigorous, within a radius of 5 meters, (*e*) the most outstanding shoots measuring 21, 24, 25, 26, 27, 28, and 37 millimeters in diameter at a height of 1.30 meters.

POPULUS CANADENSIS CULTIVAR I-154

With this famous cultivar, quite well known in Argentina as the Arnaldo Mussolini poplar, two tests were performed with a somewhat different experimental technique: cuttings and plants with roots were tried as variants, combined with variable planting depth, in three blocks with random distribution. One experiment was set up in a completely active dune, and the other 20 meters away in a portion of the same dune that had been stabilized with rye four months earlier.

The result with the rooted plants was perfect, since all the plants took root, as well in the active dune as in the part of the dune stabilized with rye. On the other hand, 90 percent of the rootless cuttings took root in the active dune and 80 percent in the area stabilized with rye, and the rate of success increased in proportion to the planting depth (100, 150, and 200 centimeters). From the first moment, our attention was drawn to the swiftness of growth and the strong system of roots, both shallow and deep, which all the I-154 poplars developed, establishing themselves perfectly in full activity (*36*).

The good response obtained with the unrooted cuttings in these two initial trials served as a basis for orienting research in that direction because it is much simpler and more economical to work with cuttings than rooted plants.

To the same degree that it happened in the stabilization experiments with pasture, an obvious difference occurred in growth of trees on the ridges and in the depressions. On the crests growth was exceptional, while in the areas excavated by the wind it was moderate and in the deepest sections it was feeble. This difference should be attributed to a difference in availability of assimilable nutrients, since the surface of the depressions corresponds to the deeper layers of the subsoil.

For growth characteristics we rely upon data from the 1960/1969 period, considering each block as a small independent wood, each of about 100 trees. On the dune stabilized with rye, the three blocks yielded these values: 94, 76, and 73 millimeters (diameter) growth in 9 years, with a general average of 81 millimeters, or a mean of 9 millimeters per year. On the active dune the three blocks gave these results: 159, 113, and 92 millimeters, with a general mean of 121 millimeters, which gives an annual average of 13.5 millimeters.

The first observation reveals the heterogeneity of the terrain with regard to topography being more important than the influence of the treatments — which fact is directly reflected in the growth of the poplar trees. As is logical, the largest trees exhibit greater differences between blocks.

The second one is that on the active dune the plants developed a trunk 50 percent larger than on the dune stabilized with pasture, which in volume indicates more than double growth. Field observations substantiate this result with conclusive figures.

The values given show that *P. canadensis* cultivar I-154 performs well in stabilizing active dunes in the semiarid region of Trenque Lauquen, which is in the transition zone to subhumid conditions. Experimenting with this cultivar was halted in Argentina because it is easier to obtain good material for actual planting from the 214 cultivar.

It should be pointed out here that during years 1960 (in which the experiment was started), 1961, and 1962 there was a drought in the area, and, as a result, the water table continually declined. According to monthly readings of the phreatometers installed within the two experimental areas, between March 1960 and March 1963 the groundwater level dropped from 660 to 796 centimeters (Table 5). Then it began to rise slowly, due to rains occurring in 1963. Nevertheless, in spite of the sharp decline (which reached 1.35 meters), during these three years the poplars exhibited no external sign of lack of water, by which fact it is thought that the plants used only the soil water stored in the subsoil.

POPULUS CANADENSIS CULTIVAR I-214

This hybrid poplar was the basic constituent used in the system of sand-dune afforestation tests, hundreds of trials having been set up in the provinces of Buenos Aires, Córdoba, and La Pampa — that is, in the whole central region of the semiarid Pampas zone. The general outcome is the same in all sectors, indicating adaptation of the tree to the critical environmental conditions. The results at the experimental areas of Trenque Lauquen, Roosevelt, Vértiz, and Caleufú deserve special notice in that the tests could be kept under close supervision, there was great continuity, and those locations exhibit all the conditions representative of the semiarid environment.

The experiments were conducted upon active dunes and upon pasture-established dunes or those occupied by spontaneous vegetation. Irrigation was never applied. Results are presented in Tables 6 and 7.

Afforestation of Stabilized Dunes

The object of this study was to confirm the possibility of afforestation of dunes already stabilized.

As can be gathered from Table 6, the percentages of sprouting were heterogeneous; the general averages fluctuated around 70 percent and were acceptable in some years, such as in 1963. Nevertheless, continued observation permitted verification that as time elapsed, the number of live plants decreased. In Trenque Lauquen, for example, on the dune stabilized with rye in March

1960, the planting of I-214 made in July 1963 yielded an average of 70 percent sprouting at the end of one year. An examination in June 1965 substantiated that the average of living plants had declined to 30 percent.

Comparing the inventory of rooted plants set in dunes four months after pasture establishment with others planted several years after dune stabilization by pasture, the following was confirmed: on dunes recently established with pasture, the poplars (I-214) succeeded in establishing themselves and overpowering the competition of herbaceous vegetation, which gradually became sparse and lost its vigor. Conversely, on sand dunes covered by vegetation more than a year before planting, the poplars did not succeed in establishing themselves permanently: the majority died and the remaining ones had very slow growth and stayed feeble. This response was observed in all places and also in all planting years. Furthermore, in afforestation experiments, where sometimes the active section was not large enough to contain all the plantings, and, thus, the test zone extended over to a contiguous area covered with vegetation, the same phenomenon then occurred: the cuttings located on the pasture either did not root, dried up in a short time, or, if they survived, were feeble. The harmful effects of adverse factors such as hares and ants increased on the weaker plants. It is obvious, then, that in the semiarid zone poplars cannot compete for water and nutrients with previously planted grassy vegetation. Only on the dunes where pasture had been established just four months earlier could the swiftness of growth of the root system of *Populus* exceed and overcome that of the rye.

Consequently, after four years of field experimenta-

Fig. 9. Cultivar I-214 and *Eragrostis curvula* after five years — near Roosevelt.

tion, it was decided to abandon the attempt to forest vegetated dunes with poplars.

Afforestation of Active Dunes with Cultivar I-214

Table 6 gives the statistics of establishment or sprouting in 42 planting experiments with hybrid I-214, performed from 1962 through 1967 in 18 locations that cover the entire central area of the semiarid Pampas region — as can be seen in Figure 1 (*35*).

Analysis of the data allows the following statements to be formulated:

a) The general averages for all years, for all locations,

Fig. 10. Eragrostis *curvula* in competition with a stand of cultivar I-214.

and for the three types of experimentation fluctuated between 75 and 90 percent. For 126 sites, only four figures appear that are below 50 percent (rate of sprouting), that is, 3 percent of the cases. If the requirement is increased to 70 percent of cuttings established, we find that there were 20 figures below. In other words, in 85 percent of the instances, the rate of establishment exceeded 70 percent. It is fitting to point out that more than half the cases of sprouting rates lower than 70 percent occurred in the areas of Intendente Alvear, Vértiz, and Caleufú, where presence of a compact calcareous layer prevented planting at the suitable depth.

b) By observing the statistics of establishment for cuttings planted at 1.50- and 2.00-meter depth, we find that in only nine cases out of 84 do the figures fall below 70 percent, and that of the nine, all except one belong to the areas where the calcareous layer did not allow planting deeply.

c) Consistent differences in establishment rates between plantings at 1.50 or 2.00 meters were not substantiated.

d) Planting at 1.00 meter merits special attention. Although the sprouting rate averages only slightly lower (8 points) than deeper planting, heterogeneity of the figures is noteworthy. Percentages vary between 30 and 100 percent, and the numbers lower than 70 percent rise to almost one-third of the total. It should be pointed out that in years with a very dry spring, rate of establishment falls below 50 percent; this is a risk that cannot be taken.

e) Influence of the compact calcareous layer (*tosca:* hardpan) near the surface (at 1 meter or less) prevents deep planting, an essential element of the system, and this influence brings about the loss of the plants.

f) Size of the cutting is an aspect vital to the success of the deep planting method. When the large-scale experiments began (in 1960), long (4 to 5 meters) cuttings were used, and excellent results were obtained. When it was established that their weight hindered handling and transportation, the size was gradually reduced, and in 1965 the cuttings used measured between 2.50 and 3.00 meters. The spring of 1965 was very dry, and the two factors together effected an almost total failure: rooting indices declined in 1965 to 20 to 30 percent. The tests were eliminated because a sufficient stand was not observed. Subsequently, 4- to 5-meter-long cuttings were used again and the high rates of rooting were once more observed.

g) The only important adverse factor was the hare (*Lepus europaeus*), which gnaws the trunks of young trees. When growth is normal, the plant overcomes injury and very few losses occur. Conversely, if the plant is weak (as on pasture-established dunes) the losses are severe. In the case of *Populus deltoides,* the damage observed was much greater than in *P. canadensis.*

h) Toward the western part of the semiarid Pampas, where precipitation decreases, percentages of establishment are reduced, but only slightly (by 10 or 20 percent), making the need for deeper planting very important.

Fig. 11. Cultivar I-214, eight months after planting.

As conclusions on establishment, the following can be stated:

a) *P. canadensis* I-214 combines good qualities for adaptation to the semiarid Argentine area.

b) Planting material should be cuttings (terminal shoots, *guías*) one or two years old and 4 to 5 meters long.

c) Planting depth should average close to 150 centimeters.

d) Where there are compact strata that prevent going deep, planting shouldn't be done.

e) Planting should be done only on fully active sand dunes.

f) There should be a good wire fence to prevent entry by livestock.

Growth of Cultivar I-214 on an Active Dune

As was mentioned in a previous paragraph, competition is quite vigorous on the active dune, and poplars hinder the spread of the grass cover. Beneath the forest of *Populus,* herbaceous vegetation advances very slowly and never succeeds in covering the whole surface. On the other hand, the poplars progress rapidly.

In order to confirm the growth, at the moment of planting the diameter of all specimens is measured and is marked at a height of 1.3 meters. Nonetheless, the matter is not simple, since while the plants begin their growth during the first two to three years, erosive movements are occurring that change the height of the reference level and in some ways alter conditions of growth of the various individuals. The mobility of the sand surface creates a problem. For the purpose of relying upon a measurable basis for establishing growth, tree diameter is measured annually at the height of the mark in some trials in representative locations, and the differences are given in millimeters. The readings are taken in the season of winter plant dormancy.

Table 7 facilitates consideration of the increase in diameter generated between planting and winter, 1968.

In this regard, the following observations may be made:

a) Great variability between sites and between years is evidenced, which indicates that the treatments do not create wide and consistent differences, as happens with sprouting.

b) Notwithstanding (*a*), there is a trend in the sense that plantings at 1-meter depth produce smaller trees, but there are various contradictory cases.

c) In the matter of spacing between plants, if one experiment with contradictory values is disregarded, in the remaining five trials there is a definite tendency for the diameters to increase slightly as distances increase.

d) Undoubtedly the difficulty cited (internal dynamics of the dune in the first few years) prevents us from obtaining conclusive data and from attempting a statistical analysis of the values measured.

Bearing of Dune Topography on Growth of Cultivar I-214

As was previously mentioned, there exists a generally known difference in growth of trees in different parts of the active sand dune. It has been substantiated visually on all the forested dunes: the trees are large on crests, moderate on level areas (playas) and small in depressions. Data from Trenque Lauquen and Paso confirm these facts. In Paso recorded growths were as follows:

Average of 100 plants located on crests: 134 millimeters in 6 years.

Average of 170 plants located in nearly level areas: 113 millimeters in 6 years.

Average of 170 plants located for the most part in depressions: 35 millimeters in 5 years.
At Trenque Lauquen the following figures were obtained:

Average of 55 plants grown on ridges: 84 millimeters in 6 years.

Average of 150 plants grown in nearly level areas: 56 millimeters in 6 years.

From all the data, the great differences recorded in Paso between trees grown largely in depressions and those grown on ridges is worthy of note. On the ridges, there were plenty of specimens with more than 150 millimeters growth in 6 years. Moreover, just as at Trenque Lauquen, a notable number of shoots ("offspring") emerged around the mother plant on the active crests — mainly on the steep slip face (leeward slope) next to the crest. These shoots grow perfectly well.

AVAILABILITY OF SOIL MOISTURE IN ACTIVE DUNES

The good response of the poplars, which have high water requirements, merits some remarks on humidity of the dunes.

The principles that govern storage and conservation of rainwater explain the moisture availability. Due to their textural characteristics and their structure of individual grains, the active dunes offer ideal conditions for captation, deep penetration, and conservation of rainwater. Dunes allow excellent infiltration, very rapid internal drainage, and, at the same time, low evaporation. If the factor of total absence of vegetation that transpires water is added, it should follow that the dune is the ideal fallow ground of the highest efficiency, since it stores practically all water that falls upon its surface and loses very little. For that reason, even in periods of prolonged drought, at a few centimeters below the surface one finds moist sand, except on the edges of the crests where the motion created by the wind forms a narrow band of dry sand.

Naturally, moisture capacity of sand is low. According to conclusions from a dune at Intendente Alvear, equivalent moisture varies between 4 and 5 percent, but hygroscopic humidity reaches only 0.80 to 0.90 percent and, above all, depth of the sandy strata is great. At Trenque Lauquen 24 hours after a rainfall, humidity between 10 and 12 percent was recorded at 30 centimeters and approximately 6 percent at 60 and 90 centimeters. The percentage of total sand that is classified fine and very fine at Alvear varies between 80 and 90 percent. Medium-sized grains make up about 7 to 8 percent, organic matter content is at 0.05 percent and pH is between 7.5 and 8.3. To estimate the change of moisture reserves beneath the dune while the trees are growing, the experiments with distances between plants were conceived. According to the results, moisture reserves do not seem to be dwindling, since the oldest plantings of poplar (9 years) do not exhibit to date signs of wilting even in the most restricted spacings (5 meters x 5 meters).

We should point out that the water supply of the plants depended solely upon rainfall, since irrigation was not applied. There was a single exception: the eucalyptus received an initial irrigation.

The observations outlined were the basis of the afforestation experiments on active dunes with more water-demanding species like willows (*Salix* sp), which will be discussed below.

AFFORESTATION WITH WILLOWS (*SALIX* SP)

Three willows were tried: *Salix interior, Salix* 131/25 (hybrid) and *Salix* 131/27 (hybrid). All the tests conducted with willows were directed toward determining response, by means of planting at different depths with three repetitions.

Salix Interior

The experiment was performed in the experimental area at Trenque Lauquen, and sprouting at a 15 percent average took place at the three depths. The best establishment rate was obtained with planting at a depth of 2 meters — achieving a 25 percent rate. Comparison with the neighboring experiment with *Salix* 131/25 led to the elimination of *Salix interior* Rowl., owing to its lack of adaptation to the semiarid medium.

Salix 131/25

Table 8 presents percentages of sprouting of the cuttings planted in 10 trials in the vicinities of Trenque Lauquen, Roosevelt, and Vértiz. Given the high water requirements of the willow, the experiments were located in the eastern, more humid half of the semiarid zone. The predictions were confirmed by the results.

The statistics for this INTA-developed willow reveal an obvious sensitivity to planting depth. Establishment average for the 10 tests yields only 51 percent for 1-meter depth. In all instances, sprouting was greater at 150 and 200 centimeters. Although in some cases there is a difference in favor of 200 centimeters, in other instances planting at a depth of 150 centimeters yields figures equal to or better than those at 200 centimeters, which indicates that good establishment is obtained at depths of 150 centimeters and below.

With regard to location, there is also a clear finding: in the more humid sector of the semiarid zone (Trenque Lauquen and Roosevelt) the best establishment rates were obtained; moderate rates resulted in the transition region, and, toward the west, rates were more haphazard.

Growth of hybrid willows was good in the areas of Trenque Lauquen and Roosevelt, as can be seen in Table 8. At the end of 5 years, the *Salix* 131/25 displayed an average diameter of 140 millimeters at Trenque Lauquen, with a slight tendency toward better development at the two deeper levels. The trees planted in 1965 — a drier year — exhibited smaller diameters; in four years the average diameter was only 50 millimeters. Conversely, trees planted in 1967 reached an average diameter of 40 millimeters in only two years.

In Table 9 growth of *Salix* 131/25 during the 1968/69 biological cycle is evaluated, confirming the tendency of slower growth in planting at a depth of 100 centimeters. Averages in the table indicate a trend between the growth of willows planted at a depth of 100 centimeters (11 millimeters in diameter per year) and those planted at 150 and 200 centimeters (14 millimeters annual increase). Likewise, the averages disclose that as the tree attains greater size, the annual increase in diameter is also larger.

The same thing occurs with willows as with poplars in regard to the size of cuttings. In a year in which small cuttings were used, many dried up, except those planted at a depth of 2 meters. The little trees, due to their small diameter, broke off very easily, although they then put forth new shoots. Thus it is necessary to use cuttings 3 to 5 meters long.

Fig. 12. Willows after two years, in a playa area. Note the roots exposed by the wind.

Fig. 13. Willows after five years — near Trenque Lauquen.

In 1968 a detailed observation was made in the season for sprouting (middle of September or the beginning of spring) in the area of Trenque Lauquen, comparing poplars and willows:

a) The majority of poplars located on the crests had put forth shoots.

b) In the depressions sprouting of poplars was scanty; the phenomenon could be owing to the fact that in the low areas cold air accumulates and there is less fertility.

c) The difference occurred as much with the 1964 poplars as with those of 1968.

d) The influence of individual characteristics became obvious, because plants of similar size on the same ridge exhibited noticeable differences among themselves.

e) Willows sprouted before the poplars and did not show differences.

Salix 131/27

This hybrid willow developed by INTA was tested in a system of nine trials of planting depth in the same locations as 131/25.

The establishment statistics, which can be evaluated in Table 8, allow formulation of these observations:

a) General averages are slightly higher (10 percent) than *Salix* 131/25.

b) Good establishment percentages are recorded at Trenque Lauquen and Roosevelt, those at Vértiz being contradictory.

c) The good results of deep planting (150 and 200 centimeters) are confirmed, planting at 100 centimeters being acceptable.

d) Even in 1965 when there was a drought, the outcome was good. Hybrid 131/27 therefore gives assurance of establishment in the more humid sector of the semiarid strip, as four years of testing prove.

The growth also shows a tendency to be less in trees planted at a depth of 100 centimeters. In four years the willows planted in 1965 achieved a trunk diameter of 100 millimeters, on the average, and the 3-year-olds achieved 35 millimeters. Thus, we observe that growth is not consistent through the years, the same as in the case of hybrid 131/25.

Pasture Establishment or Afforestation?

This chapter synthesizes the results obtained from many years of work (*37*) and discusses the techniques utilized in pasture establishment and afforestation with the objective of dune stabilization. At this point, it is fitting to inquire: Which system is more suitable?

The reply depends upon the purposes and feasibilities. Pasture establishment by means of protected seedlings allows a transition in the brief time span of 6 months (and in favorable circumstances in only 3 months) from the fully active sand dune to a firm meadow on which all wind erosion has ceased. Needed for achieving this is a combination of men, machines, and sealing materials, welded into an efficient plan and work force. The transformed dune, with care, can be used as a forage reserve.

Afforestation succeeds in fixing securely a sandy area in two or three years, depending upon environment and the weather. Subsequently, it offers a small forest which, at the most, can provide protection and some firewood, lumber, or fenceposts for local use. On the other hand, afforestation is extremely simple to implement. Planting can be done by one man with no technical or other preparation, furnished with a soil auger which costs about $10 (U.S.); if the necessary cuttings are available, he can work all winter, gradually achieving afforestation of the active dunes. In addition, the cost is much lower, perhaps 40 to 50 percent less than that of pasture establishment, if the cuttings are grown in the same semiarid region.

Choice of the method of stabilization depends then upon the means of each rural settlement and on the suitability of major land use. When the farmer cannot or does not want to do otherwise, at least he should forest his dunes.

SUMMARY AND CONCLUSIONS

Inland dunes (médanos), composed of accumulations of the sandy constituents of soils, are a spectacular product of wind erosion. The grains that form the dunes are not capable of cohesion, which makes them extremely mobile and encroaching. From this emanates their dangerous quality. Upon taking form as dunes and expanding, they destroy intact soils, disrupt rail lines and highways, damage or bury crops, cover wire fences, ranch watering places, and other farm structures, and they even encroach upon homes and villages. It is thus important and urgent to halt dune movement by means of stabilization.

The problem reaches great proportions in the Argentine Republic, where 75 percent of the land has some type of aridity problem. Immense expanses of the semiarid Pampas (more than 22 million hectares) undergo an acute erosion process, including formation of dunes, when highly susceptible land is used without attention to proper conservation practices.

Since the end of the last century technical experts have been concerned with the problem, but only in 1950 could concerted action be begun in the central region of the semiarid Pampas.

The region in which the investigations and experiments were performed is located between 35° and 37° S and 62° and 65° W. *Rainfall* varies from 750 millimeters in the east to 450 millimeters in the west, 75 percent occurring in the summer half year. At the end of winter and beginning of spring (August-October) the normal winter dry spell is at its height and the winds increase, this being the critical period for wind erosion. The water balance indicates an annual deficit fluctuating between 100 and 300 millimeters.

Three *great soil groups* are represented in the region: chestnut, brown, and regosolic — the predominant textures of the surface horizon being loamy-sand and sandy-loam, the sand content increasing with depth. The investigation sites are typical regosolic accumulations.

Natural vegetation consists of characteristic pampean rangelands among which it is important to mention "psammophytic steppes"; to the west the xerophytic woodlands forest appears, the most typical representative of which is *Prosopis calden.*

Present land use is for agricultural and livestock production, cattle-raising being of greatest importance, with sheep-raising in second place. Major crops are cereals: wheat, barley, and rye, and grain-bearing sorghum. Main perennial forage crops are alfalfa (*Medicago sativa*), black sorghum (*Sorghum almun*) and lastly, weeping lovegrass (*Eragrostis curvula*) and tall wheatgrass (*Agropyrum elongatum*). Among annual forage crops, rye (the most important) and forage sorghum are foremost. Summer fallowing is quite extensive, and the one-way disc is in general use. In the last few years the duck-foot cultivator and the rotary rod weeder are becoming widely used.

The *Pampas dunes* in full activity have quite varied dimensions, although the most general heights and depressions vary between 2 and 5 meters. The surface area which they encompass is likewise very diverse — some of 50 hectares being not uncommon, but the most numerous active nuclei measure between 2 and 10 hectares. Sand dunes do not move great distances in massive form (for example, hundreds of meters), but the sand that is carried away lights the flame of destruction in the adjoining area in the lee of prevailing winds (toward the south in Argentina); the effects spread, like an illness, through contagion.

According to observations (experiments) in the areas in the process of afforestation, the *internal mobility of the sand surface* can be summarized thusly:

a) Within a year's time 80 percent of the surface shows accumulations higher than 5 centimeters.

b) Approximately 25 percent records accumulations higher than 30 centimeters.

c) More than 15 percent gives negative variations (depressions) greater than 30 centimeters.

d) The figures of accumulation are always larger than those of depression.

e) In some instances variations of the relief approach amounts of 1 meter.

This list points up the impossibility of carrying out pasture establishment (or afforestation with small plants) without some type of herbaceous mulch.

In the main area of investigation (Trenque Lauquen) the water table, measured monthly for three years, varied from 5 to 10 meters in depth, according to place. Under poplar plantings a 6-to-7-meter depth was found.

The major investigations of *pasture establishment of dunes* performed between 1950 and 1960 resulted in the following conclusions:

a) Before sowing with forage mixtures, it is necessary to smooth out the terrain.

b) The smoothing should be made only to such a degree that sowing with a cereal seeder is possible.

c) Smoothing is achieved by three principal methods: (*1*) with wind channels (wind intensifiers and dispersers of sand), (*2*) by going over terrain with graders, and (*3*) by going over dunes with two or three rails or beams that topple the sand to leeward, breaking up the sharp crests and the steep slip face; the grader and the rails (either rails or the grader or both can be used) should be drawn by a caterpillar or track-type tractor or by animal — wheeled tractors cannot travel on high dunes.

d) Sowing of a forage mixture with the cereal or forage seeder: (*1*) In fall (March-April) a mixture of 60 kilograms per hectare of rye with 5 kilograms per hectare of *Agropyrum elongatum* and 5 kilograms of *Bromus brevis* is sown, although rye alone can stabilize the dune. (*2*) In spring (October-November) a forage mixture of 5 kilograms of each of the following species is used: maize, *Panicum miliaceum, Sorghum caffrorum, S. sudanense,* and *S. almun;* 1 kilogram of *Eragrostis curvula* should be added; if these seeds are not available, some species may be omitted (essential ones are: *P. miliaceum, Sorghum,* and *E. curvula*).

e) While seeding, one covers the dune with a thin layer of stubble mulch in spring and "chala" (sorghum straw) in fall.

f) The dune is gone over with a simple disc harrow to anchor the mulch somewhat, so that the wind does not blow it away.

With this method, in 3 to 6 months a transition is made from a fully active sand dune to a stable pasture. Substitution of asphalt binding material for the protective covering instead of herbaceous mulch gave good results. In spite of its being an acceptable technique, use of this method is not very widespread, since more sophisticated machinery for applying the binding material is not commonplace in rural areas and the cost is greater than that of the stubble mulch.

Investigations of *afforestation of sand dunes* were carried out by experiments to test species, varieties, planting depths, size of cuttings to plant, and distances between plants. The tests always had at least three repetitions, and distribution of plots was in random blocks or a Latin square. In all instances the experiments were situated on typical dunes of the most varied shape, size, and condition in fields belonging to farmers; the experimental plots were fenced off with wire around the perimeter.

In dune afforestation the following species have been tested: *Ulmus pumila* variety *arborea, Eucalyptus viminalis, Populus nigra* cultivar *italica, Populus deltoides, Populus canadensis* cultivars 154 and 214.

U. pumila and *E. viminalis* should be planted on dunes previously stabilized with pasture in order to obtain good results. *E. viminalis* gives a very fine response in the eastern half of the region.

P. deltoides does not thrive under conditions of the semiarid zone. *P. nigra* gives a mediocre response, very inferior to hybrids 154 and 214.

Populus canadensis cultivars 154 and 214 have shown good adaptation to the environment, particularly cultivar 214.

Conclusions about *P. canadensis* cultivar 214 can be summarized thusly:

a) Plantings were done in all instances without irrigation.

b) For all ials, all years, and all locations, the percentage of sprouting or establishment varied between 75 and 90 percent.

c) Best results were obtained in the more humid east.

d) Considering only the results of 150 and 200 centimeters planting depth, practically all figures exceed 80 percent for establishment or sprouting; there are no consistent differences between these two depths, in view of which we can advise planting at 150 centimeters.

e) Presence of a compact layer of calcium carbonate at a depth of 100 centimeters or less prevents deep planting and brings about the loss of many trees.

f) Planting material should be "cuttings" (terminal shoots) 1 or 2 years old, from suckers or rooted plants, and 4 to 5 meters long.

g) Planting should be done only on active dunes, since on pasture-covered dunes poplars do not tolerate competition and they die.

h) There should be a good fence to protect against livestock.

i) Spacings of 4 by 4 meters and 5 by 5 meters square have shown good results.

j) Poorest growth is noted in the depressions (basins), so it is advisable to begin planting in high and middle areas.

k) General average growth in the eastern sector of the semiarid zone was 56 millimeters in diameter in 6 years; in the high areas (ridges) the average rose to 84 millimeters for 6 years.

In the eastern portion of the semiarid zone, plantings with *hybrid willows* developed by INTA were made, excellent rooting and growth being attained in areas with annual rainfall between 600 and 700 millimeters, with results similar to those obtained with poplars regarding rooting, and better for growth. On the other hand, in the western sector (between 500 and 600 millimeters of rainfall), where the poplars continue to show good responses, the hybrid willows are not thriving.

In conclusion it should be pointed out that pasture establishment is more complicated to carry out than afforestation but in 3 to 6 months it stabilizes the dune. Conversely, afforestation is very simple (it requires only drilling a hole 150 centimeters deep and 7 centimeters in diameter), but it takes 2 to 3 years to halt the shifting of the sand completely.

With regard to *cost,* in addition to the costs of materials, pasture establishment requires 10 working days per hectare and afforestation with poplar or willow cuttings 4 to 5 days per hectare.

REFERENCES AND NOTES

1. PREGO, A. J., AND OTHERS
   1957   Utilización y conservación del suelo en la Argentina. IDIA X(114):1–46 / INTA, Instituto de Suelos y Agrotecnia, Publicación 56.

2. BONFILS, C. G.
   1966   Rasgos principales de los suelos pampeanos. INTA, Instituto de Suelos y Agrotecnia, Publicación 97. 62 p.

3. BELLÓN, C. A., AND OTHERS
   1955   Conservación del suelo y del agua. Argentina, Ministerio de Agricultura y Ganadería, Publicación Miscelánea 416:118.

4. BOVET, P. A.
   1910   El problema de los médanos en nuestro país. Argentina, Dirección de Enseñanza Agrícola, Buenos Aires. 155 p.

5. DESIMONE, A. J.
   1918   Los médanos en la región pampeana. Estación Agropecuaria, Pergamino, Publicación Miscelánea 10.

6. ISSOURIBEHERE, P. J.
   1901   Fijación de las arenas por la plantación de árboles. Argentina, Ministerio de Agricultura, Boletín Oficial 1901:182–183.

7. MIATELLO, H.
   1915   Problemas agrícolas, fijación de los médanos en el país. Argentina, Ministerio de Agricultura, Boletín 19:167–176.

8. OLIVERA, L.
   1941   El primer trabajo sobre fijación de médanos realizado en el país (1879). Anuario Rural 9:97–98.

9. VELASCO, R.
   1912   Los médanos de la provincia de Córdoba. Boletín de Agricultura (Córdoba) 1: 153–173.

10. VIDAL, J. J.
   1948   Dunas y médanos. J. J. Vidal, La Plata, Argentina. 305 p.

11. TALLARICO, L. A.
   1948   Trabajos previos a la fijación de médanos. Argentina, Ministerio de Agricultura y Ganadería, Almanaque 25:279–282.

12. PREGO, A. J., AND OTHERS
   1963   Estabilización de médanos en la región pampeana semiárida. In Conferencia Latinoamericana para el Estudio de las Regiones Aridas, Buenos Aires, Sept. 16–21, 1963, Comunicaciones y Resumenes de Trabajos: 73–74.

13. RAGONESE, A. E.
   1959   Sistema original de plantación de álamos utilizados en Italia. IDIA 144:1–3.

14. FOOD AND AGRICULTURE ORGANIZATION OF THE UNITED NATIONS
   1961   Plantación de álamos en dunas de arena. FAO, Forestry Equipment Notes, A. 2 Tec. 61.

15. PREGO, A. J., and J. E. CALCAGNO
   1960   Técnica rapida de fijación de médanos mediante siembras protegidas I (Trabajo resumido). IDIA, Suplemento 1:208–210.

16. Research for the present contribution was conducted at the Instituto Nacional de Tecnología Agropecuaria (INTA), Argentine Republic.

17. WEBER, T. F. A.
   1947   El clima de General Pico vinculado a la erosión eólica. INTA, Instituto de Suelos y Agrotecnia, Publicación 5:72.

18. PROHASKA, F. J.
    1960 El problema de las sequías en la región semi-
          árida pampeana y la sequía actual. IDIA,
          155:54–67.
19. ———.
    1961 Las características de las precipitaciones en
          la región semiárida pampeana. Revista de
          Investigaciones Agrícolas 15(2):199–232 /
          INTA, Instituto de Suelos y Agrotecnia, Pub-
          licación 172.
20. ———.
    1952 Regímenes estacionales de precipitación de
          Sudamérica y mares vecinos (desde 15°S
          hasta Antártica). Meteoros 2(1–2):66–100.
21. CHEPIL, W. S.
    1945 Dynamics of wind erosion: I. Nature of move-
          ment of soil by wind. Soil Science 60(4):305–
          420.
22. BURGOS, J. J., AND J. HOFFMAN
    1963 Las tierras áridas y semiáridas de la Repúb-
          lica Argentina, Informe Nacional. II Estruc-
          turas naturales, A–1: Clima. In Conferencia
          Latinoamericana para el Estudio de las Re-
          giones Aridas, Buenos Aires, 1963, p. 6–14.
23. THORNTHWAITE, C. W., AND J. R. MATHER
    1957 Instructions and tables for computing poten-
          tial evapotranspiration and the water balance.
          Drexel Institute of Technology, Laboratory
          of Climatology, Centerton, New Jersey, Pub-
          lications in Climatology 10 (3):183–311.
24. BURGOS, J. J., AND A. L. VIDAL
    1951 Los climas de la República Argentina según
          la nueva clasificación de Thornthwaite.
          Meteoros 1(1):3–32.
25. BONFILS, C. G., AND OTHERS
    1960 Suelos y erosión en la región pampeana semi-
          árida. Revista de Investigaciones Agrícolas
          XIII(4):321–328 / INTA, Instituto de Suelos
          y Agrotecnia, Publicación 65. 82 p.
26. RAGONESE, A. E.
    1966 Vegetación y ganadería en la República
          Argentina. INTA, Colección Científica 5.
27. CABRERA, A. L.
    1953 Esquema fitogeográfico de la República Ar-
          gentina. La Plata, Argentina, Universidad Na-
          cional, Facultad de Ciencias Naturales,
          Museo, Revista (Nueva Serie) VIII (Botán-
          ica 33):87–168.
28. PREGO, A. J.
    1952 El agua y la fertilidad del suelo pampeano.
          Revista Argentina de Agronomía 19(2):76–
          83 / INTA, Instituto de Suelos y Agrotecnia,
          Publicación 24.

29. ———.
    1961 La erosión eólica en la República Argentina.
          Ciencia e Investigación 17(8):307–324 /
          INTA, Instituto de Suelos y Agrotecnia, Pub-
          licación 78.
30. PREGO, A. J., AND F. J. PROHASKA
    1960 Dinámica de los médanos en la región pam-
          peana semiárida, I. IDIA, Suplemento 1:30–
          33.
31. PROHASKA, F. J., AND A. J. PREGO
    1963 Movimiento de la capa freática en la región
          pampeana semiárida, II. In Conferencia
          Latinoamericana para el Estudio de las Re-
          giones Aridas, Buenos Aires, Sept. 16–21,
          1963. Comunicaciones y Resúmenes de Tra-
          bajos: 41.
32. IPUCHA AGUERRE, J.
    1964 La primera visita de Bennett. IDIA, Suple-
          mento 13:32–35.
33. PROHASKA, F. J., A. J. PREGO, AND H. F. PETERS
    1962 Utilización de substancias aglutiantes para la
          fijación de médanos. Reunión Argentina de la
          Ciencia del Suelo, 2ª, April, 1962, Mendoza,
          Argentina. Resumen de trabajos y comunica-
          ciones: 110–111.
34. FERRANDO, J. C.
    1965 Praderización de médanos y dunas con el
          empleo de productos derivados del petróleo.
          Reunión Nacional de Regiones Aridas y Semi-
          áridas, 2ª, Actas. IDIA, Suplemento 19:45–
          47.
35. PREGO, A. J., AND OTHERS
    1966 Estabilización de médanos mediante foresta-
          ción en la región pampeana semiárida. IDIA,
          Suplemento Forestal 2:75–92 / INTA, Insti-
          tuto de Suelos y Agrotecnia, Publicación 100.
36. PREGO, A. J., F. RIAL ALBERTI, AND F. J. PROHASKA
    1964 Forestación de médanos en la región pam-
          peana semiárida, I. INTA, Instituto de Suelos
          y Agrotecnia, Publicación 92.

37. Performance of the series of tests, both pasture establishment and afforestation, has been possible thanks to
the close cooperation of the extension agencies of
INTA. Those agencies located in Trenque Lauquen,
General Pico, and General Villegas, and their managers, agricultural engineer F. Rojo and agronomists
H. F. Peters, U. Catalani, and O. Traverzaro, as well
as assistant E. Reinhardt, merit special recognition. As
direct collaborators we thank agricultural engineers R.
A. Alonso and agronomy students R. Mon and R.
Michelena. Lastly, we acknowledge the help of agricultural engineers A. E. Ragonese and A. Marzocca.

Effect of Insecticides on an Ecosystem
in the Northern Chihuahuan Desert

HOWARD G. APPLEGATE
Department of Civil Engineering
University of Texas at El Paso
El Paso, Texas, U.S.A.

WAYNE HANSELKA
Department of Wildlife Science
Texas A&M University
College Station, Texas, U.S.A.

DUDLEY D. CULLEY, JR.
School of Forestry and Wildlife Management
Louisiana State University
Baton Rouge, Louisiana, U.S.A.

ABSTRACT

THE EFFECTS OF DISTRIBUTION of DDT and methyl parathion from cotton fields to the surrounding area were studied in the Presidio Basin in a portion of the Chihuahuan Desert. The chemicals broke down more rapidly there than in mesic zones. They persisted for only a few weeks in the summer. The primary environmental factor inducing the breakdown was high surface-soil temperature. Ultraviolet radiation played an important but secondary role. The alkaline milieu also contributed to the rapidity of breakdown. Bacteria were isolated that could metabolize DDT. Their importance in breaking down DDT under natural conditions cannot be assessed.

The concentrations and distribution of the insecticides in birds, lizards, rodents, jackrabbits, cotton, and leatherstem plants were determined through a 15-kilometer transect that originated in the cotton fields. Although there did not appear to be any direct insecticide-mortality relationship, there may have been an indirect relationship.

Most insecticides entered the fauna via the food and were transported to muscle tissue and fat, including lipoid materials in the brain. During periods of stress the muscle fat was metabolized and its insecticides mobilized. A portion of the mobilized compounds was then stored in the lipids of the brain. Insecticides stored in the brain lipids appeared to be very immobile.

We hypothesize that the insecticides in the brain led either to impaired motor ability or to behavioral changes. Rodents carrying high body burdens of methyl parathion were recaptured more often than rodents containing less residue. The jackrabbit population did not have normally distributed age-classes. Juveniles having high methyl parathion concentrations in the brain died.

Insecticide uptake by desert plants was closely correlated with rainfall but showed little correlation with insecticide applications to cotton plants. Dormant plants apparently did not take up insecticides to any great degree. One desert plant, leatherstem, apparently stored methyl parathion in its rhizomes.

Several breakdown products have been identified in the arid Presidio Basin and correspond to those from mesic zones. At least one product, apparently unique to arid areas, is first detected in late summer and disappears rapidly with onset of cooler weather.

The ultimate fate of insecticides applied to cotton plants in the Presidio Basin depends on the time of application and on the weather.

393

EFFECT OF INSECTICIDES ON AN ECOSYSTEM IN THE NORTHERN CHIHUAHUAN DESERT

Howard G. Applegate, Wayne Hanselka, and Dudley D. Culley, Jr.

There was a time when meadow, grove, and stream,
The earth, and every common sight
To me did seem
Apparell'd in celestial light,
The glory and the freshness of a dream.

THE FRESHNESS OF Wordsworth's dream appears to be a nightmare to many biologists today. The heavy demands for food, clothing, and shelter made by our expanding population have changed the meadow, grove, and stream beyond recognition. Until recently, the arid and semiarid areas of the world were little touched by this demand. Modern technology, however, can now furnish water to formerly inaccessible places. More and more of these areas are being used for production of food and fibers. The use of pesticides (herbicides, fungicides, insecticides, and nematicides) is an integral part of modern agriculture. Studies are scanty on the effect of these compounds on an arid ecosystem. Such studies are urgently needed if we are to use desert lands to produce food and fiber without destroying the flora and fauna of adjacent areas. This is a report of a study on the effect of *insecticides* on an arid ecosystem.*

THE GEOGRAPHICAL AREA

The Presidio Basin is the northernmost extension of the Chihuahuan Desert in Texas (Fig. 1). The two towns in the basin — Presidio (United States) and Ojinaga (Mexico) — are at an elevation of 790 meters. Floodplains under intensive cultivation extend along the two rivers in the basin — the Rio Conchos and the Rio Grande. The floodplains vary in width from 1 to 6 kilometers. A series of terraces extends back from the floodplains for 40 to 60 kilometers. The terraces succeed each other at distances of roughly 6 to 8 kilometers. In the

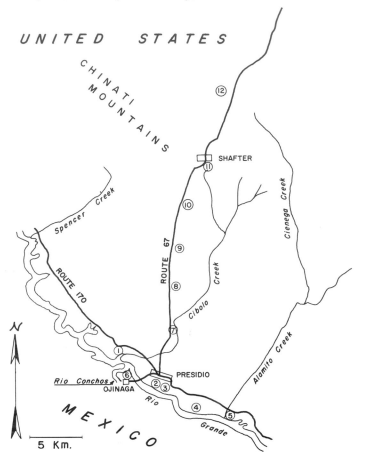

Fig. 1. Map of the sampling area in the Presidio Basin. Sites 1, 2, 3, and 4 are in the cotton fields; sites 5 and 6 are water-sampling areas; sites 7, 8, 9, and 10 are in the desert and aerodynamically part of the Presidio Basin; site 11 is in the Chinati Mountains; site 12 is north of the Chinati Mountains.

United States 40 kilometers from the Rio Grande, the Chinati Mountains (north and west) and the Torneros Mountains (east) rise to 2,380 and 1,530 meters respectively. In Mexico 60 kilometers from the Rio Grande, La Sierra Rica (south and east) and La Sierra Chinati (west) rise to 3,000 meters.

The Rio Grande is usually a dry stream west of Presidio-Ojinaga. At that point the Rio Conchos, which arises in La Sierra Madre Occidental, flows into the Rio Grande. It furnishes most of the irrigation water in the basin. No other perpetually running stream enters the basin.

The basin contains approximately 156,000 hectares of land. Of this, approximately 2,600 hectares are cultivated. Over 90 percent of the cultivated land is devoted to cotton. Agricultural soil is classed as a silty clay; desert soil, a gravelly loam.

---

* Abbreviations used in this chapter:

ppm: parts per million
DDT: 1, 1, 1-trichloro-2,2-bis(p-chlorophenyl)ethane
DDD: 1,1-dichloro-2,2-bis(p-chlorophenyl)ethane
DDE: 1,1-dichloro-2,2-bis(p-chlorophenyl)ethylene
BHC: 1,2,3,4,5,6-hexachlorocyclohexane, mixed isomers
MP: Methyl parathion — 0,0-dimethyl 0-p-nitrophenyl phosphorothioate
P: Ethyl parathion — 0,0-diethyl 0-p-nitrophenyl phosphorothioate
LD$_{50}$: Lethal dose to 50 percent of the population

NOTE: This chapter is an adaptation of a presentation at "Arid Lands in a Changing World," an international conference sponsored by the Committee on Arid Lands, American Association for the Advancement of Science, June 3–13, 1969, at the University of Arizona, Tucson, Arizona, U.S.A.

Since it is a basin, this area is relatively isolated both aerodynamically and hydrodynamically from other pesticide-using areas. The nearest major pesticide use is in the vicinity of Toyahvale (110 miles to the north in the United States) and Camargo (150 miles to the south in Mexico). The basin has a climate much different from those on the other sides of the mountains. Obviously, weather fronts originating in other areas do affect the basin. In general, however, the basin's climate does exhibit aspects differing from the surrounding macroclimate. It is justifiable, therefore, to conclude that the basin has its own distinct mesoscale patterns.

Since the Rio Grande is usually dry west of the Presidio Basin, only the Rio Conchos could bring pesticides into the basin via water. Data presented later in this chapter show that only negligible amounts are brought in by the river.

MATERIALS AND METHODS

Insecticides are applied by airplane in the Presidio Basin. The applicator for United States fields is based at Presidio, while a pilot from Chihuahua applies chemicals to the Mexican fields. In general, the same kinds of chemicals are applied in both countries. The kinds and amounts of insecticides used in the United States are shown in Table 1; only fragmentary records are available relative to applications in Mexico.

TABLE 1

Insecticides Applied at Presidio
(kilograms of pure chemical)

	1965	*1968*
DDT	9,432	20,549
MP	7,230	18,900
Sevin	1,180	0
BHC	1,175	0
P	910	500
Endrin	90	0

There is a decided trend away from all insecticides except DDT and methyl parathion (MP). The first application is usually made in the middle of June and the last in the middle of September. The frequency of application varies with the insect populations but usually is at 4- to 6-day intervals. In addition to use of the compounds shown in Table 1, a joint spraying program to control boll weevils (*Anthonomus grandis* Boh.) was undertaken by the governments of the United States and Mexico. In 1965, 7 sprays (3 in late September and 4 in October) of low-volume, high-concentrate malathion were applied. A total of 8,000 kilograms of malathion (active ingredient bases) were applied in the Presidio Basin. In 1966, guthion was applied under the same program, at approximately the same time of year and in the same concentrations.

During the five years that this study has been underway, various methods have been used in isolating and quantifying the insecticides (*1-7*). Essentially, all isolations involved partitioning between solvents of differing polarity. Just prior to processing, each sample was thoroughly mixed (soil) or diced (flora and fauna). The samples were then weighed into two equal portions. One portion was spiked by adding 10 milliliters of hexane containing known concentrations of the insecticides that were expected to be found; 10 milliliters of hexane were added to the remaining portion. Both portions were then treated identically.

Extraction procedures were as follows:

Soil: 10 grams were placed in a flask and 50 milliliters of hexane:acetone solution (9:1 v/v) was added; after one hour on a wrist shaker, the solution was filtered off and the soil reextracted with another 25 milliliters of the solution; the solutions were bulked, concentrated in a Kuderna-Danish concentrator, cleaned up by sweeping through a Kontes codistiller, and then made to volume (1 milliliter of liquid = 10 grams of sample). Extraction efficiency based on spiked samples ranged (in percentages) from 81 to 92 for DDT; 90 to 95 for DDE; 85 to 94 for DDD; 87 to 95 for MP; 88 to 93 for P.

Meat: The tissues were ground with 10 grams of anhydrous sodium sulfate; the mash was extracted first with benzin and then with acetonitrile; the benzin layer was drawn off and the mash extracted three more times with acetonitrile (the bottom layer was drawn off each time and bulked with the benzin); the bulked extract was extracted three times with 2 percent sodium chloride; the benzin layer was then filtered through anhydrous sodium sulfate, concentrated in a Kuderna-Danish concentrator, cleaned up by passing through a Kontes codistiller, and then made up to volume. Extraction efficiency based on spiked samples ranged (in percentages) from 85 to 90 for DDT; 89 to 95 for DDE; 84 to 94 for DDD; 85 to 95 for MP; 85 to 93 for P.

Leaves: 10 grams of leaves were blended with ethyl acetate, filtered through glass wool, and the filtrate treated with 5 grams of Nuchar; after filtering, crystalline sodium chloride was added to the filtrate and the top layer concentrated in a Kuderna-Danish concentrator, cleaned up by sweeping through a Kontes codistiller, and then made up to volume (1 milliliter of liquid = 10 grams of sample). Extraction efficiency based on spiked samples ranged (in percentages) from 93 to 95 for DDT; 94 to 95 for DDD; 90 to 94 for DDE; 91 to 95 for MP; 89 to 93 for P.

Solvents were cleaned up by the methods of Burke and Giuffrida (*8*). All values reported in this chapter were corrected for percentage loss during extraction and cleanup.

Three gas chromatographs were used in this study. The MicroTek 2500R was equipped with a nickel-63 detector. The operating parameters were: nitrogen gas flow — 50 ml/min; temperatures (°C) — column 190, inlet 200, outlet 230, detector 270; power supply — voltage, 54 v; pulsewidth, 9 μsec; pulse rate, 30 μsec. The column was 6.35 mm x 182.88 cm glass packed with a 50:50 (w/w) mixture of 7 percent OV-17 and 9 percent QF-1 on 80/100 mesh chromosorb W, AW, DMCS, HP.

The Barber Colman was equipped with an automatic injection device. The device and all operating conditions have been described previously by Applegate and Chittwood (9). The Aerograph 680 chromatograph operating conditions have been described in previous papers (1-7). Quantification for all three instruments was by use of a digital integrator.

Samples to be analyzed by mass spectrometry were collected off the gas chromatograph. After retention times were established for the compounds, the column was detached from the detector. A sample was injected and, at the appropriate time, a glass wool collecting device placed on the open end according to the method of Mumma and Kantner (10). The samples were washed from the glass wool with hexane. The hexane was evaporated to near dryness and collected in capillary tubes. A model 21-110B Consolidated Electrodynamics Corporation mass spectrometer was used for analysis of standard and collected samples.

Thin-layer chromatography was used as an additional confirmational tool. Extracts were spotted or streaked on silica gel HF 254 and hexane:ether (100:1 v/v) was used as the solvent. Compounds were located under ultraviolet radiation, and their R_f's were compared to those of standards on the same plate. Silica gel containing the sample was scraped off, eluted with hexane, and the eluate injected both alone and spiked with standards into the gas chromatographs.

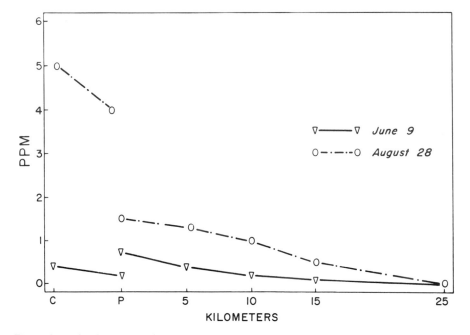

Fig. 2. Organic phosphates (methyl and ethyl parathion) in leatherstem and cotton plants. Only cotton plants are in the center of the field (C), while cotton plants and leatherstem both are at the periphery of the field (P). Only leatherstem is in the desert.

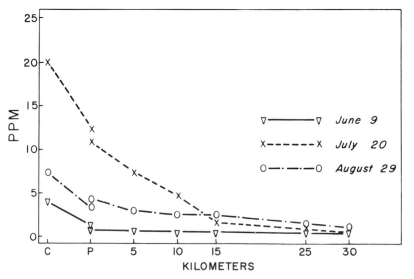

Fig. 3. DDT concentrations in leatherstem and cotton plants. Only cotton plants are in the center of the field (C), while cotton plants and leatherstem both are at the periphery (P). Only leatherstem is in the desert.

INSECTICIDE CONCENTRATIONS

Vegetation

Due to the widespread distribution in the desert of leatherstem (*Jatropha dioica* var. *dioica* Sesse *ex* Cerv.), we collected and used its leaves as a measure of insecticide drift. The plant is an important year-round food supply for desert mammals. Many humans of the Presidio Basin make a boiled extract of the plant's rhizomes for use as a medicine for various ills. Cotton leaves (*Gossypium hirsutum* L.) were also sampled.

Insecticide concentrations found in the plants for 1965 are shown in Figures 2 and 3. Essentially no halogenated insecticides drifted further than 15 kilometers from the cotton fields, while phosphorylated insecticides drifted no further than 10 kilometers. An unexpected finding was that each summer, prior to any application of phosphorylated insecticides, leatherstem leaves on plants growing near the cotton fields contained 0.5 to 1.0 parts per million of the compounds. This finding suggested that these plants could store phosphorylated insecticides in their rhizomes during winter dormancy. In the spring, when growth started, the compounds were mobilized and translocated to the new leaves.

An experiment was set up to determine if leatherstem could store MP. Preliminary data indicated that the compound was readily absorbed and could be stored in the rhizomes of dormant plants. When new growth was initiated, MP was translocated into the leaves. Unfortunately, the laboratory containing the plants was destroyed by fire before conclusive data were obtained.

Cotton leaves also contained small but consistent amounts (0.1 - 0.3 ppm) of phosphorylated insecticides in the spring, just prior to any application of the compounds. In contrast to the leatherstem, which is a perennial, cotton plants are annuals. We suspect that the compounds were on the previous year's plants, which were plowed under the previous fall. The time interval between the last application in the fall and our sampling (6 to 7 months) was not sufficient to allow complete breakdown of the compounds under environmental conditions in the Presidio Basin.

Prior to 1965, most spray applicators applied what is known as a high-volume, low-concentrate mixture; that is, the active ingredients were less than 20 percent by weight of the liquid volume. Midway through 1965, several growers in the Presidio Basin started to use a low-volume, high-concentrate mixture, in which the active ingredients were more than 90 percent by weight of the liquid volume.

A few hours prior to the application of a low-volume, high-concentrate spray of MP to cotton plants, leaf samples were taken. At various time-intervals after the application, further sampling was done. The data are shown in Figure 4. Concentrations of 66 ppm MP were found four hours after spraying. The concentrations dropped rapidly; at the end of three days they were down to 12 ppm and after 14 days to 5 to 7 ppm.

There was a high correlation between spray applications and insecticide concentrations in leaves of cotton plants. We could discover no such correlation between spraying and insecticide concentrations in leaves of leatherstem and other desert plants. There was, however, a correlation between flushes of growth of the desert plants and insecticide uptake. When, due to rain, the desert plants were growing rapidly, insecticide uptake

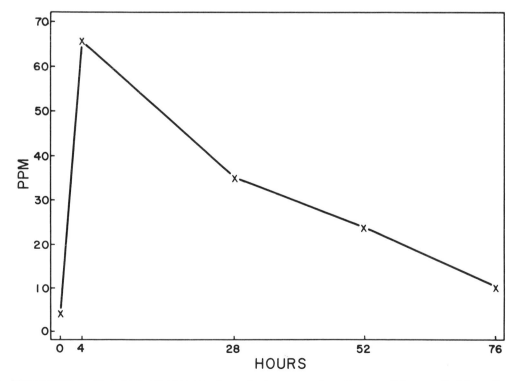

Fig. 4. Disappearance of a high-concentrate, low-volume spray of methyl parathion applied to cotton plants.

was great in ephemerals, annuals, and perennials. When perennials were in a state of dormancy, they took up a little of the insecticide.

Water

Wells used for drinking water in the basin vary in depth from 10 to 25 meters. At no time were any insecticides found in the well water. Insecticide concentrations found in the Rio Grande water and silt in 1965 were exceedingly high (Table 2). A special effort was made in 1966 to pinpoint the major source of this pollution.

A large cattle-holding yard is located in Mexico approximately halfway between the two water-sampling stations (Fig. 1). Prior to entry into the United States, the cattle are forced to swim through a vat containing 15.2 kiloliters of an insecticide mixture. The vat empties into a small arroyo and drains into the Rio Grande about 25 meters away. In 1966, we found that this vat was the major source of contamination of Rio Grande water. A small dam was constructed across the arroyo in 1967 to hold the vat contents in check so they had to sink into the soil and enter the Rio Grande via subsurface

flow. Thereafter insecticide concentrations at collecting site 5 dropped markedly while concentrations at site 6 (Fig. 1) remained constant. This suggests that the majority of the compounds found in the Rio Grande come not from spraying but from the cattle dipping vat and from the sewage of Ojinago (10,000 population).

Water samples were taken from the Rio Conchos where the river cuts through the La Sierra Rica and enters the Presidio Basin. This site is approximately 60 kilometers upriver from Ojinaga. Insecticide concentrations were found to be on the order of 0.1 to 1 ppm.

Soil

Soil samples were taken from both the surface and from a 12-centimeter depth. The subsurface soil at all locations showed only traces of insecticides. Concentrations found in surface soils are shown in Table 3. Only trace amounts of the compounds were found further than 15 kilometers from the cotton fields. As might be expected, concentrations increased in the desert soils during the spraying season. In contrast to the situation in the desert soils, concentrations in the cotton-field soils increased during June and July, but during August con-

TABLE 2

Insecticides in the Rio Grande (ppm)

| Year 1965 | Site 5 | | | | | | | | Site 6 | | | | | | | |
| | Silt | | | | Water | | | | Silt | | | | Water | | | |
	P	M	D	E	P	M	D	E	P	M	D	E	P	M	D	E
June 6	1.0	0.2	0.2	0.0	10.0	2.0	2.0	2.0	0.1	0.1	0.1	0.0	0.4	0.2	0.0	0.0
July 26	1.5	1.0	3.0	2.0	16.0	4.0	4.0	4.0	0.2	0.0	0.0	1.0	2.0	2.0	6.0	1.0
Aug 29	4.0	2.0	5.0	3.0	21.0	9.0	11.0	6.0	0.3	0.4	0.2	2.0	4.0	6.0	8.0	3.0

P = Ethyl parathion; M = Methyl parathion; D = DDT; E = Endrin

TABLE 3

Insecticide Concentrations in the Surface Soil in the Presidio Basin (ppm)

| | Site 1 West Cotton Field | | | | | | Site 2 Center Cotton Field | | | | | | Site 4 East Cotton Field | | | | | |
	BHC	MP	P	DDE	DDD	DDT	BHC	MP	P	DDE	DDD	DDT	BHC	MP	P	DDE	DDD	DDT
June	0.1	0.3	2.4	2.5	0.0	0.0	0.6	0.0	1.9	2.6	0.0	0.0	0.1	0.0	1.0	2.0	0.0	0.0
July	0.2	4.9	3.9	4.8	0.3	0.5	1.5	2.9	2.4	4.0	0.0	0.1	0.2	1.4	2.0	3.6	0.0	0.0
Aug	0.1	3.8	3.1	4.9	0.0	0.0	0.4	0.0	1.1	1.9	0.9	0.6	0.2	1.4	1.9	0.9	0.1	0.3

| | Site 1 West Desert Periphery | | | | | | | | | | | | Site 4 East Desert Periphery | | | | | |
	BHC	MP	P	DDE	DDD	DDT							BHC	MP	P	DDE	DDD	DDT
June	0.1	0.8	1.4	2.1	0.2	0.0							0.1	0.4	1.0	2.0	0.1	0.1
July	0.2	1.9	2.7	3.8	0.3	1.4							0.1	0.9	1.6	2.8	0.7	0.0
Aug	0.2	3.1	4.5	3.6	0.0	0.0							0.1	2.7	3.4	2.9	0.0	0.0

| | Site 7 5 Kilometer Desert | | | | | | Site 9 15 Kilometer Desert | | | | | | Site 12 50 Kilometer Desert | | | | | |
	BHC	MP	P	DDE	DDD	DDT	BHC	MP	P	DDE	DDD	DDT	BHC	MP	P	DDE	DDD	DDT
June	0.3	0.8	1.0	1.9	0.3	0.1	0.2	0.3	0.2	0.9	0.2	0.0	0.3	0.0	0.0	0.1	0.0	0.0
July	0.2	1.1	1.6	1.0	0.0	0.0	0.3	0.3	0.1	1.0	0.1	0.5	0.2	0.1	0.0	0.1	0.0	0.0
Aug	0.3	2.0	2.0	1.0	0.0	0.0	0.1	0.6	0.2	0.9	0.4	0.0	0.2	0.0	0.0	0.1	0.0	0.0

centrations decreased. We believe this decrease was due to the large, mature cotton plants shielding the soil from direct application of insecticides.

We were surprised to find so little DDT and so much DDE in the soils. Obviously the DDE arose from breakdown of DDT. Previous papers have reported DDT to remain in the soil for up to 14 years (*11*). The present data show that DDT persisted in soils at the Presidio Basin for only a few weeks. Of the various factors listed by Lichtenstein (*12*) as affecting the persistence of pesticidal residues in the soil, only soil temperature and ultraviolet radiation appear to play a major role in the Presidio Basin.

To test this hypothesis, typical soils from the basin were spiked with known amounts of DDT and MP. They were then placed in environmental chambers where the temperatures and ultraviolet radiation of a typical summer day in Presidio were reproduced. Both insecticides broke down in a matter of weeks. (This work forms part of a doctoral dissertation in preparation.)

Such conclusions are also supported by the data of Mulla (*13*), who applied a DDT-endrin mixture to a sandy soil subjected to temperatures as high as 49°C. Two days later, bioassay against mosquito larvae and eye gnats showed a loss of 18 percent of the DDT and 38 percent of the endrin. These losses probably represent minimum figures. The bioassay would indicate the DDT breakdown products DDE and DDD as DDT. Soil-surface temperatures in the Presidio Basin are over 60°C for several hours each day during June, July, and August. It is highly probable that a greater loss of DDT occurred under these conditions than occurred in Mulla's experiment. Mosier, Guenzi, and Miller (*14*) have shown that under ultraviolet radiation DDT will be 80 percent decomposed within 48 hours.

Soil Biota

Soil samples were examined for fungi, nematodes, and bacteria. Desert soils taken at a distance from plants were relatively poor in microbiota compared to soil samples from root zones. In no case could we find any correlation between insecticide concentrations and microbiota. There was a direct correlation between rainfall and the populations. In cotton fields, once again, no correlation could be found between insecticide concentrations and microbiota while there was a correlation between irrigations and populations.

We next set up an experiment to discover if soil bacteria could metabolize DDT (*7*). Soils from cotton fields were placed in foil-lined boxes and scattered through a cotton field so they would be sprayed along with the cotton plants. Once again, the bacterial population was directly correlated with rainfall. During this study the amount of DDT decreased from 20 ppm to 0.8 ppm in the first eight days. After eight days more, the concentration was 0.5 ppm. Since the foil would prevent any leaching of DDT from the boxes, the compound must have broken down. Various cultures of bacteria were isolated from the soil within the boxes. Several of the isolates could be grown in culture with DDT as the sole source of carbon.

Birds

The English sparrow (*Passer domesticus* L.) was chosen as a representative avian species in the basin. It was the most accessible, widespread, and abundant bird in the settled areas around the cotton fields. Breast muscles, brains, livers, and gizzards from birds collected in the cotton fields were analyzed for insecticide concentrations. Not enough birds were collected in the desert to furnish reliable data. In general, concentrations rose during the spraying season (July-September), remained steady in the season immediately after spraying (October-November), increased again later with no spraying (December-February), and then dropped (Table 4).

We believe the pattern reflects both the availability of insecticides and of food. We would expect the concentrations in the birds to rise during the spraying season. After spraying has stopped, food is still in plentiful supply but it is not being contaminated with new applications of the compounds. Later, during the winter, food is in short supply. The birds then metabolize their fat reserves in which the insecticides are stored. This leads to the second increase of concentrations in muscle. Finally, the winter rains lead to an increased supply of insects and seeds not contaminated with insecticides. Because of this greater intake of food, fat reserves would no longer

TABLE 4

Mean Insecticide Concentrations in English Sparrows
(Each value is for 5 or 6 birds; ppm)

Month	Breast Muscle					Brain Tissue					Liver				
	MP	P	DDE	DDD	DDT	MP	P	DDE	DDD	DDT	MP	P	DDE	DDD	DDT
June	0.8	0	5.0	0.3	1.4	0.3	0	2.9	0.1	0.9	0.1	0	3.8	0.2	0.5
July	1.2	0	6.4	1.4	2.9	0.9	0	3.4	0.9	1.9	0.2	0	4.0	0.8	0.9
Aug	4.4	0.1	7.0	1.0	3.5	1.2	0	5.0	2.8	3.0	0.5	0	5.1	1.1	2.8
Oct	4.0	1.2	7.8	0.9	4.0	1.4	0.1	4.0	3.0	3.2	0.5	0	4.8	2.4	3.7
Nov	4.2	0.2	7.9	1.4	4.7	1.9	0.1	4.0	3.1	4.2	0.4	0	5.5	2.0	4.9
Dec	5.3	0.1	9.2	0.9	4.8	2.3	0.1	5.4	1.9	5.8	0.4	0.1	5.5	3.0	4.9
Feb	5.8	0.1	9.5	0.9	4.3	2.5	0.1	5.3	1.8	4.9	0.1	0	5.0	2.9	5.7
May	1.3	0	5.1	0.5	2.5	2.0	0.1	3.1	0.4	1.7	0.1	0	3.9	1.0	2.6

TABLE 5

Mean Insecticide Concentrations in Postcoelomic Fat of Lizards
(Each sample is five or six postcoelomic fat-bodies; ppm)

Month	Site 1 West Cotton Field						Site 2 Center Cotton Field						Site 4 East Cotton Field					
	BHC	MP	P	DDE	DDD	DDT	BHC	MP	P	DDE	DDD	DDT	BHC	MP	P	DDE	DDD	DDT
June	◄——— No Data ———►						1.5	0.0	0.3	31.5	19.2	14.5	◄——— No Data ———►					
July	11.5	4.2	2.8	17.4	34.3	43.0	8.9	0.0	1.4	45.9	32.0	44.3	◄——— No Data ———►					
Aug	8.5	2.1	1.0	21.2	20.8	25.1	1.2	0.0	0.0	18.8	7.6	7.6	0.4	0.3	0.6	5.8	3.4	4.0

Month	Site 1 West Desert Periphery						Site 2 Center Desert Periphery						Site 4 East Desert Periphery					
June	◄——— No Data ———►						0.7	0.0	0.0	25.2	0.0	2.8	0.2	0.0	0.3	19.4	2.1	1.5
July	6.8	3.7	4.1	31.4	7.9	26.1	2.1	1.6	3.6	35.4	10.3	8.4	◄——— No Data ———►					
Aug	7.0	0.0	0.0	22.4	10.4	16.2	2.3	0.0	0.0	28.2	7.3	4.0	0.1	0.1	0.1	17.7	0.7	1.9

Month	Site 7 5 Kilometer Desert						Site 9 15 Kilometer Desert						Site 12 50 Kilometer Desert					
June	0.0	1.9	1.1	8.4	3.1	4.8	0.0	1.0	0.0	7.9	3.7	3.6	◄——— No Data ———►					
July	1.0	1.8	2.3	10.4	4.0	4.3	1.1	2.0	1.3	10.1	6.9	3.6	1.2	0.6	0.0	5.4	2.2	1.7
Aug	1.4	2.4	1.8	15.7	5.1	4.0	◄——— No Data ———►						0.0	0.0	0.0	9.4	1.5	0.0

be metabolized, and insecticide concentrations therefore drop.

We obtained interesting data relative to avian uptake from the Big Bend National Park, approximately 160 kilometers east of the Presidio Basin. No insecticides have been applied in the park since 1944. After we secured a collecting permit, we collected birds (other than the English sparrow) in various areas of the park. The species fell into two groups as far as insecticide concentrations were concerned: they contained either a great deal or else very little insecticide. In every case, those species containing higher insecticide concentrations were transients who spent part of each year outside of the park; those species containing little insecticide were residents. We plan to publish these data together with data from soils and mammals of the Big Bend National Park in the near future.

Lizards

Three species of lizards were studied: *Cnemidophorus tesselatus* Say, *C. tigris* Baird and Girard, and *C. inornatus* Baird and Girard. No species differences in insecticide concentrations were found. In general, concentrations increased in tail muscle, brain, and liver tissues during the spraying season. Residue concentrations fell as a function of the distance from cotton fields. Lizards collected more than 15 kilometers from the fields had essentially no insecticides.

Insecticide concentrations in the postcoelomic fat-bodies of lizards are summarized in Table 5. There was a sharp rise during June and July followed by a sharp drop in August. Even in August, however, the postcoelomic fat-bodies contained greater concentrations of chlorinated insecticides than did the muscle, liver, brain, and stomach contents.

In the latter part of July many of the female lizards contained eggs. Separate determinations were made of the gravid female muscle tissue and of the egg. In every case, the egg contained higher insecticide concentrations than did the muscle tissue (Table 6). No significant differences could be detected when concentrations in the muscles of gravid females were compared with concentrations in the muscles of nongravid females collected on the same dates and at the same sites.

TABLE 6

Mean Insecticide Concentrations in Muscle of Four Gravid and
Four nongravid Female Lizards and Eggs
(All lizards were collected in same field; ppm)

	BHC	MP	P	DDE	DDD	DDT
Females, gravid	1.1	2.5	1.4	3.4	2.8	2.1
Eggs	5.6	11.6	8.5	16.4	7.3	10.7
Females, nongravid	0.8	2.3	2.1	3.4	2.8	2.0

Lizards hibernate from October to May in the Presidio Basin. The adults of any given year have been born the previous summer (*4, 6*). Small and immature eggs appear in the latter part of July. Hahn and Tinkle (*15*) and Hoddenbach (*16*) report that the postcoelomic fat-body serves a reproductive function in *Uta stansburiana* and in *C. sexlineatus*. Our data on insecticide concentrations support the idea that the postcoelomic fat-body of *Cnemidorphorus* species in the Presidio Basin has a similar reproductive function. The compounds, after leaving the stomach, are stored in the postcoelomic fat-body. During egg development this body is metabolized and the insecticides are mobilized. They then are stored in the developing egg. There was an inverse relationship between the size of the egg and the size of the postcoelomic fat-body. Obviously, the stored insecticides had to go someplace; the eggs had to obtain their insecticides from someplace. They did not obtain it from muscle tissue. (Gravid and nongravid females had similar muscle concentrations of insecticides, Table 6.) We believe the postcoelomic fat-body supplied the insecticides to the eggs. A more intensive study of the fat-body–reproduction–insecticide relationship forms part of a doctoral dissertation now in preparation.

Mammals

Two species of pocketmice (*Perognathus pencillatus* Woodhouse and *P. merriami* Allen) and one species of kangaroo rat (*Dipodomys merriami* Mearns) were used as examples of burrowing mammals. The black-tailed jackrabbit (*Lepus californicus texianus* Waterhouse) was sampled as a nonburrowing mammal. None of these mammals was found in the cotton fields. Many were collected in the desert less than 30 meters from the fields, however. Data for the rodents will be discussed first.

The overall pattern of insecticide concentration in the rodents was similar to the pattern in the English sparrows (4, 6). Concentrations in the livers and leg muscles increased during the spraying season (June-August),

remained constant in the season immediately after spraying (September-November), and then rose later (December-February) before dropping (March-May). We believe the pattern can be explained by the same food-insecticide relationship advanced for the birds.

We made an interesting observation during the live-trapping of the rodents. Recapture occurred much more frequently in areas having high insecticide concentrations in the soil and vegetation than in areas having lesser concentrations. Rodents from these areas where concentrations were high had a higher bodyload of residues than did rodents taken where environmental concentrations were low. The various areas were examined for difference in soil temperature, rainfall, vegetation, and predators; we could detect no clearcut differences between the areas of high and low recaptures. This finding suggests that the insecticides may have been the crucial factor in leading to recaptures.

A minimum of 10 and a maximum of 50 rabbits were collected each month from June 1966 through August 1967, and from June 1968 through August 1969. The effects of insecticides on biological parameters during June 1966 through August 1967 are presented in Table 7 (18).

TABLE 7
Effects of Insecticides on Biological Parameters in Jackrabbits

	Collected more than 5 km from cotton fields		Collected less than 5 km from cotton fields		Significance per F-tests
	Mean	Standard Error	Mean	Standard Error	
Insecticide Concentrations (ppm)					
MP Brain	8.6	1.0	11.0	1.4	12.5
MP Muscle	6.4	0.6	8.0	1.7	NS
DDT Brain	1.9	0.6	2.2	0.8	NS
DDT Muscle	1.5	0.3	1.3	0.4	NS
Individual Indices					
Weight	2355 gr	57	2388 gr	44	NS
Length	501 mm	4	499 mm	3	NS
Foot	124 mm	0.7	124 mm	0.7	NS
Ear	136 mm	3	131 mm	0.7	NS
Tail	97 mm	9	88 mm	0.8	NS
Lens weight	234 mg	10	213 mg	4	NS
Reproductive Indices					
Testes Position *	1.5	0.1	1.4	0.05	NS
Testes Length	41.4 mm	1.7	35.9 mm	1.5	20.4
Testes Weight	7.05 gr	1.8	5.2 gr	0.6	NS
Uterus Condition *	4.18	0.29	4.10	0.16	NS
Mammary Condition *	2.5	0.10	2.2	0.08	NS
Ovary Weight	0.63 gr	0.05	0.78	0.08	13.5
Uterine Scars	3.8	0.3	3.4	1.4	NS
Fertility	1.66	0.22	1.74	0.13	NS
C-R † Length	49.5 mm	8.4	38.6 mm	4.2	293
Condition Index (Fat) *	2.90	0.1	3.29	0.07	8.4

* Scales were subjectively assigned for comparative purposes.
† Crown-rump.

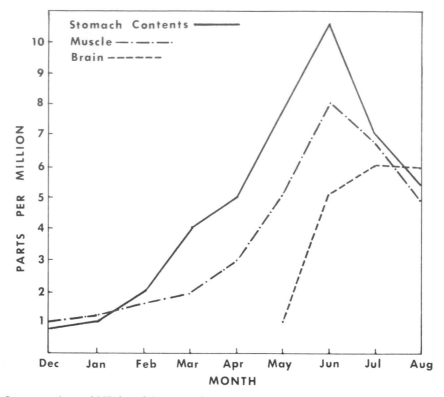

Fig. 5. Concentrations of MP (ppm) in stomach contents, muscle, and brain tissue of jackrabbits.

Insecticide levels were generally higher in specimens from the flood plain than those from the adjoining desert; however, only MP levels in the brain differed significantly between the two locations. According to the biological parameters the two populations were essentially alike, that is, equally successful. F-tests showed all but four differences to be nonsignificant. Desert rabbits had fetus crown-rump length and testes length significantly greater than did river rabbits. Animals closer to the river had ovaries significantly greater than those of desert rabbits. Jackrabbits from the desert were fatter than those closer to the river.

Since MP concentration in the brain was the only significant insecticide factor (Table 7), we took a closer look at it. The concentrations of MP in stomach contents, muscle, and brain tissues are given in Figure 5. It is apparent that a food-muscle-brain sequence occurs in jackrabbit-MP relationship. Metabolism of contaminated food allows the chemical to be stored in muscle and muscle fat. As more insecticides become available in the environment, more are stored. During periods of low food availability, fat deposits are mobilized and used. By June the reserves of fat were exhausted, and with lower concentrations in ingested food, a corresponding drop in quantity of residues in the muscle was noted.

The decline in residues located in the muscle was associated with an increase of amount in the brain. Brain tissue, because of its high lipid content, apparently was the storage point for residues mobilized from the muscle.

What effect, if any, did the MP concentrations in the brain have on an individual jackrabbit? As judged by the data in Table 7, little, if any, effect on condition or reproductive indices can be ascribed to the MP. What effect, if any, did the MP concentrations in the brain have on the jackrabbit population? We determined the age of each rabbit by means of dried eye-lens weight and the degree of epiphysial closure of long bones. Each rabbit was put in one of three classes: Class I, 2 to 9 months; Class II, 10 to 12 months; and Class III, older than one year. During 1966–1967 we determined the concentrations of MP in the brains of each age class: Class I (juveniles) had 9.5 parts per million; Class II (yearlings) had 9.7 parts per million; Class III (adults) had 5.2 parts per million. In the summer of 1966 adults made up 40 percent of the population, while in the summer of 1967 only 11 percent of the population were adults. The high percentage of young jackrabbits in 1966 contributed to the yearling class in 1967, but the 1966 yearlings failed to contribute to the adult class of 1967.

No data presented here lead inescapably to the conclusion that MP in the jackrabbit brains leads to this population shift. Results suggest this conclusion, however — particularly when considered together with the rodent data. Rodents having high body burdens of insecticides are more likely to be recaptured in live traps than rodents having low body burdens, according to our data. We propose that MP in the brain interferes either with motor ability or with behavioral patterns. A rodent having a high body burden of MP would therefore be unable to escape a trap once triggered or would walk into a trap before recognizing it. A live trap makes little ultimate difference to the rodent, that is, upon release the rodent again becomes a constituent of the ecosystem. Replace a live trap with a natural predator, and it does make a difference to the rodent. We are, of course, biasing our data. A rodent with a body burden of MP should be more liable to capture by a natural predator, while in a live trap the rodent is immune from natural predation. Selection is being made to conserve these rodents.

But what of jackrabbits? These were shot, not captured in live traps. If the MP in the brain interfered with either the motor ability of rabbits or with behavioral patterns, then these rabbits should be captured by predators more often than rabbits with unimpaired motor ability or normal behavioral patterns. We would thus tend to sample that portion of the jackrabbit population with lower concentrations of MP in the brain, while for rodents we would tend to sample that portion of the population with higher MP concentrations in the brain. The reason for the bias would be the same: MP in the brain either hindered motor ability or changed behavior patterns. We recognize that this is highly speculative and based on empirical data. Controlled experimentation is needed to confirm our speculation .

DISCUSSION

We have studied the effects of insecticides (chiefly DDT and MP) on an isolated ecosystem of the Chihuahuan Desert for 5 years. Only a small portion of our data is presented here. The sheer quantity of the data has hindered our effort to integrate a coherent picture. A broad pattern is beginning to emerge, however. The pattern will be modified and changed as further data are fitted into it. We believe the broad outlines of the pattern will not be changed by subsequent data, however.

When we started this study we expected to find that insecticides in the Presidio ecosystem would follow patterns reported from mesic and aquatic ecosystems (*17*), that is: (*a*) halogenated compounds would persist for long periods of time, (*b*) phosphorylated compounds would disappear in relatively short periods of time, (*c*) the breakdown of all insecticides would be to the same products as reported for more mesic zones, and (*d*) toxic and lethal effects would be found. None of the expectations occurred.

Water determines the rhythm of life in the desert. It also determines the fate of insecticides in the Presidio Basin. The compounds are applied in the early morning when there is little wind. Our studies have shown there is little drift during spraying operations. The insecticides fall on both soil and plants. In the early part of the growing season, when cotton plants are small, most of the insecticides reach the soil; during the growing season more and more falls on the plants and less and less on the soil. This change in initial deposition is reflected in the final deposition of the compounds.

The irrigated fields are geographically at the lowest point in the basin. During the day this low portion acts as a huge heat sink. The heated air rises and moves out over the cooler and higher desert. Cooler air moves down the arroyos to replace the hotter air. As the air moves it picks up dust particles to which insecticides are adsorbed. In the early summer much of the dust comes from the cotton fields and thus contains high concentrations of insecticides. Later, the cotton-field soils are sheltered by the plants, and most of the dust comes from roadways. This dust contains lesser amounts of insecticides. Dust devils are an everyday occurrence along the cotton fields. In the summer afternoons, dust storms commonly occur along the floodplains. This dust moves from the floodplains to the desert areas where it is deposited on plants, soil, and any exposed fauna in the desert.

Once the insecticide-carrying dust is deposited in the desert, a complex series of actions can take place. If no moisture is available, the dust remains on the soil and plant surfaces. The insecticides are degraded by high temperatures, intense ultraviolet radiation, and an alkaline soil pH. If the desert plants are actively growing and water is present either as soil moisture or dew, the plants can absorb the compounds. They then can be passed along the food chain to herbivores.

If the desert plants are not actively growing when the dust is deposited, insecticides and their breakdown products will remain on the soil and plant surfaces until a rain. During the summer, rain in the Presidio Basin is sporadic, scattered, and violent. The rain will wash the insecticide-laden dust into arroyos and thence into the Rio Grande.

But what of the compounds that do enter the food chain? We have shown that leatherstem can accumulate MP. Very few of these plants are growing along the floodplains adjacent to the cotton fields. Those plants that do grow there are sheltered by boulders and shrubs from direct drift from the cotton fields. On floodplains above and below the Presidio Basin, leatherstem is common. Is MP toxic to leatherstem? Plants in our laboratory grew well when watered with a 1,000 parts-per-million MP solution. Were the plants used by the native population for medicinal purposes? (The remaining leatherstem plants along the Presidio river bench, while sheltered from insecticide drift from the cotton fields, are not hidden from human view and thus could have been used.) Is some breakdown product or combination of breakdown products toxic to the plant? We do not know.

Let us emphasize the following fact: In all probability insecticides are not a direct factor in the mortality of organisms in the Presidio Basin. This statement is based on the comparison of our residue analysis with residue data from other laboratories for white rats, albino rabbits, and robins, and our failure to find or capture any animals showing poisoning symptoms. Do the compounds play an indirect role? This is moot. There are only suggestions that they do play such a role.

Our data indicate that very little of the DDT is absorbed by any of the organisms because it is rapidly broken down by the environment of the Presidio Basin. We have identified some of the breakdown products of DDT. The major breakdown product, as judged by peak area in gas-liquid chromatography, has not been identified. In August this unknown compound is by far the most plentiful of those detected by electron-capture techniques in all samples: soil, lizards, birds, mammals, and vegetation. It disappears during the winter months, only to reappear in July and peak in concentration in August. Until it is identified we cannot even speculate as to its role in the ecosystem.

The times reported in this chapter for insecticide breakdown and the various biological phenomena are entirely dependent upon the weather. The 42-year climatic record of the U.S. Weather Bureau for the basin was analyzed. Our 5-year climatic records for various sites within the basin were compared to the longterm record. Sites only 3 miles apart had different climates (B. Harris, unpublished thesis). In all cases, rain was the major factor in determining other weather phenomena. Air, plant, and soil temperatures, radiation intensity, and, of course, humidity were directly correlated with rain. Insecticide breakdown and the biological phenomena were advanced or delayed as much as two months

from the times given in this chapter. In some years, the advances or delays were characteristic of the entire basin; in other years only small areas in the basin differed from the times reported here. In every case, the advances or delays could be correlated with rain. As might be expected, the advances or delays were more pronounced in the desert than in the irrigated fields (19).

As is true of most experiments, ours probably raised more questions than it answered. One thing is clear: you cannot extrapolate from insecticide data gathered in mesic environments to arid environments. Obviously, much more work is needed if our children are to understand Robinson Jeffers:

The quality of these trees, green heights; of the sky, shining; of water, a clear flow; of the rock, hardness and reticence: each is noble in its quality.

REFERENCES AND NOTES

1. APPLEGATE, H. G.
   1966   Pesticides at Presidio. I: General survey. Texas Journal of Science 18:171–178.
2. APPLEGATE, H. G.
   1966   Pesticides at Presidio. II: Vegetation. Texas Journal of Science 18:266–271.
3. LAHSER, C., AND H. G. APPLEGATE
   1966   Pesticides at Presidio. III: Soil and water. Texas Journal of Science 18:386–395.
4. CULLEY, D. D., AND H. G. APPLEGATE
   1967   Pesticides at Presidio. IV: Reptiles, birds, and mammals. Texas Journal of Science 19:301–310.
5. APPLEGATE, H. G.
   1967   Pesticides at Presidio. V: Synopsis. Texas Journal of Science 19:353–361.
6. CULLEY, D. D., AND H. G. APPLEGATE
   1967   Insecticide concentrations in wildlife at Presidio, Texas. Pesticides Monitoring Journal 1:21–28.
7. TAYLOR, R. G., AND H. G. APPLEGATE
   1968   Pesticide levels and fluctuation in soil bacteria in an irrigated field and a pond in southwest Texas. Southwestern Naturalist 13:393–400.
8. BURKE, J., AND L. GUIFFRIDA
   1964   Investigations of electron capture gas chromatography for the analysis of multiple chlorinated pesticide residues in vegetables. Journal Association Agricultural Chemists 47:326–342.
9. APPLEGATE, H. G., AND G. CHITTWOOD
   1968   Automation of pesticide analysis. Bulletin Environmental Contamination and Toxicology 3:211–226.
10. MUMMA, R. O., AND T. R. KANTNER
    1966   Identification of halogenated pesticides by mass spectroscopy. Journal Economic Entomology 59:491–492.
11. NASH, R. G., AND E. A. WOOLSON
    1967   Persistence of chlorinated hydrocarbon insecticides in soils. Nature 157:924–927.
12. LICHTENSTEIN, E. P.
    1965   Persistence and behavior of pesticidal residues in soils. Archives of Environmental Health 10:825–826.
13. MULLA, M. S.
    1960   Loss of chlorinated hydrocarbon insecticides from soil surfaces in the field. Journal of Economic Entomology 53:650–655.
14. MOSIER, A. R., W. D. GUENZI, AND L. L. MILLER
    1969   Photochemical decomposition of DDT by a free-radical mechanism. Science 164:1083–1085.
15. HAHN, W. E., AND D. W. TINKLE
    1965   Fat body cycling and experimental evidence for its adaptive significance to ovarium follicle development in the lizard *Uta stansbuirana*. Journal of Experimental Zoology 158:79–86.
16. HODDENBACH, G. A.
    1966   Reproduction in western Texas of *Cnemidophorus sexlineatus* (Sauria: Teiidae). Copeia 1:110–113.
17. RUDD, R. L.
    1964   Pesticides and the living landscape. University of Wisconsin Press, Madison. 320 p.
18. HANSELLEA, C. W.
    1968   Effects of pesticides on black tailed jackrabbits in the Presidio Basin, Texas. Texas A & M University (M.S. thesis).
19. The research reported in this chapter was supported by Grant AP 28, National Air Pollution Control Administration, Consumer Protection and Environmental Health Service, Public Health Service; and Grant CC 272, National Communicable Disease Center, Public Health Service. Some of the data were gathered by the following students as part of their graduate research: R. G. Taylor (bacteria), J. Saxon (lizards), B. Harris and D. Ellermeier (meteorology). Drs. J. Griffiths and W. H. Thames have aided greatly in their discussions of the problem with us. This is Technical Article No. 8154, Texas Agricultural Experiment Station, College Station.

Part Five
INFORMATION SOURCES

Arid-Lands Information in United States Government-Sponsored Indexing Tools: What? Where? When? How?

PATRICIA PAYLORE
Office of Arid Lands Studies
University of Arizona
Tucson, Arizona, U.S.A.

ABSTRACT

SEVERAL UNITED STATES GOVERNMENT-SPONSORED indexing and abstracting tools (USGRDR, STAR, TT, and their antecedents and indexes) particularly useful to arid-lands research are described and their development since the end of World War II traced. Case histories illustrate their use in such bibliographic searches as those for translations, retrospective subject, and technical report number information. The paper is intended for the arid-lands scientist without the services of skilled bibliographers, and for the librarian unfamiliar with the format and coverage of these complex indexes.

ARID-LANDS INFORMATION IN UNITED STATES
GOVERNMENT-SPONSORED INDEXING TOOLS:
WHAT? WHERE? WHEN? HOW?

Patricia Paylore

INFORMATION EXPLOSION is as common a contemporary cliché as population explosion, but its very commonality is an index to the alarm with which it is viewed by the scientific community and the source of the frenetic attempts to bring it under bibliographic control. There is nothing new about this observation. The literature explosion on the information explosion is itself creating something approaching a crisis. The number of societies and organizations that have sprung up to bring together the documentalists and information specialists, with anxious librarians crowding the fringes, poses budget problems of priority for the average man who, though now thinking of himself as an exlibrarian, nevertheless is still reluctant to forsake his former library affiliations. Still, lemming-like, the computer-oriented professional bibliographer, till now with a foot in each camp, seems to be moving inexorably into the sea of information science, hopefully not to drown there. Two questions come to mind: Is it too late? Is it worthwhile?

The computer will doubtless save us. Though the hour is late, many are giving their all for the chance to demonstrate that it is not too late. The tremendously impressive computer-produced indexing and abstracting tools proliferating throughout the world are striking evidence that we are succeeding. The United States government has fostered and supported and disseminated much of the intellectual effort that has been marshalled to this end, and it is the purpose of what follows here to describe in some specificity the historical development and present status of those government publications in this field that constitute the most useful services to the arid-lands scientist. The cooperative alliances, the shifts in emphases in response to new developments, and the expanded services provided by many of these publications during the last twenty-five years are this country's best answer to date to the question, "Is it too late?"

Is it worthwhile? Yes, and no. Yes, if we assume that all we are indexing and abstracting and describing and formatting and retrieving is worthwhile. If we acknowledge, however, that a large percentage of the information that is being manipulated by our expensive programmers for our expensive hardware contributes, at best, nothing earth-shaking to society except "more and more about less and less," or, at worst, perpetuation of actual misinformation, the answer might be no. Historically it has not been the bibliographer's function to evaluate. On the horizon, however, or nearer still, and already somewhat larger than a man's hand, is the shape of things to come, the subject bibliographer, working within the framework of an information analysis center, who brings to bear on his bibliographical investigations the scientific judgment to evaluate as well. The eventual result, theoretically, might be a slowdown in the mass of information being disseminated through abstracting and indexing. One distinguished scientist has already pronounced the ultimate caveat when he says, "We must find a way to throw out, for good and for all, a good fraction of the published scientific literature, and after a short time virtually all of the unpublished report literature" (*1*). To the cynical information specialist, gloomily contemplating the outpouring of undigested "information" in whatever form, no matter how slickly picked up by an appropriate specialized tool of the trade, such a euphoric prospect seems beyond the realm of possibilities. Other things being equal, though they are not actually so, we shall be obliged to continue exploring the means to maintain bibliographic order while we seek the wisdom and knowledge to discriminate.

The logical and ideal extension of such bibliographic phenomena might be a move to provide scientists with their own personal bibliographers, skilled in the intricacies of the very tools designed to make the search easier, but at the same time technically oriented toward the subject need of the scientist and/or his department. The need to train scientific interpreters qualified to employ subject competence in providing such services is now widely acknowledged, most articulately, perhaps, by Kieffer (*2*) when he wrote: "Scientific scholarship of the highest order is required for this work, but at present [physical] scientists do not regard evaluation of the work of others as an acceptable scientific career in itself." Can a second order of scientific personnel, the scientific bibliographers, then, bridge the gulf that presently does and will increasingly separate the scientist from the literature of his field? To fail to provide by some acceptable means for this orderly transfer of information — which, incidentally, is much more complex than merely the acquisition, storage, and retrieval of documents — is to force the scientist himself into a specialization narrow enough to accommodate his personal responsibility for being informed through its literature. On several counts the implementation of such a potential development is a denial of the direction in which modern scientific man is going.

Today, without the bibliographic apparatus of the several centralized agencies of the U.S. government to handle and control the half a million or more technical reports being generated yearly by the laboratories and contractors of the various government agencies concerned with national research, the communications gap would indeed be catastrophic. But year by year, as the title histories below indicate, there has been a forward movement toward a coherent network of information resources and services in which many agencies are cooperating. The standards arrived at through the trials and shortcomings and experimentation of just the one great tool, *U. S. Government Research and Development Reports,* are now being adopted and applied throughout the world of research. The almost universal acceptance, for instance, of the COSATI list of subject categories, which led to the development of the great TEST thesaurus (constructed as a joint venture of the DoD and the EJC), provided not only a common vocabulary but an intellectual brake on the almost undisciplined trend toward independent and freewheeling indexing.

Perversely enough, it is also true that now that we are disciplined, we are also seeking ways to bring a degree of specificity, within the larger framework, to indexing that the more generalized thesauri do not permit. Even here, however, the philosophy and format of vocabulary control through hierarchical and generic relationships is so firmly accepted and understood that specialized indexing becomes itself a first cousin, or perhaps even a true and natural child, of the basic concept pioneered by the government's indexing services.

THE TOOLS

What follows is intended for the individual scientist without access to the services of competent and willing librarians, but also for librarians not yet sufficiently at home with these specialized tools to make full use of the technical information they contain.

To circumscribe our undertaking to manageable limits, let us arbitrarily begin with the post World War II era. Prior to that time, technical information had achieved a reasonably stable equilibrium, with most of the variables being under fair bibliographic control. Thereafter, increasing government support of scientific research spawned a new way of packaging scientific information that was to become the librarian's bane, the technical report, that very unclassical maverick "characterized principally by its heterogeneity of style, professional stature, size and form of reproduction, and by the absence of anything like volume and number relationships" (*3*). For a generation it would be considered a menace to conventional bibliographic control.

Bibliography of Scientific and Industrial Reports (BSIR)

In the summer of 1945, the Office of the Publication Board, operating under the Department of Commerce, was created by executive order to "disseminate unclassified scientific and technical information obtained from enemy sources and Government-sponsored research." To fulfill this goal, on January 11, 1946, volume I, number 1, of the *Bibliography of Scientific and Industrial Reports* (BSIR) appeared, to be issued weekly for the next two years and nine volumes. It was arranged first in 15 subject categories, relating largely to products and processes, many of which might be the subject of U. S. patents. Within each category, the arrangement was alphabetical by author, and included the title, PB (for Publication Board) number, date, pages, price, and an abstract for each entry. The issue for April 19, 1946, carried the following information:

The Reports listed have been received from civil and military agencies of the U.S. Government and from cooperating foreign governments [German, Japanese, Swiss, Dutch, Russian, Italian]. Many of the reports cover information captured in enemy countries. Secrecy restrictions on all reports have been lifted. (vol 1, no. 15)

On July 1, 1946, the Office of Technical Services (OTS) was established and assumed the functions of the old Office of the Publication Board. The PB symbol of the Board's early publications was retained (to the very day of this present writing, as a matter of information) as a designation for its reports. In February of the following year the OTS announced:

Reports listed and described in this publication have been deposited in the appropriate Federal libraries where they become part of their permanent collections. Photographic copies are prepared and furnished by the duplicating divisions of these libraries. . . . OTS can furnish information as to the depository library at which the materials can be examined. (vol 4, no. 7)

Effective with volume 10, July 1, 1948, the BSIR became a monthly publication, with its entries being made on a much more selective basis. The rather full abstracts that had characterized the original entries were dropped, a descriptive format more nearly resembling a conventional library catalog card adopted, and, in lieu of abstracts, subject headings applied. Unless we are very much mistaken, the latter were the first evidences of the tremendous vocabulary control that eventually evolved being applied to the technical-report literature. And, evidently in response to complaints about the absence of abstracts, BSIR, shortly after dropping them, resumed publication of a very few.

Bibliography of Technical Reports (BTR)

One year later, on July 1, 1949 (vol 12, no. 1), the BSIR became the *Bibliography of Technical Reports* (BTR), still published by the OTS, and, initially, with the same coverage, still furnishing only occasional abstracts.

Still somewhat oriented towards products and processes, the BTR devised a *Newsletter* to include in its monthly issue, a brief bulletin highlighting business opportunities arising from federal and other nonconfidential research, with specific emphasis upon items useful to the smaller firm. The *Newsletter* lived on through 1964, when it was finally discontinued with the December issue of USGRR.

Prepared (On-Demand) Bibliographies

Now, after little more than three years from its inception, the need for prepared bibliographies to isolate information on special topics from the mass of reports being announced confronted the OTS. It responded through the BTR with perceptibly lengthening lists of such prepared bibliographies that are still available, with the concurrent evolution of the present on-demand bibliographies offered by all the government documentation centers. A typical output of the on-demand type would be the following:

Deserts, A Report Bibliography, requested by Dr. George Howe, October 1966. Prepared by Defense Documentation Center. Unclassified. ARB-No. 059393.

This particular report bibliography itself was cited in the bibliography accompanying U. S. Army Natick Laboratories Technical Report 69-38-ES (*4*). The report bibliography was produced by a computer search of the DDC's data bank and appears in the familiar computer format.

In January 1950 (vol 13, no. 1) the BTR offered, for the first time as far as could be determined, information on translation services available commercially from the Special Libraries Translations Pool (see below for detailed development of translations services).

U.S. Government Research Reports (USGRR)

With the issue for October 15, 1954 (vol 22, no. 4), the BTR became *U. S. Government Research Reports* (USGRR), with its format unchanged. The OTS stated in the first issue under the new title that it "describes more accurately the function of the publication: the monthly listing of from 300 to 400 technical reports from Government-sponsored research made available by OTS to industry and the general public."

By July 1, 1961 (vol 36, no. 1), USGRR was being issued twice monthly, reflecting the quantum jump in the number of reports being announced. At the same time, a cooperative program for such announcement of research reports was arranged by ASTIA and OTS to make DoD research information more widely available to the American public. ASTIA's TAB (not covered in this present chapter, because of TAB's partially restricted status) released to OTS its "white" unrestricted section for inclusion in the latter's USGRR, where the section was reproduced in identical text and format. Thereafter, until July 10, 1966 (vol 41, no. 13), when these two sections were merged in USGRDR, that reporting service was divided into two parts: (*a*) technical announcement bulletin (i.e., the white section of TAB, listing mostly unrestricted ADs); and (*b*) nonmilitary and older military research reports (mostly PBs from AEC, Office of Saline Water, and reports of other civilian agencies, plus military reports not found in TAB). For this period, July 1961–July 1966, this agreement between ASTIA and OTS substantially increased the number of government reports available to science and industry. The merging of information in 1966 did not lessen this increased coverage but simply enabled users to search each issue once instead of twice.

During this decade or more of development, other more specialized tools were being refined: *Nuclear Science Abstracts* (NSA, not covered in this paper because of limited subject application to arid-lands research), *Scientific and Technical Aerospace Reports* (STAR), and *Technical Translations* (TT), to name only three. Because of the degree of cooperation that had evolved among these government documentation centers, the USGRR in August 1964 announced that it would henceforth contain listings and abstracts of only that material not listed and abstracted in the other specialized journals abstracting documents available from the OTS, namely NSA, STAR, and TT.

The Clearinghouse

Though a White House announcement on February 28, 1964, called upon the Department of Commerce to strengthen and broaden the functions of the OTS, "to make the agency more effectively a clearinghouse for federal scientific and technical information," the actual imprint of this newly named agency that supplanted certain functions of the OTS did not appear until January 1965, when the Clearinghouse was established

to serve as the central source for Government research data in the physical sciences and engineering ... and the document distribution program of the OTS ... [and] as the single agency through which unclassified technical reports and translations generated by all Government agencies are uniformly indexed and made available to the public; provide information on Federal research in progress; and operate a referral service to sources of specialized technical expertise in the Government. ... It is a part of the new Institute for Applied Technology in the National Bureau of Standards. ... In addition to providing research information to industry, the Clearinghouse is designed to reduce duplication in both industry and Government in research and information processing. An agreement with the Department of Defense designates to the Clearinghouse the task of processing all unclassified/unlimited DoD research reports, as well as reproduction and distribution of these documents to both the public and DoD agencies and contractors. (USGRDR, vol 40, no. 1, Jan. 5, 1965)

National Technical Information Service

On September 2, 1970, the Secretary of Commerce established the *National Technical Information Service* (NTIS) "to simplify and improve public access to Federal publications and data files of interest to the business, scientific, and technical communities." The Clearinghouse was at that time transferred from the National Bureau of Standards to the NTIS, and its functions merged with the broader mission stated for the latter agency. No deviation from the format or coverage of previous Clearinghouse announcements is yet visible. The nomenclature change appears at this writing to have achieved nothing more than to add to our acronym list and further complicate our understanding of these tools and their relationships.

U.S. Government Research and Development Reports (USGRDR)

Concurrent with the 1965 announcement of the establishment of the Clearinghouse, USGRR became *U. S. Government Research and Development Reports* (USGRDR), with the issue for January 5, 1965 (vol 40, no. 1), and there, until early 1971 when it became *Government Reports Announcements* (GRA), it remained, having come to maturity as an eminently useful and admirably organized searching tool for this most difficult of all publication formats.

For a single year, 1966 (vol 41), a feature called "Current r & d Projects" was carried in each issue of USGRDR, designed to "promote active interchange between research workers in the same or related fields by providing current awareness information on who is doing what research where." Information was based on data supplied by the Science Information Exchange (SIE) of the Smithsonian Institution, to which federal agencies reported their new projects. The feature was discontinued, however, at the end of the year 1966.

On September 10, 1967, USGRDR stated that it would announce all DoD unclassified-unlimited AD reports formerly announced in the white section of TAB, as well as AEC, NASA, etc, reports, all of which would be indexed in Government-Wide Index (see below for details of indexing developments). DDC users were informed that they might obtain AD documents from the DDC without cost by using Form #1. Under this rearrangement, non-AD documents are not available from the DDC and cannot be ordered on DDC Form #1, but must be purchased from the NTIS or obtained from the source indicated in the announcement of availability. At this same time, it was announced that USGRDR would be distributed to registered users of DDC in lieu of the white section of DDC's TAB.

On January 10, 1968, USGRDR assumed reporting of technical translations, following the demise of *Technical Translations* (see below), and on that date it picked up for listing in its "Report Locator List" (which itself first appeared in January 1967) the JPRS and TT numbers. As far as could be established, the first JPRS number to be listed in USGRDR was #42786, the first TT number #66-59006/12. It would seem, therefore, to be futile to search in USGRDR for earlier numbers in these two translations series. At the same time, USGRDR announced it would no longer carry abstracts for AEC or NASA reports listed in USGRDR, but instead refer the user to the abstracting journal of the original announcement (for example, STAR).

Selected Water Resources Abstracts

One of the more specialized government-sponsored abstracting and indexing tools is *Selected Water Resources Abstracts* (SWRA), issued since volume 1,

number 1, January 1968, by the Water Resources Scientific Information Center (WRSIC), Office of Water Resources Research, U. S. Department of the Interior. This Center was designated by the Federal Council for Science and Technology as the national center for water resources scientific and technical information activities. It seeks to improve and expedite the exchange of scientific and technical information among authors and users in water resources activities by "initiating efforts to coordinate and complement existing Federal and Federally supported technical information services; providing central operation of such water resources technical information services as can be best accomplished on a nationwide level; and by conducting studies as a basis for compatibility standards and new methods for processing water resources information, and for identifying current and changing user needs."

SWRA, the tool developed to achieve these goals, started out as a monthly, but in September 1968 became a semimonthly publication that includes abstracts of current and earlier pertinent monographs, journal articles, reports, and other publication formats. Selective acquisition by technical specialists is carried on through assigned subject areas by seven "centers of competence" as well as by the Water Resources Research Institutes throughout the country. This information base provides for a variety of information services, over and beyond SWRA, such as bibliographies, specialized indexes, literature searches, and state-of-the-art reviews. Because of its single discipline-oriented coverage, WRSIC developed its own specialized thesaurus of terms (5) used to classify the input of this system and adopted in lieu of the COSATI subject categories those fields and groups established by the Committee on Water Resources Research of the Federal Council for Science and Technology.

While SWRA is published by the NTIS in the same format as USGRDR announcements, WRSIC does not furnish either loan or retention copies of the publications described in SWRA. Some entries abstracted can be purchased from the NTIS, and when available in this way, these documents are priced with availability information (usually documents abstracted by other agencies for different purposes, and listed in USGRDR as PBs). Each issue of SWRA includes (besides a subject index based on only those descriptors applied to each document with an asterisk) an author index, an organizational index, an accession number index (the W- series only, as applied to all documents in SWRA whether or not a PB number may have been assigned by another agency for an abstract appearing in a different tool), and abstract sources. There are annual subject, and author/organizational/ accession number indexes, as well.

SWRA is an excellent example of a specialized tool, based on national standards, compatible with other informational services, yet using in-depth analysis to furnish access in a detail that would not be possible in a tool of greater coverage. The expectation is that this is the direction in which information science will take us in future.

But within this diversity, there will be increasing need to observe the discipline of compatibility.

Scientific and Technical Aerospace Reports (STAR)

Like SWRA, *Scientific and Technical Aerospace Reports* (STAR), issued by the National Aeronautics and Space Administration (NASA), is a specialized government-sponsored tool, with its comprehensive subject indexing based on its own thesaurus of terms (6). Its history dates back to 1915, however, when NASA was called NACA (National Advisory Committee for Aeronautics), whose publications are described in a long series of indexes: 1915/1949, plus seven biennials and annuals from 1949/51 through 1957/58 when the name of the agency was changed to its present nomenclature. During this period only NACA publications were indexed, based on its *Research Abstracts and Reclassification Notices*. This publication was superseded on November 14, 1958, by number 1 of *Publications Announcements* (later *Technical Publications Announcements*), which lasted through volume 2, number 19, December 20, 1962, and was itself succeeded on January 8, 1963, by STAR. STAR's coverage includes scientific and technical reports of NASA and its contractors as well as those of other government agencies, universities, and industrial and independent research organizations in the United States and abroad; and those prepared by NASA and contractor authors appearing in learned and technical journals. It is oriented to the science and technology of aeronautics and space. It carries abstracts but no keywords, and users should be on guard to the fact that index references are usually to NOCs rather than to titles. Although STAR's classification scheme is quite detailed for its specialized coverage, it does at least refer to the COSATI subject category list when citing a report that is also covered in USGRDR. Such references, for example, look like: CSCL 04 F. NASA's worldwide information exchange brings many foreign documents under its purview, and its format (similar to USGRDR except for the lack of keywords) combined with this coverage makes it readily useful to a searcher accustomed to the standard appearance of NTIS indexes.

The American Institute of Aeronautics and Astronautics also issues its own reporting publication, *International Aerospace Abstracts* (IAA), a commercially sponsored tool and hence not covered in this chapter, that deals with books and journals worldwide and uses STAR's subject categories in its indexing. It should be used with STAR to be assured of comprehensive access to current literature on aerospace science and technology.

Because of the characteristics of extraterrestrial environments known to us, much reporting to NASA does have bearing on arid-lands research, and the current interest in sensing techniques for the identification of arid Earth terrain features further emphasizes the use, albeit limited, of STAR to our purposes.

In the foregoing an attempt has been made to spell out what a searcher might expect to find in various gov-

ernment-sponsored indexing and abstracting tools: what subjects are likely to be covered during what time periods, their frequency of publication, how they merged, and separated, and took on new coverage. The arid-lands scientist, working on retrospective searches, will find the usefulness of these tools increasing in direct proportion to the currency of their publication. And perhaps more subtly, the more contemporary the tool the more likelihood of its research coverage being more basic, more interdisciplinary, more applicable to a broader range of subject interests. The user, it is hoped, will find such tools as revealing of hitherto unknown sources of information as the author does, once the hurdle of their immediate incomprehensibility is overcome.

There is one further step, nevertheless, in this self-education process that we must understand, namely, the indexes to the various services discussed at the beginning of this section on The Tools.

THE INDEXES TO THE TOOLS

The dictionary defines an index as "a guide for facilitating reference; something that serves as a pointer or indicator." The exercises undertaken to display the information in what follows may seem to result in insultingly elementary instructions, yet they were designed deliberately to enable the user to leapfrog over the trial-and-error manipulations that might otherwise be required. These indexes to the indexing and abstracting services are indeed the indispensable guides to facilitate the reference work entailed in information searches.

For the earliest of our tools, for instance, the BSIR, indexes were issued for each volume, providing a key to material contained in reports of the OTS, which, as we will recall, was formerly the Office of the Publication Board. In its indexes the reference was to PB numbers and the page in the bibliography itself where the abstract of the report might be found. These early indexes are in great detail and quite extensive, that for volume 1, as an example, running to 169 pages. A spot check of random arid-lands references (see Case 1, below), though nothing appeared under the heading "arid lands," showed:

> Arizona — climate (5)
> desert survival (1)
> Libya — climate (3)
> Libya — maps (1)
> water supply — Egypt (1)

In 1949 the Special Libraries Association compiled what was to become the first of a series of useful indexes to arrive at numerical information, so often the only significant clue in a bibliographic search. This first one, entitled "Numerical Index to the Bibliography of Scientific and Industrial Reports, v. 1-10, 1946-1948," facilitated identification and acquisition of reports mentioned by PB numbers only. A later cumulative numerical and correlation index for volumes 11-22, 1949-1954, was issued by Technical Information Service, Washington, D. C. The latest of such indexes known here is that produced as document AD-823 394, entitled "PB Number to AD/ATI Number Correlation Index" (7), superseding a similar report dated January 1967, AD-806 759. Its listing provides AD or ATI numbers for over 22,000 documents when only the PB number is known.

Aside from these cumulated indexes to document numbers, the earliest index reference to an AD number discovered was that for AD-2135, appearing in volume 32, number 4, October 16, 1959, of USGRR, but a lower-numbered AD document issued in 1952, AD-307, was indexed as late as 1960. (These references are more convenient, obviously, than the cumulated index references described above, useful as they are, since the description of the document is already in hand and requires no further identification in another source.) And finally, also in 1959, there appeared an index first to PB numbers, then toward the end of the year to all report numbers announced, alphabetically by prefix (AD, AFCRL, PB, etc). The cumulative number index to volume 32, July-December 1959, of USGRR, for instance, was the first to carry other than PB and AEC numbers. TTs and JPRSs, however, still were not included until 1968.

In 1954, the USGRR indexing began including the following categories of information: (*1a*) numerical index to PB reports, (*1b*) series and number of selected AEC documents, (2) correlations with PB numbers, (3) cooperating research laboratories, (4) author index, and (5) list by subject classification; also included was a key to subject classification code number, and a key of abbreviations (acronyms) of more than 400 government document series found in the correlations. A later publication of this nature was issued separately by the Special Libraries Association (8). The list by subject classification, item 5 above, is very broad indeed, with reference to PB number only. A sample check under "meteorology and climatology," for instance, revealed several hundred PB numbers, each one of which would have to be located in the bibliography itself to determine its usefulness to the particular search.

In 1959, with volume 31, the USGRR index added "sources" to its array of information, and also enlarged its subject categories to include earth sciences, which took under its umbrella the former single headings of "climatology and meteorology" and, later, "geography," signalling a breakaway from the initial emphasis on products and processes. At this same time, it also began issuing indexes to each issue, referring from the subject heading being consulted to the PB number.

On February 15, 1965, volume 1, number 1 of *Government-Wide Index to Federal Research and Development Reports* (GWI) appeared, providing indexing to NSA, STAR, TAB, and USGRDR through five keys: subject, personal author, corporate source, report number, and accession number. USGRDR continued its own issue-by-issue indexes, however, until July 1966. In July 1968, the GWI became *U. S. Government Research and Development Reports Index* (USGRDR-I), using the same format and key approach as the GWI. Since

July 1966, the GWI and later the USGRDR-I have been issued twice monthly. Beginning in 1968, mercifully, there are quarterly and annual cumulations of the five keys. From 1965 through 1967, however, each individual monthly and semimonthly index for each of the five keys (color-coded to help) must be scanned separately, a tedious task despite the detailed coverage. With the issue for March 25, 1971 (vol 71, no. 6), this index became known as *Government Reports Index* (GRI), corresponding with the change in name of USGRDR to *Government Reports Announcements* (GRA).

Independent indexing to STAR's announcements began with the dissolution of GWI, 1968, and its present indexing corresponds to that of USGRDR-I except that it uses its own thesaurus (6) as the authority for the vocabulary that appears in the subject index. It, too, like USGRDR-I, comprises five sections: subject, corporate source, personal author, report number, and accession number, with separate cumulative indexes published quarterly, semiannually, and annually.

TRANSLATIONS

For a quarter of a century, or perhaps longer, there have been determined efforts to provide information on available translations of technical reports and other types of publication. The greatest impetus followed the end of World War II when, as we learned from the brief history of the beginnings of BSIR, captured documents from enemy countries began pouring into U. S. government repositories. At the same time, several friendly governments also arranged to furnish us with technical literature.

Translation Monthly

The most ambitious of these efforts in the United States was initiated in 1955 by the Special Libraries Association Translation Pool (later called Center) in its *Translation Monthly,* published for four years, 1955-1958, for the Association by the John Crerar Library, Chicago. It was an author list only, with name indexes to each of the first two volumes.

Beginning in 1957, the SLA Translations Center assumed the functions and collections formerly provided by the Science Translation Center of the Library of Congress, whose monthly *Bibliography of Translations from Russian Scientific and Technical Literature* (prepared under the sponsorship of the NSF and the AEC) lasted through 39 numbers (October 1953–December 1956). All its Russian translations were subsequently added to the SLA Translations Center. Prior to this transfer of responsibility, SLA had referred all its Russian translations to the Library of Congress which, in turn, had sent all its non-Russian items to SLA.

Also beginning in 1957, with volume 3, SLA's *Translation Monthly* began reporting translations by subject categories, including an author index to each issue, followed by an annual cumulation. In May 1957, TM began including translations from other sources (i.e., sources of translations not held by the SLA Translations Center), each symbolized by source agency (for example, UKSM for United Kingdom Scientific Mission; or CTT for Columbia Technical Translations). This list of other sources grew longer and longer, until the final issue of *Technical Translations* (TM's successor) in December 1967, when 74 translating agencies operating in 18 different countries were cited.

This evolution of the original SLA project was accelerated by the 1957 Executive Order designating the Department of Commerce as a collection and distribution center for the publication, in cooperation with the SLA Translations Center, of TM, now to include not only listing but abstracts of translated technical literature available from OTS, LC, SLA, cooperating foreign governments, commercial translators and publishers, universities, and other sources.

Technical Translations

In 1959, TM became *Technical Translations* (TT), with new voluming, and was now issued twice monthly. It featured translations in progress by U. S. government agencies, a list of periodicals translated cover-to-cover (see Current Ordering Information [list in this chapter] for topics presently covered by cover-to-cover serials published by the NTIS and available on subscription), and author, subject, journal, and translation number indexes, with a semiannual cumulative index. With the issue for January 15, 1964 (vol 11, no. 1), TT began reproducing contents pages of journals recently translated cover-to-cover, and at the same time discontinued abstracts and/or descriptors. In addition, the subject index appearing in individual issues was dropped. To compensate for the loss of abstracts, entries now carried the notation "abstract available" following the end of the translation note. Availability refers to the scheme OTS worked out in March 1958 to provide subscriptions to TM in card form, so that to obtain an abstract, after abstracts were dropped from TT in 1964, one would request it by TT number, presumably, and receive it in card-catalog form.

When the Clearinghouse was established in 1965, all translations received with AD or NASA numbers were announced under those numbers rather than under the TT reference; nor did the GWI include TT numbers unless such numbers were also cited in any of the four journals indexed in the GWI. In the summer of 1965 with the issue of volume 14, number 1, the subject-category arrangement was brought into conformity with the COSATI subject category list.

Finally, with the December 1967 issue, *Technical Translations* was discontinued (vol 18, no. 12). After that date, only government-sponsored technical translations appeared in USGRDR, the first such issue carrying these being that for January 10, 1968 (vol 68, no. 1). Though it is hard to accept the justification for this decapitation of a valuable searching tool, it was meant; in the Clearinghouse's words, "to make a more comprehensive announcement journal covering all U. S. Government-sponsored research and development reports and translations," all under one cover, with one index (in five sections!).

For a decade before TT's demise, the Joint Publications Research Service (JPRS) provided a parallel service. Established in 1957 as an agency to translate under contract, for all federal agencies, foreign documents and journals in all fields of science, the JPRS was originally placed under the OTS, later under the Clearinghouse, and finally under NTIS, which still distributes its translations. JPRS's own report series, *USSR Scientific Abstracts and East European Scientific Abstracts,* was expanded in 1964 to include abstracts formerly published separately by OTS under the title *Current Review of the Soviet Technical Press,* which was discontinued with the issue dated October 9, 1965. JPRS also expanded its services to include a staggering number of translations in the social sciences, announcement and distribution being assumed by the Clearinghouse in October 1965. As pointed out above, JPRS reports are now carried in the report number index to USGRDR.

National Translations Center

Inevitably, these measures to insure reporting of translations were considered less than totally satisfactory, for about the time TT was discontinued, the National Translations Center (NTC), formerly the SLA's Translations Center, still at the John Crerar Library in Chicago, undertook the publication of *Translations Register-Index* (TR-I), while abroad a *World Index of Scientific Translations,* under the administration of the European Translations Centre in Delft, began also in 1967. The latter is beyond the scope of this chapter, but full information about its coverage and purpose may be found in any issue of this published *World Index.*

When the SLA relinquished its sponsorship of the National Translations Center to the Crerar in 1968, the latter continued the services as direct recipient of support from the National Science Foundation, which agency had supported the former SLA Translations Center through grants from 1956 forward (*9*). The present NTC now characterizes itself as "the principal U. S. depository and information center for unpublished translations into English from the world literature of the natural, physical, medical, and social sciences." Its purposes are to help eliminate costly duplication of translations effort, to disseminate information on available translations, and to provide copies of translations on file or to refer inquiries to other known sources. Its files constitute the holdings of the Center from its beginnings in the mid-1950s, through all the shifts in sponsorship, and since 1968 also include translations listed in USGRDR as well as all those reported to the Center directly by commercial translating agencies and pools.

Translations Register-Index

The TR-I is arranged by (*a*) a Register Section announcing new accessions of the National Translations Center, including prices, listed by COSATI subject categories; and (*b*) an Index Section including the Center's accessions described in the Register Section, but also items listed in USGRDR and those translations reported by commercial translators and other sources but which may not necessarily be available from the Center, and which are not reported in the Register Section. The U. S. government-sponsored translations so indexed are available from the Clearinghouse as indicated by a specific symbol. The full citation of these appears in USGRDR, and may be located therein from the index entry in TR-I which carries the Accession/Report Number section of USGRDR-I.

The National Translations Center has its counterpart in some respects in many other countries, reporting its accessions to the European Translations Centre in Delft, which in turn "provides an instrument through which all organizations active in the field of translations can coordinate their policy with regard to translations," in a veritable joint European-American translation network for non-Western scientific and technical literature.

What emerges from this rather murky development appears to be a comprehensive source of information about scientific and technical translations in our National Center, but with the source of the translations themselves being fractured among many agencies: the Center, the Clearinghouse, the NTIS, the European Translations Centre, or any one of a variety of commercial, private, or institutional agencies. Again, it requires a certain familiarity with the intricacies of coverage of the many overlapping services to get the maximum usefulness from each and all, and one is often left with the feeling of being strangled in this web of possibilities.

HYPOTHETICAL CASES

Case 1

Let us suppose a client has asked us to provide information in technical reports on physiological responses of humans to heat stress, including any investigations the government may have made based on World War II experiences in the North African desert. Suppose the client has worked his way through conventional library materials such as books and journal articles, but he is baffled and impatient before the technical-report literature. We embark bravely, at the beginning, with the index to volume 1 of BSIR, and find:

a) Nothing under physiology, heat, stress, aridity, arid lands.

b) Under deserts, however, we find a reference on "desert survival" citing PB 5143. On the page indicated, we find:

DILL, DAVID B.

1942   Desert water requirements. Army Air Forces, Experimental Engineering Center.

with an adequate abstract. With the PB number, this report can be ordered from the Clearinghouse.

c) Emboldened by this success, we try other descriptors, one of which, Arizona — climate, brings to light a series of Army Air Forces "Studies on local forecasting"

including several in the arid parts of Arizona, such as Douglas, Tucson, Phoenix, Marana, and Kingman, each of which "discusses geographical and topographical considerations, climatic and synoptic considerations, peculiarities of local weather, and contains topographic maps and climatic statistics." Each of these also carries PB numbers for ordering purposes. If our library has these, we can scan for pertinence to our search; if not, we can order, or refer to our client.

d) We move on to indexes to other volumes of BSIR; and in volume 2, under "desert" we find PB 37883, a reference to War Department Field Manual 31-25, 1942, Desert Operations, 72 p. The abstract says it is based "on experience in the Western Desert, Africa." We accept this as relevant to our search.

e) Index to volume 3, under "desert deterioration," brings to light a technical report on aircraft maintenance in the desert, and how to minimize the destructive effects of sand, dust, and heat on machines, not necessarily people. We will show it to our client, but only to let him know we are trying.

f) Now we hit a streak of bad luck, and find nothing more until the index to volume 28, where, under "desert —humidity — effect of irrigation" we find:

OHMAN, H. L. AND R. L. PRATT

1956   Daytime influence of irrigation upon desert humidities. 34 p. U.S. Quartermaster Research and Engineering Center, Natick. PB 125047.

Since humidity is a factor in heat stress, we take this.

g) In the index to volume 32 of USGRR, we encounter the series of other Technical Reports prepared by the Army Quartermaster Research and Engineering Center in Natick, dealing specifically with "deserts, physiological effects":

BAKER, P. T.

1958   Theoretical model for desert heat tolerance. 33 p. Rept. EP-96. PB 135 853

FREGLEY, H. J. AND P. F. IAMPIETRO

1959   Dietary potassium supplementation and performance in the desert. 21 p. Rept. EP-109. PB 142 310.

There are others, all of whose abstracts indicate pertinence to our search.

h) In the index to volume 33, we find index terms to "deserts, Arizona," "deserts, geophysical factors," and "deserts, survival factors," of which only the last is suitable:

POND, A. W.

1956   Afoot in the desert, a contribution to basic survival. Rev. (from 1951 publication). Arctic, Desert, Tropic Information Center, Maxwell Air Force Base, Alabama. ADTIC Publ. D-100. 57 p. PB 143 817.

However, Dodd and McPhilmy's 1959 report on "Yuma summer microclimate," 1959, 39 p. (Army Engineer Waterways Experiment Station Technical Report Ep-120) looks like a possibility, so we report it for client decision.

i) As we proceed methodically through the whole sequence of indexes we find occasional references relating specifically to our search, and by volume 38 we are finding descriptors, starred (for indexing) and nonstarred when examining the actual references with their abstracts, which give us clues to further analysis of the indexes. For instance, for

TALSO, P. J. AND R. W. CLARKE

> 1948   Observations on physiological problems in desert heat, Task Force Furnace, Yuma, Arizona. Report on the physiological effects of high temperatures. Army Medical Research Laboratory, Ft. Knox, Kentucky. Report 14. 25 p. PB 162 575 AD-62 767.

we find descriptors for "desert tests" as well as "adaptation (physiology)," "survival," and "heat tolerance." We may use these terms, then, in future searching, as well as the more limited terminology we have been employing, to broaden our options.

j) By the time we reach the 1968 annual subject index to USGRDR, we find under "desert adaptation" the following:

> Diet and feeding habits of Sahara nomadic tribe and adaptation to desert climate
> N 68-22210   6 P   U 68 14

We know that U 68 14 refers to USGRDR, volume 68, number 14; within that number, we locate field 6 (Biological and Medical Sciences), group P (Physiology), and begin our search for the reference either under "Diet and feeding habits . . ." or for the report number, N68-22210. Because this wording is not necessarily the actual title of the report but merely a "notation of content" (NOC) used to provide a more exact description of the subject matter, the word search may not be successful. In this case it is not, so we conduct a report-number search, and find:

> Alimentary behavior and organic adaptation to desert zones   N68-22210
> For abstract, see STAR 06 12

In volume 6, number 12 of STAR, we search its index (or the annual cumulated index if it is available) under either the same subject heading we searched in the subject index to USGRDR, "desert adaptation," or the report number N68-22210. Either of these will refer to p. 1823 of volume 6, number 12. On this page, we search among the dozen or so abstracts until we find N68-22210. There we find complete information about N68-22210, including its true title, "Alimentary behavior . . ." and other items such as corporate source, date, citation, and an abstract.

This so-called retrospective search through one tool for this topic brought to light about 25 pertinent references and took a day's time of a skilled bibliographer — an expensive undertaking. Was it worth it to our client? He thought so. Could it have been done "for free" by his friendly librarian? Hard-pressed librarians have less and less time to conduct such specialized personal searches. Could it have been done by computer? Yes, if

our mythical client understood the resources and services of the Clearinghouse and the NTIS. In this case, he might have gotten a certain percentage of "garbage" but the time involved in a manual sort to eliminate irrelevant items would undoubtedly be less than was consumed by the equivalent decisions made under our own search.

If you are saying now, "Then why have you taken my time to tell me how to use these tools, if I can get the same information without having to know how?" you might consider some of the by-products of this know-how:

We uncovered 19 references not relevant to our client's need but highly significant for our master arid-lands bibliography, maintained in the Office of Arid Lands Studies; our client learned of otherwise unknown colleagues working in the same field of his interest with whom he can now communicate, cooperate, and perhaps collaborate. Furthermore, he knows of the existence of this great bibliography, and its satellites, and he no longer feels alien toward its format, even though he himself may not have gone through the search personally. And when the day comes that he, his students, or his coworkers come face to face with an office console, he will understand better how to establish a dialogue with the total resource, whether it be the great NTIS or the European Translations Centre or the University of Arizona's Arid Lands Information Center.

Case 2

Suppose we have no more than a vague reference to CRES-37-1 in a very bad bibliography we have been checking. From the context, we will assume it is a current publication. This search could develop as follows:

a) Select the 1968 annual accession and number index to USGRDR.

b) Search the number index for CRES-37-1.

c) Find CRES-37-1 refers to N68-22044 9 F U68 14.

d) At this point, we have two options:

(*1*) If we do not understand the significance of N68-, we can proceed directly to USGRDR, volume 68, no. 14, field 9 (Electronics and Electrical Engineering), group F (Telemetry), and search for and find N68-22044 as:

> A study of earth radar returns from Alouette satellite
> NASA-CR-94294, CRES-37-1, Grant NSG-477.

We can be satisfied with the information found here, which in addition to title, "A study of earth radar . . .," will include authors, corporate source (Center for *Re*search in *Engineering* and *Science* of the University of Kansas), date, and several keyword descriptors, but no abstract; or,

(*2*) if we know that N68- will ultimately produce an abstract in the 1968 volume of STAR, and we decide we want to read the abstract before proceeding to locate the actual document in our library, we can proceed through the steps set forth under (*j*), as detailed under Case 1, that is, search STAR's number index for N68-

22044, either in the annual index if it has been cumulated by this time, or if it has not, through the several individual issues of STAR for 1968, each of which has a number index. When found, the index reference will be to the exact page of STAR on which the citation is abstracted.

We should be mindful that where USGRDR references include an array of descriptors (of which each descriptor starred [*] is announced in the subject index), these references carry no abstracts when the reference is to STAR (N- numbers), NSA, or TT. STAR, on the other hand, gives an abstract but no descriptors, though the same starred descriptors applied to the reference as it appears in USGRDR also are used as subject keys to STAR's indexing. In Case 2, above, for instance, USGRDR used "*Alouette satellites" as a descriptor; the same term was used in the STAR index, referring to the abstract on this subject appearing on page 1843 of that volume of STAR.

Case 3

Let us suppose our great and good friend M. P. Petrov has sent us an inscribed copy of his book whose transliterated title, as we make it out from our scanty knowledge of the Russian alphabet, is "Pustyni Tsentralnoi Azii," dated 1966. If we would like to know if this work has been translated, we could proceed as follows:

a) Go first to *Technical Translations* for 1966, where we will find nothing; go on to the cumulative index for January-June 1967 of this same source, and in the author list find:

PETROV, M. P.   04 8F   TT-66-35568

b) Proceed to no. 4 (04) of volume 17 (the volume indexed in the cumulative index we are examining), look under field 8 (Earth sciences), group F (Geography), then scan for TT-66-35568. We find:

PETROV, M. P.
    The deserts of Central Asia, vol. 1: The Ordos, Alashan and Peishan. 16 Dec. 1966. 362 p., refs. JPRS-39145. Order from CFSTI: HC $3.00 MF 65¢ as TT66-35568. Trans. of mono. Pustyni Tsentralnoi Azii, Moscow, 1966. 273 p.

We note that there is no abstract, but if we wish to acquire a copy of the translation of the work Petrov has sent us, we can order a "hard" or paper copy (HC) from the NTIS for $3.00 or a microfiche (MF) for 65¢.* We can order under either the TT or the JPRS number, and to avoid any uncertainty, can submit both numbers as well as the author and title. If we have a handy supply of coupons from the NTIS (issued in $3.00 denominations), we can simply record the order (report) number on the coupon, together with the name and address to which the copy is to be sent, and send directly off. For foreign customers of NTIS, these prepaid coupons are the simplest and most convenient way to order NTIS documents, since they eliminate the use of an exchange bank and avoid fluctuations in exchange rates.

---

* Beginning January 1971, 95¢.

WHAT OF THE FUTURE?

H. Bentley Glass, in a troubled look ahead (*10*), cited as one of five limiting factors to the growth of science, which merits our study, "the sheer volume of scientific information that already exists without adequate codification and storage and that is so rapidly accumulating. . . . Perhaps a majority of scientists have frankly and openly abandoned the effort to cope with the flood of knowledge, even in their own specialties. They read little, especially in the foreign literature, and depend mainly upon word-of-mouth reports from fellows of the 'invisible college' to know what is going on. . . ."

The government has addressed itself to this, as we have seen, for almost a generation now, refining, perfecting, cooperating, supporting, searching. And at the end of September 1969, the DDC announced (*11*):

At almost the same instant one day last week, Air Force and Navy specialists sat down at typewriter-like consoles at their own installations and held "conversations" with the computer at the DDC. This dialog between people and machine took place while DDC's own employees busily utilized the same computer for other data processing chores.

This was a milestone event . . . in an experimental program to assess the economic and operational feasibility of providing key user organizations access to the DDC bank of computer-stored management and technical information. The on-line retrieval capability means that with associated terminal equipment the user may obtain without human intervention at DDC material stored in the Center's huge computer at Cameron Station. . . .

It is done by special computer programming and a "retrieval manual" worked out by DDC data processing experts and information specialists. With only a few minutes of checking out at the console, the user is able to ask the computer for certain information and in return the computer talks back, even giving directions for proceeding from one step to another. . . .

Whereas the conventional method requires 8 steps that may take from one to four days, the new on-line system has only four that may happen in one to two minutes.

Will Dr. Glass's scientists, the ones who have "frankly and openly abandoned the effort to cope with the flood of knowledge," be willing to sit down at their consoles and ask the computer what they are unable or unwilling to ask their librarians, or those personal bibliographers we talked about earlier? Considering the massive infusion of federal funds into the bewildering plethora of projects and studies to make this latest dialogue feasible for the government's own bibliographical resource, one would certainly hope so — or in the parlance of the day, "You'd better believe it!" The younger scientist, computer-oriented from childhood, will find this growing capability, not only between himself and government-sponsored information bases and/or institutions but even between himself and his own institutional libraries and information centers, a natural and an easy manipulation.

If the efforts of the NTIS and the several agencies whose reporting services the NTIS now handles can be developed in an orderly way to provide the specificity required to satisfy the scientist, the feverish activity in which we are now employed may subside to a normal degree insofar as it will be reflecting an acknowledgment on the part of the entire scientific community that such networks, national and international, must, can, are, and will standardize, cooperate, share, and disseminate, to the mutual benefit of all involved.

Certainly arid-lands information, now needed so desperately to insure orderly, civilized, and quality development of the world's deserts, to preserve this delicately poised environment, so that we may benefit from it in aesthetic as well as in economic ways, will be one of the vital elements in such an all-encompassing network. The time *is* short, yes, but the stakes are high: survival, in a word, bibliographic as well as environmental.

REFERENCES AND NOTES

1. BRANSCOMB, L. M.
   1968   Is the literature worth reviewing? Scientific Research 3(11):49–56.

2. KIEFFER, L. J.
   1969   The information analysis center and the creation of reliable data. In COSATI Panel on Information Analysis Center, The Information Analysis Center, seven background papers, Report 69-6, pp. 19–27.

3. GRAY, D. E.
   1951   Controlling the Technical Report. Bibliography of Technical Reports 16(5):141–142.

4. HOWE, G. H., AND OTHERS
   1968   Classification of world desert areas. U.S. Army Natick Laboratories, Technical Report 69-38-ES. 104 p.

5. U.S. OFFICE OF WATER RESOURCES RESEARCH
   1966   Water resources thesaurus; a vocabulary for indexing and retrieving the literature of water resources research and development. U.S. Government Printing Office, Washington, D.C. 237 p.

6. U.S. NATIONAL AERONAUTICS AND SPACE ADMINISTRATION
   1967   NASA thesaurus, preliminary edition. 3 vols. NASA SP-7030. Clearinghouse, or U.S. Government Printing Office, Washington, D.C.

7. BROWN, K. R.
   1967   PB Number to AD/ATI number correlation index. Garrett Corporation, Los Angeles, California. AiResearch Manufacturing Division. 202 p. AD-823 394.

8. REDMAN, H. F., AND L. E. GODFREY (EDS)
   1962   Dictionary of report series codes. Special Libraries Association, N.Y. 648 p.

9. NATIONAL SCIENCE FOUNDATION, OFFICE OF SCIENCE INFORMATION SERVICE
   1969   Annual report, fiscal year 1969. 19 p.
   "Fiscal Year 1969 witnessed a significant change in the management and a broadening of the former SLA Translations Center at the John Crerar Library into a National Translations Center (NTC) and the beginning of the development of Federal policies governing various aspects of Federal translations activities. The Center became the focal point in the U.S. for the handling of translations while the Clearinghouse assumed a lesser role as the central depository and distributor of U.S. Government-sponsored translations. The Panel on International Information Activities of the Committee on Scientific and Technical Information accepted a Foundation-prepared statement of "Policies Governing the Announcement and Dissemination of Translations by Agencies of the U.S. Federal Government." The adoption of this policy statement will facilitate continuing cooperation among federal agencies in making their translations readily available through the CFSTI and NTC," pp. 11–12.

10. GLASS, H. B.
    1969   Letter from the President. AAAS Bulletin 14(3):1–2.

11. DEFENSE DOCUMENTATION CENTER
    1969   On-line retrieval system being tested at DDC. DDC Digest 39:1–3.

ACRONYMS

AD	Originally referred to ASTIA documents; designation continued by DDC
AEC	Atomic Energy Commission
AFCRL	Air Force Cambridge Research Laboratory
ASTIA	Armed Services Technical Information Agency; became DDC
BSIR	Bibliography of Scientific and Industrial Reports
BTR	Bibliography of Technical Reports
CFSTI	Clearinghouse for Federal Scientific and Technical Information
COSATI	Committee on Scientific and Technical Information (of the Federal Council on Science and Technology)
CRES	Center for Research in Engineering and Science (University of Kansas)
DDC	Defense Documentation Center
DoD	Department of Defense
EJC	Engineers Joint Council
ETC	European Translations Centre

GRA	Government Reports Announcements
GRI	Government Reports Index
GWI	Government-Wide Index; superseded by USGRDR–I
IAA	International Aerospace Abstracts
JPRS	Joint Publications Research Service (when followed by a number, applies to translations prepared by this agency)
LC	Library of Congress
NACA	National Advisory Committee for Aeronautics
NASA	National Aeronautics and Space Administration
NOC	Notation of content
NSA	Nuclear Science Abstracts
NSF	National Science Foundation
NTC	National Translations Center
NTIS	National Technical Information Service
OTS	Office of Technical Services (of the Department of Commerce)
PB	Publication Board (i.e., Office of the Publication Board, Department of Commerce); used to designate its reports and continued by OTS and later the Clearinghouse, both of which succeeded the now defunct Publication Board
SIE	Smithsonian Information Exchange
SLA	Special Libraries Association
STAR	Scientific and Technical Aerospace Reports
SWRA	Selected Water Resources Abstracts
TAB	Technical Abstract Bulletin
TEST	Thesaurus of Engineering and Scientific Terms
TM	Translation Monthly
TR–I	Translations Register-Index
TT	Technical Translations
USGRDR	U.S. Government Research and Development Reports
USGRDR–I	U.S. Government Research and Development Reports Index
USGRR	U.S. Government Research Reports
WIST	World Index of Scientific Translations

CURRENT ORDERING INFORMATION

National Technical Information Service
Springfield, Va. 22151

USGRDR subscriptions (send to NTIS):
annual, $30.00 ($37.50 foreign)
single copies, $3.00
domestic and foreign air mail charges quoted on request
Index to USGRDR, $22.00 ($27.50 foreign)

STAR subscriptions (send to U.S. Government Printing Office,
Washington, D.C., 20402):
annual, $54.00 ($68.25 foreign)
single, $2.25 domestic (plus 25 percent extra for foreign mailing)
Index to STAR, $30.00 ($35.00 foreign)
UNESCO coupons accepted for subscriptions

SWRA subscriptions (send to NTIS):
annual, $22.00 ($27.50 foreign)

TR–I subscriptions (send to National Translations Center,
John Crerar Library, 35 West 33rd St., Chicago, Illinois 60616):
annual, $30.00

WIST subscriptions (send to European Translations Centre,
        Delft, The Netherlands):
        annual, $25.00

IAA subscriptions (send to Institute of Aerospace Sciences,
        750 Third Ave., New York, N.Y. 10017):
        annual, $54.00 ($68.25 foreign)
        Index to IAA, $30.00 ($35.00 foreign)

COVER-TO-COVER serial subscriptions (send to NTIS):
        Meteorology and Hydrology, $12.00
        Space Biology and Medicine, $12.00
        USSR Academy of Sciences, Vestnik, $18.00

A Regional Bibliography of Calcrete

ANDREW S. GOUDIE
Departmental Demonstrator
School of Geography
University of Oxford
Oxford, England

ABSTRACT

CALCRETE APPEARS to be the most suitable international term for materials formed by the cementation and/or alteration of a pre-existing soil or rock by dominantly calcium carbonate. Variously called caliche, kunkur (or kankar), kalkruste, croûte calcaire, tosca, and so on, calcrete is highly variable in nature and extremely extensive in arid and semiarid areas of the world. Its significance lies in its great areal extent, its geomorphic importance, its effect on hydrology, its uses, its importance for mineral prospecting, and its possible palaeoclimatic meaning.

Some 450 citations, with full bibliographical information, are given by regions: the Americas, Australia, East Africa, North Africa, southern Africa, India, and miscellaneous (including Europe and the Middle East). The arrangement is alphabetical by author within the regional groups.

A REGIONAL BIBLIOGRAPHY OF CALCRETE

Andrew S. Goudie

Calcrete, a term coined by Lamplugh for cemented Pleistocene debris around Dublin Bay (Lamplugh, 1902), was later used by the same geologist for indurated calcareous deposits in southern Africa (Lamplugh, 1907). He proposed the terms "silcrete" and "ferricrete" for analogous siliceous and iron-rich formations. Calcrete is equivalent to various local terms, including:

Caliche ....United States
Croûte calcaire ...France and North Africa
Kalkruste ....Germany and South West Africa
Kunkur, kankar ..East Africa, India, Australia
Nari ....Israel
Kafkalla and
   havara ....Cyprus
Tosca ....Argentina and Spain

Because the term caliche is frequently used to denote non-calcareous deposits like the soda nitrates of Chile and Peru, and because the term kunkur simply means "gravel" in Hindi practice, calcrete is probably the most suitable international term for materials formed by the cementation and/or alteration of a pre-existing soil or rock by (dominantly) calcium carbonate, with silica and magnesium carbonate playing a minor role.

Calcretes are highly variable in nature, ranging from soft powder, through nodular and honeycomb forms, to extremely tough hardpan or boulder forms. Classifications on the basis of such physical properties have been made in America (Price 1933; Gile 1961), in North Africa (Durand 1949, 1963; Wilbert 1962; Ruellan 1967) and in southern Africa (Netterberg 1967).

World calcrete chemistry is known from more than 280 partial analyses obtained from the published and unpublished literature (Table 1). Complete analyses are rare, though some detailed work has been done by Aristarain (1970) in New Mexico. The calcium carbonate generally occurs as calcite, not aragonite. Magnesium carbonate has a mean ratio to calcium carbonate of 1:14, but locally becomes a major constituent of some calcretes, notably in Ovamboland-Etoscha Pan area of South West Africa (Gevers 1930), at Keimoes in South Africa (von Backstrom 1964), in parts of Australia (Johns 1963), and in Hawaii (Sherman and Ikawa 1958). In such cases the calcretes are sometimes referred to as "dolocretes" or "magnesicretes." Gypsum contents are usually low but are important in the coastal zone of the Namib Desert, while water-soluble salts become significant only in areas of impeded drainage. Amorphous silica is present in large quantities in certain Kalahari "cal-silcretes," in some central Australian valley calcretes (Mabbutt 1967) and in some American calcretes (Sidwell 1943; Price 1933; Swineford and Franks 1959). The dominant clastics are quartz grains, while the dominant clay minerals in calcrete hardpan layers are frequently attapulgite and sepiolite (Van den Heuvel 1966; Millot, Paquet, and Ruellan 1969).

The processes involved in the formation of calcrete deposits are still imperfectly understood, though detailed reviews have appeared in Bretz and Horberg (1949), Brown (1956), Durand (1959) and Ruellan (1967). Older hypotheses involving deposition by capillary rise *per ascensum* from groundwater (Blake 1902; Pomel 1871), lacustrine deposition, and sheetflood deposition, have been largely rejected in favor of various pedogenic hypotheses relevant to the conditions of semiaridity under which most calcretes form (Gile, Peterson, and Grossman 1966).

Calcretes are extremely extensive in semiarid areas with slopes of around 2.5 m/km. They are a major component of the High Plains of Texas and New Mexico, and are also widespread in Kansas, Oklahoma, Arizona and California. In Argentina, tosca, early recognized by Charles Darwin, underlies large stretches of the Pampas (Siragusa 1964) and is found in neighboring Missiones Province, and in the dry intermontane basins around Mendoza and elsewhere. Calcrete is extensive in the semiarid Sao Francisco basin of northeastern Brazil, in the coastal lowlands of northwest Venezuela, and in Central America.

Many of the hamada surfaces of North Africa are underlain by croûte calcaire, and the literature for the former French territories is large. Knowledge of Libyan calcretes is more restricted, but it is clear that they are a major element of the Gefara Plain, the Calanschio sandsea, and the inland margin and wadi basins of the Cyrenaican limestone mountains.

TABLE 1

The Gross Chemistry of Calcrete

	%	Ratio of CaO	Range
CaCO₃	79.28	. .	−99.40
SiO₂	12.30	3.47	0.34−
MgO	3.05	13.96	. .
Al₂O₃	2.12	20.02	0.04− 7.40
Fe₂O₃	2.03	21.02	0.17−15.97
CaO	42.62	. .	. .
MgCo₃	. .	. .	0.11−45.41

Calcretes exist also in the northern Mediterranean lands, particularly eastern Spain, Provence in France, and Apulia in Italy. In the eastern Mediterranean, areas with less than 500 mm of precipitation *per annum* in Syria, Jordan, Lebanon, Israel and Cyprus have calcretes associated with terra rossas and brown earths, and some calcretes, locally called "gatch" in Kuwait, are present around the Persian Gulf.

In India, where some of the earliest work on calcretes was undertaken by British geologists in the middle of the nineteenth century, deposits are reported from Rajasthan, Gujarat, the Punjab, United Provinces, and the Madras area. The Indian terminology was transferred to East Africa, where calcretes occur widely in the Wajir and El Wak depressions of northern Kenya, in the Serengeti and Masai Steppes of Tanzania, and in the plateau areas of Somalia.

The same terminology has been used in Australia though silcretes and ferricretes are more dominant in the semiarid heart of Australia than is calcrete. In southern Africa, where the term originated, calcretes have a spectacular development, reaching thicknesses of as much as 60 m in the vicinity of the Molopo Valley, and being a frequently encountered layer within the terrestrial sediments of the Kalahari Basin in South Africa and Botswana. Large expanses are also present in the Cape Flats (near Cape Town), in the wetter margins of the Namib Desert both in South West Africa and Angola, in the Etoscha Pan region, and in the drier parts of Mozambique and Madagascar.

The calcretes of the world's semiarid regions have a wide significance because of their:

1) Great areal extent.

2) Geomorphic importance as scarp formers (Elias 1948), as protectors of old geomorphic surfaces such as pediments (Dumas 1969), as an influence on desert depression morphology (Judson 1950), as a cause of rock disintegration (Price 1925; Jennings and Sweeting 1961; Young 1964), as a cause of karstic development (Price 1940), and as an accelerator of soil erosion (Beaudet, Martin, and Maurer 1964).

3) Effect on hydrology and their use for groundwater location (Cronin 1964; Vegter 1953).

4) Use for road construction in otherwise sandy areas (Caiger 1968; Weinert 1968; Gillette 1934; Netterberg 1967).

5) Importance for mineral prospecting, geological interpretation (Weeks and Eargle 1963; Bosazza 1965; Cuyler 1930), and cement manufacture (Green 1964).

6) Restriction on root development which necessitates expensive treatment with explosives or bulldozers for farming and afforestation purposes, and results in a highly characteristic natural vegetation assemblage (Shreve and Mallery 1933; Leistner 1959).

7) Possible palaeoclimatic significance and stratigraphic indication (Dunham 1969; Mason, Brink, and Knight 1959; Burgess 1961; Nagtegaal 1969).

THE AMERICAS

ALBRITTON, C. C.
  1958 Quaternary stratigraphy of the Guadiana Valley, Durango, Mexico. Geological Society of America, Bulletin 69:1197–1216.

ARELLANO, A. R. V.
  1953 Barrilaco Pedocal — a stratigraphic marker, ca. 5000 B.C. and its climatic significance. International Geological Congress, 19th, Algiers, 1952, Proceedings 7:53–76.

ARISTARAIN, L. F.
  1962 Caliche deposits of New Mexico. Harvard University (unpublished Ph.D. thesis). 292 p.
  1970 Chemical analyses of caliche profiles from the High Plains, New Mexico. Journal of Geology 78:201–212.

BAKER, C. L.
  1933 Reynosa problem of south Texas and origin of caliche. American Association of Petroleum Geologists, Bulletin 17:1534.

BENNETT, H. H.
  1924 The soils of Central America and northern South America. American Soil Survey Association, Bulletin 6(1):69–81.

BISSELL, H. J.
  1963 Lake Bonneville: Geology of southern Utah Valley, Utah. U.S. Geological Survey, Professional Paper 257-B:101–130.

BLAKE, W. P.
  1902 The caliche of southern Arizona: an example of deposition by the Vadose Circulation. American Institute of Mining and Metallurgical Engineers, Transactions 31:220–226.

BLANK, H. R., AND E. W. TYNES
  1965 Formation of caliche *in situ*. Geological Society of America, Bulletin 76:1387–1392.

BLOUNT, D. N., AND C. H. MOORE
  1969 Depositional and non-depositional carbonate breccias, Chiantla Quadrangle, Guatemala. Geological Society of America, Bulletin 80:429–442.

BRANNER, J. C.
  1911 Aggraded limestone plains of the interior of Bahia and the climate changes suggested by them. Geological Society of America, Bulletin 22:187–206.

BREAZEALE, J. F., AND H. V. SMITH
  1930 Caliche in Arizona. Arizona Agricultural Experiment Station, Bulletin 131:419–441.

BRETZ, J. H., AND L. HORBERG
  1949 Caliche in S.E. New Mexico. Journal of Geology 57:491–511.

BROWN, C. N.
  1956 The origin of caliche in the north east Llano Estacado, Texas. Journal of Geology 64(1):1–15.

BRYAN, K.
  1929 Solution-facetted limestone pebbles. American Journal of Science 18:193–208.

BUOL, S. W.
1965 Present soil forming factors and processes in arid and semi-arid regions. Soil Science 99: 45–49.

BUOL, S. W., AND M. S. YESILSOY
1964 A genesis study of a Mohave sandy loam profile. Soil Science Society of America, Proceedings 28:254–256.

CAMPBELL, L. C.
1929 Caliche as a surfacing material. Western Highways Builder, April, pp. 20–22.

CARTER, W. T.
1932 Some relationships of native vegetation to soils and climate in Texas. American Soil Survey Association, Bulletin 13:1–5.

CLARK, J. G., J. R. BEERBOWER, AND K. KIETZKE
1967 Oligocene sedimentation, stratigraphy, palaeoecology and palaeoclimatology in the Big Badlands of S. Dakota. Fieldiana: Geology Memoir 5. 158 p.

CRONIN, J. G.
1964 A summary of the occurrence and development of groundwater in the southern High Plains of Texas. U.S. Geological Survey, Water Supply Paper 1963. 88 p.

CUYLER, R. F.
1930 Caliche as a fault indicator. Geological Society of America, Bulletin 41(1):109. (Abstr.)

DARWIN, C.
1890 On the formations of the Pampas. In Geological Observations on South America, Chap. 4, p. 350–381. Minerva Library Edition, London.

DAY, R.
1955 Long caliche base laid in west Texas. Contractors and Engineers 52(5):116–119.

DELANEY, P. J.
1965 Fisigraphia e geologia de superficie da planicie Costeire de Rio Grande do Sul. Porto Alegre, Brazil, Universidade do Rio Grande do Sul, Escola de Geologia, Publicacao Especial 6:1–105.

DOLLFUS, O., AND J. TRICART
1959 Note sur les périodes froides dans les Andes centrales Péruviennes (Région de la Oroya). Société Géologique de France, Compte Rendu Sommaire, sér. 7, 1(8):236–238.

DUNHAM, R. J.
1965 Vadose pisolite in the Capitan Reef. American Association of Petroleum Geologists, Bulletin 49:338. (Abstr.)

1969 Vadose pisolite in the Capitan Reef (Permian), New Mexico and Texas. In Depositional Environments in Carbonate Rocks, Society of Economic Paleontologists and Mineralogists, Special Publication 14: 182–191.

ELIAS, M. K.
1931 Geology of Wallace County, Kansas. Kansas Geological Survey, Bulletin 18. 254 p.

1948 Ogallala and post Ogallala sediments. Geological Society of America, Bulletin 59:609–612.

FAIRBRIDGE, R. W. (ed)
1969 Encyclopedia of Geomorphology. Reinhold, New York. 1295 p. (See, for example, p. 15, 108, 113.)

FLACH, K. W., W. D. NETTLETON, L. H. GILE, AND J. G. CADY
1969 Pedocementation: induration by silica, carbonates and sesquioxides in the Quaternary. Soil Science 107(6):442–453.

FRYE, J. C., AND A. B. LEONARD
1957 Studies of cenozoic geology along the eastern margins of the Texas High Plains, Armstrong to Howard counties. University of Texas, Bureau of Economic Geology, Report of Investigations 32. 62 p.

1959 Correlation of the Ogallala formation (Neogene) in West Texas with type localities in Nebraska. University of Texas, Bureau of Economic Geology, Report of Investigations 39. 46 p.

1965 Quaternary of the southern Great Plains. In H. E. Wright and D. G. Frey (eds), The Quaternary in the United States, p. 203–216. Princeton University Press, Princeton, N.J.

1967 Buried soils, fossil mollusks, and Late Cenozoic paleoenvironments. In Essays in Palaeontology and Stratigraphy. University of Kansas, Department of Geology, Special Publication 2: 429–444.

GARDNER, L. R.
1968 The Quaternary geology of the Moapa Valley, Clark County, Nevada. Pennsylvania State University (Ph.D. thesis). 236 p.

GILE, L. H.
1961 A classification of Ca horizons in soils of a desert region, Doña Ana County, New Mexico. Soil Science Society of America, Proceedings 25:52–61.

1966 Cambic and certain non-cambic horizons in desert soils of southern New Mexico. Soil Science Society of America, Proceedings 30: 773–781.

1967 Soils of an ancient basin floor near Las Cruces, New Mexico. Soil Science 103(4):265–276.

1970 Soils of the Rio Grande border in southern New Mexico. Soil Science Society of America, Proceedings 34(3):465–472.

GILE, L. H., AND R. B. GROSSMAN
1968 Morphology of the argillic horizon in desert soils of southern New Mexico. Soil Science 106(1):6–16.

GILE, L. H., AND J. W. HAWLEY
1966 Periodic sedimentation and soil formation on an alluvial-fan piedmont in southern New Mexico. Soil Science Society of America, Proceedings 30:261–268.

GILE, L. H., F. F. PETERSON, AND R. B. GROSSMAN
1965 The K horizon: A master soil horizon of carbonate accumulation. Soil Science 99(2): 74–82.

1966 Morphologic and genetic sequences of carbonate accumulation in desert soils. Soil Science 101(5):347–360.

GILLAM, W. S.
1937 The formation of lime concretions in the Moody and Crofton Series. Soil Science Society of America, Proceedings 2:471–477.

GILLETTE, H. S.
1934 Soil tests useful in determining quality of caliche. Public Roads 15(10):237–240.

1951 Co-operative study research study of methods of preparing base course materials for the disturbed soil indicator tests. Texas Highway Department, mimeographed report. 20 p.

HASSAN, A. A.
1969 Vertical distribution of exchangeable cations in soils of arid zones as an indication to the depth of water infiltration. *In* Water in the Unsaturated Zone, Proceedings of the Wageningen Symposium, International Association of Scientific Hydrology and UNESCO, Vol 2, p. 650–658.

HAVENS, J. S.
1966 Recharge studies on the High Plains in northern Lea County, New Mexico. Geological Survey, Water Supply Paper 1819-F. 52 p.

HAWKER, H. W.
1927 Soils of Hidalgo County, Texas. Soil Science 23:475–485.

HAWLEY, J. W., AND L. H. GILE
1968 Caliche development related to the geomorphic evolution of the Rio Grande desert. Geological Society of America, Special Paper 121:130. (Abstr.)

HAWORTH, E.
1897 Physical properties of the Tertiary. Kansas Geological Survey, Report 2:251–284.

HUBBS, C. L., G. S. BIEN, AND H. E. SUESS
1965 La Jolla radiocarbon measurements, 4. Radiocarbon 7:66–117. (*See* p. 101.)

HUNT, C. B.
1967 Physiography of the United States. Freeman, San Francisco. 480 p.

JENNY, H.
1941 Factors of soil formation. McGraw-Hill, New York. 281 p.

JOHNSON, D. L.
1967 Caliche on the Channel Islands. California Division of Mines and Geology, Mineral Information 120(12):151–158.

JUDSON, S.
1950 Depressions of the northern portion of the southern High Plains of eastern New Mexico. Geological Society of America, Bulletin 61: 253–274.

LEE, W. T.
1905 Underground waters of Salt River Valley, Arizona. U.S. Geological Survey, Water Supply Paper 136. 196 p.

LEVINGS, W. S.
1951 Late cenozoic erosional history of the Raton Mesa region. Colorado School of Mines, Quarterly 46(3). 111 p.

LIDDLE, R. A.
1946 The geology of Venezuela and Trinidad. 2nd ed., rev. and enl. Paleontological Research Institution, Ithaca, New York. 890 p.

LUGN, A. L.
1948 The Pleistocene of the Great Plains. Geological Society of America, Bulletin 59:625.

MARBUT, C. F.
1923 Soils of the Great Plains. Association of American Geographers, Annals 13(2):41–66.

MATHEWS, H. L., G. PRESCOTT, AND S. S. OBENSHAIN
1965 The genesis of certain calcareous floodplain soils of Virginia. Soil Science Society of America, Proceedings 29:729–732.

MAZZA, C. A.
1967 Estudio de un sequencia de suelos isohumicos con costra calcárea del sur de la Provincia de Buenos Aires. Universidad Nacional de la Plata, Facultad de Agronomía, Revista 43: 215–239.

McDOWELL, C.
1966 Comparison of AASHO and Texas Test Methods and Specifications for flexible base materials. Texas Highways Departmental Research Report 48-1F. 67 p.

MERRIAM, D. F., AND J. C. FRYE
1954 Additional studies of the Cenozoic of western Kansas. Kansas Geological Survey, Bulletin 109:49–64.

MOORE, F. E.
1959 The geomorphic evolution of the east flank of the Laramie Range, Colorado and Wyoming. University of Wyoming (unpublished thesis).

MOTTS, W. S.
1958 Caliche genesis and rainfall in the Pecos Valley area of southeastern New Mexico. Geological Society of America, Bulletin 69:1737. (Abstr.)

1965 Hydrologic types of playas and closed valleys and some relations of hydrology to playa geology. U.S. Air Force Cambridge Research Laboratories, Environmental Research Paper 96:73–104.

MUIR, J.
1936 Geology of the Tampico Region, Mexico. American Association of Petroleum Geologists, Tulsa, Oklahoma; T. Murby and Co., London. 280 p.

NELSON, H. F.
1959 Deposition and alteration of the Edwards Limestone, Central Texas. University of Texas Publication 5905, Bureau of Economic Geology. 235 p. (*See* p. 40, 67.)

NEWMAN, A. L.
1964 Soil survey of Cochran County, Texas, U.S. Department of Agriculture, Soil Survey Series 1960(17). 80 p.

NICHOLSON, A., AND A. CLEBSCH
1961 Geology and ground-water conditions in southern Lea County, New Mexico. New Mexico State Bureau of Mines and Mineral Resources, Ground Water Report 6. 123 p.

NIKIFOROFF, C. C.
1941 Hardpan and microrelief in certain soil complexes of California. U.S. Department of Agriculture, Technical Bulletin 745. 46 p.

PANOS, V., AND O. STELCL
1968 Physiographic and geologic control in development of Cuban mogotes. Zeitschrift für Geomorphologie 12(2):117–164.

PFEFFER, K. H.
1969 Kalkrusten und kegelkarst. Erdkunde 23(3): 230–236.

PRICE, W. A.
1925 Caliche and pseudo-anticlines. American Association of Petroleum Geologists, Bulletin 9: 1009–1017.

1933 The Reynosa problem of South Texas and the origin of caliche. American Association of Petroleum Geologists, Bulletin 17(5):488–552.

1940 Caliche karst. Geological Society of America, Bulletin 51:1938. (Abstr.)

1940 Origin of caliche. Geological Society of America, Bulletin 51:1939. (Abstr.)

1958 Sedimentology and Quaternary geomorphology of south Texas. Gulf Coast Association of Geological Societies, Transactions 8:41–75.

PUTZER, H.
1958   Quartäre krustenbildungen im tropischen Süd-Amerika. Geologisches Jahrbuch 76:37–52.
1962   Geologie von Paraguay. Gebrüder Borntraeger, Berlin. 184 p.

REEVES, C. C.
1966   Pluvial lake basins of W. Texas. Journal of Geology 74:269–291.
1968   Introduction to palaeolimnology. Elsevier, Amsterdam. 228 p.
1970   Caliche. In R. W. Fairbridge (ed), Encyclopedia of the Earth Sciences, vol 4. Reinhold, New York.
1970   Origin, classification, and geologic history of caliche on the southern High Plains, Texas and eastern New Mexico. Journal of Geology 78(3):352–362.

REEVES, C. C., AND J. D. SUGGS
1964   Caliche of central south Llano Estacado. Journal of Sedimentary Petrology 34(3):669–672.

RIGHTMIRE, C.
1967   A radiocarbon study of the age and origin of caliche deposits. University of Texas (unpublished M.S. thesis).

RODRIGUEZ, E. J.
1966   Estudio hidrogeológico de sector nordeste de la provincia de Mendoza. Asociación Geológica Argentina, Revista 21(1):39–60.

ROTHROCK, E. P.
1925   On the force of crystallization of calcite. Journal of Geology 33:80–82.

RUHE, R. V.
1967   Geomorphic surfaces and surficial deposits in southern New Mexico. New Mexico State Bureau of Mines and Mineral Resources, Memoir 18. 66 p.

RUSSELL, W. L.
1929   Stratigraphy and structure of the Smoky Hill Chalk in western Kansas. American Association of Petroleum Geologists, Bulletin 13:595–604.

SAYRE, A. N.
1937   Geology and groundwater resources of Duval County, Texas. U.S. Geological Survey, Water Supply Paper 776. 116 p.

SCHOFF, S. L.
1939   Geology and groundwater resources of Texas County, Oklahoma. Oklahoma Geological Survey, Bulletin 59. 248 p.

SCHULTZ, G. E.
1969   Geology and paleontology of a late Pleistocene basin in southwest Kansas. Geological Society of America, Special Paper 105. 85 p.

SHANTZ, H. L.
1923   The natural vegetation of the Great Plains region. Association of American Geographers, Annals 13:81–107.

SHERMAN, G. D., AND H. IKAWA
1958   Calcareous concretions and sheets in soils near south point, Hawaii. Pacific Science 12:255–257.

SHREVE, F., AND T. MALLERY
1933   The relation of caliche to desert plants. Soil Science 35:99.

SIDWELL, R.
1943   Caliche deposits of the southern High Plains, Texas. American Journal of Science 241:257–261.

SIRAGUSA, A.
1964   Contribución al conocimiento de las toscas de la República Argentina. Gaea 12:123–148.

STAPPENBECK, R.
1926   Geologie und grundwasserkunde der Pampa. Stuttgart. 409 p.

STUART, D. M., M. A. FOSBERG, AND G. C. LEWIS
1961   Caliche in south west Idaho. Soil Science Society of America, Proceedings 25:132–135.

SWEETING, M. M.
1969   Karstic morphology of the Yucatan. In University of Edinburgh Report, "Expedition to British Honduras-Yucatan 1966," section 4: 37–40.

SWINEFORD, A., AND P. C. FRANKS
1959   Opal in the Ogallala Formation in Kansas. Society of Economic Palaeontologists and Mineralogists, Special Publication 7:111–120.

SWINEFORD, A., A. B. LEONARD, AND J. C. FRYE
1958   Petrology of the Pliocene pisolitic limestone in the Great Plains. Kansas Geological Survey, Bulletin 130:97–116.

THEIS, C. V.
1936   Possible effect of groundwater on the Ogallala formation of the Llano Estacado. Washington Academy of Sciences, Journal 26(9):390–392.

THOMAS, C. M.
1965   Origin of pisolites. American Association of Petroleum Geologists, Bulletin 49:360. (Abstr.)

THORNBURY, W. D.
1965   Regional geomorphology of the United States. Wiley, New York. 609 p.

TOLMAN, C. F.
1909   Erosion and deposition in the southern Arizona bolson region. Journal of Geology 17: 136–163.

TROEH, F. R.
1969   Noteworthy features of Uruguayan soils. Soil Science Society of America, Proceedings 33(1):125–128.

TROWBRIDGE, A. C.
1926   Reynosa formation in Lower Rio Grande region, Texas. Geological Society of America, Bulletin 37:455–462.

UDDEN, J. A.
1923   The "Rim-Rock" of the High Plains. American Association of Petroleum Geologists, Bulletin 7:72–74.

VALASTRO, S., E. MOTT DAVIS, AND C. T. RIGHTMIRE
1968   University of Texas at Austin Radiocarbon dates VI. Radiocarbon 10(2):384–401.

VAN DEN HEUVEL, R. C.
1966   The occurrence of sepiolite and attapulgite in the calcareous zone of a soil near Las Cruces, New Mexico. National Conference on Clays and Clay Minerals, 13th, 1964, Proceedings, pp. 193–207.

VAN SICLEN, D. C.
1957   Cenozoic strata of the southwestern Osage Plains of Texas. Journal of Geology 65:47–60.

VAN TUYL, F. M., AND W. S. LEVINGS
1949   Pliocene Ogallala algal limestone in Union County, New Mexico. American Association of Petroleum Geologists, Bulletin 33:1429–1430.

VEDDER, J. G., AND R. M. NORRIS
  1963   Geology of San Nicolas Island, California. U.S. Geological Survey, Professional Paper 369. 65 p.
WANLESS, H. R.
  1922   Lithology of White River sediments. American Philosophical Society, Proceedings 61:184–203.
WARD, W. C., R. L. FOLK, AND J. L. WILSON
  1969   Surficial alteration of Pleistocene(?) limestone adjacent to saline lakes, Isla Mujeres, Quintana Róo, Mexico. American Association of Petroleum Geologists, Bulletin 53(3):748–749. (Abstr.)
  1970   Blackening of eolianite and caliche adjacent to saline lakes, Isla Mujeres, Quintana Róo, Mexico. Journal of Sedimentary Petrology 40(2):548–555.
WEEKS, A. D., AND D. H. EARGLE
  1963   Relation of diagenetic alteration and soil forming processes to the uranium deposits of the southern Texas coastal plain. National Conference on Clays and Clay Minerals, 10th, 1961, Proceedings, p. 23–41.
WEEKS, A. W.
  1933   Lissie, Reynosa, and upland terrace deposits of the coastal plain of Texas between Brazos River and Rio Grande. American Association of Petroleum Geologists, Bulletin 17(5):453–487.
WHITE, W. N., W. L. BROADHURST, AND J. W. LANG
  1946   Groundwater in the High Plains of Texas. U.S. Geological Survey, Water Supply Paper 889-F:381–420. (*See* p. 386–387.)
YOUNG, R. G.
  1964   Fracturing of sandstone cobbles in caliche cemented terrace gravels. Journal of Sedimentary Petrology 34:886.

AUSTRALIA

AITCHISON, G. D., R. C. SPRIGG, AND G. W. COCHRANE
  1954   The soils and geology of Adelaide and suburbs. Geological Survey of South Australia, Bulletin 32. 126 p.
BLACKBURN, G., R. D. BOND, AND A. R. P. CLARKE
  1965   Soil development associated with stranded beach ridges in south east South Australia. CSIRO, Australia, Soil Publication 22. 66 p.
CASEY, J. N., AND A. T. WELLS
  1964   The geology of the north east Canning Basin, Western Australia. Australia, Bureau of Mineral Resources, Geology, and Geophysics, Report 49. 61 p.
CONNOLLY, J. R.
  1968   Submarine canyons of the continental margins, East Bass Strait. Marine Geology 6:449–461.
CRAWFORD, A. R.
  1965   The geology of Yorke Peninsula. Geological Survey of South Australia, Bulletin 39. 96 p.
CROCKER, R. L.
  1941   Notes on the geology and physiography of south east South Australia with reference to late climatic history. Royal Society of South Australia, Transactions 65:103–107.
  1944   Soil and vegetation relationships in the lower south east of South Australia — a study in ecology. Royal Society of South Australia, Transactions 68:144–172.

  1946   Post Miocene climatic and geologic history and its significance in relation to the genesis of the major soil types of South Australia. CSIR, Australia, Bulletin 193. 56 p.
DAVIDSON, W. A.
  1968   Millstream hydrogeological investigation. Western Australia, Department of Mines, Annual Report, p. 57–62.
FAIRBRIDGE, R. W., AND C. TEICHERT
  1953   Soil horizons and marine bands in the coastal limestones of Western Australia. Royal Society of New South Wales, Journal and Proceedings 86:68.
FENNER, C.
  1927   Adelaide, South Australia, a study in human geography. Royal Society of South Australia, Transactions 51:193–256.
FIRMAN, J. B.
  1967   Stratigraphy of late Cenozoic deposits in South Australia. Royal Society of South Australia, Transactions 91:165–178.
GLAESSNER, M. F., AND L. W. PARKIN (eds)
  1958   Geology of South Australia. Geological Society of Australia, Journal 5(2). 163 p.
HOWCHIN, W.
  1918   The geology of South Australia. Government Printer, Adelaide. 543 p.
  1923   A geological sketch-section of the sea-cliffs on the eastern side of Gulf St. Vincent, from Brighton to Sellicks Hill with descriptions. Royal Society of South Australia, Transactions 47:279–315.
JACKSON, E. A.
  1967   Soil features in arid regions with particular reference to Australia. Australian Institute of Agricultural Science, Journal 25:196–208.
JENNINGS, J. N., AND M. SWEETING
  1961   Caliche pseudo-anticlines in the Fitzroy Basin, Western Australia. American Journal of Science 259:635–639.
JESSUP, R. W.
  1967   Soils and eustatic sea level fluctuations in relation to Quaternary history and correlation in South Australia. *In* R. B. Morrison and H. E. Wright (eds), Quaternary Soils, International Congress on Quaternary, VII, Denver and Boulder, 1965, Proceedings 9:191–204. University of Nevada, Desert Research Institute, Center for Water Resources Research, Reno.
JOHNS, R. K.
  1963   Limestone, dolomite and magnesite resources of South Australia. Geological Survey of South Australia, Bulletin 38. 100 p.
LINK, A. G.
  1967   Late Pleistocene-Holocene climatic fluctuations; possible solution pipe-Foiba relationships and the evolution of limestone cave morphology. Zeitschrift für Geomorphologie 11(2):117–145.
LITCHFIELD, W. H.
  1969   Soil surfaces and sedimentary history near the Macdonnell Ranges, N.T. CSIRO, Australia, Soil Publication 25. 45 p.
LOWERY, D. C.
  1967   The origin of Easter Cave doline, Western Australia. Australian Geographer 10(4):300–302.

MABBUTT, J. A.
1967   Denudation chronology in Central Australia. *In* J. N. Jennings and J. A. Mabbutt (eds), Landform Studies from Australia and New Guinea, p. 144–181. Cambridge University Press, London. 434 p.

MABBUTT, J. A., AND OTHERS
1963   General report on lands of the Wiluna-Meekatharra area, Western Australia, 1958. CSIRO, Australia, Land Research Series 7. 215 p.

MCLEOD, W. N.
1966   The geology and iron deposits of the Hammersley range area, Western Australia. Geological Survey of Western Australia, Bulletin 117. 170 p.

NORRISH, K., AND L. E. R. ROGERS
1956   The mineralogy of some terra rossas and rendzinas of South Australia. Journal of Soil Science 7(2):294–301.

PRESCOTT, J. A.
1931   The soils of Australia in relation to vegetation and climate. CSIR Bulletin 52.

REYNOLDS, M. A.
1953   The Cainozoic succession of Maslin and Aldinga Bays, South Australia. Royal Society of South Australia, Transactions 76:114–140.

SANDERS, C. C.
1968   Hydrogeological reconnaissance of calcrete areas in the East Murchison and Mount Margaret goldfields. Western Australia, Department of Mines, Annual Report, pp. 54–57.

SOFOULIS, J.
1963   The occurrence and hydrological significance of calcrete deposits in Western Australia. Geological Survey of Western Australia, Annual Report, pp. 38–42.

STACE, H. C. T., AND L. E. R. ROGERS
1954   Morphological, chemical and mineralogical data for some South Australian terra rossa and rendzina soils. CSIRO, Australia, Division of Soils, Report 10/54. 43 p.

VAN ANDEL, T. H., AND OTHERS
1967   Late Quaternary history, climate and oceanography of the Timor Sea, N.W. Australia. American Journal of Science 265(9):737–758.

WARD, W. G.
1966   Geology, geomorphology and soils of the south-western part of County Adelaide, South Australia. CSIRO, Australia, Soil Publication 23. 115 p.

WATKINS, J. R.
1967   The relationship between climate and the development of landforms in the Cainozoic rocks of Queensland. Geological Society of Australia, Journal 14(1):153–168.

WELLS, A. T., D. J. FORMAN, AND L. C. RANFORD
1964   Geological reconnaissance of the Rawlinson and Macdonald 1:250,000 sheet areas, Western Australia. Australia, Bureau of Mineral Resources, Geology and Geophysics, Report 65. 35 p.

WELLS, A. T., A. J. STEWART, AND S. K. SKWARKO
1966   Geology of the south-eastern part of the Amadeus Basin, Northern Territory. Australia, Bureau of Mineral Resources, Geology and Geophysics, Report 88. 59 p.

WHITEHOUSE, F. W.
1940   Studies in the late geological history of Queensland. University of Queensland, Department of Geology, Paper n.s. 2(5):1–79.

WOOLNOUGH, W. G.
1930   Influence of climate and topography in the formation and distribution of products of weathering. Geological Magazine 67:123–132.

EAST AFRICA

ANDERSON, E. D., AND L. M. TALBOT
1965   Soil factors affecting the distribution of the grassland types and their utilization by wild animals in the Serengeti Plains, Tanganyika. Journal of Ecology 53:33–56.

BAKER, B. H.
1958   Geology of the Magadi area. Geological Survey of Kenya, Report 42. 81 p.

BAKER, B. H., AND E. P. SAGGERSON
1958   Geology of the El Wak — Aus Mandula area. Geological Survey of Kenya, Report 44. 48 p.

BESTOW, T. T.
1953   Report on the geology and hydrology of Wajir District. Geological Survey of Kenya, unpublished report.

BORNHARDT, W.
1900   Zur oberflachengestaltung und geologie Deutsch-Ostafrikas. Dietrich Reimer, Berlin. 595 p. (*See* p. 24, 25, and 291.)

BUTZER, K. W., AND D. L. THURBER
1969   Some late Cenozoic sedimentary formations of the Lower Omo Basin. Nature 222:1138–1143.

CECIONI, G.
1941   Il crostone selcioso di Bur-Uen in Somalia. Società Geologica Italiana, Bollettino 58:235–241.

COMBE, A. D.
1942   Limestone deposits, S.W. Toro, South West Uganda. Uganda Protectorate, Geological Survey Department, Annual Report for the year ending December 31, 1942, Field Work Report, p. 13.

DIXEY, F.
1948   Geology of northern Kenya. Geological Survey of Kenya, Report 15. 43 p.

HAY, R. L.
1963   Stratigraphy of Beds I through IV, Olduvai Gorge, Tanganyika. Science 139:829–833.

HUNT, J. A.
1958   Report on the geology of the Adadleh area, Hargeisa and Berbera districts. Somaliland Protectorate, Geological Survey, Report 2. 16 p.

JOUBERT, P.
1963   Geology of the Wajir-Wajir Bor area. Geological Survey of Kenya, Report 57.

LEAKEY, L. S. B., R. PROTSCH, AND R. BERGER
1968   Age of Bed V, Olduvai Gorge, Tanzania. Science 162:559–560.

MACFADYEN, W. A.
1952   Water supply and geology of parts of British Somaliland. Government of Somaliland Protectorate. 184 p.

OATES, F.
1933   The limestone deposits of Tanganyika Territory. Geological Survey of Tanganyika, Bulletin 4. 118 p.

PALLISTER, J. W.
  1963 Notes on the geomorphology of the northern region, Somali Republic. Geographical Journal 129(2):184–187.

RIX, P.
  1964 Geology of the Enyali-Ndiandaza area. Geological Survey of Kenya, Report 68.

SAGGERSON, E. P.
  1952 Geology of the Kisumu District. Geological Survey of Kenya, Report 21. 86 p.
  1966 Geology of the Loita Hills area. Geological Survey of Kenya, Report 71. 60 p.

WILLIAMS, M. A. J.
  1968 A dune catena on the clay plains of the west central Gezira, Republic of the Sudan. Journal of Soil Science 19(2):367–378.

NORTH AFRICA

AGAFONOFF, V.
  1935 Sols types de Tunisie. Service Botanique et Agronomique de Tunisie, Annales 12–13: 43–414.
  1936 Les sols bruns et rouge a croûte carbonatée en Tunisie. Académie des Sciences, Paris, Comptes Rendus 202:1597–1599.

AUBERT, G.
  1948 Les sols a croûtes calcaires. Conférence Internationale de Pédologie Mediterranéenne, Algér-Montpellier, 1947, Comptes Rendus pp. 330–332.

AWAD, H.
  1963 Some aspects of the geomorphology of Morocco related to the Quaternary climate. Geographical Journal 129(2):129–139.

BEAUDET, G., J. MARTIN, AND G. MAURER
  1964 Remarques sur quelques facteurs de l'érosion des sols. Revue de Géographie Marocain 6: 65–72.

BEAUDET, G., G. MAURER, AND A. RUELLAN
  1967 Le quaternaire marocain; observations et hypothèses nouvelles. Révue de Géographie Physique et Géologie Dynamique, sér. 2, 9(4): 269–310.

BEHRMANN, W.
  1932 Beobachtungen am Rande der Wüste. Geographische Zeitschrift 38:321–334.

BELLAIR, P.
  1949 Le quaternaire de Tejerhi (Fezzan). Société Géologique de France, Comptes Rendus pp. 161–162.

BOULAINE, J.
  1958 Sur la formation des carapaces calcaires. Travaux des Collaborateurs Service de la Carte Géologique de l'Algérie, Bulletin 20:7–19.
  1960 Sur quelques sols rouges a carapace calcaire. Association Française pour l'Etude du Sol, Bulletin 3:130–134.
  1966 Sur le rôle de la végétation dans la formation des carapaces calcaires mediterranéennes. Académie des Sciences, Paris, Comptes Rendus 253:2568–2570.
  1966 Sur les relations entre les carapaces calcaires et les sols isohumiques de climat xérothèrique. Science du Sol 1:3–16.

BOULAINE, J., AND J. P. L'HERMITTE
  1963 Sur les sols noirs a carapaces calcaires formes sous climat xérothèrique. Algérie, Centre de Recherche et d'Experimentation Forestières, Annales 2:5–8.

BUTZER, K. W.
  1964 Environment and archaeology. University of Wisconsin Press, Madison. 525 p.
  1968 Desert and river in Nubia. University of Wisconsin Press, Madison. 562 p.

CASTANY, G.
  1966 Le Tyrrhénien de la Tunisie. Quaternaria 6: 229–269.

CHARLES, G.
  1949 Sur la formation de la carapace zonaire en Algérie. Académie des Science, Paris, Comptes Rendus 228:261–263.

COMEL, A.
  1932 Richerche pedologiche sui terreni della Tripolitania. Società Geologica Italiana, Bollettino 51(2):317–342.

COMPAGNIE DES TECHNIQUES HYDRAULIQUES ET AGRICOLES, GRENOBLE
  1954 Mémoire explicatif, aménagement de l'uadi Megenin. Rapport général en application de la convention du 30 Mars, 1954.
  1956 Caam Wadi Development Scheme, Geologic and Topographic Survey for a storage dam site.

CONCARET, J., AND P. MAHLER
        Sue les paleosols du Haouz de Marrakech et leur importance agronomique. Académie Agricole Française, Comptes Rendus 46:654–658.

COQUE, R.
  1955 Morphologie et croûte dans le Sud-Tunisien. Annales de Géographie 64:359–370.
  1962 La Tunisie présaharienne, étude géomorphologique. Armand Colin, Paris. 476 p.

COQUE, R., AND A. JAUZEIN
  1966 The Quaternary of Tunisia. Quaternaria 8: 139 154.

CROFTS, R.
  1967 Raised beaches and chronology in north west Fuenaventura, Canary Islands. Quaternaria 9: 247–260.

DALLONI, M.
  1951 Sur la genèse et l'âge des "terrains a croûte" Nord Africain (On the genesis and age of the North African "crust terrains"). Centre National de la Recherche Scientifique, Colloques Internationaux 35:278–285.

DESPOIS, J.
  1954 Les croûtes calcaire et leur origine. Annales de Géographie 63:59–60.
  1955 La Tunisie orientale; sahel et basse steppe, étude géographique. Institut des Hautes Etudes de Tunis, Section des Lettres, Publication 1. 554 p.

DICÉSARE, F., A. FRANCHINO, AND C. SOMMARUGA
  1963 The Pliocene-Quaternary of the Girabub Erg region. Institut Française du Pétrole, Revue 18(10):1344–1354.

DURAND, J. H.
  1949 Essai de nomenclature des croûtes. Société des Sciences Naturelles de Tunisie, Tunis, Bulletin 3–4: 141–142.

DURAND, J. H. (cont)

1953   Étude géologique, hydrogéologique et pédologique des croûtes en Algérie. Algérie, Direction du Service de la Colonisation et de l'Hydraulique, Clairbois-Birmandreis. 209 p.

1954   Les sols d'Algérie. Algérie, Direction du Service de la Colonisation et de l'Hydraulique, Service des Etudes Scientifiques, Pédologie 2. 244 p.

1959   Les sols rouges et les croûtes calcaires en Algérie. Algérie, Direction de l'Hydraulique et de l'Equipement Rural, Clairbois-Birmandreis, Service des Études Scientifiques, Étude Général 7. 188 p.

1963   Les croûtes calcaires et gypseuses en Algérie: formation et age. Société Géologique de France, Bulletin, sér. 7, 5:959–968.

DUTCHER, L. C., AND H. E. THOMAS

1966   The occurrence, chemical quality and use of groundwater in the Tabulbah area, Tunisia. U.S. Geological Survey, Water Supply Paper 1757-E. 29 p.

FISCHER, T.

1910   Schwarzerde und Kalkruste in Marokko. Zeitschrift für Praktische Geologie 18:105–114.

FLANDRIN, J., M. GAUTIER, AND R. LAFFITTE

1948   Sur la formation de la croûte calcaire superficielle en Algérie. Académie des Sciences, Paris, Comptes Rendus 226:416–418.

FLORIDA, G. B.

1939   Osservazioni sul Miocene dei dintorni di Homs. Società Geologica Italiana, Bollettino 58:245–260.

FOURNET, A.

1960   Croûte villafranchien saumon, relation avec les sols steppiques dans la région de Mezzouna. Société des Sciences Naturelles de Tunisie, Bulletin 9/10:109–111.

FRANCHETTI, L.

1914   La missione Franchetti in Tripolitania: Il Gebel, pp. 120–122. Societa Italianà per lo Studio della Libia, Firenze. 3 vols.

FRANZ, H., AND G. FRANZ

1969   Beiträge zur kenntnis der bildung von kalkrusten in böden der warmen trockengebiete. Zeitschift für Pflanzenernährung und Bodenkunde 121(1):34–42.

GAUCHER, G.

1947   Les sols rubifiés et les sols a croûtes du Bas Chelif et des basses plaines oranaises (Régions d'Inkermann, de Relizane, de Perregaux, et de St. Denis du Sig). Académie de Science, Paris, Comptes Rendus 225:133–135.

1948   Les sols a croûtes calcaires. Conférence Internationale de Pédologie Mediterranéenne, Algér-Montpellier, 1947, Comptes Rendus pp. 334–335.

1948   Sur certains caracteres des croûtes calcaires en rapport avec leur origine. Académie de Science, Paris, Comptes Rendus 227:154–156.

1948   Sur quelques conditions de formation des croûtes calcaires. Académie de Science, Paris, Comptes Rendus 227:215–217.

1958   Les conditions géologiques de la pedogenèse nord-Africaine. Travaux des Collaborateurs Service de la Carte Géologique de l'Algérie, Alger, Bulletin 20 (n.s.):55–94.

GIGOUT, M.

1951   Etudes géologiques sur la meseta marocaine occidentale. Institut Scientifique Chérifien, Travaux, sér. Géologie et Géographie Physique 3:198–201.

1958   Sur le mode de formation des limons et croûtes calcaires du Maroc. Académie de Science, Paris, Comptes Rendus 247(1):97–100.

1960   Sur la genèse des croûtes calcaires pléistocènes en Afrique du Nord. Société Géologique de France, Compte Rendu Sommaire, sér. 7, 2(1):8–10.

GILCHRIST SHIRLAW, D. W., S. G. WILLIMOTT, J. I. CLARKE, AND M. E. FRISBY

1961   Soil survey of Tauorga, Tripolitania, Libya. University of Durham Department of Geography. 65 p.

GLINKA, K. D.

1931   Treatise on soil science. 4th ed. Translated, 1963, by Israel Program for Scientific Translations, Jerusalem. 670 p.

GOBERT, E. G.

1948   Sur le problème des croûtes et sur les sols capsiens. Société des Sciences Naturelles de Tunisie, Bulletin 1:56–65.

HAUSEN, H.

1958   On the geology of Fuerteventura. Societas Scientarum Fennica, Commentationes Physico-Mathematicae 22(1):1–211.

1958   Contribución al conocimiento de las formaciones sedimentarias de Fuerteventura. Anuario de Estudios Alanticos (Madrid-Las Palmas) 4:37–84.

HOUÉROU, H. N. LE

1960   Contribution a l'étude des sols du sud tunisien. Annales Agronomiques 11(3):241–308.

HUVELIN, P.

1965   Sols rubifiés et croûtes calcaires du piedmont septentrional du haut Atlas de Demate. Service Géologique du Maroc, Notes 25; Notes et Mémoires 185:95–97.

JARANOFF, D.

1937   Etudes de géologie dynamique au Maroc, dans les confins Algero-Marocains et en afrique occidentale française. Revue de Géographie, Physique et de Géologie Dynnamique 10:131–141.

JOFFE, J. S.

1949   Pedology. With an introduction by C. F. Marbut. 2nd ed. Pedology Publications, New Brunswick, New Jersey. 662 p. (See p. 229.)

JOLY, F.

1962   Études sur le relief du sud-est Marocain. Institut Scientifique Chérifien, Travaux, sér. Géologie et Géographie Physique 10:350–353.

KANTER, H.

1965   Die serir Kalanscho in Libyen, eine Landschaft der Vollwüste. Petermanns Geographische Mitteilungen 109:265–272.

KRECJI-GRAF, K. VON

1960   Zur geologie der Makaronesen. 4: Krustenkalke. Deutsche Geologische Gesellschaft, Zeitschrift 112:36–61.

LIPPARINI, T.

1940   Tettonica e geomorfologia della Tripolitania. Società Geologica Italiana, Bollettino 59:221–299.

MARGAT, J., R. RAYNAL, AND P. TALTASSE
  1954 Deux series d'observations nouvelles sur les croûtes au Maroc (Couloir sud-Rifain et Maroc oriental). Service Géologique du Maroc, Notes 10; Notes et Mémoires 122:26–38.

MAURER, G.
  1968 Les montagnes du Rif Central. Institut Scientifique Chérifien, Travaux, sér. Géologie et Géographie Physique 14:415–417.

MILLOT, G., H. PAQUET, AND H. RUELLAN
  1969 Néoformation de l'attapulgite dans les sols a carapaces calcaires de la basse Moulouya (Maroc Oriental). Académie des Sciences, Paris, Comptes Rendus 268(D)(23):2771–2774.

MOSELEY, F.
  1965 Plateau calcrete, calcreted gravels, cemented dunes and related deposits of the Maallegh-Bomba region of Libya. Zeitschrift für Geomorphologie 9:167–185.

MÜLLER, S.
  1954 Beobachtungen an rezenten Kalkrindenböden im nördlichen Algerien. Zeitschrift für Pflanzenernährung und Bodenkunde 65(1/3):107–117.

NEUVILLE, R., AND A. RUHLMANN
  1942 La place du paléolithique ancien dans le quaternaire marocain. Institut des Hautes Etudes Marocaines, Casablanca, Collection Hespéris. 156 p.

PASSARGE, S.
  1909 Verwitterung und abtragung in den steppen und wüsten Algeriens. Geographische Zeitschrift 15:493–510.

PERVINQUIÉRE, L.
  1903 Etude géologique de la Tunisie centrale. Rudeval, Paris. 359 p.

PIMIENTA, J.
  1950 Les nodules de tuf et la formation actuelle de croûtes calcaires dans les environs de Tunis. Société Geologique de France, Compte Rendu Sommaire, pp. 211–212.

POMEL, A.
  1871 Le Sahara, observations de géologie et de géographie physique et biologique. Société de Climatologie d'Alger, Bulletin 8:133–165.

POMEL, A., AND J. POUYANNE
  1882 Texte explicatif de la carte géologique provisoire au 1:800,000e des provinces d'Alger et d'Oran. Adolphe Jourdan, Alger. 56 p.

RENOU, E.
  1848 Description géologique de l'Algérie. Paris. 182 p.

ROSEAU, H.
  1948 Les sols a croûte calcaire. Conférence Internationale de Pédologie Méditerranéenne, Alger-Montpellier, 1947, Comptes Rendus pp. 332–333. (Abstr.)

RUELLAN, A.
  1962 Utilisation de la géomorphologie pour l'étude pédologique au 1:20,000 de la Plaine du Zebra (Basse Moulouya). Révue de Géographie du Maroc 1/2:23–30.
  1967 Les sols isohumiques subtropicaux au Maroc. Conference on Mediterranean Soils, Madrid, 1966, Transactions, pp. 81–89.

  1967 Individualisation et accumulation du calcaire dans les sols et les depôts quaternaires du Maroc. Cahiers Orstom, sér. Pédologique 5(4):421–462.
  1968 Les horizons d'individualisation et d'accumulation du calcaire dans les sols du Maroc. International Congress of Soil Science, 9th, Adelaide, 1968, Transactions 4:501–510.

SIDORENKO, A. V.
  1959 The calcareous desert crusts of Egypt (translated title). Academy of Sciences of the USSR, Doklady 128 (4):812–814.

TRICART, J., AND A. CAILLEUX
  1962– Le modelé des régions sèches. Centre de Documentation Universitaire, Paris. 2 vols. (*See* vol 1963   2, p. 147.)

VITA-FINZI, C.
  1960 Post Roman changes in Wadi Lebda. In S. G. Willimott and J. I. Clarke (eds), Field Studies in Libya. University of Durham, Department of Geography, Research Papers series 4:46–51.
  1969 The Mediterranean valleys. Cambridge University Press, London. 140 p.

WILBERT, J.
  1962 Croûtes et encroûtements calcaires au Maroc. Al Awamia 3:175–192.

WILLIAMS, G. E.
  1970 Piedmont sedimentation and late Quaternary chronology in the Biskra Region of the northern Sahara. Zeitschrift für Geomorphologie, Supplementband 10:40–63.

WILLIMOTT, S. G.
  1960 Soils of the Jefara. In S. G. Willimott and J. I. Clarke (eds), Field Studies in Libya. University of Durham, Department of Geography, Research Papers series 4:26–45.

YANKOVITCH, L.
  1935– Etude pédo-agrologique de la Tunisie. Service 1936   Botanique et Agronomique de Tunisie, Annales 12–13:417–559. (*See* Chapter 4 esp.)
  1937– Etude agrologique detaillée de la Tunisie. Ser-1938   vice Botanique et Agronomique de Tunisie, Annales 14 15:159–163.
  1947 Les sols a croûtes calcaires. Conférence Internationale de Pédologie Méditerranéenne, Alger-Montpellier, 1947, Comptes Rendus pp. 336–337 (Abstr.)

SOUTHERN AFRICA

AMDURER, S. S.
  1956 The engineering geology of the Cape Flats. University of Cape Town (unpublished Ph.D. thesis).

BESAIRIE, H.
  1948 Sheet mémoir, Carte Géologique de Reconnaissance 1:200,000 Fort Dauphin, Madagascar. Paris.

BLOOMFIELD, K., AND M. S. GARSON
  1965 Geology of the Kirk Range-Lisungwe Valley area. Malawi, Geological Survey Department, Bulletin 17. 227 p.

BOND, G. W.
  1946 A geochemical survey of the underground water of the Union of South Africa. Geological Survey of South Africa, Memoir 41. 208 p.

BOND, G. W. (cont)
  1963 Pleistocene environments in southern Africa. *In* F. C. Howell and F. Bourlière (eds), African Ecology and Human Evolution, pp. 308–334. Wenner-Gren Foundation for Anthropological Research, Inc., New York.
  1965 Quantitative approaches to rainfall and temperature changes in the Quaternary of southern Africa. *In* H. E. Wright and D. G. Frey (eds), International Studies on the Quaternary. Papers prepared on the occasion of the VII Congress of the International Association for Quaternary Research, Boulder, 1965. Geological Society of America, Special Paper 84:323–336.

BOSAZZA, V. L.
  1964– Formation of malachite and chrysocolla from
  1965   chalcopyrite in rocks of the Bushman Mine series, Northern Bechuanaland Protectorate. Institution of Mining and Metallurgy, Transactions 74(4):201–215.

BRAIN, C. K.
  1957 Observations on the recent geological history of the Kalahari Gemsbok National Park. Pretoria, mimeo.

BRINK, A. B. A.
  1962 Airphoto interpretation applied to soil engineering mapping in South Africa. Archives Internationale de Photogrammetrie 14:498–506.

CAIGER, J. H.
  1964 The use of airphoto interpretation as an aid to prospecting for road building materials in South West Africa. University of Cape Town, Department of Mechanical Engineering (unpublished Master's thesis).
  1968 Aerial photography and aerial reconnaissance applied to gravel road construction in the Okavango Territory, South West Africa. Civil Engineer in South Africa 10(5):99–106.

COERTZEE, C. B.
  1960 The geology of the Orange Free State Gold-Field. Union of South Africa, Department of Mines and Geological Survey, Memoir 49:110.

CROCKETT, R. N.
  1962 Notes on the geology of the area N.W. of Shushong Bamangwato tribal territory. Bechuanaland Protectorate, Geological Survey, unpublished report.
  1965 A provisional description of the geology of the Shashi area. Buchuanaland Protectorate, Geological Survey, unpublished report.
  1965 Geology of the country around Mahalype and Machaneng. Bechuanaland Protectorate, Geological Survey, Record 1961–1962:77–99.

CROCKETT, R. N., AND C. M. H. JENNINGS
  1965 Geology of part of the Okwa Valley, Western Bechuanaland. Bechuanaland Protectorate, Geological Survey, Record 1961–1962:101–113.

DEBENHAM, F.
  1952 The Kalahari today. Geographical Journal 118:12–23.

DEMANGEON, F.
  1906 Le Kalahari d'après le livre de M. Siegfried Passarge. Annales de Géographie 15:43–58.

DIXEY, F.
  1927 The limestone resources of Nyasaland. Nyasaland Protectorate, Geological Survey, Bulletin 3. 43 p.

DRYSDALL, A. R., AND R. K. WELLER
  1966 Karroo sedimentation in northern Rhodesia. Geological Society of South Africa, Transactions 69:39–70. (*See* p. 58.)

DU TOIT, A. L.
  1906 Geological survey of the eastern portion of Griqualand West. Geological Commission of the Cape of Good Hope, Annual Report 11th, pp. 87–176.
  1907 Geological survey of portions of Hopetown, Britstown, Prieska, and Hay. Geological Commission of the Cape of Good Hope, Annual Report, 12th, pp. 161–192.
  1915 Underground water in S.E. Bechuanaland. South African Philosophical Society, Transactions 16(3):251–262.
  1954 Geology of South Africa. 3rd ed. Oliver and Boyd, Edinburgh. 611 p.

FEIO, M.
  1964 A evoluçao da escadaria de aplanaçoes do sudoeste de Angola. Garcia de Orta (Lisboa) 12(2):323–354.

GANSSEN, R.
  1960 Landschaft und böden in südwestafrika. Die Erde 9(2):115–131.
  1960 Böden und Landschaft in südwestafrika. International Congress of Soil Science, 7th, Madison, 1960, Transactions 4:49–55.
  1963 Südwest-afrika, böden und bodenkultur, versuch einer klimapedologie warmer trockengebiete. Berlin.

GANSSEN, R., AND W. MOLL
  1961 Beiträge zur kenntnis der Böden warm-arider Gebiete, dargestellt am beispiel südwestafrika. Zeitschrift für Pflanzenernährung, Düngung und Bodenkunde 94(1):9–25.

GERRARD, I.
  1962 Physiographic note to accompany the West Tuli contour map. Bechuanaland Protectorate, Geological Survey, unpublished report.
  1963 The geology of the Foley area. Bechuanaland Protectorate, Geological Survey, Records 1959–1960:135–148.
  1965 Geology of the area West Tuli. Bechuanaland Protectorate, Geological Survey, Report 1961–1962:5–24.
  1965 The lower Karroo succession at Lelanamagadi Pan, S.W. Bechuanaland, and its correlation with the Molopo R. sequence. Bechuanaland Protectorate, Geological Survey, unpublished report.
  1965 Geology of the Ramathlabama, Pitsani Molopo, and Mabule areas. An explanation of portions of quarter degree sheets, 2524D, 2525C, 2525D. Bechuanaland Protectorate, Geological Survey, unpublished report.

GEVERS, T. W.
  1930 Terrester Dolomit in der Etoscha-Pfanne, Südwest-afrika. Zentralblatt für Mineralogie, Abt. B, 6:224–230.

GREEN, D.
  1961 The Mamabule coal area. Bechuanaland Protectorate, Geological Survey, Mineral Resources Report 2. 65 p.
  1964 Possible manufacture of portland cement at Mamabule. Bechuanaland Protectorate, Geological Survey, unpublished report.
  1965 Origin of the Lebung Gypsum deposits, some suggested principles. Bechuanaland Protectorate, Geological Survey, unpublished report.

GROVE, A. T.
1969 Landforms and climatic change in the Kalahari and Ngamiland. Geographical Journal 135(2):191–212.

HALPENNY, L. C.
1957 Development of groundwater in parts of Angola, Portuguese West Africa. 1: Inner Cunene area. World Mining Consultants, Inc., New York. 105 p.

HERMANN, P.
1908 Beiträge zur geologie von Deutsch-Südwestafrika. Deutsche Geologische Gesellschaft, Zeitschrift, Monatsberichte 11:259–270.

HOORE, J. L. D'
1964 Soil map of Africa. Scale 1:5,000,000: Explanatory monograph. Commission for Technical Co-operation in Africa, Lagos, Joint Project 11, Publication 93.

JAEGER, F.
1936 Kalkpfannen des östlichen Südwestafrikas. International Geological Congress, 16th, Washington, D.C., 1933, Report 2:741–752.
1939 Die Trockenseen der erde. Petermanns Geographische Mitteilungen, Ergänzungsheft 236. 159 p.

JAEGER, F., AND L. WAIBEL
1920 Beiträge zur Landeskunde von Südwestafrika. Mitteilungen aus den Deutschen Schutzgebieten, Erg.-Heft 14. 804 p.

JENNINGS, C. M. H.
1958 Geological note on the Lebung-Seruruma area. Bechuanaland Protectorate, Geological Survey, unpublished report.
1963 The geology of the Makhware Hills area. Bechuanaland Protectorate, Geological Survey, Records 1959–1960:23–33.
1965 Geology of the Serowe area. Bechuanaland Protectorate, Geological Survey, Records 1961–1962:61–76.

KAISER, E.
1923 Abtragung und Auflagerung in der Namib, der südwestafrikanischen küstenwüste. Geologische Charakterbilder 27/8:40.
1926 Die Diamantenwüste Südwestafrikas. Reimer, Berlin. 2 volumes.

KENT, L. E.
1947 Diatomaceous deposits in the Union of South Africa, with special reference to Kieselguhr. Union of South Africa, Geological Survey, Memoir 42. 252 p.

KNETSCH, G.
1937 Beiträge zur kenntnis von Krustenbildung. Deutsche Geologische Gesellschaft, Zeitschrift 89:177–192.

KORN, H., AND H. MARTIN
1937 Die jüngere geologische und klimatische geschichte Südwestafrikas. Zentralblatt für Mineralogie, Geologie, und Paläontologie, Abt. B, 11:456–473.

LAMPLUGH, G. W.
1902 Calcrete. Geological Magazine 9:575.
1907 Geology of the Zambezi Basin around Batoka Gorge. Geological Society of London, Quarterly Journal 63:162–216.

LEISTNER, O. A.
1959 Notes on the vegetation of the Kalahari Gemsbok National Park with special reference to its influence on the distribution of antelopes. Koedoe 2:128–151.

LIVINGSTONE, D.
1857 Missionary travels and researches in South Africa. John Murray, London. 711 p. (*See*, for example, p. 112, 162, 527.)

LOGAN, R. F.
1960 The central Namib desert, South West Africa. National Academy of Sciences/National Research Council, Publication 758. 162 p.

LOXTON, R. F.
1962 Soil survey of the Kroonstad, 1:50,000 Topo-Cadastral, sheet 2727A. South Africa, Department of Agricultural Technical Services, Technical Communication 15. 63 p., map.

MABBUTT, J. A.
1956 The physiography and surface geology of the Hopefield fossil site. Royal Society of South Africa, Transactions 35:21–58.
1957 Physiographic evidence for the age of the Kalahari sands of the southwestern Kalahari. Pan African Congress on Prehistory, 3rd, Livingstone, 1955, Proceedings pp. 123–126.

MACGREGOR, A. M.
1930 Geological notes on a circuit of the Great Makarikari Salt Pan, Bechuanaland Protectorate. Geological Society of South Africa, Transactions 33:89–102.

MARTIN, H.
1963 A suggested theory for the origin and a brief description of some gypsum deposits of South West Africa. Geological Society of South Africa, Transactions 55:345–350.

MASON, R. J., A. B. A. BRINK, AND K. KNIGHT
1959 Pleistocene climatic significance of calcretes and ferricretes. Nature 184:568.

MISSAO DE PEDOLGIA DE ANGOLA
1963 Carta geral dos solas de Angola, distrito de Mocamedes. Portugal, Junta de Investigacoes do Ultramar, Lisboa, Memorias ser. 2, 45. 192 p.

MOFFAT, R.
1858 Journey from Colesberg to Steinkopf in 1854–1855. Royal Geographical Society, Journal 28:153.

MOUNTAIN, M. J.
1967 Pedogenic materials, their engineering properties and distribution in South Africa and South West Africa. Regional Conference for Africa on Soil Mechanics and Foundation Engineering, 4th, Cape Town, Proceedings pp. 65–70.
1967 The location of pedogenic materials using aerial photographs with some examples from South Africa. Regional Conference for Africa on Soil Mechanics and Foundation Engineering, 4th, Cape Town, Proceedings pp. 35–40.

MUIR, A., AND I. STEPHEN
1957 The superficial deposits of the Lower Shire Valley. Colonial Geology and Mineral Resources 6(4):391–407.

NETTERBERG, F.
1967 Some roadmaking properties of South African calcretes. Regional Conference for Africa on Soil Mechanics and Foundation Engineering, 4th, Cape Town, Proceedings pp. 77–81.
1969 The geology and engineering properties of South African calcretes. University of Witwatersrand, Johannesburg, Department of Geology (unpublished Ph.D. thesis). 1070 p.

NETTERBERG, F. (cont)
1969 The interpretation of some basic calcrete types. South African Archaeological Bulletin 24:117–122.
1969 Ages of calcretes in Southern Africa. South African Archaeological Bulletin 24:88–92.

NEVES, FERRAO, C. A.
1961 A hidrogeologia e o problema do abastecimento de agua ao Saixo Cunene (Angola). Garcia de Orta (Lisboa) 9(3):315–338.

PARTRIDGE, T. C., AND A. B. A. BRINK
1967 Gravels and terraces of the lower Vaal River basin. South African Geographical Journal 49:21–38.

PASSARGE, S.
1897 Summary of the geological conditions of Ngamiland. Unpublished report for the Geological Exploration of Ngamiland, translated by F. D. Lugard for Royal Geographical Society Library, London.
1904 Die Kalahari. Reimer, Berlin. 822 p.
1943 Die kalkpfannen im Hereroland und in der Kalahari. Beiträge zur Kolonialforschung 5.

PEABODY, F. E.
1954 Cave deposits of the Kaap Escarpment. Geological Society of America, Bulletin 65:671–705.

PIAGET, J. E. H.
1963 Coarse fraction mineralogy and granulometry of selected soils of the western Orange Free State. University of the Orange Free State, Bloemfontein, Faculty of Agriculture (unpublished DSc thesis). 255 p.

RANGE, P.
1912 Topography and geology of the German south Kalahari. Geological Society of South Africa, Transactions 15:63–73.

RIMANN, E.
1914 Zur enstehung von Kalaharisand und Kalaharikalk, Insbesondere der Kalkpfannen. Zentralblatt für Mineralogie, pp. 394–400, 443, 448.

ROGERS, A. W.
1934 The build of the Kalahari. South African Geographical Journal 17:3–12.
1936 Surface geology of the Kalahari. Royal Society of South Africa, Transactions 24(1):57–80.
1940 Pans. South African Geographical Journal 22:55–60.

ROGERS, A. W., AND A. L. DU TOIT
1909 Report on the geology of parts of Kenhardt, Prieska and Carnarvon. Geological Commission of the Cape of Good Hope, Annual Report, 14th, pp. 8–110.

SCHOLZ, H.
1963 Studien über die bodenbildung zwischen Rehoboth und Walvis Bay. Bonn (unpublished D. Agr. dissertation). 185 p.
1968 Die böden der wüste Namib Südwestafrika. Zeitschrift für Pflanzenernährung und Bodenkunde 119(2):91–107.
1968 Die böden die Halbwüste Südwestafrikas. Zeitschrift für Pflanzenernährung und Bodenkunde 120(2):105–118.
1968 Die böden der feuchten savanne südwestafrikas. Zeitschrift für Pflanzenernährung und Bodenkunde 120(3):208–221.

SINGER, R., AND J. WYMER
1968 Archaeological observations at the Saldanha skull site in South Africa. South African Archaeological Bulletin 23(3):63–74.

SMITH, D. A. M.
1965 The geology of the area around the Khan and Swakop rivers in South West Africa. Geological Survey of South Africa, South West Africa Series, Memoir 3:75.

SOHNGE, P. G., H. D. LE ROEX, AND H. J. NEL
1948 The geology of the country around Messina. Union of South Africa, Geological Survey, Explanation of Sheet 146. 74 p.

TINLEY, K. L.
1966 An ecological reconnaissance of the Moremi Wildlife Reserve, Botswana. Okavango Wildlife Society, Johannesburg. 146 p. (See p. 15.)

TRUTER, F. C., AND OTHERS
1938 The geology and mineral deposits of the Oliphants Hoek area, Cape Province. Union of South Africa, Geological Survey, Explanation of Sheet 173. 136 p.

VAN DER MERWE, C. R.
1963 Soil groups and subgroups of South Africa. 2nd rev. ed. South Africa, Department of Agricultural Technical Services, Science Bulletin 356 (Chemistry Series 165). 355 p., map. (Original edition, 1941, issued by South Africa, Department of Agriculture, as its Science Bulletin 231.)

VAN EEDEN, O. R., AND OTHERS
1955 The geology of the E. Soutpansberg and the Lowveld to the north. South Africa, Geological Survey, Explanation of Sheet 42.

VAN STRATEN, O. J.
1959 The Morapule coalfield, Palapye area. Bechuanaland Protectorate, Geological Survey, Mineral Resources Report 1:7.

VEGTER, J. R.
1953 Underground water supplies in the crystalline complex of the Kenhardt district C.P. and the water supply of Walvis Bay. University of Pretoria, Faculty of Science (unpublished Master's thesis). (See, for example, p. 20, 39.)

VERMAAK, C. F.
1960 The geology and economic resources of the Bakgatla Reserves. Bechuanaland Protectorate, Geological Survey, unpublished report.

VISSER, H. N.
1953 Geology of the Koedoesrand Area, northern Transvaal. Union of South Africa, Geological Survey, Explanation of Sheets 35 and 36. 101 p.

VISSER, J. N. J. (COMP.)
1964 Analyses of rocks, minerals and ores. Republic of South Africa, Department of Mines, Geological Survey, Handbook 5. 409 p.

VON BACKSTRÖM, J. W.
1953 The geology of the area around Lichtenburg. Union of South Africa, Geological Survey, Explanation of Sheet 54. 61 p.
1964 The geology of an area around Keimoes. Republic of South Africa, Geological Survey, Memoir 53. 218 p.

WAGNER, P. A.
1927 The geology of the north eastern part of the Springbok flats and surrounding country. Union of South Africa, Geological Survey, Explanation of Sheet 17.

WAYLAND, E. J.
1954 Outline of the prehistory and stone age climatology in the Bechuanaland Protectorate. Institut Royal Colonial Belge, Bruxelles, Section des Sciences Naturelles et Medicales, Mémoires 25(4). 47 p.

WEINERT, H. H.
1968 Engineering petrology for roads in South Africa. Engineering Geology 2(6):363–395.

WILSON, P. T.
1961 Underground water development on Molopo Farms Settlement and Barolong Tribal Territory. Bechuanaland Protectorate, Geological Survey, unpublished report.

WRIGHT, E. P.
1957 Report on the limestone deposit at Kalusi Kankanga. Bechuanaland Protectorate, Geological Survey, unpublished report.
1957 Monthly report on Mababe depression. July, 1957. Bechuanaland Protectorate, Geological Survey, unpublished report.
1957 Geology of Kihabedum Valley and Kihabe Hill. Bechuanaland Protectorate, Geological Survey, unpublished report.
1958 Reconnaissance geology of W. Ngamiland. Bechuanaland Protectorate, Geological Survey, unpublished report.
1958 Report on a visit to the Shinamba Hills. Bechuanaland Protectorate, Geological Survey, unpublished report.
1958 The Tsodilo Hills, Ngamiland. Bechuanaland Protectorate, Geological Survey, unpublished report.
1958? Traverse 2 west of Lake Ngami: Well at Ngabanyane. Bechuanaland Protectorate, Geological Survey, unpublished report.
1958 Geology of the valley of the Schadum. Bechuanaland Protectorate, Geological Survey, unpublished report.
1958 Geology of the area south of Lake Ngami. Bechuanaland Protectorate, Geological Survey, Records 1956:29–35.

WYBERGH, W.
1918– The limestone resources of the Union. Union
1920 of South Africa, Geological Survey, Memoir 11. 2 vols.
1919 The coastal limestones of the Cape Province. Geological Society of South Africa, Transactions 22:46–67.

INDIA

ABBOTT, J.
1845 On kunker formations with specimens. Asiatic Society of Bengal, Journal 14:442–444.

ABICHANDANI, C. T., AND B. B. ROY
1966 Rajasthan Desert, its origin and amelioration. Indian Geographical Journal 41(3–4):35–43.

AIYENGAR, N. K. N.
1964 Minerals of Madras. Madras, Department of Industries and Commerce. 210 p.

AUDEN, J. B.
1952 Some geological and chemical aspects of the Rajasthan salt problem. Proceedings of the Symposium on the Rajputana Desert. National Institute of Sciences of India, Bulletin 1: 53–67.

BLAGOVESCHENSKIY, E. N.
1968 The dry savannah of northwest India (Thar Desert). Soviet Geography 9(6):519–537.

CAMPBELL, J.
1843 Suggestions regarding the probable origin of some kinds of kunkar and the influence of deliquescent salts on vegetation. Calcutta Journal of Natural History 3:25–28.

CARTER, H. J.
1881 On the kunkur formation of the alluvium in India compared with the flint formation of the chalk of England. Annals and Magazine of Natural History, ser. 5, 7:308–312.

GORE, J. E.
1873 Note on a bed of fossiliferous "kunkur" in the Punjab. Royal Geological Society of Ireland, Journal n.s. 4:9–10.

JONES, W.
1829 Remarks on cancar (kankar). Gleanings of Science 1:365–367.

LA TOUCHE, T. H. D.
1911 Geology of western Rajputana. Geological Survey of India, Memoir 35(1). 116 p.

NARASIMHAN, A. S.
1968 Limestone and kankar deposits, east of Ariyalur, Tiruchirapalli District, Madras. Geological Survey of India, Records 95(2):415–417.

NEWBOLD, T. J.
1844 Note on a recent fossil fresh-water deposit in southern India, with a few remarks on the origin and age of the kunker, and on the supposed decrease of thermal temperature in India. Asiatic Society of Bengal, Journal 13: 313–318.

NIELLY, A.
1872 Essay on the geology of kankar. Professional Papers on Indian Engineering, ser. 2, 1:598–603.
1873 Report on experiments made on kunkur mortars and concrete. Professional Papers on Indian Engineering, ser. 2, 2:115–140.
1877 Kunkar limes and cements, Bari Doab Canal. Professional Papers on Indian Engineering, ser. 2, 6:378–389.

PRASAD, M., M. LAL, AND B. L. DHAWAN
1965 Use of kankar lime in soil estabilisation for roads. Indian Roads Congress, Journal 29(1): 121–135.

RAO, R.
1966 Nature and origin of calcareous kankar in the granitic terrain of Hyderabad. Geological Mining, and Metallurgical Society of India, Quarterly Journal 38(3):135–138.

SHERWILL, R. S.
1846 Note on shelly kunkur from Benares. Asiatic Society of Bengal, Journal (proceedings section) 15:14–15.

SINGH, D., AND G. LAL
1946 Kankar composition as an index of the nature of the soil profile. Indian Journal of Agricultural Science 16:328–342.

SMITH, E.
1833 Notes on the specimens of the kankar formation, and on fossil bones collected on the Jamna. Asiatic Society of Bengal, Journal 2: 622–631.

THOMSON, M.
 1872   Kunkur and mortar analysis. Professional Papers on Indian Engineering, ser. 2, 1:491–496.

UPPAL, H. L., AND R. L. NANDA
 1968   Survey of hidden calcrete in alluvial plains of India by airphoto interpretation. Australian Road Research Board, Conference, 4th, Proceedings 4(2):1677–1682.

VERSTAPPEN, H. T., B. GHOSE, AND S. PANDEY
 1969   Land forms and resources in central Rajasthan (India). International Institute for Aerial Survey and Earth Sciences (ITC), Delft, Publications ser. B, 51.

WYNNE, A. N.
 1872   Geology of Kutch. Geological Survey of India, Memoir 9(1):293.

MISCELLANEOUS (INCLUDING EUROPE AND THE MIDDLE EAST)

BARTLETT, H. H.
 1951   Radiocarbon datability of peat, marl, caliche and archaeological materials. Science 114:55–56.

BEAR, L. M.
 1960   Geology and mineral resources of the Akaki-Lythrodondha area. Cyprus, Geological Survey Department, Memoir 3. 122 p.

BLANC, J. J.
 1959   Recherches géologiques et sédimentologiques. Institut Océanographique, Annales 34:157–211.

BLANCK, E., S. PASSARGE, AND A. REISER
 1926   Uber krustenböden und krustenbildungen wie auch roterden, insbesondere ein Beitrag zur kenntnis der bodenbildungen Palastinas. Chemie der Erde 2:348–395.

BOWEN JONES, H., AND J. H. STEVENS
 1967   Mileiha development project, Hamraniyah development area. Trucial States Council, Survey of Soils and Agricultural Potential in the Trucial States. 54 p.

BURGESS, I. C.
 1961   Fossil soils of the Upper Old Red Sandstone of South Ayrshire. Geological Society of Glasgow, Transactions 24(2):138–153.

BUTZER, K. W.
 1963   Climatic-geomorphologic interpretation of Pleistocene sediments in the Eurafrican subtropics. In F. C. Howell and F. Bourlière (eds), African Ecology and Human Evolution, pp. 1–27. Wenner-Gren Foundation for Anthropological Research, Inc., New York.

CAILLEUX, A.
 1965   Quaternary secondary chemical deposition in France. Geological Society of America, Special Paper 84:125–139.

CARR, J. M., AND L. M. BEAR
 1960   Geology and mineral resources of the Peristerona-Lagoudhera area. Cyprus, Geological Survey Department, Memoir 2. 79 p.

CERECEDA, J. D.
 1916   Acerca de la costra caliza superficial en los suelos aridos de Espagna. R. Sociedad Española de Historia Natural, Boletín 16:305–311.

CLARE, K. E., AND P. J. BEAVEN
 1961   Road making gravels and soils of Nigeria. Great Britain, Department of Scientific and Industrial Research, Road Research Laboratory, Research Note 3914.

DAN, J.
 1962   Disintegration of nari lime crust in relation to relief, soil and vegetation. Archives Internationales de Photogrammetrie 14:189–194.

DEGENS, E. T., AND E. RUTTE
 1960   Geochemische untersuchungen eines kalkrustenprofils von Altkorinth-Griechenland. Neues Jahrbuch für Geologie und Paläontologie, Monatsheft (6):263–276.

DE LABURU, C. R., AND J. M. O. SANCHEZ
 1967   Une forme d'accumulation des carbonates calcique et magnesique en bandes horizontales et "grillages" sous climat semiaride mediterranean. Conference on Mediterranean Soils, Madrid, 1966, Transactions pp. 465–471.

DE VAUMAS, M. E.
 1967   Phenomènes karstiques en Mediterranée orientale. Centre de Recherches et Documentation Cartographiques et Géographiques, Mémoires et Documents 4 (n.s.):197–282.
 1968   Questions de géomorphologie en Israel. Association de Géographes Français, Bulletin 362–363:167–178.

DOBROVOL'SKIY, V. V.
 1961   Typomorphic neoformations in Quaternary deposits of the USSR desert belt. Soviet Soil Science 1961 (10):1085–1097.

DUCLOZ, C.
 1967   Les formations quaternaires de la région de Klepini (Chypre), et leur place dans la chronologie du quaternaire mediterranéen. Archives des Science (Génève) 20(2):123–197.

DUMAS, M. B.
 1969   Glacis et croûtes calcaires dans le levant espagnol. Association de Géographes Français, Bulletin 375–376:553–561.

EVERARD, C. E.
 1964   Climatic change and man as factors in the evolution of slopes. Geographical Journal 130(1):65–69.

FEDOROFF, N.
 1961   Les croûtes et les encroûtements calcaires dans le midi mediterranéen Français. Révue de Géographie Physique et de Géologie Dynamique 4(1):43–50.
 1967   Sols rouges a la limite nord du Bassin Mediterranéan (Haute Province). Conference on Mediterranean Soils, Madrid, 1966, Transactions pp. 443–450.

FENELON, P.
 1967   Vocabulaire français des phénomènes karstiques. Centre des Recherches et Documentation Cartographiques et Geographiques, Mémoires et Documents 4(n.s.):13–68.

GÉZE, B.
 1956   Carte de reconnaissance des sols du Liban au 1/200,000e, Notice explicative. Republique Libanaise, Station Agronomique Libano-Français. 52 p.

GIBB, SIR ALEXANDER, AND PARTNERS
1969   Water resources survey, interim report. Abu Dhabi, Department of Development and Public Works. 121 p.

GLINKA, K. D.
1927   The great soil groups of the world and their development. Translated from the German by C. F. Marbut. Edwards Brothers. 235 p.

GOLDBERG, A. A.
1959   The development of Nari. Research Council of Israel, Bulletin 8G:219.

HALCROW, SIR W., AND PARTNERS
1969   Report on the water resources of the Trucial States. Trucial States Council, Water Resources Survey, Dubai.

HEATH, G. R.
1966   Carbonate nodules formed in soil profiles: the nomenclature problem. Australian Journal of Science 28(10):395.

KRUMBEIN, W. E.
1968   Geomicrobiology and geochemistry of the "Nari-lime-crust" (Israel). *In* G. Miller and G. M. Friedman (eds), Recent Developments in Carbonate Sedimentology in Central Europe, pp. 138–147. Springer, Heidelberg.

KUBIENA, W. L.
1953   The soils of Europe. Murby, London. 318 p. (*See* pp. 157–158.)

KULIK, N.
1959   Lime concretions in semidesert soils. Soviet Soil Science 1959 (1):126.

LAMPLUGH, G. W.
1903   The geology of the country around Dublin. Geological Survey of Ireland, Memoir, Explanation of Sheet 112. 160 p.

MARCELIN, P.
1947   Observations sur des terres et des sols en région Méditerranéenne. I: Terres et sols en Costière. Impr. de Chastanier et Alméras, Nimes 147 p.

MILLER, R. P.
1937   Drainage lines in bas-relief. Journal of Geology 45:432–438.

NAGTEGAAL, P. J. C.
1969   Microtextures in recent and fossil caliche. Leidse Geologische Mededelingen 42:131–142.

NETTERBERG, F.
1967   Discussion on the nomenclature of soil carbonates. Australian Journal of Science 29(7):224.

ONUR, A.
1966   Eshabülkehf tepesinde bazi cŏgrafi müsahedeler. Cografaya Araştirmalari Dergisi 1:159–177.

OSMUND, D. A.
1954   Report on some Cyprus soils. Colonial Office, London. 81 p.

OSMUND, D. A., AND I. STEPHEN
1957   The micropedology of some red soils from Cyprus. Journal of Soil Science 8:19–26.

PANTAZIS, T. M.
1967   The havara and kafkalla deposits of Cyprus. Cyprus, Ministry of Agriculture and Natural Resources. 32 p.

1967   The geology and mineral resources of the Pharmakas-Kalavasos area. Cyprus, Geological Survey Department, Memoir 8.

PEREL'MAN, A. I.
1967   Geochemistry of epigenesis. Translated from the Russian by N. N. Kohanowski. Plenum Press, New York. 266 p.

QUIGLEY, R. M., AND F. J. HEFFERMAN
1968   Swelling sand (gatch) from Kuwait. Engineering Geology 2(5):351–356.

RUTTE, E.
1958   Kalkruste in Spanien. Neues Jahrbuch für Geologie und Paläontologie, Abhandlungen 106(1):52–138.

1960   Kalkrusten im ostlichen Mittelmeergebiet. Deutsche Geologische Gesellschaft, Zeitschrift 112(1):81–90.

SANCHEZ, J. A., AND F. ARTES
1967   Sols sur croûtes calcaires dans les zones cotières du sud est de l'Espagne. Conference on Mediterranean Soils, Madrid, 1966, Transactions pp. 331–340.

SCHOELLER, A.
1944   La croûte calcaire des Pouilles (Italie). Société Géologique de France, Compte Rendu Sommaire sér. 5, 14:181–183.

SOLE SABARIS, L.
1962   Le quaternaire marin des Baléares et ses rapports avec les côtes mediterranéennes de la peninsule Ibérique. Quaternaria 6:309–342.

SOTERIADES, C. G., AND C. KOUDOUNAS
1968   Soils Memoirs of Paphos Sheet No. 51. Republic of Cyprus, Soils Series Report 1.

STRAKHOV, N. M.
1970   Principles of lithogenesis. Vol 3. Plenum, New York.

TRICART, J.
1967   Certaines formes de sedimentation calcaire sont-elles dues a des déséquilibres géochemiques périodiques? Chemical Geology 2:233–248.

WOLFART, R.
1967   Geologie von Syrien und dem Libanon. Gebrüder Borntraeger, Berlin. 326 p.

ACKNOWLEDGMENTS

I should like to acknowledge the assistance of the following in the preparation of this bibliography, which was undertaken while the author was in receipt of a research studentship from the Natural Environmental Research Council:

C. C. Reeves (Texas Technological University, Lubbock, Texas).

A. T. Grove (Department of Geography, Cambridge).

C. Vita-Finzi (University College, London).

The Chief of Soils (CSIRO, Adelaide).

J. P. Walsh (Geological Survey, Nairobi).

The Directors of the Geological Surveys of Botswana, South Africa, South West Africa, and Zambia.

The Librarian of the Geological Survey of India.

F. Netterberg (National Institute for Road Research, Pretoria).

National Agricultural Settlement Authority (Tripoli, Libya).

Helga Besler (University of Stuttgart).

H. Scholz (Windhoek, South West Africa).

C. G. Soteriades (Director of Soil Survey, Cyprus).

Patricia Paylore (University of Arizona).